Channel Coding for Telecommunications

Channel Coding for Telecommunications

Martin Bossert

University of Ulm, Germany

JOHN WILEY & SONS, LTD

Chichester · New York · Weinheim · Brisbane · Singapore · Toronto

Other Wiley Editorial Offices

John Wiley & Sons, Inc., 605 Third Avenue,
New York, NY 10158-0012, USA

WILEY-VCH Verlag GmbH, Pappelallee 3,
D-69469 Weinheim, Germany

Jacaranda Wiley Ltd, 33 Park Road, Milton,
Queensland 4064, Australia

Singapore • Toronto

Library of Congress Cataloging-in-Publication Data

Bossert, Martin, 1955-
[Kanalcodierung. English]
Channel coding for telecommunications / Martin Bossert.
p. cm.
Includes bibliographical references.
ISBN 0-471-98277-6 (alk. paper)
1. Coding theory. I. Title.
TK5102.92.B67 1999
003'. 54—dc21 99-32323
CIP

British Library Cataloguing in Publication Data
A catalogue record for this book is available from the British Library

ISBN 0-471-98277-6

Produced from PostScript files supplied by the author.
Printed and bound in Great Britain by Bookcraft (Bath) Limited.
This book is printed on acid-free paper responsibly manufactured from sustainable forestry, in which at least two trees are planted for each one used for paper production.

Contents

Preface

This book is a translation of the second German edition of my book *Kanalcodierung*, which was published in 1998 by B.G.Teubner, Stuttgart, Germany. The German edition was based on lectures held at the Universities of Karlsruhe and Ulm from 1987 to 1998. The lectures were intended for students majoring in communications engineering, computer science and mathematics. The scope of the book is also intended to be such that it can be used by researchers working in the field of reliable data transmission.

This book is more than a direct translation. Some modifications have been made to clarify more advanced topics, and in particular the section on error location codes has been restructured. This edition also includes an appendix dealing with the popular topic of serial and parallel (turbo) code concatenation, which uses iterative decoding. Here, examples are given that clearly illustrate the principles behind the code construction. Selected simulation results are then presented showing some of the advantages and limitations of the code.

I am aware of the fact that the success of a book greatly depends on the environment in which it was created, and, furthermore, it is a measure of it. Therefore my first thanks are to my teacher Viktor Zyablov from the Institute of Problems of Information Transmission (IPPI) of the Academy of Science, Moscow. Next many thanks to my colleagues Sergo Shavgulidze from the Georgian Technical University in Tblisi, Georgia and Valodja Sidorenko also from IPPI. From numerous discussions and projects spanning many years, I have learned a lot in the field of science as well as in social interaction.

My thanks go also to my research/teaching assistants Markus Breitbach, Hans Dieterich, Adrian Donder, Jürgen Freudenberger, Thomas Frey, Helmut Grießer, Günther Haas, Armin Häutle, Ralph Jordan, Rainer Lucas, Paul Lusina, Johannes Maucher, Ramon Nogueroles, and Walter Schnug. Without the work of these people, this book would not have been possible. In addition I would like to thank the guests who frequently visited my group, in particular Ernst Gabidulin and Stefan Dodunekov, and the engineering students from Ulm: Christian Paintz,

Achim Fahrner and Stephan Leuschner.

I would especially like to thank Paul Lusina for the careful translation and frequent discussions. Further thanks should also be given to Helmut Grießer for the many hours spent working on the book.

Finally I would like to translate the last sentence of the German preface, as follows: I always wondered why the authors of books thank their family members. After writing this book I know why and I would like to thank my beloved family for their understanding, my wife Inge and my children Marie-Luise, Magdalena and Sebastian.

MARTIN BOSSERT

Ulm, Germany, July 1999

For comments/errors: Martin.Bossert@e-technik.uni-ulm.de
Errata list for the book: http//it.e-technik.uni-ulm.de/~boss

Introduction

The term 'coding' is generally associated with the mapping of information to a set of symbols or numbers. In this way, different goals are achieved. Encryption or authentication encoding is termed cryptography, and is often used for information transmission to protect information from misuse. Source coding techniques aim to compress information, while channel coding seeks to make information immune from random distortion.

A pervasive example for channel coding is speech or written text. Language contains redundancy, and this can be used to correct corrupted transmission (printing errors, dyslexia, etc.). The redundancy is illustrated by the fact that the reordering of 7 letters does not always produce a valid word from a particular language. The valid words are the codewords while decoding corresponds to finding a valid codeword that is as close as possible to the received set of symbols.

In order to explain to my wife and children what channel coding is, we carried out the following experiment. I selected a saying while the children played the role of the channel. One of the 26 letters of the alphabet was selected. Dice were then thrown to determine which position in the saying was replaced by the selected letter. The result was as follows:

'A trap df ond milrian cileq begens wifh a sinqle soep.' *Chinese saying*

My wife carried out the decoding by trying to find a valid word for every corrupted word (at the same time, the meaning and structure of the entire sentence is implicitly used to find a valid word). She could correct all the errors. The meaning of the saying gives new courage to those attempting to solve what is at first glance a very complex problem (on a par with writing or studying from a textbook).

The origin of information theory and channel coding began with work by C. E. Shannon [Sha48] in 1948. Shannon proved the theoretical limitations of channel coding, which have yet to be achieved. Similarly to language, channel coding first adds redundancy to the information stream, which is later used to detect or correct errors due to transmission distortion. This redundancy can be determined by using two different code classes. The first class comprises block codes, where independent blocks (codewords) of symbols are constructed with a constant mapping between information and redundancy. The second method is convolutional encoding, where the redundancy is continuously calculated for the code sequence using a combination (convolution) of successive information symbols.

Channel coding is a very young field. However, it has gained increasing importance in communication systems, which are almost inconceivable without channel coding.

This book presents the basic principles, methods and techniques used for channel coding. The choice of a code is critically dependent on the channel. An important point for the practical use of codes, which should be heavily emphasized, is the decoding. Furthermore, in communication systems, the powerful concepts of code concatenation and coded modulation are becoming of increasing importance, and are treated in chapter 10. Elementary knowledge of probability theory and communications theory is required to understand these concepts.

In chapter 1, several elementary principles, models, concepts and definitions are given. A calculation of the achievable parameters of block codes results in an upper bound for their performance. Possible decoding principles are clarified, and decoding error probability is defined. In chapter 2, the theory of Galois fields is presented. This topic includes prime fields and extension fields. Often, the greatest common divisor of two numbers is required for a coding calculation. This can be solved using the Euclidean algorithm. Elementary algebraic principles are used to simplify code constructions, and later allow for the description of cyclic codes, which are also labeled CRC codes (*cyclic redundancy check*).

Reed–Solomon (RS) codes are introduced in chapter 3. For two reasons, we present RS codes only with respect to a prime field, though the theory is unchanged for extension fields. The first reason is that additional calculations in an extension field can make understanding of the theory more complicated. The second reason is that Reed–Solomon codes can be understood only on the basis of prime field theory. Several algebraic decoding techniques, based on the solution of a special set of equations, are presented. These include the Berelkamp–Massey and Euclidean algorithms. We go on to show the equivalence of these algorithms.

For the explanation of BCH codes in chapter 4, we require the theory of extension fields. We concentrate on the extension of the basic binary field. The connection between Reed–Solomon and BCH codes is shown, and decoding can also be performed based on algebraic techniques.

Additional code classes that are of practical interest are presented in chapter 5. They include codes related to PN sequences (*pseudonoise*), and orthogonal and biorthogonal signals. In particular, these codes include Simplex, Reed–Muller and Hamming codes, as well as Walsh and Hadamard sequences. In addition, the binary quadratic residue codes, are defined in section 5.4. These codes are among the set of best codes known today.

In chapter 6, the general properties of block codes are developed, and are used in chapter 7 to derive other decoding techniques. We also include the trellis description of block codes, which can be an important aid for decoding. We give details of the fundamental principles, definitions and properties of trellises.

The decoding of block codes is dealt with extensively in chapter 7. Different non-algebraic decoding methods are presented. These decoding methods are also advantageous because their decoding speed may be faster and because of their ability to incorporate reliability information in the decoding. This last point is generally not possible for algebraic codes. The ability to accurately recover a signal lies not only in making a binary decision, but also in deciding how reliable the decision is. How this information is employed in a decoding method is described in section 7.4. Decoding methods described include permutation decoding, threshold decoding, list decoding, iterative symbolwise decoding and majority logic decoding. We explain the basic difference between optimal and suboptimal decoding algorithms. Finally, we consider the problem of decoding simplification. On this basis, codes must be chosen that can correct as many errors as possible for the particular channel, and still have a tractable decoding complexity.

The elementary principles of convolutional codes are presented in chapter 8. Here we also give the classical description of convolutional codes, and the algebraic representation. The most important practical attribute of convolutional codes is their ability to be decoded using the Viterbi algorithm. We explain the application of this algorithm to convolutional codes, although the topic was already presented with respect to block codes in chapter 7. We describe sequential decoding of convolutional codes by means of the Fano and Zigangirov–Jelinek (ZJ) algorithm. Decoding that outputs a reliability in addition to the decided symbol is gaining increasing importance. The *puncturing* of convolutional codes is possible in order to construct codes with different levels of error protection. At the end of this chapter, tables of good convolutional codes are given.

In chapter 9, the principle of generalized code concatenation is explained. This is a very powerful principle, which comes from the idea of concatenating codes (inner and outer codes) together. Generalized concatenation is described by the partitioning of the inner code and the use of multiple outer codes. A description of this theory has for a long time only been available in Russian textbooks. For this reason, the theory of this topic is comprehensively presented in chapters 9 and 10 and expanded with examples of its application. A clear advantage of generalized concatenation is the possibility of constructing codes with unequal error protection. Such codes can simultaneously correct burst and individual errors. Another important fact is that many *short* codes can be used to construct *long* codes that have tractable decoding complexities. We next focus on the generalized concatenation of block and convolutional codes. We stress that this subject lies largely in the realm of research, and there remain a large number of unanswered problems.

With the help of generalized concatenation, it is easy to describe the class of Reed–Muller codes. This description has produced new decoding techniques that allow the use of the aforementioned reliability information. These new decoding techniques are presented in section 9.5.1. In comparison with traditional known decoding techniques for Reed–Muller codes, these new techniques have a much smaller decoding complexity, i.e. the decoding requires fewer computer operations.

The theory of generalized concatenation (GC) can be applied to the special case of modulation, which is described in chapter 10. In this approach, the modulation is considered to be the inner code of the GC encoder. Therefore the principles from chapter 9 are for the most part required for the understanding of chapter 10. Coded modulation is at present an important research focus in communications theory. Generalized code concatenation allows this area to be abstractly described, allowing for construction methods for coded modulation systems. Coded modulation is often referred to as trellis coded modulation. The use of multidimensional spaces allows for a more flexible use of the applied techniques for coded modulation systems. Multidimensional signal constellations are also known as 'lattices', and several are applied to generalized concatenated codes.

Iterative decoding in combination with serial and parallel concatenation of convolutional codes will also be explained in the last part of the book. This construction method has resulted in codes with the best known decoding results to date.

Over 200 examples serve to demonstrate the theoretical principles presented. The examples are referenced in order to clarify concepts presented later. At the end of chapters 1–8, practice problems are encoded and the solutions are included in appendix D. The problems aid in deepening understanding, but are not required for the contents of subsequent chapters.

Readers of this book will have different focuses and interests. The following sections are recommended for readers interested only in a particular topic:

- The coding and decoding of BCH and RS codes are presented in chapters 1–4.
- The decoding of binary linear block codes is described in chapters 1, 6 and 7.
- The theory of convolutional codes and their decoding is found in chapter 8. In addition, sections 7.1 and 7.2 presents the required background for channel models, metrics and reliability information.
- The description of block codes using a trellis and the theory of minimal trellises are presented in section 6.8. Here the concepts from chapter 1 are assumed.
- Iterative decoding and serial and parallel concatenated codes are considered in chapter 1, section 8.8 and appendix A.

Several themes, for example Viterbi decoding or reliability information values, are dealt within the context of block or convolutional codes. Repetition, however, is necessary because of the different perspectives afforded by the different techniques. With respect to notation, I have focused on popular publications and sought to make this notation easier and more unified. The notations for block and convolutional codes are somewhat different, and this is also noted in this book.

It is clear that such a book can not be completely error-free. In view of the presented theme it is hoped that enough redundancy is contained in the material such that the corresponding errors can be detected and corrected.

1

Fundamentals

The goal of channel coding can be explained using the model depicted in figure 1.1. The information vector $\mathbf{i}$ is transformed into the coded bit vector $\mathbf{c}$. Transmission over the channel introduces additive noise, which results in the received bit vector $\mathbf{r}$. $\mathbf{r}$ is decoded to produce the information bit vector $\hat{\mathbf{i}}$, which is designed to have a reduced probability of error due to the coding technique implemented. In order to correct or identify errors caused during transmission, it is equivalent to calculate an estimate of the codeword $\hat{\mathbf{c}}$ or the error vector $\hat{\mathbf{f}}$ from the received vector $\mathbf{r}$.

The channel can be modeled to take into account different factors, for example reading to and writing from memory, corruption from a channel, manufacturing tolerances, crosstalk interference, etc. In order to simulate distortion, an appropriate channel model must be used. If the channel is memoryless then the conditional probability may be used based on given probabilities associated with the system:

$$P(r|c).$$

The above expression is the probability that the vector $\mathbf{r}$ has been received if the codeword $\mathbf{c}$ was transmitted. In figure 1.2, a basic model of a binary symmetric channel (BSC) is presented.

In the next sections, we will consider only binary information sequences. In a BSC, the event that a transmitted 0 is received as a 1 at the receiver occurs with probability p. The event that the transmitted symbol is not altered by the channel occurs with a probability of $1-p$. These probability transitions are the same in the case of a transmitted 1. In order to guarantee

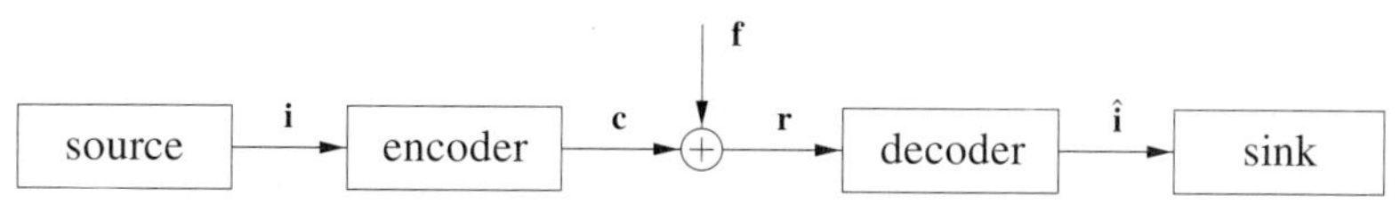

Figure 1.1 Digital transmission system.

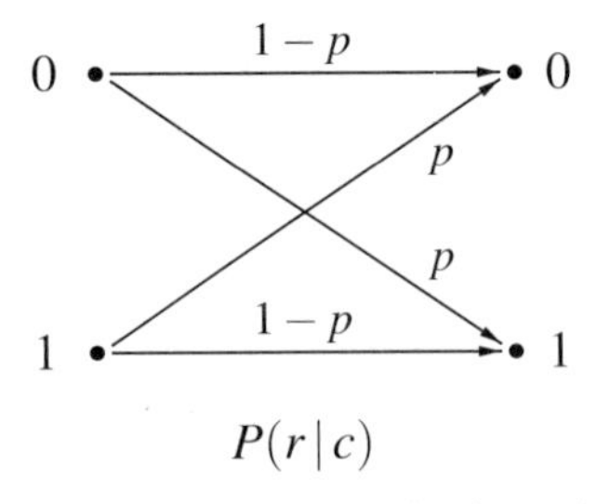

Figure 1.2 Binary symmetric channel model.

that the calculation of the received vector produces binary values of 0 and 1, we define the modulo operation:

Definition 1.1 (Modulo calculation) *The modulo operation is defined for the term $a = cb + d$, such that*

$$a = d \mod b,$$

where $a, c \in \mathbb{Z}$, $b \in \mathbb{N}$, and $d \in \mathbb{N}_0$, (spoken as 'a congruent d modulo b').

In the above definition $\mathbb{Z}, \mathbb{N}$, and $\mathbb{N}_0$ are the sets of integer numbers, natural numbers, and natural numbers including zero, respectively. When a modulo b operation is performed on a number, values that are a multiple of b cannot be uniquely recovered. This can be seen in the following equation:

$$cb = 0 \mod b.$$

Example 1.1 (Modulo calculation) The following are calculation examples for the modulo operation: $71 = 1 \bmod 7$, $-13 = 2 \bmod 5$, $30 = 6 \bmod 8$, $(25 \cdot 31) = 4 \cdot 3 = 5 \bmod 7$. ◇

Block code A block code C uniquely maps a block of information symbols of length k: $i_0, i_1, \ldots, i_{k-1}$ to a codeword with the length n: $c_0, c_1, \ldots, c_{n-1}$. The number of redundancy symbols is $n - k$. The ratio k/n is defined as the *code rate*. If a binary information bit stream is divided into independent blocks of bits for encoding then the coding technique is termed a binary block code, where the symbols 0 and 1 make up the code alphabet.

Example 1.2 (Single parity check code, PC code) The coding technique for a single PC code of length n is a block of k information bits $i_0, i_1, \ldots i_{k-1}$ to a codeword of length $n = k+1$, $c_0, c_1, \ldots c_{n-1}$ by equating $c_0 = i_0, c_1 = i_1, \ldots c_{k-1} = i_{k-1}$. The last coordinate of the codeword c_{n-1} satisfies the equation $\sum_{j=0}^{n-2} i_j + c_{n-1} = 0 \bmod 2$. For the case where $n = 3$, the following table shows the codewords:

i_0	i_1	c_2
0	0	0
0	1	1
1	0	1
1	1	0

◇

Example 1.3 (Repetition code, RP code) A binary repetition code of length n consists of two codewords: the all-zero word $c_0 = c_1 = \ldots = c_{n-1} = 0$ and the all-one word $c_0 = c_1 = \ldots = c_{n-1} = 1$. ◇

Definition 1.2 (Binary addition of codewords) *The addition of two codewords* $\mathbf{c}+\mathbf{a}$ *is defined through the addition of the j-th coordinates* $c_j + a_j$, $j = 0, 1, \ldots, n-1$, *with the* mod 2 *operation performed on the sum of every coordinate.*

Definition 1.3 (Linear code) *A code C is linear, if all linear combinations (e.g. addition) of codewords are again codewords.*

Linear codes will be the main focus of this book. For information on nonlinear codes, [McWSl] may be consulted. Single parity check codes (example 1.2) and repetition codes (example 1.3) are linear codes.

A binary bit stream can be represented as a vector with binary components in the binary vector space $\mathbb{F}_2^n$ of all possible vectors of length n.

Definition 1.4 (Scalar product) *The scalar (inner) product is defined for* $\mathbf{a}, \mathbf{b} \in \mathbb{F}_2^n$ *as*

$$\langle \mathbf{a}, \mathbf{b} \rangle = \sum_{i=0}^{n-1} a_i b_i \mod 2 .$$

1.1 Weight, distance

We will now define the Hamming metric. This will be the first metric considered in our investigation of coding theory and will be sufficient for our investigation of codes up until chapter 7. In chapter 7 and appendix B, other metrics will be presented.

Definition 1.5 (Hamming weight) *The (Hamming) weight of a vector* $\mathbf{c}$ *is defined as the number of non-zero vector coordinates. This number ranges from a minimum value of zero to the length n of the vector* $\mathbf{c}$.

$$\mathrm{wt}(\mathbf{c}) = \sum_{j=0}^{n-1} \mathrm{wt}(c_j), \quad \textit{where} \quad \mathrm{wt}(c_j) = \begin{cases} 0, & c_j = 0 \\ 1, & c_j \neq 0 \end{cases} .$$

Definition 1.6 (Hamming distance) *The (Hamming) distance between two vectors* $\mathbf{a}$ *and* $\mathbf{c}$ *is the number of coordinates where* $\mathbf{a}$ *and* $\mathbf{c}$ *differ:*

$$\begin{aligned} \mathrm{dist}(\mathbf{a}, \mathbf{c}) &= \sum_{j=0}^{n-1} \mathrm{wt}(a_j + c_j), \quad \textit{where} \quad \mathrm{wt}(a_j + c_j) = \begin{cases} 0, & c_j = a_j, \\ 1, & c_j \neq a_j, \end{cases} \\ \mathrm{dist}(\mathbf{a}, \mathbf{c}) &= \mathrm{wt}\,(\mathbf{a} + \mathbf{c}) . \end{aligned}$$

Definition 1.7 (Weight enumeration) *The j-th coordinate w_j of the weight enumeration vector $W = (w_0, w_1, \ldots, w_n)$ of a code C of length n is defined as the number of codewords with weight j. W can also be written as a polynomial in x, giving* $W(x) = \sum_{j=0}^{n} w_j x^j$.

A linear code C always has the zero codeword: thus $W_0 = 1$. The zero codeword is necessary due to the linearity of the code: $(\mathbf{c} + \mathbf{c} = \mathbf{0} \in C)$.

Example 1.4 (Weight enumeration) The weight enumeration vector of the single parity check code of length $n = 3$ from example 1.2 is

$$W = (1, 0, 3, 0) .$$

The weight enumeration vector of the repetition code of length n from example 1.3 is

$$w_0 = 1;\ w_n = 1;\ w_j = 0,\ j = 1, \ldots, n-1 .$$ ◇

1.1.1 Minimum distance, and error correction

Definition 1.8 (Minimum distance) *The minimum distance d of a code C is the minimum distance between two different codewords:*

$$d = \min_{\substack{\mathbf{a},\mathbf{c} \in C \\ \mathbf{a} \neq \mathbf{c}}} \{\mathrm{dist}(\mathbf{a}, \mathbf{c})\} .$$

For linear codes, the minimum distance is equal to the minimum weight

$$d = \min_{\substack{\mathbf{a},\mathbf{c} \in C \\ \mathbf{a} \neq \mathbf{c}}} \{\mathrm{wt}(\mathbf{a} + \mathbf{c})\} = \min_{\substack{\mathbf{c} \in C \\ \mathbf{c} \neq 0}} \{\mathrm{wt}(\mathbf{c})\} .$$

The property that the *minimum distance equals the non-zero minimum weight* is a very important tool for designing codes. The minimum distance among codewords specifies the error detection and error correction capabilities of a code. Calculation of the minimum distance through the minimum weight is much easier to compute than calculating the minimum distance directly.

A question that naturally arises from knowledge of the minimum distance is: how many errors can one correct using the code C?

We compare two codewords having minimum distance d. When a single error in a codeword occurs, the received codeword has a distance of one from the transmitted codeword. Naturally, an n-dimensional space cannot be depicted on paper; however, we shall consider an illustrative example to clarify the concept of error correction.

Two neighboring points have a Hamming distance of 1. Using this measure, the condition for error correction is shown in figure 1.3.

In the left-hand diagram of figure 1.3, we assume a code with minimum distance $d = 3$, which is equal to the Hamming distance between vectors $\mathbf{a}$ and $\mathbf{c}$. According to the definition of the minimum distance, the Hamming distance between any two codewords must be ≥ 3. If we surround each codeword within a *sphere* of Hamming distance 1, we observe that the spheres do not overlap (figure 1.3). Therefore each vector with a Hamming distance ≤ 1 from any codeword can be uniquely mapped to a codeword.

In the right-hand diagram of figure 1.3, the minimum distance is set to $d = 4$. If a received vector has a distance of two from a codeword, it may be equidistant from two different codewords. This makes a unique mapping of the received vector to a codeword impossible.

The vector with the Hamming distance of 2 between $\mathbf{a}$ and $\mathbf{c}$ cannot be mapped unambiguously to codeword $\mathbf{a}$ or $\mathbf{c}$. Therefore we can only correct vectors with a Hamming distance ≤ 1 in this example.

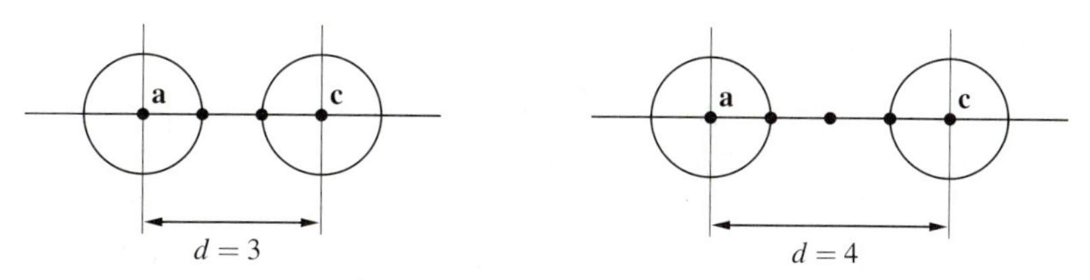

Figure 1.3 Hamming distances.

We shall now generalize the concept. A received vector $\mathbf{r} = \mathbf{c} + \mathbf{f}$, ($\mathbf{c} \in C$, $\mathbf{f}$ an error vector), can only be corrected if the distance between it and any other codeword $\mathbf{a} \in C$ satisfies

$$\mathrm{dist}(\mathbf{c}, \mathbf{c} + \mathbf{f}) < \mathrm{dist}(\mathbf{a}, \mathbf{c} + \mathbf{f})$$

or

$$\mathrm{wt}(\mathbf{f}) < \mathrm{wt}(\mathbf{a} + \mathbf{c} + \mathbf{f}).$$

If this result holds for the whole set of errors $\{\mathbf{f}\}$ having a certain weight then we get

$$\mathrm{wt}(\mathbf{f}) \leq \left\lfloor \frac{d-1}{2} \right\rfloor,$$

where $\lfloor x \rfloor$ is the largest integer number that is less than or equal to x. When d is odd, $\lfloor \frac{d-1}{2} \rfloor = \frac{d-1}{2}$, and when d is even, the result is $\lfloor \frac{d-1}{2} \rfloor = \frac{d-2}{2}$.

We assume that fewer errors are more probable. This is the rationale for mapping the received vector to the nearest codeword. Minimum distance corrections correspond to the error event with the fewest errors, and therefore the greatest likelihood of occurring.

Example 1.5 (Minimum distance) The single parity check code from example 1.2 has minimum distance $d = 2$, which gives $\lfloor \frac{d-1}{2} \rfloor = 0$. With a single PC code, error correction is impossible; however, an odd number of received errors can be detected because the parity check sum is $\neq 0$. ◇

Example 1.6 (Minimum distance) The repetition code from example 1.3 has a minimum distance of $d = n$; therefore

$$\left\lfloor \frac{d-1}{2} \right\rfloor = \left\lfloor \frac{n-1}{2} \right\rfloor = \begin{cases} \frac{n-1}{2}, & n \text{ odd} \\ \frac{n-2}{2}, & n \text{ even}. \end{cases}$$

In the case of a BSC (figure 1.2), the following decoding method for a repetition code can be performed. Count the number of zeros in the received word. If this total is larger than $\lfloor \frac{n-1}{2} \rfloor$ then decode $\mathbf{0}$, otherwise $\mathbf{1}$. For the case of equality (i.e. $\frac{n}{2}$), the decision is arbitrary (for the case of even n). ◇

If the vector $\mathbf{c} + \mathbf{f}$ is not a codeword then the error can be detected. However, for any $\mathbf{f} \neq 0$ with $\mathrm{wt}(f) \leq d$, $\mathbf{c} + \mathbf{f}$ cannot be a codeword. We may now state the following proposition:

Linear block code, error correction capability For any code $C(n,k,d)$, up to $e = \lfloor \frac{d-1}{2} \rfloor$ errors can always be corrected, or up to $d - 1$ errors can always be detected.

1.1.2 Hamming bound

The question concerning the number of codewords that can exist with a given minimum distance d and a length n is a central problem in channel coding. There are two possible questions that can be asked. For a given set of code parameters, what are the *largest* and *smallest* numbers of codewords that can exist? These limits correspond to the upper and lower bounds of a code's parameters respectively. The Hamming bound is an upper bound. For a vector $\mathbf{c} \in C(n,k,d)$, there exist $\binom{n}{1}$ vectors at a distance of 1, $\binom{n}{2}$ vectors at a distance of 2, and so on. The binomial coefficients $\binom{n}{t}$ are defined as

$$\binom{n}{t} = \frac{n(n-1)\cdots(n-t+1)}{t(t-1)\cdots 1}.$$

All together, there are 2^n binary vectors, where n is the codeword length.

Theorem 1.9 (Hamming bound) *For a binary code $C(n,k,d)$, the following inequality holds:*

$$2^k\left(1+\binom{n}{1}+\cdots+\binom{n}{e}\right) \leq 2^n, \quad \text{where} \quad e = \left\lfloor \frac{d-1}{2} \right\rfloor.$$

The Hamming bound may be interpreted as follows. A codeword can be bounded within a correction sphere with the largest possible radius such that no two spheres overlap. This maximum diameter corresponds to the minimum distance among codewords (see figure 1.3). All vectors within the correction sphere can be uniquely mapped to the codeword located at the center of the sphere.

Definition 1.10 (Perfect code) *A perfect code is a code $C(n,k,d)$ that satisfies the Hamming bound with equality.*

Only a few perfect codes exist. In particular, repetition codes with odd length, single error correction Hamming codes and the Golay code (see exercise 1.7 c) are all binary perfect codes that can exist.

The Hamming bound compares the number of vectors contained inside the correction spheres with the total number of vectors in the space $\mathbb{F}_2^n$. A conceptual interpretation of what the Hamming bound measures is based on *how good the vector space is covered by the correction spheres*. With a perfect code, the correction spheres cover the entire space; therefore all vectors lie within a correction sphere and can be unambiguously mapped to a codeword. Based on this conceptualization, the Hamming bound is also termed the sphere packing bound.

Example 1.7 (Hamming bound for the binary single parity check code) For the binary single parity check code from example 1.2, the Hamming bound is calculated: $k = n-1$, $e = 0$,

$$2^{n-1}(1) < 2^n;$$

therefore the code is not perfect. ◇

Example 1.8 (Hamming bound for the repetition code) For a repetition code of length $n = 3$ (example 1.3), the parameters are $k = 1$, $e = 1$,

$$2^1\left(1+\binom{3}{1}\right) = 2^1 4 = 8 = 2^3;$$

therefore the code is perfect. ◇

Theorem 1.11 (Perfect repetition codes) *All binary repetition codes of odd length are perfect.*

Proof A repetition code of odd length n has the parameters $C(n,1,n)$. Using the equalities

$$\sum_{j=0}^{n}\binom{n}{j} = 2^n \qquad \text{and} \qquad \binom{n}{j} = \binom{n}{n-j},$$

we calculate the Hamming bound:

$$2\left(1+\binom{n}{1}+\ldots+\binom{n}{\frac{n-1}{2}}\right) = 1+\binom{n}{1}+\cdots+\binom{n}{\frac{n-1}{2}}+\binom{n}{\frac{n-1}{2}}+\ldots+\binom{n}{1}+1$$

$$= 1+\binom{n}{1}+\cdots+\binom{n}{\frac{n-1}{2}}+\binom{n}{\frac{n+1}{2}}+\ldots+\binom{n}{n-1}+\binom{n}{n} = \sum_{j=0}^{n}\binom{n}{j} = 2^n . \qquad \square$$

1.2 Parity check matrix and syndrome

Codewords $\mathbf{c} = (c_0, c_1, \ldots, c_{n-1})$ of a linear block code can be defined by the following equation:

$$\mathbf{H}\cdot\mathbf{c}^T = \mathbf{0} \quad \text{or} \quad \mathbf{c}\cdot\mathbf{H}^T = \mathbf{0}$$

$\mathbf{H}$ is the parity check matrix. For the case of a binary code with length n and dimension k, a binary $(n-k)\times n$ matrix results. Matrix multiplication of $\mathbf{H}$ with a vector $\mathbf{c}^T$ produces a vector where the values are the inner products of the rows of $\mathbf{H}$ and $\mathbf{c}^T$. The fact that the number of rows is equal to the number of parity checks will be apparent through the systematic representation of the parity check matrix $\mathbf{H}$ (see definition 1.12). With the definition of a code in terms of $\mathbf{H}$ and a minimum distance of d, any $d-1$ columns from $\mathbf{H}$ must be linearly independent, and some d columns must be linearly dependent.

Example 1.9 (Parity check matrix) The parity check code from example 1.2 has the following parity check matrix:

$$\mathbf{H} = (1, 1, \ldots, 1)\,,$$

$n-k = 1$, e.g. a $1\times n$ matrix.

A repetition code of length n has the following parity check matrix $\mathbf{H}$:

$$\mathbf{H} = \begin{pmatrix} 1 & 1 & & & \\ 1 & & 1 & & \\ \vdots & & & \ddots & \\ 1 & & & & 1 \end{pmatrix} \qquad \text{(free space corresponds to 0)}\,.$$

$\mathbf{H}$ is an $(n-1)\times n$ matrix. ◇

The scalar product of a row of the parity check matrix $\mathbf{H}$ with a valid transposed codeword must be zero. The linear combination of rows of the parity check matrix $\mathbf{H}$ can produce a parity check matrix $\mathbf{H}'$ that corresponds to the same code.

Furthermore, codes are called equivalent if one can be constructed from the other through permutations of columns and/or the linear combination of rows (see section 1.7).

Definition 1.12 (Systematic coding) *The mapping of information symbols into a codeword is systematic if the k information symbols are unchanged within the n codeword coordinates. Therefore the information bits and the redundancy bits are at distinct positions. The parity check matrix has the following form:*

$$\mathbf{H} = (\,\mathbf{A} \mid \mathbf{I}\,), \quad \textit{where } \mathbf{I} \textit{ is the } (n-k) \times (n-k) \textit{ identity matrix.}$$

Every linear block code can be encoded in systematic form. A systematic representation of $\mathbf{H}$ is not restricted to the form $(\,\mathbf{A} \mid \mathbf{I}\,)$. For an equivalent code, the $n-k$ identity matrix columns can be distributed throughout any of the n columns (see also section 1.8).

Definition 1.13 (Syndrome) *The transposed syndrome* $\mathbf{s}$ *is defined as the multiplication of a transposed received word* $\mathbf{r} = \mathbf{c} + \mathbf{f}$, *with the parity check matrix, where* $\mathbf{c} \in C$ *and* $\mathbf{f}$ *is an error vector:*

$$\mathbf{s}^T = \mathbf{H} \cdot \mathbf{r}^T = \mathbf{H} \cdot (\mathbf{c}^T + \mathbf{f}^T) = \mathbf{H} \cdot \mathbf{f}^T \; .$$

The syndrome depends only on the error, and not the codeword. This arises from the property that $\mathbf{H} \cdot \mathbf{c}^T = \mathbf{0}$ *for* $\mathbf{c} \in C$.

Decoding can be divided into two steps. The first involves determining the syndrome of the received codeword. The second step lies in determining the most probable error that gives the syndrome.

1.3 Decoding principles

When a received word is decoded, the output is correct or false, or no decision can be made. These outputs may be defined as correct decoding, false decoding and decoding failure respectively.

Possible decoding results

- *Correct decoding:* The transmitted codeword is equal to the decoded codeword. The decoder has removed the errors introduced by the channel.
- *False decoding:* The transmitted codeword is different than the decoded codeword. In this case the decoder has calculated an error that does not correspond to the true channel error. For example, if the error is a valid codeword, the received vector will also be a valid codeword due to the linearity of the code.
- *Decoding failure:* The decoder can find no solution (e.g. no valid codeword). This result occurs for certain specific decoding methods.

Many different decoding principles can be implemented when a given codeword from $C(n,k,d)$ is corrupted to produce a received vector $\mathbf{r} \notin C$. The word $\mathbf{r}$ is defined as $\mathbf{r} = \mathbf{c} + \mathbf{f}$ ($\mathbf{c} \in C$, $\mathbf{f}$ error). The decoder makes an estimate of the received codeword $\hat{\mathbf{c}}$ and of the corresponding error $\hat{\mathbf{f}}$ that would produce the received vector $\mathbf{r}$:

$$\mathbf{c} + \mathbf{f} = \mathbf{r} = \hat{\mathbf{c}} + \hat{\mathbf{f}} \, .$$

In the case of correct decoding, $\mathbf{c} = \hat{\mathbf{c}}$ (and $\mathbf{f} = \hat{\mathbf{f}}$), and in the case of incorrect decoding, $\mathbf{c} \neq \hat{\mathbf{c}}$ (or $\mathbf{f} \neq \hat{\mathbf{f}}$).

Note The following decoding methods are applied exclusively to the BSC, based on the Hamming metric. In section 7.1, other channels and metrics will be studied in detail.

Possible decoding principles

- *Error detection:* This decoding method verifies whether the received word $\mathbf{r}$ is a valid codeword. Therefore the decoder only tests whether $\mathbf{r} \in \mathcal{C}$. Errors are detected for the case when $\mathbf{r} \notin \mathcal{C}$. For the case where $\mathbf{f} = 0$, $\mathbf{r}$ is likewise correctly identified (no error in the received vector). The case where $\{\mathbf{f} \in \mathcal{C}, \mathbf{f} \neq \mathbf{0}\}$ results in an incorrect decoding of $\mathbf{r}$. The error cannot be detected due to the linearity of the code. A decoding failure is not possible using error detection.

- *Maximum-likelihood decoding (ML):*[1] The probability that a received word $\mathbf{r}$ corresponds to a codeword $\hat{\mathbf{c}}$ is calculated. The codeword with the largest probability is selected as the transmitted codeword:

$$P(\mathbf{r}|\hat{\mathbf{c}}) = \max_{\mathbf{c} \in \mathcal{C}} P(\mathbf{r}|\mathbf{c}) \; .$$

  In the case where more than one codeword results with the same maximum probability, a random decision among these codewords is made. For a binary symmetric channel, the decoder selects the codeword that has the smallest Hamming distance from the received codeword $\mathbf{r}$. A decoding failure is impossible for a ML decoder. Only correct or false decoding results.

- *Symbolwise maximum a posteriori decoding (s/s-MAP):* Each element c_i of a codeword is examined. The probability that a 1 or a 0 was transmitted is calculated. This results in an independent decoding decision for each element of the received codeword. When all n elements are decoded, the vector $(\hat{c}_0, \hat{c}_1, \ldots, \hat{c}_{n-1})$ – in contrast to ML decoding – may or may not be a codeword. When the decoded word is not a codeword, we consider whether the code is in systematic form. Using systematic coding, we can extract the information bits from the received codeword through knowledge of the location of the elements which contain the information bits. For a systematic encoder, a decoding failure will not occur; however, false decoding may result if an error occurred in an information coordinate during transmission. When a code is not in systematic form, a decoding failure results.

- *Bounded minimum-distance decoding (BMD):* A decoding decision is made only in the case where $\mathbf{r}$ lies within a correction sphere of radius $\left\lfloor \frac{d-1}{2} \right\rfloor$. In this case, all three possible decoding results can occur: correct decoding, false decoding and decoding failure.

- *Decoding beyond half the minimum distance:* A decoding decision can be attempted for the case where $\mathbf{r}$ lies outside a BMD decoding sphere. The decoding method is not ML. In this case, the decoding sphere is made larger than the BMD radius of $\left\lfloor \frac{d-1}{2} \right\rfloor$. The spheres now overlap, but will in turn cover more of the vector space. A decoding failure is still possible. Clearly, all decoding results can occur: correct decoding, false decoding and decoding failure.

[1] For the first chapters of this book, the definition of MAP and ML decoding are sufficient. For an exact mathematical definition, see sections 7.2 and 8.4

Comment The principle of error detection is often used in radio systems to reduce the impact of errors. False information is identified, but not used (hidden). This is termed *error concealment*. Error concealment can be used in combination with other coding techniques, for example, two codes can be used in succession. In the transmitter, the outer algorithm encodes the information for the purpose of error detection and the inner algorithm encodes with the goal of error correction. In the receiver, the first algorithm decodes the inner code of the transmitter, and corrects errors, while the second algorithm, which corresponds to the outer code of the transmitter, detects errors which remain. One also has the flexibility of only using part of a decoder's ability.

BMD decoding means that errors are only corrected when the codeword $\mathbf{c}$ exists such that

$$\mathrm{dist}(\mathbf{c},\mathbf{r}) \leq e = \left\lfloor \frac{d-1}{2} \right\rfloor .$$

This means that, according to the BMD decoding rule, all vectors that do not lie within a correction sphere cannot be decoded. This number corresponds exactly to the difference between the left and the right side of the Hamming bound (theorem 1.9).

Often, many vectors exist that can be uniquely decoded that do not lie within a correction sphere of half the minimum distance. It is often desirable to decode at least a few of these vectors as codewords. Examples of such techniques include the case of a decoding failure for a binary linear code (section 7.3.3) and for generalized concatenated codes (section 9.2.4). Non-ML decoding techniques do not completely exhaust the decoding ability of a particular code.

The decoding of a repetition code in example 1.6 is an ML decoding technique. For any code, an ML decoder is implemented when all codewords are compared with the received word. In general, this can be carried out in the following manner:

Standard array decoding (ML) A coset for a linear block code $C(n,k,d)$ is defined as the addition of any vector $\mathbf{b} \in \mathbb{F}_2^n$ to all codewords of the code C. Therefore

$$[C]_{\mathbf{b}} = \{\mathbf{b}+C\} = \{\mathbf{b}+\mathbf{c},\mathbf{c} \in C\}.$$

Cosets have the following properties:

- Each coset consists of 2^k vectors (which corresponds to the number of codewords in C).
- There exist $b_1, b_2, ..., b_{2^{n-k}-1}$ such that $C \cup \{\mathbf{b}_1 + C\} \cup \{\mathbf{b}_2 + C\} \cup \ldots \cup \{\mathbf{b}_{2^{n-k}-1} + C\}$ includes all of the 2^n vectors in the vector space.
- Two cosets are either equal or disjoint (partial overlapping is impossible).

It is clear that $\mathbf{b}_i$ must not be an element of C in order to generate all cosets of C (excluding $\mathbf{b}_i = 0$). If $\mathbf{b}_i \in C$ then all vectors are in C.

Decoding a received vector using cosets is performed in the following manner. The cosets are ordered such that the vector with the smallest weight is placed in the first column. If two or more vectors within one coset have the same weight, their order of representation is randomly determined. The coset leader is defined as the first vector in a coset. The decoding of a received vector $\mathbf{r} = \mathbf{c} + \mathbf{f}$ consists of finding the coset that contains the vector $\mathbf{r}$. The coset leader of this set corresponds to the error with the smallest weight (and largest probability) that can be added to a codeword to give the received error vector. This method of decoding is called standard array decoding, and is clearly an ML decoding method. Standard array decoding can only be implemented practically for short codes.

Example 1.10 (Standard array) We are given a code with the four codewords

$$(0000) \quad (0011) \quad (1100) \quad (1111)$$

from a (4,2,2) code. The standard array consists of four cosets

$$\begin{array}{llllll} \mathbf{b} = (0000): & \{(0000), & (0011), & (1100), & (1111)\} \\ \mathbf{b} = (1000): & \{(1000), & (1011), & (0100), & (0111)\} \\ \mathbf{b} = (0010): & \{(0010), & (0001), & (1110), & (1101)\} \\ \mathbf{b} = (1001): & \{(1001), & (1010), & (0101), & (0110)\} & . \end{array}$$

The minimal-weight coset leaders used to generate the cosets in this example are not unique. This is expected because the minimum distance of this code is 2 (no error correction). If $\mathbf{b} = (0100)$ is added to C, we obtain the coset $\{(0100)\ (0111)\ (1000)\ (1011)\}$. This generates the same coset as from the vector $\mathbf{b} = (1000)$. ◇

Based on BMD decoding, we can now correct every received vector $\mathbf{r}$ that belongs to a coset having an unique coset leader satisfying $\leq \left\lfloor \frac{d-1}{2} \right\rfloor$. The Hamming bound gives us information on the difference between BMD and ML decoding. In a perfect code, each coset has a coset leader of weight $\leq e$.

1.4 Error probability

Definition 1.14 (Error probability) *The block error probability P_{Block} gives the probability that a transmitted codeword does not correspond to the decoded codeword (false decoding and decoding failure). Similarly, the bit error probability P_{Bit} specifies with what probability a transmitted information bit is incorrect.*

In order to calculate the error probability, three decoding cases must be identified: corrected, uncorrected, and falsely corrected. We have already described that false decoding occurs when $\mathbf{c}$ is transmitted, and $\text{dist}(\mathbf{c}+\mathbf{f},\mathbf{c}) > \text{dist}(\mathbf{c}+\mathbf{f},\mathbf{b})$ where $\mathbf{c},\mathbf{b} \in C$ and $\mathbf{f}$ is the error. This can occur when $\mathbf{f} \in C$, $\mathbf{f} \neq \mathbf{0}$.

We now wish to determine the block error probability for the case of the binary symmetric channel. This will be accomplished using figure 1.2 and the code $C(n,k,d)$ for the three decoding principles (error detection, BMD and ML decoding).

For decoding over more than half the minimum distance, the error probability is dependent on the decoding method. The same is true for s/s-MAP decoding.

For a BSC channel, the probability that exactly t errors occur during the transmission of a binary codeword consisting of n bits is

$$p(t) = p^t(1-p)^{n-t}\ .$$

There exist $\binom{n}{t}$ different vectors of weight t. The probability for any t number of errors in n coordinates is

$$\binom{n}{t} p(t) = \binom{n}{t} p^t(1-p)^{n-t}\ .$$

BMD decoding method If more than e errors occur then

$$P_{Block} = \sum_{j=e+1}^{n} \binom{n}{j} p^j (1-p)^{n-j} ,$$

is the block error probability using the BMD-decoding method. This method can only correct errors with a weight of at most $e = \left\lfloor \frac{d-1}{2} \right\rfloor$.

In practice, the algorithm for the BMD decoding method can in certain cases correct more errors. Therefore P_{Block} is an upper bound on the error probability. An equivalent representation of P_{Block} is

$$P_{Block} = 1 - \sum_{j=0}^{e} \binom{n}{j} p^j (1-p)^{n-j} .$$

Error detection In this case, a decoding error only occurs when the error results in a valid codeword. For the case where W is the weight distribution of the code

$$P_{EBlock} = \sum_{j=1}^{n} w_j p^j (1-p)^{n-j} .$$

For $p = \frac{1}{2}$,

$$P_{EBlock} = \sum_{j=1}^{n} w_j \left(\frac{1}{2}\right)^n = \left(\frac{1}{2}\right)^n \sum_{j=1}^{n} w_j = \left(\frac{1}{2}\right)^n (2^k - 1) \approx \frac{1}{2^{n-k}} .$$

If the vector $\mathbf{f}$ is randomly chosen from the vector space $\mathbb{F}_2$, the probability that this vector is a codeword results in the error probability P_{EBlock}. When $p = \frac{1}{2}$, then $P_{EBlock} \leq \frac{1}{2^{n-k}}$ can be used as a practical upper bound for the block error.

Example 1.11 (Block error probability) A repetition code of length $n = 3$ can correct one error. Using this parameter, the block error probability for a BSC with error probability p is

$$P_{Block} = 1 - \sum_{j=0}^{1} \binom{3}{j} p^j (1-p)^{3-j} = 1 - (1-p)^3 - 3p(1-p)^2 .$$

◇

Example 1.12 (Error detection) The weight components of the parity check codes of the length $n = 3$ are calculated in example 1.4 to be $W = (1,0,3,0)$. The block error detection probability in this case is

$$P_{EBlock} = \sum_{j=1}^{3} w_j p^j (1-p)^{n-j} = 3p^2(1-p) .$$

◇

Maximum-likelihood decoding (ML) P_{MBlock} is calculated by

$$P_{MBlock} = 1 - \sum_{j=0}^{n} \alpha_j p^j (1-p)^{n-j} , \quad \alpha_0 = 1 ,$$

where α_j is the number of the coset leader that corresponds to the weight j. The distribution of α_j can only be determined for short codes. In the case of long codes, no practical implementation of the ML decoding method can be given.

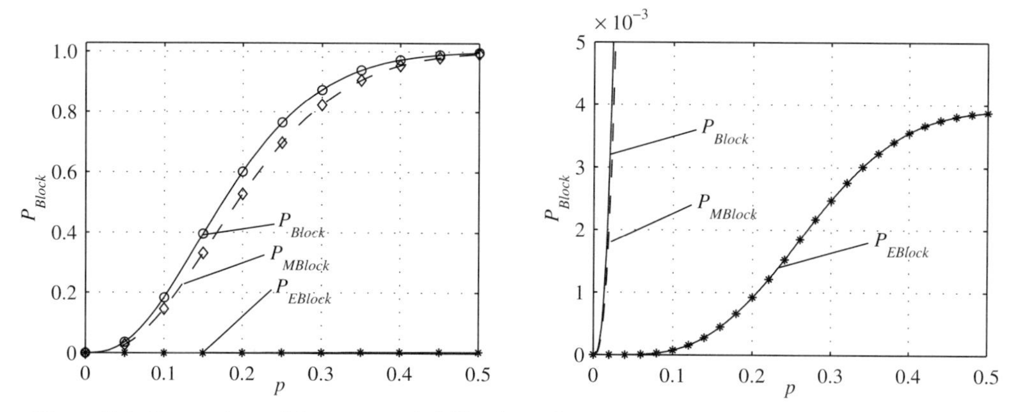

Figure 1.4 Comparison of the error probability through error detection, and BMD and ML decoding.

Example 1.13 (Error probability) For a code with the parameters (15,7,5), we can calculate the error probability and can plot the block decoding performance of different methods as a function of the error probability p. This can be seen in figure 1.4. The weight distribution of the code is

$$A(x) = 1 + 18x^5 + 30x^6 + 15x^7 + 15x^8 + 30x^9 + 18x^{10} + x^{15} .$$

The right-hand side of the graph uses a different scaling of the P_{Block} axis to clarify the progression of the error probability through error detection. ◇

1.5 Hamming codes

Definition 1.15 (Hamming code) *The parity check matrix* $\mathbf{H}$ *of a Hamming code consists of all different* $2^h - 1$ *columns that are from* $\mathbb{F}_2^h$*, excluding the all-zero column.*

Theorem 1.16 (Parameters of the Hamming code) *The parameters of the Hamming code are*

$$\begin{aligned} \textit{Length:} \quad & n = 2^h - 1, \\ \textit{Dimension:} \quad & k = n - h, \\ \textit{Minimum distance:} \quad & d = 3 . \end{aligned}$$

Proof The parameters n and k are apparent. Any two columns are linearly independent, and there are three columns that are linearly dependent. The minimum distance is $d = 3$. □

We now wish to show that Hamming codes are perfect, or in other words satisfy the Hamming bound with equality (see theorem 1.9).

Theorem 1.17 (Hamming codes are perfect) *All single error correcting Hamming codes defined by definition 1.16 are perfect.*

Proof A Hamming code has parameters $n = 2^h - 1,\ k = n - h,\ e = 1 = \left\lfloor \frac{d-1}{2} \right\rfloor$. Substituting into the equation from theorem 1.9,

$$\begin{aligned} 2^k\left(1+\binom{n}{1}\right) &\leq 2^n, \\ 1+n &\leq 2^{n-k}, \\ 1+2^h-1 &= 2^h . \end{aligned}$$

The result is equality; therefore all Hamming codes are perfect. □

Example 1.14 (Hamming code and decoding) We construct the Hamming code for the case where $h = 3$. The parity check matrix has $n = 2^3 - 1 = 7$ columns and $n - k = h = 3$ rows: We write this in such a way that the binary representation of the column number defines the column:

$$\mathbf{H} = \begin{pmatrix} 0 & 0 & 0 & 1 & 1 & 1 & 1 \\ 0 & 1 & 1 & 0 & 0 & 1 & 1 \\ 1 & 0 & 1 & 0 & 1 & 0 & 1 \end{pmatrix}, \qquad \mathbf{H} \cdot \mathbf{c}^T = \mathbf{0}\,,\ \mathbf{c} \in C\,.$$

For the minimum distance to be 3, any two columns must be linearly independent. This condition is satisfied because there are no two columns that are identical. Furthermore, there must be three columns that are linearly dependent. For example column 1 + column 2 + column 3 $= \mathbf{0}$. Therefore $\mathbf{c} = (1110000)$ is a codeword of minimum weight:

$$\mathbf{H} \cdot \mathbf{c}^T = \begin{pmatrix} 0 \\ 0 \\ 1 \end{pmatrix} + \begin{pmatrix} 0 \\ 1 \\ 0 \end{pmatrix} + \begin{pmatrix} 0 \\ 1 \\ 1 \end{pmatrix} = \mathbf{0}\,.$$

We will now consider the case that an error occurs during the transmission of the codeword. The received vector is described by $\mathbf{r} = (1110010) = \mathbf{c} + \mathbf{f}$, where $\mathbf{f} = (0000010)$:

$$\mathbf{H} \cdot \mathbf{r}^T = \begin{pmatrix} 0 \\ 0 \\ 1 \end{pmatrix} + \begin{pmatrix} 0 \\ 1 \\ 0 \end{pmatrix} + \begin{pmatrix} 0 \\ 1 \\ 1 \end{pmatrix} + \begin{pmatrix} 1 \\ 1 \\ 0 \end{pmatrix} = \begin{pmatrix} 1 \\ 1 \\ 0 \end{pmatrix} = \mathbf{s}^T = \mathbf{H} \cdot \mathbf{f}^T$$

(compare with definition 1.13). The decoding result gives the syndrome that is the binary representation of the column where the error occurred. ◇

We shall now transform the Hamming code from example 1.14 into systematic form.

Example 1.15 (Systematic encoded Hamming code) Through the corresponding linear combination of the rows, one can construct a systematic representation of the parity check matrix for the Hamming code with length $n = 7$:

$$\mathbf{H} = \left(\begin{array}{cccc|ccc} 0 & 1 & 1 & 1 & 1 & 0 & 0 \\ 1 & 0 & 1 & 1 & 0 & 1 & 0 \\ 1 & 1 & 0 & 1 & 0 & 0 & 1 \end{array}\right).$$

◇

1.6 Generator matrix

We should now like to consider how to construct the codeword $\mathbf{c} = (c_0, c_1, \ldots, c_{n-1})$ from the information vector $\mathbf{i} = (i_0, i_1, \ldots, i_{k-1})$. Because a linear code is defined in a linear vector space, there exists a basis for this space. The rows of $\mathbf{G}$ are the k linearly independent basis vectors of the vector space $\mathcal{C}$. The matrix $\mathbf{G}$ is termed the generator matrix, and each codeword $\mathbf{c}$ can be constructed by

$$\mathbf{c} = \mathbf{i} \cdot \mathbf{G} .$$

The generator matrix is a $k \times n$ matrix.

For a given generator matrix $\mathbf{G}$, one can construct a generator matrix $\mathbf{G}'$ through the permutation and addition of rows. This new matrix will result in the generation of the same code; however, the mapping of the information sequence to the code sequence will be different as a result of the permutation of rows (see problem 1.14).

For the special case of a systematic representation of the generator matrix, the parity check matrix can be calculated. We are given $c_0 = i_0,\ c_1 = i_1, \ldots, c_{k-1} = i_{k-1}$, and, using this

$$\mathbf{H} \cdot \mathbf{c}^T = (\ \mathbf{A} \mid \mathbf{I}\ ) \cdot \mathbf{c}^T = \mathbf{0},$$

$$\begin{pmatrix} c_k \\ \vdots \\ c_{n-1} \end{pmatrix} = -\mathbf{A} \cdot \begin{pmatrix} c_0 \\ \vdots \\ c_{k-1} \end{pmatrix} = -\mathbf{A} \cdot \begin{pmatrix} i_0 \\ \vdots \\ i_{k-1} \end{pmatrix},$$

$$\mathbf{c}^T = \left(\frac{\mathbf{I}'}{-\mathbf{A}} \right) \cdot \mathbf{i}^T,$$

$$\mathbf{c} = \mathbf{i} \cdot [\ \mathbf{I}' \mid -\mathbf{A}^T\ ] = \mathbf{i} \cdot \mathbf{G} .$$

Here $\mathbf{I}$ is the $(n-k) \times (n-k)$ identity matrix and $\mathbf{I}'$ is the $k \times k$ identity matrix.

Example 1.16 (Generator matrix of the Hamming code) The generator matrix of the Hamming code from example 1.14 can be calculated from the parity check matrix:

$$\mathbf{H} = \left(\begin{array}{cccc|ccc} 0 & 1 & 1 & 1 & 1 & 0 & 0 \\ 1 & 0 & 1 & 1 & 0 & 1 & 0 \\ 1 & 1 & 0 & 1 & 0 & 0 & 1 \end{array} \right) = (\ \mathbf{A} \mid \mathbf{I}\ ),$$

$$\mathbf{G} = (\ \mathbf{I}' \mid -\mathbf{A}^T\ ) = \left(\begin{array}{cccc|ccc} 1 & 0 & 0 & 0 & 0 & 1 & 1 \\ 0 & 1 & 0 & 0 & 1 & 0 & 1 \\ 0 & 0 & 1 & 0 & 1 & 1 & 0 \\ 0 & 0 & 0 & 1 & 1 & 1 & 1 \end{array} \right).$$

◇

In order to obtain the codeword $\mathbf{c} = (1110000)$ from example 1.14, we select the information vector $\mathbf{i} = (1110)$:

$$\mathbf{c} = \mathbf{i} \cdot \mathbf{G} = (1000011) + (0100101) + (0010110) = (1110000) .$$

With the aid of the generator matrix, we can show the following:

Theorem 1.18 (Frequency of 0 and 1 in a linear code) *The frequencies of a one or a zero for each codeword coordinate c_i, $i = 0 \ldots, n-1$, from the set of all codewords of a linear code C are the same.*

Proof Let $\mathbf{i}$ be a binary information vector of length k and let $\mathbf{G}$ be the $k \times n$ generator matrix of the code C with length n. The addition of rows of the generator matrix for the code C does not change the codeword set; it only changes the mapping of information vectors to codewords. By addition of different rows, we can put the generator matrix in a form such that jth column contains a single 1 value at row i, i.e. $g_{ij} = 1, i \in 1 \ldots k$, $g_{m,j} = 0$, $m \in 1 \ldots k, m \neq i$. When the codeword is calculated, the element c_j will be identically 1 if the information vector has a 1 in the coordinate i, i.e. $i_i = 1$. The generation of information bits is defined as random, with an equal likelihood of a 1 or a 0. This results in an equally likely zero or a one occurring in the coordinate c_j. □

1.7 Code, encoding and and equivalent codes

A code is a set of codewords. The encoding is the mapping from the set of information vectors to the set of codewords. In general there exist $2^k!$ different mappings for a binary code. For the first information vector, we may chose one of the 2^k possible codewords, for the second information vector, we may chose one of $2^k - 1$ code vectors, and so on. Of course, in this case, we allow the all zero information vector to be mapped to a non-zero codeword.

Among all possible mappings, there are systematic ones. Clearly, all mappings where the corresponding information vector of a codeword appears unchanged at some k positions of the codeword is called systematic. Any linear block code can be encoded in systematic form.

Given a generator matrix $\mathbf{G}$, a particular mapping (encoding) of information vectors onto codewords is defined. Obviously, if we obtain a generator matrix $\mathbf{G}'$ by performing row operations on $\mathbf{G}$, we change this mapping, but get the same set of codewords, and therefore the same code.

A so-called equivalent code is a code that can be obtained by permuting columns of the generator matrix $\mathbf{G}$. Clearly, permutation of columns generates another set of codewords, and thus another code. However, the properties of the code, namely rate, minimum distance and weight distribution, remain unchanged as compared with the original code.

1.8 Cyclic codes

A cyclic shift of a codeword $\mathbf{c} = (c_0, c_1, \ldots, c_{n-1})$ over i coordinates results in the following vector: $(c_i, c_{i+1}, \ldots, c_{n-1}, c_0, \ldots, c_{i-1})$.

Definition 1.19 (Cyclic code) *A code is cyclic if each cyclic shift of the coordinates in a codeword results in a codeword.*

Cyclic codes can be represented by polynomials, and are interesting for practical purposes. This is because the multiplication and division of polynomials can be realized through the use of a shift register (see problem 3.2). We shall later define cyclic codes in terms of a polynomial representation.

1.9 Dual codes

The generator matrix of a code can be used as the parity check matrix for an associated code and vice versa. These codes are termed dual codes.

Definition 1.20 (Dual code) *Define the code $\mathcal{C}^{\perp}$ as having a generator matrix $\mathbf{G}^{\perp}$ that is also the parity check matrix $\mathbf{H}$ for the code $\mathcal{C}$. This code is termed the dual code $\mathcal{C}^{\perp}$ of the code $\mathcal{C}$.*

The inner product in definition 1.4 of dual codes results in

$$\mathbf{c} \in \mathcal{C},\, \mathbf{b} \in \mathcal{C}^{\perp} \mid \langle \mathbf{c} \cdot \mathbf{b}^T \rangle = 0\,.$$

Because each (transposed) row of the parity check matrix must result in a zero when multiplied by the generator matrix, the same must be true for all linear combinations of the rows.

The length of a dual code $\mathcal{C}^{\perp}$ is equal to the length of the code $\mathcal{C} : n^{\perp} = n$. The dimension is calculated as $k^{\perp} = n - k$. There is no easy method of calculating the minimum distance of $\mathcal{C}^{\perp}$ directly.

Example 1.17 (Dual code) The generator matrix $\mathbf{G}^{\perp}$ and the parity check matrix $\mathbf{H}^{\perp}$ of the dual of the Hamming code from example 1.14 are calculated below:

$$\mathbf{G}^{\perp} = \mathbf{H} = \left(\begin{array}{cccc|ccc} 0 & 1 & 1 & 1 & 1 & 0 & 0 \\ 1 & 0 & 1 & 1 & 0 & 1 & 0 \\ 1 & 1 & 0 & 1 & 0 & 0 & 1 \end{array} \right) = (\,\mathbf{A} \mid \mathbf{I}\,)\,,$$

$$\mathbf{H}^{\perp} = \mathbf{G} = \left(\begin{array}{cccc|ccc} 1 & 0 & 0 & 0 & 0 & 1 & 1 \\ 0 & 1 & 0 & 0 & 1 & 0 & 1 \\ 0 & 0 & 1 & 0 & 1 & 1 & 0 \\ 0 & 0 & 0 & 1 & 1 & 1 & 1 \end{array} \right).$$

The inner product of the first row of $\mathbf{H}$ with the second row of $\mathbf{G}$ results in

$$0 \cdot 0 + 1 \cdot 1 + 1 \cdot 0 + 1 \cdot 0 + 1 \cdot 1 + 0 \cdot 0 + 0 \cdot 1 = 2 = 0 \mod 2.$$

◇

1.10 Shortening and extension of codes

The shortening of codes can be carried out through one of the following two methods:

(i) Instead of an information vector $\mathbf{i}$ of k symbols, we select only the first $k - m$ symbols. Therefore both the dimension and the length of the code are reduced by m, which gives a code rate of $\frac{k-m}{n-m}$. However, the minimum distance of the new code is greater than or equal to that of the original code.

(ii) Certain m coordinates of each codeword can be punctured (the coordinate can be a 0 or a 1). Through the puncturing, the minimum distance may become smaller, and the code rate is $\frac{k}{n-m}$.

Both techniques of code shortening result in a linear code. For cyclic codes, the shortening method is defined in section 4.3. The puncturing technique is primarily used in the case of convolutional codes to create a desired code rate (see section 8.1.10).

Example 1.18 (Shortening of codes) We use the generator matrix $\mathbf{G}$ from example 1.16. In order to shorten the code according to (i), we set the first two information columns to zero and then delete these elements from the codeword.

Using the codeword from example 1.14, the following result is obtained for the information vector $\mathbf{i} = (0010)$:

$$\mathbf{c} = \mathbf{i} \cdot \mathbf{G} = (0010110) \Longrightarrow (10110) .$$

This gives a codeword of length $n = 5$.

Through puncturing according to (ii), we discard, for example, the third element of all codewords:

$$(0010110) \Longrightarrow (000110) . \qquad \diamond$$

For code *extension*, an element (either 0 or 1) is added to a code to give an even Hamming weight. This technique is only meaningful in terms of codes that have odd minimum weight. The minimum distance increases by a value of 1.

Theorem 1.21 (Lengthening by adding an overall parity check) *Let the code $C(n,k,d)$ (d odd) have an $(n-k) \times n$ parity check matrix $\mathbf{H}$. We can construct an $(n-k+1) \times (n+1)$ parity check matrix through code extension by adding a row of ones, and a column in the form of $(100\ldots0)^T$. The code is extended, giving the parity check matrix $\mathbf{H}^{ex}$ with the parameters $(n+1,k,d+1)$:*

$$\mathbf{H}^{ex} = \begin{pmatrix} 1 & 1 & \ldots & 1 & 1 \\ & & & & 0 \\ & & \mathbf{H} & & \vdots \\ & & & & 0 \end{pmatrix} .$$

Proof The length and the dimension are defined by the matrix parameters. Because the code is linear, the minimum distance equals the minimum weight. Take $\mathbf{c} \in C$ with $\text{wt}(\mathbf{c}) = d$. For the extended codeword $\mathbf{c}^{ex}$, the weight becomes $\text{wt}(\mathbf{c}^{ex}) = d+1$. $\square$

Example 1.19 (Code lengthening by adding an overall parity check) A one will be appended to the codeword from example 1.18:

$$(0010110) \Longrightarrow (00101101) .$$

The corresponding parity check matrix that results is

$$\mathbf{H} = \left(\begin{array}{cccc|cccc} 1 & 1 & 1 & 1 & 1 & 1 & 1 & 1 \\ 0 & 1 & 1 & 1 & 1 & 0 & 0 & 0 \\ 1 & 0 & 1 & 1 & 0 & 1 & 0 & 0 \\ 1 & 1 & 0 & 1 & 0 & 0 & 1 & 0 \end{array}\right) . \qquad \diamond$$

If a parity check coordinate is added to change an even-weight code to an odd-weight code, the weight is increased but the distance remains the same; this produces a nonlinear code:

$$\mathbf{0} \longrightarrow 00\ldots01, \quad (\mathbf{a}+\mathbf{a}) = \mathbf{0} \notin C, \quad \mathbf{a} \in C .$$

1.11 Channel capacity and the coding theorem

The channel capacity defines how much information can be sent over a given channel. This phenomenon was investigated in 1948 by C. E. Shannon in his groundbreaking work in information theory. Shannon's channel coding theorem proves that with channel coding, the bit error probability can be made arbitrarily small if the code rate is smaller than the channel capacity. This means that through code design, almost error-free information transmission can be achieved. Unfortunately, the proof is based on probability theory, and does not result in a method for the construction of the corresponding code.

The goal of channel coding is to find a code that is easy to encode and decode, and at the same time gives a high code rate for the largest minimum distance. We have already defined an upper bound, (the Hamming bound, see theorem 1.9), and other coding bounds will be defined in sections 6.4 and 6.5.

In order to explain the theoretical basis for channel capacity and coding, we shall present some of the basic concepts of information theory without the use of elaborate derivations. For an in depth proof, a variety of textbooks may be referenced, for example [PeWe] or [WoJa].

Self-information The self-information I of a number x which has a probability $p(x)$ is defined as

$$I := \mathrm{ld}\,\frac{1}{p(x)},$$

where ld is the logarithm to base 2. The self-information measures the amount of information contained in a number. A *non-random* number, where $p(x) = 1$, has the self-information value $I = 0$. It can be seen that the more improbable a number is, the greater the self-information that it contains: $I \longrightarrow \infty$ for $p(x) \longrightarrow 0$.

Entropy Let X be a set of symbols $x_i, i = 1, \ldots, n$, with a probability distribution $p(x_i)$, where $\sum_{i=1}^{n} p(x_i) = 1$. The entropy $H(X)$ of the set X is defined as the average of the self-information:

$$H(X) := \sum p(x_i)\,\mathrm{ld}\,\frac{1}{p(x_i)}.$$

For equally probable binary symbols $p(0) = \frac{1}{2}$, $p(1) = \frac{1}{2}$, the entropy is equal to $H = \frac{1}{2}\ \mathrm{ld}\ 2 + \frac{1}{2}\ \mathrm{ld}\ 2 = 1$. In general, the entropy is maximized when symbols occur with equal probability.

The conditional probability $P(y_j|x_i)$ represents the probability of the event y_j occurring if the event x_i is known to have occurred. The joint probability $P(x_i, y_j)$ is a measure of the likelihood that x_i and y_i occur at the same time. With these measures, one can define the joint entropy and the conditional entropy:

$$H(X,Y) = \sum_i \sum_j P(x_i, y_j)\,\mathrm{ld}\,\frac{1}{P(x_i, y_j)}\,,$$

$$H(Y|X) = \sum_i \sum_j P(x_i, y_j)\,\mathrm{ld}\,\frac{1}{P(y_j|x_i)}\,,$$

$$H(X|Y) = \sum_i \sum_j P(x_i, y_j)\,\mathrm{ld}\,\frac{1}{P(x_i|y_j)}\,,$$

$$H(X,Y) = H(Y) + H(X|Y) = H(X) + H(Y|X)\,.$$

For the statistically independent case ($P(x_i,y_j) = p(x_i) \cdot p(y_j)$) we have

$$H(X,Y) = H(X) + H(Y), \quad H(X|Y) = H(X), \quad H(Y|X) = H(Y) .$$

The **mutual information** $I(X;Y)$ is defined as

$$I(X;Y) = H(X) - H(X|Y) = H(Y) - H(Y|X)$$

Definition 1.22 (Channel capacity) *The channel capacity is defined as the maximum value for the mutual information I(X;Y):*

$$C = \max_{p(x_i)} \{H(X) - H(X|Y)\} = \max_{p(x_i)} \{H(Y) - H(Y|X)\} .$$

In the following example, we calculate the channel capacity of a binary symmetric channel.

Example 1.20 (Channel capacity for a BSC) For a binary symmetric channel with the error probability p, we have

$$\begin{aligned} P(1|1) &= P(0|0) = 1-p, & P(1,1) &= P(0,0) = \tfrac{1}{2}(1-p), \\ P(1|0) &= P(0|1) = p, & P(0,1) &= P(1,0) = \tfrac{1}{2}p . \end{aligned}$$

This results in ($\max_{p(x_i)} \{H(Y)\} = 1$ when every element x_i is equally probable):

$$H(Y|X) = p \operatorname{ld} \frac{1}{p} + (1-p) \operatorname{ld} \frac{1}{1-p} .$$

The channel capacity for C depends only on the channel, and is shown for the BSC to be

$$C(p) = 1 - H(p) = 1 - p \operatorname{ld} \frac{1}{p} - (1-p) \operatorname{ld} \frac{1}{1-p} .$$

In figure 1.5, the channel capacity C for a binary symmetric channel is shown as a function of the error probability p. ◇

The channel coding theorem states that the block error probability through the selection of a particular code can be made arbitrarily small. Hence, also the bit error probability can be made arbitrarily small. Of course, the channel capacity is a theoretical bound, which in practice can only be achieved at great expense.

Theorem 1.23 (Channel coding theorem) *For every real number $\varepsilon > 0$ and every code rate R smaller than the channel capacity ($R < C$), there exists a code C of length n (with rate $R = \frac{k}{n}$), for an n sufficiently large, such that the block error probability after decoding is smaller than ε.*

Instead of a full proof, we present a principle for understanding the theorem that is valid for small error probabilities. Choose a code with length n that is sufficiently large, for example large enough that the errors in a codeword can almost always be corrected given that the expected value of errors is smaller than the number of correctable errors. If it is possible to bound these numbers using probability calculations, the channel coding theorem can be proved.

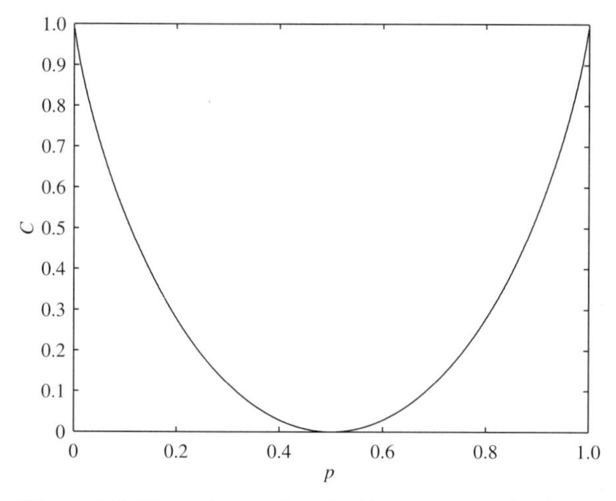

Figure 1.5 Channel capacity of a binary symmetric channel.

1.12 Summary

The basic theory for block codes was developed by M. J. E. Golay [Gol49] and R. W. Hamming [Ham50]. Shannon knew of the existence of Hamming codes before the publication of his fundamental work on information theory (1948). By 1949, Golay had designed the only non-trivial binary, perfect, multiple error correcting code that can exist based on combinatorial considerations. Work on the group structure of codes was pioneered by D. Slepian, and we shall use his notation in the following chapters.

In this first chapter, we have introduced a few of the basic concepts of linear block codes and their properties The error correction capability of a code depends on the minimum distance, which for a linear block code is equivalent to the minimum weight. A result of this is that for linear block codes, the *distance distribution* is equal to the *weight distribution*. The channel model we have used is the basic binary symmetric channel (BSC). One central problem for channel coding is the calculation of the various parameters (length, dimension and minimum distance) of a code. The significance of the Hamming bound has also been presented.

The description of a linear block code through the parity check matrix **H** and the generator matrix **G** has been given. We have emphasized the difference between code and encoding (the mapping of information vectors to codewords). A systematic encoder can be found for any linear block code. Two codes are equivalent if they are constructed through the permutation of columns of the generator matrix. The parity check (PC) code and the repetition code (RP) has been used as examples to demonstrate the function of the parity check and generator matrix. We have defined cyclic codes, and outlined the shortening and extension of codes. The parity check matrix multiplied by the received vector results in the syndrome. The syndrome only produces a zero result for the case when the transmission error is equal to 0 or another codeword.

Many different decoding principles can be selected: error detection, maximum-likelihood decoding (ML), bounded minimum-distance decoding (BMD), symbol-by-symbol MAP decoding (s/s-MAP) and decoding over more than half the minimum distance. In the case of a BSC, the resulting block error probability can be calculated. In the general case, only the block error probability for the bounded minimum distance method can be calculated. This is the case because in general, the weight distribution of a code is not known. The standard array decoding is an example of a ML decoding technique, which cannot be practically implemented for *long codes*.

We have defined the information theory concept of entropy and channel capacity. We have also mentioned the channel coding theorem, which states that the error probability can be made arbitrarily small after decoding, for the case when the code rate is chosen to be smaller than the channel capacity.

Further information concerning the description of the elementary principles behind coding theory can be found in the introductory chapters from [McWSl], [LCF] and [Bla], as well as in [ClCa] and [Gal].

The description of codes through the parity check and the generator matrix has practical drawbacks. In the next chapter, we shall provide the necessary mathematical basis to describe codes using polynomials over a finite field.

1.13 Problems

Problem 1.1
In the usual representation of integers to base 10, a number consists of digits $(0\ldots 9)$. Show that the sum of digits of an integer modulo 9 is equal to the complete number modulo 9. Show first that:

$$(a+b) \mod 9 = ((a \mod 9) + (b \mod 9)) \mod 9$$

and

$$(a \cdot b) \mod 9 = ((a \mod 9) \cdot (b \mod 9)) \mod 9.$$

Problem 1.2
You are given the parity check matrix of a binary linear block code C with length $n = 7$:

$$\mathbf{H} = \begin{pmatrix} 0 & 0 & 0 & 1 & 1 & 1 & 1 \\ 0 & 1 & 1 & 0 & 0 & 1 & 1 \\ 1 & 0 & 1 & 0 & 1 & 0 & 1 \end{pmatrix}$$

(a) Is $\mathbf{c}_1 \in \mathbb{F}_2^7, \mathbf{c}_1 = (0,1,0,1,0,1,0)$ a codeword of the code C?

(b) Determine the parameters of the code, and the code rate R.

(c) Take $\mathbf{c}_2 = (1,1,0,1,1,0,1) \in \mathbb{F}_2^7$. Give three possible error vectors $\mathbf{f}$ such that the addition $\mathbf{c}_2 + \mathbf{f}$ is a codeword from C.

(d) Decode $\mathbf{c}_3 = (0,1,1,1,0,1,0)$ with respect to the code C.

Problem 1.3

(a) How many different vectors $\mathbf{h}$ belong to $\mathbb{F}_2^m$?

(b) Using all vectors from $\mathbb{F}_2^m$ that are not equal to the $\mathbf{0}$, a matrix can be constructed in which the vectors $\mathbf{h}^T$ are the column vectors. This matrix is the parity check matrix of a Hamming code. Determine the length n, dimension k and code rate R of a Hamming code that depends on m.

(c) Carry out the calculation for $m = 4$. How many errors can this code correct?

Problem 1.4

(a) Calculate the probability $p(e)$ that the transmission of n bits over a binary symmetric channel (BSC) with error probability p gives exactly e false symbols.

(b) Which condition must n, e and p ($e < \frac{n}{2}$) satisfy to give

$$P(e+1) < P(e) \quad ?$$

Determine the validity of this condition for the parameters $n = 7$, $e = 1$ and $p = 10^{-1}$.

Problem 1.5
Calculate a systematic form of the parity check matrix **H** and the corresponding generator matrix **G** of a Hamming code with length $n = 15$ in such a way that the first 11 coordinates in each codeword coincide with the 11 information bits.

Problem 1.6
A systematic binary channel has error probability p. Calculate the expected value $E(n)$ for the number of errors in the block consisting of n binary coordinates.

Hint: Use the binomial theorem

$$(x+y)^n = \sum_{i=0}^{n} \binom{n}{i} x^i y^{n-i}.$$

Problem 1.7

(a) A binary code C has the parameters $n = 15$, $k = 7$, $d = 5$. Is the Hamming bound satisfied? What is the difference between the left and right side of the Hamming bound?

(b) Can a code with the parameters $n = 15$, $k = 7$ and $d = 7$ exist?

(c) Determine whether a code with the parameters $n = 23$, $k = 12$ and $d = 7$ can exist.

Problem 1.8
Calculate the standard array for a $(4,1,4)$ code.

Problem 1.9
Consider the code with the four codewords

$$(0000)\ (0011)\ (1100)\ (1111).$$

We receive (1010), which is transmitted over a time-varying channel. The first and fourth symbols are transmitted over a channel with error probability $p = 0.3$, and the second and third symbols are transmitted over a channel with the error probability $p = 0.1$. What is the decision of a MAP decoder?

Problem 1.10
Calculate the block error probability of a $(23,12,7)$ code using the BMD decoding method over a binary symmetric channel with error probabilities $p = 0.05$, $p = 0.02$, $p = 0.01$ and $p = 0.005$.

Problem 1.11
Calculate the error probability when a $(7,4,3)$ Hamming code is used for the purpose of error detection over a binary symmetric channel as a function of the error probability p.

Problem 1.12
Consider the parity check matrix of a $(7,4,3)$ Hamming code:

$$\mathbf{H} = \begin{pmatrix} 0 & 1 & 1 & 1 & 1 & 0 & 0 \\ 1 & 0 & 1 & 1 & 0 & 1 & 0 \\ 1 & 1 & 0 & 1 & 0 & 0 & 1 \end{pmatrix}.$$

(a) Show that all elements of the coset of any vector **b** result in the same syndrome.

(b) Create a table of all correctable error patterns and the corresponding syndrome.

(c) Use the coset table to explain why only errors with a weight of one can be corrected, and why the code can detect errors of weight 2.

Problem 1.13
You are given the codewords (1111), (0011) and (0101) of a linear cyclic code. Determine all the missing codewords and give the code parameters.

Problem 1.14
The mapping from information bits to codewords can be carried out through different methods. Using the $(7,4,3)$ Hamming code:

(a) Give a generator matrix of the systematic code.

(b) Modify the generator matrix in such a way that the information vector (1010) results in the codeword (0101011).

(c) How many different possibilities exist for the mapping of the 16 information vectors to codewords?

(d) The exchange of columns of a code is called a permutation, and gives an equivalent code. How many permutations are possible? How many possibilities exist where exactly two columns are exchanged?

2

Galois fields

'God created integers, everything else is the work of man.' Leopold Kronecker

Rings and fields are sets of elements with two operations, addition and multiplication, that satisfy a number of specific axioms. Rational, real and complex numbers are well-known examples of infinite fields. For coding applications, a finite field is needed. Such a field is called a Galois field. In this chapter, we shall proceed by defining a ring and then by defining the rules for the construction of a Galois field. Three representations of a Galois field will be presented.

2.1 Groups

Group A non-empty set $\mathcal{A}$ of elements is called a group under the operation $*$ if the following axioms are satisfied:

I. Closure: $\forall a, b \in \mathcal{A}: \ a * b \in \mathcal{A}$.
II. Associativity: $\forall a, b, c \in \mathcal{A}: \ a * (b * c) = (a * b) * c$.
III. The existence of a neutral element e: $\exists e \in \mathcal{A} : \forall a \in \mathcal{A}: \ a * e = a$.
IV. Inverse element: $\forall a \in \mathcal{A} : \exists b = a^{-1} \in \mathcal{A}: \ a * b = e$.

Commutative, or Abelian, groups satisfy the additional axiom

V. Commutativity: $\forall a, b \in \mathcal{A} : a * b = b * a$.

Example 2.1 (Group) The finite set of integers $\{0, 1, 2, 3\}$ is a group under modulo 4 addition. I: $(a+b) \mod 4 \in \{0, 1, 2, 3\}$, II: $(a+b)+c = a+(b+c) \mod 4$, III: $a+0 = a$, IV: $a+(4-a) = 4 = 0 \mod 4$. ◇

2.2 Rings, fields

Ring A *ring* $\mathcal{R}$ is defined as a set $\mathcal{A}$ with two operations (addition [+] and multiplication [·]) if the following axioms are satisfied:

I. $\mathcal{A}$ is an Abelian group under addition.
II. Closure: under multiplication: $\forall a,b \in \mathcal{A}: \ a \cdot b \in \mathcal{A}$.
III. Associativity: $\forall a,b,c \in \mathcal{A}: \ a \cdot (b \cdot c) = (a \cdot b) \cdot c$.
IV. Distributivity: $\forall a,b,c \in \mathcal{A}: \ a \cdot (b+c) = a \cdot b + a \cdot c$.

Note [+] and [·] have the typical meaning for addition and multiplication respectively. What is important is the connection of these operations through the distributive property.

Definition 2.1 (Integer ring $\mathbb{Z}_m$) *The computation modulo m (definition 1.1) satisfies the axioms of a ring and is called the ring of residue classes* $\mathbb{Z}_m = \{[0]_m, [1]_m, \dots, [m-1]_m\}$, *where* $[0]_m = 0, m, 2m, 3m, \dots$. *The set of elements* $\{0, \dots, m-1\}$ *are the representative elements of the residue classes.* $\mathbb{Z}_m$ *is also called an integer ring.*

The elements from $\mathbb{Z}_m$ do not necessarily have an inverse element under the operation of multiplication. The property

$$a \in \mathbb{Z}_m \quad \Longrightarrow \quad \exists a^{-1} \in \mathbb{Z}_m, \quad \text{with} \quad a^{-1} \cdot a = 1 \mod m,$$

need not be satisfied for all $a \in \mathbb{Z}_m$.

Theorem 2.2 (Invertible elements) *An element* $a \in \mathbb{Z}_m$ *is invertible if and only if* $\gcd(a,m) = 1$. *The set of invertible elements in* $\mathbb{Z}_m$ *is an Abelian group under multiplication.*

Proof See the proof of theorem 2.12 on page 30. □

Definition 2.3 (Euler $\Phi(m)$ function) *For* $m \in \mathbb{N}$, *the* Euler $\Phi(m)$ *function is defined for* $m \in \mathbb{N}$ *as the number of elements i,* $1 \le i < m$, *such that* $\gcd(i,m) = 1$:[1]

$$\Phi(m) = |\{i \mid \gcd(i,m) = 1\}|, \qquad 1 \le i < m.$$

By definition, $\Phi(1) = 1$.

The Euler $\Phi(m)$ function also represents the exact number of invertible elements that are contained in the ring $\mathbb{Z}_m$.

Theorem 2.4 (Euler $\Phi(m)$ function of a prime number) *For every prime integer p,* $\Phi(p) = p-1$ *(definition 2.3).*

Proof $\gcd(1,p) = 1$. Furthermore, a prime number is not divisible by any integer $1 < i < p$; therefore

$$\gcd(i,p) = 1 \quad \text{for all } 1 \le i < p\,.$$ □

Theorem 2.5 (Euler/Fermat theorem) *(see [HaWr]) Let* $m \in \mathbb{N}$, $a \in \mathbb{Z}_m$ *and* $\gcd(a,m) = 1$. *Then*

$$a^{\Phi(m)} = 1 \mod m.$$

[1] $|\cdot|$ is defined as the cardinality of the set.

Example 2.2 (Integer ring) For $\mathbb{Z}_6$

$$
\begin{array}{lllll}
1\cdot 1=1 \mod 6, & & & & \\
1\cdot 2=2 \mod 6, & 2\cdot 2=4 \mod 6, & & & \\
1\cdot 3=3 \mod 6, & 2\cdot 3=0 \mod 6, & 3\cdot 3=3 \mod 6, & & \\
1\cdot 4=4 \mod 6, & 2\cdot 4=2 \mod 6, & 3\cdot 4=0 \mod 6, & 4\cdot 4=4 \mod 6, & \\
1\cdot 5=5 \mod 6, & 2\cdot 5=4 \mod 6, & 3\cdot 5=3 \mod 6, & 4\cdot 5=2 \mod 6, & 5\cdot 5=1 \mod 6.
\end{array}
$$

The elements 1 and 5 are invertible, and the elements $2,3,4$ are not invertible (0 is not invertible):

$$\Phi(6)=2, \quad 5^2=1 \mod 6, \quad 1^2=1 \mod 6.$$

◇

Field A set $\mathcal{A}$ with two operations $(+,\cdot)$ is called a *field*, if the following axioms are satisfied:

I. $\mathcal{A}$ is an Abelian group under addition.
II. $\mathcal{A}$ (without the null element) is an Abelian group under multiplication.
III. Distributive law: $\forall a,b,c \in \mathcal{A} : a(b+c) = a\cdot b + a\cdot c$.

2.3 Prime fields

Definition 2.6 (Galois fields and prime fields) *A Galois field is defined as any finite set satisfying the axioms of a field, and is denoted by $GF(q)$, where $q \in \mathbb{N}$. A prime field $GF(p)$ has the additional condition that $p \in \mathbb{N}$ is prime. The set of integers $(0,\ldots,p-1)$ satisfies the axioms of a field under the operations $(+,\cdot)$ mod p.*

Because p is prime, $\gcd(a,p)=1 \ \forall a \in \mathbb{Z}_p \backslash \{0\}$. According to theorem 2.2, $a \in \mathbb{Z}_p$ is invertible if $\gcd(a,p)=1$. Therefore every non-zero element in the set $GF(p)$ is invertible, which gives an Abelian group under the operation of multiplication. Definition 1.1 shows the calculation of a number using the modulo operation with respect to the prime integer p. We shall now construct the prime field $GF(5)$:

Example 2.3 (Prime field, $p=5$) The prime field $GF(5)$ consists of five elements, $\mathcal{A} = \{0,1,2,3,4\}$. The addition and multiplication tables are presented below. For example $4+3 = 7 = 5+2 = 2 \bmod 5$, $2\cdot 4 = 8 = 5+3 = 3 \bmod 5$.

+	0	1	2	3	4
0	0	1	2	3	4
1	1	2	3	4	0
2	2	3	4	0	1
3	3	4	0	1	2
4	4	0	1	2	3

·	0	1	2	3	4
0	0	0	0	0	0
1	0	1	2	3	4
2	0	2	4	1	3
3	0	3	1	4	2
4	0	4	3	2	1

The set must be an Abelian group with respect to addition. One can see from the above table that the condition of closure is satisfied. The commutativity property results from the symmetry of the calculations about the diagonal. The associativity property is satisfied because $a+(b+c)=(a+b)+c$ is true for all numbers in the set. The inverse element for a is $p-a=5-a$, because $a+p-a=p=0 \bmod p$, where 0 is the neutral element under addition.

The set must be an Abelian group with respect to multiplication (excluding the 0 element).

The commutativity and closure properties can be read from the multiplication table results. The associativity property is fulfilled with respect to multiplication because $a \cdot (b \cdot c) = (a \cdot b) \cdot c$ is true for all numbers in the set. The neutral element is $e = 1$. The inverse element for 2 is 3 because $2 \cdot 3 = 1 \mod 5$, and vice versa due to the commutativity property. The inverse element for 4 is 4, because $4 \cdot 4 = 1 \mod 5$; therefore 4 is self-inverse.

The last axiom requires that the distributive property be valid. This is an integer property. ◇

Based on these results, $GF(5)$ satisfies all requirements of a field. The prime field $GF(2)$ has already been used in the first chapter.

2.3.1 Primitive elements

Definition 2.7 (Primitive element) *The multiplicative group of a prime field $GF(p)$ is a cyclic group. This means that an element α exists such that any non-zero element of the field can be represented as some power of α.*

Because a Galois field contains only a finite number of elements $a \in GF(p)$, the elements $a^i, i = 1, 2, 3, \ldots$, modulo p must be repeated. Therefore there must exist some integer i_r such that i_r: $a^{i_r} = a \mod p$. For all elements $\beta \in GF(p)$, $\beta^p = \beta \mod p$ and $\beta \neq 0$, $\beta^{p-1} = 1 \mod p$ according to theorem 2.5 (where $\Phi(p) = p-1$). For a primitive element $\alpha \in GF(p)$, the smallest number $0 < n < p-1$ that gives $\alpha^n = 1$ is $p-1$. This implies that all $p-1$ non-zero elements of $GF(p)$ can be generated from the element α. In the case where $a^{i_r} = a \mod p$, we have $a^{i_r+j} = a^{1+j} \mod p$.

Definition 2.8 (The order of an element) *The order of an element $a \in GF(p)$, $a \neq 0$, is the smallest exponent $r > 0$, such that $a^r = 1 \mod p$. An element is primitive when $r = p-1$.*

In a ring, it should be noted that there may not exist an exponent $r > 0$ for which $a^r = 1$ for an element a. In this case, the order of the element is not defined.

Example 2.4 (The order of an element) The order of the elements for the field $\mathbb{Z}_5$ are calculated:

$$1^1 = 1 \mod 5,\ 2^4 = 1 \mod 5,\ 3^4 = 1 \mod 5,\ 4^2 = 1 \mod 5\ .$$

The element 1 has order 1, the element 4 has order 2, and the elements 2 and 3 have order $4 = 5-1$. Therefore 2 and 3 are primitive elements. ◇

Theorem 2.9 (The existence of a primitive element) *(see [HaWr]) Any Galois field contains at least one primitive element.*

If α is a primitive element from $GF(p)$ then all numbers α^j with $\gcd(j, p-1) = 1$ are also primitive elements of $GF(p)$. There are $\Phi(p-1)$ primitive elements.

Theorem 2.10 (Orders must be divisors of $p-1$) *The order of the elements of a Galois field $GF(p)$ must be a factor of $p-1$.*

Proof An element b has order r; therefore $b^r = 1 \bmod p$. Any element b can be written as a power of the primitive element α^k. We consider an integer $r \not| (p-1)$ ($a \not| b$ means a does not divide b):

$$p-1 = jr+i, \quad \text{remainder } 0 < i < r.$$

Using the Euler/Fermat theorem (theorem 2.5),

$$b^{p-1} = 1 = b^{jr+i} = (b^r)^j b^i = 1b^i = b^i \, .$$

But according to definition 2.8, r is the smallest number for which $b^r = 1$. Therefore r must divide $p-1$. □

Example 2.5 (Primitive element) The element 2 is a primitive element of $GF(5)$:

$$2^1 = 2, \ 2^2 = 4, \ 2^3 = 8 = 3, \ 2^4 = 16 = 1 \quad \bmod 5.$$

The element 3 is likewise a primitive element:

$$3^1 = 3, \ 3^2 = 9 = 4, \ 3^3 = 27 = 2, \ 3^4 = 81 = 1 \quad \bmod 5.$$

The element 4 has order 2 (2 is a factor of $4 = p-1$):

$$4^1 = 4, \ 4^2 = 16 = 1 \quad \bmod 5.$$

All of the elements from $GF(5)$ can be generated by raising either 2 or 3 to some integer power. The null element is generated by the exponent $-\infty$. When two elements are multiplied, the two exponents are added modulo $p-1$.

$$2 \cdot 4 = 8 = 3 = 2^1 \cdot 2^2 = 2^3 = 3 \quad \bmod 5 \, .$$

We have two possible ways to represent the elements in a prime field: either as a primitive element raised to an exponent or as an integer; these are called the exponential and component representations respectively. ◇

2.3.2 Euclidean algorithm

The Euclidean algorithm is an important algorithm both for natural numbers and numbers in a Galois field:

Theorem 2.11 (Euclidean algorithm) *The Euclidean algorithm finds the greatest common divisor* $\gcd(a,b)$ *of two integers* $a, b \neq 0$*, where* $a < b$*. This is achieved through integer division, which iteratively calculates the remainder* $r > 0$*, as follows (the remainder is always non-negative and smaller than the divisor):*

$$\begin{aligned} b &= q_1 a + r_1 \\ a &= q_2 r_1 + r_2 && \text{with } a = r_0, \ b = r_{-1}: \\ r_1 &= q_3 r_2 + r_3 && r_{j-1} = q_{j+1} \cdot r_j + r_{j+1}, \ j = 0, \ldots, l+1 \\ &\vdots \\ r_l &= q_{l+2} r_{l+1} + 0 \end{aligned}$$

$$\gcd(a,b) = r_{l+1}, \qquad \text{where } 0 < r_{j+1} < r_j \, .$$

Proof r_{l+1} divides r_l, and therefore divides r_{l-1} ($r_{l-1} = q_{l+1}r_l + r_{l+1}$) and $r_{l-2},\dots,$ which includes the integers a and b. On the other hand, if $t \in \mathbb{N}$, $t > r_{l+1}$ divides a and b , it could also be factored by $r_1, r_2, \dots, r_{l+1}$. Based on this fact, r_{l+1} is the gcd. □

Theorem 2.12 (Euclidean algorithm) *For any integer numbers $a, b \neq 0$, there exist integers v_j and w_j such that every remainder r_j can be decomposed as*

$$r_j = aw_j + bv_j, \quad j = -1, 0, 1, \dots l+2.$$

This gives the following equation: $\gcd(a,b) = aw_{l+1} + bv_{l+1}$.

Proof The coefficients v_j and w_j are calculated recursively by starting with the integers $b, a, r_1, \dots$:

$$\begin{array}{rcrcrl} b & = & v_{-1}b & + & w_{-1}a, & \\ a & = & v_0 b & + & w_0 a, & \\ r_1 & = & v_1 b & + & w_1 a, & \\ & \vdots & & & & \\ r_j & = & v_j b & + & w_j a, & \\ & \vdots & & & & \\ r_{l+2} & = & v_{l+2}b & + & w_{l+2}a & = 0\,. \end{array}$$

The recursive formula for determining v_j and w_j is

$$\begin{array}{rcl} v_{-1} & = & 1 \\ v_0 & = & 0 \\ v_1 & = & v_{-1} - q_1 v_0 \\ & \vdots & \\ v_j & = & v_{j-2} - q_j v_{j-1} \end{array} \qquad \begin{array}{rcl} w_{-1} & = & 0, \\ w_0 & = & 1, \\ w_1 & = & w_{-1} - q_1 w_0, \\ & \vdots & \\ w_j & = & w_{j-2} - q_j w_{j-1}\,. \end{array}$$

□

Now we can prove the statement of theorem 2.2. If $\gcd(a,m) = 1$ for $a \in \mathbb{Z}_m$ then, using the above decomposition of the remainder in terms of the integers v and w, an inverse element can be shown to exist such that

$$va + wm = 1 \quad \Longrightarrow \quad va = 1 \mod m;$$

therefore $a^{-1} = v \bmod m$.

Example 2.6 (Euclidean algorithm) We shall calculate $\gcd(18,30)$:

$$\begin{array}{lll} 30 : 18 = 1 & \text{remainder } 12, & 30 = 1 \cdot 18 + 12, \\ 18 : 12 = 1 & \text{remainder } 6, & 18 = 1 \cdot 12 + 6, \\ 12 : \ 6 = 2 & \text{remainder } 0, & 12 = 2 \cdot \ 6 + 0, \end{array}$$

$$\gcd(18,30) = 6,$$
$$6 = 1 \cdot 18 - 1 \cdot 12,$$
$$12 = 1 \cdot 30 - 1 \cdot 18,$$
$$6 = 1 \cdot 18 - 1 \cdot (1 \cdot 30 - 1 \cdot 18) = 2 \cdot 18 - 1 \cdot 30\,.$$

◇

Example 2.7 (Euclidean algorithm) We shall calculate $\gcd(24,42)$:

$$\begin{array}{lll} 42:24=1 & \text{remainder } 18, & 42=1\cdot 24+18, \\ 24:18=1 & \text{remainder } 6, & 24=1\cdot 18+6, \\ 18:\ 6\ =3 & \text{remainder } 0, & 18=3\cdot\ 6\ +0, \end{array}$$

$$\gcd(24,42)=6\ .$$

Now we shall generate the numbers 42 and 24 through the remainders 18, 6 and 0. In order to calculate the values v_i and w_i,

$$\begin{array}{ll} v_{-1}=\ 1, & w_{-1}=\ 0, \\ v_0=\ 0, & w_0=\ 1, \\ v_1=v_{-1}-q_1v_0=\ 1, & w_1=w_{-1}-q_1w_0=-1, \\ v_2=\ v_0-q_2v_1=-1, & w_2=\ w_0-q_2w_1=\ 2, \\ v_3=\ v_1-q_3v_2=\ 4, & w_3=\ w_1-q_3w_2=-7. \end{array}$$

The result is

$$\begin{array}{rcrcl} 42 & = & 1\cdot 42 & - & 0\cdot 24, \\ 24 & = & 0\cdot 42 & + & 1\cdot 24, \\ 18 & = & 1\cdot 42 & - & 1\cdot 24, \\ 6 & = & -1\cdot 42 & + & 2\cdot 24, \\ 0 & = & 4\cdot 42 & - & 7\cdot 24\ . \end{array}$$

◇

Properties of the gcd

This subsection presents some of the properties of the gcd that are useful for calculation and analysis:

(1) $\gcd(0,b)=b,\ b>0$;

(2) $\gcd(a,b)=\gcd(a+ib,b),\ i\in\mathbb{Z}$;

(3) $\gcd(a,b)=vb+wa,\ v,w\in\mathbb{Z}$ (*Note:* v,w are not unique!);

(4) $\gcd(a,b)=\gcd(a,c)=1\implies\gcd(a,b\cdot c)=1$;

(5) (a) a,b even, $\gcd(a,b)=2\cdot\gcd\left(\frac{a}{2},\frac{b}{2}\right)$,
(b) a even, b odd, $\gcd(a,b)=\gcd\left(\frac{a}{2},b\right)$,
(c) a,b odd, $\gcd(a,b)=\gcd\left(\frac{a-b}{2},b\right)$;

(6) $a\mid bc$ and $\gcd(a,b)=1\implies a\mid c$ ($a\mid c$ means a divides c).

Proof of the properties of $\gcd(a,b)$ Properties (1), (2), (5a) and (5b) are evident. Property (3) has already been proved in theorem 2.12. We shall next present the proofs of properties (4), (5c) and (6).

(4) If a does not have a common factor with either b or c then it also does not have a common factor with bc.

(5c) For property (2) we can write $\gcd(a,b) = \gcd(a+b,b)$. Because $a+b$ is even, property (5b) results. With $\frac{1}{2}(a+b) = \frac{a-b}{2} + b$ and using property (2), the result follows.

(6) We write $1 = vb + wa$. Multiplying by c, we obtain $c = vbc + wac$. It follows from this that a must divide c. □

Definition 2.13 (Coprime numbers) *Two integers $a, b \neq 0$ are said to be relatively prime or coprime numbers if* $\gcd(a,b) = 1$.

The existence of an inverse element According to the definition of a prime number p, the element $\alpha < p$ has no common divisor with p. Therefore, for all $\alpha \in GF(p)$,

$$\gcd(\alpha, p) = 1 \ .$$

It follows from theorem 2.12 that two numbers a and b exist for which

$$a\alpha + bp = 1.$$

This can also be written as

$$\begin{aligned} a\alpha &= 1 \mod p, \\ a &= \alpha^{-1} \ . \end{aligned}$$

We have shown that for each prime field $GF(p)$, and for every element in $GF(p)$, where $\alpha \neq 0$, there exists an inverse element α^{-1}.

The number of primitive elements in $GF(p)$ The number of primitive elements in $GF(p)$ is $\Phi(p-1)$.

Let α be a primitive element of $GF(p)$ and let s be the order of α^l. Assume that $\gcd(l, p-1) = 1$. Then there exist integers v and w such that $1 = vl + w(p-1)$, or $s = svl + sw(p-1)$. From theorem 2.10 we have $p-1 \mid sl$. With this we obtain $s = p-1$, since $p-1$ divides $sw(p-1)$ and slv, it must divide also s. Clearly, there exist $\Phi(p-1)$ integers l with $\gcd(l, p-1) = 1$.

At this point, enough background has been presented to allow the reader to proceed to chapter 3 on Reed–Solomon codes.

2.3.3 Gaussian integers

In this section, we shall present an illustrative example of the prime field $GF(p)$ represented as Gaussian integers, which will deepen the understanding of the concept of a field. A Gaussian number z is a complex number with integer values for the real and imaginary components [Hub94]:

$$\zeta = u + \mathrm{j}v, \quad u, v \in \mathbb{Z} \text{ and } \mathrm{j} = \sqrt{-1} \ .$$

It is known from number theory that every prime number p of the form $p = 4a+1$ (or $p = 1 \mod 4$) can be written as

$$p = (u + \mathrm{j}v)(u - \mathrm{j}v) = u^2 + v^2 \ .$$

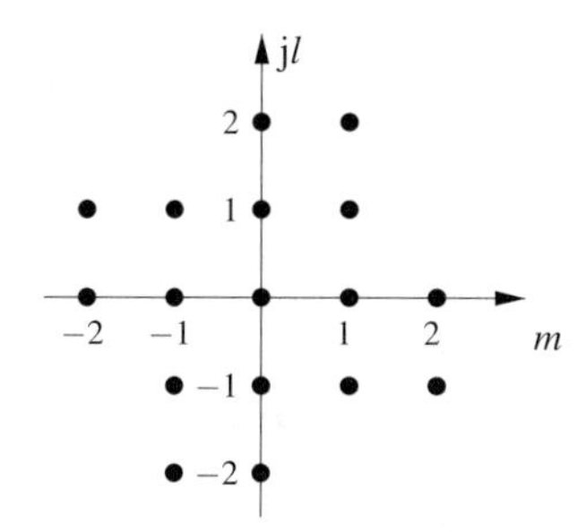

Figure 2.1 Gaussian field $\mathbb{G}_{4+\mathrm{j}}$.

The number $\Pi = u + \mathrm{j}v$ will be defined as a Gaussian prime number, and the calculation modulo Π is defined as $(\Pi^* = u - \mathrm{j}v)$

$$\zeta = w \mod \Pi = w - \left[\frac{w\Pi^*}{\Pi\Pi^*}\right]\Pi ,$$

where $[\cdot]$ performs the operation of rounding to the nearest Gaussian number. The possible prime numbers that we do *not* want to consider are $1+\mathrm{j},\ -1-\mathrm{j},\ -1+\mathrm{j},\ 1-\mathrm{j}$ and the numbers $l,\ -l,\ \mathrm{j}l,\ -\mathrm{j}l$, where l is prime with the property $l = 3 \bmod 4$.

Theorem 2.14 (Gaussian number field $\mathbb{G}_\Pi$) *Let $p = 4a+1,\ a \in \mathbb{Z}$, a prime number with $p = (u+\mathrm{j}v)(u-\mathrm{j}v) = \Pi\Pi^*$. Then the Gaussian numbers modulo Π form a field $\mathbb{G}_\Pi = \{0, \zeta_1, \zeta_2, \ldots, \zeta_{p-1}\}$:*

$$\zeta_i = m + \mathrm{j}l \mod \Pi, \quad m, l \in \mathbb{Z} .$$

Proof $GF(p)$ is a field. Each element $b \in GF(p)$ can be uniquely inverted from an element $\zeta \in \mathbb{G}_\Pi$ through

$$b \mod \Pi = \zeta = b - \left[\frac{m\Pi^*}{\Pi\Pi^*}\right]\Pi, \qquad b = (\zeta b\Pi^* + \zeta^* t\Pi) \mod p ,$$

where t and b are defined by $1 = t\Pi + b\Pi^*$. Furthermore, for all $b_1, b_2 \in GF(p)$ and $\zeta_1, \zeta_2 \in \mathbb{G}_\Pi$,

$$\begin{aligned} b_1 + b_2 \mod p \quad &\longleftrightarrow \quad \zeta_1 + \zeta_2 \mod \Pi , \\ b_1 b_2 \mod p \quad &\longleftrightarrow \quad \zeta_1\zeta_2 \mod \Pi . \end{aligned}$$

Note that

$$\begin{aligned} \zeta_1 + \zeta_2 \mod \Pi &= \zeta_1 + \zeta_2 - \left[\frac{(\zeta_1+\zeta_2)\Pi^*}{\Pi \cdot \Pi^*}\right]\Pi , \\ \zeta_1\zeta_2 \mod \Pi &= \zeta_1\zeta_2 - \left[\frac{(\zeta_1\zeta_2)\Pi^*}{\Pi\Pi^*}\right]\Pi . \end{aligned}$$

□

Example 2.8 (Gaussian field $\mathbb{G}_{4+\mathrm{j}}$) The prime number $p = 17 = 1 \bmod 4$ can be represented as $17 = (4+\mathrm{j})(4-\mathrm{j})$. The Gaussian number $13 + 11\mathrm{j}$ under the operation $\bmod \Pi = 4 + \mathrm{j}$ produces

$$
\begin{aligned}
13+11\mathrm{j}-\left[\frac{(13+11\mathrm{j})(4-\mathrm{j})}{(4+\mathrm{j})(4-\mathrm{j})}\right](4+\mathrm{j}) &= 13+11\mathrm{j}-\left[\frac{63+31\mathrm{j}}{17}\right](4+\mathrm{j}) \\
&= 13+11\mathrm{j}-[3,7+1,8\mathrm{j}](4+\mathrm{j}) \\
&= 13+11\mathrm{j}-(4+2\mathrm{j})(4+\mathrm{j}) \\
&= 13+11\mathrm{j}-(14+12\mathrm{j}) = -1-\mathrm{j}.
\end{aligned}
$$

In a similar way, we calculate the numbers 9 and 10 mod Π:

$$
\begin{aligned}
9 \mod \Pi &= 9-\left[\frac{9(4-\mathrm{j})}{(4+\mathrm{j})(4-\mathrm{j})}\right](4+\mathrm{j}) = 9-(2-\mathrm{j})(4+\mathrm{j}) = 9-(9-2\mathrm{j}) = 2\mathrm{j}, \\
10 \mod \Pi &= 1+2\mathrm{j}.
\end{aligned}
$$

Calculating the numbers $\{1,2,\ldots,16\}$ mod Π gives $\mathbb{G}_\Pi = \{1,2,-1-\mathrm{j},-\mathrm{j},1-\mathrm{j},2-\mathrm{j},-1-2\mathrm{j},-2\mathrm{j},2\mathrm{j},1+2\mathrm{j},-2+\mathrm{j},-1+\mathrm{j},\mathrm{j},1+\mathrm{j},-2,-1\}$. We should now like to check whether $9\cdot 10 = 90 = 5$ mod 17 also corresponds to $2\mathrm{j}(1+2\mathrm{j}) = 1-\mathrm{j}$ mod Π. We compute

$$
\begin{aligned}
2\mathrm{j}(1+2\mathrm{j})-\left[\frac{2\mathrm{j}(1+2\mathrm{j})(4-\mathrm{j})}{(4+\mathrm{j})(4-\mathrm{j})}\right](4+\mathrm{j}) &= (-4+2\mathrm{j})-\left[-\frac{14}{17}+\frac{12}{17}\mathrm{j}\right](4+\mathrm{j}) \\
&= -4+2\mathrm{j}-(-1+\mathrm{j})(4+\mathrm{j}) \\
&= (-4+2\mathrm{j})-(-5+3\mathrm{j}) = 1-\mathrm{j}.
\end{aligned}
$$

The elements from the field $\mathbb{G}_{4+\mathrm{j}}$ are shown in figure 2.1. ◇

2.4 Extension fields

Up until now we have only considered the prime fields $GF(p)$. However, there exist fields $GF(q)$ where q is the power of a prime p, $q = p^m$. For other numbers, no finite fields exist. The fields $GF(q)$ with $q = p^m, m > 1$ are called extension fields. In this section we shall present the techniques to construct extension fields. The case frequently arises where a prime field cannot be efficiently represented in terms of bits. For example $p = 71$ requires 7 digits, $2^7 = 128$. In practice, extension fields from $GF(2)$ are almost always used, for among other things, to better exploit the dual representation of elements of a Galois field.

Note Another notation for the ring $\mathbb{Z}_m$ is $\mathbb{Z}/(m)$. Let us define $\mathbb{Z}_m[x]$ to be the set of polynomials with coefficients from $\mathbb{Z}_m$ where $/(m)$ represents the computation modulo m. The set $\mathbb{Z}_m[x]$ is an infinite ring under the operations $(+,\cdot)$. If the modulo $p(x)$ operation is performed on the set $\mathbb{Z}_m[x]$, we obtain the finite ring $\mathbb{Z}_m[x]/p(x)$. The modulo $p(x)$ operation produces the remainder of $f(x)$ when $f(x)$ is divided by the polynomial $p(x)$. We shall show that if m is a prime, $m = p$ and $p(x)$ is a so-called primitive polynomial of degree s, then $\mathbb{Z}_p[x]/p(x) = GF(p)[x]$ is a finite Galois field $GF(q)$, where $q = p^i$.

2.4.1 Irreducible polynomials

Definition 2.15 (Irreducible polynomial) *A polynomial $p(x)$ with coefficients from $GF(p)$ is irreducible if there are no non-zero polynomial factors of smaller degree that also have coefficients from $GF(p)$.*

An irreducible polynomial cannot have roots from the base field $GF(p)$. However, this is not sufficient, because if $p_1(x)$ and $p_2(x)$ have no roots from $GF(p)$ then the product $p(x) = p_1(x)p_2(x)$ also has no roots from $GF(p)$, but is not irreducible. If $p(x)$ is an irreducible polynomial then $ap(x)$ is also irreducible. We shall consider only polynomials where the coefficient of the term with the highest power is equal to one. Such polynomials are called monic.

Example 2.9 (Irreducible polynomial) Consider the polynomial $p(x) = x^4 + x + 1$ with coefficients from $GF(2)$. To test whether $p(x)$ is irreducible, we can check whether a polynomial $f(x)$ with $\deg f(x) < \deg p(x)$ exists that gives a remainder of zero.

For $p(x)$ with order m, there exist 2^m (in general p^m) polynomials $f(x)$ with $\deg f(x) < m$ (including $f(x) = 0$). Therefore, when $m = 4,\ p = 2 \Longrightarrow 16$ polynomials exist:

$$0, 1, x, x^2, x^3, 1+x, \ldots, 1+x+x^2+x^3.$$

For all $f(x) \neq 0, 1$, we must check if there is a polynomial that divides $p(x)$ and gives a zero remainder:

$$\left.\begin{array}{rcll} x^i & \not| & p(x) & i = 1,2,3 \\ (1+x^i) & \not| & p(x) & i = 1,2,3 \\ & \vdots & & \\ (1+x+x^2+x^3) & \not| & p(x) & \end{array}\right\} \Longrightarrow p(x) \text{ is irreducible with respect to } GF(2).$$

◇

Theorem 2.16 (Existence of an inverse polynomial) *Let an irreducible polynomial $p(x)$, $\deg p(x) = m$, have coefficients $p_i \in GF(p)$. Each of the $p^m - 1$ (excluding 0) polynomials $b(x)$ with $\deg b(x) < m$ and coefficients $b_i \in GF(p)$ has a unique inverse polynomial $b^{-1}(x)$ mod $p(x)$. This means that*

$$b(x)b(x)^{-1} = 1 \mod p(x)\ .$$

Proof

$$\forall b(x) : b(x) \not| \ p(x) \Longrightarrow \gcd(b(x), p(x)) = 1\ .$$

Using the Euclidean algorithm, we can write

$$b(x)a(x) + c(x)p(x) = 1,$$

which gives

$$a(x) = b^{-1}(x) \mod p(x)\ .$$

□

Example 2.10 (Inverse polynomial) The following calculation can be performed for the polynomial $p(x) = x^4 + x + 1,\ p_i \in GF(2)$:

$$\begin{array}{rcll} p(x) & = & 0 & \mod p(x), \\ x^4+x+1 & = & 0 & \mod p(x), \\ x^4+x & = & 1 & \mod p(x), \\ x\cdot(x^3+1) & = & 1 & \mod p(x), \end{array}$$

$$x^{-1} = 1 + x^3\ .$$

With the aid of the Euclidean algorithm, an inverse $b^{-1}(x)$ can be calculated for each polynomial $b(x)$ where $\deg b(x) < 4$, $b_i \in GF(2)$.

◇

A ring can be constructed for every polynomial $p(x), \deg p(x) = m$ (not necessarily irreducible) with coefficients from $GF(p)$, through the addition and multiplication operations $\bmod\, p(x)$ of polynomials $b(x)$ with $\deg b(x) < m$. When $p(x)$ is irreducible, the ring also satisfies the requirements of a field. This is proved in theorem 2.16, which states that each polynomial of smaller order than the irreducible polynomial $p(x)$ has an inverse. In the following section, we shall consider the extension field of polynomials based on the field generated by the irreducible polynomial $p(x)$. The extension field has $p^{\deg p(x)}$ elements.

2.4.2 Primitive polynomials and roots of polynomials

Let us review how to generate complex numbers from real numbers. The equation $x^2 + 1 = 0$ has no solution $\in \mathbb{R}$, but it can be solved using the set of complex numbers $\mathbb{C}$, with the definition $\mathrm{j}^2 + 1 = 0$, $\mathrm{j} = \pm\sqrt{-1}$, $\mathrm{j} \in \mathbb{C}$. Therefore j is an element of the extension field. In the following, we shall use the same idea in order to construct an extension field.

Definition 2.17 (Roots of a polynomial) *Let $p(x), p_i \in GF(p)$, be an irreducible monic polynomial. We introduce an element α as a formal root of $p(x)$:*

$$p(\alpha) = 0.$$

α is from the extension field $GF(p^{\deg p(x)})$ that is to be constructed.

Definition 2.18 (Extension field) *Let $p(x)$ be an irreducible polynomial over $GF(p)$ and let $\alpha \notin GF(p)$ be a root of $p(x)$, where $\deg p(x) = m$. The extension field $GF(p^m)$ is the smallest field that contains α and $GF(p)$.*

Example 2.11 (Roots) According to example 2.9, $p(x) = x^4 + x + 1$ is irreducible in the field of $GF(2)$. When α is a root of $p(x)$,

$$\alpha^4 + \alpha + 1 = 0\,. \qquad \diamond$$

Definition 2.19 (Primitive polynomial) *$p(x)$ is an irreducible polynomial with $\deg p(x) = m$ and coefficients $p_i \in GF(p)$. An element $\alpha \in GF(p^m)$ is called a primitive element if $\alpha^i \bmod p(\alpha)$ can produce all $p^m - 1$ elements of the extension field $GF(p^m)$ (excluding the zero element). The polynomial $p(x)$ is called primitive if it has a primitive element as a root (see definition 2.7).*

Elements of an extension field An element of an extension field $GF(p^m)$ is defined as the root α of a primitive polynomial $p(x)$, namely $p(\alpha) = 0$. The powers of the element $\alpha^i \bmod p$, $i \in \mathbb{N}$, produce the extension field excluding the element $\{0\}$. We can represent the elements of the extension field through two methods.

Exponential representation According to definition 2.19, we can represent every non-zero element of an extension field as a power of the primitive element α. This is called the *exponential representation*.

Table 2.1 Galois field $GF(2^4)$ (logarithm table).

Exp.	Polynomial					Calculation
$-\infty$				0	0000	
0				1	0001	
1			α		0010	
2		α^2			0100	
3	α^3				1000	
4			α	$+1$	0011	$\alpha^4 = \alpha+1$
5		α^2	$+\alpha$		0110	$\alpha^5 = \alpha\alpha^4 = \alpha^2+\alpha$
6	α^3	$+\alpha^2$			1100	$\alpha^6 = \alpha\alpha^5 = \alpha^3+\alpha^2$
7	α^3		$+\alpha$	$+1$	1011	$\alpha^7 = \alpha\alpha^6 = \alpha^4+\alpha^3 = \alpha^3+\alpha+1$
8		α^2		$+1$	0101	$\alpha^8 = \alpha\alpha^7 = \alpha^4+\alpha^2+\alpha = \alpha^2+1$
9	α^3		$+\alpha$		1010	$\alpha^9 = \alpha\alpha^8 = \alpha^3+\alpha$
10		α^2	$+\alpha$	$+1$	0111	$\alpha^{10} = \alpha\alpha^9 = \alpha^4+\alpha^2 = \alpha^2+\alpha+1$
11	α^3	$+\alpha^2$	$+\alpha$		1110	$\alpha^{11} = \alpha\alpha^{10} = \alpha^3+\alpha^2+\alpha$
12	α^3	$+\alpha^2$	$+\alpha$	$+1$	1111	$\alpha^{12} = \alpha\alpha^{11} = \alpha^3+\alpha^2+\alpha+1$
13	α^3	$+\alpha^2$		$+1$	1101	$\alpha^{13} = \alpha\alpha^{12} = \alpha^3+\alpha^2+1$
14	α^3			$+1$	1001	$\alpha^{14} = \alpha\alpha^{13} = \alpha^4+\alpha^3+\alpha = \alpha^3+1$
15				1	0001	$\alpha^{15} = \alpha\alpha^{14} = \alpha^4+\alpha = 1$

Additive or polynomial representation The *additive or polynomial representation* is defined by the representation of the elements of an extension field using a polynomial $f(\alpha)$ of $\deg f(x) \leq \deg p(x) - 1$ over $GF(p)$ and that satisfies the field operations $(+,\cdot)$ mod $p(x)$. One writes the polynomial $f(\alpha)$ (according to definition 2.19) as

$$f_{m-1}x^{m-1} + f_{m-2}x^{m-2} + \ldots + f_0.$$

This allows the polynomial to be defined through its unique coefficient values:

$$f_{m-1}f_{m-2}\cdots f_1 f_0\,, \qquad f_i \in GF(p)\,.$$

The zero element $00\ldots0$ is written in exponential form as $\alpha^{-\infty}$. Multiplication is carried out through the addition of the exponents modulo $p^m - 1$. The addition of polynomials is carried out through the addition of the polynomial coefficients in $GF(p)$ mod p (vector addition).

As an example, we shall construct the extension field for $GF(2^4)$ using the primitive polynomial $p(x)$ and the primitive element α from example 2.11.

Example 2.12 (Galois field $GF(2^4)$) The Galois field $GF(2^4)$ with the primitive polynomial $p(x) = x^4 + x + 1$, $p_i \in GF(2)$, and the primitive element α, $p(\alpha) = 0$, is presented in table 2.1.

Multiplication: $\alpha^9\alpha^7 = \alpha^{16 \bmod 15} = \alpha$, $\quad (\alpha^{i \bmod 15})$.

Addition: $\alpha^4 + \alpha^5 = 0011 \oplus 0110 = 0101 = \alpha^8$ ($\oplus$ represents the addition of coefficients in the base field $GF(p)$). ◇

We have seen that in order to describe an extension field $GF(p^m)$, one needs a primitive polynomial $p(x)$, $\deg p(x) = m$, $p_i \in GF(p)$. It is therefore useful to know of the existence of a primitive polynomial.

Theorem 2.20 (Existence of a primitive polynomial) *(see [Art]) For every field $GF(p)$ and every number $m \in \mathbb{N}$, there exists at least one primitive polynomial $p(x) = p_0 + p_1x + \ldots + p_mx^m$, $p_i \in GF(p)$.*

Table 2.2 Primitive polynomials.

m	Primitive polynomial $p(x)$	m	Primitive polynomial$p(x)$
1	$x+1$	9	x^9+x^4+1
2	x^2+x+1	10	$x^{10}+x^3+1$
3	x^3+x+1	11	$x^{11}+x^2+1$
4	x^4+x+1	12	$x^{12}+x^7+x^4+x^3+1$
5	x^5+x^2+1	13	$x^{13}+x^4+x^3+x+1$
6	x^6+x+1	14	$x^{14}+x^8+x^6+x+1$
7	x^7+x+1	15	$x^{15}+x+1$
8	$x^8+x^6+x^5+x^4+1$	16	$x^{16}+x^{12}+x^3+x+1$

Note In the search for a primitive polynomial, it is not sufficient to only find an irreducible polynomial; therefore

not every irreducible polynomial is a primitive polynomial.

We show this through a counterexample: $q(x) = x^4+x^3+x^2+x+1$ is irreducible over $GF(2)$. It is not primitive, because

$$q(\alpha) = 0 : \qquad \alpha^4 = \alpha^3+\alpha^2+\alpha+1,$$
$$\alpha^5 = \alpha\cdot\alpha^4 = \alpha^4+\alpha^3+\alpha^2+\alpha = 1 = \alpha^0 .$$

There exists different primitive polynomials, which result in an isomorphic extension field; however, it is sufficient to only know *one*. The advantage of knowing more than one primitive polynomial comes from the fact that some polynomials have an easier shift register implementation. Table 2.2 shows primitive polynomials for the construction of extension fields $GF(2^m)$ for $m \leq 16$.

This method of extension field representation is often called a *circular field* due to its similarity to $(e^{\mathrm{j}\frac{2\pi}{N}})^N = 1$. In the following section, we shall investigate other properties of extension fields.

2.4.3 Properties of extension fields

The complex numbers are an extension field of the real numbers. We consider the complex-conjugate numbers $-\mathrm{j}$, j, which result in

$$(x-\mathrm{j})(x+\mathrm{j}) = x^2+1 .$$

The product of two polynomials with complex-conjugate roots ($\in \mathbb{C}$) gives a polynomial with coefficients from $\mathbb{R} \subseteq \mathbb{C}$. This is exactly analogous to the extension fields produced through $GF(p^m)$.

Theorem 2.21 (Complex-conjugate roots) *The polynomial $p(x)$, $\deg p(x) = m$, $p_i \in GF(p)$, has a root $\alpha \in GF(p^m)$ such that*

$$\alpha^p, \alpha^{p^2}, \ldots, \alpha^{p^{m-1}}$$

are also roots from $p(x)$ mod $p(x)$. These roots are called complex-conjugate roots.

Proof See theorem 4.1. □

With this result, one can examine the polynomial $p(x)$ in terms of linear factors:

$$p(x) = (x-\alpha)(x-\alpha^p)\cdots(x-\alpha^{p^{m-1}}) .$$

The product of all complex-conjugate linear factors (elements from $GF(p^m)$) gives a polynomial with coefficients from $GF(p)$. We have already defined the order of elements $\alpha \in GF(p)$. The definition of the order of the extension field corresponding to definition 2.6 is presented below:

Definition 2.22 (The order of elements) *The order n of the element β, $\beta \in GF(p^m)$, is defined as $\beta^n = 1$, where n is the smallest number that satisfies this equation.*

When $n = p^m - 1$, β is a primitive element. In this case, β is often called the nth primitive root of unity. It follows from definition 2.22 that all β^i, $0 \leq i < n$, are different.

Theorem 2.23 (Product of irreducible polynomials) *(see [Art]) $x^{p^m} - x$ is the product of all irreducible polynomials in $GF(p)$ having degree s such that $s \mid m$, $1 \leq s < m$.*

This theorem will become meaningful after the definition of cyclotomic cosets in the next section. From theorems 2.23 and 2.10, it follows that

$$\beta^{p^m-1} = 1 \quad \text{for all } \beta \in GF(p^m),\ \beta \neq 0.$$

Theorem 2.24 (Divisors of the order) *The order n of the element β, $\beta \in GF(p^m)$ must divide $p^m - 1$, $n \mid p^m - 1$. If n can be written as*

$$n = p^s - 1 \quad \textit{then} \quad s \mid m .$$

Proof If $n \nmid p^m - 1$ then

$$p^m - 1 = xn + r, \quad 0 < r < n,$$
$$\beta^{p^m-1} = \beta^{xn}\beta^r \neq 1 .$$

This is a contradiction to theorem 2.23.

The second statement that can be made is that $p^s - 1 \mid p^m - 1$ if and only if $s \mid m$:

$$m = xs + r, \quad 0 \leq r < s .$$

The following results:

$$\frac{p^m-1}{p^s-1} = p^r\frac{p^{xs}-1}{p^s-1} + \frac{p^r-1}{p^s-1}$$

and

$$p^s - 1 \mid p^{xs} - 1, \quad \frac{p^r-1}{p^s-1} < 1 .$$

Since r must be an integer, the only value of r that satisfies this equation is $r = 0$.

Note: $p^s - 1 \mid p^{xs} - 1$ results from the property that when an element β has order $p^s - 1$, $\beta^{p^s-1} = 1$. We can write the element β as a power of a primitive element α: $\beta = \alpha^a$, and

$$\alpha^{p^{xs}-1} = 1 = \beta^{p^s-1} = \alpha^{a(p^s-1)},$$

which implies $p^s - 1 \mid p^{xs} - 1$. □

The extension field $GF(p^s)$ for the case when β has order n is the smallest field that contains β. One calls $GF(p^s)$ a subfield and the corresponding $GF(p^m)$ an extension field ($\beta \in GF(p^s)$, $\beta \in GF(p^i)$. All subfields of $GF(p^m)$, described as $GF(p^i)$ with $i \mid m$, $1 \leq i \leq m$, can be related through $l = \gcd(s,m)$ to produce

$$GF(p^l) = GF(p^m) \cap GF(p^s).$$

Example 2.13 (Subfield) We wish to find the subfield for the element $\beta = \alpha^5 \in GF(2^4)$ from example 2.12. The order β is equal to 3 according to Definition 2.22. Therefore $n = p^s - 1 \mid p^m - 1 (3 \mid 15)$ and $s \mid m (2 \mid 4)$. The result is the field $GF(2^2)$ with four elements:

$$\begin{array}{rl} -\infty, 0, 5, 10 & \text{in exponential form,} \\ 0000, 0001, 0110, 0111 & \text{in polynomial form.} \end{array}$$

◇

2.5 Cyclotomic cosets

The introduction of cyclotomic cosets allows us to more easily compute the complex-conjugate elements of an extension field (in coefficient representation) according to theorem 2.21. Cyclotomic cosets will be used in chapter 4 in order to define the class of BCH codes.

Definition 2.25 (Cyclotomic coset) *The cyclotomic cosets K_i corresponding to a number $n = q^m - 1$ are*

$$K_i := \{ i q^j \mod n, \ j = 0, 1, \ldots, m-1 \},$$

where i is the smallest element of the set in K_i.

One observes that the number q can be generated from a prime number; therefore $q = p^l, l \geq 1$, p is prime. Furthermore, $GF(q)$ is a subfield of $GF(q^m)$. This definition is also applicable to $n \mid q^m - 1$.

The cyclotomic coset K_i has the following properties:

$$\begin{aligned} |K_i| &\leq m, \\ K_i \cap K_j &= \emptyset \quad \text{(empty set) if } i \neq j, \\ K_0 &= \{0\}, \\ \bigcup_i K_i &= \{0, 1, \ldots, n-1\}. \end{aligned}$$

Example 2.14 (Cyclotomic coset) We shall now determine the cyclotomic coset with respect to the number 15 (with reference to example 2.13):

$$n = 15 = 2^4 - 1, \quad GF(2),$$

$$\begin{array}{ll} K_0 = \{0\}, & K_5 = \{5, 10\}, \\ K_1 = \{1, 2, 4, 8\}, & K_7 = \{7, 11, 13, 14\}. \\ K_3 = \{3, 6, 9, 12\}, & \end{array}$$

◇

If n is a prime, all cyclotomic cosets K_i, $i > 0$, have the same number of elements.

Note The product $\prod_{j\in K_i}(x-\alpha^j)$ corresponds to an irreducible polynomial $m_i(x)$. The product of all irreducible polynomials (excluding $x-\alpha^{-\infty}$) gives x^n-1, which satisfies theorem 2.23. p can be replaced by $q=p^l$.

Definition 2.26 (Trace function) *The trace function is a mapping of elements $\beta \in GF(q^m)$ to $GF(q)$, where $q=p^l$, $l\geq 1$, p is a prime number. It is defined through*

$$\mathrm{tr}(\beta)=\beta+\beta^q+\beta^{q^2}+\ldots+\beta^{q^{m-1}}=a\in GF(q)$$

(see theorems 2.21 and 4.1).

Because we can express an element β through a primitive element α, it is also possible to substitute $\beta=\alpha^i$ into the definition of the trace function. This allows us to write the trace function in terms of the cyclotomic coset in the following way:

$$\mathrm{tr}(\beta)=\sum_{j=0}^{m-1}\beta^{q^j}=\sum_{j\in K_i}\alpha^j\ .$$

An apparent property of the trace function is

$$\mathrm{tr}(\beta+\gamma)=\mathrm{tr}(\beta)+\mathrm{tr}(\gamma)\ .$$

The property that the trace function is a mapping to the base field $GF(q)$ is not apparent, and will be proved in theorem 4.1.

2.6 Quadratic residues

Definition 2.27 (Quadratic residues) *The set $\mathbb{M}_Q$ of quadratic residues for p prime is*

$$\mathbb{M}_Q:=\left\{i^2 \mod p,\ i=1,\ldots,\frac{p-1}{2}\right\}\ .$$

Using this gives

$$|\mathbb{M}_Q|=\frac{p-1}{2}\ .$$

Example 2.15 (Quadratic residue) We shall determine the set $\mathbb{M}_Q$ of quadratic residues for $p=17$:

$$\begin{array}{ccccccccc}
1^2 & = & 1 & & & = & 16^2, & & \\
2^2 & = & 4 & & & = & 15^2, & & \\
3^2 & = & 9 & & & = & 14^2, & & \\
4^2 & = & 16 & & & = & 13^2, & & \\
5^2 & = & 25 & = & 8 & = & 12^2, & & \\
6^2 & = & 36 & = & 2 & = & 11^2, & & \\
7^2 & = & 49 & = & 15 & = & 10^2 & = & 100, \\
8^2 & = & 64 & = & 13 & = & 9^2 & = & 81,
\end{array}$$

$$\mathbb{M}_Q=\{1,2,4,8,9,13,15,16\},\quad |\mathbb{M}_Q|=\frac{17-1}{2}=8\ . \qquad \diamond$$

Quadratic residues will be used in section 5.4 to define quadratic residual codes. As with cyclotomic cosets, the definition of quadratic residues allows a more convenient description of the corresponding code class.

2.7 Summary

At the age of 20, on the eve of a duel (most likely because of a love affair in 1832), E. Galois wrote a friend a letter that outlined the basic principles of Galois theory. Upon his ideas, the fundamentals of algebraic coding theory have been constructed.

In this chapter, we have defined the Galois field and used it to demonstrate a few of the properties and calculation rules for prime and extension fields. Only the most important properties of Galois fields have been presented. A more complete study of Galois field theory may be found in [McWSl] and [Art].

We have defined groups, rings and fields. The Gaussian number field modulo Π has been presented as an example of a prime field. The Euler $\Phi(m)$ function gives the number of invertible elements for integer rings. Furthermore, we have introduced the concept of element order and the primitive element, which we have used to compute a Galois field and its subfields. With irreducible and primitive polynomials, we have defined extension fields.

The extension field $GF(2^m)$ has almost exclusively been used for the application of codes. This is because the exponential representation can also be easily represented in polynomial form in hardware. The advantages of both the binary exponential and the polynomial notation have been presented. Using the polynomial form, the addition of two elements is accomplished through modulo 2 addition of the coefficients, and multiplication is calculated through modulo $2^m - 1$ addition of the exponents.

We have described the Euclidean algorithm, which is extremely useful for theoretical problems and for decoding. In chapter 3 we shall introduce the Euclidean algorithm for polynomials, used to calculate the greatest common divisor of two polynomials. In the case that the greatest common divisor is 1, the number or the polynomial are defined as relatively prime.

Lastly, we have defined the cyclotomic coset and quadratic residues. The cyclotomic coset will be used in chapter 4 to define BCH codes, and quadratic residues are used in section 5.4 to define quadratic residue codes.

We shall use the following notation in the remaining sections of the book: p will represent a prime number, and q will be an integer power of p.

The Reed–Solomon codes will be presented in the next chapter from the point of view of prime fields, and therefore results presented in this chapter form a basis for their understanding. The prime field and the extension field can be used interchangeably, and calculations in either field give the same final result.

2.8 Problems

Problem 2.1

(a) Is the set of all vectors $a \in \mathbb{F}_2^n$ a group with respect to addition $\bmod 2$?

(b) With respect to which operation, addition or multiplication, is the set $\mathbb{Z}$ of integer numbers a group?

Problem 2.2

You are given the set $\mathcal{M} = \{a, b, c, d\}$ and the addition and multiplication table for the elements from $\mathcal{M}$:

+	a	b	c	d
a	a	b	c	d
b	b	a	d	c
c	c	d	a	b
d	d	c	b	a

Addition table

·	a	b	c	d
a	a	a	a	a
b	a	b	c	d
c	a	b	d	c
d	a	d	b	c

Multiplication table

Is the set $\mathcal{M}$ with the defined operations a field?

Problem 2.3
You are given the set $\mathcal{M}_q := \{0, 1, \dots, q-1\}$ and the two operations:

- Addition modulo q,
- Multiplication modulo q.

Calculate the operation table for the elements when $q = 2, 3, 4, 5, 6$.

Problem 2.4
Determine $\Phi(70)$ and $\Phi(288)$.

Problem 2.5
Determine the addition and multiplication table for $(\mathbb{Z}_4, +)$, $(\mathbb{Z}_4, \cdot)$, $(\mathbb{Z}_{11}, +)$, $(\mathbb{Z}_{11}, \cdot)$.

(a) To which algebraic structure[2] do the sets belong?

(b) What is the difference between $(\mathbb{Z}_4 \setminus \{0\}, \cdot)$ and $(\mathbb{Z}_{11} \setminus \{0\}, \cdot)$?

(c) What algebraic structure results for $(\mathbb{Z}_4, +, \cdot)$ and $(\mathbb{Z}_{11}, +, \cdot)$?

(d) Determine the order of the element $p = 7$ in $(\mathbb{Z}_{11}, +, \cdot)$.

(e) How many primitive elements exist in $(\mathbb{Z}_{11}, +, \cdot)$?

Problem 2.6
Find the $\gcd(294, 816)$ using the Euclidian algorithm. Show that every remainder can be represented as a combination of 816 and 294.

Problem 2.7

(a) Determine the greatest common divisor $\gcd(585, 1768)$.

(b) Consider the polynomials

$$u(x) = x^{12} + x^{10} + x^7 + x^4 + x^3 + x^2 + x + 1,$$
$$v(x) = x^{11} + x^9 + x^7 + x^6 + x^5 + x + 1$$

with coefficients from $GF(2)$. Determine the greatest common divisor

$$\gcd\left(u(x), v(x)\right).$$

Problem 2.8
Determine x under the conditions:

(a) $6x = 47 \mod 127$,

[2] A set under operation $*$ is called a commutative semigroup if all group axioms are fulfilled except that at least one element has no inverse (see [Art]).

(b) $7^x = 5 \mod 17$.

Problem 2.9
Is the polynomial $p(x) = 4 + 4x + 2x^2 + x^3$ with coefficients from $GF(5)$ a primitive polynomial?

Problem 2.10
Calculate the cyclotomic coset corresponding to $n = 3^4 - 1$.

Problem 2.11
Determine the extended field for $GF(3^2)$.

(a) Give the addition and multiplication tables in polynomial and exponential form. What is the order of the element 11? *Hint:* Use $x^2 + 2x + 2$ as the primitive polynomial.

(b) Determine the trace function of the element.

3

Reed–Solomon codes

Reed–Solomon (RS) codes are a class of codes that are often used in practical systems. In this chapter, we shall investigate in detail the properties of RS codes. Transformation techniques will be used to represent the codes so that we may employ 'engineering paradigms' in our discussions, which may be more familiar to the reader. We shall define the extension of RS codes by one and two coordinate positions. The decoding of RS codes is also a central point in this chapter, which solves the practical problem of estimating the information contained in a received vector. The solution of the key equation is the central point in the decoding method. Two efficient methods for this will be presented: the Euclidean and the Berlekamp–Massey algorithms.

We shall consider only prime Galois fields in our examples; however, all arguments and algorithms are also applicable for extension fields. It should be mentioned that the reader need not be familiar with the concept of extension fields (section 2.4) in order to understand our presentation of RS codes. In practice, extension fields to base two are often used because they can be easily implemented in real systems.

3.1 Definition of RS codes

The following algebraic theorem may be applied to Galois fields.

Theorem 3.1 (Fundamental theorem of algebra) *A polynomial* $A(x) = A_0 + A_1x + A_2x^2 + \ldots + A_{k-1}x^{k-1}$ *of degree* $k-1$ $(A_{k-1} \neq 0)$ *with coefficients* $A_i \in GF(p)$ *has at most* $k-1$ *roots* $\alpha_j \in GF(p)$.

Proof $\alpha \in GF(p)$ is a root from $A(x)$, which implies that $x-\alpha$ is a linear factor, i.e. $(x-\alpha)|A(x)$. We may write $A(x) = (x-\alpha)A^*(x)$, where $\deg(A(x)) = \deg(A^*(x)) + 1$ holds. This argument can be applied $\deg(A(x))$ times, which gives at most $k-1$ linear factors of $A(x)$. □

Theorem 3.2 (Evaluation of $A(x)$) *Consider n different non-zero elements $(\alpha_0, \ldots, \alpha_{n-1})$ from the Galois field $GF(p)$, $n \leq p-1$. Let $A(x)$ be a polynomial of degree $k-1$ with coefficients from $GF(p)$. Suppose $k-1 \leq n-d$, where d is a given integer. The weight of a vector $\mathbf{a} = (a_0, a_1, \ldots, a_{n-1})$ with $a_i = A(\alpha_i)$, $i = 0, 1, \ldots, n-1$, is greater than or equal to d:*

$$\mathrm{wt}(\mathbf{a}) \geq d \ .$$

Proof By theorem 3.1, $A(x)$ has at most $k-1$ roots. Hence $\mathbf{a}$ has at least $n-k+1 \geq d$ elements that are not roots. □

With this theorem, we can construct a vector with a weight larger than a specified value, which in turn determines the minimum weight and the minimum distance of the code (see section 1.1).

Definition 3.3 (RS code) *Let $\alpha \in GF(p)$ be an element of order n. The RS code of length n is defined by the set of polynomials $A(x)$ of degree less than k such that $A(x) = A_0 + A_1 x + A_2 x^2 + \ldots + A_{k-1} x^{k-1}$, $A_i \in GF(p)$, $k \leq n$. The codewords $\mathbf{a} = (a_0, a_1, \ldots, a_{n-1})$ are generated by $a_i = A(\alpha^i)$:*

$$C := \left\{ \mathbf{a} \mid a_i = A(\alpha^i),\ i = 0, 1, \ldots, n-1,\ \deg A(x) < k \right\} \ .$$

The minimum distance is $d = n-k+1$, and the dimension of the code is k.

An element $\alpha \in GF(p)$ of order n satisfies definition 2.8:

$$\alpha^n = 1 \mod p \ .$$

However, all powers $\alpha^0, \alpha^1, \ldots, \alpha^{n-1}$ represent n different elements from $GF(p)$. When α is a primitive element, $n = p-1$. The equation $x^n - 1 = 0$ is valid for all powers of the element $\alpha \in GF(p)$ of order n. This includes all elements α^i, $i = 0 \ldots, n-1$. Therefore the polynomial $x^n - 1$ has exactly n different linear factors $x - \alpha^i$, which generate the polynomial:

Theorem 3.4 (Linear factors of $x^n - 1$) *The linear factors of the polynomial $x^n - 1$ can be written as*

$$x^n - 1 = \prod_{i=0}^{n-1} (x - \alpha^i) \ .$$

$\alpha \in GF(p)$ has order n.

Note For a primitive element $\alpha \in GF(p)$, $n = p-1$ and

$$\alpha^{p-1} = 1 = \alpha^0 \ .$$

The exponent of α is calculated with respect to modulo $p-1$. The elements or coefficients are calculated with respect to modulo p.

For all elements, we get

$$\alpha^n = 1 \longrightarrow x^n - 1 = 0 \ .$$

Therefore the polynomial $x^n - 1$ can be used as a modulo divider for any polynomial. For the coefficients of a polynomial with degree greater than $n-1$, the modulo $x^n - 1$ operation results in the addition of the coefficient at $in + j$, $i = 1, 2, \ldots$, to the coefficient of x^j. For $n = 6$,

$$1 + x^3 + x^{12} + x^{29} = 2 + x^3 + x^5 \mod (x^6 - 1) \ .$$

Multiplication with $x^i \mod (x^n - 1)$ results in a cyclic shift of the polynomial coefficient by i coordinates.

Example 3.1 (Reed–Solomon code) We should like to construct the RS code of length 6 with minimum distance 5 over GF(7). We must first find a primitive element for this field, or, in other words, an element with the order 6. We must also choose all polynomials $A(x)$ with degree $k - 1 \leq n - d$. This results in $k = 2$.

We first verify that $\alpha = 5$ is a primitive element from $GF(7)$:

$$5^1 = 5, \ 5^2 = 25 = 4, \ 5^3 = 25 \cdot 5 = 4 \cdot 5 = 20 = 6 \ .$$

At this point, we can conclude that 5 is a primitive element because, according to theorem 2.10, the order of any element in $GF(p = 7)$ must divide $p - 1 = 6$. Therefore the possible orders of the elements are $2, 3$ and 6. Since we know that the order of 5 is greater than 3, it must have order 6, and is therefore a primitive element. The complete calculation gives

$$5^4 = 30 = 2, \ 5^5 = 10 = 3, \ 5^6 = 15 = 1 \ .$$

At this point, it is instructive to show that if the exponent is calculated $\mod (p-1)$, this will produce the same answer as the integer calculation:

$$\begin{aligned} 5^9 = 5^4 \cdot 5^5 = 2 \cdot 3 = 6 = 5^{9 \bmod 6} &= 5^3 = 6 \\ \neq 5^{9 \bmod 7} &= 5^2 = 4 \ . \end{aligned}$$

The codewords $\mathbf{a}$ of a RS code can be obtained through the formula $a_i = A(\alpha^i)$, $A(x) = A_0 + A_1 x$ with $A_0, A_1 \in GF(7)$. There exist $p^k = 7 \cdot 7 = 49$ different $A(x)$ and therefore 49 codewords. The calculation of the codeword for the information polynomial $A(x) = 5 + 3x$ gives

$$\begin{array}{lllllll} a_0 &= A(\alpha^0) &= A(1) &= 5 + \ \ 3 &= 1 & \mod 7, \\ a_1 &= A(\alpha^1) &= A(5) &= 5 + 3 \cdot 5 &= 6 & \mod 7, \\ a_2 &= A(\alpha^2) &= A(4) &= 5 + 3 \cdot 4 &= 3 & \mod 7, \\ a_3 &= A(\alpha^3) &= A(6) &= 5 + 3 \cdot 6 &= 2 & \mod 7, \\ a_4 &= A(\alpha^4) &= A(2) &= 5 + 3 \cdot 2 &= 4 & \mod 7, \\ a_5 &= A(\alpha^5) &= A(3) &= 5 + 3 \cdot 3 &= 0 & \mod 7, \end{array}$$

$$\Longrightarrow \quad \mathbf{a} = (1, 6, 3, 2, 4, 0) \ . \qquad \diamond$$

3.1.1 Discrete Fourier transform (DFT)

We now ask the question: Can one calculate $A(x)$ from $\mathbf{a}$? In doing so, we first represent the vector $\mathbf{a}$ by the polynomial $a(x) = a_0 + a_1 x + a_2 x^2 + \ldots + a_{n-1} x^{n-1}$. Our task is to find a transformation from $a(x)$ to $A(x)$. The forward and reverse transformations have been performed in a variety of ways by different authors. In [McWSl], the transformation is defined as the Mattson–Solomon polynomial, and in [Bla] it is referred to as the discrete Fourier transform.

Definition 3.5 (Discrete Fourier transform) *The DFT is defined as* $(i, j \in 0 \ldots n-1)$

$$\text{DFT} \quad \boxed{a_i = A(\alpha^i)}$$

$$a(x) \circ\!\!-\!\!\bullet A(x)\,.$$

$$\text{inverse DFT} \quad \boxed{A_j = n^{-1} \cdot a(\alpha^{-j})}$$

$\alpha \in GF(p)$ *is an element of order n, and* $a(x)$ *and* $A(x)$ *are polynomials of degree* $\leq n-1$ *with coefficients from* $GF(p)$.

Let $A(x) = A_0 + \ldots + A_{n-1}x^{n-1}$ and $B(x) = B_0 + \ldots + B_{n-1}x^{n-1}$ be two polynomials from $GF(p)$. Then the cyclic convolution $A(x) * B(x)$ is defined as a polynomial $C(x) = C_0 + \ldots + C_{n-1}x^{n-1}$, where $C_j = \sum_{i=0}^{n-1} A_i B_{j-i}$, $j = 0, 1, \ldots, n-1$. All indices are calculated modulo n.

Note For extension fields $GF(2^m)$, $n^{-1} = 1$. The calculation is done in the basic field, where $(2^m - 1) \cdot 1 = 1 \bmod 2$.

Multiplication of two polynomials $a(x)$ and $b(x)$ can be related in the transform domain through the convolution theorem:

Theorem 3.6 (Convolution theorem) *The multiplication of two polynomials* $\bmod x^{n-1}$ *corresponds to cyclic convolution:*

$$a(x) \circ\!\!-\!\!\bullet A(x), \qquad b(x) \circ\!\!-\!\!\bullet B(x),$$

$$c_i = a_i b_i, \qquad c(x) \circ\!\!-\!\!\bullet C(x) = A(x)B(x) \mod (x^n - 1),$$

$$C_i = A_i B_i, \quad C(x) \bullet\!\!-\!\!\circ c(x) = \frac{1}{n} a(x)b(x) \mod (x^n - 1)\,.$$

Proof The proof of the convolutional equations is based on definition 3.5:

$$C_j = \frac{1}{n} c(\alpha^{-j}) = \frac{1}{n} \sum_{i=0}^{n-1} \alpha^{-ij} c_i = \frac{1}{n} \sum_{i=0}^{n-1} \alpha^{-ij} a_i b_i$$

$$= \frac{1}{n} \sum_{i=0}^{n-1} \alpha^{-ij} a_i \left(\sum_{k=0}^{n-1} \alpha^{-ik} B_k \right) = \sum_{k=0}^{n-1} B_k \left(\frac{1}{n} \sum_{i=0}^{n-1} \alpha^{-i(j-k)} a_i \right) = \sum_{k=0}^{n-1} B_k A_{j-k}.$$

Therefore

$$C(x) = A(x)B(x) \mod (x^n - 1).$$

□

3.1.2 Generator polynomial

The RS code of length n, dimension k and minimum distance $d = n - k + 1$ can also be defined by all (information) polynomials $i(x)$ of degree $< k$. The codewords are calculated through multiplication by the generator polynomial:

$$a(x) = i(x)g(x), \quad \deg g(x) = n - k\,.$$

Each codeword $a(x)$ must be divisible by $g(x)$. The RS code may be defined through the set of polynomials $A(x)$ of degree $\leq k-1$. Therefore $A_i = 0$ for $k \leq i \leq n-1$. This corresponds to the transformation from definition 3.5:

$$A_i = n^{-1} a(\alpha^{-i}) = 0 \quad \text{for } k \leq i \leq n-1\,.$$

In each codeword $a(x)$, the elements $x = \alpha^i$ must be roots. These corresponds to the linear factors $x - \alpha^{-i}$. The product of these linear factors results in the *generator polynomial* $g(x)$ of degree $n-k$:

$$g(x) = \prod_{i=k}^{n-1}(x - \alpha^{-i}) .$$

This is valid, since the convolution theorem (theorem 3.6) gives $a(x) = i(x)g(x) \circ\!\!-\!\!\bullet\, IG = A$, with $G_i = 0$ for $k \leq i \leq n-1$. The RS code is cyclic, therefore, if $c(x) \in C$,

$$xc(x) = xi(x)g(x) = i'(x)g(x) \mod (x^n - 1) \in C.$$

Example 3.2 (Generator polynomial) The generator polynomial of the RS code from example 3.1 is calculated as follows ($\alpha = 5$: a primitive element from $GF(7)$, $A_i = 0,\ i = 2,3,4,5$):

$$\begin{aligned}
g(x) &= \prod_{i=2}^{5}(x - \alpha^{-i}) \\
&= (x-\alpha^{-2})(x-\alpha^{-3})(x-\alpha^{-4})(x-\alpha^{-5})\,\Big|_{\text{mod 6 (for exp.)}} \\
&= (x-\alpha^{4})(x-\alpha^{3})(x-\alpha^{2})(x-\alpha) \\
&= (x^2 - (\alpha^4+\alpha^3)x + \alpha^7)(x^2 - (\alpha^2+\alpha)x + \alpha^3) \\
&= (x^2 - x + 5)(x^2 - 2x + 6) \\
&= x^4 - x^3 + 5x^2 - 2x^3 + 2x^2 - 10x + 6x^2 - 6x + 30\,\Big|_{\text{mod 7 (for coef.)}} \\
g(x) &= x^4 + 4x^3 + 6x^2 + 5x + 2 .
\end{aligned}$$

The codeword **a** from example 3.1 must be divisible by $g(x)$. Therefore $a(x)/g(x) = i(x)$:

$$(4x^4 + 2x^3 + 3x^2 + 6x + 1)/(x^4 + 4x^3 + 6x^2 + 5x + 2) = 4$$

Therefore the polynomial is $a(x) = 4g(x)$, $i(x) = 4 + 0x$. ◇

3.1.3 Parity check polynomial

The parity check polynomial $h(x)$ is defined as

$$a(x)h(x) = 0 \mod (x^n - 1) \text{ for all } a(x) \in C .$$

The corresponding definition in the transform domain based on definition 3.5 is

$$A_i H_i = 0,\ \ i = 0,1,\ldots n-1, \qquad h(x) \circ\!\!-\!\!\bullet\, H(x).$$

H_i must be zero when the coefficients A_i are not equal to zero. The parity check polynomial is calculated through a similar equation as for the generator polynomial:

$$h(x) = \prod_{i=0}^{k-1}(x - \alpha^{-i}) ,\ \ \deg h(x) = k.$$

The roots of $g(x)$ and $h(x)$ are disjoint. All possible roots α^i are included in the two polynomials. Therefore, according to theorem 3.4, the product of $g(x)$ and $h(x)$ gives

$$g(x)h(x) = \prod_{i=0}^{n-1}(x-\alpha^{-i}) = x^n - 1.$$

Example 3.3 (Parity check polynomial) The parity check polynomial $h(x)$ of the RS code from example 3.1 is

$$h(x) = \prod_{i=0}^{1}(x-\alpha^{-i}) = (x-\alpha^0)(x-\alpha^{-1}) = (x-1)(x-3) = x^2+3x+3.$$

The product with the codeword $a(x) = 4x^4+2x^3+3x^2+6x+1$ gives

$$a(x)h(x) = 0 \mod (x^6-1),$$

$$(4x^4+2x^3+3x^2+6x+1)(x^2+3x+3) =$$

$$\begin{array}{rccccccc}
 & 4x^6 + & 2x^5 + & 3x^4 + & 6x^3 + & x^2 & & \\
 & & 12x^5 + & 6x^4 + & 9x^3 + & 18x^2 + & 3x & \\
 & & & +\ 12x^4 + & 6x^3 + & 9x^2 + & 18x + & 3 \\
\hline
\text{mod } 7: & * & 14 & 21 & 21 & 28 & 21 & * \\
 & & \| & \| & \| & \| & \| & \\
 & & 0 & 0 & 0 & 0 & 0 &
\end{array}$$

$*$ For $GF(7)$: $4x^{6 \bmod (n-1)} + 3x^0$, where $n = 7$ gives $(4+3)x^0 = 7 \mod 7 = 0$. ◇

3.1.4 Encoding

At this point, we introduce RS code generation using four different methods. Two techniques provide systematic encoding where information and parity check symbols are separated. All methods employ the same code, but a particular information vector is mapped, in general, to a different codeword, depending on the encoding technique. A summary of these methods is presented below.

Method 1 (non-systematic) The k information digits are the coefficients of the polynomial $A(x) = A_0 + A_1x + \ldots + A_{k-1}x^{k-1}$; the codeword $a(x)$ is given by the inverse transformation.

Method 2 (non-systematic) The k information digits are the coefficients of the polynomial $i(x) = i_0 + i_1x + \ldots + i_{k-1}x^{k-1}$; the codeword $a(x)$ results from the multiplication by the generator polynomial $a(x) = i(x)g(x)$.

Method 3 (systematic) The k information coordinates are $a_{n-k}, a_{n-k+1}, \ldots, a_{n-1}$; the $n-k$ parity check coordinates are calculated in the following manner:

$$\left(a_{n-1}x^{n-1} + \ldots + a_{n-k}x^{n-k}\right) : g(x) = i(x),\ \mathit{remainder}(x)$$

$$a(x) = a_{n-1}x^{n-1} + \ldots + a_{n-k}x^{n-k} - \mathit{remainder}(x).$$

Method 4 (systematic) The k information coordinates are $a_{n-k}, a_{n-k+1}, \ldots, a_{n-1}$; the $n-k$ parity check coordinates are calculated in the following manner:

$$a_j = -\frac{1}{h_0}\sum_{i=1}^{k} a_{n-i+j}h_i, \quad j = 0, 1, \ldots, n-k-1 \; .$$

The index $n-i+j$ is calculated $\bmod n$, and $h(x)$ is the parity check polynomial.

In section 3.2, where we present the decoding, we shall see that for decoding it does not matter which encoding method was used. However, it is apparent that the information can only be recovered when the coding method is known. This principle is true for all types of codes. The symbol error rate of the information depend on the encoding transformation from the information vector to the code vector but the block error does not.

3.1.5 General definition of RS codes

We may generate an equivalent code with the same minimum distance as the original code by taking an element α of order n and multiplying the coordinates a_i of a codeword by $\alpha^{ib}, b \in \mathbb{N}_0$:

$$a_i\alpha^{ib} = 0 \iff a_i = 0.$$

For $A(x) \bullet\!\!-\!\!\circ\, a(x)$,

$$a_i\alpha^{ib} \circ\!\!-\!\!\bullet\, x^bA(x) \mod (x^n - 1).$$

The multiplication by x^b corresponds to a cyclic shift of the polynomial $A(x)$.

For $U(x) = x^bA(x) \bmod (x^n - 1)$,

$$A_i = U_{i+b} \; .$$

The coordinates $A_i = 0$ are cyclicly shifted (the index $i+b$ is calculated $\bmod n$).

Definition 3.7 (RS code) *A RS code with dimension k, length n and minimum distance* $d = n-k+1$ *is defined as* $u(x) \circ\!\!-\!\!\bullet\, U(x)$, *with* $U(x) = x^bA(x) \bmod (x^n - 1)$ *and* $\deg A(x) < k$:

$$C := \{u(x) \mid u(x) \circ\!\!-\!\!\bullet\, U(x)\}$$

The transformation of all codewords in a RS code results in $d-1$ consecutive zeros, which implies a minimum distance of d. The indices of α for 0 and n are equivalent; therefore successive coordinates can also be interpreted in a cyclic manner:

$$n-i, n-i+1, \ldots, n-1, 0, 1, \ldots, j \; .$$

3.1.6 Generalized RS codes and extended RS codes

Choose n different non-zero elements $\alpha_0, \alpha_1, \ldots, \alpha_{n-1}$ from $GF(p)$. [1] Choose n (not necessarily different) non-zero elements $\beta_i \in GF(p), i = 0, \ldots, n-1$. A generalized Reed–Solomon code (GRS code) is defined by

$$C_{GRS} = \{a(x) \mid a_i = \beta_iA(\alpha_i),\ \beta_i \neq 0,\ i = 0, 1, \ldots, n-1\} \; ,$$

where $A(x) = A_0 + A_1x + \ldots + A_{k-1}x^{k-1}$, $A_i \in GF(p)$.

[1] A special case is to take an element α of order n; then $\alpha^i, i = 0, \ldots, n-1$, gives n different elements of $GF(p)$.

Theorem 3.8 (GRS code) *A generalized RS code C_{GRS} according the above definition has the same parameters as an RS code, length n, dimension k and minimum distance $d = n - k + 1$. If all $\beta_i = 1$, we get an RS code according to definition 3.7 .*

Proof According to theorem 3.1, $A(x)$ has at most $k-1$ consecutive zeros; therefore the polynomial $a(x)$ has at least $n-k+1 = d$ coordinates that are non-zero. Because the factor $\beta_i \neq 0$, the weight of $a(x)$ is not changed. □

Extension of RS codes by one coordinate

There exist RS codes $C(n,k,d)$ over $GF(p)$ that can be extended by one coordinate a_n as follows:

$$C_{erw} = \left\{ a_i,\ i = 0, 1, \ldots, n-1,\ a_n = -\sum_{i=0}^{n-1} a_i = -a(x=1) \right\},$$

where $a(x) \circ\!\!-\!\!\bullet A(x) = A_0 + A_1 x + \ldots + A_{k-1} x^{k-1}$, $A_i \in GF(p)$.

Theorem 3.9 (Simple extension of RS codes) *The extended RS code according to the above definition has length $n+1$, dimension k and minimum distance $d = n-k+2$; thus*

$$C_{erw}(n+1, k, d+1) = C_{erw}(p, k, n-k+2) .$$

Proof Take $a(x) \in C(n,k,d)$ a codeword with minimum weight and a generator polynomial constructed from the roots α^i $i \in 1 \ldots d-1$. The two cases that must be considered are when $a_n \neq 0$ and when $a_n = 0$. If $a_n \neq 0$ then the weight and therefore the minimum distance are increased by one owing to the linearity of the code. Therefore the minimum distance of the code C_{ext} is equal to $d+1$. We shall now examine the case when $a_n = 0$. By definition, $a(x) = i(x)g(x)$. In order to satisfy $a(x=1) = 0$, either $i(x=1) = 0$ or $g(x=1) = 0$, or both. According to our seletion of roots α^i, $i \in 1 \ldots d-1$, $g(x-1) \neq 0$. If $i(x=1)$ equals 0 then we may write $i(x) = (x-1)\tilde{i}(x)$, which gives

$$a(x) = i(x)g(x) = \tilde{i}(x)(x-1)g(x) .$$

We can redefine the generator polynomial of the extended code as

$$\tilde{g}(x) = (x-1)g(x)$$

The factors of the new generator polynomial $\tilde{g}(x)$ include the factor α^0, which by theorem 3.2 increases the minimum distance by one. It is clear that the code polynomial $a(x)$ is divisible by the new generator polynomial $\tilde{g}(x)$. It is therefore a codeword in $C_{ext}(n+1, k, d+1)$. □

Note that the RS codes extended by one position are not necessarily cyclic.

Using the definition of the DFT, one can also define the coordinate a_n as $-nA(x=0)$, because $A_0 = n^{-1}a(x=1)$.

Extension of RS codes by two coordinates

In the following description, knowledge of extension fields (section 2.4) and of non-primitive BCH codes (section 4.2) will be assumed.

We shall first define the non-primitive BCH code of length n over the base field $GF(q)$ where q is a prime p or a power of a prime p^m. Let $GF(q^l)$ be an extension field and let $\beta \in GF(q^l)$ be an element of order n. The cyclotomic cosets in $GF(q^l)$ are defined as

$$K_i = \{iq^j \mod n, \; j = 0, 1, \ldots, l-1\}.$$

The generator polynomial of the code $C(n,k,d)$ over $GF(q)$ is defined (see section 4.2) by

$$g(x) = \prod_{i \in \mathcal{M}} (x - \beta^i), \quad \mathcal{M} = \bigcup K_i,$$

and the maximum number $d-1$ of consecutive zeros of $g(x)$ defines the designed minimum distance d. For the dimension, we have $k = n - |\mathcal{M}|$.

A double extended RS code has the parameters $C_{ext^2}(n = q+1, k, d)$. This is possible if we choose $l = 2$ and $n = q+1$. Thus we obtain the specific cyclotomic cosets

$$K_i = \{iq^j \mod q+1, \; j = 0, 1\}.$$

Note that the calculation mod $(q+1)$ gives $aq = q - a + 1 = -a \bmod (q+1)$. We shall show in the following that each cyclotomic coset (except K_0) consists of exactly two elements. This is because for $j = 2$ and $j = 0$, we get the same element: $iq^0 = iq^2 \bmod (q+1)$, since $q^2 = (q+1)^2 - 2(q+1) + 1 = 1 \bmod (q+1)$. Thus the list of cyclotomic cosets is as follows:

$K_0 = \{0\}$;

$K_1 = \{1, q\}$, because $1q^2 : (q+1) = q$ remainder $-q$, and $-q = 1 \mod (q+1)$;

$K_2 = \{2, q-1\}$, because $2q^2 : (q+1) = 2q$ remainder $-2q$, and $-2q = 2 \mod (q+1)$;

$K_3 = \{3, q-2\}$, because $3q^2 = 3 \mod (q+1)$;

$\vdots$

If q is even then any code generated by the union of cyclotomic cosets $\mathcal{M} = K_0 \cup K_1 \cup \ldots \cup K_t$ has the designed minimum distance $d = 2t+2$. Because $2t+1$ consecutive numbers in $\mathcal{M}$ occur, namely $\mathcal{M} = \{q-t+1, \ldots, q-1, q, 0, 1, 2, \ldots, t\}$, $|\mathcal{M}| = 2t+1$. The code defined by the cyclotomic cosets $\mathcal{M} = \{K_{\frac{q}{2}}, K_{\frac{q}{2}-1}, \ldots, K_{\frac{q}{2}-t+1}\}$ has the designed minimum distance $d = 2t+1$ and $|\mathcal{M}| = 2t$. In both cases this is also the true minimum distance. For odd q, the minimum distance can be proved through similar arguments.

A polynomial defined by a single cyclotomic coset is

$$\begin{aligned} m_i(x) &= \prod_{j \in K_i} (x - \alpha^j) = (x - \alpha^i)(x - \alpha^{q-i+1}) \\ &= (x - \alpha^i)(x - \alpha^{-i}) = x^2 - (\alpha^i + \alpha^{-i}) + 1 \, . \end{aligned}$$

Clearly these factors satisfy the requirements of a generator polynomial.

Remark The sum $\alpha^i + \alpha^{-i}$ is an element of the basic field $GF(q)$ (compare with the trace function). The product over all possible $m_i(x)$ results in $x^n - 1$.

Thus we have proved the following theorem:

Theorem 3.10 (Double extended RS codes) *For RS codes $C(n,k,d)$ of length $n = 2^m - 1$, there exists a double extended cyclic code with the parameters*

$$C_{erw^2}(2^m + 1, k, n-k+3), \quad \textit{or equivalently} \quad C_{erw^2}(n+2, k, d+2) \, .$$

The following example from [McWSl] illustrates the construction of a double extension of an RS code.

Example 3.4 (Double extended RS codes) Let us construct a code of length $n = (2^3 - 1) + 2 = 9$ over the basic field $GF(2^3)$ according to theorem 3.10. To do this, we use a primitive element of an extended field $\eta \in GF(2^6)$, giving $2^6 - 1 = 63$, and $9 \mid 63$, and $7 \mid 63$. The element $\alpha = \eta^9$ has order 7 and can be used as a primitive element of $GF(2^3)$. In contrast the element $\beta = \eta^7$ has order 9 in $GF(2^6)$. We calculate the powers of the element α:

$$\alpha = \eta^9, \; \alpha^2 = \eta^{18}, \; \alpha^3 = \eta^{27}, \; \alpha^4 = \eta^{36}, \; \alpha^5 = \eta^{45}, \; \alpha^6 = \eta^{54}, \; \alpha^7 = \eta^{63} = 1 \, ,$$

and

$$\beta = \eta^7, \; \beta^9 = \eta^{63} = 1 \, .$$

The cyclotomic coset is calculated through

$$K_i = \{i \cdot 8^j \mod 9, \; j = 0, 1\}, \quad i = 0, 1, 2, 3, 4,$$

which gives

$$K_0 = \{0\}, \; K_1 = \{1, 8\}, \; K_2 = \{2, 7\}, \; K_3 = \{3, 6\}, \; K_4 = \{4, 5\}.$$

The subfield elements from $GF(2^3)$ of $GF(2^6)$ can be calculated through

$$\begin{aligned} \beta + \beta^{-1} &= \eta^7 + \eta^{56} &= \alpha^5, \\ \beta^2 + \beta^{-2} &= \eta^{14} + \eta^{49} &= \alpha^3, \\ \beta^3 + \beta^{-3} &= \eta^{21} + \eta^{42} &= \alpha^0, \\ \beta^4 + \beta^{-4} &= \eta^{28} + \eta^{35} &= \alpha^6 \, . \end{aligned}$$

Using this subfield, the polynomial $x^9 - 1$ can be factored using the coefficients from $GF(2^3)$:

$$x^9 - 1 = (x-1)(x^2 + x + 1)(x^2 + \alpha^3 x + 1)(x^2 + \alpha^5 x + 1)(x^2 + \alpha^6 x + 1) \, .$$

A polynomial $x^{2^m+1} - 1$ has root 1 and roots β^i and β^{-i}, and thus only quadratic factors.

With this, we can construct the following codes of length 9. The minimum distance is calculated by the number of consecutive roots.

Code	Generator polynomial $g(x)$
$C(9,7,3)$	$x^2 + \alpha^6 x + 1$
$C(9,6,4)$	$(x-1)(x^2 + \alpha^5 x + 1)$
$C(9,5,5)$	$(x^2 + x + 1)(x^2 + \alpha^6 x + 1)$
$C(9,4,6)$	$(x-1)(x^2 + \alpha^3 x + 1)(x^2 + \alpha^5 x + 1)$
$C(9,3,7)$	$(x^2 + x + 1)(x^2 + \alpha^3 x + 1)(x^2 + \alpha^5 x + 1)$
$C(9,2,8)$	$(x-1)(x^2 + x + 1)(x^2 + \alpha^3 x + 1)(x^2 + \alpha^5 x + 1)$

◇

Remark In contrast to single extended RS codes, double extended RS codes are cyclic.

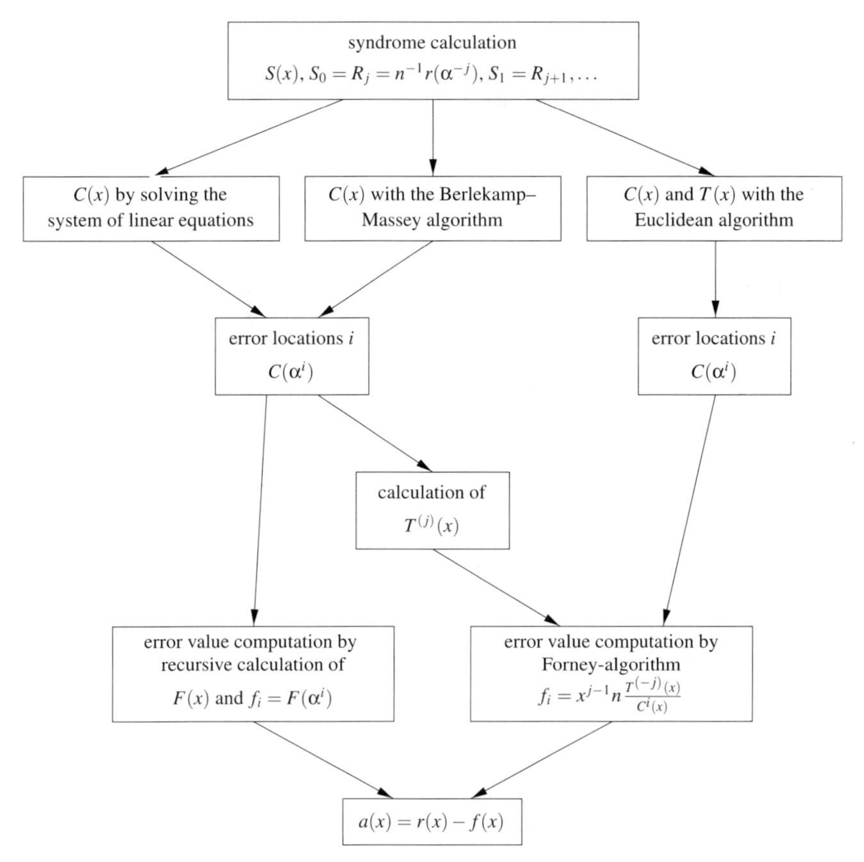

Figure 3.1 Overview of algebraic decoding.

3.2 Algebraic decoding

In figure 3.1, a flow chart for algebraic decoding is presented. In this section we describe algorithms for the particular steps. We shall assume an RS code whose codewords $a(x)$ are defined according definition 3.5 as

$$a(x) \circ\!\!-\!\!\bullet\, A(x), \ \ A_0 = A_1 = A_2 = \ldots = A_{d-2} = 0\ .$$

$A(x)$ has exactly $d-1$ successive coefficients equal to zero; hence the RS code has distance d and can correct $\lfloor \frac{d-1}{2} \rfloor$ errors.

We assume that we receive the vector $r(x) = a(x) + f(x)$, where the coefficient $f_i \in GF(q)$. The error polynomial $f(x)$ is defined as having $f_i \neq 0$ for every error location i. We shall also assume that fewer than $\lfloor \frac{d-1}{2} \rfloor$ errors have occurred during the transmission, i.e.

$$\mathrm{wt}(\mathbf{f}) \leq \left\lfloor \frac{d-1}{2} \right\rfloor .$$

The error locations can be expressed in terms of the support by

$$\mathrm{supp}(\mathbf{f}) := \{i \mid f_i \neq 0\}\ .$$

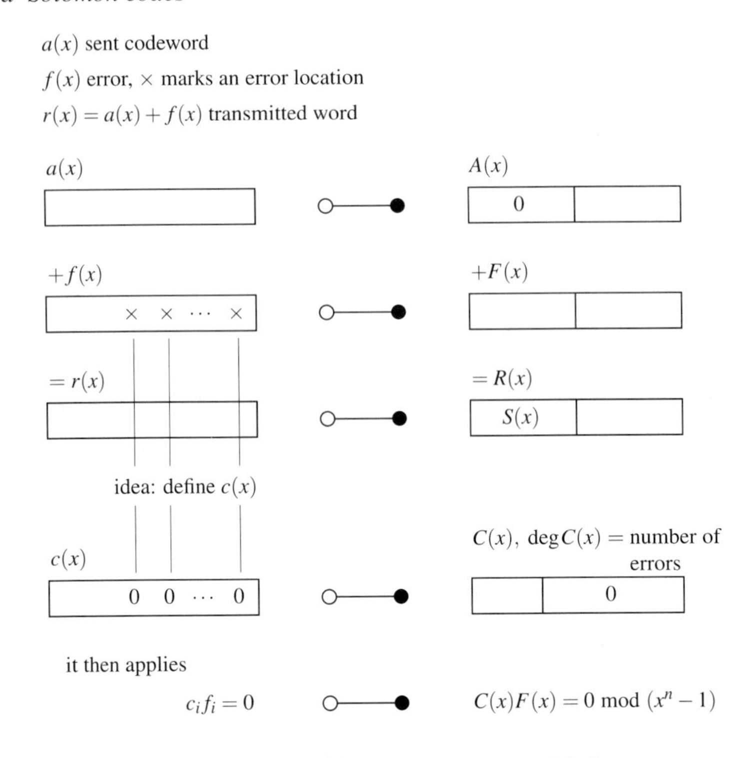

Figure 3.2 Error correction concept.

To determine whether $r(x)$ is a codeword, $r(x)$ is transformed using the Fourier transform and one checks if $R_0 = \ldots = R_{d-2} = 0$, since

$$a(x) + f(x) = r(x) \circ\!\!-\!\!\bullet R(x) = A(x) + F(x) .$$

If $f(x) = 0$ then $R(x) = A(x)$, and the coefficients $R_0, \ldots, R_{d-2}$ are equal to zero. On the other hand, if $f(x) \neq 0$ then we define the syndrome polynomial $S(x) = S_0 + S_1 x + \ldots + S_{d-2} x^{d-2}$, where

$$R_i = F_i = S_i, \;\; i = 0, 1, \ldots, d-2 .$$

This means $j = 0$ in figure 3.1. The syndrome $S(x)$ depends only on the transformed error polynomial $F(x)$ because, by definition, $A_0 = \ldots = A_{d-2} = 0$.

For the selection of a code for a given channel, the expected value for the number of errors should be smaller than half the minimum distance of the code. The decoding problem can now be formulated as finding an error vector $f(x)$ with the fewest possible number of non-zero coefficients which generate the received vector $r(x)$. This is equivalent to finding an error vector $\mathbf{f}$ with the smallest possible weight. Unfortunately, the property *smallest weight* cannot be used algebraically; therefore we consider the corresponding property *smallest degree* instead.

Error locator polynomial The *error locator polynomial* is defined such that

$$c_i = 0 \Longleftarrow f_i \neq 0, \quad \text{i. e.} \quad c_i f_i = 0 .$$

The coefficients of $c(x)$ are equal to zero at the error locations and arbitrary at the non-error locations. This defines a large class of polynomials. From this class of polynomials we select only those for which $\deg C(x), c(x) \circ\!\!-\!\!\bullet\, C(x)$ is equal to the number of errors. This implies that $c_i \neq 0 \Leftarrow f_i = 0$. With the transformation according to definition 3.5, we can map $c(x)$ to $C(x)$ as

$$C(x) := \prod_{i \in \text{supp}(\mathbf{f})} (x - \alpha^i). \tag{3.1}$$

The degree of $C(x)$ is the number of coefficients from $c(x)$ that are equal to zero, or, in other words, the number of errors in the error vector $f(x)$.

Using the Fourier transform convolution theorem, we get

$$c_i \cdot f_i = 0 \circ\!\!-\!\!\bullet\, C(x)F(x) = 0 \mod (x^n - 1) .$$

When $e \leq \left\lfloor \frac{d-1}{2} \right\rfloor$ is the actual number of errors that have occurred, $C(x)$ results in the polynomial from equation 3.1:

$$C(x) = C_0 + C_1 x^1 + \ldots + C_e x^e ,$$

This produces $e+1$ coefficients C_i, $i = 0, 1, \ldots, e$, which must be determined. Because we are interested only in the e zeros of the polynomial, any single coefficient can be freely chosen. By equation 3.1, C_e is equal to one. We can also choose $C_0 = 1$, but then get another representation of $C(x)$, as follows:

$$C(x) := \prod_{i \in \text{supp}(\mathbf{f})} (1 - \alpha^{-i} \cdot x) .$$

It is advisable to note that the normalization of $C(x)$ does not change the roots of the polynomial. The strategy of algebraic decoding is shown schematically in figure 3.2.

3.2.1 Key equation

We will now present different methods for the computation of the coefficients of error locator polynomial $C(x)$. Let us assume that $t = \left\lfloor \frac{d-1}{2} \right\rfloor$ errors have occurred. The multiplication of polynomials

$$C(x)F(x) = 0 \mod (x^n - 1) \tag{3.2}$$

can be written as a linear system of equations (LSE). We use the fact that the syndrome $S(x)$ is known and defined in the following manner:

$$F_i = S_i, \quad i = 0, 1, \ldots, d-2 .$$

This gives

$$
\begin{array}{llllllll}
 & C_0S_0 & + C_1F_{n-1} & + C_2F_{n-2} & + \ldots & + C_tF_{n-t} & = 0, & \\
 & C_0S_1 & + C_1S_0 & + C_2F_{n-1} & + \ldots & + C_tF_{n-t+1} & = 0, & \\
 & \vdots & & & & \vdots & & \\
 & C_0S_{t-1} & + C_1S_{t-2} & + C_2S_{t-3} & + \ldots & + C_tF_{n-1} & = 0, & \\
* & C_0S_t & + C_1S_{t-1} & + C_2S_{t-2} & + \ldots & + C_tS_0 & = 0, & \\
* & C_0S_{t+1} & + C_1S_t & + C_2S_{t-1} & + \ldots & + C_tS_1 & = 0, & (3.3)\\
* & & \vdots & & & & \vdots & \\
* & C_0S_{2t-1} & + C_1S_{2t-2} & + C_2S_{2t-3} & + \ldots & + C_tS_{t-1} & = 0, & \\
 & C_0F_{2t} & + C_1S_{2t-1} & + C_2S_{2t-2} & + \ldots & + C_tS_t & = 0, & \\
 & & \vdots & & & & \vdots & \\
 & C_0F_{n-1} & + C_1F_{n-2} & + C_2F_{n-3} & + \ldots & + C_tF_{n-t-1} & = 0\,. &
\end{array}
$$

The t equations marked with $*$ depend only on the known syndrome coefficients and the unknown error locator polynomial coefficients (one coefficient may be freely chosen). Because we have t equations and t unknown variables, the linear system of equations is theoretically solvable. From the product $C(x)F(x) = 0 \bmod (x^n - 1)$, we shall derive the key equation as follows. Consider the product $C(x)S(x)$:

$$
\begin{array}{rllllll}
0: & C_0S_0, & & & & & \\
1: & C_0S_1 & + C_1S_0, & & & & \\
 & \vdots & & & & & \\
t-1: & C_0S_{t-1} & + C_1S_{t-2} & + \ldots & & + C_{t-1}S_0, & \\
t: & C_0S_t & + C_1S_{t-1} & + C_2S_{t-2} & + \ldots & + C_tS_0, & \\
t+1: & C_0S_{t+1} & + C_1S_t & + C_2S_{t-1} & + \ldots & + C_tS_1, & (3.4)\\
 & \vdots & & & & & \\
2t-1: & C_0S_{2t-1} & + C_1S_{2t-2} & + C_2S_{2t-3} & + \ldots & + C_tS_{t-1}, & \\
2t: & & C_1S_{2t-1} & + C_2S_{2t-2} & + \ldots & + C_tS_t, & \\
 & \vdots & & & & & \\
3t-1: & & & & & C_tS_{2t-1}\,. &
\end{array}
$$

The coefficients $0, 1, \ldots, t-1$ produced by the product of $C(x)S(x)$ according to equations 3.3 are not relevant in the solution of the polynomial $C(x)$. These coefficients need not be zero, however, they can be zero. We shall set the coefficients $0, 1, \ldots t-1$ of the product $C(x)S(x)$ equal to $-T_0, -T_1, \ldots, -T_{t-1}$, or equivalently $-T(x) = -T_0 - T_1x - \ldots - T_{t-1}x^{t-1}$. The polynomial $T(x)$ is called the *error evaluator* polynomial and is not relevant to the solution of the error locator polynomial $C(x)$.

The coefficients $2t, 2t+1, \ldots, 3t-1$ produced by $C(x)S(x)$ from equations 3.3 are also not important for the solution of the coefficients of $C(x)$. By calculating the product $C(x)S(x)$ modulo x^{2t}, these coefficients can be easily eliminated.[2]

[2]The calculation of $x^{n+1} \bmod (x^n - 1)$ gives the value x. However, the calculation $\bmod x^{2t}$ results in $x^{2t+i} = 0$, $i = 0, 1, \ldots$.

These considerations lead to the *key equations* :

$$C(x)S(x) = -T(x) \mod x^{2t} \ . \tag{3.5}$$

As we have indicated, the necessary section part of the linear system of equations (marked by $*$) in equations 3.3 is contained in the key equations, namely

$$0 = \sum_{i=0}^{t} C_i S_{j-i}, \qquad j = t, t+1, \ldots, 2t-1 \ .$$

Both representations can be used to determining $C(x)$.

For $e < t$ errors Assume that $t-1$ errors have occurred; therefore $C_t = 0$, and it follows from the linear system of equations 3.3 that we need to determine only $t-1$ coefficients of $C(x)$ (one coefficient can be chosen freely). There still exist t equations (marked by $*$ in equations 3.3) that contain only the calculated syndrome coefficients and the unknown coefficients of $C(x)$. We also have an extra equation containing only known syndrome coefficients:

$$C_0 S_{t-1} + C_1 S_{t-2} + C_2 S_{t-3} + \ldots + C_{t-1} S_0 = 0 \ .$$

In this case, the system of linear equations is overdetermined. We have $t-1$ unknowns and $t+1$ equations.

There also exist in the key equations 3.4 an extra equation, corresponding to the coefficient $t-1$. This means that $-T_{t-1} = 0$.

In general, to determine the polynomial $C(x)$ for any number of errors $e \leq t = \left\lfloor \frac{d-1}{2} \right\rfloor$, we have to determine e coefficients from t equations. Note that the determined coefficients should satisfy all equations containing exclusively C_i and S_i. Again, we have $2t - e$ such equations. The corresponding part of the linear system of equations can be written as

$$\begin{pmatrix} S_e & \ldots & S_1 & S_0 \\ S_{e+1} & \ldots & S_2 & S_1 \\ \vdots & & & \vdots \\ \vdots & & & \vdots \\ S_{2t-1} & \ldots & S_{2t-e} & S_{2t-e-1} \end{pmatrix} \cdot \begin{pmatrix} C_0 \\ C_1 \\ \vdots \\ C_{e-1} \\ 1 \end{pmatrix} = 0 \ . \tag{3.6}$$

For the case $e < t$, the following relationship for the key equations must hold:

$$\deg T(x) < \deg C(x).$$

The the key equation gives the same solution as do equations 3.3

Definition 3.11 (Key equation). *The key equation used to calculate the error locator polynomial $C(x)$ is defined as*

$$C(x)S(x) = -T(x) \mod x^{2t}, \qquad \textit{with } \deg T(x) < \deg C(x)$$

The degree of $C(x)$ is equal to the number of errors that have occurred.

Calculation of $C(x)$ by LSE The number of errors e that have occurred in the codeword is not known. One can determine this number by trial and error. We start with $e = 1$ and if the solution satisfies all equations in 3.6 then one error has occurred. Otherwise, we set $e = 2$ and so on. A second method is to determine the rank of the matrix of syndrome coefficients in equation 3.6. The rank of the matrix corresponds to the number of errors that have occurred in the codeword.

For large e or t, the solution has a high complexity. One can take advantage of the regular structure of a LSE to reduce the complexity of determining $C(x)$ with the smallest degree possible.

Before we give an example, we reiterate that an encoded RS codeword has $A_k = \ldots = A_{n-1} = 0$ for all codewords $a(x)$. Multiplication of the equation $C(x)F(x) \bmod (x^n - 1)$ by any power of x results in a cyclic reordering of equations 3.3. Using $S_0 = R_i = F_i$, $S_1 = R_{i+1} = F_{i+1}, \ldots, S_{d-2} = R_{d-2+i} = F_{d-2+i}$ as coefficients of $S(x)$, one can see that equations 3.4 are valid for $i \in \{0, 1, \ldots, n-1\}$.

According to the definition of the key equation (definition 3.11), only the syndrome is required for the solution. However, for any location of the syndrome $S(x)$, we get the same solution for $C(x)$.

Example 3.5 (Error locations by LSE) We shall assume that the codeword **a** from example 3.1 has been transmitted. We choose $e = \lfloor \frac{d-1}{2} \rfloor = 2$. Assume that the error $f(x) = 5x^4 + 3x$ has occurred during transmission. If $a(x) = 4x^4 + 2x^3 + 3x^2 + 6x + 1$ then the received vector is

$$r(x) = a(x) + f(x) = 2x^4 + 2x^3 + 3x^2 + 2x + 1\,.$$

At the receiver, we know only $r(x)$ and the RS code used. We calculate the syndrome $S(x)$ of the received vector ($\alpha = 5$ is the primitive element):

$$\begin{aligned} S_0 = R_2 &= n^{-1}r(\alpha^{-2}) = 6r(\alpha^4) \\ &= 6(2\alpha^{16} + 2\alpha^{12} + 3\alpha^8 + 2\alpha^4 + 1) \\ &= 6(2\alpha^4 + 2\alpha^0 + 3\alpha^2 + 2\alpha^4 + 1) \\ &= 6(2 \cdot 2 + 2 \cdot 1 + 3 \cdot 4 + 2 \cdot 2 + 1) \\ &= 6(4 + 2 + 12 + 4 + 1) = 6 \cdot 23 \\ &= 5 \mod 7, \end{aligned}$$

$$\begin{aligned} S_1 = R_3 &= 6 \cdot r(\alpha^3) = 6(2 \cdot \alpha^{12} + 2\alpha^9 + 3\alpha^6 + 2\alpha^3 + 1) \\ &= 6(2\alpha^0 + 2\alpha^3 + 3\alpha^0 + 2\alpha^3 + 1) \\ &= 6(2 + 12 + 3 + 12 + 1) \\ &= 5 \mod 7, \end{aligned}$$

$$S_2 = R_4 = 3,$$

$$S_3 = R_5 = 3$$

$$\Longrightarrow \quad S(x) = 5 + 5x + 3x^2 + 3x^3\,.$$

First we assume that a single error has occurred ($e = 1$). The LSE that must be satisfied are calculated using equation 3.6, $C_0 + x = C(x)$:

$$S_1 \cdot C_0 + S_0 = 0 = 5 \cdot C_0 + 5 = 0$$
$$\Longrightarrow \quad C_0 = 6, \text{ and } 5 \cdot 6 + 5 = 0 \mod 7 \,.$$

We get $C(x) = 6 + x$ and check if the equations

$$S_2 \cdot 6 + S_1 = 0 \quad \text{and} \quad S_3 \cdot 6 + S_2 = 0$$

are also fulfilled. However, we discover that

$$S_2 \cdot 6 + S_1 = 3 \cdot 6 + 5 = 2 \mod 7.$$

Hence we must assume that two errors have occurred: $C(x) = C_0 + C_1 x + x^2$,

$$\begin{array}{rcl} S_2 C_0 + S_1 C_1 + S_0 &=& 0, \\ S_3 C_0 + S_2 C_1 + S_1 &=& 0, \end{array}$$

$$\left.\begin{array}{rcl} 3C_0 + 5C_1 + 5 &=& 0, \\ & & \\ 3C_0 + 3C_1 + 5 &=& 0, \end{array}\right\} \begin{array}{rcl} 2C_1 &=& 0, \\ C_1 &=& 0, \\ C_0 &=& 3, \end{array}$$

$$C(x) = 3 + x^2 \,.$$

The two roots of $C(x)$ are the two error locations. We try all possible elements of $GF(7)$ to find these roots:

$$\begin{array}{rlcll} 1: & 3+1 &=& 4, & \\ 2: & 3+4 &=& 0 & \mod 7 \Longrightarrow \text{1st root: } 2 = \alpha^4, \\ 3: & 3+9 &=& 5 & \mod 7, \\ 4: & 3+16 &=& 5 & \mod 7, \\ 5: & 3+25 &=& 0 & \mod 7 \Longrightarrow \text{2nd root: } 5 = \alpha^1, \end{array}$$

$$C(x) = (x-2)(x-5) = (x-\alpha^4)(x-\alpha^1) \,.$$

The errors are therefore in the first and fourth coordinates:

$$f(x) = f_4 x^4 + f_1 x \,.$$

At this point, we cannot say what the values are for f_4 and f_1 (the error value calculation is given in section 3.2.4). ◇

Finding the roots of $C(x)$ In the above example, the roots of $C(x)$ were determined by substituting all the elements of $GF(7)$ into $C(x)$. This technique is also known as the *Chien search algorithm.*

Decoding failure If the number of roots is smaller than the degree of $C(x)$ then the decoding has failed. This means that if more errors have occurred during transmission than the number of correctable errors then a decoding failure might happen.

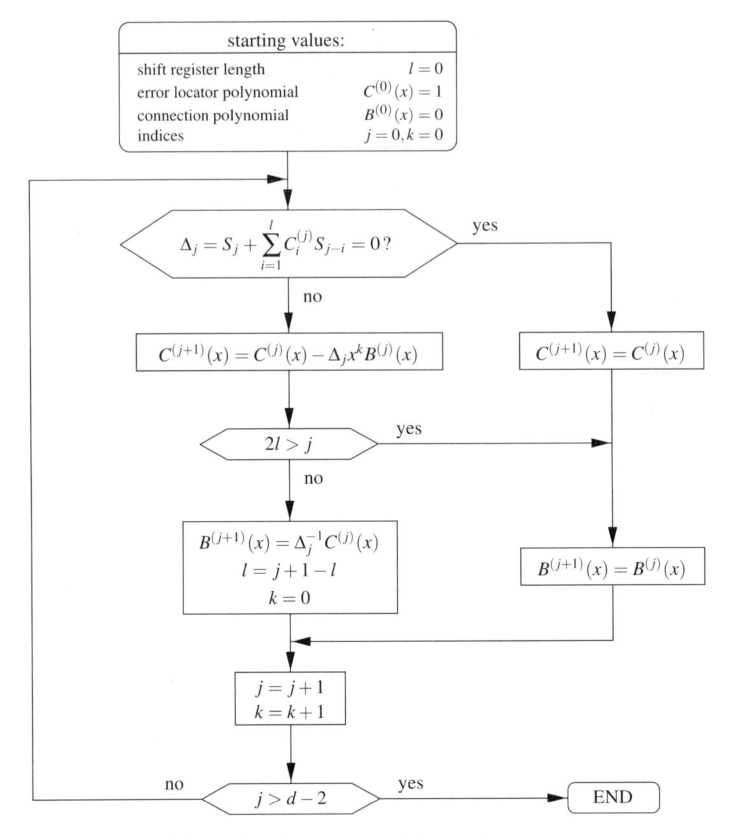

Figure 3.3 Berlekamp–Massey flow chart.

3.2.2 Berlekamp–Massey algorithm

The Berlekamp–Massey algorithm (BMA) is a very efficient method of calculating $C(x)$ that has the smallest degree satisfying the key equation 3.5. The BMA solution can also be formulated as the shortest feedback shift register with feedback coefficients C_i that can generate all syndrome coefficients. The algorithm is iterative, and begins with the error locator polynomial $C(x)$ with degree 1. The degree of $C(x)$ is increased by at most one for every iteration. The new $C_i(x)$ is calculated from the preceding polynomial $C_{i-1}(x)$.

Figure 3.3 shows a flow chart of the BMA that can be used in the solution of the key equation 3.5. We shall show that the Euclidean algorithm (section 3.2.3) can also generate a solution for $C(x)$ and that these two algorithms are equivalent (section 3.2.5).

Example 3.6 (Berlekamp–Massey algorithm) We should like to find the error locator polynomial for the syndrome $S(x) = 5 + 5x + 3x^2 + 3x^3$ from example 3.5 (see table 3.1). The error coordinates are 1 and 4.

Because the solution of the BMA always has $C_0 = 1$, it results in an equivalent solution of the LSE multiplied by a factor. The solution is shown through

$$C_{LGS}(x) = C_{0,LGS} C_{BMA}(x) .$$

Table 3.1 Calculation of the error locator polynomial with the BMA (example 3.6).

j	k	l	Δ_j	$C^{(j+1)}(x)$	$2l > j$	$B^{(j+1)}(x)$
0	0	0	$\Delta_0 = S_0 = 5$	$C^{(1)} = 1 - 5x^0 \cdot 0 = 1$	no	$B^{(1)} = 5^{-1} = 3$
1	1	1	$\Delta_1 = S_1 + C_1^{(1)} S_0 = 5$	$C^{(2)} = 1 - 5x \cdot 3 = 1 + 6x$	yes	$B^{(2)} = B^{(1)} = 3$
2	2	1	$\Delta_2 = S_2 + C_1^{(2)} S_1 = 3 + 6 \cdot 5 = 5$	$C^{(3)} = C^{(2)} - \Delta_2 x^2 B^{(2)} = 1 + 6x - 5x^2 \cdot 3 = 1 + 6x + 6x^2$	no	$B^{(3)} = 3(1+6x) = 3 + 4x$
3	1	2	$\Delta_3 = S_3 + C_1^{(3)} S_2 + C_2^{(3)} S_1 = 3 + 6 \cdot 3 + 6 \cdot 5 = 2$	$C^{(4)} = C^{(3)} - 2xB^{(3)} = 1 + 6x + 6x^2 - 2x(3+4x) = 1 + 5x^2$	yes	$B^{(4)} = B^{(3)}$

The factor does not alter the roots of $C(x)$.

The BMA is a very efficient method for the solution of the error locator polynomial. It greatly reduces the effort of solving the key equations, allowing RS codes to be used in practical situations. Nevertheless, we shall give another method for the solution of the key equation.

3.2.3 Euclidean algorithm

The Euclidean algorithm is given in theorem 2.11. It calculates the greatest common divisor of two integers or polynomials. We shall generalize the Euclidean algorithm for integers from section 2.3.2 to polynomials. Moreover, we present it in a modified form. The algorithm finds the greatest common divisor

$$\gcd(a(x), b(x)), \quad \text{with } \deg\, a(x) < \deg\, b(x) .$$

The chain of divisions

$$\begin{aligned} b(x) &= q_1(x)a(x) + r_1(x), \\ a(x) &= q_2(x)r_1(x) + r_2(x), \\ r_1(x) &= q_3(x)r_2(x) + r_3(x), \\ &\vdots \end{aligned}$$

can be described as

$$\begin{pmatrix} r_{i-2}(x) \\ r_{i-1}(x) \end{pmatrix} = \begin{pmatrix} q_i(x) & 1 \\ 1 & 0 \end{pmatrix} \cdot \begin{pmatrix} r_{i-1}(x) \\ r_i(x) \end{pmatrix}, \quad i = 1, 2, \ldots , \tag{3.7}$$

where $a(x) = r_0(x)$ and $b(x) = r_{-1}(x)$. In the same manner, we introduce recursive relations for the polynomials $v_j(x)$ and $w_j(x)$ that allow us to present every remainder as a linear combination of $a(x)$ and $b(x)$:

$$r_j(x) = v_j(x)b(x) + w_j(x)a(x) . \tag{3.8}$$

Through the relationship

$$\begin{pmatrix} w_i(x) & -w_{i-1}(x) \\ -v_i(x) & v_{i-1}(x) \end{pmatrix} = \begin{pmatrix} w_{i-1}(x) & -w_{i-2}(x) \\ -v_{i-1}(x) & v_{i-2}(x) \end{pmatrix} \cdot \begin{pmatrix} -q_i(x) & -1 \\ -1 & 0 \end{pmatrix}, \tag{3.9}$$

a recursive relationship is derived

$$\begin{aligned}
v_{-1}(x) &= 1, & w_{-1}(x) &= 0, \\
v_0(x) &= 0, & w_0(x) &= 1, \\
v_1(x) &= v_{-1}(x) - q_1(x)v_0(x), & w_1(x) &= w_{-1}(x) - q_1(x)w_0(x), \\
&\vdots & &\vdots \\
v_j(x) &= v_{j-2}(x) - q_j(x)v_{j-1}(x), & w_j(x) &= w_{j-2}(x) - q_j(x)w_{j-1}(x).
\end{aligned}$$

With the recursion equations 3.7 and 3.9, we can give an elegant formulation showing that every remainder can be represented in the form of equation 3.8:

$$\begin{aligned}
&\begin{pmatrix} w_i(x) & -w_{i-1}(x) \\ -v_i(x) & v_{i-1}(x) \end{pmatrix} \\
&= \begin{pmatrix} w_{i-1}(x) & -w_{i-2}(x) \\ -v_{i-1}(x) & v_{i-2}(x) \end{pmatrix} \cdot \begin{pmatrix} q_i(x) & 1 \\ 1 & 0 \end{pmatrix} \cdot (-1) \\
&= \begin{pmatrix} w_{i-2}(x) & -w_{i-3}(x) \\ -v_{i-2}(x) & v_{i-3}(x) \end{pmatrix} \cdot \begin{pmatrix} q_{i-1}(x) & 1 \\ 1 & 0 \end{pmatrix} \cdot \begin{pmatrix} q_i(x) & 1 \\ 1 & 0 \end{pmatrix} \cdot (-1) \cdot (-1) \\
&= \begin{pmatrix} w_{i-3}(x) & -w_{i-4}(x) \\ -v_{i-3}(x) & v_{i-4}(x) \end{pmatrix} \cdot \begin{pmatrix} q_{i-2}(x) & 1 \\ 1 & 0 \end{pmatrix} \cdot \begin{pmatrix} q_{i-1}(x) & 1 \\ 1 & 0 \end{pmatrix} \cdot \begin{pmatrix} q_i(x) & 1 \\ 1 & 0 \end{pmatrix} \cdot (-1)^3 \qquad (3.10) \\
&\vdots \\
&= \begin{pmatrix} 1 & 0 \\ 0 & 1 \end{pmatrix} \cdot \begin{pmatrix} q_1(x) & 1 \\ 1 & 0 \end{pmatrix} \cdot \begin{pmatrix} q_2(x) & 1 \\ 1 & 0 \end{pmatrix} \cdots \begin{pmatrix} q_i(x) & 1 \\ 1 & 0 \end{pmatrix} \cdot (-1)^i .
\end{aligned}$$

Another equivalent representation of equation 3.7 is

$$\begin{aligned}
\begin{pmatrix} r_{i-2}(x) \\ r_{i-1}(x) \end{pmatrix} &= \begin{pmatrix} q_i(x) & 1 \\ 1 & 0 \end{pmatrix} \cdot \begin{pmatrix} r_{i-1}(x) \\ r_i(x) \end{pmatrix}, \qquad (3.11) \\
\begin{pmatrix} r_{i-3}(x) \\ r_{i-2}(x) \end{pmatrix} &= \begin{pmatrix} q_{i-1}(x) & 1 \\ 1 & 0 \end{pmatrix} \cdot \begin{pmatrix} q_i(x) & 1 \\ 1 & 0 \end{pmatrix} \cdot \begin{pmatrix} r_{i-1}(x) \\ r_i(x) \end{pmatrix}, \\
\begin{pmatrix} r_{i-4}(x) \\ r_{i-3}(x) \end{pmatrix} &= \begin{pmatrix} q_{i-2}(x) & 1 \\ 1 & 0 \end{pmatrix} \cdot \begin{pmatrix} q_{i-1}(x) & 1 \\ 1 & 0 \end{pmatrix} \cdot \begin{pmatrix} q_i(x) & 1 \\ 1 & 0 \end{pmatrix} \cdot \begin{pmatrix} r_{i-1}(x) \\ r_i(x) \end{pmatrix}, \\
&\vdots \\
\begin{pmatrix} r_{-1}(x) \\ r_0(x) \end{pmatrix} &= \begin{pmatrix} q_1(x) & 1 \\ 1 & 0 \end{pmatrix} \cdot \begin{pmatrix} q_2(x) & 1 \\ 1 & 0 \end{pmatrix} \cdots \begin{pmatrix} q_i(x) & 1 \\ 1 & 0 \end{pmatrix} \cdot \begin{pmatrix} r_{i-1}(x) \\ r_i(x) \end{pmatrix} .
\end{aligned}$$

We insert equation 3.10 in 3.11, giving

$$\begin{pmatrix} r_{-1}(x) \\ r_0(x) \end{pmatrix} = (-1)^i \cdot \begin{pmatrix} w_i(x) & -w_{i-1}(x) \\ -v_i(x) & v_{i-1}(x) \end{pmatrix} \cdot \begin{pmatrix} r_{i-1}(x) \\ r_i(x) \end{pmatrix} . \qquad (3.12)$$

To inverse the matrix in equation3.12, we need the determinant. This can be determined based on the equation representation of 3.10:

$$\begin{vmatrix} q_i(x) & 1 \\ 1 & 0 \end{vmatrix} = -1.$$

Therefore the determinant is 1, which gives

$$\begin{pmatrix} r_{i-1}(x) \\ r_i(x) \end{pmatrix} = \begin{pmatrix} v_{i-1}(x) & w_{i-1}(x) \\ v_i(x) & w_i(x) \end{pmatrix} \cdot \begin{pmatrix} r_{-1}(x) \\ r_0(x) \end{pmatrix} .$$

This representation may be used to present a few properties of the Euclidean algorithm.

Properties of the Euclidean algorithm polynomials

(1) $v_i(x)$ and $w_i(x)$ are relatively prime, $\gcd(v_i(x), w_i(x)) = 1$, which gives

$$\tau(x)v_i(x) + \delta(x)w_i(x) = 1 .$$

(2) $\deg r_i(x) < \deg r_{i-1}(x)$.

(3) $\deg w_i(x) < \deg r_{-1}(x)$.

(4) $\deg r_i(x) < \deg r_{-1}(x) - \deg w_i(x)$.

(5) For the case where $\kappa(x)$ and $\lambda(x)$ are relatively prime, and where $\deg(\kappa(x)r_0(x) + \lambda(x)r_{-1}(x)) < \deg r_{-1}(x) - \deg \kappa(x)$, there exists a factor c (from $GF(p)$) and a number i such that

$$\kappa(x) = cw_i(x) \quad \text{and} \quad \lambda(x) = cv_i(x) .$$

This means that the solutions of the polynomials $w_i(x)$ and $v_i(x)$ through the Euclidean algorithm are unique to within a factor.

Proof of the properties

(1) Consider equation 3.10. The determinant of the right-hand side is 1. We equate this to the formula for the determinant of the left-hand side to obtain

$$w_i(x)v_{i-1}(x) - w_{i-1}(x)v_i(x) = 1 .$$

Therefore, according to the Euclidean algorithm, the gcd is equal to 1. □

(2) This property follows directly from the division of equation 3.7, since the degree of the remainder polynomial is smaller than the degree of the dividing polynomial. □

(3) and (4) The degree of the polynomial $w_i(x)$ is equal to the degree of the products of the polynomials $q_1(x)$ to $q_i(x)$:

$$\deg\ w_i(x) = \deg\left(\prod_{j=1}^{i} q_j(x)\right) = \sum_{j=1}^{i} \deg q_j(x) .$$

If we multiply $r_{i-1}(x)$ by the polynomial $w_i(x)$, we get the degree of $r_{-1}(x)$. To verify this, one can recursively apply equation 3.7. We get

$$\deg\ r_{i-2}(x) = \deg\ q_i(x) + \deg\ r_{i-1}(x) ;$$

in other words,

$$\deg r_{-1}(x) = \sum_{j=1}^{i} \deg q_j(x) + \deg r_{i-1}(x) .$$

This gives us the equation

$$\deg w_i(x) = \sum_{j=1}^{i} \deg q_j(x) = \deg r_{-1}(x) - \deg r_{i-1}(x) < \deg r_{-1}(x) ,$$

which is exactly property 4. □

(5) We shall prove this property only with respect to a special case that will be needed later. We define

$$\begin{aligned} r_{-1}(x) &= x^{d-1}, \\ r_0(x) &= S_0 + S_1 x + S_2 x^2 + \ldots + S_{d-2} x^{d-2}, \\ r_i(x) &= r_0 + r_1 x + r_2 x^2 + \ldots + r_{d-e-2} x^{d-e-2}, \\ w_i(x) &= w_0 + w_1 x + w_2 x^2 + \ldots + w_e x^e, \\ v_i(x) &= v_0 + v_1 x + v_2 x^2 + \ldots + v_{e-1} \cdot x^{e-1} . \end{aligned} \tag{3.13}$$

From properties 3 and 4, we know that the degree from $r_i(x)$ is at most $d-e-2$, because it must be smaller than $\deg r_{-1}(x) - \deg w_i(x) = d-1-e$. We want to calculate the specific coefficients of the equation

$$r_i(x) = w_i(x) r_0(x) + v_i(x) r_{-1}(x)$$

The following system of equations is generated. Each row corresponds to a coefficient of the variable x. The degree of x is indicated by the first column:

$$\begin{array}{rlllll}
0: & w_0 S_0, & & & & \\
1: & w_0 S_1 & + w_1 S_0, & & & \\
 & \vdots & & & & \\
d-e-2: & w_0 S_{d-e-2} & + w_1 S_{d-e-1} & + \cdots & + w_e S_{d-2e-2}, & \\
d-e-1: & w_0 S_{d-e-1} & + w_1 S_{d-e-2} & + \cdots & + w_e S_{d-2e-1} & = 0, \\
d-e: & w_0 S_{d-e} & + w_1 S_{d-e-1} & + \cdots & + w_e S_{d-2e} & = 0, \\
 & \vdots & & & & \\
d-2: & w_0 S_{d-2} & + w_1 S_{d-3} & + \cdots & + w_e S_{d-e-2} & = 0, \\
d-1: & v_0 & + w_1 S_{d-2} & + \cdots & + w_e S_{d-e-1} & = 0, \\
d: & v_1 & + w_2 S_{d-2} & + \cdots & + w_e S_{d-e} & = 0, \\
d+1: & v_2 & + w_3 S_{d-2} & + \cdots & + w_e S_{d-e+1} & = 0, \\
 & \vdots & & & & \\
d+e-4: & v_{e-3} & + w_{e-2} S_{d-2} & + w_{e-1} S_{d-3} & + w_e S_{d-4} & = 0, \\
d+e-3: & v_{e-2} & + w_{e-1} S_{d-2} & + w_e S_{d-3} & & = 0, \\
d+e-2: & v_{e-1} & + w_e S_{d-2} & & & = 0.
\end{array} \tag{3.14}$$

The e rows from $d-e-1$ to $d-2$ contain only coefficients of $w_i(x)$, and the e rows from $d-1$ to $d+e-2$ contain both coefficients of $w_i(x)$ and $v_i(x)$. At this point, we have $2e$ equations for the $(e+1)+e$ coefficients of $w_i(x)$ and $v_i(x)$. Furthermore, the coefficients of $v_i(x)$ may be found recursively if $w_i(x)$ is known.

The relations

$$v_{e-i} = -\sum_{j=1}^{i} w_{e-j+1} S_{d-i-2+j}, \quad i = 1, 2, \ldots, e , \tag{3.15}$$

can be derived from the last e rows of equations 3.14. Suppose there exist two additional polynomials

$\kappa(x)$ and $\lambda(x)$ that are relatively prime and such that

$$\kappa(x) \neq cw_i(x) \quad \text{and} \quad \lambda(x) \neq cv_i(x) .$$

Remember that the following equation must hold: $\deg(\kappa(x)r_0(x) + \lambda(x)r_{-1}(x)) < \deg r_{-1}(x) - \deg \kappa(x)$. From equations 3.14, we see that we have two solutions for the e rows from $d-e-1$ to $d-2$, namely $\kappa(x)$ and $w_i(x)$. We know that $w_i(x)$ is a solution and that only one of the coefficients can be chosen freely. Hence each solution can be represented as $cw_i(x)$. This means that $\kappa(x) \neq cw_i$ and $\lambda(x) \neq cv_i(x)$ cannot be solutions to equations 3.14. □

In order to use the Euclidean algorithm, we shall first modify equation 3.2 in the following manner:

$$C(x)F(x) = 0 \mod (x^n - 1) = T(x)(x^n - 1),$$

with

$$F(x) = S_0 + S_1x + \ldots + S_{d-2}x^{d-2} + F_{d-1}x^{d-1} + \cdots + F_{n-1}x^{n-1} = S(x) + F^*(x) .$$

We can write

$$C(x)F(x) = C(x)S(x) + C(x)F^*(x) = -T(x) + x^nT(x) .$$

The following must hold according to the results of the key equation (definition 3.11):

$$\deg T(x) \leq e - 1, \ \deg C(x) = e .$$

The Euclidean algorithm can be used in two different ways to find the error coordinates of a received codeword.

Euclidean algorithm I (EAI)

$$\begin{aligned} \text{Initialization:} &\quad r_0(x) = a(x) = S(x),\ r_{-1}(x) = b(x) = x^{d-1},\ d \text{ odd}. \\ \text{Stop condition:} &\quad \deg(r_t) < \frac{d-1}{2} \text{ and } \deg(r_{t-1}) \geq \frac{d-1}{2}. \\ \text{Result:} &\quad -T^{\mathrm{I}}(x) = r_t(x),\ C(x) = w_t(x) . \end{aligned}$$

The following relation is valid:

$$\overbrace{C(x)S(x)}^{\deg = e+d-2} + \overbrace{C(x)F^*(x)}^{\text{coef.} \geq d-1} = -T(x) + x^nT(x) .$$

The difference $C(x)F^*(x) - x^nT(x)$ has coefficients equal to 0 for the coordinates $0, 1, \ldots,$ $d-2$; therefore one can represent them as any polynomial $(\ldots)$ times x^{d-1}, because this polynomial is not important for determining the error locations.

When using the Euclidean algorithm to determine the gcd of x^{d-1} and $S(x)$, we can represent the remainder polynomial $r_t(x)$ as

$$r_t(x) = v_t(x)x^{d-1} + w_t(x)S(x) .$$

This allows us to write

$$\begin{array}{rcccl} -T^{\mathrm{I}}(x) & = & (\ldots)x^{d-1} & + & C(x)S(x) \\ r_t(x) & = & v_t(x)x^{d-1} & + & w_t(x)S(x)\,, \end{array}$$

where $(\ldots)$ corresponds to $v_t(x)$, and $-T^{\mathrm{I}}(x)$ and $C(x)$ correspond to $r_t(x)$ and $w_t(x)$ respectively. When the solution satisfies $\deg r_t(x) < \frac{d-1}{2}$, $r_t(x)$ is equal to $-T^{\mathrm{I}}(x)$ and $C(x)$ is equal to $w_t(x)$.

If we assume that no more than $\lfloor \frac{d-1}{2} \rfloor$ errors have occurred, we can use the Euclidean algorithm to solve the system of equations 3.14 and thereby calculate the error locator polynomial $C(x)$. Property 5 of the Euclidean algorithm ensures that the roots of the solution are unique.

Euclidean algorithm II (EAII)

$$\begin{array}{rl} \text{Conditions:} & \text{RS code with } \tilde{a}(x),\ \deg(\tilde{A}(x)) < k,\ u(x) \text{ received.} \\ \text{Initialization:} & r_0(x) = a(x) = U_{n-1}x^{n-1} + U_{n-2}x^{n-2} + \ldots, \\ & r_{-1}(x) = b(x) = x^n - 1. \\ \text{Stop condition:} & \deg(r_t) < n - \dfrac{d-1}{2} \text{ and } \deg(r_{t-1}) \geq n - \dfrac{d-1}{2}. \\ \text{Result:} & -T^{\mathrm{II}}(x) = v_t(x),\ C(x) = w_t(x)\,. \end{array}$$

Here $\tilde{a}(x)$ and $\tilde{A}(x)$ are the transmitted codeword polynomials in time and frequency respectively.

$F(x)$ has roots in the non-error locations j, $C(x)$ has roots in the error location i, and the polynomial $x^n - 1$ has roots in all coordinates l. The product $\prod(\alpha^j - x)$ over all non-error coordinates j is the gcd of $F(x)$ and $x^n - 1$. At the same time, this product can also be represented as the product $\prod(\alpha^l - x)$ over all coordinates l divided by the product $\prod(\alpha^l - x)$ over all error coordinates i. This corresponds to $(x^n - 1)$ divided by $C(x)$. This gives

$$\gcd(F(x), x^n - 1) = \frac{x^n - 1}{C(x)}\,,$$

or equivalently

$$\begin{array}{rcrcl} (\sim) & = & -T^{\mathrm{II}}(x)(x^n - 1) & + & C(x)S(x)x^{n-d+1}, \\ r_t(x) & = & v_t(x)(x^n - 1) & + & w_t(x)S(x)x^{n-d+1}\,, \end{array}$$

where $(\sim)$, $-T^{II}(x)$ and $C(x)$ correspond to $r_t(x)$, $v_t(x)$ and $w_t(x)$ respectively. From $F(x)$, we only need to know the coefficients of the syndrome $S(x)$, $S_{d-2} = F_{n-1} = U_{n-1}$, $S_{d-3} = F_{n-2} = U_{n-2}, \ldots$, which are known, since $U(x)$ is the received vector.

As with the EAI, the Euclidean property 5 can be employed to show that the roots of the polynomial $w_t(x)$ are unique and $w_t(x)$ corresponds to the error locator polynomial $C(x)$.

Example 3.7 (Euclidean algorithm) We shall again use the syndrome $S(x) = 3x^3 + 3x^2 + 5x + 5$ from example 3.5 to calculate $C(x)$ using EAI and the transform of the transmitted codeword $U(x) = 3x^5 + 3x^4 + 5x^3 + 5x^2 + \sim$ for the solution using EAII.

Calculation with the EAI

$$a(x) = 3x^3 + 3x^2 + 5x + 5, \qquad b(x) = x^{d-1} = x^4,$$

$$\begin{array}{rllllll}
b : a = & x^4 & & & & & : 3x^3 + 3x^2 + 5x + 5 = 5x + 2, \\
& -(x^4 + & x^3 + & 4x^2 + & 4x) & & \\
\hline
& & 6x^3 + & 3x^2 + & 3x & & \\
& & -(6x^3 + & 6x^2 + & 3x & + 3) & \\
\hline
& & & 4x^2 & & + 4 &
\end{array}$$

$$b = q_1 a + r_1, \qquad q_1 = 5x + 2, \qquad r_1 = 4x^2 + 4,$$

$$\begin{array}{rlllll}
a : r_1 = & 3x^3 + & 3x^2 + & 5x & + 5 & : 4x^2 + 4 = 6x + 6, \\
& -(3x^3 & & + 3x) & & \\
\hline
& & 3x^2 + & 2x & + 5 & \\
& & -(3x^2 & & + 3) & \\
\hline
& & & 2x & + 2 &
\end{array}$$

$$a = q_2 \cdot r_1 + r_2, \qquad q_2 = 6x + 6, \qquad r_2 = 2x + 2\,.$$

Here we stop because the degree of $r_2(x) = 2x + 2$ is less than $\frac{d-1}{2} = 2$. Hence

$$-T^{\mathrm{I}}(x) = r_2(x) = 2x + 2, \;\; T^{\mathrm{I}}(x) = 5x + 5\,.$$

The calculation of $C^{\mathrm{I}}(x) = w_2(x)$ gives

$$\begin{array}{lllllll}
w_{-1} & = & 0, & & & & \\
w_0 & = & 1, & & & & \\
w_1 & = & w_{-1} - q_1 w_0 & = & -(5x+2) \cdot 1 & = & 2x + 5, \\
w_2 & = & w_0 - q_2 w_1 & = & 1 - (6x+6)(2x+5) & & \\
& = & 1 - (5x^2 + 2) & = & 2x^2 + 6 & = & C^{\mathrm{I}}(x).
\end{array}$$

$C^{\mathrm{I}}(x)$ is equal to $6C(x)$, where $C(x)$ was calculated in example 3.6. Therefore $C^{\mathrm{I}}(x)$ has the same roots. If no more than $\frac{d-1}{2}$ errors have occurred, the solution of the EAI is equivalent to that of the BMA according to

$$C^{\mathrm{I}}(x) = (\text{constant})C(x)\,,$$

Calculation with the EAII

$$a(x) = 3x^5 + 3x^4 + 5x^3 + 5x^2 + \sim, \qquad b(x) = x^6 - 1,$$

$$\begin{array}{rlllllll}
b : a = & x^6 - & 1 & & & & & : 3x^5 + 3x^4 + 5x^3 + 5x^2 + \sim = 5x + 2 \\
& -(x^6 + & x^5 + & 4x^4 + & 4x^3 + & 5x \cdot \sim) & & \\
\hline
& & 6x^5 + & 3x^4 + & 3x^3 + \sim & & & \\
& & -(6x^5 + & 6x^4 + & 3x^3 + & 3x^2 & + 2 \cdot \sim) & \\
\hline
& & & 4x^4 & & + \sim & &
\end{array}$$

$$b = q_1 a + r_1, \qquad q_1 = 5x + 2, \qquad r_1 = 4x^4 + \sim,$$

$$
\begin{array}{ll}
a : r_1 = & 3x^5 + 3x^4 + 5x^3 + 5x^2 + \sim \; : \; 4x^4 + \sim = 6x + 6, \\
& \underline{-\ (3x^5 \qquad + 6x \cdot \sim)} \\
& \qquad 3x^4 + \sim \\
& \qquad \underline{-\ (3x^4 \qquad + 6 \cdot \sim)} \\
& \qquad\qquad \sim
\end{array}
$$

$$
a = q_2 \cdot r_1 + r_2, \qquad q_2 = 6x + 6, \qquad r_2 = \sim .
$$

At this point, we stop because the degree of $r_2(x) = \sim$ is less than $n - \frac{d-1}{2} = 6 - 2 = 4$; $C^{\text{II}}(x) = 2x^2 + 6$ because q_1 and q_2 are the same as from EAI. The calculation of $T^{\text{II}}(x)$ gives

$$
\begin{aligned}
v_{-1} &= 1, \\
v_0 &= 0, \\
v_1 &= v_{-1} - q_1 v_0 = 1, \\
v_2 &= v_0 - q_2 v_1 = -(6x+6) \cdot 1 = x + 1 ,
\end{aligned}
$$

or

$$
T^{\text{II}}(x) = 6x + 6 .
$$

◇

3.2.4 Calculation of the error values

So far, we have studied different methods for solving the key equation

$$
C(x)S(x) = -T(x) \mod x^{2t}, \quad t = \left\lfloor \frac{d-1}{2} \right\rfloor, \quad \deg C(x) > \deg T(x) .
$$

The coefficients $S_0, S_1, \ldots, S_{2t-1}$ correspond to the coordinates of the transform of the received polynomial $r(x)$, which depends only on the error polynomial $f(x)$:

$$
A_i = A_{i+1} = \ldots = A_{i+2t-1} = 0 \text{ for all } A(x) \bullet\!\!-\!\!\circ\, a(x) \in C
$$

(compare with section 3.1.5). For the transformation of the received errors, the syndrome is

$$
S(x) = S_0 + S_1 x + \ldots + S_{2t-1} x^{2t-1}, \text{ with } S_0 = F_i,\ S_1 = F_{i+1}, \ldots, S_{2t-1} = F_{i+2t-1} .
$$

The Euclidean algorithm II is an exception, since the syndrome positions are fixed.

By finding $C(x)$, we know the error locations in the received codeword. Note that in the case of a binary code, we also know the error value, specifically $C(\alpha^i) = 0 \Longrightarrow f_i = 1$. In the case of a non-binary code, we must calculate the error values $f_i \neq 0$. Two different techniques are considered. The first possibility is to determine $F(x)$ completely. The error values can then be calculated through

$$
f_i = F(\alpha^i).
$$

The coefficients for $F(x)$ can be determined recursively from the equation

$$
C(x)F(x) = 0 \mod (x^n - 1).
$$

$C(x)F(x) = 0 \bmod (x^n - 1)$ corresponds to the system of equations

$$0 = \sum_{j=0}^{e} C_j F_{n-j+l}, \qquad l = 0, 1, \ldots, n-1 ,$$

where $C(x) = C_0 + C_1 x + \ldots + C_e x^e$ and $S(x) = F_i + F_{i+1}x + \ldots + F_{i+2t-1}x^{2t-1}$ are known. The remaining coefficients from $F(x)$ are calculated recursively by

$$F_{2t+i+l} = -\frac{1}{C_0} \sum_{j=1}^{e} C_j F_{2t+i-j+l} \text{ for } l = 0, 1, \ldots, n-1-2t .$$

Now that $F(x)$ is known, the error values f_i for the error coordinates $C(\alpha^i = 0)$ can be calculated using $f_i = F(\alpha^i)$

The second solution is much less complex, and is known as the Forney algorithm. For the implementation of this algorithm, we need to introduce the variable l, which corresponds to a cyclic shift of $F(x)$:

$$F^{(l)}(x) = x^l F(x) \mod (x^n - 1).$$

The variable l is chosen such that

$$F_0^{(l)} = S_0,\ F_1^{(l)} = S_1, \ldots, F_{2t-1}^{(l)} = S_{2t-1} .$$

As before, we have the relationship

$$C(x)F^{(l)}(x) = 0 \mod (x^n - 1),$$

which can also be formulated as

$$\begin{aligned} C(x)F^{(l)}(x) &= T^{(l)}(x)(x^n - 1) \\ &= -T^{(l)}(x) + x^n T^{(l)}(x), \end{aligned}$$

with

$$\deg T^{(l)}(x) \le e - 1,\ \deg C(x) = e .$$

This gives

$$\begin{aligned} T_0^{(l)} &= -C_0 F_0^{(l)} &&= -C_0 S_0, \\ T_1^{(l)} &= -C_0 F_1^{(l)} - C_1 F_0^{(l)} &&= -C_0 S_1 - C_1 S_0, \\ T_2^{(l)} &= -C_0 S_2 - C_1 S_1 - C_2 S_0, \\ &\vdots \\ T_{e-1}^{(l)} &= -C_0 S_{e-1} - C_1 S_{e-2} - \ldots - C_{e-1} S_0 . \end{aligned}$$

$T^{(l)}(x)$ is called the *error evaluator polynomial*, and is calculated by multiplication of the error locator polynomial $C(x)$ and the syndrome $S(x)$. The error values f_i can be calculated by

$$f_i = F(\alpha^i) = x^{-l} F^{(l)}(x) \Big|_{x=\alpha^i} = x^{-l} \frac{T^{(l)}(x)(x^n - 1)}{C(x)} \Big|_{x=\alpha^i} .$$

With $C(\alpha^i) = 0$ for the error coordinates and $x^n - 1 = 0$ for all elements of the Galois field, we get

$$\frac{T^{(l)}(x)(x^n-1)}{C(x)}\Big|_{x=\alpha^i} = \frac{0}{0}.$$

Using de L'Hôpital's rule, we can use the derivatives of the numerator and denominator of the above fraction to solve for the coefficients of $f(x)$:

$$\begin{aligned}(T^{(l)}(x)(x^n-1))' &= T^{(l)\prime}(x)(x^n-1) + T^{(l)}(x)nx^{n-1} \\ &= T^{(l)}(x)nx^{-1}, \ \forall\, \alpha^i \in GF(p).\end{aligned}$$

Error value computation

$$f_i = x^{-l}nx^{-1}\frac{T^{(l)}(x)}{C'(x)}\Big|_{x=\alpha^i}.$$

The integer l is defined by

$$l: \qquad F_0^{(l)} = S_0,\ F_1^{(l)} = S_1,\ \ldots, F_{2t-1}^{(l)} = S_{2t-1};$$

therefore

$$F^{(l)}(x) = x^l F(x) \mod (x^n - 1)\ ,$$

and $T^{(l)}(x)$ is defined by

$$T_j^{(l)} = -\sum_{i=0}^{j} S_{j-i}C_i,\ j = 0,1,\ldots,e-1, \quad \text{with } \deg C(x) = e\ .$$

Remarks

- For the EAII, the shift is $l = 0$. This is because the syndrome already occupies the highest-degree coordinates of the FFT transformed error polynomial $F(x)$. Therefore $T^{(0)}(x) = T^{\text{II}}(x)$.
- If the representation $A_i, A_{i+1}, \ldots, A_{i+k-1}$ is the information polynomial of the codeword $a(x) \circ\!\!-\!\!\bullet A(x)$, the complete error polynomial $F(x)$ must be determined. Because $A(x) = R(x) - F(x)$, and the received polynomial is $r(x) \circ\!\!-\!\!\bullet R(x)$, $r(x)$ must also be calculated.
- If the extension field $GF(2^m)$ is used then $n = 2^m - 1$ and $n^{-1} = 1$. A simplification can be used when the derivative of a polynomial with coefficients from $GF(2^m)$ is taken. In this case, the terms in the polynomial with even exponents vanish, and in the terms with odd exponents i, each i is set to $i-1$ (0 is even). For example, taking α as a primitive element from $GF(2^m)$,

$$\begin{aligned}C(x) &= \alpha^{i_0} + \alpha^{i_1}x + \alpha^{i_2}x^2 + \alpha^{i_3}x^3, \\ C'(x) &= \alpha^{i_1} + \alpha^{i_3}x^2\ .\end{aligned}$$

Example 3.8 (Error value computation) We shall now calculate the error values of the received polynomial from example 3.5. The syndrome was calculated in example 3.7 to be $S(x) = 5 + 5x + 3x^2 + 3x^3$, with $S_0 = F_2, \ldots, S_3 = F_5$. The error locator polynomial was calculated to be $C(x) = 5x^2 + 1$ using the BMA algorithm, and $C^{\mathrm{I}}(x) = C^{\mathrm{II}}(x) = 2x^2 + 6$ using the EAI and EAII algorithms. For the BMA case, we must calculate the error evaluator polynomial $T^{(-2)}(x)$. The EAI and EAII gave error evaluator polynomials $T^{\mathrm{I}}(x) = 5x + 5$ and $T^{\mathrm{II}}(x) = 6x + 6$ respectively. According to equation 3.5, we get

$$\left.\begin{array}{lllll} T_0^{(-2)} & = & -C_0S_0 & = & 2, \\ T_1^{(-2)} & = & -C_0S_1 - C_1S_0 & = & 2 \end{array}\right\} \Longrightarrow T^{(-2)}(x) = 2x + 2\,.$$

We shall now calculate the error values for f_1 and f_4:

$$\text{BMA:} \quad C'(x) = 3x, \qquad T^{(-2)}(x) = 2x + 2,$$

$$f_{1,4} = x^{2-1} n \frac{T^{(-2)}(x)}{C'(x)}\Bigg|_{\substack{x=5=\alpha^1 \\ x=2=\alpha^4}} = \begin{cases} 5 \cdot 6 \cdot \frac{5}{1} & = 3, \\ 2 \cdot 6 \cdot \frac{6}{6} & = 5, \end{cases}$$

$$\text{EAI:} \quad C^{\mathrm{I}'}(x) = 4x, \qquad T^{\mathrm{I}(-2)}(x) = 5x + 5,$$

$$f_{1,4} = xn \frac{T^{\mathrm{I}(-2)}(x)}{C'(x)}\Bigg|_{\substack{x=5 \\ x=2}} = \begin{cases} 5 \cdot 6 \cdot \frac{2}{6} & = 3, \\ 2 \cdot 6 \cdot \frac{1}{1} & = 5, \end{cases}$$

$$\text{EAII:} \quad C^{\mathrm{II}'}(x) = 4x, \qquad T^{\mathrm{II}}(x) = 6x + 6,$$

$$f_{1,4} = x^{-1} n \frac{T^{\mathrm{II}}(x)}{C^{\mathrm{II}'}(x)}\Bigg|_{\substack{x=5 \\ x=2}} = \begin{cases} \frac{1}{5} \cdot 6 \cdot \frac{1}{6} & = 3, \\ \frac{1}{2} \cdot 6 \cdot \frac{4}{1} & = 5\,. \end{cases}$$

◇

3.2.5 Equivalence of the Euclidean and Berlekamp–Massey algorithms

In this section we shall show the relation between the Euclidean algorithms and the BMA (see [Dorn87]). The first step is to consider the system of equations 3.14 and interpret it as an inversion:

$$x^e w_i\left(\frac{1}{x}\right) x^{d-2} r_0\left(\frac{1}{x}\right) + x^{e-1} v_i\left(\frac{1}{x}\right) x^{d-1} r_{-1}\left(\frac{1}{x}\right)\,.$$

Clearly an inversion means

$$\begin{array}{rcl} w_i(x) & = & w_0 + w_1x + w_2x^2 + \ldots + w_ex^e, \\ x^e w_i(\frac{1}{x}) & = & w_e + w_{e-1}x + w_{e-2}x^2 + \ldots + w_0x^e\,. \end{array}$$

This results in the following system of equations:

$$
\begin{array}{rlllll}
d+e-2: & w_0S_0, & & & & \\
d+e-3: & w_0S_1 & +\ w_1S_0, & & & \\
 & \vdots & & & & \\
d-2e-2: & w_0S_{d-e-2} & +\ w_1S_{d-e-1} & +\ \ldots & +\ w_eS_{d-2e-2}, & \\
d-2e-1: & w_0S_{d-e-1} & +\ w_1S_{d-e-2} & +\ \ldots & +\ w_eS_{d-2e-1} & =0, \\
d-2e: & w_0S_{d-e} & +\ w_1S_{d-e-1} & +\ \ldots & +\ w_eS_{d-2e} & =0, \\
 & \vdots & & & & \\
e: & w_0S_{d-2} & +\ w_1S_{d-3} & +\ \ldots & +\ w_eS_{d-e-2} & =0, \\
e-1: & v_0 & +\ w_1S_{d-2} & +\ \ldots & +\ w_eS_{d-e-1} & =0, \\
e-2: & v_1 & +\ w_2S_{d-2} & +\ \ldots & +\ w_eS_{d-e} & =0, \\
e-3: & v_2 & +\ w_3S_{d-2} & +\ \ldots & +\ w_eS_{d-e+1} & =0, \\
 & \vdots & & & & \\
2: & v_{e-3} & +\ w_{e-2}S_{d-2} & +\ w_{e-1}S_{d-3} & +\ w_eS_{d-4} & =0, \\
1: & v_{e-2} & +\ w_{e-1}S_{d-2} & +\ w_eS_{d-3} & & =0, \\
0: & v_{e-1} & +\ w_eS_{d-2} & & & =0.
\end{array}
\tag{3.16}
$$

The equations corresponding to the rows 0 to $d-2e-1$ can be understood as follows:

$$x^e w_i\left(\frac{1}{x}\right)x^{d-2}S\left(\frac{1}{x}\right) = x^{e-1}v_i\left(\frac{1}{x}\right) \quad \bmod x^{2e+1}\ .$$

If one uses the BMA to solve the inverted syndrome, the intermediary solution is shown to be the inverted polynomials $w_i(x)$ and $v_i(x)$ of the Euclidean algorithm.

For the stop condition $\deg r_i(x) < (d-1)/2$, therefore, $d-e-2 < (d-1)/2$.

3.2.6 Erasure correction

Consider the situation where we receive a vector and know in which coordinates the errors have occurred, but do not know the values of these errors. Such a coordinate is called an erasure. The existence of erasures will be important in chapter 7.

For RS codes, erasures can be corrected because one knows the error location, and need only calculate the value of the error. Erasure corrections can be incorporated into algebraic decoding in the following manner.

Assume that one of the coordinates of the received vector $r(x)$ is an erasure. We must first define the multiplication and addition of a Galois field element $\alpha^i \in GF(p)$ with an erasure:

$$
\begin{aligned}
\alpha^i + \otimes &= \otimes, \\
\alpha^i \cdot \otimes &= \begin{cases} 0, & \text{where} \quad \alpha^i = 0, \\ \otimes, & \text{where} \quad \alpha^i \neq 0\,. \end{cases}
\end{aligned}
$$

We can now represent our received vector $r(x)$ as

$$r(x) = a(x) + v(x) + f(x)\ .$$

$a(x) \in C$, $f(x)$ is the error and $v(x)$ is the erasure; therefore $v_i \in \{0, \otimes\}$. We now define a polynomial $b(x)$ whose coefficients b_i are equal to zero where the erasures have occurred (this is a similar definition as used for the error locator polynomial $c(x)$):

$$b(x) \circ\!\!-\!\!\bullet B(x) = \prod_{i:v_i=\otimes} (x - \alpha^i) \; .$$

The degree of $B(x)$ is equal to the number of erasures:

$$v_i = \otimes \implies b_i = 0 \qquad \text{and} \qquad v_i = 0 \implies b_i \neq 0 \; .$$

We multiply the coefficients of $r(x)$ by those from $b(x)$ to obtain the polynomial $\tilde{r}(x)$:

$$\tilde{r}_i = b_i r_i, \;\; i = 0, 1, \ldots, n-1 \; .$$

The coefficients of $\tilde{r}(x)$ come from the Galois field, thus allowing the transformation. We get

$$\begin{array}{rcl} \tilde{r}(x) & \circ\!\!-\!\!\bullet & \tilde{R}(x), \\ (a_i + f_i + v_i)b_i = (a_i + f_i)b_i & \circ\!\!-\!\!\bullet & (A(x) + F(x))B(x) \; . \end{array}$$

We further assume that $\deg A(x) \leq n - d - 1$. This makes it apparent that certain syndrome coefficients $S(x)$ are no longer exclusively dependent on the error, but also on the product $A(x)B(x)$. Because we do not know $A(x)$, we cannot use the affected syndrome coefficients in the error location calculation. This situation is depicted schematically in figure 3.4. In practice, this shortens the number of syndrome coefficients by one for each erasure. The number of errors e and the erasures t that can be corrected with respect to the minimum distance d is equal to

$$2e + t < d \; .$$

With the shortened syndrome, we can use the Berlekamp–Massey algorithm to determine the error locator polynomial $\tilde{C}(x)$. The error values can then be calculated through

$$C(x) = B(x)\tilde{C}(x).$$

The transmitted codeword $a(x)$ is calculated by

$$a(x) = \tilde{r}(x) - f(x).$$

This means that $\tilde{r}(x)$ must be calculated explicitly.

The effectiveness of the above technique is connected to the occurrence of errors and erasures during transmission. In order to efficiently correct errors in a codeword with erasures by using known algorithms, it would be simpler if the erasures were given known values. In other words, the receiver knows the location of the error and need only calculate the Galois field elements that must be added to the received vector in order to correct the error.

- Determine the polynomial $r^*(x)$ using

$$r_i^* = \begin{cases} r_i, & r_i \neq \otimes, \\ 0, & r_i = \otimes. \end{cases}$$

  Therefore the erasure coordinates in $r(x)$ are set to zero (in principle, to any Galois field element).

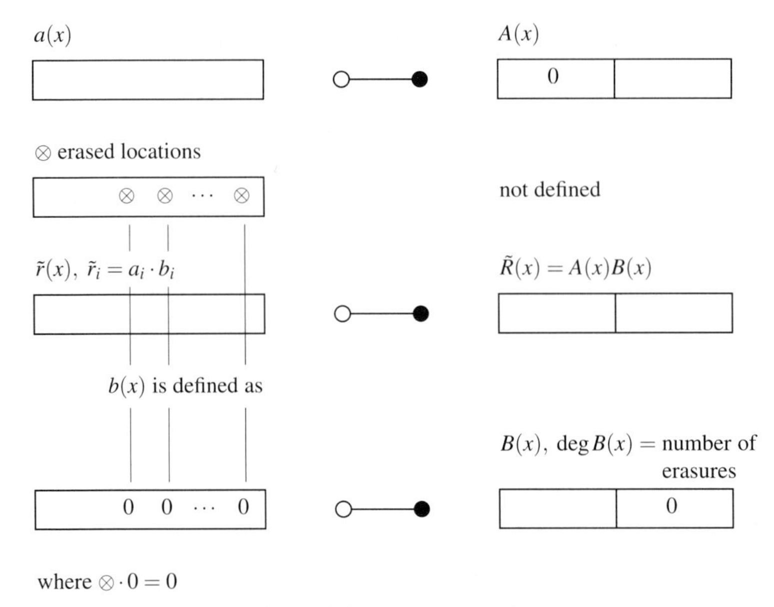

Figure 3.4 Erasure correction.

- Calculate $C_{\otimes}(x)$ using

$$C_{\otimes}(x) = \prod_{i:r_i=\otimes} (x - \alpha^i) .$$

- Calculate the syndrome $S(x)$ as a part of $R^*(x)$.
- Start the Berlekamp–Massey algorithm with $C_{\otimes}(x)$.

This method adapts the algebraic decoding of errors so that it may take advantage of erasures. This technique can be used in many cases to improve decoding results. In particular, erasure decoding is needed for concatenated codes, which are presented in chapter 9.

Erasure correction is also possible for other decoding methods that will be introduced in chapter 7. The problem is always an unknown value, and a known location. For the binary case (e.g. with BCH codes), the value of an erasure is either 0 or 1. An investigation of when erasures are relevant in binary cases can be found in [BZ95].

3.3 Summary

The class of RS codes introduced by Reed and Solomon in [RS60] have been described. An efficient decoding of these codes can be implemented using the BM algorithm. A means of solving the key equation using shift registers was presented by Massey; see [Mas69]. The decoding of RS codes using the Euclidean algorithm was developed by Sugiyama et al. in [SKHN75]. A continuous fraction solution has been developed that is equivalent to the Euclidean algorithm [WS79]. The equivalence of the Euclidean algorithm and the BMA has been proved based on work by Dornstetter [Dorn87]. Sorger [Sor93] developed the Newton interpolation algorithm for the solution of the system of equations. Another use of the Euclidean algorithm is in the decoding of Goppa codes [Man77].

In this chapter, RS codes, their generalized representation and extension by one or two coordinates have been developed. Four different encoding methods have been introduced. An algebraic decoding technique based on an error locator polynomial has been defined, and this polynomial can be calculated through the solution of the key equation. To this end, the syndrome of the transformed codeword must be calculated. We have shown how to calculate the error locator polynomial $C(x)$ using the Euclidean algorithm and the BMA. When the error coordinates are known, or, in other words, the roots of $C(x)$ have been found, the error values can be calculated recursively using $F(x)$ and through the inverse transformation. A more efficient procedure, however, is the Forney algorithm. This algorithm can directly calculate the error values. It is important to know which coordinates in the transform domain generate the syndrome.

The decoding possibilities are summarized in figure 3.1. If, however, the k information coordinates correspond to the k coefficients from $A(x) \bullet\!\!-\!\!\circ\, a(x) \in \mathcal{C}$ then $R(x)$ can be calculated and $A(x)$ can be found through $A(x) = R(x) - F(x)$. We have also introduced erasure correction, which will be used in later chapters.

There exist many illustrative examples of applications of RS codes. One such is for error correction of music CDs [HTV82] based on the RS codes $\mathrm{RS}(32,28,5)$ and $\mathrm{RS}(28,24,5)$. These codes are both shortened codes constructed from $\mathrm{RS}(255,251,5)$ over $GF(2^8)$. The serial implementation of both codes gives a collective rate of $3/4$ ($\frac{28}{32} \cdot \frac{24}{28} = \frac{3}{4}$; see chapter 9). These codes serve mainly to correct fabrication defects; however they also correct scratches and other damage.

The well-known ESA/NASA standard for satellite data transmission uses a $(255,223,33)$ RS code [WHPH87].

Before the specification of the GSM radio communication system in 1985, several test systems were considered, one of which (CD 900) used an RS code of length 63 for error correction.

In the ETSI standard on *digital broadcasting systems for television, sound and data services*, one of the two codes used is a RS code with parameters $(255,239,17)$, which is shortened to the parameters $(204,188,17)$. This code can correct eight symbol errors.

3.4 Problems

Problem 3.1

Consider the Galois field $GF(7)$, with the primitive element $\alpha = 5$:

(a) Determine the generator polynomial $g(x)$ of a RS code $\mathcal{C}$ of length $n = 6$ that can correct one error. Choose the highest-degree coefficients in the transformation domain as zero.

(b) Determine the parity check polynomial $h(x)$ such that $h(x)g(x) = x^n - 1$.

(c) Is the polynomial $c(x) = 6 + 4x + 6x^2 + x^3$ a codeword of $\mathcal{C}$?

(d) Design the following encoding schemes for $\mathcal{C}$:

- non-systematic coding through multiplication with the generator polynomial;
- systematic coding with the generator polynomial;
- systematic coding with the parity check polynomial.

Problem 3.2
For each case, give a shift register circuit that corresponds to the Galois field $GF(7)$:

(a) a polynomial multiplied by $f(x) = 3 + x + 2x^2$;

(b) calculated by modulo $f(x) = x^4 + 2x + 1$.

Problem 3.3
Consider the Galois field $GF(7)$, with the primitive element $\alpha = 5$ and the generator polynomial $g(x) = x^2 + 5x + 6$ of an RS code C of length $n = 6$. Take $c(x) \in C$, which is a codeword with $c(x) = i(x)g(x)$. Is $x^i c(x) \mod (x^n - 1)$, $i = 1, \ldots, 5$, a codeword?

Problem 3.4
Consider the RS code C of length $n = 6$ over the Galois field $GF(7)$. The primitive element is $\alpha = 5$. The code can correct two errors.

(a) Determine the codeword $b \in C$, $b \circ\!\!-\!\!\bullet B = (0,0,0,0,2,5)$:

- through the inverse transform,
- through the use of the convolution theorem of the DFT.
  It is known that $A = (2,5,0,0,0,0) \bullet\!\!-\!\!\circ a = (0,6,1,4,5,3)$.

(b) The received word is $\mathbf{r} = (3,5,3,4,3,5)$. Carry out the decoding and determine the transmitted codeword $\mathbf{c} \in C$ under the condition that $e \leq 2$ errors in $\mathbf{r}$ have occurred. Solve the problem using the following procedures:

- syndrome calculation;
- key equation;
- solution of the key equations through
  - a linear system of equations,
  - the Berlekamp-Massey algorithm,
  - the Euclidean algorithm;
- error locator polynomial;
- error location calculation (Chien search);
- error value calculation (Forney algorithm).

Problem 3.5
In this problem, we must first build the extension field $GF(2^4)$, from which a three-error correction RS code can be constructed that can be decoded by means of the Berlekamp-Massey algorithm.

(a) Let α be a root of the primitive polynomial $x^4 + x^3 + 1$. Create a table of all powers of α ($\alpha^i, i = 0, \ldots, n-1$; $n = 2^4 - 1$) in component form (logarithm table).

(b) Determine the generator polynomial $g(x)$, where

$$g(x) \circ\!\!-\!\!\bullet \sum_{i=0}^{8} G_i x^i \,.$$

(c) You receive

$$r(x) = \alpha^6 + \alpha^2 x + \alpha^7 x^2 + \alpha^2 x^3 + \alpha^9 x^4 + \alpha^{12} x^5 + x^6 + \alpha^9 x^{11} \,.$$

Determine the transmitted codeword $a(x)$ by means of algebraic decoding. Find the solution of the key equation using the Berlekamp-Massey algorithm.

Table 3.2 Logarithm table for $GF(2^8)$.

GF(2^8), $p(x) = x^8 + x^6 + x^5 + x^4 + 1$									
Exp.	Comp.	Exp.	Comp.	Exp.	Comp.	Exp.	Comp.	Exp.	Comp.
0	00000001	51	01001101	102	01100111	153	10111101	204	10010110
1	00000010	52	10011010	103	11001110	154	00001011	205	01011101
2	00000100	53	01000101	104	11101101	155	00010110	206	10111010
3	00001000	54	10001010	105	10101011	156	00101100	207	00000101
4	00010000	55	01100101	106	00100111	157	01011000	208	00001010
5	00100000	56	11001010	107	01001110	158	10110000	209	00010100
6	01000000	57	11100101	108	10011100	159	00010001	210	00101000
7	10000000	58	10111011	109	01001001	160	00100010	211	01010000
8	01110001	59	00000111	110	10010010	161	01000100	212	10100000
9	11100010	60	00001110	111	01010101	162	10001000	213	00110001
10	10110101	61	00011100	112	10101010	163	01100001	214	01100010
11	00011011	62	00111000	113	00100101	164	11000010	215	11000100
12	00110110	63	01110000	114	01001010	165	11110101	216	11111001
13	01101100	64	11100000	115	10010100	166	10011011	217	10000011
14	11011000	65	10110001	116	01011001	167	01000111	218	01110111
15	11000001	66	00010011	117	10110010	168	10001110	219	11101110
16	11110011	67	00100110	118	00010101	169	01101101	220	10101101
17	10010111	68	01001100	119	00101010	170	11011010	221	00101011
18	01011111	69	10011000	120	01010100	171	11000101	222	01010110
19	10111110	70	01000001	121	10101000	172	11111011	223	10101100
20	00001101	71	10000010	122	00100001	173	10000111	224	00101001
21	00011010	72	01110101	123	01000010	174	01111111	225	01010010
22	00110100	73	11101010	124	10000100	175	11111110	226	10100100
23	01101000	74	10100101	125	01111001	176	10001101	227	00111001
24	11010000	75	00111011	126	11110010	177	01101011	228	01110010
25	11010001	76	01110110	127	10010101	178	11010110	229	11100100
26	11010011	77	11101100	128	01011011	179	11011101	230	10111001
27	11010111	78	10101001	129	10110110	180	11001011	231	00000011
28	11011111	79	00100011	130	00011101	181	11100111	232	00000110
29	11001111	80	01000110	131	00111010	182	10111111	233	00001100
30	11101111	81	10001100	132	01110100	183	00001111	234	00011000
31	10101111	82	01101001	133	11101000	184	00011110	235	00110000
32	00101111	83	11010010	134	10100001	185	00111100	236	01100000
33	01011110	84	11010101	135	00110011	186	01111000	237	11000000
34	10111100	85	11011011	136	01100110	187	11110000	238	11110001
35	00001001	86	11000111	137	11001100	188	10010001	239	10010011
36	00010010	87	11111111	138	11101001	189	01010011	240	01010111
37	00100100	88	10001111	139	10100011	190	10100110	241	10101110
38	01001000	89	01101111	140	00110111	191	00111101	242	00101101
39	10010000	90	11011110	141	01101110	192	01111010	243	01011010
40	01010001	91	11001101	142	11011100	193	11110100	244	10110100
41	10100010	92	11101011	143	11001001	194	10011001	245	00011001
42	00110101	93	10100111	144	11100011	195	01000011	246	00110010
43	01101010	94	00111111	145	10110111	196	10000110	247	01100100
44	11010100	95	01111110	146	00011111	197	01111101	248	11001000
45	11011001	96	11111100	147	00111110	198	11111010	249	11100001
46	11000011	97	10001001	148	01111100	199	10000101	250	10110011
47	11110111	98	01100011	149	11111000	200	01111011	251	00010111
48	10011111	99	11000110	150	10000001	201	11110110	252	00101110
49	01001111	100	11111101	151	01110011	202	10011101	253	01011100
50	10011110	101	10001011	152	11100110	203	01001011	254	10111000

Problem 3.6

Consider the extension field $GF(2^m)$. Construct a two-error correcting RS code of length $n = 2^m - 1$.

(a) How many binary information coordinates has this code?

(b) How many binary errors can this code correct?

Problem 3.7

(a) You are given the codeword $c(x) = \alpha^6 + \alpha^{11}x + \alpha^7x^2 + \alpha^2x^3 + x^4 + \alpha^{12}x^5 + x^6$ of an RS code of length $n = 15$. Determine the corresponding codeword for the extension of the code by one coordinate.

(b) Determine the code parameters and the generator polynomial of the two-coordinate extension code built from the RS code of length $n = 15$.

Note: Use the logarithm table from problem 3.5 for calculations in $GF(2^4)$ and table 3.2 for calculations in $GF(2^8)$.

4

BCH codes

Binary BCH codes are an important class of cyclic codes. We fix the alphabet and then construct codes of arbitrary length. We begin this chapter by defining the primitive BCH code construction and continue with the representation of non-primitive BCH codes, including the Golay code. We shall also clarify the extension and shortening of BCH codes. This method can be applied to any type of code. Non-binary BCH codes will be defined and their connection with RS codes will be given. We shall study the decoding of BCH codes and the properties of very long BCH codes.

4.1 Primitive BCH codes

In this section, we introduce the construction of primitive BCH codes based on different definitions.

4.1.1 Definition based on cyclotomic cosets

First, we should like to prove that a polynomial having all complex-conjugate elements of an extension field as roots is irreducible with respect to the base field, and all the coefficients of this polynomial are elements of the base field.

Theorem 4.1 (Minimal polynomial) *Let K_i be the cyclotomic class (definition 2.25) corresponding to the number $n = p^m - 1$ and let α be a primitive element from $GF(p^m)$. The minimal polynomial $m_i(x)$ is defined by*

$$m_i(x) = \prod_{j \in K_i} \left(x - \alpha^j\right),$$

and has only coefficients from $GF(p)$ and is irreducible over $GF(p)$.

Proof First, we raise the cyclotomic polynomial m_i to the pth power:

$$(m_i(x))^p = \prod_{j \in K_i} \left(x - \alpha^j\right)^p .$$

For each of the factors, we have

$$\left(x - \alpha^l\right)^p = x^p - \binom{p}{1} \alpha^l x^{p-1} + \ldots - \ldots = x^p - \alpha^{pl} ,$$

because $\binom{p}{i} = 0 \bmod p, i = 1, \ldots, p-1$.

We raise the element α^i to the power $p^l, l = 0, 1, 2, \ldots$, and obtain the elements of a cyclotomic coset as follows:

$$\alpha^i \longrightarrow \alpha^{ip} \longrightarrow \alpha^{ip^2} \longrightarrow \ldots \longrightarrow \alpha^{ip^{m-1}} \longrightarrow \alpha^i \longrightarrow \alpha^{ip} \longrightarrow \ldots ,$$

which implies that we can write

$$\prod_{j \in K_i} (x - \alpha^j)^p = m_i(x^p).$$

From the relation

$$(m_i(x))^p = m_i(x^p),$$

it follows that the coefficients of the polynomial $m_i(x)$ must satisfy

$$\left(m_{ij}\right)^p = m_{ij} ,$$

which can only be true for $m_{ij} \in GF(p)$, since for any element of the field $GF(p)$ the exponent $p-1$ produces 1. Because the polynomial $m_i(x)$ is made up of linear factors, with roots only from the irreducible polynomial corresponding to the cyclotomic coset K_i, it also satisfies the definition of a irreducible polynomial over $GF(p)$. □

Definition 4.2 (Primitive BCH code) *Let K_i be the cyclotomic coset corresponding to $n = 2^m - 1$ (definition 2.25). Let α be a primitive element of $GF(2^m)$ and let $\mathcal{M}$ be the union of any number of cyclotomic cosets ($\mathcal{M} = K_{i_1} \cup K_{i_2} \ldots$). A primitive BCH code of length n and dimension $k = n - |\mathcal{M}|$ is defined by the generator polynomial*

$$g(x) := \prod_{i \in \mathcal{M}} (x - \alpha^i), \qquad g_i \in GF(2) .$$

The designed distance is d if a set of $d-1$ successive integers exist in $\mathcal{M}$. The true minimum distance is $\delta \geq d$.

The generator polynomial $g(x)$ has only binary coefficients according to theorem 4.1. To avoid the search for $d-1$ successive numbers in $\mathcal{M}$, we shall show how to directly determine the designed distance based on an appropriate grouping of the cyclotomic cosets.

4.1.2 Designed distance

Sort the cyclotomic cosets K_i to obtain

$$K_{i_0}, K_{i_1}, \ldots, K_{i_s} \quad \text{with } i_0 = 0 < i_1 = 1 < i_2 < \ldots < i_s$$

(i_j is the smallest number of the set K_{i_j}). The designed distance is $d = i_{s+1}$ if $\mathcal{M}$ is constructed from the union of cyclotomic cosets $\mathcal{M} = K_{i_1} \cup K_{i_2} \ldots \cup K_{i_s}$ and $d = i_{s+1} + 1$ for $\mathcal{M}_0 = \mathcal{M} \cup K_0$. The number i_{s+1} is obviously the smallest number that is not contained in the set $\mathcal{M}$.

Example 4.1 (Generator polynomial of a BCH code) We shall construct a two-error correcting BCH code of length $n = 15 = 2^4 - 1$. In order to correct two errors, the designed distance d must satisfy $d \geq 5$. The cyclotomic coset K_i has been calculated in example 2.14. We choose $\mathcal{M} = K_1 \cup K_3$ such that the designed distance is $d = 5$. This is due to the fact that the numbers $1, 2, 3, 4$ are included in $\mathcal{M}$:

$$\mathcal{M} = \{1, 2, 3, 4, 6, 8, 9, 12\} .$$

This gives a dimension

$$k = 15 - |\mathcal{M}| = 15 - 8 = 7 .$$

The generator polynomial $g(x)$ is given by

$$\begin{aligned} g(x) &= \prod_{i \in \mathcal{M}} (x - \alpha^i) \\ &= (x-\alpha)(x-\alpha^2)(x-\alpha^3)(x-\alpha^4)(x-\alpha^6)(x-\alpha^8)(x-\alpha^9)(x-\alpha^{12}) \\ &= x^8 + g_7 x^7 + \ldots + g_1 x + g_0 . \end{aligned}$$

In order to calculate the coefficients g_i of $g(x)$, we use theorem 4.1. A polynomial $m_i(x)$ has coefficients from $GF(2)$, when we use the complex-conjugate roots of a cyclotomic coset K_i as linear factors of this polynomial. With the primitive element α from example 2.12, we first calculate $m_1(x)$:

$$\begin{aligned} m_1(x) &= (x-\alpha)(x-\alpha^2)(x-\alpha^4)(x-\alpha^8) \\ &= (x^2 - (\alpha+\alpha^2)x + \alpha\alpha^2)(x^2 - (\alpha^4+\alpha^8)x + \alpha^4\alpha^8) \\ &= x^4 - (\alpha+\alpha^2)x^3 + \alpha\alpha^2 x^2 - (\alpha^4+\alpha^8)x^3 \\ &\quad + (\alpha+\alpha^2)(\alpha^4+\alpha^8)x^2 - \alpha\alpha^2(\alpha^4+\alpha^8)x \\ &\quad + \alpha^4\alpha^8 x^2 - (\alpha+\alpha^2)\alpha^4\alpha^8 x + \alpha\alpha^2\alpha^4\alpha^8, \end{aligned}$$

$$\begin{aligned} x^3: &\quad (\alpha+\alpha^2) + (\alpha^4+\alpha^8) = \alpha^5 + \alpha^5 = 0, \\ x^2: &\quad \alpha\alpha^2 + (\alpha+\alpha^2)(\alpha^4+\alpha^8) + \alpha^4\alpha^8 = \alpha^3 + \alpha^5\alpha^5 + \alpha^{12} = 0, \\ x\ : &\quad \alpha^3(\alpha^4+\alpha^8) + \alpha^{12}(\alpha+\alpha^2) = \alpha^3\alpha^5 + \alpha^{12}\alpha^5\alpha^8 + \alpha^2 = 1, \\ x^0: &\quad \alpha\alpha^2\alpha^4\alpha^8 = \alpha^{15} = 1 . \end{aligned}$$

Therefore the polynomial $m_1(x)$ is

$$m_1(x) = x^4 + x + 1 .$$

Similarly,

$$m_3(x) = x^4 + x^3 + x^2 + x + 1 .$$

The generator polynomial is therefore

$$g(x) = m_1(x) m_3(x) = x^8 + x^7 + x^6 + x^4 + 1 . \qquad \diamond$$

4.1.3 Definition by the DFT

BCH codes can also be described using the discrete Fourier transform (DFT), which was defined in section 3.1.1.

Definition 4.3 (Binary primitive BCH code by the DFT) *The binary primitive BCH code C of length $n = 2^m - 1$ and the designed distance d are defined as*

$$C := \{a(x) \mid a(x) \circ\!\!-\!\!\bullet A(x),\ A_{n-1} = \ldots = A_{n-d+1} = 0,\ \forall_i : A_i^2 = A_{2i}\}$$

(compare with definition 3.3 for RS codes). The condition $A_i^2 = A_{2i}$, according to theorem 4.1, guarantees that the coefficients a_i are binary. The dimension k can be found by determining the number of possibilities that exist for the coefficients A_i.

The coefficients A_i of the polynomial $A(x)$ are from $GF(2^m)$ and the coefficients of $a(x)$ are from $GF(2)$. This condition is satisfied if $A_i^2 = A_{2i}$ is valid for all i. Using this, the symbols from $GF(2^m)$ are transformed to symbols in $GF(2)$. The computations are done in $GF(2^m)$.

Example 4.2 (BCH code via DFT) We shall construct the same BCH code as in example 4.1:

$\mathcal{A} =$	A_0	A_1	A_1^2	0	A_1^4	A_5	0	0	A_1^8	0	A_5^2	0	0	0	0
	0	1	2	3	4	5	6	7	8	9	10	11	12	13	14

The zero coordinates must satisfy the relation $A_i^2 = A_{2i}$. Therefore the zero at coordinate 14 implies zeros at coordinates 11, 13 and 7. Similarly, the zero at coordinate 12 forces zeros at 6, 9 and 3. Thus we have four consecutive zeros at 11, 12, 13 and 14, and hence a designed minimum distance of 5.

The next task is to determine to which subfield of $GF(2^4)$ the non-zero coordinates of $A(x)$ belong by using the coefficient relationship $A_i^2 = A_{2i}$. The coordinate A_0 corresponds to no other coordinate. Therefore the only subfield that satisfies the coefficient relation $A_0 = A_0^2$ is $GF(2)$. For A_5 and A_{10}, we can only chose elements of the field $GF(2^2)$ (compare with example 2.13). For A_1, we may chose elements from $GF(2^4)$, and the associated coordinates A_2, A_4, A_8 depend on the choice of A_1.

The cardinality of the code can be determined based on the number of possible independent coefficients we may choose for the polynomial $A(x)$. The independent coordinates include A_0, A_1 and A_5. Based on the subfields for these coordinates, the number of possible polynomials (or codewords) we may generate is

$$|C| = \underset{\substack{\uparrow \\ A_0}}{2} \cdot \underset{\substack{\uparrow \\ A_5}}{4} \cdot \underset{\substack{\uparrow \\ A_1}}{16} = 128, \quad k = 7 .$$

Table 4.1 Parameters for primitive BCH codes.

n	k	δ	n	k	δ	n	k	δ	n	k	δ
7	4	3	127	120	3	255	247	3	255	91	51
15	11	3		113	5		239	5		87	53*
	7	5		106	7		231	7		79	55
	5	7		99	9		223	9		71	59*
31	26	3		92	11		215	11		63	61*
	21	5		85	13		207	13		55	63
	16	7		78	15		199	15		47	85
	11	11		71	19		191	17		45	87*
	6	15		64	21		187	19		37	91*
63	57	3		57	23		179	21*		29	95
	51	5		50	27		171	23		21	111
	45	7		43	31**		163	25*		13	119
	39	9		36	31		155	27		9	127
	36	11		29	43*		147	29*			
	30	13		22	47		139	31			
	24	15		15	55		131	37*			
	18	21		8	63		123	39*			
	16	23					115	43*			
	10	27					107	45*			
	7	31					99	47			

** $\delta = d + 2$ * lower bound

For the base field $GF(2)$, the cardinality satisfies $k = \text{ld}|C|$. The generator polynomial is obtained using the DFT as follows:

$$\begin{aligned} g(x) &= (x-\alpha^{-3})(x-\alpha^{-6})(x-\alpha^{-7})(x-\alpha^{-9})(x-\alpha^{-11})(x-\alpha^{-12})(x-\alpha^{-13})(x-\alpha^{-14}) \\ &= g(x) \quad \text{(from example 4.1)} . \end{aligned}$$

◇

4.1.4 Properties of primitive BCH codes

We can construct BCH codes with length $n = 2^m - 1$. Table 4.1 shows the parameters for the dimension k and the true minimum distance δ for primitive BCH codes of length $n = 7, 15, 31, 63, 127$ and 255.

For the true minimum distance the following statement holds.

Theorem 4.4 (True minimum distance) *(without proof) Primitive BCH codes of length $n = 2^m - 1$ and the designed distance $d = 2^h - 1$ have the true minimum distance $\delta = d$.*

Two-error correcting BCH codes are distinguished through the following special property:

Definition 4.5 (Quasiperfect code) *An l-error correction code is quasiperfect if every vector of weight $t \geq l + 1$ may result in a codeword when added to a vector of weight $t \leq l + 1$. The result of the addition is a codeword.*

Remark The significance of a quasiperfect l error correcting code is that there is no coset leader with weight $t > l + 1$ (see section 1.3).

Theorem 4.6 (Quasiperfect BCH codes) *(see [McWSl]) All two-error correcting primitive BCH codes are quasiperfect.*

Conceptually, this means that if we consider a radius-three correction sphere around each codeword then all vectors in the space are within at a sphere. For a perfect code, each vector can be associated with one correction sphere while for a quasiperfect code, one vector may lie within more than one correction sphere. This theorem is illustrated by the following example.

Example 4.3 (Quasiperfect BCH code) For the code from example 4.1, we obtain the Hamming bound through

$$2^7\left(1+\binom{15}{1}+\binom{15}{2}\right) \leq 2^{15},$$
$$1+15+15\cdot 7 < 256\ .$$

However,

$$2^7\left(1+\binom{15}{1}+\binom{15}{2}+\binom{15}{3}\right) > 2^{15},$$
$$1+15+15\cdot 7+35\cdot 13 > 256\ .$$

◇

4.1.5 Calculation of the generator polynomial

The calculation of the generator polynomial can be simplified by calculating the irreducible polynomial $m_i(x)$ associated with each cyclotomic coset K_i that belongs to the code and then multiplying these polynomials together:

$$g(x) = m_{i_1}(x)\dots m_{i_s}(x).$$

In table 4.2, the irreducible polynomials are listed (in octal form). The entry marked with the asterisk in table 4.2 is

$$3 \quad 1\ 2\ 7\ .$$

This means that the cyclotomic coset K_3 with regard to the number $2^6-1=63$ gives the polynomial $m_3 \mathrel{\widehat{=}} 127$ in octal representation. The cyclotomic coset for K_3 is obtained through

$$K_3 = \{3,6,12,24,48,33\},$$

and the corresponding polynomial is

$$\begin{array}{llccc} m_3(x) = \prod\limits_{i\in K_3}(x-\alpha^i) & \widehat{=} & 1 & 2 & 7 \\ & \widehat{=} & 0\ 0\ 1 & 0\ 1\ 0 & 1\ 1\ 1 \\ & & \quad\ \ \updownarrow & \ \ \updownarrow & \updownarrow\ \updownarrow\ \updownarrow \\ & = & x^6+ & x^4+ & x^2+x+1 \end{array}\ .$$

When the polynomial $m_j(x)$ corresponding to K_j is not contained in table 4.2, it may be calculated as follows. From

$$K_j = \{j\cdot 2^k \mod n,\ k=0,\dots m-1\},$$

Table 4.2 Irreducible polynomials[1] of degree ≤ 11.

Degree						
degree 2	1 7					
degree 3	1 13					
degree 4	1 23	3 37	5 07			
degree 5	1 45	3 75	5 67			
degree 6	1 103	∗3 127	5 147	7 111	9 015	11 155
	21 007					
degree 7	1 211	3 217	5 235	7 367	9 277	11 325
	13 203	19 313	21 345			
degree 8	1 435	3 567	5 763	7 551	9 675	11 747
	13 453	15 727	17 023	19 545	21 613	23 543
	25 433	27 477	37 537	43 703	45 471	51 037
	85 007					
degree 9	1 1021	3 1131	5 1461	7 1231	9 1423	11 1055
	13 1167	15 1541	17 1333	19 1605	21 1027	23 1751
	25 1743	27 1617	29 1553	35 1401	37 1157	39 1715
	41 1563	43 1713	45 1175	51 1725	53 1225	55 1275
	73 0013	75 1773	77 1511	83 1425	85 1267	
degree 10	1 2011	3 2017	5 2415	7 3771	9 2257	11 2065
	13 2157	15 2653	17 3515	19 2773	21 3753	23 2033
	25 2443	27 3573	29 2461	31 3043	33 0075	35 3023
	37 3543	39 2107	41 2745	43 2431	45 3061	47 3177
	49 3525	51 2547	53 2617	55 3453	57 3121	59 3417
	69 2701	71 3323	73 3507	75 2437	77 2413	83 3623
	85 2707	87 2311	89 2327	91 3265	93 3777	99 0067
	101 2055	103 3575	105 3607	107 3171	109 2047	147 2355
	149 3025	155 2251	165 0051	171 3315	173 3337	179 3211
	341 0007					
degree 11	1 4005	3 4445	5 4215	7 4055	9 4015	11 7413
	13 4143	15 4563	17 4053	19 5023	21 5623	23 4757
	25 4577	27 6233	29 6673	31 7237	33 7335	35 4505
	37 5337	39 5263	41 5361	43 5171	45 6637	47 7173
	49 5711	51 5221	53 6307	55 6211	57 5747	59 4533
	61 4341	67 6711	69 6777	71 7715	73 6343	75 6227
	77 6263	79 5235	81 7431	83 6455	85 5247	87 5265
	89 5343	91 4767	93 5607	99 4603	101 6561	103 7107
	105 7041	107 4251	109 5675	111 4173	113 4707	115 7311
	117 5463	119 5755	137 6675	139 7655	141 5531	147 7243
	149 7621	151 7161	153 4731	155 4451	157 6557	163 7745
	165 7317	167 5205	169 4565	171 6765	173 7535	179 4653
	181 5411	183 5545	185 7565	199 6543	201 5613	203 6013
	205 7647	211 6507	213 6037	215 7363	217 7201	219 7273
	293 7723	299 4303	301 5007	307 7555	309 4261	331 6447
	333 5141	339 7461	341 5253			

one can calculate

$$K_i = \{-j \cdot 2^k \mod n,\ k = 0, \ldots m-1\}.$$

The polynomial $m_j(x)$ is the *reciprocal* polynomial of $m_i(x)$. This means

$$m_j(x) = x^{\deg m_i(x)} m_i\left(\frac{1}{x}\right) .$$

Example 4.4 (Use of table 4.2) For the case $n = 127 = 2^7 - 1$, the cyclotomic coset K_{15} is

$$K_{15} = \{15, 30, 60, 71, 99, 113, 120\} .$$

[1] This table was reproduced from [PeWe, pp. 476–492]

The polynomial m_{15} is not contained in table 4.2. We calculate (mod127): $-15 = 112$, $-30 = 97$, $-60 = 67$, $-71 = 56$, $-99 = 28$, $-113 = 14$, $-120 = 7$. This corresponds to the cyclotomic coset K_7. From table 4.2, we select

$$m_7 \mathrel{\widehat{=}} 367 \mathrel{\widehat{=}} 011110111 = x^7 + x^6 + x^5 + x^4 + x^2 + x + 1 \; .$$

The *reciprocal* polynomial is

$$m_{15} = x^7 + x^6 + x^5 + x^3 + x^2 + x + 1 \; . \qquad \diamond$$

4.2 Non-primitive BCH codes

Definition 4.7 (Non-primitive BCH codes) *Let $\beta \in GF(2^m)$ be an element of order $n < 2^m - 1$. Consider the cyclotomic coset K_i with respect to n and let $\mathcal{M}$ be the union of any number of cyclotomic cosets. This generates a non-primitive BCH code with length n and the generator polynomial*

$$g(x) := \prod_{i \in \mathcal{M}} (x - \beta^i) \; .$$

The designed distance is d if $d - 1$ successive numbers exist in the set $\mathcal{M}$. The true minimum distance is $\delta \geq d$.

A frequently used example to illustrate non-primitive BCH codes is the perfect Golay code $\mathcal{G}_{23}$.

Example 4.5 (Golay code) The Golay code $\mathcal{G}_{23}$ has length $n = 23$ and dimension $k = 12$. We require an element of order 23. According to theorem 2.22,

$$n \mid 2^m - 1 \; .$$

$23 \nmid 31$, $23 \nmid 63$, $23 \nmid 127$, ..., $23 \nmid 1023$, $23 \mid 2047$, $23 \cdot 89 = 2047$. Therefore the element $\alpha^{89} \in GF(2^{11})$ has order 23. We next determine the cyclotomic coset with respect to 23. This gives

$$K_1 = \{1, 2, 4, 8, 16, 9, 18, 13, 3, 6, 12\} \; .$$

No further cyclotomic coset is needed, because the dimension $k = n - |K_1| = 23 - 11 = 12$ is already satisfied. The designed distance for $\mathcal{G}_{23}$ is $d = 5$ because K_1 contains 4 successive integers. The true minimum distance is 7. The generator polynomial $g(x)$ is obtained using $(\beta = \alpha^{89} \in GF(2^{11}))$:

$$g(x) = (x-\beta)(x-\beta^2)(x-\beta^3)(x-\beta^4)(x-\beta^6)(x-\beta^8)(x-\beta^9) \cdot \\ \cdot (x-\beta^{12})(x-\beta^{13})(x-\beta^{16})(x-\beta^{18}) \; .$$

From table 4.2, we select $m_{89}(x) \mathrel{\widehat{=}} 5343$:

$$\begin{array}{rcl} m_{89}(x) & \widehat{=} & 5 \quad\quad 3 \quad\quad 4 \quad\quad 3 \\ & \widehat{=} & 1\;0\;1\quad 0\;1\;1\quad 1\;0\;0\quad 0\;1\;1 \\ & = & x^{11} + \; x^9 + \; x^7 + x^6 + \; x^5 + \; x + 1 \; . \end{array}$$

Therefore the generator polynomial for $\mathcal{G}_{23}$ is

$$g(x) = x^{11} + x^9 + x^7 + x^6 + x^5 + x + 1 \; . \qquad \diamond$$

4.3 Shortening and extending BCH codes

Shortening BCH codes

Definition 4.8 (Shortened BCH code) *Let C be a BCH code of length n, dimension k, designed minimum distance d and generator polynomial $g(x)$ (definitions 4.2 and 4.3):*

$$C := \{i(x)g(x) \mid \deg i(x) < k\} .$$

The shortening of the BCH code C^* is obtained by using the information polynomial $i(x)$ with degree $\deg i(x) < k^* < k$. This shortened code has length $n^* = n - (k - k^*)$, dimension k^* and designed distance d.

In other words, the shortening of BCH codes is achieved by using a subset of the original code, namely the codewords having zeros as coefficients in the upper $k - k^*$ information positions. This technique of shortening is also possible for RS codes. An additional method of shortening a code is by puncturing. This technique removes one or more coordinates from a codeword. In contrast to the shortening technique described in definition 4.8, the minimum distance is often decreased while the dimension of the code remains the same. Shortening and puncturing can be applied to all classes of codes. Once again, puncturing keeps the number of codewords constant and reduces the minimum distance, while shortening preserves the minimum distance and reduces the dimension.

Using these methods, we can construct BCH codes of arbitrary length.

Extending BCH codes

Definition 4.9 (Extended BCH code) *Let C be a binary BCH code. The extended code $\hat{C}$ has length $n+1$ and is defined by the addition of one coordinate with the value 0 or 1 such that all codewords $\mathbf{a} \in \hat{C}$ have even weight.*

In principle, any code can be extended. An extended code remains linear, but is no longer cyclic. (It should be noted, though, that the double extension of RS codes outlined in section 3.1.6 remains cyclic.)

Minimum distance of extended codes If the minimum distance d of a codeword is odd then the minimum distance $\hat{d}$ of an extension code $\hat{C}$, as proved in theorem 1.21, is

$$\hat{d} = d + 1 .$$

4.4 Non-binary BCH codes

Primitive as well as non-primitive BCH codes can be non-binary. We modify the corresponding definitions such that, instead of using $GF(2)$, we use $GF(q)$, where q can be either a prime p or a power of a prime p^l. The order n of the elements from $GF(q^m)$ must satisfy either $n = q^m - 1$ or $n < q^m - 1$. This generates either a primitive or non-primitive BCH code respectively. The cyclotomic cosets can be calculated through

$$K_i = \{iq^j \mod n,\ j = 0, 1, \ldots, m-1\} .$$

The generator polynomial, the minimum distance and the dimension are obtained through analogous arguments to those for binary primitive BCH codes. A double extended RS code, as described in section 3.1.6, can be classified as a non-primitive BCH code.

4.5 Relationship between BCH and RS codes

The statement is often made that RS codes comprise a subset of BCH codes, and occasionally that BCH codes comprise a subset of RS codes. Which statement is true? The answer is both.

First we consider BCH codes as subfield subcodes of RS codes. This is done using theorem 4.1, which states that a generator polynomial has only coefficients from $GF(q)$ if all complex-conjugate roots of $GF(q^m)$ are used. It can be shown that a BCH code is a subfield subcode of an RS code:

$$\text{BCH}(n \leq q^m - 1, k = n - |\mathcal{M}|, d \leq \delta) \text{ over } GF(q) \subset$$
$$\text{RS}(n \leq q^m - 1, k, d = n - k + 1) \text{ over } GF(q^m).$$

For the BCH code, we choose only the codewords from the RS code that have coefficients from $GF(q)$. The BCH code uses more roots in its generator polynomial from $GF(q^m)$ than necessary in order to achieve a specified minimum distance (i.e. not all roots of a BCH code are in series). This results in loss of the MDS property of RS codes.

On the other hand, it can be shown that if we limit the length of a BCH code with the parameters $(n \leq q^m - 1, k, d)$ over $GF(q)$ to length $n \leq q - 1$ then we get an RS code with the parameters $(n \leq q - 1, k, d = n - k + 1)$. Thus we have

$$\begin{array}{ll}
\text{RS} & C(n \leq q^m - 1, k, d = n - k + 1) \text{ over } GF(q^m) \\
\Downarrow & \text{subfield subcode} \\
\text{BCH} & C(n \leq q^m - 1, k = n - |\mathcal{M}|, d \leq \delta) \text{ over } GF(q) \\
\Downarrow & \text{bounded length} \\
\text{RS} & C(n \leq q - 1, k, d = n - k + 1) \text{ over } GF(q)\,.
\end{array}$$

Example 4.6 (Relationship of RS and BCH codes) We construct a $(n = 5^3 - 1 = 124, k = 100, d = 124 - 100 + 1 = 25)$ RS code over $GF(5^3)$. The subfield subcode is a $(n = 5^3 - 1, k = 66, d = 25)$ BCH code over $GF(5)$. By bounding the length of this code, we obtain the RS code with the parameters $(n = 5 - 1 = 4, k < 4, d = n - k + 1)$. ◇

4.6 Asymptotic behavior of BCH codes

BCH codes with a *very* great length are poor. This means that as the length goes to infinity, there exist no BCH codes where the code rate $R = k/n > 0$ and the normalized minimum distance $d/n > 0$ simultaneously. This effect is called *asymptotically poor* . This statement is not valid for short BCH codes up to length $\sim 2^{13}$. Within this set, there exist many good codes. BCH codes offer the following properties: they are easy to decode, they can be designed for any code length, and they have good coding characteristics. As a result of these characteristics, BCH codes are extensively used in practice.

Theorem 4.10 (BCH codes are asymptotically poor) *There exist no binary primitive BCH code for which $d/n > \varepsilon$ and simultaneously $k/n > \varepsilon$ for $n \to \infty$.*

Proof For a detailed proof, [McWSl, theorem 13, p. 269] is recommended. We must show that for $n \longrightarrow \infty$ that either d/n approaches 0, for the case where k/n does not approach 0 and/or k/n approaches 0 for the case d/n does not approach 0:

$$\text{(i)} \quad \text{from } \frac{k}{n} > 0 \quad \text{it follows that} \quad \frac{d}{n} \longrightarrow 0\,,$$
$$\text{(ii)} \quad \text{from } \frac{d}{n} > 0 \quad \text{it follows that} \quad \frac{k}{n} \longrightarrow 0\,.$$

We shall only prove the simpler condition (i). For primitive BCH codes, we may write

$$k = n - |\mathcal{M}|, \quad |\mathcal{M}| \le m(d-1)\,,$$

where m defines the size of the field $GF(2^m)$.

The inequality $|\mathcal{M}| \le m(d-1)$ can be obtained through the following consideration. The cyclotomic coset K_1 for $n = 2^m - 1$ will always result in a minimum distance of $d = 3$ because $\{1,2\} \in K_1$. The increase in the minimum distance is clearly larger than 1 for any additional cyclotomic coset. In this way, we obtain the upper bound for $|\mathcal{M}|$, which gives

$$k \ge n - m(d-1)\,.$$

From 'k/n does not go to 0' it follows that

$$1 - \frac{m(d-1)}{n} > 0 \Longrightarrow \frac{m(d-1)}{n} < 1\,;$$

therefore

$$d < \frac{n}{m} + 1\,,$$

resulting in

$$\frac{d}{n} < \frac{1}{n}\left(\frac{n}{m} + 1\right) = \frac{1}{m} + \frac{1}{n} \xrightarrow[n \longrightarrow \infty]{} 0\,,$$

where $n = 2^m + 1$. □

The result that BCH codes are asymptotically poor is dependent on the estimation of the minimum distance of a BCH code. However, in [BJ74], a cyclic BCH code construction is given for which an improved distance estimation is recognized (compare with section 9.2.6).

4.7 Decoding of BCH codes

The algebraic decoding method (using the Berlekamp–Massey or Euclidean algorithm for the solution of the key equations) that was explained in section 3.2 may be used for the decoding of BCH codes. In addition to the $2t$ successive *roots* in the transformation domain, BCH codes result in an additional *zeros* for the complex-conjugate positions (see example 4.2). These *roots* cannot be used by the decoding method, because they are dependent (see the conditions explained in section 4.1.3). For the decoding of binary BCH codes, the calculation of the error value polynomial is not necessary, because the error value is always equal to 1.

Because the algebraic decoding method can only use successive *roots*, we see that error correction will only work for errors with half the weight of the designed minimum distance ($\lfloor \frac{d-1}{2} \rfloor$), even when the true minimum distance is larger.

For the decoding of BCH codes, we can made the following observations:

- For a BCH code, the decoding is identical to that for RS codes. The solution of the error value polynomial is not necessary for binary BCH codes.
- The *roots* in the complex-conjugate coordinates of the transformed codewords cannot be used to improve decoding.
- For the shortening (RS and BCH) codes, the decoding methods are the same. This is not the case for the puncturing of a code.
- In the case where more than $\lfloor \frac{d-1}{2} \rfloor$ errors occur, the calculation of the error locator polynomial $C(x)$ results in a degree larger than the number of roots in $C(x)$. In this case, a decoding failure is possible.

4.8 Summary

Binary BCH codes were developed independently by Bose and Ray-Chaudhuri [BRC60a, BRC60b] and Hocquenghem [Hoc59]. Decoding methods were developed by Peterson [Pet60]. Non-binary BCH codes and their decoding were published by Gorenstein and Zierler in [GZ61]. Because the algebraic decoding can be applied to both RS and BCH codes, important developments in this area are also mentioned in the summary of chapter 3 (section 3.3).

We have defined binary primitive BCH codes (length $n = 2^m - 1$) using two different approaches. The first uses cyclotomic cosets, while the second uses the DFT. A codeword $a(x)$ of a binary BCH code has coefficients from $GF(2)$ and the transformed codeword $A(x)$ has coefficients from $GF(2^m)$. This results in the restriction $A_{2i} = A_i^2$, which implies that some of the coefficients of the transformed codeword $A(x)$ must belong to a subfield of $GF(2^m)$. As for RS codes, the algebraic decoding using the key equation can be applied to BCH codes. Both the BM and the Euclidean algorithm can be used to compute the error locator polynomial. This requires calculations to be done in the Galois field $GF(2^m)$, despite the fact that the code is binary. The BCH codes can only decode error vectors having a Hamming weight not more than half the designed distance, even if the true minimum distance is larger. The solution of the error value polynomial is not necessary for binary BCH codes, because the error value is always one.

Two properties of primitive BCH codes have been explained. When the designed distance $d = 2^h - 1$, it is also equal to the true minimum distance δ. Two-error correcting BCH codes are quasiperfect. Quasiperfect codes violate the Hamming bound if the correction sphere is increased by one. This is not a sufficient condition, because it is also true for three-error correcting codes that are not quasiperfect.

The class of non-primitive BCH codes has also been defined. An example of such a code is the perfect Golay code $\mathcal{G}_{23}$ [Gol49, Gol54]. Non-primitive BCH codes have length $n < 2^m - 1$. They are constructed using an element of order n, i.e. a non-primitive element from $GF(2^m)$. The element may be a primitive element of a subfield.

Shortening and extending of BCH codes has been presented. Shortening of a code can also be achieved through puncturing, where the original designed distance is often reduced.

Table 4.3 Logarithm table for $GF(2^5)$.

$GF(2^5)$ with $(\alpha) = \alpha^5 + \alpha^2 + 1 = 0$							
Comp.	Exp.	Comp.	Exp.	Comp.	Exp.	Comp.	Exp.
00000	$-\infty$	00101	7	11111	15	11110	23
10000	0	10110	8	11011	16	01111	24
01000	1	01011	9	11001	17	10011	25
00100	2	10001	10	11000	18	11101	26
00010	3	11100	11	01100	19	11010	27
00001	4	01110	12	00110	20	01101	28
10100	5	00111	13	00011	21	10010	29
01010	6	10111	14	10101	22	01001	30

Binary and non-binary BCH codes can be considered as subfield subcodes from RS codes. On the other hand, RS codes can also be considered as subcodes of non-binary BCH codes over $GF(q)$.

The binary cyclic BCH codes have a wide range of applications in the form of CRC (*cyclic redundancy check*) codes. These codes are used mainly for error detection. The ISDN protocol LAPD uses a CRC code according to the ITU-T recommendation X.25 having the generator polynomial $g(x) = x^{16} + x^{12} + x^5 + 1$. The same code is also used in the DAB standard for digital broadcasting. In the DAB standard, the generator polynomials used for error detection are $g(x) = x^{16} + x^{15} + x^2 + 1$ and $g(x) = x^8 + x^4 + x^3 + x^2 + 1$. The DECT (Digital Enhanced Cordless Telecommunications) system, cyclic codes are also employed. In this case, the ETSI standard uses a cyclic binary code to protect signaling data for digital cordless telephones.

The ETSI standard for GSM (Global System for Mobile Communications) uses binary cyclic codes for the purpose of error detection in a variety of signaling channels. In the synchronization channel, the generator polynomial $g(x) = x^{10} + x^8 + x^6 + x^5 + x^4 + x^2 + 1$ is used, and for initial access, the polynomial $g(x) = x^6 + x^5 + x^3 + x^2 + x^1 + 1$ is used. The polynomial $g(x) = x^{40} + x^{26} + x^{23} + x^{17} + x^3 + 1$ is employed for error detection in all protocols. This is a burst error correcting Fire code [LiCo, pp. 261–267]. Furthermore, the detection of errors in the speech channel is achieved using $g(x) = x^3 + x + 1$.

In ERMES, the ETSI standard of the European digital paging service, a type of shortened $(31,20,6)$ BCH code is used that has a generator polynomial $g(x) = x^{12} + x^{11} + x^9 + x^7 + x^6 + x^3 + x^2 + 1$. This code is used for both error correction and error detection.

4.9 Problems

Problem 4.1
Calculate the generator polynomial $g(x)$ of a single-error correcting BCH code with length 31. Use the cyclotomic coset K_1 and table 4.3 as a logarithm table.

Problem 4.2
Show that the binary single-error correction primitive BCH codes of length $n = 2^m - 1$ are equal to Hamming codes.

Problem 4.3
Determine the dimension k_i of the i-error correcting BCH code with length $n = 31$ for $i = 1,2,3,5,7$.

Problem 4.4
Consider a BCH code (15,7,5) with the generator polynomial $g(x) = x^8 + x^7 + x^6 + x^4 + 1$. We receive $r(x) = x^{10} + x^8 + x^6 + x^2 + 1 = c(x) + e(x)$. Calculate the transmitted code word $c(x)$ under the assumption that no more than two errors have occurred.

Hint: In calculations in $GF(2^4)$ use the primitive polynomial $p(x) = x^4 + x + 1$, or use the logarithm table from example 2.12.

5

Other classes of codes

In addition to RS and BCH codes, there are other classes of codes that are important from a practical and theoretical standpoint. In this chapter, a few of the more important classes will be defined and discussed. The first class of codes to be considered are built on the premise of orthogonal transmission (e.g. using biorthogonal signals). This class includes the first-order Reed–Muller codes (RM codes), simplex, Hamming and Hadamard codes, and Walsh sequences. The close interrelationships among these codes will be explored. Another construction that arises from the simplex code structure is the pseudonoise sequence (PN sequence). The properties of PN sequences will also be investigated.

Next, higher-order RM codes will be introduced. These codes will again be described in chapter 9 as generalized concatenated codes. The binary quadratic residue codes (QR codes) include some of the best known codes, and will also be discussed. We shall return to the subject of cyclic codes and generalize its definition, which will allow us to evaluate the consta- and negacyclic codes and i-cyclic codes. These techniques are used for algebraic implementations of higher-rate modulation methods.

Finally, we will present the binary representation of q-ary codes. Some interesting examples of this representation include the nonlinear Kerdock and Preparata codes and the Nordstrom–Robinson code. These codes all correspond to a binary interpretation of linear codes over the ring $\mathbb{Z}_4$.

5.1 First-order Reed–Muller codes, simplex codes and Walsh sequences

There exists a close connection among the Hamming, simplex, RM and Hadamard codes, and Walsh sequences. This relationship is depicted schematically in figure 5.1 and explained in this section. In particular, there exists a class of codes in which any one code from this class

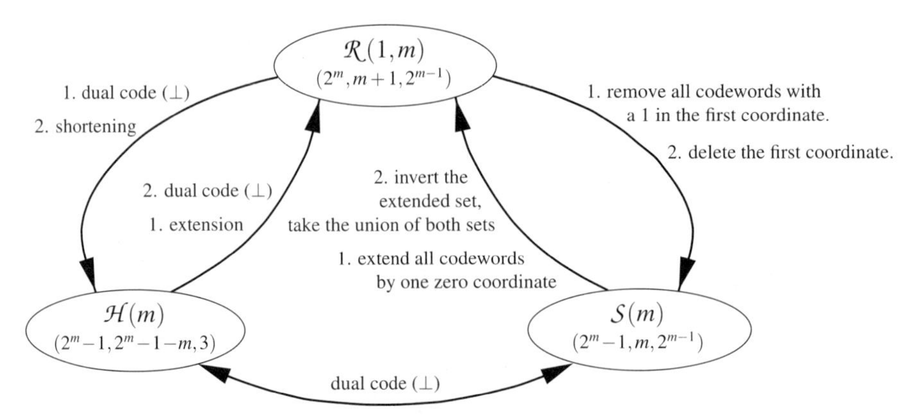

Figure 5.1 Interdependence among Hamming, simplex and first-order Reed–Muller codes.

can generate the other codes through a simple transformation operation. We shall begin our investigation with the RM code definition.

Definition 5.1 (First-order RM codes) *The generator matrix $\mathbf{G}_{\mathcal{R}}$ of a first-order RM code is constructed as follows. The top row of the matrix is the all-ones vector of length 2^m. The submatrix below this row contains as columns all 2^m vectors of length m from $\mathbb{F}_2^m$. This code is denoted by $\mathcal{R}(1,m)$, and has length $n = 2^m$, dimension $k = m+1$ and minimum distance 2^{m-1}. The $\mathcal{R}(1,m)$ code is a binary $(2^m, m+1, 2^{m-1})$ code.*

In general the generator matrix is a $k \times n$ matrix of rank k. It is clear by construction of the matrix $\mathbf{G}_{\mathcal{R}}$ that it consists of 2^m columns and $m+1$ rows. The proof for the minimum distance is given in theorem 5.2.

Example 5.1 (First-order RM codes) The generator for the RM code with $m = 4$ is

$$\mathbf{G}_{\mathcal{R}} = \begin{pmatrix} 1111 & 1111 & 1111 & 1111 \\ 0000 & 0000 & 1111 & 1111 \\ 0000 & 1111 & 0000 & 1111 \\ 0011 & 0011 & 0011 & 0011 \\ 0101 & 0101 & 0101 & 0101 \end{pmatrix} .$$

$\mathcal{R}(1,4)$ is a binary $(16,5,8)$ code. ◇

RM codes can be recursively calculated, as shown in the following theorem:

Theorem 5.2 (Recursive construction of first-order RM codes) *RM codes of length $2n$ can be constructed from first-order RM codes of length n as follows:*

$$\mathcal{R}(1,m+1) = \{|\mathbf{a}|\mathbf{a}+\mathbf{b}|,\ \mathbf{a} \in \mathcal{R}(1,m),\ \mathbf{b} = (0\ldots0) \text{ or } (1\ldots1)\} .$$

Proof We begin by considering $\mathcal{R}(1,2)$, or in other words a $(4,3,2)$ parity check (PC) code. The code words of the new code are constructed from the eight original code words as follows:

$$|\mathbf{a}|\mathbf{a}| \text{ or } |\mathbf{a}|\mathbf{a}+(1111)| .$$

The length of the new code is $n = 2^3$ and the dimension is $k = 3+1 = 4$. Three information coordinates are used to select the vector $\mathbf{a}$, and one coordinate is used to decide if $\mathbf{a}$ should be inverted.

The code is linear, and therefore the distance can be found as the minimum weight of codewords. There are two cases to consider. One arises when $\mathbf{a}$ is the minimum codeword from $\mathcal{R}(1,2)$ and $\mathbf{b}$ is the all zero vector. This gives $\{\mathbf{a}_{min}|\mathbf{a}_{min}\}$ with a weight of $2\,\mathrm{wt}(\mathbf{a}_{min})$. The second case arises when $\mathbf{b}$ is the all-ones vector of length 2^m. In this case, assume that $\mathrm{wt}(\mathbf{a}) = a$; then $\mathrm{wt}(|\mathbf{a}|\mathbf{a}+(1111)|) = a+2^m-a = 2^m$. The minimum distance is the minimum of these two cases. Therefore $d = \min\{2\cdot 2, 4\}$ for this example.

The new code is $\mathcal{R}(1,3)$, or similarly a $(8,4,4)$ code. Recursion can be used to find the parameters for m and $m+1$. It follows from this, that $\mathcal{R}(1,m)$ has minimum distance 2^{m-1}. □

Theorem 5.3 (Weight distribution of $\mathcal{R}(1,m)$) *The weight distribution of the $\mathcal{R}(1,m)$ code based on definition 1.7 is*

$$W(x) = 1 + (2^{m+1}-2)x^{2^{m-1}} + x^{2^m}.$$

Proof All codewords excluding $(0\ldots 0)$ and $(1\ldots 1)$ have the same weight. This follows directly from the recursion technique used in theorem 5.2. If this statement is valid for $\mathcal{R}(1,m)$ then it is also valid for $\mathcal{R}(1,m+1)$, since the construction

$$|\mathbf{a}|\mathbf{a}| \text{ and } |\mathbf{a}|\mathbf{a}+(1\ldots 1)|$$

doubles the weight of all codewords, excluding $(0\ldots 0)$ and $(1\ldots 1)$. □

5.1.1 Reed–Muller and Hamming codes

The dual code (definition 1.20) of an $\mathcal{R}(1,m)$ code is an extended Hamming code (see theorem 1.21 on extended codes).

Let $\mathbf{H}_{\mathcal{H}}(m)$ be an $(n-k)\times n$ parity check matrix of a Hamming code $\mathcal{H}(m)$ (see theorem 1.16). Then the $((n-k)+1)\times n$ parity check matrix of an extended Hamming code $\mathcal{H}^{ex}(m)$ can be constructed by adding the all-ones row and an extra column $(10\ldots 0)^T$. Hence $\mathcal{H}^{ex}(m)$ has parameters $(2^m, 2^m-m-1, 4)$. It can be seen that the parity check matrix of the extended Hamming code is identical to the generator matrix of a first-order RM code.

Example 5.2 (Connection between RM and Hamming codes) We select $m = 3$ and construct the generator and parity check matrices for the codes with length 8:

$$\mathbf{G}_{\mathcal{R}} \quad \overset{\perp}{\Longleftrightarrow} \quad \mathbf{H}^{ex}_{\mathcal{R}} \quad \overset{ext.}{\Longleftrightarrow} \quad \mathbf{H}_{\mathcal{R}}$$

$$\begin{pmatrix} 1111 & 1111 \\ 0000 & 1111 \\ 0011 & 0011 \\ 0101 & 0101 \end{pmatrix} \qquad \begin{pmatrix} 1111 & 1111 \\ 0001 & 1110 \\ 0110 & 0110 \\ 1010 & 1010 \end{pmatrix} \qquad \begin{pmatrix} 000 & 1111 \\ 011 & 0011 \\ 101 & 0101 \end{pmatrix}.$$

By carrying out a cyclic shift of coordinates, $\mathbf{G}_{\mathcal{R}}$ can be transformed to $\mathbf{H}^{ex}_{\mathcal{R}}$. In $\mathbf{G}_{\mathcal{R}}$, coordinate $n-1$ is shifted to coordinate 0, and so on. Finally, the first row and the last column of the new matrix are deleted to produce $\mathbf{H}_{\mathcal{R}}$. ◇

5.1.2 Hamming and simplex codes

The dual code (definition 1.20) of a binary Hamming code is known as the simplex code.

Definition 5.4 (Simplex code) *The generator matrix of a simplex code $\mathcal{S}(m)$ is the parity check matrix of a Hamming code. This gives a code with length $n = 2^m - 1$, dimension $k = m$, and minimum distance $d = 2^{m-1}$. With the exception of the zero codeword, all codewords have the same weight, which is also the minimum distance.*

Example 5.3 (Simplex code) The generator matrix of a $(7,3,4)$ simplex code is

$$\mathbf{G}_{\mathcal{S}(3)} = \begin{pmatrix} 000 & 1111 \\ 011 & 0011 \\ 101 & 0101 \end{pmatrix} .$$

◇

Orthogonal Walsh sequences A so-called Walsh sequence $\mathcal{W}(m)$ can be constructed by extending all the codewords of a $\mathcal{S}(m)$ simplex code by one zero coordinate, and by carrying out the following mapping:

$$\phi : \begin{cases} 0 \rightarrow 1, \\ 1 \rightarrow -1 . \end{cases}$$

The length of a Walsh sequence $\mathcal{W}(m)$ is 2^m. Let $\mathbf{a}, \mathbf{b} \in \mathcal{S}(m)$ be different codewords and let $\mathbf{x} = \phi(\mathbf{a})$ and $\mathbf{y} = \phi(\mathbf{b})$. Then $\mathbf{x}, \mathbf{y} \in \mathcal{W}(m)$ and

$$\sum_{i=0}^{2^m-1} x_i y_i = 0 .$$

Example 5.4 (Walsh sequence) The simplex code $\mathcal{S}(2)$ has the following four codewords: (000), (011), (101), (110). These codewords are extended by adding the zero coordinate to the beginning of each and then transforming them according to the mapping $\phi : 0 \rightarrow 1, 1 \rightarrow -1$:

$$\begin{matrix} 0 & 0 & 0 \\ 0 & 1 & 1 \\ 1 & 0 & 1 \\ 1 & 1 & 0 \end{matrix} \quad \longrightarrow \quad \begin{matrix} 0 & 0 & 0 & 0 \\ 0 & 0 & 1 & 1 \\ 0 & 1 & 0 & 1 \\ 0 & 1 & 1 & 0 \end{matrix} \quad \longrightarrow \quad \begin{matrix} 1 & 1 & 1 & 1 \\ 1 & 1 & -1 & -1 \\ 1 & -1 & 1 & -1 \\ 1 & -1 & -1 & 1 \end{matrix} .$$

Any two rows are orthogonal to one another. For instance, the dot product of rows 2 and 3 is

$$\sum_{i=0}^{3} x_i y_i = 1 \cdot 1 + 1 \cdot (-1) + (-1) \cdot 1 + (-1) \cdot (-1) = 0 .$$

◇

Remark There is a difference between the definition of duality and orthogonality of codes. Duality means that the inner product (definition 1.4) is zero,

$$\mathbf{a} \in \mathcal{C},\ \mathbf{b} \in \mathcal{C}^{\perp} :\ \langle \mathbf{a}, \mathbf{b} \rangle = 0 .$$

Orthogonality is defined for values -1 and 1, and is calculated for x and y as

$$\sum_{i=0}^{2^m-1} x_i y_i = 0 .$$

Let $\mathbf{a} = \phi^{-1}(\mathbf{x})$ and $\mathbf{b} = \phi^{-1}(\mathbf{y})$. Then it follows that if $\mathbf{x}$ and y are orthogonal,

$$\sum_{i=0}^{2^m-1} a_i + b_i = 0 \mod 2 .$$

When $\mathbf{x}$ is orthogonal to $\mathbf{y}$, this does not imply that $\langle \mathbf{a}, \mathbf{b} \rangle = 0$. For example, $\mathbf{x} = (1,1,-1,-1)$ is orthogonal to $\mathbf{y}(1,-1,1,-1)$. The inner product is determined using the inverse transformation $\phi^{-1}(x)$: $\langle \mathbf{a}, \mathbf{b} \rangle = \langle \phi^{-1}(\mathbf{x}), \phi^{-1}(\mathbf{y}) \rangle = \langle (0,0,1,1), (0,1,0,1) \rangle = 1 \neq 0$.

Similarly, if $\langle \mathbf{a}, \mathbf{b} \rangle = 0$, this does not imply that $\mathbf{x}$ is orthogonal to $\mathbf{y}$. This can be seen for the case when $\langle \mathbf{a}, \mathbf{b} \rangle = \langle (1,1,1,1), (1,1,1,1) \rangle = 0$.

5.1.3 Simplex codes and binary pseudonoise (PN) sequences

$\mathcal{S}(m)$ is often called a *maximal-length feedback shift register code*. All non-zero codewords can be represented as sequences of maximal length generated by a shift register of length m (see also [Gol]). The codewords are also called m sequences or pseudonoise sequences (PN).

We shall first explain the concept of randomness, using the definition of a pseudonoise sequence from [Gol]. For this, we need some notation and definitions.

Definition 5.5 (Sequence, period, periodic sequence, j-shift) *Let $\mathbf{c}_\infty$ be a semi-infinite binary sequence*

$$\mathbf{c}_\infty = \{c_0, c_1, c_2, \ldots\} .$$

The period of this sequence is the smallest value n such that

$$c_i = c_{i+n} = c_{i+2n} = \ldots = c_{i+j\cdot n} = \ldots, \quad i = 0, 1, \ldots .$$

All elements with indices in the same coset modulo n must be identical. Therefore we need only consider the sequence

$$\mathbf{c} = (c_0, c_1, c_2, \ldots, c_{n-1}) .$$

A cyclic j-shift of this sequence is defined by

$$\mathbf{c}_j = (c_j, c_{j+1}, \ldots, c_{n-1}, c_0, \ldots, c_{j-1}) .$$

Definition 5.6 (i-run) *A run of length i consists of i successive symbols with the same value between two symbols of another value, i.e.*

$$c_{j-1} \neq c_j = c_{j+1} = \ldots = c_{j+i-1} \neq c_{j+i} .$$

Definition 5.7 *The periodic autocorrelation function of a sequence of length n is defined as*

$$\varphi_{\mathbf{c}}(k) = \frac{n - 2\,\mathrm{dist}(\mathbf{c}, \mathbf{c}_k)}{n}, \quad k = 0, 1, \ldots, n-1 .$$

The periodic autocorrelation coefficients $\varphi_{\mathbf{c}}(k)$ have the following bounds:

$$-1 \leq \varphi_{\mathbf{c}}(k) \leq 1 \quad \textit{and} \quad \varphi_{\mathbf{c}}(0) = 1.$$

We are now ready to give Golomb's three postulates for a binary pseudonoise sequence $\mathbf{c}$ with period n.

Golomb's postulates

(1) The numbers of zeros and ones in a sequence are the same for even n, and differ by one for odd n:

$$\mathrm{wt}(\mathbf{c}) = \begin{cases} \frac{n}{2} & \text{for } n \text{ even,} \\ \frac{n \pm 1}{2} & \text{for } n \text{ odd .} \end{cases}$$

Goal: The probability for 0 and 1 should in every case be equal to $1/2$.

(2) The following relations should be valid for a period of length n:

half $(1/2)$ of all runs have length 1 (half of which are zero);

a quarter $(1/4)$ of the runs have length 2 (half of which are zero);

an eighth $(1/8)$ of the runs have length 3 (half of which are zero);

$\vdots$

$1/2^i$ of all runs have length i (half of which are zero)

Goal: This distribution guarantees that the appearance of a zero or one after a subsequence of length i is equiprobable.

(3) The periodic autocorrelation must be constant:

$$\varphi_{\mathbf{c}}(k) = \text{const}, \quad 0 < k < n\ .$$

Goal: The correlation value should not vary over one period. This protects the period from being known through correlation except when $k = 0$ or for multiples of n.

Theorem 5.8 (Autocorrelation value) *Let* $\mathbf{c}$ *be a sequence of period n satisfying Golomb's postulates. Then, for* $0 < j < n$, *we have*

$$\varphi_{\mathbf{c}}(j) = \begin{cases} -\frac{1}{n-1} & \textit{for } n \textit{ even,} \\ -\frac{1}{n} & \textit{for } n \textit{ odd .} \end{cases}$$

Proof We shall prove this theorem using a standard method from combinatorics namely, calculating a sum in two different ways. We represent a sequence and all of its $n-1$ cyclic shifts as an array:

$$\begin{array}{rcccc|l}
\mathbf{c}_0: & c_0 & c_1 & \dots & c_{n-1} & \\
\mathbf{c}_1: & c_1 & c_2 & \dots & c_0 & s = n\varphi_{\mathbf{c}}(j) \\
 & \vdots & & & & \\
\mathbf{c}_j: & c_j & c_{j+1} & \dots & c_{j-1} & s = n\varphi_{\mathbf{c}}(j) \\
 & \vdots & & & & \\
\mathbf{c}_{n-1}: & c_{n-1} & c_0 & \dots & c_{n-2} & s = n\varphi_{\mathbf{c}}(j) \\
\hline
 & s' & s' & & s' & \Sigma = n(n-1)\varphi_{\mathbf{c}}(j)
\end{array}$$

Let s be the number of identical coordinates minus the number of different coordinates between the first row $\mathbf{c}_0$ and the selected row $\mathbf{c}_j$. According to the third postulate, this number is $n\varphi_{\mathbf{c}}(j)$. The number of different coordinates is equal to the distance between $\mathbf{c}$ and $\mathbf{c}_j$. The number of identical coordinates is $n - \mathrm{dist}(\mathbf{c}, \mathbf{c}_j)$. According to the definition of the autocorrelation function, we obtain

$$s = n - 2\ \mathrm{dist}(\mathbf{c}, \mathbf{c}_j)\frac{n}{n} = n\varphi_{\mathbf{c}}(j)\ .$$

The sum over all the rows gives $\Sigma = n(n-1)\varphi_{\mathbf{c}}(j)$.

Next, we assume n to be even, and we calculate s' for each column. The columnwise calculation is performed by summing the number elements in a column that are identical to the first element, and subtracting the number that are unlike the first element. According to the first postulate (that the numbers of zeros and ones are the same),

$$s' = \frac{n}{2} - 1 - \frac{n}{2} = -1\ .$$

The sum over all column coordinates is therefore $-n$, and must be equal to Σ. Hence $\Sigma = n(n-1) \cdot \varphi_{\mathbf{c}}(j) = -n$, or $\varphi_{\mathbf{c}}(j) = -\frac{1}{n-1}$. A similar proof may be carried out for odd n. □

Cyclic representation of PN sequences A Hamming code can be represented as a cyclic primitive BCH code of length $2^m - 1$ and $d = 3$ with generator polynomial $g_{BCH}(x)$ (see definition 4.2). Therefore, the cyclic simplex code has as a generator polynomial that is a parity check polynomial of a BCH code, i.e.

$$g_S(x) = \frac{x^n - 1}{g_{BCH}(x)} = h_{BCH}(x),$$

where

$$g_{BCH}(x) = \prod_{i=0}^{m-1}(x - \alpha^{2^i}), \quad \alpha \text{ is a primitive element.}$$

In other words, using the cyclotomic coset K_1 (definition 2.25), we get

$$g_S(x) = \prod_{i \mid i \notin K_1}(x - \alpha^i), \quad \alpha \text{ is a primitive element.}$$

Similarly, the parity check polynomial of a simplex code is the generator polynomial of a BCH Hamming code, and is a primitive polynomial (definition 2.15). Such a code is termed minimal (see [McWSl, p. 219]).

Example 5.5 (Simplex code) The generator polynomial of a $(7,4,3)$ BCH code can be found in table 4.2 as a polynomial having degree 7 and the representation $13 = (001011)$:

$$g_{BCH}(x) = x^3 + x + 1 \longrightarrow g_S(x) = x^7 - 1 : x^3 + x + 1 = x^4 + x^2 + x + 1\ .$$

From this polynomial we can create 8 codewords for the cyclic simplex code $\mathcal{S}(3)$:

0000000
0010111
0101110
1011100
0111001
1110010
1100101
1001011 .

◇

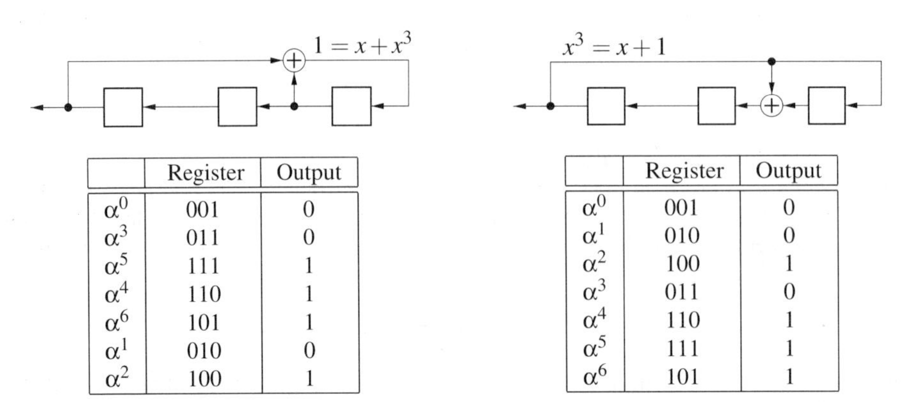

	Register	Output
α^0	001	0
α^3	011	0
α^5	111	1
α^4	110	1
α^6	101	1
α^1	010	0
α^2	100	1

	Register	Output
α^0	001	0
α^1	010	0
α^2	100	1
α^3	011	0
α^4	110	1
α^5	111	1
α^6	101	1

Figure 5.2 Two discrete LTI systems described by primitive polynomials.

There is a relation between the Galois field $GF(2^m)$ and PN sequences. When one writes all the $2^m - 1$ elements of the field (excluding the zero element) in component representation as rows of a matrix, every column in this matrix is a PN sequence. This can be seen in example 2.12. This result implies that a PN sequence of length $n = 2^m - 1$ can be generated by a shift register. The feedback coefficients are identical to the coefficients of the reciprocal primitive polynomial. Such a shift register can produce the longest possible run with a length m. A length-m run of ones occurs in the PN sequence, but there is no such run of zeros.

Example 5.6 (PN sequences and Galois fields) The primitive polynomial $p(x) = x^3 + x + 1$ gives the feedback coefficients for two discrete LTI (linear, time invariant) systems shown in figure 5.2. After initializing the registers, both systems generate all the elements (excluding the zero element) from GF(2^3) that can be represented as all powers of α^i, $i = 0, \dots, 6$. Both systems produce a PN sequence at the output (from a shift register) with period 7. $\diamond$

5.1.4 Reed–Muller and simplex codes

The codewords of a simplex code $\mathcal{S}(m) = (2^m - 1, m, 2^{m-1})$ can be found by selecting only the codewords of the corresponding RM code $\mathcal{R}(1,m) = (2^m, m+1, 2^{m-1})$ having a zero in the first coordinate, and then by removing this coordinate. According to theorem 1.18 in any linear code exactly half of all codewords have a zero in the first coordinate. This reduces the dimension and the length of the code by one.

Conversely, all 2^{m+1} codewords of a RM code can be constructed from a simplex code by extending the simplex codewords by the coordinate 0 and then taking the inverse of these vectors. The union of the inverted and non-inverted vectors gives all the codewords of the RM code.

Example 5.7 (Relation between RM and simplex codes) The four codewords of a simplex code $\mathcal{S}(2)$ are extended by one coordinate with the value zero:

$$\begin{matrix} 0 & 0 & 0 \\ 0 & 1 & 1 \\ 1 & 0 & 1 \\ 1 & 1 & 0 \end{matrix} \longrightarrow \begin{matrix} 0 & 0 & 0 & 0 \\ 0 & 0 & 1 & 1 \\ 0 & 1 & 0 & 1 \\ 0 & 1 & 1 & 0 \end{matrix} \longrightarrow \begin{matrix} 0 & 0 & 0 & 0 \\ 0 & 0 & 1 & 1 \\ 0 & 1 & 0 & 1 \\ 0 & 1 & 1 & 0 \end{matrix} \quad \text{and inv.} \quad \begin{matrix} 1 & 1 & 1 & 1 \\ 1 & 1 & 0 & 0 \\ 1 & 0 & 1 & 0 \\ 1 & 0 & 0 & 1. \end{matrix}$$

This set of vectors is inverted to give the eight codewords of the $\mathcal{R}(1,2)$ RM code, which is also a parity check code. ◇

If one applies the mapping ϕ to the 0 extended codewords of the simplex code, the set of orthogonal sequences is produced that were presented in section 5.1.2. If one inverts the set of orthogonal sequences, this produces the set of biorthogonal sequences. Using the construction of RM codes from simplex codes as described above, the set of biorthogonal sequences is equivalent to the ϕ transform of the codewords of the RM code. Therefore the set of biorthogonal sequences corresponds to the codewords of a RM code of order 1.

Example 5.8 (The relation of biorthogonal sequences to RM codes) The biorthogonal sequences of length 4 can be constructed by applying the ϕ transform to the codewords of the $\mathcal{R}(1,2)$ RM code from example 5.7:

$$\begin{matrix} 0 & 0 & 0 & 0 \\ 0 & 0 & 1 & 1 \\ 0 & 1 & 0 & 1 \\ 0 & 1 & 1 & 0 \end{matrix} \;\; \text{inv.} \;\; \begin{matrix} 1 & 1 & 1 & 1 \\ 1 & 1 & 0 & 0 \\ 1 & 0 & 1 & 0 \\ 1 & 0 & 0 & 1 \end{matrix} \;\; \xrightarrow{\phi} \;\; \begin{matrix} 1 & 1 & 1 & 1 \\ 1 & 1 & -1 & -1 \\ 1 & -1 & 1 & -1 \\ 1 & -1 & -1 & 1 \end{matrix} \;\; \text{inv.} \;\; \begin{matrix} -1 & -1 & -1 & -1 \\ -1 & -1 & 1 & 1 \\ -1 & 1 & -1 & 1 \\ -1 & 1 & 1 & -1 . \end{matrix}$$

◇

The inner product (correlation) of biorthogonal sequences produce the following values:

$$\sum_{i=0}^{n-1} x_i y_i = \begin{cases} 0 & \text{for } \mathbf{x} \neq \mathbf{y}, \mathbf{x} \neq -\mathbf{y}, \\ +n & \text{for } \mathbf{x} = \mathbf{y}, \\ -n & \text{for } \mathbf{x} = -\mathbf{y} . \end{cases}$$

These results can be used for data transmission applications. Correlations with the 2^m orthogonal sequences allow the extraction of $m+1$ information bits. The orthogonal sequences can be enumerated by m bits, and the sign corresponds to an extra bit.

Hadamard matrices

Definition 5.9 (Hadamard matrix) *A Hadamard matrix* $\mathbf{H}_n$ *of order n is an* $n \times n$ *matrix with elements 1 and* -1 *satisfying*

$$\mathbf{H} \cdot \mathbf{H}^T = n\mathbf{I},$$

where $\mathbf{I}$ is the identity matrix. The Hadamard matrix of order $n = 2^m$ can be constructed as follows:

$$\mathbf{H}_{2n} = \begin{pmatrix} \mathbf{H}_n & \mathbf{H}_n \\ \mathbf{H}_n & -\mathbf{H}_n \end{pmatrix}, \quad \mathbf{H}_1 = (1) .$$

These matrices are also called Sylvester matrices. Their construction is identical to that of first-order RM codes (see theorem 5.2). The rows of the matrix correspond to the zero expansion of the codewords of the simplex codes, and thus result in a set of orthogonal sequences. Equivalent relationships among simplex, Hamming and first-order RM codes exist.

Example 5.9 (Hadamard matrices of orders 2, 4 and 8) We shall construct the fourth-order Hadamard matrix by using the second-order Hadamard matrix. We shall represent a 1 by +

and a -1 by $-$.

$$\mathbf{H}_2 = \begin{pmatrix} + & + \\ + & - \end{pmatrix} \Longrightarrow \mathbf{H}_4 = \begin{pmatrix} \mathbf{H}_2 & \mathbf{H}_2 \\ \mathbf{H}_2 & -\mathbf{H}_2 \end{pmatrix} = \left(\begin{array}{cc|cc} + & + & + & + \\ + & - & + & - \\ \hline + & + & - & - \\ + & - & - & + \end{array}\right).$$

$\mathbf{H}_8$ can be similarly constructed:

$$\mathbf{H}_8 = \begin{pmatrix} \mathbf{H}_4 & \mathbf{H}_4 \\ \mathbf{H}_4 & -\mathbf{H}_4 \end{pmatrix} = \left(\begin{array}{cccc|cccc} + & + & + & + & + & + & + & + \\ + & - & + & - & + & - & + & - \\ + & + & - & - & + & + & - & - \\ + & - & - & + & + & - & - & + \\ \hline + & + & + & + & - & - & - & - \\ + & - & + & - & - & + & - & + \\ + & + & - & - & - & - & + & + \\ + & - & - & + & - & + & + & - \end{array}\right).$$

◇

Remark Hadamard matrices do not exist for the case when $4 \not| \, n$ (with the exception of $n = 2$). An unsolved problem is whether Hadamard matrices exist for all values $n = 4i$, $i \in \mathbb{N}$.

5.2 Reed–Muller codes of higher order

The construction principle of RM codes of order 1 can be extended to give RM codes of higher order. This extension gives the class of binary RM codes. In order to illustrate the technique, we use the definition of the Boolean function [McWSl], which will be used in the following example.

Example 5.10 (Boolean function) We shall give an example in order to define the Boolean functions of order $0, 1 \ldots, m$. For $m = 4$, we have the following:

$$\begin{array}{rcllll}
\mathbf{1} & = & 1111 & 1111 & 1111 & 1111 \\
\mathbf{v}_4 & = & 0000 & 0000 & 1111 & 1111 \\
\mathbf{v}_3 & = & 0000 & 1111 & 0000 & 1111 \\
\mathbf{v}_2 & = & 0011 & 0011 & 0011 & 0011 \\
\mathbf{v}_1 & = & 0101 & 0101 & 0101 & 0101 \\
\mathbf{v}_3 \cdot \mathbf{v}_4 & = & 0000 & 0000 & 0000 & 1111 \\
\mathbf{v}_2 \cdot \mathbf{v}_4 & = & 0000 & 0000 & 0011 & 0011 \\
\mathbf{v}_1 \cdot \mathbf{v}_4 & = & 0000 & 0000 & 0101 & 0101 \\
\mathbf{v}_2 \cdot \mathbf{v}_3 & = & 0000 & 0011 & 0000 & 0011 \\
\mathbf{v}_1 \cdot \mathbf{v}_3 & = & 0000 & 0101 & 0000 & 0101 \\
\mathbf{v}_1 \cdot \mathbf{v}_2 & = & 0001 & 0001 & 0001 & 0001 \\
\mathbf{v}_2 \cdot \mathbf{v}_3 \cdot \mathbf{v}_4 & = & 0000 & 0000 & 0000 & 0011 \\
\mathbf{v}_1 \cdot \mathbf{v}_3 \cdot \mathbf{v}_4 & = & 0000 & 0000 & 0000 & 0101 \\
\mathbf{v}_1 \cdot \mathbf{v}_2 \cdot \mathbf{v}_4 & = & 0000 & 0000 & 0001 & 0001 \\
\mathbf{v}_1 \cdot \mathbf{v}_2 \cdot \mathbf{v}_3 & = & 0000 & 0001 & 0000 & 0001 \\
\mathbf{v}_1 \cdot \mathbf{v}_2 \cdot \mathbf{v}_3 \cdot \mathbf{v}_4 & = & 0000 & 0000 & 0000 & 0001
\end{array}.$$

We start with the generator matrix of $\mathcal{R}(1,m)$. The all-ones vector has order 0, $\mathbf{v}_1$, $\mathbf{v}_2$, $\mathbf{v}_3$, $\mathbf{v}_4$ have order 1, all possible combinations of products of two vectors have order 2, and so on. The product $\mathbf{v}_1 \cdot \mathbf{v}_2 \cdot \mathbf{v}_3 \cdot \mathbf{v}_4$ has order 4. ◇

Definition 5.10 (RM code) *The generator matrix $\mathbf{G}_{\mathcal{R}}$ of an RM code $\mathcal{R}(r,m)$ of order r is constructed from the matrix of Boolean functions of orders $0,1\ldots,r$. The code has length $n = 2^m$, dimension $k = 1 + \binom{m}{1} + \binom{m}{2} + \ldots + \binom{m}{r}$, and minimum distance $d = 2^{m-r}$. $\mathcal{R}(r,m)$ is a binary $(2^m, k, 2^{m-r})$ code.*

The generator matrix is a $k \times n$ matrix. The code construction technique produces 2^m columns and k rows. The minimum distance can be found in theorem 5.11.

Example 5.11 ($(16,11,4)$ RM code) The $\mathcal{R}(2,4)$ is a binary $(16,11,4)$ code with generator matrix

$$\mathbf{G} = \begin{pmatrix} 1111 & 1111 & 1111 & 1111 \\ 0000 & 0000 & 1111 & 1111 \\ 0000 & 1111 & 0000 & 1111 \\ 0011 & 0011 & 0011 & 0011 \\ 0101 & 0101 & 0101 & 0101 \\ 0000 & 0000 & 0000 & 1111 \\ 0000 & 0000 & 0011 & 0011 \\ 0000 & 0000 & 0101 & 0101 \\ 0000 & 0011 & 0000 & 0011 \\ 0000 & 0101 & 0000 & 0101 \\ 0001 & 0001 & 0001 & 0001 \end{pmatrix}.$$

◇

A method of constructing a new code from a given code uses the $|\mathbf{u}|\mathbf{u}+\mathbf{v}|$ construction. This technique was developed by Plotkin [Plo60]. The RM codes can be calculated recursively using this technique.

Theorem 5.11 ($|\mathbf{u}|\mathbf{u}+\mathbf{v}|$ construction) *An RM code of order $r+1$ with length $2n = 2 \cdot 2^m = 2^{m+1}$ can be constructed from an RM code of order r and $r+1$ with length $n = 2^m$ as follows:*

$$\mathcal{R}(r+1, m+1) = \{|\mathbf{u}|\mathbf{u}+\mathbf{v}| : \mathbf{u} \in \mathcal{R}(r+1,m),\ \mathbf{v} \in \mathcal{R}(r,m)\}\ .$$

Proof We shall present the so-called $|\mathbf{u}|\mathbf{u}+\mathbf{v}|$ construction. Let

$$\mathcal{C}_u(n, k_u, d_u) \text{ and } \mathcal{C}_v(n, k_v, d_v)$$

be two binary codes. The code

$$\mathcal{C} := \{|\mathbf{u}|\mathbf{u}+\mathbf{v}| : \mathbf{u} \in \mathcal{C}_u,\ \mathbf{v} \in \mathcal{C}_v\}$$

has length $2n$ and dimension $k = k_u + k_v$. The minimum distance is

$$\operatorname{dist}(\mathcal{C}) = \min\{2d_u, d_v\}.$$

We calculate two different codewords $\mathbf{a} \neq \mathbf{b}$ from $\mathcal{C}$:

$$\mathbf{a} = |\mathbf{u}|\mathbf{u}+\mathbf{v}| \text{ and } \mathbf{b} = |\mathbf{u}'|\mathbf{u}'+\mathbf{v}'|.$$

Case 1. $\mathbf{v} = \mathbf{v}'$:

$$\operatorname{dist}(\mathbf{a},\mathbf{b}) = 2\operatorname{dist}(\mathbf{u},\mathbf{u}') \geq 2d_u.$$

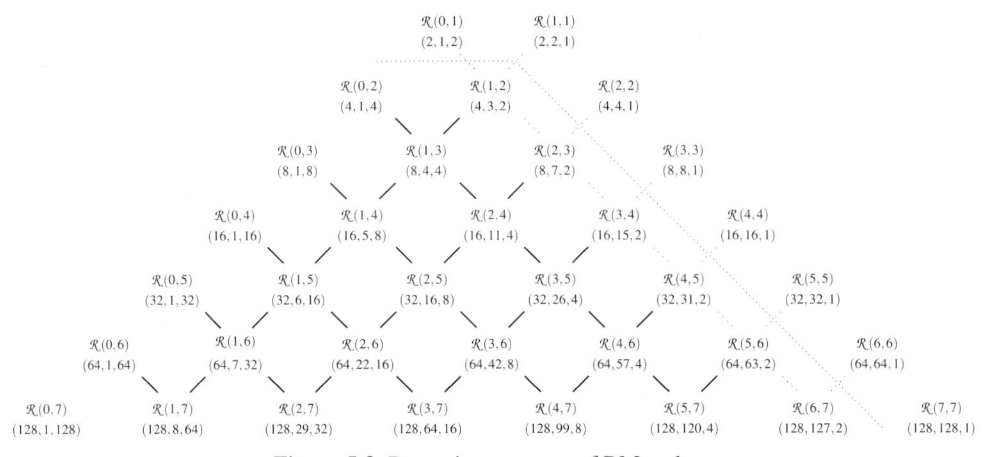

Figure 5.3 Recursive structure of RM codes.

Case 2. $\mathbf{v} \neq \mathbf{v}'$:

$$\begin{aligned}
\text{dist}(\mathbf{a},\mathbf{b}) &= \text{dist}(\mathbf{u},\mathbf{u}') + \text{dist}(\mathbf{u}+\mathbf{v},\mathbf{u}'+\mathbf{v}') \\
&= \text{wt}(\mathbf{u}-\mathbf{u}') + \text{wt}(\mathbf{u}-\mathbf{u}'+\mathbf{v}-\mathbf{v}') \\
&\geq \text{wt}(\mathbf{u}-\mathbf{u}') - \text{wt}(\mathbf{u}-\mathbf{u}') + \text{wt}(\mathbf{v}-\mathbf{v}') = \text{wt}(\mathbf{v}-\mathbf{v}') \\
&\geq d_v \, .
\end{aligned}$$

Note that both bounds are tight.

In the case when $C_u = \mathcal{R}(r+1,m)$ and $C_v = \mathcal{R}(r,m)$, we get for the minimum distance

$$d_u = 2^{m-r-1}, \quad d_v = 2^{m-r} \implies d = 2^{m-r} \, .$$

The length is $2 \cdot 2^m = 2^{m+1}$. The dimension $k = k_u + k_v$ is verified using the relation

$$\binom{m}{r+1} + \binom{m}{r} = \binom{m+1}{r+1} \, .$$

The parameters of the resulting code C correspond to the RM code $\mathcal{R}(r+1,m+1)$. In fact, the resulting code is identical to the RM code $\mathcal{R}(r+1,m+1)$. This follows from the possible construction of the generator matrix of a double-length RM code as

$$\mathbf{G}_{|\mathbf{u}|\mathbf{u}+\mathbf{v}|} = \begin{pmatrix} \mathbf{G_u} & \mathbf{G_u} \\ \mathbf{0} & \mathbf{G_v} \end{pmatrix} \, .$$

$\mathbf{G_u}$ and $\mathbf{G_v}$ are both generator matrices of RM codes, and, by linearity, $|\mathbf{u}|\mathbf{u}+\mathbf{v}|$ is also an RM codeword. □

RM codes can be represented as concatenated codes (section 5.1) due to their recursive nature, which results in efficient methods of decoding. RM code construction can be depicted as a tree diagram made up of RM codes. Such a tree diagram is shown in figure 5.3.

Theorem 5.12 (Dual RM code) *(without proof) The dual of an RM code is an RM code with parameters*

$$\mathcal{R}^{\perp}(r,m) = \mathcal{R}(m-r-1,m) \, .$$

The class of RM codes can be described using many different methods. We have limited our presentation to Boolean functions and the recursive $|\mathbf{u}|\mathbf{u}+\mathbf{v}|$ construction. For more construction techniques, see [McWSl], [WLK+94] and [For88b].

5.3 q-ary Hamming codes

In this section, we shall define the q-ary Hamming codes, which are generalizations of the binary Hamming codes introduced in section 1.5. Let q be a power of a prime. We construct a parity check matrix that has pairwise linearly independent columns of a vector in $GF(q)^h$, $\mathbf{h}_i = (h^i_0, h^i_1, \ldots, h^i_{h-1})^T$, with $h^i_j \in GF(q)$.

Definition 5.13 **(q-ary Hamming codes)** *Consider $GF(q)$ and an integer $h \geq 1$. The q-ary Hamming code is defined by the parity check matrix having $n = (q^h - 1)/(q-1)$ pairwise linearly independent columns from $GF(q)^h$.*

The parameters of the Hamming code are $(n, n-h, 3)$.

The integer $q^h - 1$ is divisible by $q - 1$, since

$$(q-1)(q^{h-1} + q^{h-2} + \ldots + 1) = q^h - 1.$$

The maximum number of possible linear independent columns in $GF(q)^h$, and therefore the maximum length of the code, is $n = (q^h - 1)/(q-1)$. Indeed, the number of non-zero vectors in $GF(q)^h$ is equal to $q^h - 1$. Each non-zero vector can produce $q - 1$ pairwise linearly dependent vectors through multiplication by the non-zero elements in $GF(q)$. The dimension $k = n - h$ is given by the rank h of the parity check matrix. Since any two columns are linearly independent, it follows that $d \geq 3$. According to the following theorem, which states that these codes are perfect, we can conclude that $d \leq 3$. Thus the minimum distance is $d = 3$.

Theorem 5.14 **(q-ary Hamming codes are perfect)** *Given $GF(q)$ and an integer $h \geq 1$, the q-ary $(n, n-h, 3)$ Hamming codes over $GF(q)$ are perfect, where $n = (q^h - 1)/(q-1)$.*

Proof The Hamming bound is satisfied with equality, since

$$q^{n-h} \sum_{i=0}^{1} \binom{n}{i} (q-1)^i = q^{n-h}(1 + n(q-1)) = q^n . \tag{5.1}$$

□

Example 5.12 **(q-ary Hamming code)** Let $q = 5$ and $h = 2$. Then the maximum length is $n = (5^2 - 1)/(5-1) = 6$. A possible parity check matrix is

$$\mathbf{H} = \begin{pmatrix} 0 & 1 & 1 & 1 & 1 & 1 \\ 1 & 0 & 1 & 2 & 3 & 4 \end{pmatrix} .$$

The dimension is 4 and the minimum distance is 3, since

- any two columns are linearly independent;
- three columns exist that are linearly dependent, e.g.

$$\begin{pmatrix} 0 \\ 1 \end{pmatrix} + \begin{pmatrix} 1 \\ 0 \end{pmatrix} + 4 \cdot \begin{pmatrix} 1 \\ 1 \end{pmatrix} = \begin{pmatrix} 5 \\ 5 \end{pmatrix} = \begin{pmatrix} 0 \\ 0 \end{pmatrix} \quad \text{mod } 5 .$$

◇

Construction as a non-primitive BCH code BCH codes can be defined as subfield subcodes over $GF(q^h)$ which implies that the codeword coefficients belong to $GF(q \geq 2)$, (section 4.5). We need an element of order n from $GF(q^h)$. We choose $n = (q^h - 1)/(q-1)$. Note that $n \mid (q^h - 1)$, and this means that there exists an element of order $n = (q^h - 1)/(q-1)$ in $GF(q^h)$.

A q-ary BCH code can be constructed from the first cyclotomic coset:

$$K_1 = \{1, q, q^2, \ldots, q^{h-1}\} \mod n .$$

We have $q^h = n(q-1) + 1 = 1 \mod n$. Therefore the generator polynomial has degree h, and the code has dimension $k = n - h$. Next we determine the minimum distance. A parity check matrix $\mathbf{H}$ can be obtained by h cyclic shifts of the parity check polynomial $h(x) = (x^n - 1)(1/g(x))$. Any two columns of this matrix are linearly independent. This proves that the minimum distance $d \geq 3$. However, the Hamming bound is satisfied with equality by equation 5.1, which implies that the minimum distance must be $d \leq 3$. Thus $d = 3$, and we have proved the following theorem:

Theorem 5.15 (q-ary Hamming code as a non-primitive BCH code) *Let $\alpha \in GF(q^h)$, $h \geq 1$, be an element of order $n = (q^h - 1)/(q-1)$. The code with the generator polynomial*

$$g(x) = (x - \alpha)(x - \alpha^q)(x - \alpha^{q^2}) \ldots (x - \alpha^{q^{h-1}})$$

is a perfect $(n, n-h, 3)$ Hamming code over $GF(q)$.

Example 5.13 (q-ary Hamming codes as non-primitive BCH codes) Using the same parameters from example 5.12, we can write

$$n = (5^2 - 1)/(5-1) = 24/4 = 6.$$

A primitive polynomial over $GF(5)$ of degree 2 is $p(x) = 2 + x + x^2$. The cyclotomic cosets are

$$K_i = \{iq^j \mod n, \ j = 0, 1, \ldots, h-1\}$$

$$\begin{aligned} K_0 &= \{0\}, & K_2 &= \{2, 4\}, \\ K_1 &= \{1, 5\}, & K_3 &= \{3\}, . \end{aligned}$$

K_1 corresponds to a code C_1 with the designed minimum distance $d = 2$ because there are no successive roots in the cyclotomic coset. If we consider the union of the $K_0 \cup K_1 = \{0, 1, 5\}$, we obtain a code C_2 with the designed minimum distance $d = 4$ ($5 = -1 \mod 6$). C_2 is generated by including the factor $x - 1$ in the generator polynomial, which is equivalent to the elimination of all codewords with uneven Hamming weight in C_1. Therefore K_0 can only increase the minimum distance by 1 (see theorem 1.21). We can deduce that the true minimum distance of K_1 is $\delta = 3$, which satisfies the distance requirement for a Hamming code. ◇

Example 5.14 (5-ary Hamming code of length 31) We choose $q = 5$ and $h = 3$ to obtain the length

$$n = \frac{5^3 - 1}{5 - 1} = 5^2 + 5 + 1 = 31 ,$$

and the above construction gives a $(31, 31-3=28, 3)$ code. The cyclotomic cosets are

$$K_i = \{iq^j \mod n,\ j = 0, 1, \ldots, h-1\}$$

$$\begin{array}{lll} K_0 = \{0\}, & K_4 = \{4,20,7\}, & K_{12} = \{12,29,21\}, \\ K_1 = \{1,5,25\}, & K_6 = \{6,30,26\}, & K_{16} = \{16,18,28\}, \\ K_2 = \{2,10,19\}, & K_8 = \{8,9,14\}, & K_{17} = \{17,23,22\}, \\ K_3 = \{3,15,13\}, & K_{11} = \{11,24,27\}\ . & \end{array}$$

If we construct the generator polynomial using K_8, we obtain a code with the designed distance equal to the true distance of $d = 3$. This BCH code satisfies the Hamming code definition. $\diamond$

5.4 Quadratic residue codes

Quadratic residue codes (QR codes) are among the best known codes. We shall only focus on the binary QR codes, which can be defined as follows.

Definition 5.16 (QR code) *Let $p = 8m \pm 1$ be a prime and let $\mathcal{M}_Q$ be the set of quadratic residues* $\bmod p$ *(section 2.6). Let $\beta \in GF(2^l)$ be an element of order p (definition 2.22). Then a QR code is defined by the generator polynomial*

$$g(x) := \prod_{i \in \mathcal{M}_Q} (x - \beta^i)\ .$$

Theorem 5.17 (Parameters of QR codes) *(without proof) The length of a QR code is $n = p$, the dimension is $k = \frac{p+1}{2}$, and the minimum distance d satisfies the following two inequalities: $d^2 \geq p$ for $p = 8m+1$ and $d^2 - d + 1 > p$ for $p = 8m-1$.*

QR codes are cyclic codes. We define another generator polynomial using the complementary set $\overline{M_Q}$:

$$\overline{\mathcal{M}_Q} := \{i \mid i \notin \mathcal{M}_Q,\ i = 1, 2, \ldots, p-1\}\ ,$$

We thereby get an equivalent QR code with the generator polynomial

$$\overline{g(x)} := \prod_{i \in \overline{\mathcal{M}_Q}} (x - \beta^i)\ .$$

It can be shown that

$$g(x)\overline{g(x)}(x-1) = x^p - 1.$$

As with the BCH codes, the factor $x - 1$ can be used to modify the generator polynomial of QR codes. This gives $g(x)(x-1)$ and $\overline{g(x)}(x-1)$ as generator polynomials. This construction allows one to select only the even-weight codewords generated by $g(x)$ and $\overline{g(x)}$. Therefore QR codes can be extended and shortened based on the same principle as BCH codes (section 4.3).

Example 5.15 (Golay code) The Golay code $\mathcal{G}_{23}$ from example 4.5 is a QR code, since $23 = 3 \cdot 8 - 1$ is a prime and the set of quadratic residues

$$\mathcal{M}_Q = \{1,4,9,16,2,13,3,18,12,8,6\}$$

is identical to the cyclotomic coset K_1 from example 4.5. The dimension is $k = \frac{p+1}{2} = 12$. We estimate the minimum distance by theorem 5.17 to give

$$d^2 - d + 1 \geq p = 23 \Longrightarrow d \geq 6.$$

The element of order p is calculated as in example 4.5 and is equal to $\beta = \alpha^{89} \in GF(2^{11})$. The generator polynomial is

$$g(x) = x^{11} + x^9 + x^7 + x^6 + x^5 + x + 1\ .$$

Next, we calculate the generator polynomial $\overline{g(x)}$ for the equivalent code:

$$\begin{aligned}
\overline{g(x)} &= (x^{23} - 1)/((x-1) \cdot g(x)) \\
&= (x^{23} - 1)/(x^{12} + x^{11} + x^{10} + x^9 + x^8 + x^5 + x^2 + 1) \\
&= x^{11} + x^{10} + x^6 + x^5 + x^4 + x^2 + 1\ .
\end{aligned}$$

The generator polynomial $\overline{g(x)}$ is the reciprocal of $g(x)$ (see section 4.1.5). Therefore $\overline{g(x)}$ corresponds to $m_{1958}(x)$, because $-89 = 1958 \bmod (2^{11} - 1)$. ◇

The algebraic decoding method presented in chapter 3 does not allow one to decode QR codes up to half the true minimum distance, only up to half the designed minimum distance. In chapter 7 a decoding method will be introduced that will be able to decode QR codes up to half the true minimum distance δ.

QR codes can be extended. The extension of the Golay code $\mathcal{G}_{23}$ by one coordinate gives the code $\mathcal{G}_{24}$ with length $n = 24$, dimension $k = 12$ and minimum distance $d = 8$. This code is not perfect, but does posses some important symmetric characteristics that can be exploited. For example, the symmetry of $\mathcal{G}_{24}$ can be used for signal modulation design, which gives an optimal partitioning of the signal space ([CoSl]).

5.5 Consta- and negacyclic codes

Constacyclic codes are a generalization of cyclic codes. Before presenting constacyclic codes, we shall list the properties of ordinary cyclic codes.

The set of all polynomials with coefficients from $GF(q)$ of degree less than n forms a ring $\mathcal{R} = GF(q)[x]/(x^n - 1)$, where all polynomials are calculated modulo $(x^n - 1)$. All the codewords from cyclic RS and BCH codes are elements of the ring $\mathcal{R}$ with primitive length $n = q^m - 1$. We can use the following interpretation based on section 4.5. The parameters of the RS code are $q = p^s$ and $m = 1$. For a BCH code, the parameters are $q = p$ and $m > 1$. Multiplication in $\mathcal{R}$ is carried out modulo $(x^n - 1)$, giving $x^n = 1$. A cyclic shift of the codewords $x(x)$ results in a codeword $c^*(x)$:

$$\begin{aligned}
c^*(x) &= x(c_0 + c_1x + \ldots + c_{n-1}x^{n-1}) \\
&= c_0x + \ldots + c_{n-2}x^{n-1} + c_{n-1}x^n \\
&= c_{n-1} + c_0x + \ldots + c_{n-2}x^{n-1}\ .
\end{aligned}$$

For every $\beta \in GF(q^m)\backslash\{0\}$, $\beta^n = 1$. The roots of the polynomial $x^n - 1$ are precisely all the non-zero elements of the field $GF(q^m)$. The polynomial $x^n - 1$ is the product of minimal polynomials with coefficients from $GF(q)$. When $m = 1$, the minimal polynomials are linear factors of the form $x - \alpha$. The generator polynomial of a cyclic code is a product of some minimal polynomials, and is therefore a factor of the polynomial $x^n - 1$.

Similarly, the codewords of a constacyclic code are elements from the polynomial ring $GF(q)[x]/(x^N - \xi)$, where N is defined as a divisor of $q^m - 1$ and $\xi \in GF(q)\backslash\{0\}$ is an element of order $r = \frac{q^m-1}{N}$. We have $x^N \equiv \xi$, and therefore a constant cyclic shift of a codeword $c(x)$ to a codeword $c^*(x)$ is as follows:

$$\begin{aligned} c^*(x) &= x(c_0 + c_1x + \ldots + c_{N-1}x^{N-1}) \\ &= c_0x + \ldots + c_{N-2}x^{N-1} + c_{N-1}x^N \\ &= \xi c_{N-1} + c_0x + \ldots + c_{N-2}x^{N-1} \quad . \end{aligned}$$

When α is a primitive element in $GF(q^m)$, and $\xi = \alpha^{bN}$, the set of roots of the polynomial $x^N - \xi$ is

$$\mathcal{N} = \{\alpha^{b+ir} \mid i \in \{0, \ldots, N-1\}\} \ .$$

As before, the polynomial $x^N - \xi$ is the product of minimal polynomials with roots in $\mathcal{N}$. A generator polynomial of a constacyclic code is the product of some minimal polynomials, and therefore is a divisor of the polynomial $x^N - \xi$.

When $\xi = -1$, the code is called a negacyclic code . We shall clarify this definition in the following two examples.

Example 5.16 (Negacyclic code) Consider a prime $p \geq 2$. Then $GF(p)$ is isomorphic to the modular ring $\mathbb{Z}_p = \{-\frac{p-1}{2}, \ldots, 0, \ldots, \frac{p-1}{2}\}$. The length of a negacyclic code is $N = \frac{p-1}{2}$. This corresponds to the element $\xi = \alpha^N \equiv -1$, which has order 2, and the roots of the polynomial $x^N + 1$ are all elements of the form α^{1+2i}, $i = 0, \ldots, N-1$. For example, if $p = 11$ then the polynomial

$$g(x) = (x - \alpha)(x - \alpha^3)(x - \alpha^5)$$

is the generator polynomial of a negacyclic code $\mathcal{C}(5, 2, 4)$. The minimum Lee distance is $d_L = 8$ (see appendix B, section B.1, for the definition of the Lee metric). The important properties of this code come from the existence of the algebraic decoding method based on the Lee metric. If the transmission method uses q-PSK modulation, both the modulation and the code can be decoded simultaneously using the same algebraic technique. ◇

Example 5.17 (Constacyclic code) A constacyclic code can be constructed with the length $N = \frac{2^4-1}{5} = 3$ using the symbol alphabet $GF(2^4)$. The order of the corresponding element ξ must equal 5, $\xi = \alpha^3$. The roots of the polynomial $x^3 + \alpha^3$ consist of all the elements of the form α^{1+5i}, $i = 0, \ldots, 2$. For example,

$$g(x) = (x - \alpha)(x - \alpha^6)$$

is the generator polynomial of a constacyclic code $\mathcal{C}(3, 1, 3)$ over $GF(2^4)$. An interesting property of constacyclic codes over $GF(p^s)$ is that their image in $GF(p)$ corresponds to a subcode of the shortened BCH codes over $GF(p)$. In this case, the Hamming distance can be larger than that of the constacyclic code. The binary image of the constacyclic code $\mathcal{C}(3, 1, 3)$ in this example is a shortened BCH code $\mathcal{C}(12, 4, 5)$. ◇

There are many possibilities for the construction of consta- and negacyclic codes. These possibilities will not be covered further in this book, but a further study of this subject is presented in [Ber].

5.6 Binary interpretation of codes over $GF(q = 2^m)$ and $\mathbb{Z}_4$

The binary interpretation of codes over $GF(q = 2^m)$ provides interesting possibilities for designing binary codes. In particular, binary interpretations of RS codes over $GF(2^m)$ have been investigated in [VB91]. In [Nec91], [HKC+94] and [CMKH96], some famous nonlinear binary codes were described as binary interpretation of linear codes over the ring $\mathbb{Z}_4$. These codes include Nordstrom–Robinson, Preparata and Kerdock codes. The goal of this section is to reveal the concepts behind the binary interpretation of codes so that yet another coding 'treasure' may be revealed.

Galois ring Analogous to the extension field of prime fields, we can define a a Galois ring $\mathbb{Z}_a^m$ of polynomials with degree m and coefficients from $\mathbb{Z}_a$. The special case $\mathbb{Z}_4$ is treated, for example, in [HKC+94]. For any primitive polynomial $p(x)$ with coefficients from $GF(2)$ and degree m (which defines a field $GF(2^m)$), there exists a primitive polynomial $p_4(x)$. We can construct $p_4(x)$ from $p(x)$ using the following method. Let $p(x)$ be represented as $p(x) = u(x) + v(x)$, where $u(x)$ and $v(x)$ are polynomials containing only even and odd powers of x respectively. We define $p_4(x^2) = \pm(u^2(x) - v^2(x))$; then $p_4(x) = p(x) \bmod 2$. This allows us to define an element $p_4(\zeta) = 0$, $\zeta \in \mathbb{Z}_4$, which has the order $n = 2^m - 1$.

Example 5.18 (Galois ring) The polynomial $p(x) = x^3 + x + 1$ is a primitive polynomial defining the field $GF(2^3)$. In this case, $u(x) = 1$ and $v(x) = x^3 + x$. This gives

$$\begin{aligned} u^2(x) &= 1, \\ v^2(x) &= x^6 + 2x^4 + x^2, \\ p_4(x^2) &= -x^6 - 2x^4 - x^2 + 1, \\ p_4(x) &= x^3 + 2x^2 + x - 1. \end{aligned}$$

$p_4(\zeta) = 0$, and ζ has order 7 ($7 \mid 63$, $63 = 4^3 - 1$). ◇

We know that $p_4(x) \mid (x^n - 1)$, and can define

$$g^{rec}(x) = (x^n - 1)/(p_4(x)(x-1)), \quad \text{of degree } n - m - 1,$$

with coefficients g_i^{rec}, and the corresponding reciprocal polynomial

$$g(x) = (4 - g_{n-m-1}^{rec}) + (4 - g_{n-m-2}^{rec})\,x + \ldots + (4 - g_0^{rec})\,x^{n-m-1}.$$

Definition 5.18 (Cyclic codes over $\mathbb{Z}_4$) *We select the generator polynomial $g(x)$ and $p_4(x)$ to obtain the dual codes with the length $2^m - 1$ over $\mathbb{Z}_4$. These codes can be denoted by C_K and C_P. The numbers of codewords are 4^{m+1} and 4^{n-m} respectively.*

Correspondence between $\mathbb{Z}_4$ and $(GF(2))^2$ In order to map $\mathbb{Z}_4$ onto $(GF(2))^2$, we use the following table. We can now give a binary interpretation of codes according to definition 5.18.

$\mathbb{Z}_4$	$(GF(2))^2$
0	00
1	01
2	11
3	10

Example 5.19 (Nordstrom–Robinson code) We extend the code C_K for $m = 3$ from definition 5.18 by one parity check coordinate from $\mathbb{Z}_4$, and map the coordinates of the code according to the previous table. The resulting code is binary and nonlinear and has length 16, and $4^4 = 256$ codewords, with a minimum distance of 6.

This code contains more codewords than any corresponding linear code of length 16 and minimum distance 6. It should be noted that if we choose the code C_P from definition 5.18, we also get the Nordstrom–Robinson code. ◇

Example 5.20 (Kerdock code) We extend the code C_K having the parameters $m \geq 3$, $n = 2^m - 1$ by one parity check coordinate from $\mathbb{Z}_4$. The binary interpretation of the $n+1$ code coordinates produces the binary Kerdock code with

$$\begin{aligned} \text{length} \quad & 2n+2 = 2^{m+1}, \\ \text{number of codewords} \quad & 4^{m+1}, \\ \text{minimum distance} \quad & 2^m - 2^{\frac{m-1}{2}}, \; m \text{ odd}. \end{aligned}$$

When $m \geq 2$ is even, the binary interpretation has the same length, the same number of codewords and a minimum distance of $d = 2^m - 2^{m/2}$. In general, Kerdock codes are of little practical interest, because double-error correcting BCH codes have better parameters. Note that the Nordstrom–Robinson code from example 5.19 is a Kerdock code having the parameter $m = 3$. ◇

Example 5.21 (Preparata code) For $m \geq 3$ and $n = 2^m - 1$, let us extend the code C_P from definition 5.18 by one parity check coordinate from $\mathbb{Z}_4$. The binary interpretation of the $n+1$ code coordinates results in the binary Preparata code with

$$\begin{aligned} \text{length} \quad & 2n+2 = 2^{m+1}, \\ \text{number of codewords} \quad & 4^{n-m}, \\ \text{minimum distance} \quad & 6. \end{aligned}$$
◇

The binary interpretation of this code may be advantageous for decoding, since errors corrupt only individual bits, not the whole symbol. Furthermore, the binary minimum distance can be larger than the symbol minimum distance. For further study of this topic, see [VB91] and [HKC+94].

5.7 Summary

The class of RM codes [Reed54, Mul54] have been defined and Plotkin's $|\mathbf{u}|\mathbf{u}+\mathbf{v}|$ construction [Plo60] (introduced in 1951 as a research report in Russian) has been described for recursive construction of RM codes. RM codes are to date the source of many publications which stem from the different possible interpretations of their construction, as well as different possible decoding methods. It is interesting to note that the Hamming codes [Ham50] which have been known since the foundation of information theory (before 1950), still play an important role.

QR codes have been thoroughly investigated by Assmus and Mattson in a series of publications in 1963 [AM63] (for further information, see [McWSl]).

Generalizations of cyclic codes with respect to classes of consta- and negacyclic codes are presented in [Ber]. Nechaev [Nec91] in 1985 began the first investigation of binary interpretation of the ring $\mathbb{Z}_4$. Further investigations have been performed by Hammons et al. [HKC+94], Helleseth and Calderbank [CMKH96], Kolev [Kol96] and Nechaev and Kuzmin [NK96].

In this chapter, some classes of codes have been defined and investigated. First, we have shown the connections among RM, simplex, Hadamard and Hamming codes and the orthogonal Walsh Hadamard sequences, biorthogonal sequences and pseudonoise sequences. The dependences result in simple calculations, transformations and constructions of these sequences and codes. We have defined the q-ary Hamming code and presented its interpretation as a BCH code. The RM codes will be represented as generalized concatenated codes in chapter 9. This results in an extremely efficient decoding algorithm.

The binary QR codes are extremely good codes, which unfortunately cannot be decoded with ordinary algebraic decoding methods. In chapter 7, we shall present a decoding algorithm that can correct some error patterns that are larger than half the minimum distance δ.

The concept of cyclic codes has been generalized, and the classes of consta- and negacyclic codes have been introduced. This definition gives an interesting generalization of algebraic codes. For more information on this topic, see Berlekamp [Ber]. We have presented the binary non-linear Kerdock and Preparata codes and the Nordstrom–Robinson code as linear codes over the ring $\mathbb{Z}_4$. Only a few of the various different possible construction methods for new codes could be given here. Many codes exist that we have not presented (e.g. the *double circulant codes*). To include all of these classes would burst the covers of this book. Some of these codes could, however, reveal surprising applications.

The first-order RM code of length 32 was used for data transmission from the satellite *Mariner 9* at the beginning of the 1970s. This code was one of the first practically implemented codes ever to be used. (See [Mas92], where J. Massey gives an anecdote that explains the first two codes used for satellite transmission.)

PN sequences are used by the GPS (*Global Positioning System*) for different purposes. There is one short sequence for public use that can be correlated to measure distance. There is another extremely long encryption sequence that has military applications. This sequence is so long that one period extends over an entire year. The public PN sequence is generated by the polynomial $x^{10}+x^3+1$ or by $x^{10}+x^9+x^8+x^6+x^3+x^2+1$. The military encryption sequence is a concatenation of 4 primitive polynomials with degree 12 [Kap].

5.8 Problems

Problem 5.1

The following questions are to be answered with respect to the binary QR code with length $n=31$:

(a) Find the quadratic residues modulo $31 = 4\cdot 8-1$.

(b) What is the minimum distance of the QR code of length 31?

(c) How many errors can be corrected using the BMA algebraic decoding method for the QR code of length 31?

Problem 5.2

Construct a Hadamard matrix of order 12. First construct a matrix from the cyclic shifts of the Legendre sequence $L_{11} = (+-+++---+--)$. Extend the matrix in the appropriate manner.

Problem 5.3

Consider the Reed-Muller code $\mathcal{R}(0,2)$. Determine the cosets of this code in such a way that a $\mathcal{R}(1,2)$ Reed-Muller code can be constructed. The coset leaders result in another linear code. Give the parameters of this code.

Problem 5.4

Consider the following generator matrix for a Hamming code over $GF(5)$:

$$\mathbf{G} = \begin{pmatrix} 1 & 0 & 0 & 0 & 4 & 1 \\ 0 & 1 & 0 & 0 & 4 & 2 \\ 0 & 0 & 1 & 0 & 4 & 4 \\ 0 & 0 & 0 & 1 & 4 & 3 \end{pmatrix}.$$

The vector $\mathbf{r} = (1\,4\,2\,1\,1\,4)$ is received. Decode $\mathbf{r}$ under the assumption that no more than one coordinate has an error.

Problem 5.5

A channel code must be determined for binary transmission over a channel producing burst errors. It is known that an error burst has a maximum length of 20 bits. It is also known that, after an error burst, at least 500 error-free bits are transmitted. As a choice of codes, there is a BCH code with length 127, a BCH code with length 255 and a (binary interpreted) RS code with the length 31. Determine the most favourable parameters for each code. All codes are able to correct the burst errors. Compare the final code rates.

Problem 5.6

Determine if the following three cases represent code words of a simplex code of length 15:

(a) (1 1 1 1 0 1 0 1 1 0 0 1 0 0 0)

(b) (1 1 1 1 1 0 0 1 1 0 0 1 0 0 0)

(c) (1 1 1 1 0 1 0 1 1 0 0 1 1 0 0)

Problem 5.7

How many code words does the Reed-Muller code $\mathcal{R}(5,8)$ have?

Problem 5.8

Given the Hadamard matrix

$$\mathbf{H}_8 = \begin{pmatrix} + & + & + & + & + & + & + & + \\ + & - & + & - & + & - & + & - \\ + & + & - & - & + & + & - & - \\ + & - & - & + & + & - & - & + \\ + & + & + & + & - & - & - & - \\ + & - & + & - & - & + & - & + \\ + & + & - & - & - & - & + & + \\ + & - & - & + & - & + & + & - \end{pmatrix},$$

as well as the received vector $\mathbf{h} = (+--++-++)$. Find the most probable transmitted row from $\mathbf{H}_8$.

6

The trellis representation and properties of block codes

In this chapter, the general properties of arbitrary codes are presented. We have already investigated one such property, namely the Hamming bound. Now we shall investigate dual codes and codeword weight distribution. We shall define a special mapping (automorphism) of codewords that will be used in the next chapter on decoding methods. After we introduce the Gilbert–Varshamov bound, we describe the (*maximum distance separable*, MDS) property which, is a characteristic of RS codes.

Next, we introduce the concept of a code trellis, and apply this concept to linear block codes. We investigate properties of block code trellises, and then present minimum trellis construction methods for block codes. The minimal trellis representation of block codes brings to light further interesting properties and applications.

6.1 Cyclic dual codes

The definition of a dual code was introduced in section 1.9, and is extended in this section for cyclic codes.

Definition 6.1 (Dual code) *Let C be a cyclic code of length n with codewords $a(x)$, $a_i \in GF(q^m)$. For each code C, the dual code $C^\perp$ with codewords $b(x)$, $b_i \in GF(q^m)$, is defined as*

$$C^\perp := \{b(x) | \forall a(x) \in C \;\; a(x)b(x) = 0 \mod (x^n - 1)\} .$$

The parity check polynomial $h(x)$ (see the parity check matrix $\mathbf{H}$; section 1.2) of a code is the generator polynomial $g(x)$ (from the generator matrix $\mathbf{G}$) of the dual code. This gives

$$g(x)h(x) = x^n - 1 = 0 \mod (x^n - 1) ,$$

and:

$$(i(x)g(x))(j(x)h(x)) = (i(x)j(x))(g(x)h(x)) = 0 \mod (x^n - 1) ,$$

where $i(x)$ and $j(x)$ are the information polynomials of C and $C^{\perp}$ respectively.

Note concerning representation A generator matrix is constructed from a generator polynomial $g(x) = g_0 + g_1 x + \ldots + g_{n-k} x^{n-k}$ as follows:

$$\mathbf{G} = \begin{pmatrix} g_0 & g_1 & \ldots & g_{n-k} & 0 & \ldots & 0 \\ 0 & g_0 & g_1 & \ldots & g_{n-k} & 0 & \ldots \\ \vdots & & \ddots & \ddots & & \ddots & \\ 0 & \ldots & 0 & g_0 & g_1 & \ldots & g_{n-k} \end{pmatrix} .$$

The corresponding parity check matrix constructed from the parity check polynomial $h(x) = h_0 + h_1 x + \ldots + h_k x^k$ is

$$\mathbf{H} = \begin{pmatrix} h_k & h_{k-1} & \ldots & h_0 & 0 & \ldots & 0 \\ 0 & h_k & h_{k-1} & \ldots & h_0 & 0 & \ldots \\ \vdots & & \ddots & \ddots & & \ddots & \\ 0 & \ldots & 0 & h_k & h_{k-1} & \ldots & h_0 \end{pmatrix} .$$

This means that the proper matrix representation of a dual code is given by a generator polynomial of the form

$$g^{\perp}(x) = x^k h\left(\frac{1}{x}\right) .$$

Using the polynomial representation of a code, one can use $h(x)$ as the generator polynomial of a dual code. The polynomials $h(x)$ and $g^{\perp}(x)$ are equivalent generator polynomials in the sense that both define equivalent codes.

The dimension of a dual code is

$$k^{\perp} = n - k .$$

For the minimum distance $d^{\perp}$, the following holds:

RS codes The minimum distance of a dual RS code is

$$d^{\perp} = n - k^{\perp} + 1 = n - (n - k + 1) + 2 = n - d + 2 .$$

This represents the number of successive numbers of the dual parity check polynomial.

$g(x) \circ\!\!-\!\!\bullet\, G(x)$	$=$		0
$h(x) \circ\!\!-\!\!\bullet\, H(x)$	$=$	0	

BCH codes In general, $d^{\perp}$ cannot be determined exactly. The designed minimum distance of a dual code is obtained by the calculation of the number of successive roots in $\overline{\mathcal{M}}$ (see section 4.1.1).

QR codes For QR codes, the polynomial $\overline{g(x)}(x-1)$ is the generator polynomial for a dual code when the original code has the polynomial $g(x)$ (see section 5.4). Because of the correspondence between $g(x)$ and $\overline{g(x)}$, it follows that $d^{\perp} = d+1$, when d is odd. The factor $x-1$ removes codewords with odd weight.

In the transform domain, we get, for $a(x) \in C$ and $b(x) \in C^{\perp}$,

$$a(x)b(x) = 0 \mod (x^n - 1) \circ\!\!-\!\!\bullet\, A_i B_i = 0 \mod q,\ i = 0, \ldots, n-1 .$$

Definition 6.1 is valid for cyclic codes. For non-cyclic codes, the following relation can be used:

$$C^{\perp} := \left\{ \mathbf{b} \in GF(q)^n \mid \forall \mathbf{a} \in C : \sum_{i=0}^{n-1} a_i b_i = 0 \right\} ,$$

where the sum is performed over the field.

Example 6.1 (Dual code) Given a code C of length $n = 3$, minimum distance $d = 2$, and dimension $k = 2$, the set of codewords is

$$C = \{(0,0,0),\ (0,1,1),\ (1,0,1),\ (1,1,0)\} .$$

The dual code $C^{\perp}$ has dimension $k^{\perp} = n - k = 1$, and consists of the following two codewords:

$$C^{\perp} = \{(0,0,0),\ (1,1,1)\} .$$

The minimum distance of $C^{\perp}$ is $d^{\perp} = 3$. The generator polynomial of C is $g(x) = 1 + x$, and the parity check polynomial $h(x)$ is calculated by

$$(x^3 - 1) : (x+1) = x^2 + x + 1 ,$$

which is the generator polynomial for the dual code $C^{\perp}$. ◇

In the following section, we define a function that groups vectors from $GF(2^n)$ as codewords of C or $C^{\perp}$.

Definition 6.2 (Dual classification function) *Let C and $C^{\perp}$ be a linear binary code and its dual code respectively. Let $(\mathbf{a}, \mathbf{b}) = \sum_{i=1}^{n} a_i b_i$ be the inner product of vectors $\mathbf{a}, \mathbf{b} \in GF(2)^n$. Define the function*

$$f(\mathbf{b}) = \sum_{\mathbf{a} \in C} (-1)^{(\mathbf{a},\mathbf{b})} . \tag{6.1}$$

Then

$$f(\mathbf{b}) = \begin{cases} |C| & \text{if } \mathbf{b} \in C^{\perp}, \\ 0 & \text{if } \mathbf{b} \notin C^{\perp}. \end{cases}$$

If $\mathbf{b} \in C^{\perp}$ then, by the definition of the dual code, $(\mathbf{a}, \mathbf{b}) = 0$ and $f(\mathbf{b}) = |C|$. If $\mathbf{b} \notin C^{\perp}$ then, at least for one $\mathbf{a} \in C$, we have $(\mathbf{a}, \mathbf{b}) = 1$. In fact, exactly one half of the code vectors from C will have $(\mathbf{a}, \mathbf{b}) = 1$, and the remaining vectors from C will have $(\mathbf{a}, \mathbf{b}) = 0$. This gives $f(\mathbf{b}) = 0$.

6.2 MacWilliams identity

Definition 6.3 (Weight distribution) *The weight distribution polynomial of a code C is defined as follows:*

$$W_C(x,y) = w_0x^n + w_1yx^{n-1} + w_2y^2x^{n-2} + \ldots + w_ny^n .$$

The coefficient w_i is equal to the number of codewords with (Hamming) weight i. The x exponent corresponds to the number of zero coordinates in the codeword, while the y exponent corresponds to the number of non-zero coordinates.

Note The weight distribution polynomial $W_C(x,y)$ can be written as

$$W_C(x,y) = \sum_{\mathbf{a}\in C} x^{n-\text{wt}(\mathbf{a})}y^{\text{wt}(\mathbf{a})},$$

where $\text{wt}(\mathbf{a})$ means the Hamming weight of $\mathbf{a}$.

If a code C has minimum distance d then

$$w_1 = w_2 = \ldots = w_{d-1} = 0,$$

and $w_0 = 1$ for all linear codes.

Example 6.2 (Weight distribution) The code $C = \{(0,0,0),\ (0,1,1),\ (1,0,1),\ (1,1,0)\}$ from example 6.1 has the weight distribution polynomial

$$W_C(x,y) = x^3 + 3xy^2 .$$

The dual code $C^\perp = \{(0,0,0),\ (1,1,1)\}$ has the weight distribution

$$W_{C^\perp}(x,y) = x^3 + y^3 .$$

◇

If a code C is used only to detect errors at the output of a binary symmetric channel then the error probability P_{FBlock} can be calculated exactly when one knows the weight distribution of the code C. A decoding error occurs only if the error vector is another codeword. When p is the error probability of a BSC, the probability that the error vector is a codeword is

$$P_{FBlock} = \sum_{i=d}^{n} w_ip^i(1-p)^{n-i} \quad \text{(compare with section 1.4)},$$

where d is the minimum distance and n is the code length.

For short codes, the weight distribution can be calculated by listing all of the codewords. For long codes, the weight distribution can be calculated for only a few codes, including RS codes [Bla]. There is no general method for calculating a code's weight distribution. Nevertheless, if the weight distribution of a binary code is known, we can use this information to calculate the weight distribution of the corresponding dual code. This can be done using the MacWilliams identity, which shows the relationship between weight distribution polynomials of a code C and its dual code $C^\perp$.

Theorem 6.4 (MacWilliams identity) *The MacWilliams identity for a binary linear code is*

$$W_{C^\perp}(x,y) = \frac{1}{|C|}W_C(x+y,x-y) .$$

Proof Consider the sum

$$\omega = \sum_{\mathbf{b} \in GF(2)^n} f(\mathbf{b}) x^{n-\mathrm{wt}(\mathbf{b})} y^{\mathrm{wt}(\mathbf{b})}.$$

We calculate this sum two different ways.

Method 1:

$$\begin{aligned}
\omega &= \sum_{\mathbf{b} \in GF(2)^n} f(\mathbf{b}) x^{n-\mathrm{wt}(\mathbf{b})} y^{\mathrm{wt}(\mathbf{b})} \\
&= \sum_{\mathbf{b} \in C^\perp} f(\mathbf{b}) x^{n-\mathrm{wt}(\mathbf{b})} y^{\mathrm{wt}(\mathbf{b})} + \sum_{\mathbf{b} \notin C^\perp} f(\mathbf{b}) x^{n-\mathrm{wt}(\mathbf{b})} y^{\mathrm{wt}(\mathbf{b})} \\
&= \sum_{\mathbf{b} \in C^\perp} |C| \, x^{n-\mathrm{wt}(\mathbf{b})} y^{\mathrm{wt}(\mathbf{b})} = |C| W_{C^\perp(x,y)}.
\end{aligned}$$

Method 2:

$$\begin{aligned}
\omega &= \sum_{\mathbf{b} \in GF(2)^n} f(\mathbf{b}) x^{n-\mathrm{wt}(\mathbf{b})} y^{\mathrm{wt}(\mathbf{b})} = \sum_{\mathbf{b} \in GF(2)^n} \sum_{\mathbf{a} \in C} (-1)^{(\mathbf{a},\mathbf{b})} x^{n-\mathrm{wt}(\mathbf{b})} y^{\mathrm{wt}(\mathbf{b})} \\
&= \sum_{\mathbf{a} \in C} \left(\sum_{\mathbf{b} \in GF(2)^n} (-1)^{(\mathbf{a},b)} x^{n-\mathrm{wt}(\mathbf{b})} y^{\mathrm{wt}(\mathbf{b})} \right).
\end{aligned}$$

Represent $(-1)^{(\mathbf{a},\mathbf{b})} x^{n-\mathrm{wt}(\mathbf{b})} y^{\mathrm{wt}(\mathbf{b})}$ as

$$(-1)^{(\mathbf{a},\mathbf{b})} x^{n-\mathrm{wt}(\mathbf{b})} y^{\mathrm{wt}(\mathbf{b})} = \left[(-1)^{a_1 b_1} x^{1-b_1} y^{b_1}\right] \left[(-1)^{a_2 b_2} x^{1-b_2} y^{b_2}\right] \cdots \left[(-1)^{a_n b_n} x^{1-b_n} y^{b_n}\right],$$

where $\mathbf{a} = (a_1, a_2, \ldots, a_n)$ and $\mathbf{b} = (b_1, b_2, \ldots, b_n)$. Then

$$\begin{aligned}
&\sum_{\mathbf{b} \in GF(2)^n} (-1)^{(\mathbf{a},\mathbf{b})} x^{n-\mathrm{wt}(\mathbf{b})} y^{\mathrm{wt}(\mathbf{b})} \\
&\quad = \sum_{b_1 \in GF(2)} \sum_{b_2 \in GF(2)} \cdots \sum_{b_n \in GF(2)} (-1)^{(\mathbf{a},\mathbf{b})} x^{n-\mathrm{wt}(\mathbf{b})} y^{\mathrm{wt}(\mathbf{b})} \\
&\quad = \left[\sum_{b_1=0}^{1} (-1)^{a_1 b_1} x^{1-b_1} y^{b_1}\right] \left[\sum_{b_2=0}^{1} (-1)^{a_2 b_2} x^{1-b_2} y^{b_2}\right] \cdots \left[\sum_{b_n=0}^{1} (-1)^{a_n b_n} x^{1-b_n} y^{b_n}\right] \\
&\quad = [x + y(-1)^{a_1}][x + y(-1)^{a_2}] \cdots [x + y(-1)^{a_n}] = (x+y)^{n-\mathrm{wt}(\mathbf{a})} (x-y)^{\mathrm{wt}(\mathbf{a})}.
\end{aligned}$$

Hence we obtain

$$\omega = \sum_{\mathbf{a} \in C} (x+y)^{n-\mathrm{wt}(\mathbf{a})} (x-y)^{\mathrm{wt}(\mathbf{a})} = W_C(x+y, x-y),$$

or $|C| W_{C^\perp}(x,y) = W_C(x+y, x-y)$. This proves the theorem. □

Example 6.3 (MacWilliams identity) The code from example 6.1 produces the codewords

$$C = \{(0,0,0),\ (0,1,1),\ (1,0,1),\ (0,1,1)\},$$

with a weight distribution (example 6.2) of

$$W_C = x^3 + 3xy^2.$$

The dual code weight distribution is given by

$$\begin{aligned}W_{C^{\perp}}(x,y) &= \frac{1}{|C|}W_C(x+y,x-y) = \frac{1}{4}((x+y)^3+3(x+y)(x-y)^2)\\ &= \frac{1}{4}(x^3+3x^2y+3xy^2+y^3+3(x^3-x^2y-xy^2+y^3))\\ &= \frac{1}{4}(4x^3+4y^3) = x^3+y^3.\end{aligned}$$

◇

Theorem 6.5 (MacWilliams identity for non-binary codes) *(see [McWSl]). The MacWilliams identity for codes over $GF(q)$ is*

$$W_{C^{\perp}}(x,y) = \frac{1}{|C|}W_C(x+(q-1)y,x-y)\ .$$

6.3 Automorphism

Definition 6.6 (Automorphism) *An automorphism ϕ is a permutation of the coordinates of a code C that gives a linear one-to-one mapping of a code C onto C:*

$$\phi(C) = C, \quad \phi(\mathbf{a}_1+\mathbf{a}_2) = \phi(\mathbf{a}_1)+\phi(\mathbf{a}_2), \quad \mathbf{a}_1,\mathbf{a}_2 \in C\ .$$

For a cyclic code, a cyclic shift of a codeword is an automorphism. For BCH and RS codes, the squaring function is another automorphism. Indeed, a codeword $i(x)g(x)$ is mapped to the word $(i(x)g(x))^2 = (i(x)^2g(x))g(x)$ mod (x^n-1), which is also a codeword. Moreover, if $i_1(x) \neq i_2(x)$ then $(i_1(x)g(x))^2 \neq (i_2(x)g(x))^2$. Otherwise, we would have $(i_1(x)g(x))^2 - (i_2(x)g(x))^2 = ((i_1(x)-i_2(x))g(x))^2 = 0$. Hence squaring is a one-to-one mapping.

Remark The cubic function is not an automorphism although it maps one codeword to another codeword. An example can always be found where $i_1(x) \neq i_2(x)$ such that $(i_1(x)g(x))^3 = (i_2(x)g(x))^3$. Hence the cubic function is not a one-to-one mapping.

Consider the mapping

$$a(x) \to a(x)^{2^m} = a(x^{2^m}) \mod (x^n-1)\ .$$

For a QR code, this implies that, for all $a(x) \in C$ (mod (x^p-1)),

$$a(x^i) \in C, \quad i \in \mathcal{M}_Q\ ,$$

where $\mathcal{M}_Q$ is the quadratic residue set (see section 5.4).

Example 6.4 The generator polynomial for $\mathcal{G}_{23}$ is

$$g(x) = x^{11}+x^9+x^7+x^6+x^5+x+1\ .$$

The polynomial $g(x^{13})$ must be a codeword, because $13 \in \mathcal{M}_Q$. Therefore $g(x) \to g(x^{13})$ is an automorphism of $\mathcal{G}_{23}$:

$$\begin{aligned}g(x^{13}) &= x^{143}+x^{117}+x^{91}+x^{78}+x^{65}+x^{13}+1\\ &= x^5+x^2+x^{22}+x^9+x^{19}+x^{13}+1\ .\end{aligned}$$

To test whether $g(x^{13})$ is a codeword, we divide it by the generator polynomial $g(x)$:

$$(x^{22}+x^{19}+x^{13}+x^9+x^5+x^2+1)/(x^{11}+x^9+x^7+x^6+x^5+x+1)$$
$$=x^{11}+x^9+x^8+x+1\ .$$

$g(x^{13})$ is therefore a codeword. ◇

6.4 Gilbert–Varshamov bound

The Hamming bound is an upper bound that corresponds to the rate of a code with a specific minimum distance. The Gilbert–Varshamov bound is, in contrast, a lower bound, which states that a 'good' code exists having code length n, dimension k and minimum distance d. It is apparent that there are 'poor' codes that do not satisfy the Gilbert–Varshamov bound. The Gilbert–Varshamov bound can be formulated for non-binary codes in the Galois field $GF(q)$.

There exist two bounds that were independently developed by Gilbert and Varshamov. T. Ericson [EEI+89] showed that there was only a slight difference between the two bounds. In effect, both inequalities are based on the same idea, but are satisfied by different descriptions. The most significant difference between the two bounds is that the Varshamov bound is satisfied by a *linear* code, whereas the Gilbert bound includes linear and nonlinear codes. Therefore the Varshamov bound is stronger because it consists of a smaller subset of codes. This fact is not mentioned in most channel coding books.

In the following proof, we shall only focus on the solution presented by Varshamov.

Theorem 6.7 (Varshamov bound) *There exists a binary linear code with length n, dimension k and minimum distance of at least d that satisfies*

$$2^k \geq \frac{2^{n-1}}{\sum_{i=0}^{d-2}\binom{n-1}{i}}\ .$$

Proof A binary code of length n has minimum distance d if any $d-1$ columns of the parity check matrix $\mathbf{H}$ are linearly independent.

We construct a parity check matrix $\mathbf{H}$ with $n-k$ rows and n columns. The first column can be any non-zero vector of length $n-k$. Assume that we select j columns such that $d-1$ columns are linearly independent. $d-2$ columns or fewer can be combined to produce a unique vector of dimension $n-k$. There are

$$\binom{j}{1}+\ldots+\binom{j}{d-2}$$

different possible linear combinations. If the this sum is smaller than the total number of non-zero vectors $2^{n-k}-1$ then j can be extended by at least one more column. This can be repeated as long as

$$1+\binom{j}{1}+\ldots+\binom{j}{d-2}<2^{n-k}\ .$$

In the last iteration, we obtain $j+1=n$, which is substituted into the above equation. A code is optimized by making the value $n-k$ as small as possible with respect to the parameters n and d. Therefore

$$2^{n-k-1}\leq\sum_{i=0}^{d-2}\binom{n-1}{i}\ .$$

This result produces the Varshamov bound. □

Theorem 6.8 (Varshamov bound for non-binary codes) *(without proof) There exists a linear code with length n, dimension k and minimum distance of at least d that satisfies*

$$q^k \geq \frac{q^{n-1}}{\sum_{i=0}^{d-2} \binom{n-1}{i} (q-1)^i} .$$

We compare this bound with the Gilbert bound:

Theorem 6.9 (Gilbert bound) *(without proof) The Gilbert bound for non-binary codes over $GF(a)$ is*

$$q^k \geq \frac{q^n}{\sum_{i=0}^{d-1} \binom{n}{i} (q-1)^i} .$$

In comparing the two bounds, it can be seen that the Varshamov bound is always larger than the Gilbert bound. What is more important, though, is that the Varshamov bound is satisfied by a linear code.

Example 6.5 (Varshamov bound) The Hamming code of length $n = 15$ has dimension $k = 11$ and minimum distance $d = 3$. When these parameters are substituted into the Varshamov bound (theorem 6.7), we get

$$\binom{14}{0} + \binom{14}{1} \geq 2^{15-11-1} = 8 , \quad 1 + 14 \geq 2^3 = 8 .$$

◇

6.5 Singleton bound (MDS)

Theorem 6.10 (Singleton bound) *For a code with length n, dimension k and minimum distance d, the following relation holds:*

$$n - k \geq d - 1 .$$

Proof Let C be a code with length n, dimension k and minimum distance d. By removing the first $d-1$ coordinates from each codeword, a new code C' is obtained of length $n-(d-1)$ with the same cardinality $|C'| = |C| = q^k$. Since the distance of C is d, the codewords in C' are different. Hence $|C'| \leq q^{n-d+1}$, and the bound follows. □

Definition 6.11 (Maximum distance separable, MDS) *A code is called MDS when*

$$n - k = d - 1 .$$

An MDS code satisfies the Singleton bound with equality. The dual code of an MDS code is also an MDS code.

Theorem 6.12 (Binary MDS codes) *Only trivial examples exist of linear binary MDS codes. They include the repetition code ($k = 1$, $d = n$), the single parity check codes ($k = n-1$, $d = 2$) and codes without redundancy ($k = n$, $d = 1$).*

Proof Consider a binary linear (n,k) MDS code. The code has $d = n - k + 1$; hence any $n - k$ columns of an $(n-k) \times n$ parity check matrix $\mathbf{H}$ of the code must be linearly independent. Denote $\mathbf{H} = (\mathbf{h}_1, \mathbf{h}_2, \ldots, \mathbf{h}_n)$. We now show that

$$\forall i > (n-k) : \mathbf{h}_i = \mathbf{h}_1 + \mathbf{h}_2 + \ldots + \mathbf{h}_{(n-k)} \,. \tag{6.2}$$

The columns $\mathbf{h}_1, \mathbf{h}_2, \ldots, \mathbf{h}_{(n-k)}$ form a basis of $n-k$ vectors, and $\mathbf{h}_i$ is equal to a linear combination of these vectors. If one coefficient of the set of basis vectors equals zero then there exist $\leq (n-k)$ linearly dependent columns in $\mathbf{H}$. The contradiction proves the expression 6.2.

It therefore follows that

$$\mathbf{h}_{(n-k)+1} = \mathbf{h}_{(n-k)+2} = \mathbf{h}_{(n-k)+3} = \ldots = \mathbf{h}_{(n)} \,. \tag{6.3}$$

We first consider $n-k = 1$. This gives the unique $(n, n-1)$ single parity check code. Next we consider $n-k > 1$. It follows from equation 6.3 that binary MDS codes do not exist for $n > n-k+1$ (i.e. $k > 1$). For $k = 0$ and 1, we get the trivial code and the repetition code respectively. □

RS codes are non-binary MDS codes. If one knows any k coordinates of a codeword, the rest of the codeword can be calculated. The Hamming weight distributions of MDS codes can be calculated, which implies that the Hamming weights of RS codes can also be calculated.

6.6 Reiger bound (burst error correction)

In practical systems, errors may occur in bursts. Therefore errors (made up of zero and non-zero values) tend to occur in groups of t consecutive codeword coordinates. The rest of the sequence is error-free. This is defined as a burst error of length t. We shall deal with this error event in section 7.1.3 by defining a special channel that models burst behavior. In section 9.2.4, special codes are constructed to deal with burst errors.

In this section, we shall derive a bound for burst errors for block codes. Assume that all errors occur over t consecutive coordinates of a codeword. What is the minimum number of redundancy coordinates of a block code that uniquely corrects all the errors in a burst of length t? The Reiger bound[1] solves this problem.

Theorem 6.13 (Reiger bound) *A linear block code that can correct a burst error of length t needs at least $n-k \geq 2t$ redundant coordinates.*

Proof First we prove that a non-zero codeword must have a burst length greater than $2t$. Assume that two burst errors of length t are summed to give a burst error of length $2t$. Therefore a codeword cannot exist such that $n-2t$ consecutive coordinates are zero, otherwise a codeword plus a correctable burst error could equal another codeword plus a correctable burst error.

From code linearity, the difference between two vectors in the same coset will produce a codeword. If we subtract any two vectors with a burst error pattern of $2t$ in the first coordinates, and zero otherwise, we obtain a codeword where $n-2t$ consecutive coordinates are zero if the vectors lie in the same coset. This contradicts our earlier statement. Therefore vectors with a burst error pattern of length $2t$ must lie in different cosets. Therefore q^{2t} different coset are required. The number of possible cosets is based on the code redundancy $n-k$, which gives q^{n-k} cosets. Therefore $2t \leq n-k$ for a t-error correcting code. □

[1]Because of a printing error in a standard reference book, this bound is often referred to as the Rieger bound

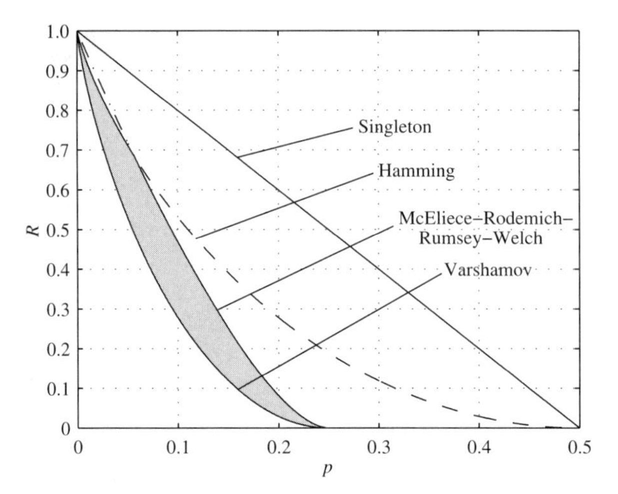

Figure 6.1 Asymptotic upper and lower bounds.

6.7 Asymptotic bounds

In this section, we asymptotically describe the code rate R that satisfies the channel coding theorem from section 1.11 based on different upper and lower bounds. We consider only the case of a BSC (figure 1.2). np is the expected number of errors in a codeword of length n transmitted over a BSC with the error probability p. The expected number of errors must be smaller than the correctable number of errors. Therefore

$$np \leq \left\lfloor \frac{d-1}{2} \right\rfloor \longrightarrow \frac{d}{n} \approx 2p \,.$$

We shall use (without proof)

$$\sum_{i=0}^{np} \binom{n}{i} \leq 2^{nH(p)} \,,$$

where $H(p)$ is the binary entropy (see section 1.11):

$$H(p) = -p\,\mathrm{ld}\,p - (1-p)\,\mathrm{ld}(1-p) \,.$$

For $n \longrightarrow \infty$ and $0 \leq p \leq \frac{1}{2}$,

$$\sum_{i=0}^{np} \binom{n}{i} = 2^{nH(p)} \,.$$

We may apply this approximation to the following bounds:

Singleton bound (upper bound)

$$n-k \geq d-1 \text{ or } 1-\frac{k}{n} \geq \frac{d}{n} - \frac{1}{n}.$$

For $n \longrightarrow \infty$, we get

$$R \leq 1-2p \,.$$

Hamming bound (upper bound)

$$\sum_{i=0}^{t} \binom{n}{i} < 2^{n-k} .$$

For $n \to \infty$, we get

$$R \leq 1 - p \text{ ld } \frac{1}{p} - (1-p) \text{ ld } \frac{1}{1-p} = 1 - H(p).$$

This corresponds to the channel capacity.

McEliece–Rodemich–Rumsey–Welch bound (upper bound) This bound [MRRW77] is the best known upper bound, and is shown only for an asymptotic value of n:

$$R \leq H\left(\frac{1}{2} - \sqrt{2p(1-2p)}\right) .$$

Varshamov bound (lower bound)

$$\sum_{i=0}^{d-2} \binom{n-1}{i} \geq 2^{n-k-1} .$$

For $n \to \infty$, we get

$$R \geq 1 - 2p \text{ ld } \frac{1}{2p} - (1-2p) \text{ ld } \frac{1}{1-2p} = 1 - H(2p).$$

In figure 6.1, all of the asymptotic bounds mentioned so far are presented. There exist good codes that can meet the Varshamov bound. No codes can exist in the region above the upper bounds, and in the gray region we do not know if codes exist.

6.8 Minimal trellis of linear block codes

In this section, we describe the trellis representation of block codes. The advantage of this representation is the possibility of performing maximum-likelihood decoding of codewords using the Viterbi algorithm, which is presented in sections 7.3.4 and 7.4.2. The disadvantage of the trellis representation is that the complexity usually increases exponentially as the code length increases. Therefore the trellis representation is only practical for short codes. Decoding using the Viterbi algorithm consists of addition and comparison operations. In order to minimize the decoding complexity, an appropriate trellis for the code must be constructed. We first define a minimal trellis, and then in section 6.8.3 we shall describe its properties. We show that a *unique* minimal trellis exists for any linear code. A possible construction algorithm based on the parity check matrix of a code is described in section 6.8.1, and a second construction technique based on the generator matrix is presented in section 6.8.2.

Definition 6.14 (Code trellis) *A trellis $T = (\mathcal{V}, \mathcal{E})$ of length n is an edge-labeled directed graph with a set of vertices (nodes or states) $\mathcal{V}$ and a set of edges (branches) $\mathcal{E}$.*

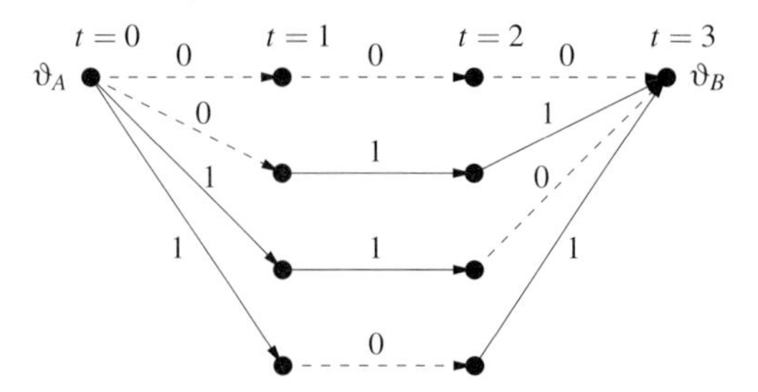

Figure 6.2 Trivial trellis of the $(3,2,2)$ code.

The set $\mathcal{V}$ can be partitioned into $n+1$ subsets:

$$\mathcal{V} = \bigcup_{t=0}^{n} \mathcal{V}_t \, ,$$

where $\mathcal{V}_t$ is a set of vertices having level t.

An edge e is a connection from a vertex with level t to a vertex with level $t+1$. Each edge e is labeled by a symbol $c(e)$ from the alphabet $GF(q)$. The vertices at level $t=0$ and $t=n$ are special cases defined as the root $\mathcal{V}_0 = \{\vartheta_A\}$ and the goal $\mathcal{V}_n = \{\vartheta_B\}$ respectively. This means that ϑ_A is the first vertex of the trellis and ϑ_B is the last vertex.

A path in the trellis is a sequence of edges $\mathbf{e} = (e_1, e_2, \ldots, e_m)$ that connects one vertex ϑ with another ϑ'. Each path $\mathbf{e}$ corresponds to a vector $\mathbf{c}(\mathbf{e}) = (c(e_1), \ldots, c(e_m)) \in GF(q)^m$, which corresponds to the collection of individual edges.

Consider $C(n,k,d)$, a linear block code of length n described over the alphabet $GF(q)$ with $M = q^k$ codewords $\{\mathbf{c}_1, \mathbf{c}_2, \ldots, \mathbf{c}_M\}$. A trellis $T(C)$ is called a *code trellis* if the following conditions hold. Each path $\mathbf{e}$ from ϑ_A to ϑ_B has a corresponding vector $\mathbf{c}(\mathbf{e})$ that belongs to the code C. Furthermore, for each codeword $\mathbf{c}$, there exists at least one path from ϑ_A to ϑ_B. There exist many different possible trellis representations for a given code C.

An important property of code trellises is that each edge is a part of a trajectory from ϑ_A to ϑ_B. We can assume, without loss of generality, that there exists no other parallel trajectory with the identical set of labels from ϑ_A to ϑ_B. We proceed by presenting the trivial trellis representation of a code.

Example 6.6 (Trivial trellis of a $(3,2,2)$ code) Given the parity check code

$$C(3,2,2) = \{(000), (011), (110), (101)\},$$

the trivial trellis is obtained when each codeword is mapped to a unique path as shown in figure 6.2. The trellis has $|\mathcal{V}| = 10$ vertices and $|\mathcal{E}| = 12$ edges.

Another possible trellis of this code is shown in figure 6.3, with $|\mathcal{V}| = 6$ vertices and $|\mathcal{E}| = 8$ edges. We shall show later that this trellis is minimal and therefore that no other trellis exists with a fewer number of vertices and edges. $\diamond$

Definition 6.15 (Minimal trellis) *The minimal trellis of a code $C(n,k,d)$ is a trellis with a minimum number of vertices $|\mathcal{V}|$.*

The Viterbi algorithm requires $|\mathcal{E}|$ additions and $|\mathcal{E}| - |\mathcal{V}| + 1$ comparisons. The value $Z(T) = |\mathcal{E}| - |\mathcal{V}| + 1$ is defined as the *cycle rank* of the graph T. To minimize the complexity

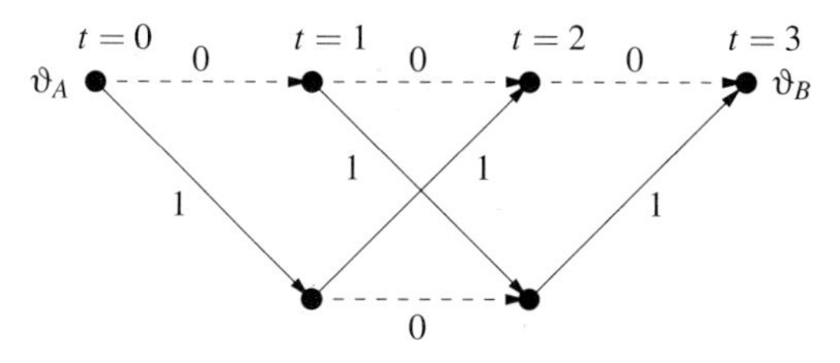

Figure 6.3 Minimal trellis of the $(3,2,2)$ code.

of the Viterbi decoding algorithm, we must use a code trellis having a minimum number of edges $|\mathcal{E}|$ and a minimum cycle rank $Z(T)$. It is not immediately apparent if this follows from a minimum number of vertices $|\mathcal{V}|$.

In section 6.8.3, it is shown that a minimal trellis of a linear code minimizes simultaneously $|\mathcal{V}|$, $|\mathcal{E}|$ and $Z(T)$.

6.8.1 Construction with the aid of the parity check matrix

Given $\mathbf{H} = (\mathbf{h}_1, \mathbf{h}_2, \dots, \mathbf{h}_n)$, the parity check matrix of a code $\mathcal{C}$, where $\mathbf{h}_i$ is the ith column of the matrix $\mathbf{H}$, each codeword $\mathbf{c} \in \mathcal{C}$ satisfies the parity check equation

$$c_1\mathbf{h}_1 + \ldots + c_n\mathbf{h}_n = \mathbf{0} \,. \tag{6.4}$$

One can divide the vector $\mathbf{c}$ with respect to the parameter t into a head (future) section $\mathbf{c}_t^A = \mathbf{a}$ and tail (past) section $\mathbf{c}_t^B = \mathbf{b}$; therefore $\mathbf{c} = (\mathbf{a}|\mathbf{b})$:

$$\mathbf{c} = (c_1, c_2, \dots, c_n) = (\mathbf{c}_t^A|\mathbf{c}_t^B) = (\mathbf{a}|\mathbf{b}) = (\{c_1, \dots, c_t\}|\{c_{t+1}, \dots, c_n\}) \,.$$

This corresponds to the partition of $\mathbf{H} = (\mathbf{H}_t^A|\mathbf{H}_t^B)$.

Syndrome trellis We label the vertices (states) of a trellis $\sigma_t(\mathbf{c})$ by the partial syndrome

$$\forall \mathbf{c} \in \mathcal{C} : \sigma_t(\mathbf{c}) = c_1\mathbf{h}_1 + \ldots + c_t\mathbf{h}_t = \mathbf{c}_t^A \cdot (\mathbf{H}_t^A)^T, \quad t = 1, 2, \dots, n-1. \tag{6.5}$$

Therefore the number of vertices (states) $|\mathcal{V}_t|$ with level t is equal to the number of different vectors $\sigma_t(\mathbf{c})$. It is apparent that one must calculate all of the codewords to find this number.

Example 6.7 (Syndrome trellis) A $(5,3,2)$ code is represented by the parity check matrix

$$\mathbf{H} = \begin{pmatrix} 1 & 1 & 0 & 1 & 0 \\ 0 & 1 & 1 & 0 & 1 \end{pmatrix} . \tag{6.6}$$

When the level equals $t = 0$, the trellis has only one vertex $\sigma_0(\mathbf{0})$. This is an $n-k$-tuple consisting of all zeros.

For each $t = 1, \dots, n-1$, all of the vertices $\mathcal{V}_{t+1}$ of level $t+1$ from the vertex $\mathcal{V}_t$ can be obtained through

$$\sigma_{t+1} = \sigma_t + \alpha\mathbf{h}_{t+1}, \quad \forall \alpha \in GF(q), \quad \forall \sigma_t \in \mathcal{V}_t \,. \tag{6.7}$$

◇

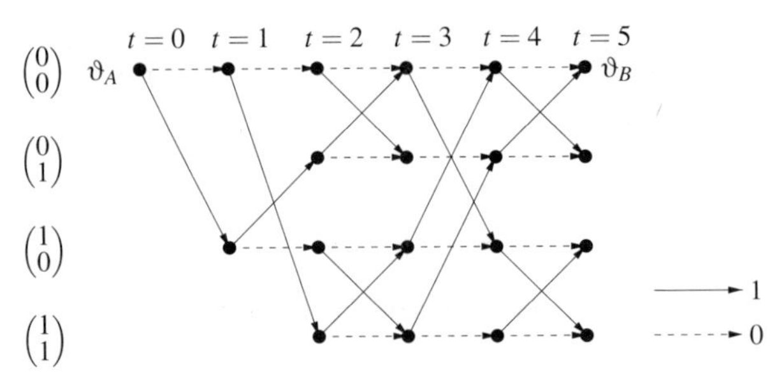

Figure 6.4 Trellis for a binary $(5,3,2)$ code before elimination.

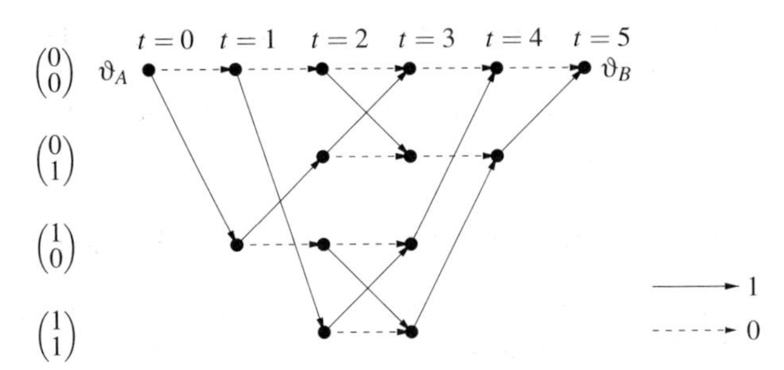

Figure 6.5 Syndrome trellis of the $(5,3,2)$ code.

The vertex σ_t is connected with σ_{t+1} by an edge that is labeled α. For the matrix in equation 6.6, we obtain the trellis in figure 6.4. It follows from equations 6.4 and 6.5 that $\sigma_n(\mathbf{c}) = \mathbf{0}, \quad \forall \mathbf{c} \in \mathcal{C}$. Every codeword $\mathbf{c}$ must have a path from ϑ_A to ϑ_B in the trellis. Each vertex that does not have a path to ϑ_B can be deleted along with all corresponding edges. Through this method, we obtain the unique syndrome trellis (figure 6.5) of the code.

The Wolf upper bound can be derived from the construction of a trellis. It proves that the number of vertices $|\mathcal{V}_t|$ of level t in a minimal trellis of a linear (n,k,d) code cannot be larger than the number of possible syndromes (q^{n-k}), or the number of possible codewords (q^k). There exist codes that have fewer vertices for a specified level. We obtain the upper bound

$$|\mathcal{V}_t| \quad \leq \quad q^{\min\{k,n-k\}}. \tag{6.8}$$

Theorem 6.16 (Syndrome trellis) *The syndrome trellis of a linear code is minimal.*

To prove this theorem, we require the following lemma on code separability. First we define the tail set $B_t(\mathbf{a})$ as the set of all possible tails $\mathbf{b} \in B_t(\mathbf{a})$ of a head $\mathbf{a}$, such that arbitrary $(\mathbf{a}|\mathbf{b}) \in \mathcal{C}$, therefore

$$B_t(\mathbf{a}) \quad = \quad \left\{\mathbf{c}_t^B \mid \mathbf{c} = (\mathbf{c}_t^A|\mathbf{c}_t^B) \in \mathcal{C},\ \mathbf{c}_t^A = \mathbf{a}\right\}. \tag{6.9}$$

Similarly, the set $A_t(\mathbf{b})$ is defined as the set of all possible heads corresponding to a tail $\mathbf{b}$ that result in codewords.

Lemma 6.17 (Separability) *Consider* $\mathbf{a}$ *and* $\tilde{\mathbf{a}}$, *two different heads of a linear code* $\mathcal{C}$. *The sets* $B_t(\mathbf{a})$ *and* $B_t(\tilde{\mathbf{a}})$ *are either identical, if* $\sigma_t(\mathbf{a}) = \sigma_t(\tilde{\mathbf{a}})$, *or disjoint, if* $\sigma_t(\mathbf{a}) \neq \sigma_t(\tilde{\mathbf{a}})$. *A similar result holds for the sets* $A_t(\mathbf{b})$ *and* $A_t(\tilde{\mathbf{b}})$.

Proof Let $\mathcal{C}_t^B$ be a code of length $n-t$ with parity check matrix $\mathbf{H}_t^B$. Then $(\mathbf{a}|\mathbf{b})\cdot\mathbf{H}^T=\mathbf{0}$, which gives

$$B_t(\mathbf{a})=\left\{\mathbf{b}\mid\mathbf{b}\cdot\left(\mathbf{H}_t^B\right)^T=-\mathbf{a}\cdot\left(\mathbf{H}_t^A\right)^T\right\}.$$

$B_t(\mathbf{a})$ is, by definition, a coset (see section 1.3) of the code $\mathcal{C}_t^B$. The cosets $B_t(\mathbf{a})$ and $B_t(\tilde{\mathbf{a}})$ are either identical, if $\mathbf{a}\cdot\left(\mathbf{H}_t^A\right)^T=\tilde{\mathbf{a}}\cdot\left(\mathbf{H}_t^A\right)^T$, i.e. $\sigma_t(\mathbf{a})=\sigma_t(\tilde{\mathbf{a}})$, or are disjoint, if $\sigma_t(\mathbf{a})\neq\sigma_t(\tilde{\mathbf{a}})$. □

Proof of theorem 6.16 For a given level t, M_t is the number of different sets $B_t(\mathbf{a})$ corresponding to lemma 6.17. For level t, denote by $|\tilde{\mathcal{V}}_t|$ the number of vertices of a minimal trellis. Two codewords $\mathbf{c}$ and $\tilde{\mathbf{c}}$ having disjoint tail sets ($B_t\left(\mathbf{c}_t^A\right)\cap B_t\left(\tilde{\mathbf{c}}_t{}^A\right)=\emptyset$), cannot pass through a common vertex at level t; hence $|\tilde{\mathcal{V}}_t|\geq M_t$. On the other hand, the construction of the syndrome trellis produces exactly M_t vertices at level t; hence $|\tilde{\mathcal{V}}_t|\leq M_t$. From this, it can be concluded that $|\tilde{\mathcal{V}}_t|=M_t$. The syndrome trellis minimizes $|\mathcal{V}_t|$ and hence $|\mathcal{V}|$. □

Another definition is given for the matrix rank $Z(\mathbf{H})$, where $\mathbf{H}$ is the parity check matrix used to generate a particular code trellis. The number of vertices $|\mathcal{V}_t|$ with level t can be found from the following theorem:

Theorem 6.18 (Number of vertices $|\mathcal{V}_t|$with level t for a minimal trellis) *Given a linear code $\mathcal{C}$ with parity check matrix $\mathbf{H}$, the number of vertices $|\mathcal{V}_t|$ in the minimal trellis with level t is*

$$|\mathcal{V}_t|=q^{Z\left(\mathbf{H}_t^A\right)+Z\left(\mathbf{H}_t^B\right)-Z(\mathbf{H})}\,, \tag{6.10}$$

where $Z(\mathbf{H})$ is $\mathrm{rank}(\mathbf{H})$.

Proof Let $\mathcal{C}_t^A$ be a code of length t having parity check matrix $\mathbf{H}_t^A$ and let $\mathcal{C}_t^B$ be a code of length $n-t$ having parity check matrix $\mathbf{H}_t^B$. Since all minimal trellises are isomorphic (see theorem 6.24), we can prove the theorem using any minimal trellis. We use the syndrome trellis. From equations 6.4 and 6.5, it follows that a codeword $\mathbf{c}=(\mathbf{a}|\mathbf{b})$ passes through a vertex ϑ with level t if and only if $\mathbf{a}\left(\mathbf{H}_t^A\right)^T=\sigma$ and $\mathbf{b}\left(\mathbf{H}_t^B\right)^T=-\sigma$. This result shows that $\mathbf{a}$ is a coset from $\mathcal{C}_t^A$ and $\mathbf{b}$ is a coset from $\mathcal{C}_t^B$. It can be seen that the number of codewords that pass through ϑ is equal to $|\mathcal{C}_t^A|\cdot|\mathcal{C}_t^B|$. The result $|\mathcal{C}_t^A|\cdot|\mathcal{C}_t^B|\cdot|\mathcal{V}_t|=|\mathcal{C}|$ proves equation 6.10. □

6.8.2 Construction with the aid of the generator matrix

The construction of a minimal trellis through the use of the generator matrix $\mathbf{G}$ of a linear code is based on the *code sum*. This produces the *Shannon product* of trellises.

Definition 6.19 (Code as a sum of codes) *Consider two linear codes $\mathcal{C}$ and $\hat{\mathcal{C}}$ of length n. The code $\mathcal{C}+\hat{\mathcal{C}}$ is the sum of all possible combinations $\mathbf{c}+\hat{\mathbf{c}}$ of codewords $\mathbf{c}\in\mathcal{C}$ and $\hat{\mathbf{c}}\in\hat{\mathcal{C}}$:*

$$\mathcal{C}+\hat{\mathcal{C}} \;=\; \left\{\mathbf{c}+\hat{\mathbf{c}}:\mathbf{c}\in\mathcal{C},\hat{\mathbf{c}}\in\hat{\mathcal{C}}\right\}.$$

With this definition, we can describe a code $\mathcal{C}$ as the sum of row codes. Let

$$\mathbf{G}=\begin{pmatrix}\mathbf{g}_1\\ \vdots\\ \mathbf{g}_k\end{pmatrix}$$

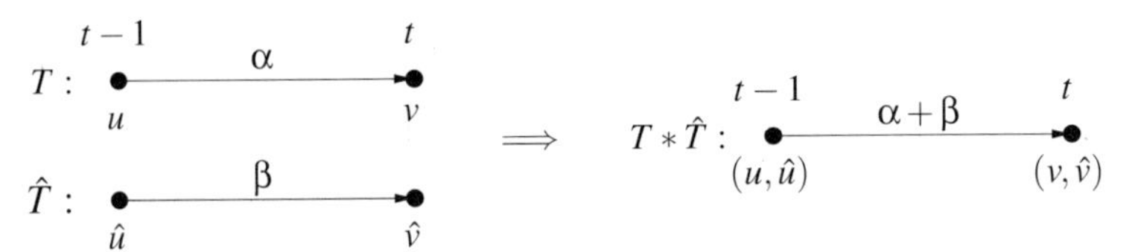

Figure 6.6 The Shannon product.

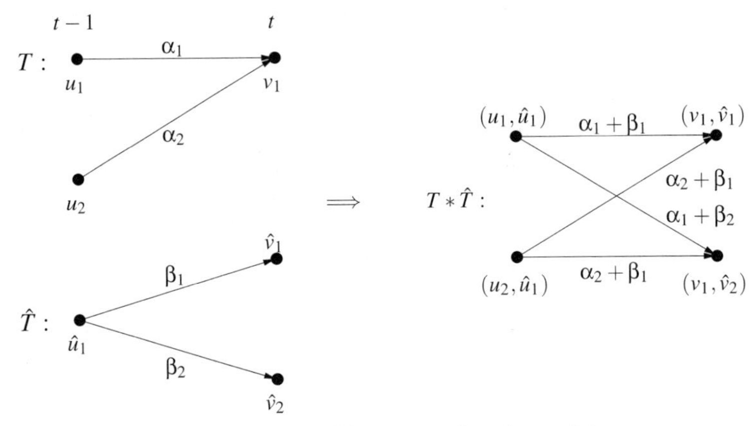

Figure 6.7 The Shannon product (example).

be a generator matrix of the codes C, and take C_i to be the code that is generated by a row vector $\mathbf{g}_i$ of the generator matrix $\mathbf{G}$; therefore $\mathbf{G}_i = (\mathbf{g}_i)$. It is then apparent that

$$C \quad = \quad C_1 + C_2 + \ldots + C_k. \tag{6.11}$$

Definition 6.20 (Shannon product) *Given trellises T and $\hat{T}$ of length n, the Shannon product $T * \hat{T}$ is a trellis that has $|\mathcal{V}_t| \cdot |\hat{\mathcal{V}}_t|$ vertices at level t. These vertices are labeled by $(\vartheta, \hat{\vartheta})$, $\vartheta \in \mathcal{V}_t$, $\hat{\vartheta} \in \hat{\mathcal{V}}_t$. Two vertices $(u, \hat{u})$ and $(v, \hat{v})$ of adjacent levels $u \in \mathcal{V}_{t-1}$, $\hat{u} \in \hat{\mathcal{V}}_{t-1}$, $v \in \mathcal{V}_t$, $\hat{v} \in \hat{\mathcal{V}}_t$, are connected by an edge with the label*

$$\alpha(u,v) + \beta(\hat{u},\hat{v})$$

if and only if the trellis T has an edge (u,v) labeled by $\alpha(u,v)$, and $\hat{T}$ has an edge $(\hat{u},\hat{v})$ labeled by $\beta(\hat{u},\hat{v})$ (figure 6.6).

Other examples of the Shannon product are presented in figure 6.7.

Theorem 6.21 (Code trellis from the Shannon product) *Let T be a trellis of code C and $\hat{T}$ a trellis of code $\hat{C}$. The Shannon product $T * \hat{T}$ is a trellis of the code $C + \hat{C}$.*

The proof is based on the one-to-one correspondence between the codewords $\mathbf{c} + \hat{\mathbf{c}}$ and the trellis path in $T * \hat{T}$.

Example 6.8 (Shannon product and minimal trellises) Given the generator matrix of a binary code

$$\mathbf{G} = \begin{pmatrix} 1 & 1 & 0 \\ 0 & 1 & 1 \end{pmatrix} = \begin{pmatrix} \mathbf{g}_1 \\ \mathbf{g}_2 \end{pmatrix}.$$

◇

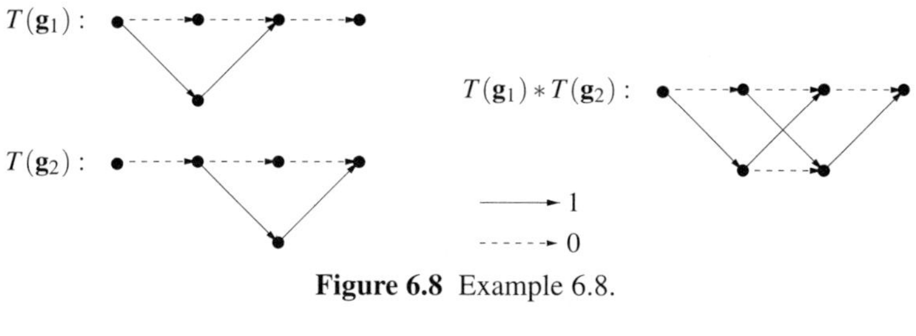

Figure 6.8 Example 6.8.

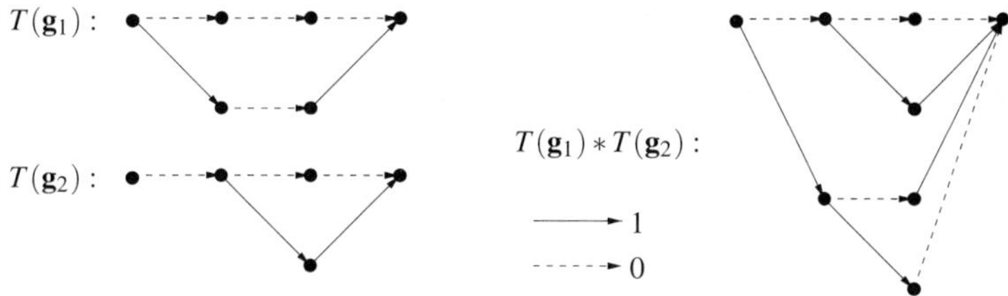

Figure 6.9 Example 6.9.

$T(\mathbf{g}_i)$ is the minimal code trellis generated by $\mathbf{G} = (\mathbf{g}_i)$. The trellises $T(\mathbf{g}_1)$ and $T(\mathbf{g}_2)$ are depicted in figure 6.8. A trellis $T(C)$ of the code C can be constructed from the Shannon product $T(C) = T(\mathbf{g}_1) * T(\mathbf{g}_2)$. This trellis is minimal.

The following example shows that the Shannon product $T(C) = T(C_1) * T(C_2)$ does not always produce a minimal trellis of the codes $C_1 + C_2$, even if the trellises $T(C_i)$, $i = 1, 2$, are minimal.

Example 6.9 (Shannon product producing a non-minimal trellis) Consider a different generator matrix of the code presented in example 6.8:

$$\mathbf{G} = \begin{pmatrix} 1 & 0 & 1 \\ 0 & 1 & 1 \end{pmatrix} = \begin{pmatrix} \mathbf{g}_1 \\ \mathbf{g}_2 \end{pmatrix} .$$

◇

The Shannon product produces the trellis shown in figure 6.9. This trellis is non-minimal.

What properties must a generator matrix have such that a minimal trellis is produced by the Shannon product $T(\mathbf{g}_1) * \cdots * T(\mathbf{g}_k)$? In order to answer this question, we must first introduce some new notation.

The *span* of a vector $\mathbf{g}$ is defined as an interval from the first to the last non-zero coordinates, and is denoted by $\text{span}(\mathbf{g}) = [i, j]$. For example, $\text{span}\big((0,1,0,1,0,0)\big) = [2,4]$. We also denote $\text{L}(\mathbf{g}) = i$ and $\text{R}(\mathbf{g}) = j$ as the left and right indices respectively. The vector $\mathbf{g}$ is described as being *t-active* for the level $t \in [i, j-1]$.

LR properties: A set of *vectors* $\{\mathbf{g}_1, \ldots, \mathbf{g}_k\}$ has the left–right (LR) properties if

$$\text{L}(g_i) \neq \text{L}(g_j) \quad \text{and} \quad \text{R}(g_i) \neq \text{R}(g_j)$$

are satisfied for all $i \neq j$. A *matrix* $\mathbf{G}$ of non-zero columns satisfies the LR property if the set of all rows satisfies the LR property. The matrix $\mathbf{G}$ from example 6.8 satisfies the LR property, while the matrix $\mathbf{G}$ from example 6.9 does not. A matrix with the LR property is termed *trellis-oriented*.

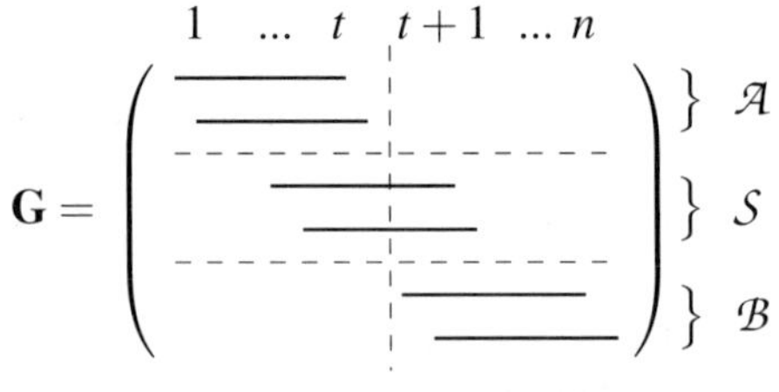

$$\mathbf{G} = \begin{pmatrix} \text{rows with span in } [1,t] \\ \hline \text{rows crossing } t \mid t+1 \\ \hline \text{rows with span in } [t+1,n] \end{pmatrix} \begin{matrix} \} \ \mathcal{A} \\ \} \ \mathcal{S} \\ \} \ \mathcal{B} \end{matrix}$$

Figure 6.10 Subset partitioning of the generator matrix.

Theorem 6.22 (Minimal trellis from the Shannon product) *Let the generator matrix of the code C be represented as*

$$\mathbf{G} = \begin{pmatrix} \mathbf{g}_1 \\ \vdots \\ \mathbf{g}_k \end{pmatrix}.$$

Every row $\mathbf{g}_i$, $i = 1, \ldots, k$, of the generator matrix $\mathbf{G}$ produces a minimal trellis $T(\mathbf{g}_i)$. The Shannon product $T = T(\mathbf{g}_1) * \cdots * T(\mathbf{g}_k)$ gives the minimal trellis of the code C if and only if $\mathbf{G}$ satisfies the LR property.

Proof The trellis $T(\mathbf{g}_i)$ of a q-ary code has q vertices at level t if $\mathbf{g}_i$ is t-active; otherwise there is only one vertex. From its construction, the trellis $T = T(\mathbf{g}_1) * \cdots * T(\mathbf{g}_k)$ has $|\mathcal{V}_t| = q^s$ vertices at level t. s is the number of t-active row vectors of the generator matrix $\mathbf{G}$. It must now be shown that $|\mathcal{V}_t| = q^s$ satisfies equation 6.10 with equality when $\mathbf{G}$ is trellis-oriented (an LR matrix).

Assume the matrix $\mathbf{G} = (\mathbf{g}_1, \ldots, \mathbf{g}_k)^T$ is trellis-oriented. It can therefore be partitioned into the following groups (figure 6.10):

$$\mathcal{A} = \{\mathbf{g}_i \in \mathbf{G} \mid \text{span}(\mathbf{g}_i) \in [1,t]\};$$
$$\mathcal{B} = \{\mathbf{g}_i \in \mathbf{G} \mid \text{span}(\mathbf{g}_i) \in [t+1,n]\};$$
$$\mathcal{S} = \mathcal{G} \setminus \mathcal{A} \setminus \mathcal{B}, \quad \mathcal{G} = \{\mathbf{g}_i \in \mathbf{G}\}.$$

Only rows from the subset $\mathcal{S}$ are t-active, which gives $|\mathcal{V}_t| = q^{|\mathcal{S}|}$.

Denote by C_t^A the subcode of C that consists of codewords from C having zeros at positions $t+1, \ldots, n$. The subcode has dimension $\dim(C_t^A) = t - Z(\mathbf{H}_t^A)$. We now show that C_t^A is generated by all rows of the matrix $\mathcal{A}$. It is enough to show that, for all $\mathbf{g}_{i_j} \in \mathbf{G}$,

$$\mathbf{c} = \alpha_1 \mathbf{g}_{i_1} + \ldots + \alpha_m \mathbf{g}_{i_m} \in C_t^A \iff \mathbf{g}_{i_j} \in \mathcal{A}, \quad j = 1, \ldots, m. \tag{6.12}$$

$\Rightarrow$ (An LR matrix gives the minimal trellis)
Assume that among the $\mathbf{g}_{i_1}, \ldots, \mathbf{g}_{i_m}$ rows there are generators that do not belong to $\mathcal{A}$. In particular, there exists a generator having a maximum right index $\mathrm{R}(\mathbf{g}_e) > t$, since $\mathbf{g}_e \notin \mathcal{A}$. Because $\mathbf{G}$ has the LR property and, according to equation 6.12, the codeword $\mathbf{c}$ has a non-zero coordinate at position $\mathrm{R}(\mathbf{g}_e) > t$, $c_{\mathrm{R}(\mathbf{g}_e)} \neq 0$. We may conclude that $\mathbf{c} \notin C_t^A$. This contradiction proves the condition 6.12. We may now write $|\mathcal{A}| = dim(C_t^{\mathcal{A}}) = t - Z(\mathbf{H}_t^A)$. Similarly,

$$|\mathcal{B}| = \dim(C_t^B) = n - t - Z(\mathbf{H}_t^B),$$

Algorithm 6.1 LR algorithm (for binary codes).

LR algorithm: as long as (LR property fails to hold)

$$
\begin{aligned}
&\{ \\
&\quad \text{find a pair } (i,j), \text{ so that} \\
&\qquad \left(\mathrm{L}(\mathbf{g}_i) = \mathrm{L}(\mathbf{g}_j) \quad \text{and} \quad \mathrm{R}(\mathbf{g}_i) \le \mathrm{R}(\mathbf{g}_j)\right) \\
&\qquad \text{or} \\
&\qquad \left(\mathrm{R}(\mathbf{g}_i) = \mathrm{R}(\mathbf{g}_j) \quad \text{and} \quad \mathrm{L}(\mathbf{g}_i) \ge \mathrm{L}(\mathbf{g}_j)\right) \\
&\quad \mathbf{g}_i = \mathbf{g}_j + \mathbf{g}_i \\
&\}
\end{aligned}
$$

and we obtain

$$
\begin{aligned}
|\mathcal{S}| &= |\mathcal{G}| - |\mathcal{A}| - |\mathcal{B}| \\
&= k - t + Z(\mathbf{H}_t^A) - n + t + Z(\mathbf{H}_t^B) \\
&= Z(\mathbf{H}_t^A) + Z(\mathbf{H}_t^B) - Z(\mathbf{H}).
\end{aligned}
$$

Therefore $|\mathcal{V}_t|$ satisfies equation 6.10, and T is minimal.

$\Leftarrow$ (The minimal trellis obtained from an LR matrix)

Let the trellis $T(\mathbf{G})$ be minimal and assume that $\mathbf{G}$ is not an LR matrix. Without loss of generality, $\mathrm{L}(\mathbf{g}_i) = \mathrm{L}(\mathbf{g}_j)$ and $\mathrm{R}(\mathbf{g}_i) < \mathrm{R}(\mathbf{g}_j)$. We multiply $\mathbf{g}_i$ by $\alpha \in GF(q)$ such that the addition of $\alpha\mathbf{g}_i + \mathbf{g}_j$ over the field produces zero in the coordinate $\mathrm{L}(\mathbf{g}_j)$.

If the row $\mathbf{g}_j$ of the generator matrix $\mathbf{G}$ is replaced by $\alpha\mathbf{g}_i + \mathbf{g}_j$ then we obtain $\mathbf{G}'$. The trellis $T(\mathbf{G}')$ has fewer than $|\mathcal{V}_t''|$ vertices at level $\mathrm{L}(\mathbf{g}_i)$. The construction proves the $\Rightarrow$ statement. □

In example 6.8, it was shown that the LR matrix $\mathbf{G}$ from example 6.8 results in a minimal trellis. In contrast, the matrix $\mathbf{G}$ from example 6.9 did not produce a minimal trellis. An LR matrix can be constructed from any generator matrix with the help of algorithm 6.1.

Example 6.10 (Construction of an LR matrix) Consider the $(5,3,2)$ code from example 6.7. This code has a parity check matrix

$$\mathbf{H} = \begin{pmatrix} 1 & 1 & 0 & 1 & 0 \\ 0 & 1 & 1 & 0 & 1 \end{pmatrix}.$$

A possible generator matrix of this code is

$$\mathbf{G} = \begin{pmatrix} 1 & 0 & 0 & 1 & 0 \\ 0 & 1 & 0 & 1 & 1 \\ 0 & 0 & 1 & 0 & 1 \end{pmatrix}.$$

This matrix does not satisfy the LR properties. In this case, we apply the LR algorithm to the matrix $\mathbf{G}$:

$$\mathbf{G}_1 = \begin{pmatrix} 1 & 0 & 0 & 1 & 0 \\ 0 & 1 & 1 & 1 & 0 \\ 0 & 0 & 1 & 0 & 1 \end{pmatrix}.$$

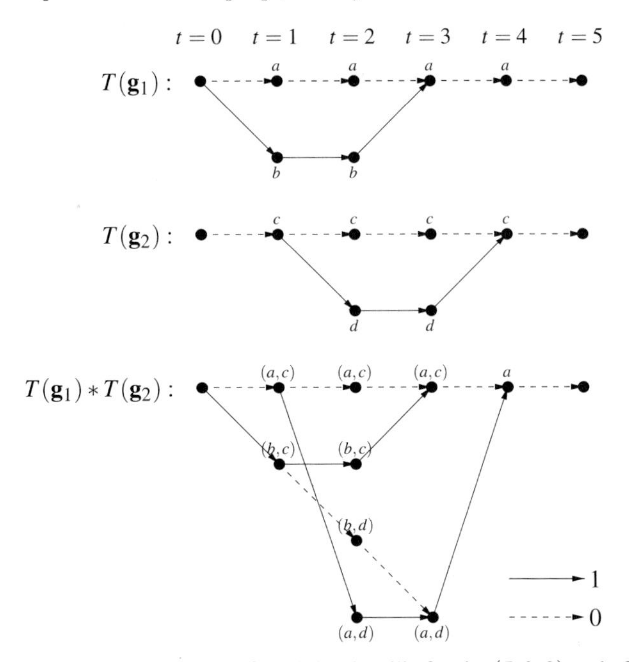

Figure 6.11 First step for the calculation of a minimal trellis for the $(5,3,2)$ code from example 6.10.

After the second step, we obtain an LR-generator matrix:

$$\mathbf{G}_{LR} = \begin{pmatrix} 1 & 1 & 1 & 0 & 0 \\ 0 & 1 & 1 & 1 & 0 \\ 0 & 0 & 1 & 0 & 1 \end{pmatrix} = \begin{pmatrix} \mathbf{g}_1 \\ \mathbf{g}_2 \\ \mathbf{g}_3 \end{pmatrix}.$$

Theorem 6.22 states that the minimum trellis of a code can be produced from the Shannon product

$$T(C) = T(\mathbf{g}_1) * T(\mathbf{g}_2) * T(\mathbf{g}_3)$$

(compare with figures 6.11 and 6.12).

One recognizes that in this example the minimal trellis of the $(5,3,2)$ code is the same as in figure 6.5. ◇

6.8.3 Properties of a minimal trellis

Now we consider some properties of a code trellis. Since tail sets of a linear code are separable (see lemma 6.17), only codewords $\mathbf{c}$ that have the common tail set (or state $\sigma(\mathbf{c})$) can pass through a common vertex ϑ. We call $\sigma(\mathbf{c})$ *the state of the vertex* ϑ, $\sigma(\vartheta) \mathrel{\widehat{=}} \sigma(\mathbf{c})$. Notice that the state $\sigma(\mathbf{c})$ can be computed from either the head $\mathbf{c}_t^A$ or the tail $\mathbf{c}_t^B$.

Two vertices ϑ_1 and ϑ_2 at level t are called *equivalent* if $\sigma(\vartheta_1) = \sigma(\vartheta_2)$. The merging of two equivalent vertices of a code trellis $T(C)$ leads to another code trellis of the same code (when vertices ϑ_1 and ϑ_2 are merged, all edges connected to ϑ_1 and ϑ_2 must be connected with the common vertex ϑ; parallel edges are deleted).

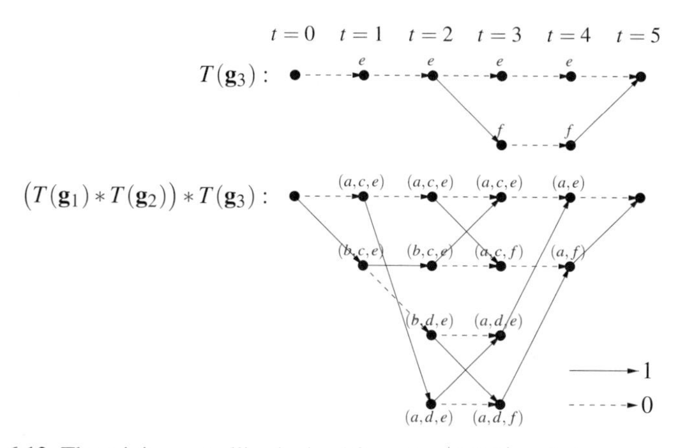

Figure 6.12 The minimum trellis obtained from the $(5,3,2)$ code from example 6.10.

A code trellis that has no equivalent vertices is called a *canonical*. Hence a canonical trellis of a linear code is a trellis where all codewords $\mathbf{c}$ with state $\sigma_t(\mathbf{c})$ at level t pass through a common vertex. Thus a canonical trellis of a linear code coincides with the syndrome trellis up to a reordering of the vertices.

It is apparent that a canonical trellis of a linear code can be constructed from any code trellis through the merging of equivalent vertices. The canonical trellis has a minimum number of vertices $|\mathcal{V}|$. The reverse is also true, namely that a trellis with minimal $|\mathcal{V}|$ is canonical.

Theorem 6.23 (Canonical trellis) *A code trellis of a linear code is minimal if and only if it is canonical.*

As a result of this statement, we again get that the syndrome trellis is minimal since it is canonical. Since all canonical trellises of a given linear code are isomorphic, we obtain the following from theorem 6.23:

Theorem 6.24 (Uniqueness of a minimal trellis) *All minimal trellises of a linear code are isomorphic.*

In the following, we show that the minimal trellis of a linear code contains not only the fewest number of vertices $|\mathcal{V}|$, but also the fewest number of edges $|\mathcal{E}|$ and the smallest cycle rank $Z(T) = |\mathcal{E}| - |\mathcal{V}| + 1$. Therefore the decoding complexity of a code using the Viterbi algorithm with a minimal trellis is minimal.

Theorem 6.25 (Minimal cyclic rank) *Consider the minimal (canonical) trellis $\tilde{T} = (\tilde{\mathcal{V}}, \tilde{\mathcal{E}})$ of a linear code $\mathcal{C}$ and $T = (\mathcal{V}, \mathcal{E})$, which is a non-minimal trellis of the code $\mathcal{C}$. The following inequalities hold:*

$$(i)\ |\tilde{\mathcal{V}}| < |\mathcal{V}|, \quad (ii)\ |\tilde{\mathcal{E}}| < |\mathcal{E}| \quad and \quad (iii)\ Z(\tilde{T}) \leq Z(T).$$

Before proving this theorem, we need some auxiliary results. Vertices $\vartheta', \vartheta'' \in \mathcal{V}_t$ at level t are called *relative* if connected by a common vertex ϑ at either level $t-1$ or $t+1$ by the edges $e' = \alpha$ and $e'' = \alpha$ having the same labels $c(e') = c(e'')$ (see figure 6.13). It is clear that relative vertices can be merged. In the following proof, we wish to show that a trellis without relative vertices is canonical.

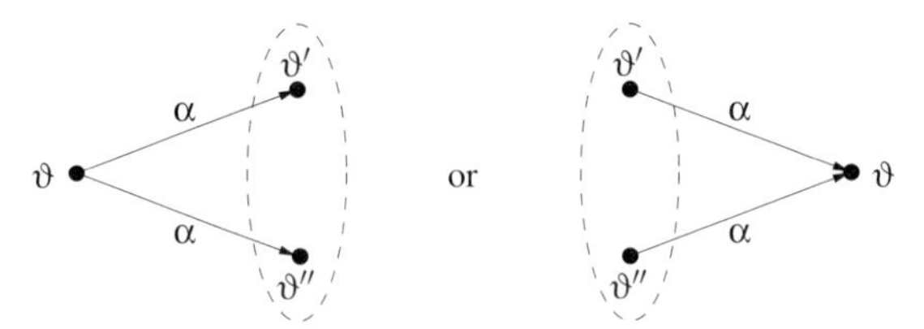

Figure 6.13 Relative vertices (Lemma 6.26).

Lemma 6.26 (Relative vertices) *A trellis of a linear code is minimal (canonical), if and only if it contains no relative vertices.*

Proof It is enough to show that every non-canonical trellis has relative vertices.

Let $\mathcal{L}(\vartheta')$ be the head set of all codewords in $T(C)$ that originate at ϑ_a and terminate at ϑ', and let $\mathcal{R}(\vartheta')$ be the set of all codewords that start from ϑ' and terminate at ϑ_b. If $T(C)$ is a non-canonical (non-minimal) trellis of a linear code C then there are at least two equivalent vertices $\vartheta_1, \vartheta_2 \in \mathcal{V}_t$ with the same states $\sigma(\vartheta_1) = \sigma(\vartheta_2) = \sigma$. Define $\vartheta_1, \ldots, \vartheta_l \in \mathcal{V}_t$ to be *all* vertices of level t that have state σ.

We prove the following property. Among $\vartheta_1, \ldots, \vartheta_l$, there are two vertices ϑ' and ϑ'' such that the following property holds:

($*$) either identical codewords exist which originate from ϑ_a and terminate at both ϑ' and ϑ'', or identical codewords exist that originate from ϑ' and ϑ'' and terminate at ϑ_b.

If this is not the case then the sets $\mathcal{L}(\vartheta_1), \ldots, \mathcal{L}(\vartheta_l)$ (and similarly $\mathcal{R}(\vartheta_1), \ldots, \mathcal{R}(\vartheta_l)$) do not intersect pairwise, and we get

$$\mathcal{L}(\vartheta_i) \cap \mathcal{L}(\vartheta_j) = \emptyset, \quad \forall i \neq j; \tag{6.13}$$

$$\mathcal{R}(\vartheta_i) \cap \mathcal{R}(\vartheta_j) = \emptyset, \quad \forall i \neq j. \tag{6.14}$$

We may merge equivalent states $\vartheta_1, \ldots, \vartheta_l$ without altering the code. We obtain the code trellis $T'(C)$ containing the codewords $\mathcal{S} = \{\mathcal{L}(\vartheta_1), \mathcal{R}(\vartheta_2)\}$. These codewords have state σ at level t, and, in the original trellis $T(C)$, they must pass trough a subset of vertices $\vartheta_1, \ldots, \vartheta_t$.

However, if equations 6.13 and 6.14 are true then it follows that all heads $(\mathcal{L}(\vartheta_1))$ of the codewords at level t can only pass through ϑ_1, and all tails $(\mathcal{R}(\vartheta_2))$ can only originate from vertex ϑ_2. Therefore there are no codewords from the set $\mathcal{S}$ contained in the trellis $T(C)$. This contradiction proves that property ($*$) must be true.

Let the same word come from ϑ_a to both ϑ' and ϑ'', i.e. let there be two paths $\mathbf{e}'$ and $\mathbf{e}''$ that come from ϑ_a to ϑ' and ϑ''respectively and such that $c(\mathbf{e}') = c(\mathbf{e}'')$. However, this condition is equivalent to having relative vertices or a non-canonical trellis. □

Proof of theorem 6.25 Let us consider an algorithm where in every step a pair of relative vertices of a trellis (when they exist) are merged, thus creating a new trellis. For the case where no relative vertices exist, the trellis is minimal according to lemma 6.26. If we assume that $T(\mathcal{V}, \mathcal{E})$ is a non-minimal trellis, there is at least one pair of relative vertices. For every step of the algorithm,

- The number of vertices is reduced by 1, therefore i) from theorem 6.25 is true.
- The number of edges is reduced by at least 1 (> 1 if parallel transitions exist), which corresponds to condition (ii) of theorem 6.25.
- The cycle rank remains the same or is reduced, corresponding to condition (iii) of theorem 6.25.

□

In [LBB96], it is shown that the following algorithm produces a minimal trellis from any arbitrary trellis.

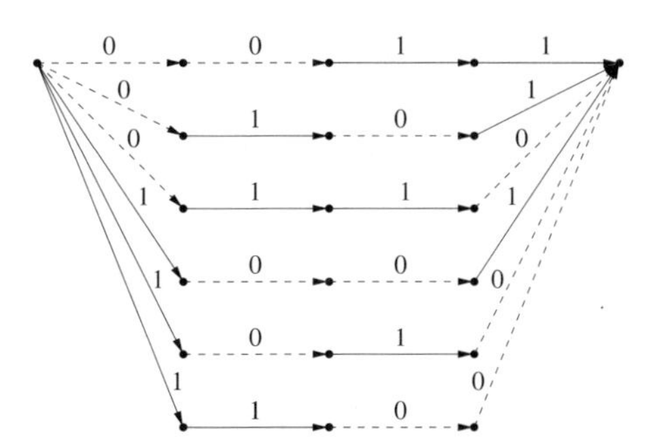

Figure 6.14 Trivial trellis.

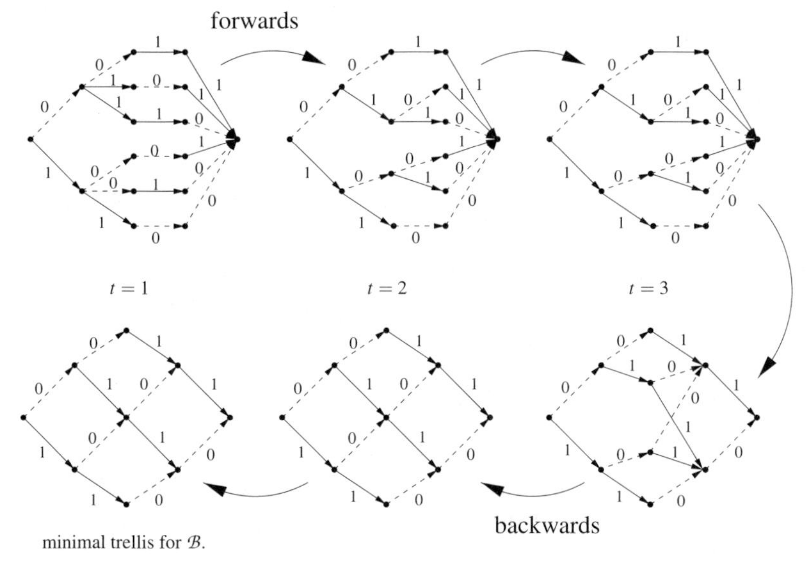

Figure 6.15 (Example 6.11) Stepwise construction of a minimal trellis from the original trivial trellis through path combination.

Algorithm for the construction of a minimal trellis Given an arbitrary trellis T of code $C(n,k,d)$,

- Merge all relative vertices $t = 1,\dots,n-1$ at level t, i.e. all edges from the vertices at level $t-1$ with the same label.
- Merge all relative vertices $t = n-1,\dots,1$ at level t, i.e. all edges from vertices $t+1$ that have the same label.
- The resulting trellis is minimal.

The following example illustrates this algorithm.

Example 6.11 (Algorithm for the construction of a minimal trellis) Six codewords of length 4 having a Hamming weight of 2 are represented by a trivial trellis. Stepwise application of the minimal trellis construction algorithm produces a minimal trellis. See figures 6.14 and 6.15.

◇

The following definition of a minimal trellis is equivalent: a minimal trellis has minimum $|\mathcal{V}|$, a minimal trellis has minimum $|\mathcal{E}|$ and a minimal trellis is canonical. These conditions can be generalized to include nonlinear codes and group codes. Therefore we next define *separable* codes.

Definition 6.27 (Separable codes) *A code is called separable if the tail set of the code is separable with respect to the level $t = 0, 1, \ldots, n-1$, i.e. if the tail corresponding to different heads is either identical or disjoint.*

Lemma 6.17 is also valid for separable codes by definition. Furthermore, theorems 6.23, 6.24 and 6.25 and lemma 6.26 can be applied to separable codes. Therefore separable codes have a unique minimal trellis (up to isomorphism) with a minimum number of vertices, edges and a minimal cycle rank.

The decoding of a separable code with the Viterbi algorithm has minimum complexity if a minimal trellis is used.

Separability of other code classes All linear codes are separable (Lemma 6.17). Furthermore, all group codes are separable. A group code is defined as follows. An alphabet Q is an additive group (possibly Abelian), and the codewords of a code C form a group (section 2.1) with respect to componentwise addition of the codewords. A general definition is such that every ith component of a codeword is from the alphabet Q_i. Furthermore, it is known [SMH96] that several famous nonlinear codes are separable. Special examples include Hadamard codes (section 5.1.4), Levenshtein codes [McWSl], Delsarte–Goethals codes, Kerdock codes and the Nordstrom–Robinson code (see [McWSl]).

Note In this section, we have examined codes with a defined order of codeword coordinates. In other words, the order of the codeword positions is fixed. Permutation of the columns of the generator or the parity check matrix of a code C creates an equivalent code $\tilde{C}$. The number of vertices $|\tilde{\mathcal{V}}|$ in the minimal trellis of $\tilde{C}$ can be smaller or larger than $|\mathcal{V}|$. This may result in a power-of-10 difference in the complexities of $|\tilde{\mathcal{V}}|$ and $|\mathcal{V}|$. An unsolved research problem for linear codes consists in finding the code permutation that produces a minimal trellis having minimum $|\tilde{\mathcal{V}}|$ for the given code. Excluding the exhaustive search method, there are many algorithms of various complexity that attempt to solve this problem; however, they do not guarantee an optimal permutation. One method is investigated in [EBMS96].

The class of RM codes in standard bit representation [KTFL93] produces an optimal trellis. It follows from this that several good permutations exist for BCH codes that may be created from shortened RM codes (see section 8.7.4).

6.9 Summary

In this chapter, we have presented several properties of codes. We have defined the dual code $C^{\perp}$ corresponding to an original code C. The generator matrix of $C^{\perp}$ is a parity check matrix of C, and vice versa. Therefore the product of any codeword from a dual code and the original code always produces zero. These properties will be used in the next chapter to define decoding methods for codewords using the dual code $C^{\perp}$.

We have defined the weight distribution of a code $\mathcal{C}$ and the MacWilliams identity, which was first published in 1962. The proof is based on [CW80]. The MacWilliams identity can be used to calculate the weight distribution of a dual code $\mathcal{C}^{\perp}$ from the weight distribution of the code $\mathcal{C}$. Furthermore, we have introduced the definition of an automorphism, where the permutation of codeword coordinates again produces a valid codeword.

The Hamming bound [Ham50] was first published in 1950, and was presented in chapter 1. This chapter has introduced other upper and lower bounds for codes. We have presented the Gilbert and Varshamov bound which are frequently presented as one bounds. Varshamov published his work [Var57] in 1957 and Gilbert in 1952 [Gil52]. These bounds prove that, given a set of parameters, certain good codes exist. The Varshamov bound is somewhat tighter, and says that *linear* codes exist that satisfy these bounds.

Next, we have presented the Singleton bound, which was developed in 1964 [Sin64]. Any code that satisfies the Singleton bound with equality is termed a *maximum distance separable* (MDS) code. This property has surprising implications. One is that the Hamming weight distribution for an MDS code can be calculated; another is that a codeword can be determined uniquely through knowledge of any k coordinates. We have shown that there are no binary MDS codes except for the trivial case. Singleton was the first to explicitly investigate MDS codes. A generalized Singleton bound for other metrics and a simple proof for nonlinear codes can be found in [BS96].

Finally, we have asymptotically compared all the aforementioned bounds as well as the McEliece-Rodemich-Rumsey-Welch bound. The asymptotic limit of the Varshamov bound results in an entropy function that shows that good codes exist. From the Hamming bound, we have obtained the channel capacity.

We have presented linear block codes using trellis diagrams, and have defined a minimal trellis and several methods for its construction. We have proved that the syndrome trellis is minimal and that the trellis-oriented generator matrix also produces a minimal trellis. Further, we have shown that a minimum trellis is composed of a minimal number of vertices $|\mathcal{V}|$ as well as a minimal number of edges $|\mathcal{E}|$ that produce a minimal cycle rank $Z(T) = |\mathcal{E}| - |\mathcal{V}| + 1$. The decoding of a block code can be achieved using a minimal trellis (see chapter 7).

We again stress that a minimal trellis can be constructed for a fixed order of coordinates of a code. If the coordinates are permuted then a new minimal trellis can be constructed, possibly with a lower degree of complexity. Out of all permutations, there exists one that produces a minimal trellis with a minimal number of vertices. This is termed an optimal trellis. Much research has been put in finding an algorithm for producing optimal trellises. References can be found in [Ksc96] and [EBMS96].

The history of the theory of minimal trellises of block codes is fairly recent. The representation of block codes by means of trellis diagrams was only introduced in 1974. Bahl, Cocke, Jelinek and Raviv [BCJR74], Wolf [Wolf78] and Massey [Mas78] introduced this concept. In 1988, interest in this representation was reawakened due to the work of Forney [For88b] and Muder [Mud88], who produced the definition of a minimum trellis. In 1993, work by Zyablov and Sidorenko [ZS94] and Kot and Leung [KL93] proved that the trellis definition of Bahl et al., Wolf and Massey is minimal for linear codes.

The use of a *trellis-oriented* generator matrix was proposed by Forney [For88b]. Kschischang and Sorokine [KS95] and Sidorenko et al. [SMH96] employed the Shannon product to construct trellises from block codes. The theory of trellis-oriented generator matrices for the construction of minimal trellises was pioneered by Kschischang and Sorokine [KS95] and McEliece [McE96]. McEliece also showed that the minimal trellis of a linear block code has a minimum number of edges $|\mathcal{E}|$. The fact that a minimal trellis produces a minimal cycle rank $|\mathcal{E}| - |\mathcal{V}| + 1$ can be found in [VK96], as well as in [Sid97]. The theory of minimal trellises with respect to group and separable codes has been investigated in [FT93, Ksc96, KS95, Sid97, SMH97] and [VK96]. Further results can be found in [DRS93] and [Sid96].

6.10 Problems

Problem 6.1
Show that the Viterbi decoding algorithm requires $|\mathcal{E}|$ additions and $|\mathcal{E}| - |\mathcal{V}| + 1$ binary comparisons for the code trellis $T = (\mathcal{V}, \mathcal{E})$.

Problem 6.2
Choose a parity check matrix **H** and a generator matrix **G**, and

(a) construct the syndrome trellis;

(b) construct the minimal trellis with the help of the code generator matrix.

Problem 6.3
Find an example of a code where the the number of nodes $|\mathcal{V}_t|$ at depth t of a minimal trellis satisfies

$$|\mathcal{V}_t| < 2^{\min\{k,n-k\}} \quad \forall t.$$

Problem 6.4
Let $T(C)$ be the minimal trellis of a linear code C and let $T\left(C^{\perp}\right) = \left(\mathcal{V}^{\perp}, \mathcal{E}^{\perp}\right)$ be the minimal trellis of the dual code $C^{\perp}$. Show that the following relation is valid:

$$|\mathcal{V}_t| = |\mathcal{V}_t^{\perp}|.$$

Problem 6.5
Consider the binary linear (n,k) code C with 2^{n-k} cosets $C_1 = C, C_2, \dots, C_{2^{n-k}}$. Propose a generalization of the Viterbi algorithm that finds in each coset $C_1 = C, C_2, \dots, C_{2^{n-k}}$ a word with the smallest Hamming distance from the received word **y** (simultaneous ML decoding of cosets).

Problem 6.6
A code trellis with edges labeled with two symbols is called a *2-sectionalized trellis*. With the help of a generator matrix, design a 2-sectionalized trellis of a $(8,4)$ Reed Muller code.

7

Decoding of block codes

In this chapter, we present decoding methods for *binary* block codes. We shall use additional channel models described in section 7.1. An important property for decoding is the distance among (code) vectors. We introduce measures of distance, or metrics, that allow us to quantify the distance between vectors. The Hamming metric as well as the Euclidean metric have already been defined. In appendix B, we give a formal definition of a metric and describe a few other metrics, including the Lee, Mannheim, Manhattan and combinatorial metrics. The metric to be used for decoding depends on the channel model.

In section 7.2, general fundamental decoding techniques will be presented, namely, *maximum-likelihood* (*ML*) and *maximum a posteriori* (*MAP*) decoding, which are applied to the binary transmission scenario. *Decision reliability* is a central point for decoding. In practice, reliability information is provided by signal estimation algorithms. We illustrate this concept with examples. Further, maximum-likelihood decoding *with* and *without* reliability information is introduced.

We consider maximum a posteriori decoding only in combination with reliability information. A disadvantage of these decoding techniques is that the decoding complexity increases exponentially with the code dimension. Therefore, in section 7.2.3, the so-called Evseev's theorem is formulated, which shows that maximum-likelihood decoding performance can be approached with a much smaller complexity. In section 7.2.4, we describe how redundancy for decoding has to be viewed from a system perspective. We define the code gain as a function of the signal energy and the information transmission rate.

Three sections are devoted to specific decoding algorithms. In section 7.3, decoding methods are described that do not use reliability information. These methods are known as *hard-decision* decoding. Methods that use reliability information are known as *soft-decision* decoding, and are described in section 7.4. In section 7.5, decoding problems including soft-decision decoding are formulated in terms of classical optimization problems, which can be solved using algorithms such as the branch-and-bound algorithm.

The Berlekamp–Massey or Euclidean algorithm can be used to solve the key equation (section 3.2.1), and are classified as hard-decision algebraic decoders. These algorithms can decode only up to the *designed* minimum distance for QR and BCH codes. We describe other algorithms that provide decoding up to half the *true* minimum distance. Also, decoding complexity and decoding speed are critical factors that may favour decoding techniques other than algebraic decoding. Practically, the calculation of the syndrome defined by a DFT requires a significant amount of decoding time. We describe *permutation decoding* and the *error trapping* technique, which is a special case of permutation decoding. A method employing *covering polynomials*, which is a modification of error trapping, is explained, as well as *majority-logic decoding*. A special case of majority-logic decoding is the *decoding algorithm* (DA), which allows us to correct some errors of weight greater than half of the minimum distance.

As far as soft-decision decoding is concerned, we introduce two different basic concepts. The first concept is based on *code symbols*, while the second relates to *codewords*.

Symbolwise decoding is illustrated by

- the *Bahl-Cocke-Jelinek-Raviv (BCJR) algorithm* of MAP decoding based on a trellis diagram;
- the *weighted majority-logic decoding*, or *APP decoding* (APP, *a posteriori probability*), proposed by Massey;
- the *iterative APP decoding* (a generalization of the previous algorithm).

All these methods provide not only hard decisions but also information on the reliability of the output. This is a very important feature for decoding of concatenated codes.

Most codeword decoding algorithms can be described in terms of list decoding. In a trivial case, the list contains all codewords. The decoder finds the most probable transmitted codeword. An example is presented in chapter 1 on page 10 using hard-decision standard array decoding. In order to calculate codeword probability, a compact code description is needed. The minimal trellis representation of a code is such a representation. Maximum-likelihood decoding using the block code trellis representation and the *Viterbi algorithm* is presented in section 7.4.2. The most successful non-exhaustive decoding techniques are based on lists that contain a reduced subset of all codewords. The list should contain the ML codeword with a high degree of probability. The crucial point lies in the selection criteria used for creating the list. We describe different algorithms, such as *generalized minimum-distance* decoding (GMD decoding), generalized Wagner decoding, algorithms by Chase and Kaneko, and also algorithms based on the concept of *ordered statistics*, developed by Fossorier and Lin. A special soft-decoding algorithm of Reed-Muller codes is given in section 9.5.

In section 7.5, we focus on decoding methods based on optimization theory, such as the *simplex algorithm*, the *branch-and-bound* and *gradient* methods. These techniques are evaluated in terms of their practical implementation. Numerical comparisons with the list decoding method are given.

7.1 Channel models and metrics

To investigate digital communication systems, we need models that accurately describe the significant components of a system. By using a good system model, one can determine by

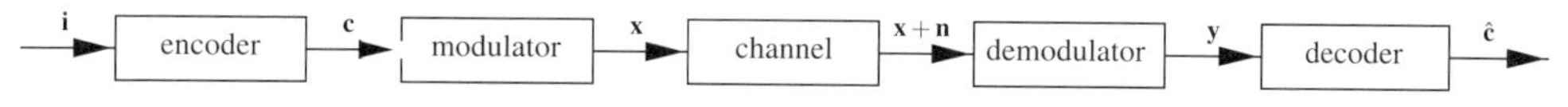

Figure 7.1 Transmission of binary symbols.

simulation optimal system design parameters. A simple block diagram of a digital data transmission channel is presented in figure 7.1. An information vector **i** is mapped by the encoder to a codeword **c**. In this chapter, we consider only block codes of length n, dimension k and minimum distance d. Code symbols are converted by a modulator into the signal vector **x**. Signals are transmitted through the channel. We assume that a channel is a physical transmission medium, for example free space, copper cables, etc. In general, a signal may be randomly distorted during transmission by the noise vector **n**. Distorted signals are received and sent to the demodulator. Here they are detected and output as the vector **y**, which is passed to the decoder. The decoder calculates the estimated codeword $\hat{\mathbf{c}}$, which is mapped back to the calculated information vector $\hat{\mathbf{i}}$. It is very important for simulations to implement adequate channel models describing real scenarios, for example satellite-earth station, ship-home port, mobile users-base station, a telephone line between modems, and so on. Both the modulator and encoder should be designed so that system requirements are satisfied as efficiently as possible. One should mathematically describe models that accurately describe the channel distortion. This allows one to implement different modulation and coding techniques and choose the best one based on system requirements. We later describe channel models that are used for the analysis of decoding techniques.

In *memoryless channels*, present events are not influenced by the past. Such examples include the q-ary symmetric channel model and the additive white Gaussian noise (AWGN) channel. More general channels with time-variant fading are described. Often these channels are used for decoding with reliability information statistics. An example of a channel with memory is the Gilbert–Elliot model, which simulates burst errors.

7.1.1 q-ary symmetric channel

The q-ary symmetric channel is described by an input alphabet consisting of q symbols (e.g. $\alpha_j \in GF(q)$), by an output alphabet consisting of Q symbols y_i, and by channel transition probabilities $P(y_i|\alpha_j)$, $i \in [1,Q]$, $j \in [1,q]$. We consider only the symmetric cases for the condition that $Q \geq q$ (i.e. the output alphabet includes the input alphabet).

'Symmetric' means that the probabilities of the correct transmission of any one symbol are identical for all symbols and that all error transition probabilities are also identical. The channel model is presented in figure 7.2. This channel is a generalization of the BSC (see figure 1.2). We distinguish two cases.

The first case deals with identical input and output alphabets ($q = Q$). The receiver gives the decoder symbols of a code alphabet. This model describes channels without reliability information. This means that the received symbol y_i is either correct or false.

In the second case, the output alphabet is greater than the input alphabet ($q < Q$). For example, the output alphabet often includes the erasure symbol ($\otimes$). Correcting erasures has already been described in section 3.2.6. By extension of the output alphabet, the reliability information of the symbols can be integrated into the channel model (see section 7.2.2). This allows us to significantly improve decoding performance.

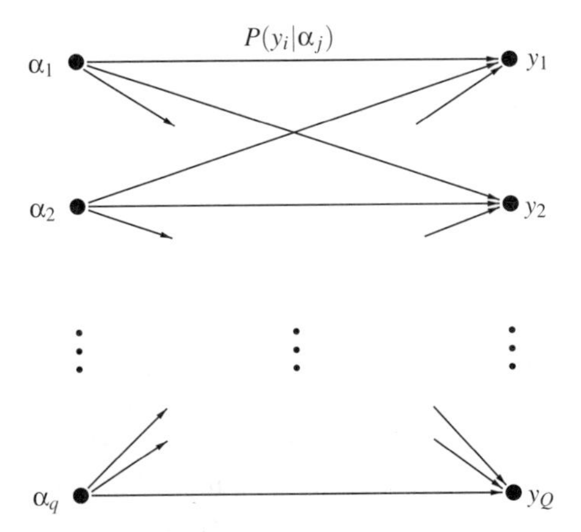

Figure 7.2 q-ary symmetric channel, where $|q| = |Q|$.

7.1.2 Additive white Gaussian noise (AWGN)

The AWGN (*additive white Gaussian noise*) model is used to describe a continuous time-invariant channel. It is by far the most important channel model used to compare the performance of different coding methods. A classical example of the use of the AWGN channel is in *deep space communication* simulations.

The application of a real channel model requires the mapping from discrete signal alphabets (e.g. Galois fields) to real numbers. With the help of the signal space model [Pro, p. 163], different discrete code symbols can be represented as points in a multidimensional real signal space ($\mathbb{R}^n$).

The channel distortion arises from the addition of zero-mean white normally distributed (Gaussian) noise with variance $\sigma^2 = N_0/2$. N_0 is the single-sided noise power spectrum density. There is no correlation among different symbols, which allows a complete description of the symbol probability distribution to be calculated.

For the case of an additive white Gaussian noise channel, the transmitted vector $\mathbf{x}$ is distorted by the random vector $\mathbf{n}$. The received signal is

$$y_i = x_i + n_i, \quad i \in [1, n] .$$

The probability distribution function of the random variable n_i is depicted in figure 7.3 for a fixed signal-to-noise ratio.

E_s is the average energy per received symbol and N_0 is the single-sided noise power spectrum density of the random process $\mathbf{n}$. The ratio E_s/N_0 is defined as the received signal-to-noise ratio. The noise has a Gaussian distribution with variance $\sigma^2 = N_0/2E_s$ and mean 0.

The AWGN channel is memoryless, and is completely described by the transition probability distribution density

$$p(\mathbf{y}|\mathbf{x}) = \prod_{l=1}^{n} p(y_l|x_l), \quad p(y_l|x_l) = \frac{1}{\sqrt{2\pi\sigma^2}} \exp\left(-\frac{(y_l \pm \sqrt{E_s}x_l)^2}{2\sigma^2}\right) . \tag{7.1}$$

The expected value of $\pm\sqrt{E_s}x_l$ for the signal component is usually normalized to 1.

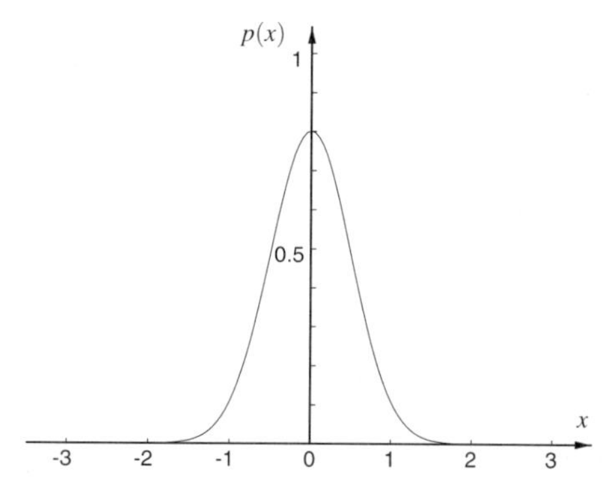

Figure 7.3 Gaussian probability distribution.

7.1.3 Time-variant channels

Channels that have a bit error probability or signal-to-noise ratio that are time-dependent are called time-variant channels. Mobile radio channels have this property. A precise model is very difficult to realize, and depends on many different parameters. Models used for channel coding employ simplified models that simulate only the most important channel parameters. These characteristics include the distribution of the receiver amplitude, and the time-dependent correlation behaviour of the channel, or, in other words, the channel memory. The amplitude distribution of the correlation is characterized by means of the model. This may be a continuous representation, as in the case of the Rayleigh channel, or a discrete model, as in the case of the Gilbert–Elliot model.

Rayleigh channel

The Rayleigh channel is an AWGN channel with time-dependent signal amplitude

$$y_i = a_i x_i + n_i \, .$$

In addition to the noise distortion n_i, there is a multiplicative distortion a_i. The Rayleigh probability distribution is

$$f_a(a) = 2a\,e^{-a^2}, \quad a > 0 \, .$$

From this, we can write the channel probability distribution as

$$p(\mathbf{y} \mid a, \mathbf{x}) = \frac{1}{\sqrt{\pi N_0}} \exp\left(-\frac{(\mathbf{y} - a\mathbf{x})^2}{N_0} \right) , \tag{7.2}$$

where the transmission probability curve depends on two parameters: the transmitted signal $\mathbf{x}$ and the channel amplitude a. Due to the time-variance of the Rayleigh channel, it is characterized by the *average* signal-to-noise ratio ($\overline{E_s}/N_0$ or $\overline{E_b}/N_0$), where $x_i = \pm 1$ is assumed. The average energy $E\{a^2\}$ for a Rayleigh process is normalized to one, and results are calculated as functions of $\overline{E_s}/N_0$ or $\overline{E_b}/N_0$, which is similar to the definition presented for the AWGN channel.

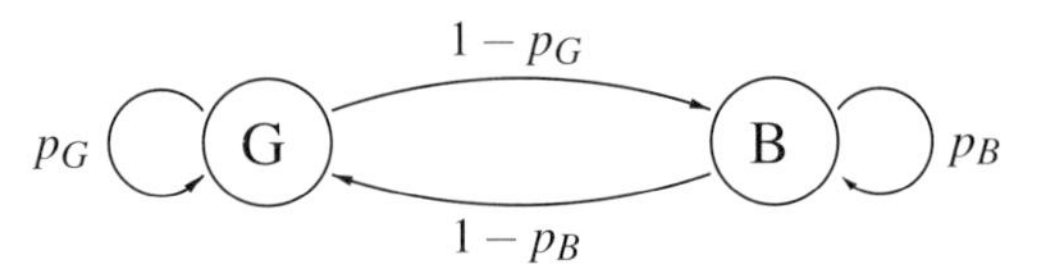

Figure 7.4 Gilbert–Elliot model.

Channels with a variable amplitude and an averaged signal-to-noise ratio, such as the Rayleigh chanel, have a poorer bit error performance than channels with a constant amplitude. The reason for this is that the bit error performance of time-variant channels is dominated by the poor sections of the received signal.

The Rayleigh channel model is often used to describe mobile radio channels. Because the variable a_i is a continuous function of time, the amplitudes of successive transmitted symbols are correlated. This correlation is generally approximated by the Jakes energy spectrum distribution [Jak]. We wish to make the coding method independent of the mobile user's velocity as well as the time-dependent channel properties (determined by the width of the Jakes spectrum). To do this, the transmitted symbols are reordered before transmission and restored after they are received. This process is called interleaving, and results in the recovered codeword's symbols $\hat{\mathbf{c}}_j$ being uncorrelated. Under this assumption, simulations can be carried out using a Rayleigh process having *uncorrelated* values a_i. This type of channel is called an *optimal interleaved Rayleigh channel* (see section 8.4.4).

The decoder must determine whether or not the channel amplitude a_i at the receiver is known. From knowledge (or estimation) of the *channel state information* (CSI), the decoder can use this information to improve the channel decoding results.

Gilbert–Elliot model

The Gilbert–Elliot model describes burst errors . The channel occupies one of two states representing good (G) and bad (B) channel conditions, as depicted in figure 7.4. The good state (G) represents a state with a relatively low bit error probability p_g. The bad channel (B) represents a state with a relatively high bit error probability p_b, which generates bursts of many successive errors. The channel remains in the good state with a probability of p_G and changes to the bad state with a probability of $1-p_G$ (figure 7.4). We define the probability that the channel is found in a good state as p_{good}, and that it is found in a bad state as p_{bad}. This gives

$$p_{good} = p_{good}p_G + p_{bad}(1-p_B) = \frac{1-p_B}{2-p_G-p_B}.$$

For the mean bit error rate of the channel, we obtain

$$p_{bit} = p_{good}p_g + (1-p_{good})p_b = \frac{p_g(1-p_B)+p_b(1-p_G)}{2-p_G-p_B}.$$

It is known that channels with memory have a larger channel capacity [Gil60] than memoryless channels with the same coded bit error probability. Therefore there exists a code that can achieve an equivalent coded bit error performance with less redundancy than an equivalent memoryless code (see [Bre97]).

7.1.4 Hamming and Euclidean metrics

An important criterion for decoding is the selected *metric*. A metric is a general measure for the distance between elements of a set. For decoding, this set is defined by the received signal. The decoding metric will be used in reference to the distance between a codeword and any other vector from the alphabet of received signals. The formal definition of a metric is given in appendix B. In this section, we shall only introduce metrics that are relevant for the decoding of binary block codes, namely the Hamming metric and the Euclidean metric. Other metrics include the Lee, Manhattan and Mannheim metrics, the combinatorial metric, and translatorial metrics such as the cyclic combining metric. These metrics and their applications are described in appendix B.

The **Hamming metric** has already been introduced in chapter 1. Here we give a formal definition. It is especially applicable for the binary signal case, where the symbols vectors from $\mathbb{F}_2^n$ are made up of elements from $GF(2)$. For the non-binary case (i.e. elements from $GF(2^m)$), the Hamming metric can be used to distinguish between an *error* and a *non-error*, even when only one of the m bits is false.

Take x and y to be two symbols from the alphabet $GF(a)$. The Hamming metric is defined as follows:

$$d_H(x,y) = \begin{cases} 0, & x = y\,, \\ 1, & x \neq y\,. \end{cases}$$

The Hamming distance between two vectors $\mathbf{x}$ and $\mathbf{y}$ is the number of different coordinates between $\mathbf{x}$ and $\mathbf{y}$:

$$d_H(\mathbf{x},\mathbf{y}) = \sum_{j=1}^{n} d_H(x_j,y_j) = w_H(\mathbf{x}-\mathbf{y})\,. \tag{7.3}$$

The Hamming weight or the Hamming norm of a vector $\mathbf{x}$ is given by the number of non-zero coordinates in $\mathbf{x}$:

$$w_H(\mathbf{x}) = \sum_{j=1}^{n} d_H(x_j,0) = d_H(\mathbf{x},\mathbf{0})\,.$$

The **Euclidean metric** is defined for $x,y \in \mathbb{R}$ by

$$d_E(x,y) = \sqrt{(x-y)^2}\,.$$

For the n-dimensional case, the Euclidean distance is given by

$$d_E(\mathbf{x},\mathbf{y}) = \sqrt{(x_1-y_1)^2 + \ldots + (x_n-y_n)^2}\,. \tag{7.4}$$

The Euclidean metric is used when each symbol of a codeword can be represented as a signal. For the norm $||\mathbf{x}||$ of $\mathbf{x}$, we get

$$w_E(\mathbf{x}) = ||\mathbf{x}|| = d_E(\mathbf{x},\mathbf{0})\,.$$

Note One observes the following characteristic of the Euclidean metric:

$$d_E(\mathbf{x},\mathbf{y}) = \sqrt{(x_1-y_1)^2+\ldots+(x_n-y_n)^2} \neq \sqrt{(x_1-y_1)^2}+\ldots+\sqrt{(x_n-y_n)^2}\,.$$

This property can cause problems if a two-dimensional space is constructed from the concatenation of two one-dimensional spaces. In this case, the quadratic Euclidean distance is used, which avoids this problem (see chapter 10). However, the quadratic Euclidean distance is not a true metric, because the triangle inequality is not satisfied (see appendix B).

As has already been mentioned, further metrics such as the Lee, Mannheim, Manhattan and combinatorial metrics, etc. are presented in appendix B.

7.2 Decoding principles, reliability, complexity and coding gain

7.2.1 Decoding principles

What is the meaning of optimal decoding?

This question does not have a general answer. The solution depends heavily on the transmission system under consideration. Nevertheless, one can differentiate between two fundamental principles:

Maximum-likelihood decoding (ML) is based on the concept of optimal codeword decoding. The codeword error probability is given by (see section 1.4):

$$P_{Block} = \sum_{\mathbf{y}} P(\hat{\mathbf{x}} \neq \mathbf{x}\,|\,\mathbf{y})\, P(\mathbf{y})\,. \tag{7.5}$$

Because $P(\mathbf{y})$ is independent of the decoding, the minimization of P_{Block} is equivalent to the minimization of $P(\hat{\mathbf{x}} \neq \mathbf{x}\,|\,\mathbf{y})$ or the maximization of $P(\hat{\mathbf{x}} = \mathbf{x}\,|\,\mathbf{y})$. Therefore the ML codeword is obtained by

$$\mathbf{x}_{opt} = \arg\left(\max_{\mathbf{x}\in C} P(\mathbf{x}\,|\,\mathbf{y})\right)\,, \tag{7.6}$$

where $\arg(\max_{\kappa} f(\kappa))$ means the argument κ of an arbitrary function $f(\kappa)$. An ML decoder determines the probability of the transmitted codeword.[1] According to equation 7.5, this decoding principle is carried out by finding the *minimal codeword error probability*.

[1] In general, we have $P(\mathbf{x}\,|\,\mathbf{y}) = P(\mathbf{y}\,|\,\mathbf{x})P(\mathbf{x})/P(\mathbf{y})$ according to Bayes' rule, where $P(\mathbf{x})$ is the a priori probability. If $P(\mathbf{x})$ is assumed to be equal for all possible values of $\mathbf{x}$, the decision is called ML. If knowledge or an estimation of $P(\mathbf{x})$ is used for decoding, the technique is called MAP. $P(\mathbf{y}\,|\,\mathbf{x})$ is given by the channel. Throughout this book, we shall use ML decoding for sequence estimation and MAP for symbol estimation. Consequently, we shall call MAP sequence estimation ML with the use of a priori probabilities, and ML symbol estimation MAP with equal a priori probabilities.

Symbolwise maximum a posteriori decoding (s/s-MAP) is the concept of optimally decoding each code symbol. The error probability of a code symbol in position i is

$$\begin{aligned} P_{Bit} &= P(\hat{x}_i \neq x_i \,|\, \mathbf{y}) = 1 - P(\hat{x}_i = x_i \,|\, \mathbf{y}) \\ &= 1 - \sum_{\mathbf{x} \in C} P(\mathbf{y} \,|\, \mathbf{x}) \, \delta_{\hat{x}_i x_i} \frac{P(\mathbf{x})}{P(\mathbf{y})} , \end{aligned} \tag{7.7}$$

where δ_{kl} is the Kronecker function. Minimization of $P(\hat{x}_i \neq x_i \,|\, \mathbf{y})$ is also equivalent to maximization of $P(\hat{x}_i = x_m \,|\, \mathbf{y})$, or, in other words, to determining the *maximum a posteriori* probability of the coordinate i. The most likely code symbol for coordinate i is obtained by

$$x_{i,opt} = \arg \left(\max_{s \in GF(q)} \{ P(x_i = s \,|\, \mathbf{y}) \} \right) . \tag{7.8}$$

If a received vector $\mathbf{y}$ is given then the most probable code symbol for each position $i, i = 1, \ldots, n$ is determined. According to equation 7.7, this decoding method gives the *minimal code symbol error probability*.

Note The output of a symbolwise MAP decoder may not produce a valid codeword. The MAP output is based on the maximum symbol probability of each coordinate given a received vector $\mathbf{y}$, and not on the maximum probability that all n coordinates produce a codeword. This may therefore give a decoding failure. In this case, the information cannot be recovered. However, in the case of a systematic code, the k information symbols can be recovered. This results directly in a corresponding information vector, and therefore a decoding failure cannot occur. This codeword is not necessarily equal to the soft-decision maximum-likelihood (SDML) codeword.

The interpretation of the above coding methods will be discussed in the next section with respect to the transmission of binary symbols.

7.2.2 Reliability and decoding principles for binary transmission

We should like to clarify the meaning of *reliability information*. We consider the simplest scenario, which is the transmission of a binary symbol over an AWGN channel (see section 7.1.2). The case of multisymbol reliability information can be obtained by a generalization of the binary case.

We assume a binary phase shift keying (BPSK or 2-PSK) method (see e. g. [Pro]), where the code symbols $c_i \in \{0, 1\}$ are mapped to the transmission symbols $x_i \in \{+1, -1\}$ corresponding to the relation

$$x_i = (-1)^{c_i}, \quad i \in [1, n] .$$

The notations $\mathbf{c}$ and $\mathbf{x}$ will be used interchangeably. After transmission over the AWGN channel described by equation 7.1, we obtain the probability distribution depicted in figure 7.5. The distribution for the received symbol y_i is given by

$$p(y_i \,|\, x_i = \pm 1) = \frac{1}{\sqrt{2\pi\sigma^2}} e^{-\frac{(y_i \mp 1)^2}{2\sigma^2}} . \tag{7.9}$$

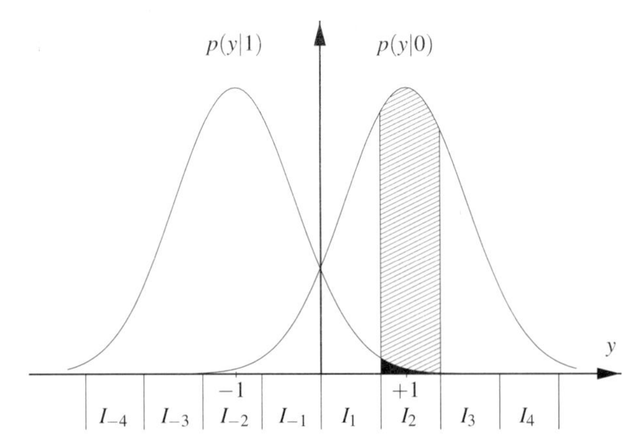

Figure 7.5 Probability distribution function for y_i.

We assume that the y_i axis in figure 7.5 is divided into intervals of width Δy_i. In practical systems, this value is often quantized.

Let us further assume that the received value of y_i lies in the interval I_2 (figure 7.5). The probability that we received this value when a $+1$ is transmitted is

$$\int_{I_2} p(y_i \,|\, x_i = +1) dy_i \;\widehat{=}\; \text{striped region in figure 7.5}\,,$$

and the probability that a -1 was transmitted is

$$\int_{I_2} p(y_i \,|\, x_i = -1) dy \;\widehat{=}\; \text{black region in figure 7.5}\,.$$

If the received value of y_i lies in the interval I_2, it is much more likely that a $+1$ was transmitted than a -1 because the integration of over I_2 gives a greater probability that $p(y_i \,|\, x_i = +1)$ occurred than $p(y_i \,|\, x_i = -1)$. To measure the *reliability* of a decision, we use the probability distribution of y_i from equation 7.9: $\ln\Big(p(y_i \,|\, x_i = +1)/p(y_i \,|\, x_i = -1)\Big) \sim sgn(y_i)|y_i|$. The magnitude of y_i can be interpreted as the reliability information, and the sign of y_i can be considered the hard-decision information. When both pieces of information are incorporated into a decoding method, this method is termed *soft-decision* decoding. This technique can considerably improve the decoding capability. The hard-decision received vector is denoted by $\mathbf{r}$ or y^H.

In addition to the channel transmission probability distribution, we can have other information about the received symbol y_i by how we deduce reliability information. We assume that the transmitted bits come from a memoryless source that produces zeros and ones with a certain probability. The a priori probability $P_a(x)$ is the probability that the symbol x is transmitted. We also know that a sequence of transmitted symbols is limited to the set of codewords of our code. This property was considered in the channel a posteriori probability $P(\mathbf{x} \,|\, \mathbf{y})$. With the help of Bayes' rule, we may write the a posteriori probability as

$$P(\mathbf{x} \,|\, \mathbf{y}) = \frac{p(\mathbf{y} \,|\, \mathbf{x}) P_a(\mathbf{x})}{p(\mathbf{y})}\,. \tag{7.10}$$

Using equation 7.8, we can obtain the MAP probability for position i through[2]

$$\max_{s \in \{+1,-1\}} \{P(x_i = s \,|\, \mathbf{y})\}.$$

The MAP probability gives the optimal reliability information for a certain position i of the received vector. The probability for multilevel modulation is outlined in section 10.2.4.

L value for BSC and AWGN channels

In many cases, it is useful to represent the above probabilities as a log likelihood ratio (here referred to as the L value). The formal calculation of the the logarithmic probability ratio is presented in appendix C based on log likelihood algebra developed by Hagenauer et. al. [HOP96]. In this section, we focus mainly on the log likelihood probability ratio of the BSC as well as for the AWGN channel.

The probability for the (information) symbol x_i before transmission is given through the a priori probability $P_a(x_i)$. Therefore the log likelihood ratio $L_a(x_i)$ of the binary symbols x_i is

$$L_a(x_i) = \ln\left(\frac{P_a(x_i = +1)}{P_a(x_i = -1)}\right). \tag{7.11}$$

The a posteriori log likelihood ratio is obtained from Bayes' rule by

$$\begin{aligned} L(\hat{x}_i) :=& \ln\left(\frac{p(x_i = +1, y_i)}{p(x_i = -1, y_i)}\right) = \ln\left(\frac{P(x_i = +1 \,|\, y_i)p(y_i)}{P(x_i = -1 \,|\, y_i)p(y_i)}\right) \\ =& \ln\left(\frac{P(x_i = +1 \,|\, y_i)}{P(x_i = -1 \,|\, y_i)}\right) = \ln\left(\frac{p(y_i \,|\, x_i = +1)}{p(y_i \,|\, x_i = -1)}\right) + \ln\left(\frac{P_a(x_i = +1)}{P_a(x_i = -1)}\right), \end{aligned}$$

or equivalently in L values

$$L(\hat{x}_i) := L(x_i, y_i) = L(x_i \,|\, y_i) = L(y_i \,|\, x_i) + L_a(x_i).$$

The sign of $L(x_i \,|\, y_i)$ corresponds to the hard decision and the absolute value $|L(x_i \,|\, y_i)|$ corresponds to the reliability of the decision. The value $L(y_i \,|\, x_i)$ is dependent on the basic channel model.

- *Binary symmetric channel (BSC)*

$$\text{for } c_i \in \{0,1\}: \quad L(y_i \,|\, c_i) = \begin{cases} +\ln\dfrac{1-p}{p} & \text{for } y_i = 0 \\ -\ln\dfrac{1-p}{p} & \text{for } y_i = 1\,. \end{cases}$$

  Here p is the probability of a transmission error.

- *AWGN channel and BPSK modulation* From equation 7.1, we obtain

$$L(y_i \,|\, x_i) = \ln\left(\frac{p(y_i \,|\, x_i = +1)}{p(y_i \,|\, x_i = -1)}\right) = \frac{2}{\sigma^2} y_i = L_{ch} y_i$$

  with $\sigma^2 = N_0/2$. N_0 is the single-sided noise power spectrum density. The term L_{ch} is a constant factor that depends only on the signal-to-noise ratio.

[2] If all $\mathbf{x}$ are assumed to be equally probable, we get an ML decision.

- *Time-variant channel* The term L_{ch} is multiplied by the time-dependent amplitude factor a_i [HOP96]:

$$L(y_i \,|\, x_i) = a_i \frac{2}{\sigma^2} y_i = a_i L_{ch} y_i \,.$$

Maximum-likelihood decoding for binary transmission

In section 7.2.1, maximum-likelihood decoding determined the most probable transmitted codeword. Here the principles for the transmission of binary symbols over a BSC as well as for the AWGN channel will be given in detail.

Hard-decision maximum-likelihood decoding (HDML) We received the vector $\mathbf{r} \in GF(2)^n$. For a BSC with the error probability p, the probability that the received vector $\mathbf{r}$ has $t = d_H(\mathbf{r}, \mathbf{c})$ errors is

$$P(\mathbf{r} \,|\, \mathbf{c}) = p^t (1-p)^{n-t} \,.$$

Here we must assume that $P(\mathbf{c})$ is equal to a constant. Because the log function is increasing and strictly monotonic, it does not shift the maximum value of the probability $P(\mathbf{r} \,|\, \mathbf{c})$. Note that $P(\mathbf{c})$ is assumed to be a constant for ML decoding.

$$\ln P(\mathbf{r} \,|\, \mathbf{c}) = -t \ln \frac{1-p}{p} + n \ln(1-p) \,.$$

From this, we find that the maximization of the log likelihood function is equivalent to the minimization of the Hamming distance between the received sequence $\mathbf{r}$ (or $\mathbf{y}^H$) and a code vector $\mathbf{c}$. Based on this result, a ML decoder for a BSC is also called a *minimum-distance* decoder.

Soft-decision maximum-likelihood decoding (SDML) We shall again investigate BPSK transmission over an AWGN channel. We assume that the transmission of all codewords have the same a priori probability. From Bayes' rule and section 7.2.2, maximization of $P(\mathbf{x} \,|\, \mathbf{y})$ based on equation 7.6 is equivalent to maximization of $p(\mathbf{y} \,|\, \mathbf{x})$. Because the AWGN channel is memoryless, we obtain

$$p(\mathbf{y} \,|\, \mathbf{x}) = \prod_{l=1}^{n} p(y_l \,|\, x_l) \,.$$

Using equations 7.1 and 7.6, we obtain the SDML codeword:

$$\begin{aligned} \mathbf{x}_{opt} &= \arg\left(\max_{\mathbf{x} \in C} \left\{ (2\pi\sigma^2)^{-\frac{n}{2}} \exp\left(-\frac{1}{2\sigma^2} \sum_{l=1}^{n} (x_l - y_l)^2 \right) \right\} \right) \\ &= \arg\left(\min_{\mathbf{x} \in C} \left\{ d_E^2(\mathbf{x}, \mathbf{y}) \right\} \right), \end{aligned} \tag{7.12}$$

where $d_E^2(\mathbf{x}, \mathbf{y}) = \sum_{l=1}^{n} |x_l - y_l|^2$ is the *quadratic Euclidean distance* between the codeword $\mathbf{x}$ and the received vector $\mathbf{y}$. Consequently, the SDML codeword has a minimal quadratic Euclidean distance with respect to the received vector. Using this fact, another approach can be taken.

For the quadratic Euclidean distance between a codeword **x** and the received vector **y**, we get

$$d_E^2(\mathbf{x},\mathbf{y}) = \sum_{l=1}^{n} x_l^2 - 2\sum_{l=1}^{n} x_l y_l + \sum_{l=1}^{n} y_l^2 = n + \text{const} - 2\sum_{l=1}^{n} x_l y_l \,. \tag{7.13}$$

From this, minimization of the quadratic Euclidean distance is equal to maximization of the scalar product of **x** and **y**. Another point of view is obtained if the scalar product is written in the following way:

$$\sum_{l=1}^{n} x_l y_l = \sum_{l=1}^{n} |y_l| - 2 \sum_{l:x_l \neq y_l^H}^{n} |y_l| \,. \tag{7.14}$$

Therefore a SDML decoder minimizes the sum of the reliability components $|y_l|$ of the coordinates to be corrected.

Symbolwise MAP decoding for BPSK transmission over an AWGN channel

In section 7.2.1, symbolwise MAP decoding determined the most probable transmitted code symbols. In the case of an AWGN channel, the MAP probability is described in the following way. We partition the code C with respect to position i,

$$C = C_i^{(+1)} \cup C_i^{(-1)} \,, \quad C_i^{(+1)} \cap C_i^{(-1)} = \emptyset \,, \quad C_i^{(\pm 1)} = \{\mathbf{x} \in C \,|\, x_i = \pm 1\} \,,$$

and we obtain the MAP probability for position i in the form of a log likelihood ratio (see appendix C):

$$\begin{aligned} L(\hat{x}_i) = L(x_i, y_i) = \ln\left(\frac{P(x_i = +1 \,|\, \mathbf{y})}{P(x_i = -1 \,|\, \mathbf{y})}\right) &= \ln\left(\frac{\sum\limits_{\mathbf{x}\in C_i^{(+1)}} \prod\limits_{l=1}^{n} p(y_l \,|\, x_l) P_a(x_l)}{\sum\limits_{\mathbf{x}\in C_i^{(-1)}} \prod\limits_{l=1}^{n} p(y_l \,|\, x_l) P_a(x_l)}\right) \\ &= \ln\left(\frac{\sum\limits_{\mathbf{x}\in C_i^{(+1)}} \prod\limits_{l=1}^{n} p(x_l, y_l)}{\sum\limits_{\mathbf{x}\in C_i^{(-1)}} \prod\limits_{l=1}^{n} p(x_l, y_l)}\right) . \end{aligned} \tag{7.15}$$

Furthermore, we observe that $P(x_i = +1 \,|\, y_i) + P(x_i = -1 \,|\, y_i) = 1$. From this, we can write

$$P(x_i = \pm 1 \,|\, y_i) = \frac{e^{-\frac{1}{2}L(x_i,y_i)}}{1 + e^{-L(x_i,y_i)}} e^{\frac{1}{2}L(x_i,y_i)x_i} \,.$$

Equation 7.15 can be divided into three mutually independent terms:

$$L(\hat{x}_i) = \ln \left(\frac{p(y_i|x_i=+1)P_a(x_i=+1) \sum\limits_{\mathbf{x}\in C_i^{(+1)}} \prod\limits_{\substack{l=1\\l\neq i}}^{n} p(x_l,y_l)}{p(y_i|x_i=-1)P_a(x_i=-1) \sum\limits_{\mathbf{x}\in C_i^{(-1)}} \prod\limits_{\substack{l=1\\l\neq i}}^{n} p(x_l,y_l)} \right)$$

$$= \ln \left(\frac{p(y_i|x_i=+1)}{p(y_i|x_i=-1)} \right) + \ln \left(\frac{P_a(x_i=+1)}{P_a(x_i=-1)} \right) + \ln \left(\frac{\sum\limits_{\mathbf{x}\in C_i^{(+1)}} \prod\limits_{\substack{l=1\\l\neq i}}^{n} p(x_l,y_l)}{\sum\limits_{\mathbf{x}\in C_i^{(-1)}} \prod\limits_{\substack{l=1\\l\neq i}}^{n} p(x_l,y_l)} \right). \quad (7.16)$$

The three terms of this equation can be interpreted in the following way:

Channel L value: $$L_{ch} y_i = \ln \left(\frac{p(y_i|x_i=+1)}{p(y_i|x_i=-1)} \right) = \frac{2}{\sigma^2} y_i, \quad (7.17)$$

A priori L value: $$L_a(x_i) = \ln \left(\frac{P_a(x_i=+1)}{P_a(x_i=-1)} \right), \quad (7.18)$$

Extrinsic L value: $$L_{ext,i}(C) = \ln \left(\frac{\sum\limits_{\mathbf{x}\in C_i^{(+1)}} \prod\limits_{\substack{l=1\\l\neq i}}^{n} p(x_l,y_l)}{\sum\limits_{\mathbf{x}\in C_i^{(-1)}} \prod\limits_{\substack{l=1\\l\neq i}}^{n} p(x_l,y_l)} \right) = \ln \left(\frac{\sum\limits_{\mathbf{x}\in C_i^{(+1)}} \prod\limits_{\substack{l=1\\l\neq i}}^{n} e^{\frac{1}{2}L(x_l,y_l)x_l}}{\sum\limits_{\mathbf{x}\in C_i^{(-1)}} \prod\limits_{\substack{l=1\\l\neq i}}^{n} e^{\frac{1}{2}L(x_l,y_l)x_l}} \right). \quad (7.19)$$

The extrinsic L value $L_{ext,i}$ is the part of the reliability information i that depends on the other coordinates.

In 1976, Hartmann and Rudolph [HR76] calculated the MAP probability of a position i by means of dual codes using the discrete Fourier transformation of equation 7.15. Hagenauer et al. [HOP96] interpreted these results as intrinsic and extrinsic information. They concluded that

$$L_{ext,i}(C^{\perp}) = \ln \left(\frac{\sum\limits_{\mathbf{b}\in C^{\perp}} \prod\limits_{\substack{l=1\\l\neq i}}^{n} \tanh\left(\frac{L(x_l,y_l)}{2}\right)^{b_l}}{\sum\limits_{\mathbf{b}\in C^{\perp}} (-1)^{b_i} \prod\limits_{\substack{l=1\\l\neq i}}^{n} \tanh\left(\frac{L(x_l,y_l)}{2}\right)^{b_l}} \right). \quad (7.20)$$

Symbolwise MAP decoding has two interesting inherent possibilities:

- After symbolwise MAP decoding, we obtain not only the decided value of a transmitted symbol, but also the reliability of the decision. This soft output decoding can be employed in concatenated coding schemes. The soft output information is used as input information for the next level of decoding. This information can result in a significant improvement in the decoding results.

- A further possible use of soft output decoding employs symbolwise iterative decoding. An example of which is presented in section 7.4.1.

The decoding complexity of the SDML as well as the symbolwise MAP decoding usually increases exponentially with the code length n. Therefore an important practical question is whether a suboptimal realizable decoding method can be found. What is the corresponding loss in bit error rate performance?

A noteworthy result based on these questions is described in the following subsection. This result of suboptimal decoding is used as motivation for the investigation of different decoding methods for block codes, and also for the motivation for the development of efficient decoding algorithms.

7.2.3 Decoding complexity and Evseev's lemma

We consider the binary transmission over a BSC where $\mathbf{c} \in C$ is transmitted and $\mathbf{r} = \mathbf{c} + \mathbf{e}$ is received. $\mathbf{e}$ is the error vector. ML decoding corresponding to section 7.2.2 is performed. We search for an error $\mathbf{f}$ with the smallest possible weight such that $\mathbf{r} + \mathbf{f} \in C$. One conceivable solution for an ML decoding algorithm is to try all possible error vectors. First we consider $\mathbf{f} = 0$, then all $\binom{n}{1}$ errors of weight 1, then all $\binom{n}{2}$ errors of weight 2, and so on. The question is, up to what error weight t must we try if we wish to obtain a ML or equivalently good decoding result? A conceivable answer was presented by Evseev [Evs83].

We define the set $\mathcal{V}_t$ as the vectors within a sphere around a codeword with radius t:

$$|\mathcal{V}_t| = \sum_{i=0}^{t} \binom{n}{i} .$$

Consider d_{VG}, the Varschamov-Gilbert minimum distance of a linear (n, k, d) code. The basis for the bound is defined by (see section 6.4)

$$d_{VG}: \quad \sum_{i=0}^{d_{VG}-1} \binom{n}{i} \geq \frac{2^n}{2^k} = 2^{n(1-R)} ,$$

where d_{VG} is the smallest integer number that gives a sum larger than $2^{n(1-R)}$. This means that $\mathcal{V}_{d_{VH}}$ is the set of vectors within a sphere centered at $\mathbf{0}$ with radius d_{VG} consisting of the vectors $|\mathcal{V}_{d_{VG}}| = \sum_{i=0}^{d_{VG}} \binom{n}{i}$.

A conceivable suboptimal decoding algorithm $\Psi(C)$ for a binary code C for the received vector $\mathbf{r}$ is defined as follows:

Algorithm 7.1 Algorithm $\Psi(C)$.

Test for all $\mathbf{f} \in \mathcal{V}_{d_{VG}}$: $\quad \mathbf{r} + \mathbf{f} \in C$?

Result: 1. Decoding failure

2. $\mathbf{r} + \mathbf{f} \in C \implies$ Decision: [a] $\hat{\mathbf{c}} = \mathbf{r} + \mathbf{f}$.

[a] For the case when more than one solution exists, the solution with the smallest weight is selected.

This algorithm is the basis for Evseev's theorem.

Theorem 7.1 (Evseev) *Consider a binary code $C(n,k,d)$ that is decoded with the algorithm $\Psi(C)$. The resulting decoding block error probability (decoding failures included) is P_Ψ:*

$$P_\Psi \leq 2P_{ML} \ ,$$

where the decoding block error probability from a ML decoder is P_{ML}.

The practical implication of Evseev's lemma is that ML decoding has, at best, half the number of block errors. In [Evs83], it is shown that, for a large number of coordinates n, the number of vectors that must be tested is $\sim 2^{n(1-R)}$. Furthermore, we know from the Wolf bound that the complexity of the ML decoding of a trellis [Wolf78] can be at most $2^{n\min\{R,(1-R)\}}$. Theorem 7.1 allows the construction of algorithms with a much lower complexity, examples of which are given in [Kro89] and [Dum96]. In the following example, a method from [Evs83] is given that has a reduced decoding complexity of $2^{nR(1-R)}$.

To simplify the description, we assume a code $C(n = lk, k, d)$ where any k successive coordinates determine the codeword (e.g. as in the case of a cyclic code). The received vector $\mathbf{r}$ can be divided into l vectors of length k, giving $\mathbf{r} = (\mathbf{r}_1, \mathbf{r}_2, \ldots, \mathbf{r}_l)$. According to theorem 7.1, we must consider all possible errors with weight d_{VG} in order to obtain a block error rate less than or equal to twice the ML block error rate. The so-called Dirichlet principle asserts that if d_{VG} errors are divided in l subgroups then there is at least one subgroup (i.e. a subvector of length k) having $\leq d_{VG}/l$ errors. This means that in every subvector l with length k, only up to d_{VG}/l errors must be tested. Because k coordinates determine the codeword, the decoding may be carried out in the following way:

$$\mathbf{r}_i + \mathbf{f}_i \implies \hat{\mathbf{c}}, \quad \mathrm{dist}(\mathbf{r}, \hat{\mathbf{c}}) \ , \quad i = 1, \ldots, l \ .$$

In using this technique, we exploit the fact that only relevant error combinations are tested that could give rise to a codeword. This approach greatly reduces the complexity of the decoding.

$$l|\mathcal{V}_{d_{VG}/l}| = \frac{1}{R}|\mathcal{V}_{d_{VG}/l}| \approx 2^{nR(1-R)} \ .$$

We have proved the following theorem:

Theorem 7.2 (Evseev's algorithm) *The decoding complexity of the algorithm $\Psi(C)$ is asymptotically $2^{nR(1-R)}$ for $n \to \infty$. The block error rate (including decoding failures) is at most twice the ML block error rate.*

This theorem and the corresponding algorithm form the basis for a series of algorithms that approach ML decoding performance with a lower complexity. Some of these related algorithms can be found in [Kro89] and [Dum96].

Note The Evseev algorithm was presented to illustrate that it is possible to develop algorithms that can approach ML decoding performance. These algorithms belong to the class of list decoding algorithms, which will be presented in detail later in this chapter.

7.2.4 Coding gain

The use of a code of rate $R = k/n$ has two consequences. First, more symbols must be transmitted (n in comparison to k), and second, more energy must be used (nE instead of kE if E is the symbol energy).

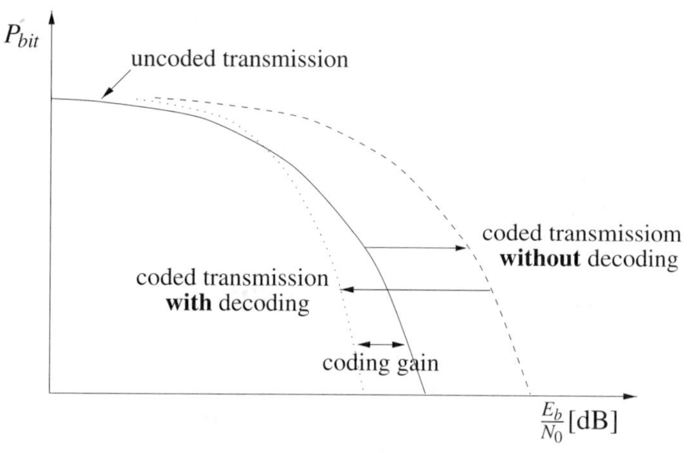

Figure 7.6 Coding gain concept.

In order to compare the performance of different coding schemes, we write the signal energy in the form of energy per information bit E_b (instead of energy per transmitted symbol E_s).

Through the normalization[3] of the energy E_b, one information bit is guaranteed a certain amount of energy, which is independent of the length and rate of the given code. This measure seeks to make a fair comparison of different encoders:

$$kE_b = nE_s \quad \Longrightarrow \quad E_b = E_s/R \, .$$

First, the use of a code increases the symbol error rate at the receiver because less energy E_s is available per symbol. In general, the decoder improves the bit error rate performance as compared with uncoded transmission, or, in other words, the information symbol error rate is smaller at the receiver. This improved error rate can also be achieved by increasing the energy E_b per information symbol in the encoded case. The required increase in E_b is called the *coding gain*. This concept is shown in figure 7.6. The asymptotic hard-decision coding gain G_{hard} is

$$G_{hard} \quad = \quad R(t+1), \tag{7.21}$$

where R is the code rate and $t = d_{min-1}/2$. A similar equation can also be calculated for soft-decision decoding, which gives

$$G_{soft} \quad = \quad Rd_{min} \, . \tag{7.22}$$

Asymptotic coding gain is again described in the context of coded modulation in section 10.2.2.

If we wish to keep the uncoded information data rate constant, then there are two possibilities:

Higher transmission rate In general, this possibility results in a larger signal bandwidth.

[3]It is also important when comparing modulation methods with different alphabets to normalize the energy per information bit.

Larger modulation alphabet If we consider transmitting 2 bits/symbol instead of 1 bit/symbol, we can use a code with rate $R = 1/2$ without an increase in bandwidth, or by decreasing the information transmission rate.

Using a code deteriorates the performance, because a larger modulation alphabet is required (see [Pro]). After decoding, there is generally an improvement or a code gain, which corresponds to the error rate in a similar way as a function of the bit energy (figure 7.6).

7.3 Decoding methods without reliability information

Decoding methods which do not use reliability information are called hard-decision decoding methods. One such example is the algebraic decoding algorithm given in chapter 3. In this section, we examine binary codes transmitted over a *binary symmetric channel* (BSC) as shown in figure 1.2. A transmitted 0 is transformed into a 1 with probability p, and is received correctly with a probability $1-p$. An identical rule holds with respect to a transmitted 1. This model can also correspond to an AWGN channel, where the additional reliability information is not used by the decoder. We show this in figure 7.5, where only the sign, i.e. $y > 0$ or $y < 0$, is used for decoding. The BSC error probability p can be calculated from the zero-mean AWGN conditional channel probability distribution $p(y|x)$ as follows:

$$p = \int_0^{\infty} p(y|-1)\,dy = \int_{-\infty}^{0} p(y|+1)\,dy.$$

Assume the codeword $\mathbf{c}$ is transmitted and the hard-decision vector $\mathbf{r}$ is received. The decoding problem is to determine either the error vector $\mathbf{f}$ or the codeword $\mathbf{c}$ based on the received vector $\mathbf{r}$. The output of the decoder is $\hat{\mathbf{c}}$. For the case $\hat{\mathbf{c}} = \mathbf{c}$, the output is correctly decoded. For the case $\hat{\mathbf{c}} \neq \mathbf{c}$, the received vector is falsely decoded. If no solution is found then a decoding failure has occurred.

7.3.1 Permutation decoding

To correct errors up to half the minimum distance d of a binary code C, the permutation decoding technique can be employed. Here, the set of automorphisms ϕ_j, $j = 1, \dots, J$ (see section 6.3) permutes $t \leq e = \left\lfloor \frac{d-1}{2} \right\rfloor$ error coordinates to the redundancy coordinate positions of the codeword. Let us assume that the code has information and redundancy coordinates that are distinct (see section 1.7).

Let the received vector $\mathbf{r}$ be permuted according to ϕ_i, which relocates all errors to the redundancy section of the codeword. The codeword $\hat{\mathbf{c}}$ is calculated based on the information coordinates of $\phi_i(\mathbf{r})$. By taking the inverse permutation, we obtain

$$\mathrm{wt}(\phi_i^{-1}(\hat{\mathbf{c}}) + \mathbf{r}) \leq e\,.$$

To describe the permutation decoding method as an algorithm, we can assume without loss of generality that the systematic parity check matrix of the binary code C is $\mathbf{H} = (\mathbf{I} \mid \mathbf{A})$ (where $\mathbf{I}$ is the identity matrix). The matrix representation is chosen because the permutation decoding algorithm can be used for decoding non-cyclic codes. We shall prove that if the information portion of a codeword is error-free then the weight of the syndrome is equal to the weight of the error.

Theorem 7.3 (Permutation decoding) *Consider the systematic encoded binary code C and the hard-decision received vector $\mathbf{r} = \mathbf{c} + \mathbf{f}$ with $\mathrm{wt}(\mathbf{f}) \le e = \lfloor \frac{d-1}{2} \rfloor$ and the parity check matrix $\mathbf{H} = (\mathbf{I} \mid \mathbf{A})$. The syndrome vector $\mathbf{s} = \mathbf{H} \cdot \mathbf{r}^T$ satisfies*

$$\mathrm{wt}(\mathbf{s}) \le e \quad \textit{and} \quad \mathrm{wt}(\mathbf{s}) = \mathrm{wt}(\mathbf{f})$$

if and only if the information symbols $r_{n-k+1}, \ldots, r_n$ are error-free. The coordinates of the errors $\mathbf{f}$ are therefore

$$f_1 = s_1, \quad f_2 = s_2, \quad \ldots, \quad f_{n-k} = s_{n-k}\,.$$

Proof

(i) If $\mathrm{wt}(\mathbf{s}) \le e$ then the information symbols are correct: for the case $f_i = 0,\ i = n-k+1, \ldots, n$,

$$\mathbf{s} = (\mathbf{I} \mid \mathbf{A}) \cdot \mathbf{f}^T = (f_1, f_2, \ldots, f_{n-k})^T.$$

Therefore $\mathrm{wt}(\mathbf{f}) = \mathrm{wt}(\mathbf{s}) \le e$.

(ii) If $\mathrm{wt}(\mathbf{s}) > e$ then at least one information symbol is false: consider two vectors

$$\mathbf{f}_1 = f_0, f_1, \ldots, f_{k-1} \quad \text{and} \quad \mathbf{f}_2 = f_{n-k}, f_{n-k+1}, \ldots, f_{n-1},$$

where $\mathbf{f}_2 \ne 0$ occupies the information coordinates. Consider the codeword

$$\mathbf{c}' = \mathbf{f}_2\mathbf{G} = \mathbf{f}_2[\mathbf{A}|\ \mathbf{I}] = [\mathbf{f}_2\mathbf{A}|\ \mathbf{f}_2\mathbf{I}] \quad \text{and} \quad \mathrm{wt}(\mathbf{c}') = \mathrm{wt}(\mathbf{f}_2\mathbf{A}) + \mathrm{wt}(\mathbf{f}_2) \ge 2t+1,$$

due to the minimum distance of the code. We next calculate the syndrome of the vector $\mathbf{f} = \mathbf{f}_1 + \mathbf{f}_2$:

$$\mathbf{s} = \mathbf{H}(\mathbf{f}_1 + \mathbf{f}_2)^T = (\mathbf{f}_1 + \mathbf{A}^T\mathbf{f}_2^T) \quad \text{and} \quad \mathrm{wt}(\mathbf{s}) = \mathrm{wt}(\mathbf{f}_1) + \mathrm{wt}(\mathbf{A}^T\mathbf{f}_2^T).$$

Using the above equations and subtracting the appropriate terms, we may write

$$\mathrm{wt}(\mathbf{s}) \ge \mathrm{wt}(\mathbf{A}^T\mathbf{f}_2^T) - \mathrm{wt}(\mathbf{f}_1) \ge 2t+1-(\mathrm{wt}(\mathbf{f}_1)+\mathrm{wt}(\mathbf{f}_2)) \ge t+1,$$

because, from the theorem, $\mathrm{wt}(\mathbf{f}_1) + \mathrm{wt}(\mathbf{f}_2) = t$. From this, we conclude that $\mathrm{wt}(\mathbf{s}) \ge \mathrm{wt}(\mathbf{A}^T\mathbf{f}_2^T) - \mathrm{wt}(\mathbf{f}_1) \ge 2e+1-\mathrm{wt}(\mathbf{f}_2) - \mathrm{wt}(f_1) \ge e+1$. □

Algorithm 7.2 describes a possible realization of a permutation decoder that is applied to the decoding of a Hamming code in the following example:

Example 7.1 (Permutation decoding) Let C be a Hamming code with length $n = 7$. The codewords have the form

$$\underbrace{c_0c_1c_2}_{\text{Redundancy}} \ \Big|\ \underbrace{c_3c_4c_5c_6}_{\text{Information}}\,.$$

The permutation set $\phi_i(\mathbf{c})$ is

$$\phi_1(\mathbf{c}) = \mathbf{c}\,, \quad \phi_2(\mathbf{c}) = x^3\mathbf{c}\,, \quad \phi_3(\mathbf{c}) = x^6\mathbf{c}\,.$$

or, explicitly,

	Redundancy	Information
$\phi_1(\mathbf{c})$:	$c_0c_1c_2$	$c_3c_4c_5c_6$
$\phi_2(\mathbf{c})$:	$c_4c_5c_6$	$c_0c_1c_2c_3$
$\phi_3(\mathbf{c})$:	$c_1c_2c_3$	$c_4c_5c_6c_0$

It is apparent that every coordinate can be relocated by one of the three permutations to one of the redundancy coordinates. We can correct one error. ◇

Algorithm 7.2 Permutation decoding.

Consider a binary code C with minimum distance d, parity check matrix $\mathbf{H} = (\mathbf{I} \mid \mathbf{A}^T)$ and a set of automorphisms $\phi_j,\ j = 1,\dots,J$, that permute any $t \leq e = \left\lfloor \frac{d-1}{2} \right\rfloor$ coordinates onto the redundancy part of the code C. Let the received vector be $\mathbf{r} = \mathbf{c} + \mathbf{f}$.

Step 1: $\mathbf{x} = \mathbf{r},\ i = 1$

Step 2: If $\mathrm{wt}(\mathbf{H} \cdot (\phi_i(\mathbf{x}))^T) \leq e$ then

$$\mathbf{s} = \mathbf{H} \cdot (\phi_i(\mathbf{x}))^T$$

goto step 4.

Step 3: $i := i + 1,\ (i > J$: failure, stop$)$ step 2.

Step 4: decode $\mathbf{r}$ as

$$\phi_i^{-1}(\phi_i(\mathbf{x}) - (s_1, s_2, \dots, s_{n-k}, 0, 0, \dots, 0)) \ .$$

Error trapping, covering polynomials: A permutation decoding method that considers only cyclic permutations is called *error trapping* . It is clear that this method is only possible for cyclic codes, and the errors must occur such that k successive error-free coordinates exist. Therefore only errors of weight $t \leq \left\lfloor \frac{n-1}{k} \right\rfloor$ can be corrected if the t errors can occur in any pattern. This coding method is favourable for the correction of burst errors (see sections 7.1.3 and 6.6).

With the *covering polynomial* approach, one can in general correct more errors than by *error trapping*. Before we proceed, we define the term 'covering polynomial'.

Definition 7.4 (Covering polynomial) *The polynomial $a(x)$ covers $b(x)$ if*

$$a(x) = \sum_{i \in \mathcal{A}} \alpha_i x^i, \qquad b(x) = \sum_{i \in \mathcal{B}} \beta_i x^i,$$

where the set $\mathcal{A} \geq \mathcal{B}$, $\alpha_i \neq 0$ and $\beta_i \neq 0$.

Covering polynomial error correction decoding is modified from the *error trapping* technique in the following way. Consider a systematic encoded binary cyclic code C, with parity check and generator polynomials $h(x)$ and $g(x)$ respectively $(h(x)g(x) = x^n - 1)$. During transmission, the error polynomial $f(x)$ is added to the codeword to give the hard-decision received polynomial $r(x) = c(x) + f(x)$.

In order to correct $t \leq e$ errors, we select J covering polynomials $Q_0(x) = 0,\ Q_1(x),\ \dots,\ Q_J(x)$ with the following property. There exists some cyclic shift x^i of the error polynomial $f(x)$ such that errors occurring in the information coordinates are covered by $Q_j(x), j \in J$. We define the syndrome

$$\begin{aligned} s^i(x) &= x^i s(x) \\ &= h(x)x^i(c(x) + f(x)) \mod (x^n - 1) \\ &= h(x)(x^i f(x)) \mod (x^n - 1) \end{aligned}$$

and $q(x) = h(x)Q_i(x) \bmod (x^n - 1)$. The received vector is correctly decoded when for $x^i r(x)$

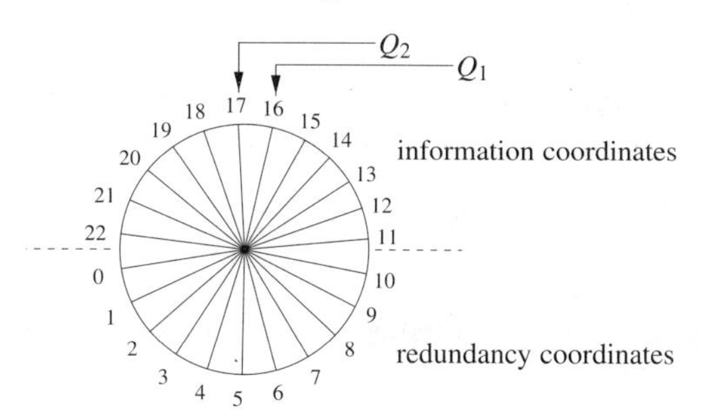

Figure 7.7 Decoding of the Golay code.

we obtain

$$\mathrm{wt}(s^i(x) + q_j(x)) \leq e - \mathrm{wt}(Q_j(x)), \qquad j = 0, \ldots, J\ .$$

The term q_j corresponds to the syndrome components that are generated when there is an error in an information coordinate. At this point we can decode $r(x)$ as

$$\hat{c}(x) = x^{n-i}(x^i r(x) + Q_j(x) + s^i(x) + q_j(x))\ .$$

For the case when $t \leq e$ errors occur, $\hat{c}(x) = c(x)$.

The proof for covering polynomial decoding follows the same logic as for theorem 7.3.

Example 7.2 (Covering Polynomials [McWSl]) With the polynomial $Q_0(x) = 0$, $Q_1(x) = x^{16}$, $Q_2(x) = x^{17}$, all errors can be corrected with weight $t \leq 3$ of the Golay code $\mathcal{G}_{23}$. We assume that codeword positions 0-10 and 11-22 correspond to the redundancy and information coordinates respectively. The covering polynomials Q_J are $Q_0(x) = 0, Q_1(x) = x^{16}, Q_2 = x^{17}$. A wheel can be used to depict the cyclic error shift. Here the top section corresponds to the information coordinates of the cyclic shift (figure 7.7). There is at least one cyclic shift x^i of errors with weight $t \leq 3$ such that all errors in the information coordinates are covered by $Q_j(x), j \in Q$ and all remaining errors are in the redundancy coordinates. We can now decode $\hat{c}$ using the covering polynomial solution. ◇

7.3.2 Majority logic decoding

One can differentiate between single-step and multistep majority logic decoding methods.

Single-step majority logic decoding The dual code (see section 6.1) can be used in the decoding of block codes. Consider a binary code $\mathcal{C}$ and the corresponding dual code $\mathcal{C}^\perp$. According to definition 1.4, the scalar product of $\mathbf{c} \in \mathcal{C}, \mathbf{b} \in \mathcal{C}^\perp$ is

$$\langle \mathbf{c}, \mathbf{b} \rangle = \sum_{i=1}^{n} c_i b_i = 0 \mod 2\ . \tag{7.23}$$

We define a vector $\mathbf{b} \in \mathcal{C}^\perp$ as a parity check vector, and the scalar product of a received vector $\mathbf{r}$ with $\mathbf{b}$ gives the parity check sum $s(\mathbf{b})$. We construct a set $\mathcal{M}_j$ made up of codewords from $\mathcal{C}^\perp$ that allow us to determine if $\mathbf{r} = \mathbf{c} + \mathbf{f}$ is a codeword from $\mathcal{C}$. Under the assumption that the vectors span the dual code (i.e. the matrix built by the vectors as rows has rank $n-k$) then all scalar products must be zero (this corresponds to multiplication by the parity check matrix $\mathbf{H}$). The set of values for the sum of all possible $\mathcal{M}_j$ parity check sums is obtained through the product $s(\mathbf{b}) = \langle \mathbf{r}, \mathbf{b} \rangle$:

$$\sum_{\mathbf{b} \in \mathcal{M}_J} s(\mathbf{b}) \in [0, J] \, .$$

Majority logic decoding is based on the premise that an erroneous coordinate is identified when a majority of the check sums equal one.

Definition 7.5 (Majority logic decision) *Consider a set of parity check vectors $\mathcal{M}_J$ of a linear binary code C, where $\mathcal{M}_J$ spans $C^{\perp}$ and position i is equal to one for every vector. The majority logic decoding of position $i \in [1, n]$ is defined by*

$$\hat{c}_i = \begin{cases} r_i \oplus 1 & \text{if } \sum_{\mathbf{b} \in \mathcal{M}_J} \langle \mathbf{r}, \mathbf{b} \rangle \bmod 2 > \left\lfloor \frac{J}{2} \right\rfloor, \\ r_i & \text{otherwise}, \end{cases} \tag{7.24}$$

where $\oplus$ is addition modulo 2.

The number of correctable errors using majority logic decoding depends on the *combinational* properties of the set $\mathcal{M}_J$. For single-step majority logic decoding, *orthogonal* parity check vectors are especially useful:

Definition 7.6 (Orthogonal parity check vectors) *A set $\mathcal{M}_J$ of parity check vectors $\mathbf{b}_j \in C^{\perp}$, $j = 1, \ldots, J$, is defined as orthogonal to a coordinate $\{i\}$ if position i is equal to 1 for every vector $\mathbf{b}_j$, $j = 1, \ldots, J$. For all other coordinates $l \neq i$, $l = 1, \ldots, n$, may only occur once in $\mathcal{M}_J$.*

The number of errors that can be corrected using single-step majority logic decoding based on orthogonal parity check vectors is given by the following theorem:

Theorem 7.7 (Majority decoding with orthogonal parity check vectors) *Consider J vectors $\mathbf{b}_j$ that are orthogonal to coordinate i and the error vector $\mathbf{f}$, $\mathrm{wt}(\mathbf{f}) \leq \left\lfloor \frac{J}{2} \right\rfloor$. The value of the error coordinate f_i is equal to the majority of parity check sums $\langle \mathbf{b}_j, \mathbf{c} + \mathbf{f} \rangle$, and $f_i = 0$ for an equal number of sums. In the case of a tie, the position i is considered to be error-free, i.e. $f_i = 0$.*

Proof If the coordinate i is correct then $f_i = 0$. This produces at least $\left\lfloor \frac{J}{2} \right\rfloor$ equations (scalar products) with the value 0, because each error coordinate affects at most one equation and there are at most $\left\lfloor \frac{J}{2} \right\rfloor$ errors by definition that can make an equation equal to one. Consider the case when the coordinate i is an error, i.e. $f_i = 1$. This coordinate affects all J equations, and the remaining errors can affect at most $\left\lfloor \frac{J}{2} \right\rfloor - 1$ equations. □

It is possible to correct a larger number of errors if more parity check vectors exist. However, the number of parity check vectors is bounded as follows:

Theorem 7.8 (Number of orthogonal parity check vectors) *Let $C^{\perp}$ be a code with codewords $\mathbf{b}_j \in C^{\perp}$ having minimum distance $d^{\perp}$. There are at most $\left\lfloor \frac{n-1}{d^{\perp}-1} \right\rfloor$ vectors $\mathbf{b}_j$ that satisfy definition 7.6.*

Proof There are $n-1$ remaining coordinate positions to choose from. For each vector, we must select $d^{\perp}-1$ different coordinates. This is only possible in $\left\lfloor \frac{n-1}{d^{\perp}-1} \right\rfloor$ different cases. □

For cyclic codes, we only require one set of orthogonal parity check vectors. The other coordinates can be checked by cyclicly shifting the received vector. This decoding method is often called *Meggit decoding*.

Multistep majority logic decoding The technique of majority decoding can be extended. We define a set of vectors that are orthogonal to a set of coordinates. Analogously to the previously technique, one can decide if the sum of a coordinate set is error-free. If an error is indicated then the parity check vectors can be subdivided into different parity check sets that correspond to different coordinate subsets. This process can be continued until the errors are found.

Example 7.3 (Majority decoding) Consider the Hamming code with length $n=7$, dimension $k=4$ and minimum distance $d=3$. The dual code $C^{\perp}$ has minimum distance $d^{\perp}=4$ and dimension $k^{\perp}=n-k=3$. According to theorem 7.8, there are at most $\left\lfloor \frac{7-1}{4-1} \right\rfloor = 2$ parity check polynomials that are orthogonal to one coordinate. The parity check matrix $\mathbf{H}$ is

$$\mathbf{H} = \begin{pmatrix} 1 & 1 & 1 & 0 & 1 & 0 & 0 \\ 0 & 1 & 1 & 1 & 0 & 1 & 0 \\ 0 & 0 & 1 & 1 & 1 & 0 & 1 \end{pmatrix} \widehat{=} \begin{pmatrix} \mathbf{h}_1 \\ \mathbf{h}_2 \\ \mathbf{h}_3 \end{pmatrix}.$$

Based on linear combinations of the vectors $\mathbf{h}_i$, $i \in [1,3]$, all codewords in $C^{\perp}$ (excluding $\mathbf{0}$) can be constructed:

$$\mathbf{h}_1 = 1110100, \quad \mathbf{h}_2 = 0111010, \quad \mathbf{h}_3 = 0011101, \quad \mathbf{h}_4 = 1001110,$$
$$\mathbf{h}_5 = 0100111, \quad \mathbf{h}_6 = 1010011, \quad \mathbf{h}_7 = 1101001,$$

where $\mathbf{h}_4 = \mathbf{h}_1 + \mathbf{h}_2$, $\mathbf{h}_5 = \mathbf{h}_2 + \mathbf{h}_3$, $\mathbf{h}_6 = \mathbf{h}_1 + \mathbf{h}_2 + \mathbf{h}_3$ and $\mathbf{h}_7 = \mathbf{h}_1 + \mathbf{h}_3$. We cannot find two parity check vectors that are orthogonal to coordinate zero. Because this code is cyclic, there are no two parity check vectors that are orthogonal to any one coordinate i. We shall, however, carry out a two-step majority decoding:

- $\mathbf{h}_1$ and $\mathbf{h}_7$ are orthogonal to coordinates 0 and 1;
- $\mathbf{h}_1$ and $\mathbf{h}_6$ are orthogonal to coordinates 0 and 2.

Therefore we can decide if the sums of coordinates 0 and 1 and/or coordinates 0 and 2 contain an error. If both sets contain an error then coordinate 0 must be erroneous. If only one error is found then the error corresponds to either coordinate 1 or 2, depending on the set. ◇

7.3.3 DA algorithm

In this section, we describe the *decoding algorithm* (DA) decoding technique, which can be applied to all binary code classes that have been mentioned up to this point. This technique even allows decoding of certain error patterns that are larger than half the minimum distance.

Algorithm 7.3 DA algorithm.

Step 1: $\mathbf{v} = \mathbf{r}$, ($\mathbf{r} = \mathbf{c} + \mathbf{f}$ received).
Calculate $X = \mathrm{WT}(\mathcal{B}, \mathbf{v})$.

Step 2: For the case $X = 0$ then step 6.

Step 3: Calculate $\varepsilon_i = \mathrm{WT}(\mathcal{B}, \mathbf{v} + \mathbf{e}_i)$, $i = 1, 2, \ldots, n$,
$\mathbf{e}_i$: ith unity vector

Step 4: Search $j \in \{1, 2, \ldots, n\}$ with $\varepsilon_j = \min_{i=1,2,\ldots,n}\{\varepsilon_i\}$.

Step 5: $\mathbf{v} = \mathbf{v} + \mathbf{e}_j$, $X = \varepsilon_j$ step 2.

Step 6: Decode $\mathbf{r}$ as $\mathbf{v}$.

Consider a binary code $C(n, k, d)$, and the corresponding dual code $C^{\perp}(n, n-k, d^{\perp})$. We define the set of decoding vectors as

$$\mathcal{B} := \{\mathbf{b} \mid \mathbf{b} \in C^{\perp}, \mathrm{wt}(\mathbf{b}) = d^{\perp}\}.$$

This gives $\langle \mathbf{c}, \mathbf{b} \rangle = 0$, $\mathbf{c} \in C$, $\mathbf{b} \in \mathcal{B} \subseteq C^{\perp}$. Furthermore, we define the syndrome weight of a vector $\mathbf{r} = \mathbf{c} + \mathbf{f}$ by

$$\mathrm{WT}(\mathcal{B}, \mathbf{r}) = 0 + \mathrm{WT}(\mathcal{B}, \mathbf{f}) = \sum_{\mathbf{b} \in \mathcal{B}} \langle \mathbf{b}, \mathbf{f} \rangle.$$

The decoding is based on the assumption that for two different errors $\mathbf{f}_1$, $\mathbf{f}_2$, with $\mathrm{wt}(\mathbf{f}_1) < \mathrm{wt}(\mathbf{f}_2)$,

$$\mathrm{WT}(\mathcal{B}, \mathbf{f}_1) < \mathrm{WT}(\mathcal{B}, \mathbf{f}_2).$$

The corresponding algorithm 7.3 describes a possible decoding implementation that uses this property.

The decoding ability of this algorithm is made plausible as follows:

Define $\mathcal{Y}_l := \{\mathbf{f} \mid \mathrm{wt}(\mathbf{f}) = l\}$ as the set of all errors of weight l. We can sum the result of the scalar products of $\mathbf{f}$ with the vector $\mathbf{b} \in \mathcal{B}$ for all vectors of weight l:

$$\sum_{\mathbf{f} \in \mathcal{Y}_l} \langle \mathbf{f}, \mathbf{b} \rangle = \binom{d^{\perp}}{1}\binom{n-d^{\perp}}{l-1} + \binom{d^{\perp}}{3}\binom{n-d^{\perp}}{l-3} + \ldots + q,$$

$$q = \begin{cases} \binom{d^{\perp}}{l}\binom{n-d^{\perp}}{1}, & l \text{ odd}, \\ \binom{d^{\perp}}{l-1}\binom{n-d^{\perp}}{1}, & l \text{ even}. \end{cases}$$

The average syndrome weight can be calculated for each error with weight l as follows:

$$\beta(n, l, d^{\perp}) = \frac{1}{\binom{n}{l}} \sum_{\mathbf{f}_j \in \mathcal{Y}_l} \langle \mathbf{f}, \mathbf{b} \rangle.$$

Therefore we can calculate the average syndrome weight for $|\mathcal{B}|$ vectors by

$$\mathrm{WD}(\mathcal{B}, l) = |\mathcal{B}| \beta(n, l, d^{\perp}).$$

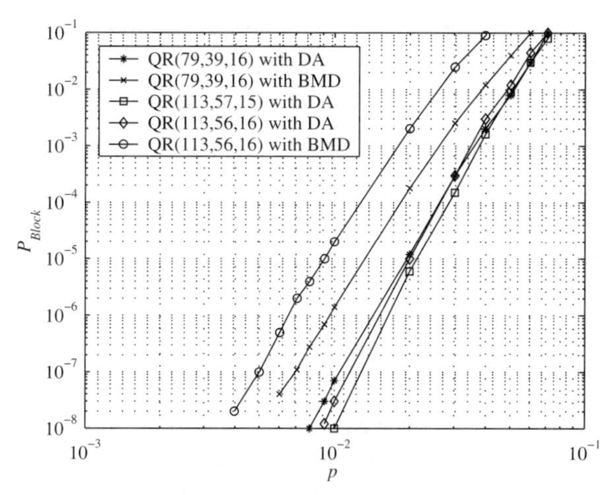

Figure 7.8 Decoding with DA.

WD increases with l for many codes and even for $l > \lfloor \frac{d-1}{2} \rfloor$. This gives the motivation for a decoding method. A necessary and sufficient condition to ensure that the DA can decode all errors up to a certain weight is given by the following theorem:

Theorem 7.9 (DA correction property) *All errors* $\mathbf{f}$ *with weight* $t \leq l$ *can be corrected with the DA if and only if the following condition is satisfied:*

$$\forall_{\mathrm{wt}(\mathbf{f}) \leq l} \; \exists_{i \in \mathrm{supp}(\mathbf{f})} : \mathrm{WT}(\mathcal{B}, \mathbf{f} + \mathbf{e}_i) < \min_{j \notin \mathrm{supp}(\mathbf{f})} \mathrm{WT}(\mathcal{B}, \mathbf{f} + \mathbf{e}_j) \; .$$

Usually, it is only possible to check the conditions of theorem 7.9 for short codes. Justification for the DA is based on the following consideration:

Theorem 7.10 (DA properties) *Consider an automorphism* ϕ *of* $C^{\perp}$. *The DA can decode* $\mathbf{f}$ *and* $\phi(\mathbf{f})$ *if the minimum in step 4 from algorithm 7.3 is unique.*

Proof

$$\mathrm{WT}(\mathcal{B}, \mathbf{f}) = \sum_{\mathbf{b} \in \mathcal{B}} \langle \mathbf{b}, \mathbf{f} \rangle = \sum_{\mathbf{b} \in \mathcal{B}} \langle \phi(\mathbf{b}), \phi(\mathbf{f}) \rangle = \sum_{\mathbf{b}' \in \phi(\mathcal{B})} \langle \mathbf{b}', \phi(\mathbf{f}) \rangle = \mathrm{WT}(\mathcal{B}, \phi(\mathbf{f})) \; .$$

Definition 6.6 ensures that the automorphism ϕ conserves the weight of the vector. Therefore $\phi(\mathcal{B}) = \mathcal{B}$. □

The DA is especially applicable to QR codes because they have a relatively large number of automorphisms. Next we present an example of the implementation of this decoding technique. We consider the QR code with length $n = 113$ and $n = 79$. No practical decoding method is known for these codes. Results are shown in figure 7.8. The decoding block error probability is plotted with respect to the bit error probability p. In comparison, we present the results of a BMD decoding method when applied to the $n = 113$ and $n = 79$ codes. Notice that the algebraic decoding method can only decode up to half the designed minimum distance.

Further investigations and examples of this technique can be found in [BH86] and [Boss87]. The Golay codes $\mathcal{G}_{23}$ and $\mathcal{G}_{24}$ can be decoded using this method up to half the design minimum distance.

Note The algorithm DA can be viewed as a *step-by-step* decoding method [Pra59]. Step-by-step decoding depends on an indication function $I(\mathbf{r})$ that can determine the number of errors in the received vector as long as this is less than or equal to t. In addition, $I(\mathbf{r})$ must be able to recognize $t+1$ errors in $\mathbf{r}$. By means of the indicator function, each position can be checked as to whether it is false, and up to t errors can be corrected in this way. This is carried out by adding a 1 to each coordinate. If the coordinate is false then recalculation of the indicator function shows one fewer error. If a 1 is added to a correct coordinate then the indicator function shows one more error. In [DN92], a step-by-step decoder for special code classes is described.

7.3.4 Hard-decision maximum-likelihood decoding: Viterbi algorithm

As outlined in chapter 6, every block code can be represented by means of a trellis $T = (\mathcal{V}, \mathcal{E})$ and can be decoded using the ML Viterbi algorithm[4] [For73]. The decoding result does not depend on whether a general code trellis (definition 6.14) or a minimal trellis (definition 6.15) is used.

Viterbi algorithm (hard decision) A codeword $\mathbf{c} \in C \subset GF(2)^n$ is transmitted and $\mathbf{r} = \mathbf{c} + \mathbf{f}$ is received. All codewords of this code are represented in a general code trellis T (see section 6.8). The algorithm uses the fact that all possible codewords follow a path through the trellis starting with vertex ϑ_A and ending with vertex ϑ_B. Based on this property, we can compare the received vector to all possible codewords that begin at ϑ_A. At each vertex of level $i = 1, 2, \ldots, n$, we can calculate the distance of the received vector to a codeword path. Assume that we arrive at a vertex with two incoming paths, and each path has a different head section and an identical tail section corresponding to a valid codeword path (see section 6.8.1). We retain only the incoming codeword path (head section) closest to the received vector. Therefore, for the considered vertex, the maximum likely (ML) head path is chosen, which is given the term *survivor*. At every vertex, only one path corresponding to the survivor, must be considered in subsequent vertices for $i = 1, 2, \ldots, n$. We formally describe this principle in the following section.

Every branch in T beginning with vertex $\vartheta \in \mathcal{V}_{i-1}$ and ending with vertex $\vartheta' \in \mathcal{V}_i$ is labeled with the symbol $c(e) \in \{0, 1\}$. The edge e is assigned a branch metric $\Lambda(e) = d_H(c(e), r_i)$ that corresponds to the Hamming distance. Each branch e is described by the 4-tuple $e = (\vartheta \in \mathcal{V}_{i-1}, c(e), \Lambda(e), \vartheta' \in \mathcal{V}_i)$. End nodes and paths are obtained using the functions $\vartheta = i(e)$ and ϑ'. Furthermore, the set $\mathcal{E}_i$ is the set of all edges with a start vertex of depth of $i-1$ and end vertex of depth of i. Therefore

$$\mathcal{E}_i = \{e \mid i(e) = \vartheta \in \mathcal{V}_{i-1},\ c(e),\ \Lambda(e) = d_H(c(e), r_i),\ f(e) = \vartheta' \in \mathcal{V}_i\}.$$

Using this relation, every vertex ϑ' is assigned a vertex metric $\alpha(\vartheta') = \alpha(f(e))$ that is calculated from the sum of $\alpha(i(e))$ and the branch metric $\Lambda(e)$. These metrics allow us to use the Viterbi algorithm (algorithm 7.4) for HDML decoding of the trellis T.

Each vertex is assigned a metric $\alpha(f(e))$ by the Viterbi algorithm that is determined by the maximum number of matching coordinates between the received vector $\mathbf{r}$, and all possible code vectors $\mathbf{c}$ passing up through the particular node. At the last vertex, the survivor is chosen with

[4] The Viterbi algorithm will be again presented in section 8.4 for convolutional codes.

Algorithm 7.4 Hard-decision Viterbi algorithm.

Initialization: $i = 1$, $\alpha(\vartheta_A) = 0$.

Step 1: $\forall e \in \mathcal{E}_i : \quad \alpha(f(e)) = \alpha(i(e)) + \Lambda(e)$.

Step 2: If more than one edge terminates at $\vartheta' \in \mathcal{V}_i$ then choose the path with maximum $\alpha(f(e))$ as the *survivor* and assign the metric to the vertex $f(e)$. (If all metric values of the incoming paths are equal, one is chosen randomly.) $i := i+1$, for the case $i \leq n$ go to step 1, otherwise go to step 3 .

Step 3: The decoded code sequence $\hat{\mathbf{x}}$ is selected from the survivor paths that terminate at the end vertex ϑ_B.

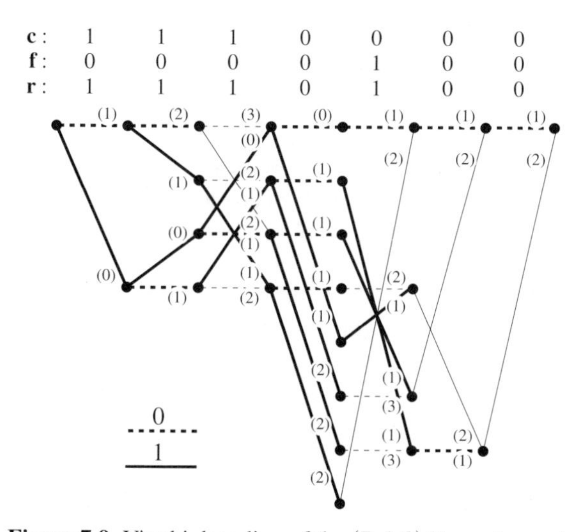

Figure 7.9 Viterbi decoding of the $(7,4,3)$ Hamming code.

the largest number of coinciding coordinates with the received vector $\mathbf{r}$. This is the codeword from C (or one of the codewords with equal distance to $\mathbf{r}$) with the smallest Hamming distance from $\mathbf{r}$ according to the ML decoding algorithm.

Example 7.4 (Viterbi decoding of the $(7,4,3)$ Hamming code) Figure 7.9 shows the minimal trellis for the $(7,4,3)$ Hamming code. The vectors $\mathbf{c}$, $\mathbf{f}$ and $\mathbf{r}$ correspond to the transmitted codeword, the error vector and the received vector respectively. The number (l) at each vertex indicates the metric value Λ_i^l. To calculate the metric value at depth $i+1$, the metric of the survivors from depth i are considered.

Note Because this code is perfect, *hard-decision* BMD decoding is equivalent to ML decoding.

Unfortunately, the usual implementation of the Viterbi algorithm can only be practically implemented for relatively short codes. Codes with a small dimension must be used, otherwise the trellis becomes too complex.

7.4 Decoding methods using reliability information

Two fundamental decoding principles will be considered here. In section 7.2.1, we consider the minimization of the codeword error probability (SDML decoding), and the minimization of the code symbol error probability (MAP decoding). Next we consider reliability information for the case of binary transmitted symbols over an AWGN channel. In this section, we present algorithms that focus on the efficient calculation of the functions (equations 7.12-7.15). In order to achieve an error probability after decoding that is as small as possible, long codewords must be used. However, for optimal decoding, the complexity increases exponentially with codeword length. What techniques can be used to achieve an exponentially decreasing bit error probability without resulting in an unrealistic complexity? This question is at the heart of research into suboptimal decoding algorithms.

We consider the representation of optimal as well as suboptimal decoding algorithms for binary block codes. The decoding algorithms are divided into two classes: those based on code symbol decoding and those based on codeword decoding. In section 7.4.1, we derive three representations for *code symbol decoding* where each element of the received vector is separately decoded. The optimal algorithm is the MAP decoder, which achieves the minimum code symbol error probability. As an example of MAP decoding, the Bahl–Cocke–Jelinek–Raviv (BCJR) algorithm is explained. Weighted majority decoding is a suboptimal symbolwise decoding method, and can be used as an approximation to MAP decoding. Iterative symbolwise decoding or iterative a posteriori probability (APP) decoding can be interpreted as generalizations of weighted majority decoding. In general, iterative decoding almost always converges to a codeword. Therefore iterative APP may be interpreted as an approximation of SDML.

Next we consider *decoding based on codewords* in which the optimization function focuses on the minimization of the quadratic Euclidean distance. Consequently, optimal decoding corresponds to SDML decoding. Such decoding methods can in general be interpreted as *list decoding*, where the output is based on a list of codewords. The construction of the list depends on the selection method of the specific algorithm. For the trivial case, the list is constructed from all codewords of the code, and this corresponds to SDML decoding. We present an example of list decoding in section 7.4.2 based on a modification of the Viterbi algorithm presented in section 7.3.4. More efficient methods consider only a subset of the decoding list. Many algorithms attempt to find the SDML codeword by eliminating codewords from the check list. An optimal implementation of this technique is the Kaneko algorithm (section 7.4.3). Several suboptimal algorithms are presented that seek to eliminate codewords from the list that can only occur with very low probabilities. The class of generalized minimum-distance (GMD) decoding algorithms introduced by Chase and decoding based on *ordered statistics* are examples of suboptimal list decoding methods. In section 7.4.5, we show that equivalent decoding algorithms exist using dual codes. These include the general Wagner algorithm, algorithms based on SDML decoding methods, and decoding based on *ordered statistics*. Figure 7.10 shows the classification of these algorithms.

7.4.1 Symbolwise soft-decision decoding

As shown in figure 7.10, this section will present optimal as well as suboptimal symbolwise decoding algorithms. The MAP decoding algorithm from section 7.2.2 is an example of optimal symbolwise decoding. The calculation of the MAP probability must be carried out for each position i for all codewords of either the code or the dual code. This is a direct evaluation

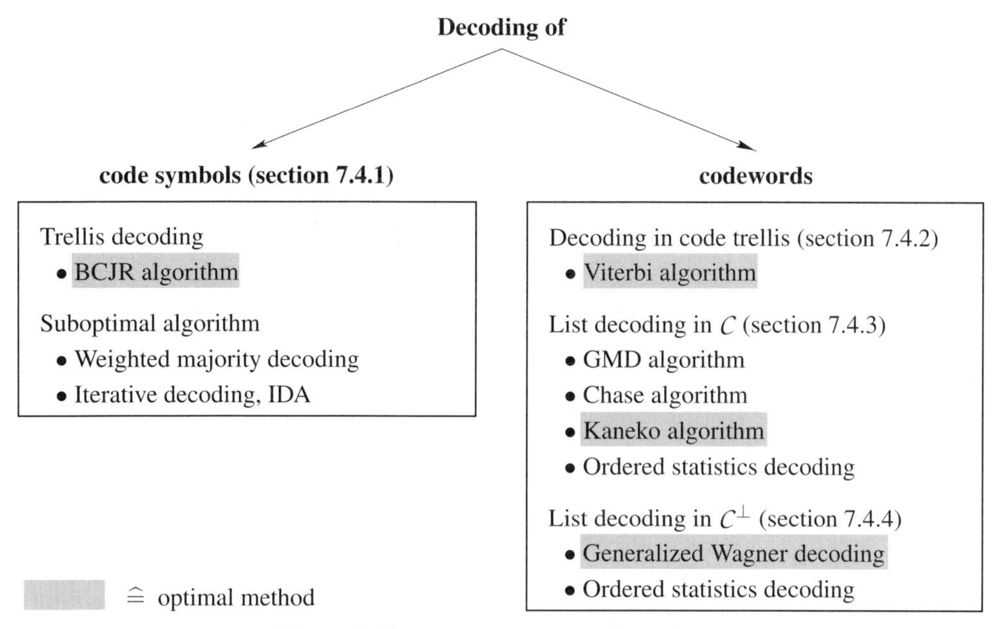

Figure 7.10 Overview of decoding algorithms.

of equation 7.16 with respect to equation 7.20 that includes intrinsic, extrinsic and channel information. Presently, such decoding can only be practically implemented for codes with $\min\{k, n-k\} \lesssim 10$. On the other hand, the trellis representation (section 6.8) states that every code can be represented by a trellis diagram. The minimal trellis corresponds to the most compact trellis, and has the least computational complexity for the evaluation of MAP decoding algorithms. Presently, MAP trellis decoding is possible for codes with $\min\{k, n-k\} \lesssim 16$.

BCJR algorithm

In 1974 Bahl, Cocke, Jelinek and Raviv [BCJR74] proposed an efficient algorithm for the calculation of the MAP probability using a trellis diagram. In order to describe this method, we must first introduce some notation.

We consider a code trellis T (see section 6.8) corresponding to the code $\mathcal{C}$. Each branch in T joins vertex $\vartheta \in \mathcal{V}_{i-1}$ with vertex $\vartheta' \in \mathcal{V}_i$, and is labeled by the symbol $x(e) \in \{+1, -1\}$. Additionally, the branch e is assigned a value based on the metric function $\Lambda(e)$, which will be presented later. Each branch e is defined by a 4-tuple according to section 7.3.4:

$$e = (\vartheta \in \mathcal{V}_{i-1}, x(e), \Lambda(e), \vartheta' \in \mathcal{V}_i) .$$

The start and end vertices with respect to a branch e are obtained from the function

$$\vartheta = i(e), \quad \vartheta' = f(e) .$$

Furthermore, the set of all branches that end at vertex ϑ are labeled $\mathcal{E}_{in}(\vartheta)$. $\mathcal{E}_{out}(\vartheta)$ is the set of all branches that begin at vertex ϑ. The set of all branches at level i having $x(e) = +1$ are labeled $\mathcal{E}_i^{(+1)}$. $\mathcal{E}_i^{(-1)}$ is defined in the same way for $x(e) = -1$.

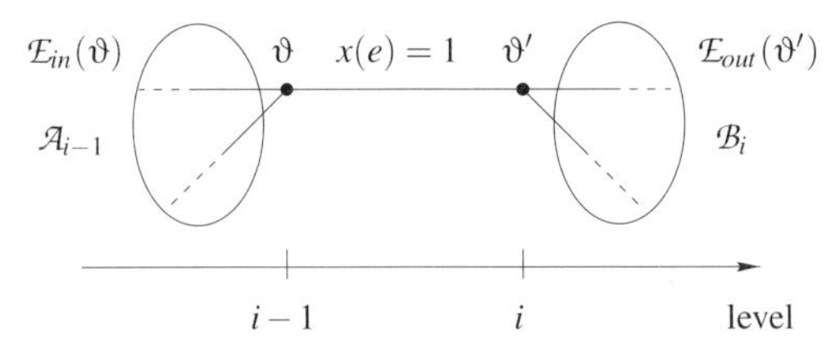

Figure 7.11 Calculation of $L_{ext,i}$.

Similar to equation 7.16, the MAP probability for position i is given in the form of a log likelihood ratio:

$$\begin{aligned} L(\hat{x}_i) &= L_a(x_i) + L_{ch}y_i + L_{ext,i}(C) \\ &= L_a(x_i) + L_{ch}y_i + \ln \left(\frac{\sum\limits_{\mathbf{x} \in C_i^{(+1)}} \prod\limits_{\substack{l=1 \\ l \neq i}}^{n} e^{\frac{1}{2}L(x_l,y_l)x_l}}{\sum\limits_{\mathbf{x} \in C_i^{(-1)}} \prod\limits_{\substack{l=1 \\ l \neq i}}^{n} e^{\frac{1}{2}L(x_l,y_l)x_l}} \right) . \end{aligned}$$

Because the a priori log likelihood value L_a and the channel log likelihood value L_{ch} are given, only the extrinsic log likelihood value $L_{ext,i}$ need be calculated.

To calculate $L_{ext,i}$ from a trellis diagram, we first assume that the trellis diagram for the code C possesses only one branch e with $x(e) = +1$ between the vertices at level $i-1$ and i, as shown in figure 7.11. All codewords with position $i = +1$ can then be written as

$$\mathbf{x} = (\mathbf{x}_{i-1}, +1, \mathbf{x}_{i+1}) \in C_i^{(+1)}, \quad \mathbf{x}_{i-1} \in \mathcal{A}_{i-1}, \quad \mathbf{x}_{i+1} \in \mathcal{B}_i,$$

where $\mathcal{A}_{i-1}$ is the set of heads and $\mathcal{B}_i$ is the set of tails (see section 6.8). For the the above special case, let the number of extrinsic L values be written as

$$\sum_{\mathbf{x} \in C_i^{(+1)}} \prod_{l=1, l \neq i}^{n} e^{\frac{1}{2}L(x_l,y_l)x_l} = \sum_{\mathbf{x}_{i-1} \in \mathcal{A}_{i-1}} \prod_{l=1}^{i-1} e^{\frac{1}{2}L(x_l,y_l)x_l} \sum_{\mathbf{x}_{i+1} \in \mathcal{B}_i} \prod_{l=i+1}^{n} e^{\frac{1}{2}L(x_l,y_l)x_l} = \alpha(i(e))\beta(f(e)) .$$

The vertex metric $\alpha(i(e))$ is assigned to the start vertex of the path e, and $\beta(f(e))$ corresponds to the end vertex.

In general, many paths exist with $x(e) = +1$, which correspond to different nodes at level $i-1$. All of these paths have associated vertex metrics α and β, which must be calculated and summed. An analogous definition holds for the paths with $x(e) = -1$.

Consider $\Lambda(e) = e^{\frac{1}{2}L(x_l,y_l)x_l}$, the branch metric for a path with a vertex of level $l-1$ connected to the vertex with level l. The extrinsic L value of position $i \in [1,n]$ can be determined as follows. The vertex metric can be calculated for $\vartheta \in \mathcal{V}_l, l \in [1, n-1]$ by forward recursion:

$$\alpha(\vartheta) = \sum_{e \in \mathcal{E}_{in}(\vartheta)} \Lambda(e)\alpha(i(e)) . \tag{7.25}$$

$\vartheta' \in \mathcal{V}_l, l \in [n-1, 1]$, can be calculated using reverse recursion:

$$\beta(\vartheta') = \sum_{e \in \mathcal{E}_{out}(\vartheta')} \Lambda(e)\beta(f(e)) , \tag{7.26}$$

with $\alpha(\vartheta_A) = \beta(\vartheta_B) = 1$. The extrinsic L value for $i \in [1,n]$ is given by

$$L_{ext,i}(C) \;=\; \ln\left(\frac{\sum\limits_{e\in\mathcal{E}_i^{(+1)}} \alpha(i(e))\beta(f(e))}{\sum\limits_{e\in\mathcal{E}_i^{(-1)}} \alpha(i(e))\beta(f(e))}\right). \tag{7.27}$$

If the path metric $\Lambda(e) = \tanh(L(x_l, y_l)/2)$ is inserted into equations 7.25 and 7.26, we obtain $L_{ext,i}$ from a trellis of the dual code according to equation 7.20, which gives

$$L_{ext,i}(C^{\perp}) = \ln\left(\frac{\sum\limits_{e\in\mathcal{E}_i^{(+1)}} \alpha(i(e))\beta(f(e)) + \sum\limits_{e\in\mathcal{E}_i^{(-1)}} \alpha(i(e))\beta(f(e))}{\sum\limits_{e\in\mathcal{E}_i^{(+1)}} \alpha(i(e))\beta(f(e)) - \sum\limits_{e\in\mathcal{E}_i^{(-1)}} \alpha(i(e))\beta(f(e))}\right). \tag{7.28}$$

This result can be compared with the BCJR algorithm for convolutional codes, which is defined in section 8.5.1.

Note Symbolwise MAP decoding is a *soft-output* decoding technique; therefore, in contrast to hard-output decision decoding, information is given concerning the reliability of the decoding result of each symbol. This is especially important for the decoding of concatenated codes, where the reliability information of previous decoding levels is used as soft input information for subsequent levels.

Threshold decoding

In 1963, Massey [Mas] extended the single-step majority decoding technique based on orthogonal parity check vectors (see section 7.3.2). This was accomplished using feedback from symbol reliability information provided by *weighted* majority decoding (also called *threshold decoding* and *a posteriori probability (APP) decoding*).

After transmission over an AWGN channel, the received vector is described as $\mathbf{y} = \mathbf{x} + \mathbf{n}$. For an orthogonal parity check vector $\mathbf{b} \in C^{\perp}$ of the binary linear code C, the syndrome $s(\mathbf{b})$ is given by

$$s(\mathbf{b}) \;=\; \langle \mathbf{y}^H, \mathbf{b} \rangle \;=\; \sum_{l=1}^{n} y_l^H b_l \;=\; \sum_{l\in\mathrm{supp}(\mathbf{b})} y_l^H \bmod 2 \;\in\; \{0,1\}\,.$$

If the hard-decision value y_l^H is replaced by $\mathrm{sign}(y_l)$, we obtain the following result:

$$s(\mathbf{b}) \;=\; \prod_{l\in\mathrm{supp}(\mathbf{b})} \mathrm{sign}(y_l) \;\in\; \{+1,-1\}\,.$$

We note that $s(\mathbf{b})$ is equal to $+1$ (-1) if the calculation with $\mathbf{b}$ produces an even (odd) number of -1 values. In order to obtain the reliability information for both of these cases, the corresponding APP probability must be found. We assume that the dual code contains only the zero codeword and the vector $\mathbf{b}$. This gives the MAP probability (equation 7.20) for the above result in the form of a log likelihood ratio:

$$L(s(\mathbf{b})) = \ln \frac{P_{even}}{P_{odd}} = \ln \left(\frac{1 + \prod\limits_{l \in \text{supp}(\mathbf{b})} \tanh\left(\frac{1}{2}L(x_l, y_l)\right)}{1 - \prod\limits_{l \in \text{supp}(\mathbf{b})} \tanh\left(\frac{1}{2}L(x_l, y_l)\right)} \right). \tag{7.29}$$

The value $L(s(\mathbf{b}))$ is a real number, which also contains the hard-decision result, while at the same time giving a measure of the reliability of the decision. The following notation is used, based on the parity check vector $\mathbf{b}_i$:

$$s_i = L^H(s(\mathbf{b}_i)) = \begin{cases} 0, & L(s(\mathbf{b}_i)) \geq 0, \\ 1, & \text{otherwise,} \end{cases} \qquad \text{and} \qquad w_i = |L(s(\mathbf{b}_i))|\,.$$

For a set $\mathcal{M}_J$ of parity check vectors that are orthogonal to position i (definition 7.6), the L value calculation of the received symbols are statistically independent according to equation 7.29. Therefore the 'L value syndrome' according to equation 7.29 can be summed for orthogonal parity check vectors. In section 7.3.2, a *weighted* majority decision can be defined for the set $\mathcal{M}_J$:

$$\hat{c}_i = \begin{cases} r_i, & \sum\limits_{l=1}^{J} w_l s_l \leq \frac{1}{2} \sum\limits_{l=1}^{J} w_l, \\ r_i \oplus 1, & \text{otherwise,} \end{cases} \tag{7.30}$$

where $r_i = y_i^H$ is the hard decision of the received value y_i.

Note If the set $\mathcal{M}_J$ is replaced with the corresponding entire dual code, and the L value is calculated with respect to equations 7.19 and 7.20, we obtain the MAP decision for position i. Using this point of view, it can be seen that the weighted majority decoding for position i based on orthogonal parity check vectors can be an approximation of the corresponding MAP probability. The calculation of this approximation is less complex, because only a very small part of the entire dual code is considered. It is clear that the error probability of this decoding result is worse compared with MAP decoding, because it considers only a subset of the dual code. Both decoding methods are equivalent for the case of parity check codes as well as repetition codes [LBB98].

Symbolwise iterative decoding

Symbolwise iterative decoding is based on algorithms that use reliability information at the input (*soft input*) and produce reliability information at the output (*soft output*). This results in an iterative decoding of the same received vector. In the first iteration, the decoder uses only the channel reliability information to calculate individual received symbols. In subsequent iterations, the new reliability information based on the extrinsic information, i.e. independent of the symbol under consideration, is incorporated by the algorithm. The process continues until the maximum allowed iteration or when a stop criterion is satisfied (see theorem 7.11). For symbolwise iterative decoding, algorithm 7.5 is considered. One observes that the decoding result $\hat{\mathbf{r}}$ need not necessarily be a codeword. Figure 7.12 illustrates the principle of symbolwise iterative decoding.

Algorithm 7.5 General definition of iterative decoding.

Initialization:
- Set $\hat{y}_i^{(0)} = L_{ch}y_i$, $L_{ext,i}^{(0)} = 0$, $i \in [1,n]$, iteration index $j = 0$.
- Define the maximum number of iterations Θ.

Step 1: For the case: $\big((\text{stop criterion 7.11 not satisfied}) \text{ and } (j < \Theta)\big)$

calculate: $L_{ext,i}^{(j)}$, $i = 1, \ldots, n$.

add: $\hat{y}_i^{(j+1)} = \hat{y}_i^{(j)} + L_{ext,i}^{(j)}$, $i = 1, \ldots, n$.

$j := j+1$, step 1.

Step 2: $\hat{\mathbf{r}} = \hat{\mathbf{y}}^{(j),H}$ (hard decision of $\hat{\mathbf{y}}^{(j)}$), stop.

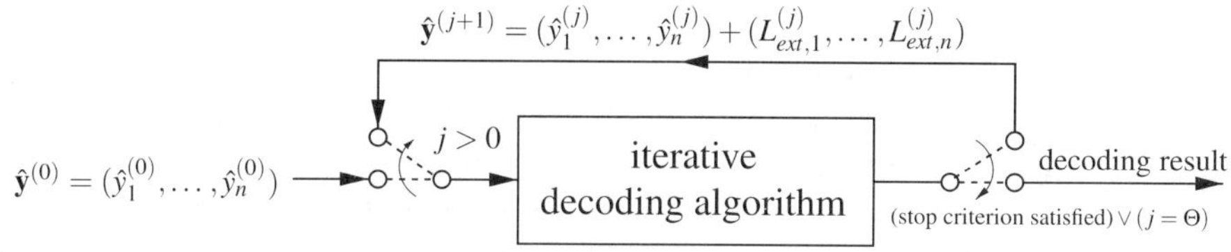

Figure 7.12 Symbolwise iterative decoder.

Let i be an incorrect error coordinate. This coordinate can be corrected by an iteration loop if $\text{sign}(\hat{y}_i^{(0)}) = -\,\text{sign}(L_{ext,i}^{(0)})$, and if $|L_{ext,i}^{(0)}| > |\hat{y}_i^{(0)}|$. The error is corrected if the sign of the incorrect symbol $\hat{y}_i^j$ changes. A similar condition is also applicable for a correct received coordinate i. In this case, the sign of $\hat{y}_i^{(j)}$ remains correct as long as in the iteration step $l \geq j$, $L_{ext,i}^{(l)}$ does not alter the sign of $\hat{y}_i^{(l)}$. We give an example to clarify this principle.

Example 7.5 (Iterative decoding) Consider the $(3,2,2)$ parity check code. It is assumed that $\mathbf{x} = (1,1,1)$ is transmitted and is received as the vector $\hat{\mathbf{y}}^{(0)} = (0.6, -0.5, 0.8)$. Clearly, the received vector $\hat{\mathbf{y}}^{(0),H} = (0,1,0)$ does not correspond to a valid codeword. Using the approximation $\tanh(y_i) = y_i$, the extrinsic information is calculated in the following manner:

$$L_{ext,1}^{(j)} = \hat{y}_2^{(j)}\hat{y}_3^{(j)}, \qquad L_{ext,2}^{(j)} = \hat{y}_1^{(j)}\hat{y}_3^{(j)}, \qquad L_{ext,3}^{(j)} = \hat{y}_1^{(j)}\hat{y}_2^{(j)} .$$

After the first iteration, we obtain

$$\hat{\mathbf{y}}^{(1)} = \hat{\mathbf{y}}^{(0)} + \mathbf{L}^{(0)} = (0.6, -0.5, 0.8) + (-0.4, 0.48, -0.3) = (0.2, -0.02, 0.5) .$$

An error has occurred in position 2, and has decreased the reliability of positions 1 and 3. On the other hand, the reliability information from positions 1 and 3 has improved the reliability information of position 2. The next iteration produces

$$\hat{\mathbf{y}}^{(2)} = \hat{\mathbf{y}}^{(1)} + \mathbf{L}^{(1)} = (0.2, -0.02, 0.5) + (-0.01, 0.1, -0.004) = (0.19, 0.08, 0.496) .$$

At this point, we obtain $\hat{\mathbf{y}}^{(2),H} = (0,0,0)$. The iterative decoder has corrected the error in position 2 and the transmitted codeword has been recovered. Further iterations do not alter the codeword, as shown in theorem 7.11. $\diamond$

Algorithm 7.6 Symbolwise iterative decoding algorithm (IDA).

Initialization:
- Set $\hat{y}_i^{(0)} = y_i$, $L_{ext,i}^{(0)} = 0, i = 1, \ldots, n$, iteration index $j = 0$.
- Define the maximum number of iterations to be Θ

Step 1: For the case $(\hat{\mathbf{y}}^{(j),H} \notin C)$ and $(j < \Theta)$):

(a) for $i = 1, \ldots, n$ calculate: $L_{ext,i}^{(j)}(\mathcal{B}_i) = \sum_{\mathbf{b} \in \mathcal{B}_i} \prod_{l \in I_i(\mathbf{b})} \tanh(\hat{y}_l^{(j)})$.

(b) for $i = 1, 2, \ldots, n$ add: $\hat{y}_i^{(j+1)} = \hat{y}_i^{(j)} + L_{ext,i}^{(j)}(\mathcal{B}_i)$.

(c) $j := j + 1$, step 1.

Step 2: $\hat{\mathbf{r}} = \hat{\mathbf{y}}^{(j),H}$, stop.

The key element of symbolwise iterative decoding is to determine the extrinsic information with respect to the *correct sign* of the symbol. The MAP algorithm produces an optimal decision with respect to the sign of a coordinate. This optimal property is exploited in the turbo decoding technique (appendix A). In general, the calculation of an optimal MAP decision is too computationally expensive. This has made the use of approximations for equations 7.19 and 7.20 necessary. Suitable approximations should satisfy several minimum requirements.

For as many cases as possible, the coordinate i must satisfy the following equation, based on decoding with parity check vectors from the set $\mathcal{B}_i \subset C^{\perp}$:

$$\mathrm{sign}(L_{ext,i}(\mathcal{B}_i)) = \mathrm{sign}(L_{ext,i}(C^{\perp})) . \tag{7.31}$$

In addition, such an approximation must reduce the calculation expense to a realizable value, and should be applicable to a large number of code classes. The requirement of equation 7.31 insures that $L_{ext,i}(\mathcal{B}_i)$ will always improve the reliability information of position i if the intrinsic information $L_{ext,i}(C^{\perp})$ of the dual code also results in an improvement. The amount of extrinsic information needed to produce optimal symbolwise iterative decoding is still an unanswered problem. It can be shown [LBB98] that iterative APP decoding can be interpreted as a generalization of weighted majority decoding. Work in [LBB98] and [BH86] shows that a set of *minimal-weight* parity check vectors can be constructed that approximate the MAP BER (see the DA algorithm, section 7.3.3). There exists a set $\mathcal{B}_i$ of *minimal weight* parity check vectors for position i, $\mathcal{B}_i = \{\mathbf{b} \,|\, i \in \mathrm{supp}(\mathbf{b}), \forall \mathbf{b} \in \mathcal{B}\}$. Let $I_i(\mathbf{b})$ be the set of coordinates that have the value 1, excluding the coordinate i itself, $I_i(\mathbf{b}) = \{l \,|\, l \in \mathrm{supp}(\mathbf{b}) \wedge l \neq i\}$. The extrinsic information can be calculated using

$$L_{ext,i}(\mathcal{B}_i) = \sum_{\mathbf{b} \in \mathcal{B}_i} \prod_{l \in I_i(\mathbf{b})} \tanh(y_l) . \tag{7.32}$$

The above equation is an approximation of $L_{ext,i}(C^{\perp})$ that satisfies the requirements for iterative decoding. From this result, we obtain the iterative decoding algorithm 7.6.

One observes that $\hat{\mathbf{r}}$ does not necessarily produce a codeword. Next we consider the stop criterion, which is based on the following theorem:

Theorem 7.11 (Stop criterion) *There exists a $\hat{\mathbf{c}} \in C$ that is the hard-decision vector $\hat{\mathbf{y}}^{(j)}$ for iteration j if $\hat{\mathbf{c}}$ remains the same for all iterations $l > j$.*

Proof Because $\hat{\mathbf{c}} \in \mathcal{C}$, we can write $\hat{\mathbf{c}} \cdot \mathbf{b} = 0 \bmod 2$ for all $\mathbf{b} \in \mathcal{B}_i$, $i \in [1,n]$. The notation $b_i \in \{+1,-1\}$ can be expanded to give $\text{sign}(\hat{y}_i^{(j)}) = \text{sign}(L_{ext,i}^{(j)}(\mathbf{b}))$, for all $\mathbf{b} \in \mathcal{B}_i$, $i \in [1,n]$. This implies that

$$\text{sign}(\hat{y}_i^{(j)}) = \text{sign}(L_{ext,i}^{(j)}(\mathcal{B}_i)),\ i \in [1,n]. \tag{7.33}$$

According to step (b), one adds $\hat{y}_i^{(j+1)} = \hat{y}_i^{(j)} + L_{ext,i}^{(j)}(\mathcal{B}_i)$, $i \in [1,n]$, and, using equation 7.33, we get for $i \in [1,n]$,

$$\text{sign}(\hat{y}_i^{(j+1)}) = \text{sign}(\hat{y}_i^{(j)}).$$

The result $\hat{\mathbf{y}}^{(j+1),H} = \hat{\mathbf{c}}$ follows, and it is apparent that $\hat{\mathbf{y}}^{(l),H} = \hat{\mathbf{c}}$, $l > j+1$. □

Remarks

- IDA can be interpreted as an iterative APP decoding algorithm that is a generalization of the algorithm developed by Gallager [Gal62] and Battail et al. [BDG79].
- The first iteration of the IDA is an approximation of an optimal MAP decision for each coordinate position.
- After the first iteration, there is a statistical dependence between the calculated a posteriori values such that an iterative decoding is only heuristically motivated.
- It can be observed that IDA usually converges to a codeword. However, values can be found that do not converge to a codeword [LBB98]. In general, the probability of these values being received is very small. Therefore the IDA algorithm produces in most cases a *suboptimal decoding*.
- A further analysis of the IDA concerning the choice of parity check vectors and convergence as well as efficient implementation can be found in [Luc97].

Example 7.6 (Simulation results and complexity for IDA) Figure 7.13 presents the bit error probability for MAP decoding and iterative decoding (IDA) for the $(15,7,5)$ and $(31,21,5)$ BCH codes. One can recognize that with IDA, the optimal bit error probability is approached for these codes. In figure 7.14, we compare the iterative decoding with the 'classic' single-step majority decoding. This shows a gain of 2.2 dB for the error probability $P_{Bit} = 10^{-3}$.
In addition, the SDML decoding results of a convolutional code with rate $2/3$ are shown in section 8.1. This code is constructed through the puncturing of the convolutional code with rate $1/2$, generator polynomial $g_1 = 561$ and $g_2 = 753$, and total memory $\nu = 9$. One observes that for this case the decoding complexities for the convolutional code and the IDA are approximately the same. Even though the $(273,191,18)$ code is suboptimally decoded, it has a better performance for $E_b/N_0 = 3\,\text{dB}$. ◇

7.4.2 List decoding using code trellises: Viterbi algorithm

In contrast to the previous methods, we now focus on soft-decision decoding not based on code symbols, but on the reliability of the entire codeword. The quadratic Euclidean distance is the metric to be minimized by the decoding algorithm. The following algorithms are classified as *list decoding* algorithms. In the most trivial case, the entire list of all codewords is considered when computing the minimum Euclidean distance to the received vector. This technique is

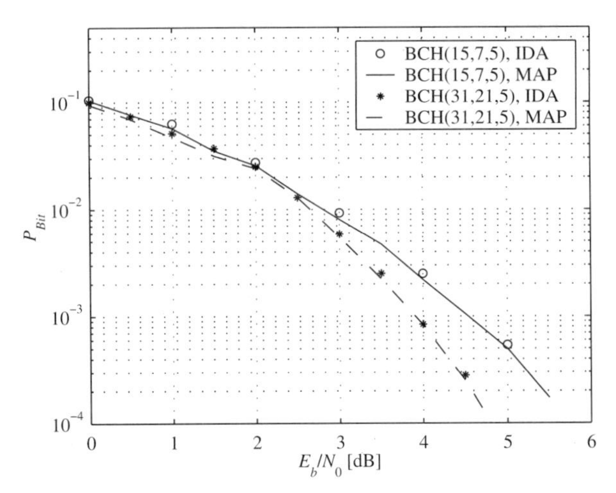

Figure 7.13 Comparison of MAP decoding and IDA for the $(15, 7, 5)$ and $(31, 21, 5)$ BCH codes.

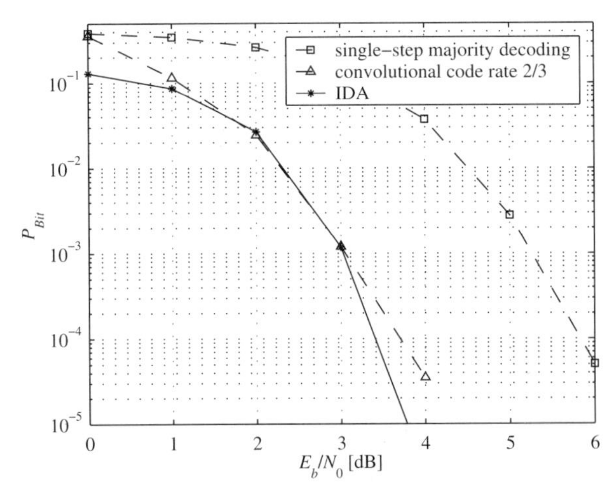

Figure 7.14 Decoding of the $(271, 191, 18)$ code with IDA in comparison with Viterbi decoding of a convolutional code of rate 2/3.

only possible for codes with a small dimension (or redundancy). Similar to symbolwise decoding, we can also employ trellis diagrams to represent the entire code in a compact form (see section 6.8). Using a trellis, the entire list can be represented and an efficient calculation of the SDML codewords can be implemented. The most famous example of this technique is the Viterbi algorithm, which has been presented in section 7.3.4 for hard-decision maximum-likelihood decoding. The only change for SDML decoding is the use of reliability information.

After the codeword $\mathbf{x}$ has been transmitted over an AWGN channel, we receive the vector $\mathbf{y} = \mathbf{x} + \mathbf{n}$, where $\mathbf{n}$ is the AWGN vector. Furthermore, we are given the code trellis T for the code $\mathcal{C}$. We again use the notation introduced for the BCJR and the hard-decision Viterbi algorithm, but with a different path metric $\Lambda(e) = y_i$. Each edge e is described by the 4-tuple $e = (\vartheta \in \mathcal{V}_{i-1}, x(e), \Lambda(e), \vartheta' \in \mathcal{V}_i)$. The set $\mathcal{E}_i$ is the set of all edges with the start vertex at

---

Algorithm 7.7 Soft-input Viterbi algorithm.

Initialization: $i = 1$, $\alpha(\vartheta_A) = 0$.

Step 1: Compute $\quad \forall e \in \mathcal{E}_i : \quad \alpha(f(e)) \; = \; \alpha(i(e)) \; + \; \Lambda(e)x(e)\,.$

Step 2: If more than one path terminates at i, choose the path with the maximum value of $\alpha(f(e))$ as the *survivor* and assign this vertex the new metric (if all metrics of incoming paths are equal, a survivor is chosen randomly).

$i := i + 1$, for the case $i \leq n$ go to step 1, otherwise go to step 3.

Step 3: Choose the code series $\hat{\mathbf{x}}$ that corresponds to the survivor path at the last vertex ϑ_B.

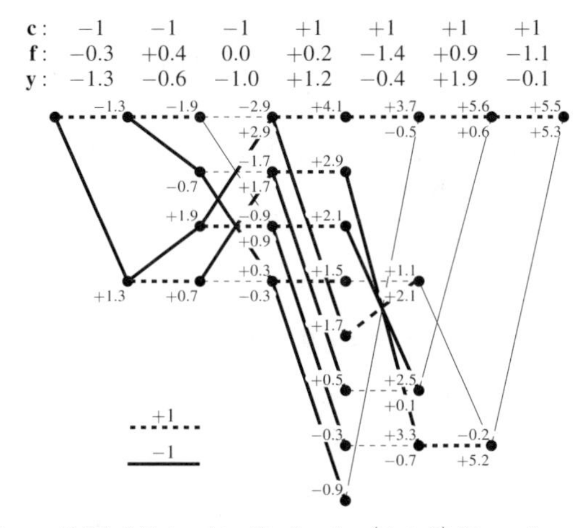

Figure 7.15 Minimal trellis for the $(7,4,3)$ Hamming code.

level $i-1$ and the end vertex at level i. This gives

$$\mathcal{E}_i \;=\; \{e \mid i(e) = \vartheta \in \mathcal{V}_{i-1}, x(e), \Lambda(e) = y_i, f(e) = \vartheta' \in \mathcal{V}_i\}\,.$$

We may formulate the SDML decoding using the code trellis T as shown in algorithm 7.7.

Example 7.7 (Viterbi decoding of the $(7,4,3)$ codes (soft-decision)) Figure 7.15 shows the minimal trellis of the $(7,4,3)$ Hamming code and the vectors **c**, **f** and **y**. The number (l) at each vertex corresponds to the metric value Λ_i^l. To calculate the metric value at level $i+1$, we use the survivor metric value at level i.

Note In the hard-decision case, if two errors occur then the received vector cannot be correctly decoded. ◇

As for hard-decision decoding, the soft-decision Viterbi algorithm is only realizable for trellises that are not too large, i.e. that correspond to relatively short codes. The difference in complexity between hard- and soft-decision decoding is in the calculation of the path metrics of the trellis.

An extension of the Viterbi algorithm that outputs a list is described in section 8.4.5. In addition to outputting the best decision, it also delivers the L next best decisions.

7.4.3 List decoding in the code space $\mathcal{C}$

Because the complexity of the SDML trellis decoding algorithm increases exponentially with $\min\{k, n-k\}$, it is necessary to consider suboptimal list decoding methods that have lower complexity and the best possible bit error rate. This technique is based on a reduced list of codewords, which should contain with a high degree of probability the ML codeword. Two basic questions arise:

1. Define the space of all possible received vectors as $\mathbb{R}^n$. Based on a received code vector, can we find a localized region in $\mathbb{R}^n$ that contains the SDML codeword without an exhaustive search of the entire vector space?

2. Can it be guaranteed that a list decoding algorithm Ψ exists that can reliably find the SDML codeword within this region?

It is also important to define a condition Ξ between the SDML codeword and the received vector that we can use to evaluate the reliability of the list algorithm Ψ.

If the codeword satisfies the requirements of Ξ then it follows that

- the SDML codeword is produced;
- the requirements of Ξ cannot be satisfied by any other codeword;
- the codeword can be found using the algorithm Ψ.

Ξ is to be used as the so-called SDML acceptance criterion. This problem was first investigated in 1966 by Forney. He proposed the *generalized minimum-distance* (GMD) decoding concept. This concept satisfied the above requirements for an algorithm Ψ and the corresponding acceptance criterion Ξ. This concept and possible extensions are presented in the following section.

Generalized minimum-distance (GMD) decoding

Consider a binary (n,k,d) block code $\mathcal{C}$. The codeword $\mathbf{x}$ is transmitted over an AWGN channel and the vector $\mathbf{y}$ is received. $\mathbf{y}^H = \mathbf{r}$ corresponds to the hard-decision received vector. The relationship between the Hamming distance and the quadratic Euclidean distance between $\mathbf{r}$ and a codeword $\mathbf{x}$ is

$$d_E^2(\mathbf{r},\mathbf{x}) = 4d_H(\mathbf{r},\mathbf{x}) .$$

Because the code $\mathcal{C}$ has minimum distance d, there is at most one codeword $\mathbf{x}_{opt}$ for which

$$d_H(\mathbf{r},\mathbf{x}_{opt}) \leq \left\lfloor \frac{d-1}{2} \right\rfloor \quad \Longleftrightarrow \quad d_E^2(\mathbf{r},\mathbf{x}_{opt}) \leq 2(d-1) . \tag{7.34}$$

The algebraic decoder Ψ is used as the decoding algorithm. t and ρ represent the number of correctable errors and erasures respectively under the condition that $2t+\rho<d$ is satisfied. This algorithm finds the codeword $\mathbf{x}_{opt}$ if equation 7.34 is satisfied. Assume that the received value is normalized as follows:

$$\tilde{y}_i = \begin{cases} +1, & y_i > +1, \\ y_i, & |y_i| \leq 1, \\ -1, & y_i < -1 . \end{cases}$$

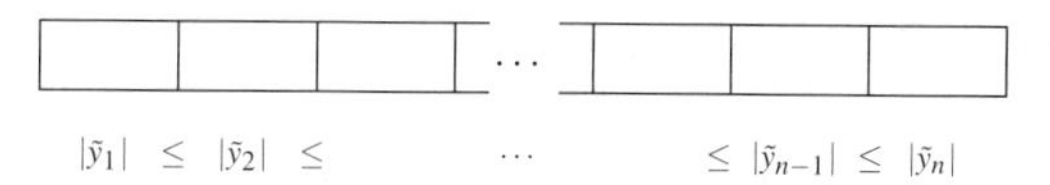

Figure 7.16 Reordering of the symbols in $\tilde{\mathbf{y}}$ according to their reliability information.

For the inequality 7.34, we can write

$$d_E^2(\tilde{\mathbf{y}}, \mathbf{x}_{opt}) \leq n + n - 2\langle \tilde{\mathbf{y}}, \mathbf{x}_{opt} \rangle \leq 2(d-1) ,$$

and thereby obtain Forney's SDML acceptance criterion.

Theorem 7.12 (SDML acceptance criterion according to Forney) *For a code C with minimum distance d and normalized received vector $\tilde{\mathbf{y}}$, $|\tilde{y}_i| \leq 1, i \in [1,n]$, there is at most one codeword $\mathbf{x}_{opt}$ that satisfies*

$$\Xi_{Forney} : \qquad \langle \tilde{\mathbf{y}}, \mathbf{x}_{opt} \rangle > n - d . \tag{7.35}$$

This codeword can always be found by an algebraic erasure decoder.

Note Forney's acceptance criterion can be interpreted as follows:

- The codeword $\mathbf{x} \in C$ is the origin of a hypercube bounded by the hyperplanes $x_i = \pm 1, i \in [1,n]$. The normalized received vector can occupy any coordinate in this hypercube.

- Ξ_{Forney} characterizes the regions within this hypercube such that for a received vector $\tilde{y}$, the SDML codeword can always be found when using an algebraic decoding algorithm. To achieve this, we define $\langle \mathbf{x}, \tilde{\mathbf{y}} \rangle = n - d$ as a hyperlevel decision that divides the decision space of $\mathbf{x}$ from other sectors of the hypercube.

- It is insured that in general, *codewords cannot occupy the same decision space* Ξ_{Forney}. If a received vector does not lie within a decision space then it will not necessarily be decoded as a SDML codeword.

A final observation focuses on the need to modify the received vector using the algorithm Ψ. By using only the decoding of $\tilde{\mathbf{r}}$, the probability is too large that either a coding failure occurs or that the recovered codeword does not satisfy the acceptance criterion of Ξ_{Forney}. In the next section, we consider how to modify the received vector by using reliability information data to remove coordinates with a low probability of being correct.

In the definition of the list decoding method based on theorem 7.12, we assume that the symbols in $\tilde{\mathbf{y}}$ are ordered with increasing information reliability, i.e. $|\tilde{y}_i| \leq |\tilde{y}_j| \, \forall \, i < j$ (see figure 7.16). This assumption does not limit the approach, because this condition can be achieved by simply renumbering the coordinate positions.

We can now consider the list decoding method within the framework of the GMD algorithm 7.8. Ψ is an algebraic decoder, and t and ρ are the number of errors and erasures respectively that can be corrected according to the equation $2t + \rho < d$.

SDML acceptance criteria for list decoding methods

It is clear that the GMD decoding performance is improved if an acceptance criterion can be found that makes the surface within a hypercube larger than that defined by Forney's criterion.

Algorithm 7.8 Generalized minimum-distance decoding algorithm.

Initialization: $\rho = 0$.

Step 1: Decode $\tilde{\mathbf{y}}$ with Ψ to obtain $\mathbf{x}_0$. If decoding failure, step 3.

Step 2: For $\langle \hat{\mathbf{x}}, \tilde{\mathbf{y}} \rangle > n - d$, set $\mathbf{x}_0 = \mathbf{x}_{opt}$ as the SDML codeword, **stop.**

Step 3: set $\rho := \rho + 2$.
For $\rho \geq d$, step 4.
for $\rho < d$, define the first ρ coordinates of $\tilde{\mathbf{y}}$ as erasures, step 1.

Step 4: Decoding failure

Therefore the number of received vectors that cannot be decoded as a SDML codeword is reduced. In the next section, we describe an extension of Forney's acceptance criterion. The following notation is introduce in order to compare the different methods. The coordinates of the received vector $\mathbf{y}$ are sorted according to their reliability information as shown in figure 7.16. The error coordinates are defined by $\mathcal{E}(\mathbf{x}, \mathbf{y})$ for any codeword $\mathbf{x}$ with respect to the received vector $\mathbf{y}$ ($x \in \mathbb{R}, \mathbf{y} \in \mathbb{R}$):

$$\mathcal{E}(\mathbf{x}, \mathbf{y}) := \{i \mid \operatorname{sign}(x_i) \neq \operatorname{sign}(y_i),\ i \in [1, n]\} . \tag{7.36}$$

We also define the set $\mathcal{T}(\mathbf{x}, \mathbf{y})$ to include the $d - |\mathcal{E}|$ most unreliable information coordinates where $\operatorname{sign}(y_i) = \operatorname{sign}(x_i)$.

We define the *weighted Hamming distance* using the set $\mathcal{E}(\mathbf{x}, \mathbf{y})$:

Definition 7.13 (Weighted Hamming distance) *The weighted Hamming distance between two vectors* $\mathbf{x}$ *and* $\mathbf{y}$ *is defined as*

$$\beta(\mathbf{x}, \mathbf{y}) := \sum_{i \in \mathcal{E}(\mathbf{x}, \mathbf{y})} |y_i| . \tag{7.37}$$

The dependence between SDML decoding and the weighted Hamming distance can be described by the following theorem:

Theorem 7.14 (SDML decoding and weighted Hamming distance) *Soft-decision maximum-likelihood decoding is equivalent to minimizing the weighted Hamming distance:*

$$\mathbf{x}_{opt} = \arg\left(\max_{\mathbf{x} \in C} p(\mathbf{y} \mid \mathbf{x})\right) = \arg\left(\min_{\mathbf{x} \in C} \beta(\mathbf{y}, \mathbf{x})\right) .$$

Proof According to equation 7.13, minimization of the quadratic Euclidean distance between a codeword $\mathbf{x}$ and a vector $\mathbf{y}$ is equivalent to maximization of their scalar product. Based on equation 7.14, we can write the scalar product as

$$\langle \mathbf{x}, \mathbf{y} \rangle = \sum_{i=1}^{n} |y_i| - 2 \sum_{i \in \mathcal{E}(\mathbf{x}, \mathbf{y})} |y_i| = \sum_{i=1}^{n} |y_i| - 2\beta(\mathbf{x}, \mathbf{y}) .$$

Because the sum over all received symbols is constant, maximization of the scalar product is equivalent to minimization of the weighted Hamming distance. □

It is assumed that $\beta(\mathbf{x}, \mathbf{y})$ is known for any codeword $\mathbf{x}$. One is interested in the smallest value of $\beta(\mathbf{x}_0, \mathbf{y})$ for one particular codeword $\mathbf{x}_0$ as compared with all $\mathbf{x} \neq \mathbf{x}_0$. This gives

$\beta(\mathbf{x}_0, \mathbf{y}) < \min_{\mathbf{x} \neq \mathbf{x}_0} \beta(\mathbf{x}, \mathbf{y})$, which indicates that $\mathbf{x}_0 = \mathbf{x}_{opt}$ is the SDML codeword. Because we must determine $\min_{\mathbf{x} \neq \mathbf{x}_0} \beta(\mathbf{x}, \mathbf{y})$ for all arbitrary codewords of the code, it is practical to find a lower bound $T_s \leq \min_{\mathbf{x} \neq \mathbf{x}_{opt}} \beta(\mathbf{x}, \mathbf{y})$ for this value. We then can deduce from $\beta(\mathbf{x}_0, \mathbf{y}) < T_s$, whether $\mathbf{x}_0 = \mathbf{x}_{opt}$ is the SDML codeword.

To define a lower bound, we assume that fewer than d errors occur, and that the codeword $\mathbf{x}_0$ can be found from the received vector $\mathbf{y}$ using the function Ψ. Based on $\mathbf{x}_0$, we find a concurrence vector $\mathbf{v}$ such that $\beta(\mathbf{x}_0, \mathbf{v}) = d$. This determines the lower bound T_s for $d_H(\mathbf{x}_0, \mathbf{v}) = d$.

In order to calculate $\mathbf{v}$, we first consider $\mathbf{y}^H = \mathbf{r}$. We have $\beta(\mathbf{r}, \mathbf{y}) = 0$ and $d_H(\mathbf{r}, \mathbf{x}_0) = |\mathcal{E}(\mathbf{x}_0, \mathbf{y})| < d$. $\beta(\mathbf{r}, \mathbf{y})$ is clearly minimal; however, $\mathbf{r}$ cannot be a concurrent vector because it cannot have a distance d from a codeword and still be correctable. In order to satisfy the definition of $\mathbf{v}$, we must change the sign of $d - |\mathcal{E}(\mathbf{x}_0, \mathbf{y})|$ coordinates of $\mathbf{y}$. Based on the reliability information, we invert the least reliable coordinates that correspond to the set $\mathcal{T}$. This produces a concurrence vector $\mathbf{v}$ for which $\beta(\mathbf{v}, \mathbf{y})$ is a minimum:

$$\beta(\mathbf{v}, \mathbf{y}) = \sum_{i \in \mathcal{E}(\mathbf{v}, \mathbf{y})} |y_i| = \sum_{i \in \mathcal{T}_{TP}(\mathbf{x}_0, \mathbf{y})} |y_i| = T_s \geq \sum_{i \in \mathcal{E}(\mathbf{x}_0, \mathbf{y})} |y_i| = \beta(\mathbf{x}_0, \mathbf{y}) \, .$$

If the weighted Hamming distance between $\mathbf{x}_0$ and $\mathbf{y}$ is smaller than that between $\mathbf{v}$ and $\mathbf{y}$ then the SDML codeword is $\mathbf{x}_0 = \mathbf{x}_{opt}$. This bound was independently proposed by Enns [Enns87] and Taipale and Pursley [TP91]. In [TP91], the following theorem is proved:

Theorem 7.15 (SDML acceptance criterion of Taipale and Pursley) *For a code $\mathcal{C}$ with minimum distance d and a received vector $\mathbf{y}$, there exists at most one codeword $\mathbf{x}_{opt}$ for which*

$$\Xi_{TP}: \qquad \beta(\mathbf{x}_{opt}, \mathbf{y}) < T_s = \sum_{i \in \mathcal{T}_{TP}(\mathbf{x}_0, \mathbf{y})} |y_i| \, . \tag{7.38}$$

This codeword can always be found with an algebraic decoding algorithm Ψ.

Remarks

- For the criterion of Taipale and Pursley, no normalization of the received vector is necessary.
- Forney's SDML acceptance criterion can be written in the following way:

$$\beta(\mathbf{x}_0, \tilde{\mathbf{y}}) + \sum_{i \notin (\mathcal{E}(\mathbf{x}_0, \tilde{\mathbf{y}}) \cup \mathcal{T}_{For}(\mathbf{x}_0, \tilde{\mathbf{y}}))} (1 - |\tilde{y}_i|) < \sum_{i \in \mathcal{T}_{TP}(\mathbf{x}_0, \mathbf{y})} |\tilde{y}_i| \, . \tag{7.39}$$

  Clearly, the criterion of Taipale and Pursley is less restrictive, which results in a larger decision space per codeword as compared with the Forney criterion. The number of received vectors that can be associated with the SDML codeword is larger using the Taipale/Pursley criterion. This vector set includes the subset of SDML vectors associated with the GMD criterion. As a result, a better bit error probability can be achieved.

The descriptions of the lower bound for $\min_{\mathbf{x} \neq \mathbf{x}_0} \beta(\mathbf{x}, \mathbf{y})$ according to Forney and Taipale-/Pursley are each based on knowledge of a single codeword $\mathbf{x}_0$ and the weighted Hamming distance to $\mathbf{y}$. An improved lower bound is achieved if $\min_{\mathbf{x} \neq \mathbf{x}_0} \beta(\mathbf{x}, \mathbf{y})$ is calculated on the basis of *two* codewords. This concept was first presented by Kaneko et al. [KNIH94] and was generalized by Kasami et al. The derivation of this criterion is omitted. Only the result is presented, and the corresponding improvement in performance is shown. We remove the requirement that fewer than d errors have occurred. The following theorem is proved in [KKT+95]:

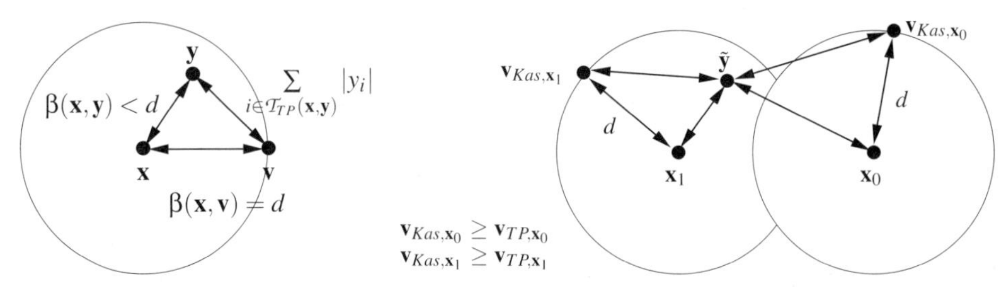

Figure 7.17 Comparison of the decision spaces for the Taipale/Pursley criterion (left) and the Kasami et al. criterion (right).

Theorem 7.16 (SDML Acceptance criterion of Kasami et al.) *Consider a binary block code $\mathcal{C}$ with the minimum distance d, and two codewords $\mathbf{x}_0$ and $\mathbf{x}_1$ from two different decoding attempts. The received vector $\mathbf{y}$ is clipped to give $\tilde{\mathbf{y}}$ such that $|\tilde{y}_i| \leq 1, i \in [1,n]$. The weighted Hamming distance to the received vector $\tilde{\mathbf{y}}$ is $n(\mathbf{x}_j) = |\mathcal{E}(\mathbf{x}_j,\tilde{\mathbf{y}})|, j \in \{0,1\}$. The cardinality of the set $\mathcal{T}_{Kas}$ is $t_{Kas} = d - \lfloor \frac{n(\mathbf{x}_0)+n(\mathbf{x}_1)}{2} \rfloor$, where $\mathcal{T}_{Kas}$ is the set of most unreliable positions with $sign(x_{i,j}) \neq sign(\tilde{y})$, $i = 0,1$ and $j = 1,\ldots,n$. The SDML codeword $\mathbf{x}_{opt}$ satisfies*

$$\Xi_{Kas}: \qquad \beta(\mathbf{x}_{opt},\tilde{\mathbf{y}}) < \sum_{i \in \mathcal{T}_{Kas},(\mathbf{x},\tilde{\mathbf{y}})} |\tilde{y}_i| \,. \tag{7.40}$$

Remarks

- It is assumed that the received vector can be decoded using an algebraic decoding algorithm to give codeword $\mathbf{x}_0$. If $\mathbf{x}_0$ satisfies the inequality 7.40 then it is the best possible choice, because $n(\mathbf{x}_0)$ is minimal.

- If only one codeword $\mathbf{x}_0$ can be found then $\mathbf{x}_0 = \mathbf{x}_1$ and therefore $n(\mathbf{x}_0) = n(\mathbf{x}_1)$. In this case, we obtain the Taipale/Pursley criterion.

- If the decoding attempt is successful then $n(\mathbf{x}_0) < n(\mathbf{x}_1)$, $\forall n(\mathbf{x}_0) \neq n(\mathbf{x}_1)$ and the cardinality of $\mathcal{T}_{Kas}(\mathbf{x}_0,\tilde{\mathbf{y}})$ in the inequality 7.40 increases. This allows us to invert more unreliable coordinates, which corresponds to an increase in the decision space as compared to the Taipale/Pursley criterion (see figure 7.17).

- The set $\mathcal{T}_{Kas}(\mathbf{x}_0,\tilde{\mathbf{y}})$ can be used to generate a correspondence vector $\mathbf{v}_{Kas,\mathbf{x}_0}$, where $|\mathcal{T}_{Kan}(\mathbf{x}_0,\tilde{\mathbf{y}})| \geq |\mathcal{T}_{TP}(\mathbf{x}_0,\mathbf{y})|$. This is analogous to the correspondence vector used in the Taipale/Pursley bound.

- The acceptance criterion of Kasami et al. can also be implemented with the use of the GMD algorithm. In general, we cannot insure that the Kasami criterion is satisfied and that the SDML codeword is recovered. A modification of this algorithm results in the Kaneko algorithm, and is presented later.

Example 7.8 (SDML acceptance criteria) Consider a $(7,4,3)$ code. The set $I = \{1,2,3,4,5,6,7\}$ describes the coordinates of the codeword vector. The received vector sorted according to the reliability information is $\mathbf{y} = (0.1, 0.3, 0.4, 0.9, 1.1, -1.4, 1.7)$. The hard-decision decoding of the received vector is $\mathbf{r} = (1,1,1,1,1,-1,1)$. The decoding of $\mathbf{r}$ using an algebraic decoder is the zero codeword $\mathbf{c} = (0, \ldots, 0)$ (corresponding to the transmitted vector $\mathbf{x}_0 = (1, \ldots, 1)$). We verify if this codeword is the SDML codeword by using the above acceptance criteria.

Taipale/Pursley criterion We consider the set $\mathcal{E}(\mathbf{x}_0, \mathbf{y}) = \{6\}$ and $\mathcal{T}_{TP}(\mathbf{x}_0, \mathbf{y}) = \{1, 2\}$. The concurrence vector with a distance d to $\mathbf{x}_0$ is therefore $\mathbf{v}_{Taip} = (-1, -1, 1, 1, 1, -1, 1)$. We now calculate the criterion:

$$\beta(\mathbf{x}_0, \mathbf{y}) = \sum_{i \in \mathcal{E}(\mathbf{x}_0, \mathbf{y})} |y_i| = 1.4 \not< 0.4 = \sum_{i \in \mathcal{T}_{TP}(\mathbf{x}_0, \mathbf{y})} |y_i| = \sum_{i \in \mathcal{E}(\mathbf{v}, \mathbf{y})} |y_i| \, .$$

The Taipale/Pursley criterion fails. Therefore we cannot decide whether $\mathbf{x}_0$ is the SDML codeword or not.

Another decoding result based on the received vector $\mathbf{y}$ gives the codeword $\mathbf{c} = (0, 1, 1, 0, 0, 1, 0)$, $(\mathbf{x}_1 = (1, -1, -1, 1, 1, -1, 1))$. We obtain the sets $\mathcal{E}(\mathbf{x}_1, \mathbf{y}) = \{2, 3\}$ and $\mathcal{T}_{TP}(\mathbf{x}_1, \mathbf{y}) = \{1\}$. From this, we calculate the Taipale/Pursley criterion based on the correspondence vector $\mathbf{v}_{TP, \mathbf{x}_1} = (-1, -1, -1, 1, 1, -1, 1)$:

$$\beta(\mathbf{x}_1, \tilde{\mathbf{y}}) = \sum_{i \in \mathcal{E}(\mathbf{x}_1, \tilde{\mathbf{y}})} |y_i| = 0.7 \not< 0.1 = \sum_{i \in \mathcal{T}_{TP}(\mathbf{x}_1, \mathbf{y})} |\tilde{y}_i| \, .$$

The codeword $\mathbf{x}_1$ has a higher probability of being a codeword than $\mathbf{x}_0$ because it has a smaller weighted Hamming distance to the received vector $\mathbf{y}$. Nevertheless, we cannot say if $\mathbf{x}_1$ is the SDML codeword.

Kasami criterion First we bound the received vector $\mathbf{y}$ according to theorem 7.16 to give $\tilde{\mathbf{y}} = (0.1, 0.3, 0.4, 0.9, -1.0, 1.0)$. The cardinality of the error sets $\mathcal{E}(\mathbf{x}_j, \tilde{\mathbf{y}})$, $j \in \{0, 1\}$, gives $n(\mathbf{x}_0) = 1, n(\mathbf{x}_1) = 2$. We first consider the codeword $\mathbf{x}_1$. The cardinality of the set $\mathcal{T}_{Kas}(\mathbf{x}_1, \tilde{\mathbf{y}})$ is given by $d - \lfloor \frac{n(\mathbf{x}_0) + n(\mathbf{x}_1)}{2} \rfloor = 2$. Based on the error set $\mathcal{E}(\mathbf{x}_1, \tilde{\mathbf{y}})$, we set $\mathcal{T}_{Kas}(\mathbf{x}_1, \tilde{\mathbf{y}}) = \{1, 4\}$. This is an extension of $\mathcal{T}_{TP}(\mathbf{x}_1, \mathbf{y})$ by the index element $\{4\}$. We may now define the concurrence vector as $\mathbf{v}_{Kas, x_1} = (-1, -1, -1, -1, 1, -1, 1)$ and compare the weighted Hamming distances of $\tilde{\mathbf{y}}$ to $\mathbf{v}_{Kas, x_1}$ and $\mathbf{x}_1$ respectively:

$$\beta(\mathbf{x}_1, \mathbf{y}) = 0.7 < 1.0 = \sum_{i \in \mathcal{T}_{Kas}(\mathbf{x}_1, \tilde{\mathbf{y}})} |y_i| \, .$$

Using the Kasami et al. criterion, we determine that $\mathbf{x}_1$ is the SDML codeword based on the codewords $\mathbf{x}_0$ *and* $\mathbf{x}_1$. The Taipale/Pursley and Kasami criteria produce identical results for the codeword $\mathbf{x}_0$ because the set $\mathcal{T}_{Kas}(\mathbf{x}_0, \tilde{\mathbf{y}}) = \mathcal{T}_{TP}(\mathbf{x}_0, \mathbf{y}) = \{1\}$. ◇

Chase algorithms

The Chase algorithms [Cha72] belong to the class of suboptimal decoding methods based on an algebraic decoding algorithm Ψ. The starting point for decoding, as for the GMD algorithm, is the calculation and sorting of the received vector $\mathbf{y}$ according to the information reliability (see figure 7.16). In contrast to the GMD decoding algorithm, the Chase algorithms calculate variants of the received vector by 'inversion' of certain coordinates instead of through erasures. This produces a set $\mathcal{T}$ of *test solutions*. The test set is based on a binary test vector $\mathbf{v} \in GF(2)^n$,

Algorithm 7.9 Chase decoding algorithm.

Initialization: Calculate the set of test vectors $\mathcal{T}$ according to variant 1, 2 or 3.
The decoding result is $\tilde{\mathbf{x}}$.
Set $i = 0$ and $\mathcal{L} = \emptyset$ and max $= 0$.

Step 1: Decode $\hat{\mathbf{r}} = \mathbf{r} \oplus \mathbf{v}_i, \mathbf{v}_i \in \mathcal{T}$, with Ψ to give $\hat{\mathbf{x}}$.
If a decoding failure is received then step 3.

Step 2: If $\langle \hat{\mathbf{r}}, \mathbf{y} \rangle > \max$, then $\max = \langle \hat{\mathbf{r}}, \mathbf{y} \rangle$ and $\mathcal{L} = \{\hat{\mathbf{x}}\}$.

Step 3: For $i < |\mathcal{T}|$, set $i = i + 1$, otherwise step 4.

Step 4: Determine the decoding result according to $\tilde{\mathbf{x}} = \begin{cases} \hat{\mathbf{x}}, & \text{if } \mathcal{L} \neq \emptyset, \\ \mathbf{r}, & \text{otherwise} . \end{cases}$

which is added to the hard-decision received vector $\mathbf{y}^H = \mathbf{r} \in \mathbb{F}_2^n$. The modified received vector $\mathbf{r} + \mathbf{v}$ is decoded using the algorithm Ψ and a list is produced of possible SDML codewords. For example, if we consider all test vectors with Hamming weight i then all codewords with Hamming distance $t + i$ from the received codeword would be decoded, where $t = \lfloor \frac{d}{2} \rfloor$. We select the codeword with the smallest Euclidean distance from the received vector, based on the the resulting list of test codewords derived from the decoding method. If the list is empty then a decoding failure is reported. The test solution should therefore correspond to the most probable error solutions. This justifies the assumption that the most unreliable positions of the received vector have the highest probability of containing errors.

In order to calculate the test solutions, Chase presented three different techniques:

Variant 1 The set of test solutions $\mathcal{T}$ is given through all binary vectors of weight $\leq \lfloor \frac{d}{2} \rfloor$ e.g. $|\mathcal{T}| = \sum_{i=0}^{\lfloor \frac{d}{2} \rfloor} \binom{n}{i}$.

Variant 2 The test solution set $\mathcal{T}$ is calculated using all binary vector combinations corresponding to the $t = \lfloor \frac{d}{2} \rfloor$ most unreliable coordinates. The cardinality of the test vector set is $|\mathcal{T}| = 2^{d/2}$. The remaining $n - \lfloor \frac{d}{2} \rfloor$ coordinates of the test vectors are set to zero.

Variant 3 The set of test solutions are chosen in an analogous way to the GMD algorithm. The test vectors are generated by setting one pair of the most unreliable coordinate positions of the received vector $\mathbf{y}$ to 1. Successive pairs of 1s are added to the previously generated test vector until the maximum number of correctable errors $t = \lfloor \frac{d}{2} \rfloor$ is reached. We have the sets $i = 0, 2, \ldots, d-1$ and $i = 1, 3, \ldots, d-1$ for even and odd d respectively. The cardinality of the test solution set is $\mathcal{T} = \lfloor \frac{d}{4} \rfloor$.

The Chase decoding algorithm can be carried out for each of the above variants. The corresponding simulation results are given in figure 7.21.

Remarks

- The Chase 1 algorithm (variant 1) gives the best bit error probability curve, because all vectors with the distance less than or equal to $t = \lfloor \frac{d}{2} \rfloor$ are considered. This also corresponds to a very high complexity. All together, $\binom{n}{0} + \binom{n}{1} + \ldots + \binom{n}{\lfloor \frac{d}{2} \rfloor}$ elements are considered in the test solution. Therefore an equal number of algebraic decoding trials must be undertaken. Variant 1 is not practical for large n and $t > 3$.

- The Chase 2 algorithm (variant 2) uses only a subset of the test solution from variant 1, and therefore only a subset of the codewords are generated as compared to variant 1. The code vectors that are not generated by variant 2 correspond to unlikely error patterns, i.e. errors in the most reliable coordinates. Therefore the actual performance compared with algorithm 1 is only slightly decreased. At the same time, the decoding complexity is reduced to $2^{\lfloor \frac{d}{2} \rfloor}$ for the algebraic decoding step.

- The Chase 3 algorithm (variant 3) is the least complex, and has a correspondingly higher bit error probability compared with the other algorithms. However, variant 3 is still better than the GMD algorithm, because, with the erasure correction method, only a subset of the codewords is found.

- In the original version of the Chase algorithms, no SDML acceptance criterion was used. The inclusion of such a criterion is useful, because it is then sometimes possible to avoid unnecessary decoding attempts if the SDML criterion is satisfied by an early trial.

Kaneko algorithm

Chase proposed three variants of his decoding algorithm with different complexities based on the generation of test vector set. The cardinality for each test solution is constant and independent of the received vector. An improvement can be made if the size of the test solution is calculated based on the received vector. Such a strategy was proposed by Kaneko et al. [KNIH94]. The modified algorithm consists of an extension of the Chase algorithm to a SDML decoding method. One notes that for the most unfavourable case, *all* codewords must be determined in order to find the SDML codeword.

In the following section, we present an adaptive calculation method for the test solution that guarantees that the SDML codeword is found.

Adaptive calculation of the test solution The Chase algorithm is not a SDML decoding algorithm, because it does not guarantee that the test solutions produce the SDML codeword. It is clear that the SDML decision can be attained if enough test solution elements are considered. In the trivial case, all 2^n possible test solutions are considered, which leads to the SDML codeword. This is clearly not a practical solution. To solve this problem, we answer the following question: Is there a way to calculate the number of necessary test solutions that must be considered to find the SDML codeword based on the received vector? Kaneko et al. found such a solution.

Ψ is an algebraic decoding algorithm that can correct $t = \lfloor \frac{d-1}{2} \rfloor$ errors. The coordinates of the received vector $\mathbf{y}$ are ordered with respect to their reliability information. Therefore $|y_i| \leq |y_j|, \forall\, i < j$ (figure 7.16). Furthermore, $\mathbf{y}^H = \mathbf{r} \in GF(2)^n$.

For the Chase algorithm, the test solutions were subdivided based on the Hamming weight. The same principle is used in the Kaneko algorithm, where all bit solutions are considered that are made up of any combination of 1s and 0s occupying the first r coordinates. The coordinates $r+1, r+2, \ldots, n$ are zero.

$$\mathcal{T}_r = \{\mathbf{v} = (v_1, \ldots, v_n) \in GF(2)^n \mid (v_1, \ldots, v_r, 0, \ldots, 0), v_i \in \{0,1\}, i \in [1,r]\} .$$

Furthermore, $\mathcal{L}_r$ is the set of all codewords that are contained in the test solution set $\mathcal{T}_r$:

$$\mathcal{T}_r \subset \mathcal{T}_{r+1}, \quad \text{and} \quad \mathcal{L}_r \subseteq \mathcal{L}_{r+1} .$$

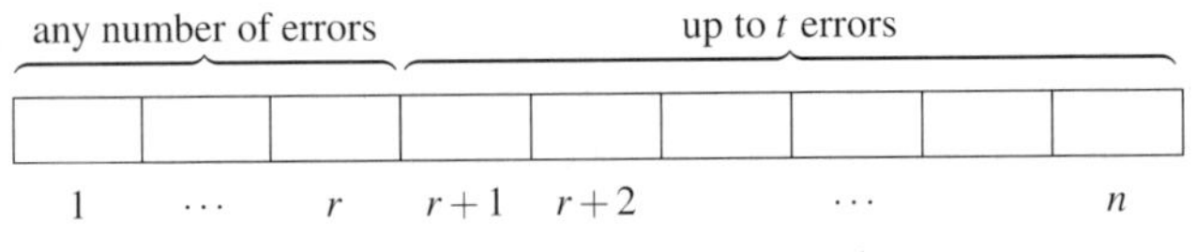

Figure 7.18 Codeword structure in $\mathcal{L}_r$.

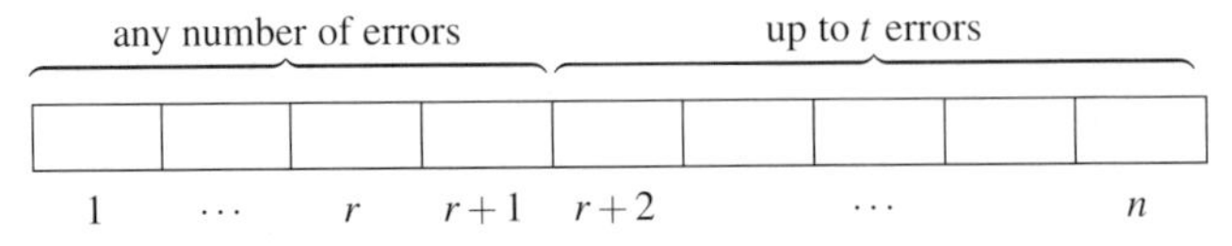

Figure 7.19 Codeword structure in $\mathcal{L}_{r+1}$.

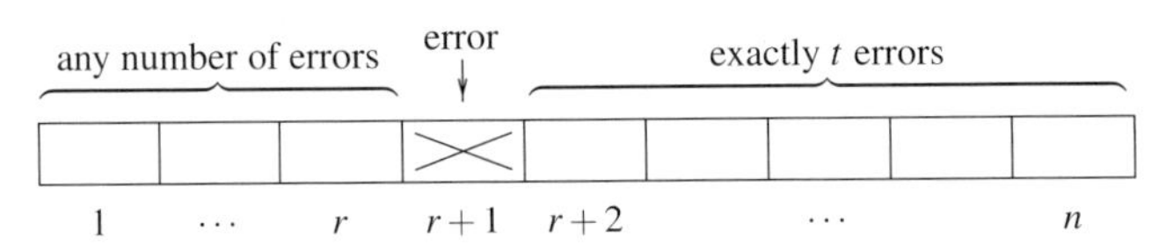

Figure 7.20 Codeword structure in $\mathcal{L}_{r+1} \setminus \mathcal{L}_r$.

The goal is to determine r and therefore the set $\mathcal{T}_r$ that contains the set $\mathcal{L}$ having the SDML codeword. We also include the condition that the cardinality of $\mathcal{T}_r$ should be as small as possible.

We assume that $\mathbf{x}$ and $\mathbf{v}, \mathbf{v} \in \mathcal{T}_r$ have at most $t = \left\lfloor \frac{d-1}{2} \right\rfloor$ different coordinates. This allows $\hat{\mathbf{r}} = \mathbf{x} \oplus \mathbf{v}$ to be decoded with the algorithm Ψ. The vector $\hat{r}$ is associated with the codeword $\mathbf{x}$.

On the other hand, if $\mathbf{v} \in \mathcal{T}_r$ is chosen such that $\mathbf{x}$ and $\mathbf{x} \oplus \mathbf{v}$ are the same for positions $1, 2, \ldots, r$ (i.e. $\mathbf{v}$ corresponds to the error coordinates in $\mathbf{x}$) then Φ decodes $\hat{\mathbf{r}}$ as $\mathbf{x}$ as long as fewer than t errors occur in positions $r+1, \ldots, n$. If this condition is not met then Φ cannot decode $\mathbf{x}$ using the solution set $\mathcal{T}_r$.

We obtain the list $\mathcal{L}_r$ of all codewords based on $\mathbf{r}$ that differ in at most t positions for the coordinates $r+1, r+2, \ldots n$. The subdivision of the coordinates is shown in figure 7.18. Correspondingly, the list $\mathcal{L}_{r+1}$ contains all codewords that differ from $\mathbf{r}$ in at most t positions within the subset $r+2, r+3, \ldots, n$. The coordinates $1, 2, \ldots, r+1$ can take any values (see figure 7.19). The additional codewords that are found using the set $\mathcal{T}_{r+1}$ instead of $\mathcal{T}_r$ corresponds to the set $\mathcal{L}_{r+1} \setminus \mathcal{L}_r$. The error pattern of this subset is shown in figure 7.20. For all $\mathbf{x} \in \mathcal{L}_{r+1} \setminus \mathcal{L}_r$, we write

$$\beta(\mathbf{x}, \mathbf{y}) \geq \sum_{i=r+1}^{r+1+t} |y_i| \,. \tag{7.41}$$

Consider the codeword $\hat{\mathbf{x}}$ that has the smallest weighted Hamming distance belonging to the test solution set $\mathcal{T}_r$ and such that $\beta(\hat{\mathbf{x}}, \mathbf{y}) < \sum_{i=r+1}^{r+1+t} |y_i|$. This condition insures that no codeword $\mathbf{x}'$ exists from the extended test solution set $\mathcal{T}_{r+1}$ that satisfies $\beta(\mathbf{x}', \mathbf{y}) < \beta(\hat{\mathbf{x}}, \mathbf{y})$. Therefore $\hat{\mathbf{x}} = \mathbf{x}_{opt}$, is the SDML codeword. If $\beta(\hat{\mathbf{x}}, \mathbf{y}) \geq \sum_{i=r+1}^{r+1+t} |y_i|$ then there is the possibility that the extended test solution set $\mathcal{T}_{r+1}$ contains a codeword that satisfies $\beta(\mathbf{x}', \mathbf{y}) \leq \beta(\hat{\mathbf{x}}, \mathbf{y})$.

We must find the smallest r that guarantees that the SDML codeword can be found in the set $\mathcal{T}_r$. This requirement is satisfied by

$$\beta(\hat{\mathbf{x}}, \mathbf{y}) < \sum_{i=r+1}^{r+1+t} |y_i| \,. \tag{7.42}$$

If a new codeword $\mathbf{x}$ is found such that $\beta(\mathbf{x}, \mathbf{y}) < \beta(\hat{\mathbf{x}}, \mathbf{y})$ then r can be recalculated with the smaller weighted Hamming distance, resulting in a smaller value for r. This results in a reduction in the number of necessary test solutions.

Assume that two codewords $\mathbf{x}_0$ ands $\mathbf{x}$ are found such that $n(\mathbf{x}_0) = |\mathcal{E}(\mathbf{x}_0, \mathbf{y})|$, $n(\mathbf{x}) = |\mathcal{E}(\mathbf{x}, \mathbf{y})|$. We can conclude that all codewords $\mathbf{x}'$ must have at least $d - \lfloor \frac{n(\mathbf{x}_0)+n(\mathbf{x})}{2} \rfloor$ errors with respect to the received vector $\tilde{y}$. From the above calculation, the necessary test solution assumes exactly $t+1$ errors. Therefore the *best possible* codeword in $\mathcal{L}_{j+1} \setminus \mathcal{L}_j$ has not only $t+1$ errors for positions $i, i+1, \ldots, i+t+1$, but also $l = d - \lfloor \frac{n(\mathbf{x}_0)+n(\mathbf{x})}{2} \rfloor - (t+1)$ errors in the unreliable information positions $1, \ldots, \mathbf{r}$. The SDML codeword is guaranteed to be in the set $\mathcal{L}_r$, where r is the smallest number satisfying the inequality

$$\beta(\mathbf{x}, \mathbf{y}) < \sum_{i=1}^{l} |y_i| + \sum_{i=r+1}^{r+t+1} |y_i| \tag{7.43}$$

This extension of the inequality 7.42 results in a reduction of the index r for certain situations, and is derived from the the triangle inequality

$$d_H(\mathbf{x}_0, \mathbf{y}^H) + d_H(\mathbf{x}, \mathbf{y}^H) \geq d_H(\mathbf{x}_0, \mathbf{x}) \geq d, \quad \mathbf{x}_0 \neq \mathbf{x} \,.$$

This can be written as $n(\mathbf{x}_0) + n(\mathbf{x}) \geq d$ and

$$d - \left\lfloor \frac{n(\mathbf{x}_0) + n(\mathbf{x})}{2} \right\rfloor - (t+1) \leq d - t - t - 1 = \begin{cases} 0, & \text{if } d \text{ is odd,} \\ 1, & \text{if } d \text{ is even.} \end{cases}$$

Consequently, the first sum for an odd value for d always gives a value of zero. The calculation of r according to a BMD decoding of $\mathbf{y}^H$ under the condition that $n(\mathbf{x}) = n(\mathbf{x}_0)$ gives a sum that decreases with t.

This method guarantees that after 2^r decoding attempts, the SDML codeword is found. The SDML acceptance criterion can also be applied in this case to eliminate unnecessary decoding attempts. This was already given in the inequality 7.40. Algorithm 7.10 describes the decoding.

The goal of this algorithm is to improve the decoding capability of a algebraic decoder based on the information reliability. Errors with weight larger than half the minimum distance in unreliable coordinates can be erased or inverted.

A $(15, 7, 5)$ BCH code is used in figure 7.21 to compare the decoding algorithms described up to this point. It can be seen that the GMD algorithm results in no significant improvement over BMD decoding. In contrast, the Chase 3 algorithm gives up to a 0.5 dB coding gain as compared with the BMD algorithm.

7.4.4 List decoding based on ordered statistics

In 1995, Fossorier and Lin [FL95] presented a new concept for soft-decision decoding of block codes. The concept is based on a generalization of the list decoding algorithms presented up

---

Algorithm 7.10 Kaneko et al. decoding algorithm.

Initialization: Set $r = n$, $j = 0$, $\beta(\mathbf{x}', \mathbf{y}) = \infty$.

Step 1: if: $(j \leq 2^r - 1)$

(a) Decode $\mathbf{y}^H \oplus \mathbf{v}_j, \mathbf{v}_j \in \mathcal{T}_r$, with Ψ to give $\mathbf{x}$.
For a decoding failure $\rightarrow$ step (c).

(b) If $\beta(\mathbf{x}, \mathbf{y}) < \beta(\mathbf{x}', \mathbf{y})$ is valid,

- set $\mathbf{x}' = \mathbf{x}$.
- If $\mathbf{x}$ satisfies the criterion Ξ_{Kan} , $\mathbf{x}_{opt} = \mathbf{x}$, **stop.**
- update r according to the inequality 7.43.

(c) $j := j + 1$, step 1.

Step 2: $\mathbf{x}_{opt} = \mathbf{x}'$.

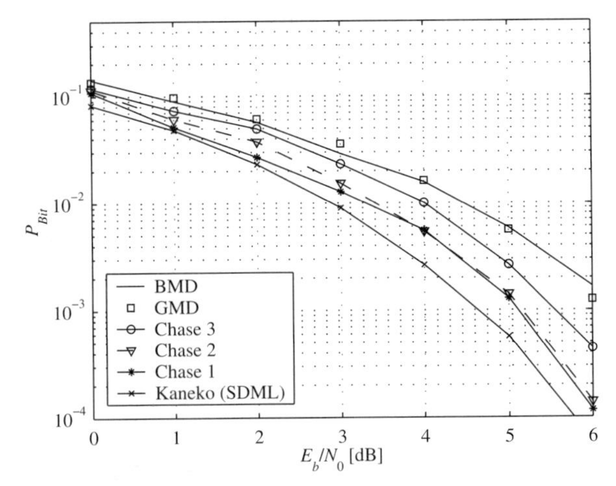

Figure 7.21 Comparison of GMD, Chase and Kaneko algorithms for the $(15, 7, 5)$ BCH code.

to this point. In this section, we present this new concept. A detailed description can be found in [FL95, FL96]. The basic idea behind this decoding method touches on the observation that errors in the reliable coordinates of a received vector are more unlikely than in unreliable positions. The concept of GMD and Chase algorithms, is to determine the test solution set from the unreliable information coordinates. An opposite approach is taken for ordered statistics decoding. The test solution set is based on the most reliable information coordinates k, and the $n - k$ remaining coordinates are recalculated.

Consider a block code C and its generator matrix $\mathbf{G}$. The starting point of the decoding begins with the received vector $\mathbf{y}$, which is permuted according to the function Φ_1, based on the reliability information values:

$$\bar{\mathbf{y}} = (\bar{y}_1, \ldots, \bar{y}_n) = \Phi_1(\mathbf{y}), \quad |\bar{y}_1| \geq \ldots \geq |\bar{y}_n| \,. \tag{7.44}$$

The vector $\bar{\mathbf{y}}$ consists of n random variables whose absolute values satisfy the ordering rule '$\geq$'. The ith coordinate of this vector is termed the ith ordered statistic [BaCo]. In contrast to the decoding algorithms presented up to this point, we permutate the vector $\bar{\mathbf{y}}$ by means of

---

Algorithm 7.11 Decoding of order l.

Initialization: Calculate $\mathbf{G}_{MRB}$ and also $\mathbf{z}_0$ and set $\hat{\mathbf{z}} = \mathbf{z}_0$.

Step 1: For the case: $1 \leq i \leq l$,

- calculate the test solution set $\mathcal{T}_i = \{\mathbf{v} \in \mathbb{F}_2^k \,|\, \mathrm{wt}(\mathbf{v}) = i\}$.
- calculate the test codewords

$$\mathcal{L}_i = \{(\mathbf{z}_A \oplus \mathbf{v}) \cdot \mathbf{G}_{LRB} \,|\, \mathbf{v} \in \mathcal{T}_i\}$$

  for every codeword $\bar{\mathbf{z}}$ that has the smallest Euclidean distance to $\mathbf{z}$.

- If $d_E(\bar{\mathbf{z}}, \mathbf{z}) < d_E(\hat{\mathbf{z}}, \mathbf{z})$ then $\hat{\mathbf{z}} = \bar{\mathbf{z}}$, goto step 1:

Step 2: $\hat{\mathbf{c}} = \Phi_1^{-1}(\Phi_2^{-1}(\hat{\mathbf{z}}))$.

---

a second permutation function Φ_2. Φ_2 calculates the vector $\mathbf{z}$ such that positions $1, \ldots, k$ are linearly independent of those coordinates that have a larger reliability information, and this can be used for systematic encoding of the code (see section 1.7):

$$\mathbf{z} = (z_1, \ldots, z_k, z_{k+1}, \ldots, z_n) = (\mathbf{z}_A | \mathbf{z}_B) = \Phi_2(\bar{\mathbf{y}}) = \Phi_2(\Phi_1(\mathbf{y}))$$
$$\text{with} \quad |z_1| \geq \ldots \geq |z_k| \wedge |z_{k+1}| \geq \ldots \geq |z_n| \,. \quad (7.45)$$

One calls the first k information coordinates of $\mathbf{z}$ the most reliable basis, MRB. The above permutations leave the code characteristics unchanged; therefore $\Phi_2(\Phi_1(\mathbf{G}))$ corresponds to an equivalent code $\mathcal{C}$. Using elementary row operations, $\Phi_2(\Phi_1(\mathbf{G}))$ can be set in systematic form:

$$\mathbf{G}_{MRB} = (\mathbf{I}_k | \mathbf{A}_{n-k}),$$

Therefore the first k coordinates of $\mathbf{z}$ can be interpreted as information positions for the calculation of the codewords. Clearly, there is a higher probability that the MRB contains fewer errors in the information coordinates than in the redundancy coordinates. Assume that the k most reliable linearly independent coordinates are error-free information symbols, resulting in $\mathbf{z}_0 = \mathbf{z}_A^H \cdot \mathbf{G}_{MRB}$. The desired codeword candidate can be recovered using a simple decoding algorithm, namely $\hat{\mathbf{x}} = \Phi_2^{-1}(\Phi_1^{-1}(\mathbf{z}_0))$ (compare with permutation decoding from section 7.3.1). In [FL95], it is shown that this results in a better bit error probability as compared with bounded minimum-distance decoding. This decoding is termed *decoding of order* 0 (*0-order decoding*). An improved decoding result is achieved if we consider the possibility of errors in the information coordinates. Similar to the previous algorithm, a test set of codewords is created in which successive coordinates of the MRB are inverted, and the corresponding codewords are constructed by matrix multiplication. Therefore algorithm 7.11, $1 \leq l \leq k$, defines the ordered statistics decoding for *decoding of order l (order-l reprocessing).*

Remarks

- In [FL95], it is shown that order-l reprocessing of $\mathbf{z}_0$ results in a drastic reduction of the search space of the SDML codeword (with higher probability) as compared with previously presented algorithms. This results in quasioptimal list decoding algorithms with reduced complexity.

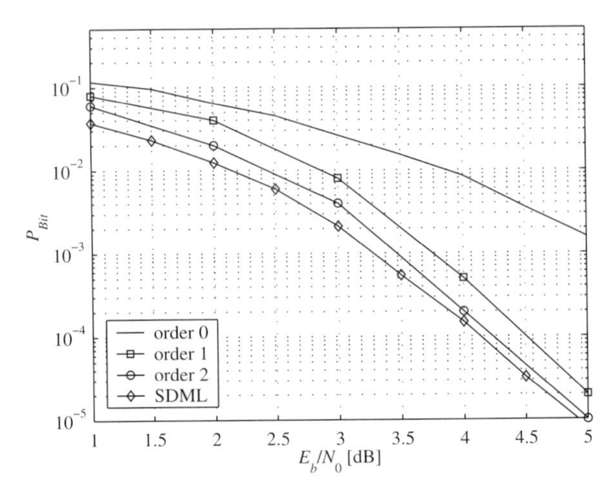

Figure 7.22 Comparison of list decoding of orders 0, 1 and 2 with SDML decoding using the $(64, 42, 8)$ RM code.

- Based on ordered statistics, the achieved bit error probability can be theoretically calculated after each decoding level (order-l reprocessing). With this information, the necessary number of decoding levels can be determined. This naturally leads to the calculation of all 2^k possible combinations of the MRB, and corresponds to SDML decoding. Because the MRB has comparably fewer errors in the information coordinates, we need to invert fewer MRB coordinate positions in order to find the SDML codeword (with larger probability). This results in fewer test codewords to consider.

- In [FL95], a special SDML acceptance criterion for order-l reprocessing is presented. In [FL96], three modifications of order-l reprocessing are given that increase the complexity while reducing the bit error probability.

- For ordered statistics decoding, an algebraic decoding algorithm is not necessary. In [FL97], it is shown that a combination of decoding of order l with an algebraic decoding algorithm results in a very good bit error probability, while having a low complexity. This is based on the introduction of a SDML acceptance criterion that can be viewed as a generalization of the Kasami criterion.

Figure 7.22 shows the decoding results for a $(64, 42, 8)$ Reed–Muller code using ordered statistics. The bit error probability approximates the theoretical SDML decoding result after order-2 reprocessing.

7.4.5 List decoding in code space $C^{\perp}$

The goal of the preceding sections on list decoding methods was to determine the most probable transmitted codeword. Therefore these algorithms dealt exclusively with the code space C. List decoding can also be carried out in the dual code space $C^{\perp}$. We consider this technique in this section.

Consider the codeword $\mathbf{c}$ and the transmitted codeword $\mathbf{x}$. $\mathbf{y}$ is the received vector. If we consider only hard-decision decoding, we obtain $\mathbf{y}^H = \mathbf{c} \oplus \mathbf{f}$. The parity check matrix of the code $\mathcal{C}$ is $\mathbf{H} = (\mathbf{h}_1^T, \ldots, \mathbf{h}_n^T)$. The syndrome $\mathbf{s}$ can be written as

$$\mathbf{s} = \mathbf{y}^H \cdot \mathbf{H}^T = (\mathbf{c} \oplus \mathbf{f}) \cdot \mathbf{H}^T = \mathbf{f} \cdot \mathbf{H}^T = \sum_{i \in \mathrm{supp}(\mathbf{f})} \mathbf{h}_i^T . \tag{7.46}$$

An error vector $\mathbf{f}$ with the same syndrome as $\mathbf{y}^H$ must be calculated, such that the addition of both vectors results in a codeword:

$$\mathbf{c} = \mathbf{y}^H \oplus \mathbf{f}, \quad \text{if} \quad \mathbf{y}^H \cdot \mathbf{H}^T = \mathbf{f} \cdot \mathbf{H}^T .$$

Theorem 7.17 (Equivalence of cosets and syndromes) *All vectors of a coset create the same syndrome according to equation 7.46.*

Because different cosets are disjoint, their vectors result in different syndromes. Each coset has exactly 2^k vectors (for a binary code). At least as many error vectors exist that when added to the $\mathbf{y}^H$ produce the 2^k codewords of the code. Hard-decision maximum-likelihood decoding is equivalent to the search for the error vector with the smallest Hamming weight. This has already been presented in chapter 1 as standard array decoding. For soft-decision maximum-likelihood decoding, we search for the most probable transmitted codeword $\mathbf{c}_{opt}$. Equivalently, this can also be viewed as the search for the most probable *error vector* $\mathbf{f}_{opt}$:

$$\mathbf{c}_{opt} = \mathbf{y}^H \oplus \mathbf{f}_{opt} .$$

The equivalence of the two statements is also evident by the following property. Over the set of error coordinates $\mathcal{E}(\mathbf{x}, \mathbf{y})$ (equation 7.36) between a codeword $\mathbf{x}$ and the received vector $\mathbf{y}$, the corresponding error vector $\mathbf{f}$ is defined in the following way:

$$\mathbf{f} := (f_1, \ldots, f_n), \quad \text{with } f_i = \begin{cases} 1, & \text{when } i \in \mathcal{E}(\mathbf{x}, \mathbf{y}), \\ 0, & \text{otherwise} . \end{cases} \tag{7.47}$$

The following is valid for the weighted Hamming distance:

$$\beta(\mathbf{x}, \mathbf{y}) = \sum_{i \in \mathcal{E}(\mathbf{x}, \mathbf{y})} |y_i| = \sum_{i \in \mathrm{supp}(\mathbf{f})} |y_i| = \beta(\mathbf{f}) .$$

Consider $\mathcal{C}_\mathbf{s}$ with the syndrome $\mathbf{s} = \mathbf{y}^H \cdot \mathbf{H}^T$. The most probable error vector is given by

$$\mathbf{f}_{opt} = \arg\left(\min_{\mathbf{f} \in \mathcal{C}_\mathbf{s}} \beta(\mathbf{f})\right) = \arg\left(\min_{\mathbf{f} \in \mathcal{C}_\mathbf{s}} \sum_{i \in \mathrm{supp}(\mathbf{f})} |y_i|\right) . \tag{7.48}$$

The calculation of the syndrome $\mathbf{s}$ can be interpreted according to equation 7.46. The values of the error vector should select columns of the parity check matrix that when summed produce the calculated syndrome. Only linearly independent columns need be considered.

Based on this consideration, the SDML decoding algorithm is formulated as in algorithm 7.12. This algorithm is named generalized Wagner decoding, and was first published in [SB89].

Algorithm 7.12 Generalized Wagner decoding.

Initialization: Calculate $\mathbf{s} = \mathbf{y}^H \cdot \mathbf{H}^T$.

Step 1: For $\mathbf{s} = \mathbf{0}$, set $\mathbf{f}_{opt} = \mathbf{0}$, step 3

Step 2: Determine a minimal set of linearly independent columns in $\mathbf{H}$ such that the sum of the columns produces the syndrome $\mathbf{s}$ and that the Hamming distance is minimal for the corresponding error vector $\mathbf{f}_{opt}$.

Step 3: Construct $\mathbf{c}_{opt} = \mathbf{y}^H \oplus \mathbf{f}_{opt}$.

In [Sny91] and [LBT93], efficient implementations of generalized Wagner decoding are given. In these techniques, the most likely error vector is calculated from a reduced error vector set. In general, the complexity increases exponentially with $\min\{k, n-k\}$. Therefore list decoding based on the dual code is also a suboptimal algorithm that has a low decoding complexity and a good bit error probability. Based on the equivalence to list decoding in code space C, many concepts derived for C can be applied to dual space decoding. It should be mentioned that in the dual space, there are no algebraic decoding methods known.

As an example of suboptimal list decoding using the dual code, we present a variation of ordered statistics decoding . The starting point for ordered statistics decoding in the dual code space is not the generator matrix of the code C but the parity check matrix $\mathbf{H}$. The received vector $\mathbf{y}$ is modified according to a permutation Φ_1 that is based on the reliability information of the coordinates. Because the parity check matrix has rank $n-k$, we apply the permutation Φ_2 to reorder the vector $\Phi_1(\mathbf{y})$ such that the linearly independent positions with the smallest reliability occupy positions $k+1, \ldots, n$:

$$\mathbf{z} = (\mathbf{z}_A | \mathbf{z}_B) = \Phi_2(\Phi_1(\mathbf{y})) = (z_1, \ldots, z_k, z_{k+1}, \ldots, z_n)$$
$$\text{with } |z_1| \geq \ldots \geq |z_k| \text{ and } |z_{k+1}| \geq \ldots \geq |z_n| \,.$$

This is called the *least reliable basis*, LRB. $\mathbf{H}_{LRB} = \Phi_2(\Phi_1(\mathbf{H})) = (A_k \,|\, I_{n-k})$ is the resulting parity check matrix. In analogy to the generator matrix $\mathbf{G}_{MRB}$, the first k columns of the parity check matrix $\mathbf{H}_{LRB}$ are information coordinates and the remaining columns are the redundant columns.

In the code space C, a list of test codewords is created based on the MRB. In contrast, the dual code space generates a list of error vectors by means of the LRB. With aid of these vectors, the test code words are calculated. Vectors are considered based on an error vector that can create the same syndrome as the hard decision permuted received vector. The syndrome is obtained by

$$\mathbf{s} = (s_1, \ldots, s_{n-k}) = \mathbf{z}^H \cdot \mathbf{H}_{LRB}^T \,.$$

$\mathbf{f}_0 = (0, \ldots, 0, s_1, \ldots, s_{n-k}) \in GF(2)^n$ is the solution of $\mathbf{f} \cdot \mathbf{H}^T = \mathbf{s} \bmod 2$. The decoding of order 0 with the dual code is defined using the constructed codewords $\mathbf{z}_0 = \mathbf{z}^H \oplus \mathbf{f}_0$ and $\hat{\mathbf{x}} = \Phi_2^{-1}(\Phi_1^{-1}(\mathbf{z}_0))$. We assume that the reliable information symbols are error-free. As an analog to decoding in code space C, we obtain an improved decoding result by considering errors in these reliable coordinate positions. In contrast to decoding in code space C, it is not enough to set the corresponding positions of $\mathbf{f}$ to 1. The error vector $\mathbf{f}$ must produce the syndrome $\mathbf{s}$. For $1 \leq l \leq k$, [FLS98] presents the decoding algorithm 7.13.

Algorithm 7.13 Decoding of order l using the dual code.

Initialization: Calculate $\mathbf{H}_{LRB}$ and $\mathbf{z}_0$ and set $\hat{\mathbf{z}} = \mathbf{z}_0$.

Step 1: For the case: $1 \leq i \leq l$,

- calculate the set of error vectors
  $\mathcal{F}_i = \{\mathbf{f} = (\mathbf{f}_k|\mathbf{f}_{n-k}) \in GF(2)^n \mid \text{wt}(\mathbf{f}_{n-k}) = i \wedge \mathbf{f} \cdot \mathbf{H}^T = \mathbf{s} \bmod 2\}$.
- Calculate from the set of codeword $\mathcal{L}_i = \{\mathbf{z}^H \oplus \mathbf{f} | \mathbf{f} \in \mathcal{F}_i\}$
  a codeword $\bar{\mathbf{z}}$ that has the smallest Euclidean distance to $\mathbf{z}$.
- If $d_E(\bar{\mathbf{z}}, \mathbf{z}) < d_E(\hat{\mathbf{z}}, \mathbf{z})$ then $\hat{\mathbf{z}} = \bar{\mathbf{z}}$.

Step 2: $\hat{\mathbf{c}} = \Phi_1^{-1}(\Phi_2^{-1}(\hat{\mathbf{z}}))$.

Remarks

- For the above decoding algorithm, [FLS98] derives special SDML acceptance criteria which are very efficient for several codes and approach the optimal decoding limit. From these, a hybrid decoding algorithm is presented that combines the MRB and the LRB techniques.

- The principle of decoding of order l using the dual code decoding algorithms is presented in [Dor74], [Omu70] and [BBLK98] as special cases.

7.5 Decoding as an optimization problem

In this section, soft-decision decoding is considered as an *optimization problem*. We follow the notation presented in [BBLK98]. It is shown that soft-decision decoding algorithms can be based on optimization theory.

The target function of the optimization (decoding) is minimization of the quadratic Euclidian distance. According to equation 7.13, this is equivalent to maximization of the inner product of a codeword $\mathbf{x}$ and the received sequence $\mathbf{y}$:

$$\max_{\mathbf{x}} \{\langle \mathbf{x}, \mathbf{y} \rangle\}, \quad \mathbf{x} \in \mathcal{C} \Leftrightarrow \mathbf{c}\mathbf{H}^T = \mathbf{0} \quad \bmod 2\,. \tag{7.49}$$

A related optimization problem, which is solved using the *simplex* algorithm (see [NeWo]), is as follows:

> Determine a vector $\mathbf{z}$ with the target function $F(\mathbf{z}) = \langle \mathbf{c}, \mathbf{z} \rangle$ that is minimized under the condition $\mathbf{z}\mathbf{A} = \mathbf{b}$, $\mathbf{z} \in \mathbb{R}_+^n$, where $\mathbf{A}$ is an $n \times m$ matrix with $m < n$.

In comparing this instruction set with the decoding instruction set (equation 7.49), the two are identical up to the modulo calculation and the requirement of a integer solution in the case of the decoding instruction (one observes that a minimization problem can be transformed into a maximization problem by a change of sign in the solution).

The idea of the simplex algorithm is to set $n - \text{rank}(\mathbf{A})$ coordinates in $\mathbf{z}$ to zero and solve $\mathbf{z} \cdot \mathbf{A} = \mathbf{b}$ with the remaining coordinates. The coordinates set to zero are called *basis variables*, the others correspond to *non-basis variables*. A basis variable is switched with a non-basis variable if the solution of $\mathbf{z}' \cdot \mathcal{A} = \mathbf{b}$ improves the target function such that $F(\mathbf{z}') < F(\mathbf{z})$. This switch is repeated until an improvement is no longer possible. One observes that the simplex algorithm gives the *global* optimal solution.

The dual code ordered statistics algorithm presented in section 7.4.5 is similar to the simplex algorithm. It has a modulo calculation and a integer solution. We again compare this algorithm with the simplex algorithm.

Consider the codeword $\mathbf{c}$ and the transmitted codeword $\mathbf{x}$. The corresponding received vector is $\mathbf{y}$ and the corresponding hard-decision vector is $\mathbf{r}$. The parity check matrix of the code $\mathcal{C}$ is $\mathbf{H}$. With $\mathbf{r} = \mathbf{c} \oplus \mathbf{f}$, we obtain the syndrome $\mathbf{s} = \mathbf{f} \cdot \mathbf{H}^T \bmod 2$. According to theorem 7.14 and equation 7.13, we can construct an equivalent target function that is the minimization of the weighted Hamming distance $\sum_{l \in \text{supp}(\mathbf{f})} |y_l|$, $\mathbf{f} \cdot \mathbf{H}^T = \mathbf{s}$. For dual code ordered statistics decoding, we assume that the coordinates of the received vector $\mathbf{y}$ are ordered such that positions $k+1, \ldots, n$ are occupied by the coordinates with the lowest reliability information values. This corresponds to the columns in the parity check matrix that are linearly independent. The corresponding parity check matrix has the form $\mathbf{H} = (\mathbf{A}_k | \mathbf{I}_{n-k})$, where $\mathbf{A}_k$ is an $(n-k) \times k$ matrix and $\mathbf{I}_{n-k}$ is the $(n-k) \times (n-k)$ unit matrix. The coordinates $n-k, \ldots, n$ can be interpreted as basis variables in analogy to the simplex algorithm. The ordered statistics decoding of order 1 calculates the vector $\mathbf{f}'$ such that all non-basis variables $e'_1, \ldots, e'_k$ are equal to zero with the exception of $e'_m = 1$, $1 \leq m \leq k$. The basis variables are subsequently calculated such that $\mathbf{f}' \cdot \mathbf{H}^T = \mathbf{s}$. After calculation of all possible vectors $\mathbf{f}'$, it is checked whether a gain in the target function is achieved, i.e. $\sum_{l \in \text{supp}(\mathbf{f}')} |y_l| < \sum_{l \in \text{supp}(\mathbf{f})} |y_l|$. One assumes that there exists at least one basis variable with $e'_l = 0$, $e_l = 1$, $k+1 \leq l \leq n$, otherwise no improvement in the target function is possible. If there are more variants that improve the target function then the variant with the largest gain is chosen. Therefore the mth and lth coordinates of $\mathbf{f}'$ and $\mathbf{y}$ are switched. Correspondingly, the lth and mth columns of the parity check matrix are exchanged to obtain the form $\mathbf{H} = (\mathbf{A}'_k | \mathbf{I}_{n-k})$. $\mathbf{f}$ is then replaced by $\mathbf{f}'$. This search is continued until an improvement in the target function is no longer possible.

Ordered statistics decoding does not always find the optimum solution due to the the nonlinear property of the $\mathbf{f} \cdot \mathbf{H}^T = \mathbf{s} \bmod 2$ calculation in the decoding method. Nevertheless, this example shows that techniques from classical optimization theory can be applied to decoding with the appropriate considerations. In the following, the classical simplex optimization algorithm is integrated as an SDML decoding algorithm.

First, the modulo calculation and the corresponding nonlinearity of the decoding technique are solved. $\mathbf{c} \cdot \mathbf{H}^T = \mathbf{0} \bmod 2$ is replaced by $\frac{1}{2}\mathbf{c} \cdot \mathbf{H}^T - \mathbf{q} = 0$, where $\mathbf{q}$ is a vector with integer coordinates of length $n-k$. The restriction that the coordinates of $\mathbf{c}$ be elements of $GF(2)$ can be written as $\mathbf{c} + \mathbf{r} = \mathbf{1}_n$, where $\mathbf{1}_n$ is the all-ones vector with length n and $\mathbf{r} = \mathbf{y}^H$ is the hard-decision received vector calculated from $\mathbf{y}$. Furthermore, c_j and r_j are non-negative, as required by the simplex algorithm. With $\mathbf{z} = (\mathbf{c}, \mathbf{q}, \mathbf{r})$ and

$$\mathbf{A} = \left(\begin{array}{c|c|c} -\frac{1}{2}\mathbf{H} & \mathbf{I}_{n-k} & \mathbf{0} \\ \hline \mathbf{I}_n & \mathbf{0} & \mathbf{I}_n \end{array} \right),$$

both restrictions can be joined such that $\mathbf{z} \cdot \mathbf{A}^T = (\mathbf{0}_{n-k}, \mathbf{1}_n)$. $\mathbf{I}_n$ and $\mathbf{I}_{n-k}$ are the n- and $(n-k)$-dimensional unit matrices. $\mathbf{0}_{n-k}$ is the zero vector with length $n-k$. The target function is given by

$$\min_{\mathbf{z}} \{ F(\mathbf{z}) = (y_0, \ldots, y_{n-1}, 0, \ldots, 0)\mathbf{z}^r \}. \tag{7.50}$$

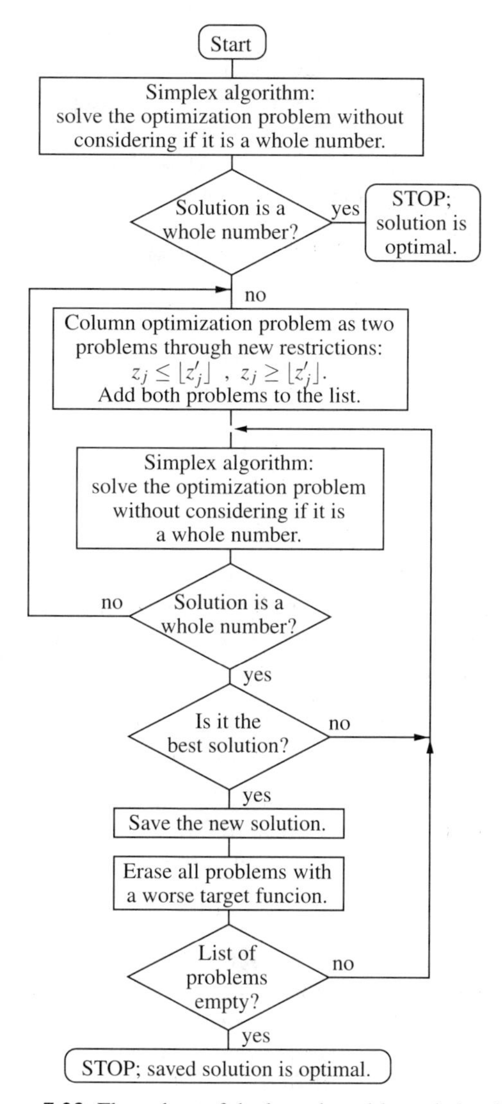

Figure 7.23 Flow chart of the branch-and-bound algorithm.

Equation 7.50 represents a classical integer optimization problem that can be solved with the *branch-and-bound* algorithm. A branch-and-bound algorithm is presented in figure 7.23.

The optimization problem is first solved without considering the integer property. If the solution $\mathbf{z}'$ is not an integer in all coordinates then the optimization problem is divided into two independent problems. The restriction $z_j \leq \lfloor z'_j \rfloor$ is the first problem and $z_j \geq \lfloor z'_j \rfloor$ is the second problem. Therefore z'_j can be any real element from $\mathbf{z}'$. Next, both problems are considered in relation to the original unsolved list of problems. If a integer solution can be found then we obtain an upper bound for the global optimum of the target function. Solutions (including real number solutions) that are above this bound are removed from the list.

These prerequisites often result in finding the global optimum. In general, the complexity in unfavorable cases can be very high. An analysis of the decoding complexity and efficient variations of the branch-and-bound technique can be found in [BBLK98].

In summary, this section shows that the adaptation of the simplex algorithm of the solution of linear and real number optimization problems can be carried out for ordered statistics decoding using dual codes. In contrast to the simplex algorithm, the ordered statistics decoding technique is suboptimal. The extension of this problem is based on a classical branch-and-bound method for the solution of the simplex algorithm. This results in an SDML decoding algorithm. The example shows that the decoding can be interpreted as a special case of an optimization problem.

7.6 Summary

The history of soft-decision decoding of block codes begins with work done by Gallager [Gal62] (*low-density parity check codes*) and Massey [Mas] (*threshold decoding*), which have been made realizable through iterative decoding. The computational capability of today's computers puts a limit on the practical performance of these algorithms. In 1966, Forney presented the GMD algorithm [For66b], which has been greatly improved since its introduction.

In the 1960s, list decoding based on algorithms from Chase [Cha72], Weldon [Wel71], and Dorsch [Dor74] were developed. The Weldon algorithm can be considered as a special case of the Chase algorithm. All three are precursors to ordered statistics decoding, where the Dorsch algorithm works with the dual of the original code. Optimal symbolwise MAP decoding was introduced by Bahl et al. [BCJR74] and Hartmann and Rudolph [HR76]. An extension in this area was presented by Battail et al. [BDG79]. They reinterpreted the ideas of Massey, and developed techniques leading to iterative decoding. Bahl et al. [BCJR74] performed MAP decoding while Wolf [Wolf78] performed SDML decoding using the Viterbi algorithm, however, both used the same trellis representation of block codes. Another successful theorem that uses graph theory is the SDML decoding using the Dijkstra algorithm [HHC93].

At the beginning of the 1980s, Evseev [Evs83] showed that the complexity of ML decoding of block codes could be significantly reduced through approximation techniques. The decoding algorithm (DA) developed by Bossert [BH86] can be considered as a MAP approximation. This approach results in a significant reduction in ML decoding complexity. Further work on soft-decision decoding and SDML acceptance criteria came from Enns [Enns87], Krouk [Kro89] and Be'ery and Snyders [BS86].

In recent years, research in the area of soft-decision decoding has generalized earlier decoding techniques as well as introducing new concepts. The work by Fossorier and Lin [FL95] on ordered statistics decoding in both the code and dual code spaces has been presented. Principles of iterative decoding from Hagenauer et al. [HOP96] and Lucas et al. [LBB96, LBBG96] have also been considered. In [KNIH94], Kaneko et al. generalized the Chase algorithm, which resulted in a ML decoding technique. How soft-decision decoding can be treated as a optimization problem is investigated by Breitbach et al. in [BBLK98].

In this chapter, we have focused on the decoding of block codes. First we have introduced several channel models, including the q-ary symmetric channel, the Gaussian channel and the Gilbert–Elliot channel model, which is useful for simulating bundle errors. We have then introduced different metrics that give us a distance measure needed for evaluation. These metrics include the Hamming and the Euclidian metric. The L value has been presented in order to deal with information reliability decisions.

Several aspects of complexity and coding gain completed the introductory sections. Next we focused on decoding techniques that do not use reliability information. For permutation decoding, a set of automorphisms with special characteristics is used. For *covering polynomial* decoding, a set of polynomials with particular properties is generated. Single-step and multi-step majority decoding requires a set of parity check vectors or codewords from the dual code that are orthogonal to a coordinate position. A special case of permutation decoding is *error trapping*, which is useful in correcting burst errors, but poor in correcting errors that have a general distribution. The DA algorithm is a method of correcting errors with weight larger than half the minimum distance.

Decoding with information reliability has been explained (soft-decision). These methods are classified as codeword or code symbol decoding and as optimal or sub-optimal. Many famous algorithms can be classified as list decoding methods. We have presented the principles of list decoding. An overview of the algorithms that have been considered here is shown in figure 7.10.

In conclusion, we have explained the relationship between optimization algorithms and decoding. The simplex algorithm and branch-and-bound algorithms have been considered.

7.7 Problems

Problem 7.1
Standard array decoding is a maximum-likelihood decoding method.

(a) Create a standard array for the repetition code with length 4.

(b) How many vectors are contained in the standard array for the Hamming code with length 7? Give all *coset leaders*.

Problem 7.2

(a) Design a permutation decoder for the cyclic single-error correcting Hamming code ($n = 15$, $k = 11$) with the generator polynomial $g(x) = x^4 + x + 1$.

(b) Decode the received polynomial $r(x) = 1 + x + x^7 + x^9 + x^{12}$.

Problem 7.3
Can the Golay code $\mathcal{G}_{23}$ be decoded using single-step majority decoding up to half the minimum distance?

Problem 7.4
Given a two-error correcting RS code $\mathcal{C}$ with length $n = 15$:

(a) How many errors and/or erasures can be corrected if two redundant coordinates are not sent?

(b) Shorten the code to $n^* = 8$ and design a permutation decoder for $\mathcal{C}^*$.

Problem 7.5
Given a two-error correcting BCH code with length $n = 15$ and parameters $k = 7$, $d = 5$. The set $\mathcal{B}$ of decoding vectors comprises all cyclic shifts of the polynomial $b(x)$. For the binary linear code, use the decoding algorithm (DA) to decode the received polynomial $r(x)$.

$$b(x) = x^7 + x^{11} + x^{13} + x^{14} \quad \text{and} \quad r(x) = x^5 + x^{10} .$$

8

Convolutional codes

In addition to block codes, convolutional codes are the other major class of codes for error correction. Convolutional codes are based on a linear mapping of a set of information words to a set of codewords. Conceptually, information and codewords are of infinite length, and therefore they are mostly referred to as information and code sequences. The existence of a maximum-likelihood decoding procedure that can be implemented with reasonable complexity is the reason for their widespread use. An important fact is that soft input can be used and soft-output can be created when decoding convolutional codes.

In section 8.1, we shall first discuss some fundamentals of convolutional codes. Different representations of convolutional codes will be introduced. The generation of a code sequence is carried out using a time-discrete and discrete-valued (digital) linear time-invariant (LTI) system: the convolutional encoder. The systems engineering representation leads to the generator matrix, i.e. the impulse response of the circuit in time domain and the system function in frequency domain. This way, we obtain a concrete mathematical description of a mapping between information and codewords. Since a digital LTI system has a finite number of states, it can also be described as a graph: the state diagram. The so-called trellis is derived from the state diagram via the code tree. The trellis is also a graph, but with a special structure. It is the backbone for ML and MAP decoding of convolutional codes. At the end of the first section, we shall have a look at the puncturing of convolutional codes. This method allows a great deal of flexibility in the selection of a rate for a convolutional code, and also allows the design of systems with *unequal error protection*.

Some aspects of algebraic description will be discussed in section 8.2. In this section, the following items are carefully defined:

- *Code:* the set of all code sequences that can be created with a linear mapping.
- *Generator matrix:* a rule for mapping information to code sequences. A given encoding structure can be realized through different mappings.

- *Encoder:* the realization of a generator matrix as a digital LTI system.

We emphasize differences among code properties, properties of the generator matrix and how the generator matrix is realized. For example, one convolutional code can be generated by several different generator matrices. These include systematic and catastrophic generator matrices among others. Hence systematic and catastrophic are properties of the generator matrix, not of the code.

The most important distance measures for convolutional codes (generator matrices) will be introduced in section 8.3.

The decoding of convolutional codes with the Viterbi algorithm (VA) will be explained in section 8.4. The representation of the code as a trellis plays a crucial role for decoding. We shall also describe the so-called soft-output Viterbi algorithm (SOVA), which provides additional information about the reliability of the output.

We shall discuss two symbol-by-symbol APP algorithms, the BCJR algorithm and the related max log algorithm, in section 8.5. These two techniques provide reliability information at the output, and therefore they are the preferred choice in concatenated and iterative systems. This form of decoding is also based on the trellis representation of the code.

In contrast to the previous two sections, a totally different method of decoding, called sequential decoding, will be discussed in section 8.6. Decoding is not performed in the trellis but in the code tree. At this point, we shall introduce the Fano and ZJ algorithms.

In section 8.7, we shall discuss a method that allows the construction of (partial) unit memory (PUM) codes, which are a special class of convolutional codes based on block code theory. A BMD decoding procedure can be used for these (P)UM codes.

Section 8.8 contains some tables of good convolutional codes.

8.1 Fundamentals of convolutional codes

A binary information sequence $\mathbf{u}$ is partitioned into blocks $\mathbf{u}_i$ with k bits. Each information block is mapped by the encoder to a code block $\mathbf{v}_i$, consisting of n bits. The ith code block $\mathbf{v}_i$ depends on the ith information block $\mathbf{u}_i$ and the m previous information blocks, i.e. the encoder has *memory* m:

$$\mathbf{u} = \mathbf{u}_0, \mathbf{u}_1, \mathbf{u}_2, \ldots \quad \longrightarrow \quad \mathbf{v} = \mathbf{v}_0, \mathbf{v}_1, \mathbf{v}_2, \ldots, \tag{8.1}$$

with $\mathbf{v}_t = \mathrm{fct}\,(\mathbf{u}_{t-m}, \ldots, \mathbf{u}_t)$ and $\mathbf{u}_t \in GF(2)^k$, $\mathbf{v}_t \in GF(2)^n$. The corresponding convolutional code with code rate $R = k/n$ is the set of all possible sequences that can be generated with a $(n,k,[m])$ convolutional encoder, and is called $C(n,k,[m])$.

8.1.1 Encoding with sequential logic

The encoder can be realized as *sequential circuit* with k inputs, n outputs and a number of memory elements. Three examples are given below.

Example 8.1 (Convolutional encoder for a $C(2,1,[2])$ code) Figure 8.1 shows a $(2,1,[2])$ encoder, consisting of a shift register of length $m = 2$, and $n = 2$ output symbols consisting of linear combinations of the memory contents of the shift registers and the current input symbol.

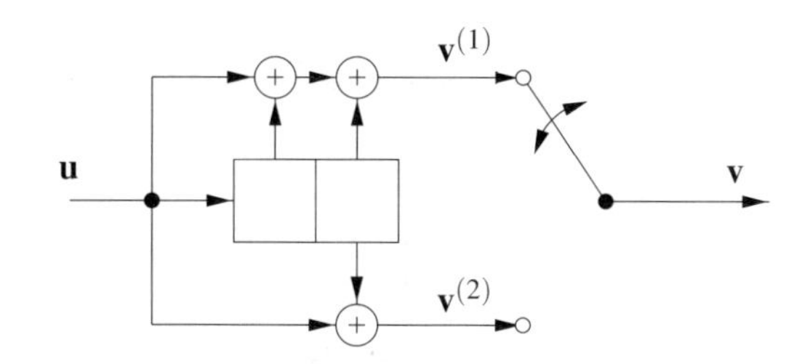

Figure 8.1 Encoder for a $C(2, 1, [2])$ convolutional code.

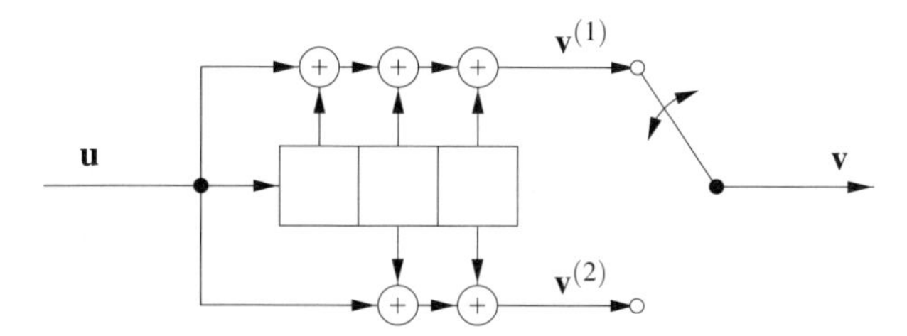

Figure 8.2 Encoder for a $C(2, 1, [3])$ convolutional code.

The information sequence $\mathbf{u} = (\mathbf{u}_0, \mathbf{u}_1, \mathbf{u}_2, \ldots)$ is shifted bitwise ($k = 1$) into the encoder, i.e. time instance t corresponds to bit u_t. The $n = 2$ output sequences

$$\mathbf{v}^{(1)} = (v_0^{(1)}, v_1^{(1)}, v_2^{(1)}, \ldots) \quad \text{and} \quad \mathbf{v}^{(2)} = (v_0^{(2)}, v_1^{(2)}, v_2^{(2)}, \ldots)$$

are generated, which gives the code sequence

$$\mathbf{v} = ((v_0^{(1)} v_0^{(2)}), (v_1^{(1)} v_1^{(2)}), (v_2^{(1)} v_2^{(2)}), \ldots) = (\mathbf{v}_0, \mathbf{v}_1, \mathbf{v}_2, \ldots). \qquad \diamond$$

Example 8.2 (Convolutional encoder for a $C(2, 1, [3])$ code) Figure 8.2 shows an encoder for a $C(2, 1, [3])$ code with memory $m = 3$. $\diamond$

In the third example, we want to have a look at an encoder with $k > 1$, i.e. more than one bit is shifted into the encoder per time instance t.

Example 8.3 (Convolutional encoder for a $C(3, 2, [1])$ code) An encoder for a $C(3, 2, [1])$ code is shown in figure 8.3. The information sequence $\mathbf{u}$ is split into blocks consisting of $k = 2$ bits. Each block element is parallelly shifted into the encoder. The following notation is used:

$$\mathbf{u}^{(1)} = (u_0^{(1)}, u_1^{(1)}, u_2^{(1)}, \ldots) \text{ and } \mathbf{u}^{(2)} = (u_0^{(2)}, u_1^{(2)}, u_2^{(2)}, \ldots).$$

Hence the information sequence is

$$\mathbf{u} = ((u_0^{(1)} u_0^{(2)}), (u_1^{(1)} u_1^{(2)}), \ldots) = (\mathbf{u}_0, \mathbf{u}_1, \mathbf{u}_2, \ldots),$$

with $\mathbf{u}_i = (u_i^{(1)} u_i^{(2)})$. Each of the two input sequences $\mathbf{u}^{(i)}$ is shifted in a register of length $m = 1$. $\diamond$

For a given circuit, the code is defined as the set of all code sequences, which is obtained when all possible information sequences are used as input. It is assumed that information as

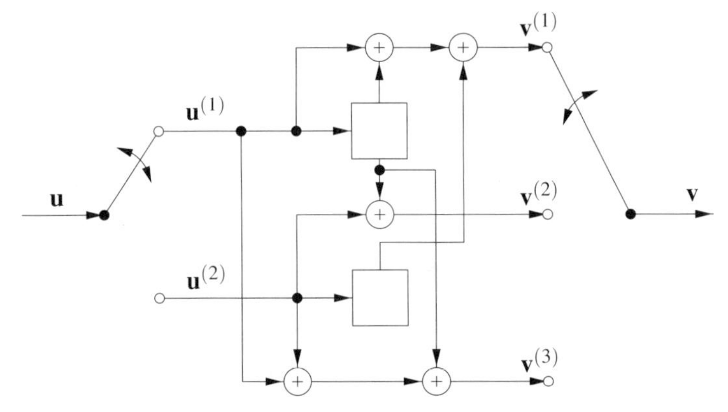

Figure 8.3 Encoder for a $\mathcal{C}(3,2,[1])$ convolutional code.

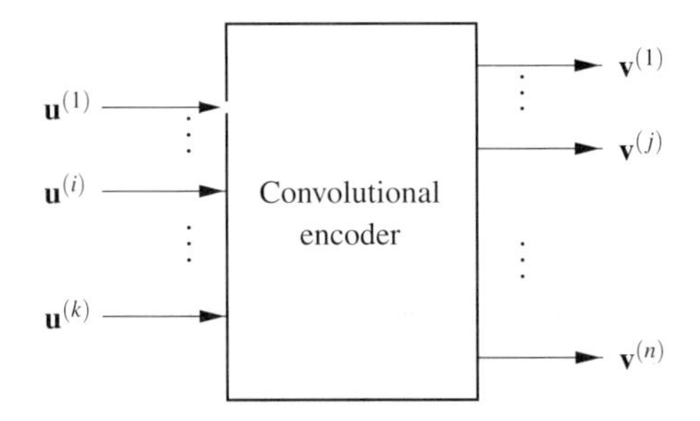

Figure 8.4 Convolutional encoder as an LTI system.

well as code sequences have infinite length. This assumption is made for all further theoretical considerations. For practical applications, one typically needs code sequences of finite length, which we shall describe in section 8.1.7.

Conventionally, all the memory elements are set to zero to initialize the encoder.

8.1.2 Impulse response and convolution

System theory can be used to analyse convolutional codes. The encoders in examples 8.1–8.3 can be represented as *linear time-invariant* (LTI) systems (see e.g. [OpWi]). In general every convolutional encoder, as shown in figure 8.4, can be described as a time-discrete LTI system with k-dimensional input and n-dimensional output.

The jth of the n output sequences $\mathbf{v}^{(j)}$ is obtained by convolving the input sequence with the corresponding system *impulse response*:

$$\mathbf{v}^{(j)} = \mathbf{u}^{(1)} * \mathbf{g}_1^{(j)} + \mathbf{u}^{(2)} * \mathbf{g}_2^{(j)} + \ldots + \mathbf{u}^{(k)} * \mathbf{g}_k^{(j)} = \sum_{i=1}^{k} \mathbf{u}^{(i)} * \mathbf{g}_i^{(j)}, \tag{8.2}$$

where $*$ is the convolution operation. This is the origin of the name *convolutional* code. The impulse response $\mathbf{g}_i^{(j)}$ of the ith input with the response to the jth output is found by stimulating the encoder with the discrete impulse $(1, 0, 0, \ldots)$ at the ith input and by observing the jth output. All other inputs are equal to the zero sequence $(0, 0, 0, \ldots)$. The impulse responses are often called *generator sequences* of the encoder.

Example 8.4 (Code sequences and convolution) We first consider the case $k = 1$. The two impulse responses (generator sequences) for the encoder in figure 8.1 are $\mathbf{g}^{(1)} = (1, 1, 1, 0, \dots)$ and $\mathbf{g}^{(2)} = (1, 0, 1, 0, \dots)$. Accordingly, the impulse responses for the encoder in figure 8.2 are $\mathbf{g}^{(1)} = (1, 1, 1, 1)$ and $\mathbf{g}^{(2)} = (1, 0, 1, 1)$. Both generator sequences $\mathbf{g}^{(1)}$ and $\mathbf{g}^{(2)}$ are of infinite length. Here and in subsequent discussions, only the first non-zero part of the sequence is considered.

The encoding equations for the two encoders are

$$\mathbf{v}^{(1)} = \mathbf{u} * \mathbf{g}^{(1)} \text{ and } ;\mathbf{v}^{(2)} = \mathbf{u} * \mathbf{g}^{(2)}.$$

In figure 8.2, the information sequence $\mathbf{u} = (1, 0, 1, 1, 1, 0, 0, \dots)$ is mapped by the convolutional encoder to the output sequences

$$\begin{aligned} \mathbf{v}^{(1)} &= (1, 0, 1, 1, 1, 0, 0, \dots) * (1, 1, 1, 1) = (1, 1, 0, 1, 1, 1, 0, 1, 0, 0, \dots), \\ \mathbf{v}^{(2)} &= (1, 0, 1, 1, 1, 0, 0, \dots) * (1, 0, 1, 1) = (1, 0, 0, 0, 0, 0, 0, 1, 0, 0, \dots), \end{aligned}$$

which form the code sequence

$$\mathbf{v} = (11, 10, 00, 10, 10, 10, 00, 11, 00, 00, \dots).$$

The discrete convolution represents a multiplication, where the components are calculated modulo 2:

$$\begin{array}{cccccccccccccc} & & & & 1 & 0 & 1 & 1 & 1 & * & 1 & 0 & 1 & 1 & = 10000001 \quad \bmod 2 \\ \hline & 1 & 0 & 1 & 1 & 1 & & & & & & & & & \\ + & & & 1 & 0 & 1 & 1 & 1 & & & & & & & \\ + & & & & 1 & 0 & 1 & 1 & 1 & & & & & & \\ \hline = & 1 & 0 & 2 & 2 & 2 & 2 & 2 & 1 & & & & & & \end{array}$$

◇

Example 8.5 (Generator sequences of the $(3, 2, [1])$ convolutional encoder) The generator sequences for the encoder in figure 8.3 can be defined similarly to the previous example. Three generator sequences ($n = 3$) exist for each of the two input sequences ($k = 2$), i.e. one sequence exists for each output:

$$\mathbf{g}_i^{(j)} = (g_{i,0}^{(j)}, g_{i,1}^{(j)}, \dots, g_{i,m}^{(j)}) \text{ for } i \in [1,2] \text{ and } j \in [1,3].$$

The generator sequences for this encoder are

$$\begin{aligned} \mathbf{g}_1^{(1)} &= (11), \quad \mathbf{g}_1^{(2)} = (01), \quad \mathbf{g}_1^{(3)} = (11), \\ \mathbf{g}_2^{(1)} &= (01), \quad \mathbf{g}_2^{(2)} = (10), \quad \mathbf{g}_2^{(3)} = (10). \end{aligned}$$

The encoding equations are

$$\begin{aligned} \mathbf{v}^{(1)} &= \mathbf{u}^{(1)} * \mathbf{g}_1^{(1)} + \mathbf{u}^{(2)} * \mathbf{g}_2^{(1)}, \\ \mathbf{v}^{(2)} &= \mathbf{u}^{(1)} * \mathbf{g}_1^{(2)} + \mathbf{u}^{(2)} * \mathbf{g}_2^{(2)}, \\ \mathbf{v}^{(3)} &= \mathbf{u}^{(1)} * \mathbf{g}_1^{(3)} + \mathbf{u}^{(2)} * \mathbf{g}_2^{(3)}. \end{aligned}$$

◇

8.1.3 Constraint length, memory and overall constraint length

The three examples of convolutional encoders that have been presented are FIR (*finite impulse response*) systems, i.e. all generator sequences have finite length, and their maximum length $m+1$ determines the memory m of the encoder.

The encoder stores m bits for the case $k=1$. The memory m is therefore equal to the number of storage elements in the encoder. Therefore m is often referred to as the memory of the encoder. This notation is a somewhat misleading for the case where $k>1$.

Constraint length We first consider only the ith input. ν_i incoming bits are stored in the encoder. The corresponding generator sequences $\mathbf{g}_i^{(j)}$, $j \in [1,n]$, have maximum length ν_i+1. ν_i is called the *constraint length* of the ith input sequence. Different input sequences can have different constraint lengths; therefore the following definitions should be used:

Memory The *memory* m of the encoder is

$$m = \max_i \nu_i, \tag{8.3}$$

and is therefore the largest of all k constraint lengths.

Overall constraint length The number of storage elements ν of an encoder is

$$\nu = \sum_{i=1}^{k} \nu_i,$$

and is called the *overall constraint length* .

The term *encoder memory* is often used ambiguously – either as the number of storage elements, i.e. the overall constraint length ν, or as the number of information blocks that affects a code block, i.e. the memory m. The overall constraint length is equal to the memory ($\nu = m$). For the case where $k=1$, no distinction is necessary. If $k>1$, the exact terminology should be used. Therefore the new notation $C(n,k,[\nu])$ will be used from now on, instead of $C(n,k,[m])$.

Example 8.6 (Constraint length ν_i of a $(4,3,[\nu=3])$ convolutional encoder) The $(4,3,[\nu=3])$ encoder[1] in figure 8.5 consists of $k=3$ shift registers with different numbers of storage elements. The constraint length of the first input sequence $\mathbf{u}^{(1)}$ is $\nu_1=0$ (hence no shift register is necessary for this input sequence!), the constraint length of the second input sequence $\mathbf{u}^{(2)}$ is $\nu_2=1$, and for $\mathbf{u}^{(3)}$ it is $\nu_3=2$. The memory of the encoder is given by the largest shift register, and is therefore $m=2$. The overall constraint length is $\nu=3$, and corresponds to the number of memory elements of the encoder. ◇

In principal, convolutional codes can also be IIR (*infinite impulse response*) systems, which have an infinite impulse response. These will be covered in section 8.1.8.

---

[1] The partitioning of the information sequence $\mathbf{u}$ in k parallel input sequences $\mathbf{u}^{(i)}$, $i \in [1,k]$, and the generation of the serial output of the code sequence $\mathbf{v}$ from the n parallel code sequences $\mathbf{v}^{(j)}$, $j \in [1,n]$, is not shown in this and all following representations.

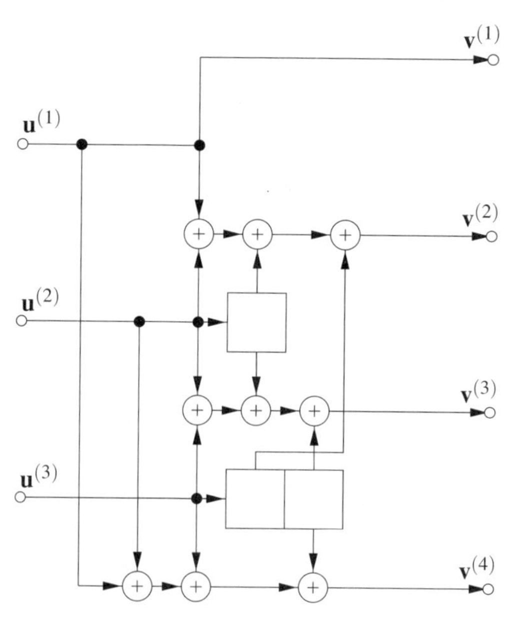

Figure 8.5 Encoder for a $C(4, 3, [\nu = 3])$ convolutional code.

8.1.4 Generator matrix in the time domain

The coding equations can be described by a matrix multiplication instead of a convolution. Then the generator sequences are organized into a semi-infinite matrix $\mathbf{G}$, which is called the *generator matrix*. With reference to the previous examples, an $(n, k, [\nu])$ convolutional encoder maps an information sequence, which is sorted in blocks with k bits,

$$\mathbf{u} = (\mathbf{u}_0, \mathbf{u}_1, \mathbf{u}_2, \ldots, \mathbf{u}_t, \ldots), \text{ with } \mathbf{u}_t = (u_t^{(1)}, u_t^{(2)}, \ldots, u_t^{(k)}) \in GF(2)^k,$$

to a code sequence consisting of blocks with n bits,

$$\mathbf{v} = (\mathbf{v}_0, \mathbf{v}_1, \mathbf{v}_2, \ldots, \mathbf{v}_t, \ldots), \text{ with } \mathbf{v}_t = (v_t^{(1)}, v_t^{(2)}, \ldots, v_t^{(n)}) \in GF(2)^n.$$

One can interpret the information sequence as being divided into k parallel input sequences, i.e.

$$\mathbf{u}^{(i)} = (u_0^{(i)}, u_1^{(i)}, u_2^{(i)}, \ldots), \quad i \in [1, k].$$

Each of these k sequences is fed into a shift register. In general, these can be of different lengths. The memory m of the convolutional encoder corresponds to the number of storage elements of the largest shift register. Based on the current bits of the input sequence and the contents of the registers, n parallel output sequences

$$\mathbf{v}^{(j)} = (v_0^{(j)}, v_1^{(j)}, v_2^{(j)}, \ldots), \quad j \in [1, n],$$

are generated using linear circuit logic. The serial output is the code sequence. The mapping from information to code sequence can be defined by multiplication by the generator matrix:

$$\mathbf{v} = \mathbf{u} \cdot \mathbf{G}.$$

The the generator matrix looks like

$$\mathbf{G} = \begin{pmatrix} \mathbf{G}_0 & \mathbf{G}_1 & \mathbf{G}_2 & \ldots & \mathbf{G}_m & & & \\ & \mathbf{G}_0 & \mathbf{G}_1 & \ldots & \mathbf{G}_{m-1} & \mathbf{G}_m & & \\ & & \mathbf{G}_0 & \ldots & \mathbf{G}_{m-2} & \mathbf{G}_{m-1} & \mathbf{G}_m & \\ & & & \ddots & & & & \ddots \end{pmatrix},$$

with the $k \times n$ submatrices

$$\mathbf{G}_l = \begin{pmatrix} g_{1,l}^{(1)} & g_{1,l}^{(2)} & \ldots & g_{1,l}^{(n)} \\ g_{2,l}^{(1)} & g_{2,l}^{(2)} & \ldots & g_{2,l}^{(n)} \\ \vdots & \vdots & & \vdots \\ g_{k,l}^{(1)} & g_{k,l}^{(2)} & \ldots & g_{k,l}^{(n)} \end{pmatrix}, \quad \text{for} \quad l \in [0, m].$$

The elements $g_{i,l}^{(j)}$, for $i \in [1,k]$ and $j \in [1,n]$, are obtained from the impulse responses (equation 8.2) of the ith input to the jth output:

$$\mathbf{g}_i^{(j)} = (g_{i,0}^{(j)}, g_{i,1}^{(j)}, \ldots, g_{i,l}^{(j)}, \ldots, g_{i,m}^{(j)}).$$

Example 8.7 (Generator matrix of a $(2, 1, [3])$ encoder) When considering the encoder from figure 8.2, the generator sequences are $\mathbf{g}^{(1)} = (1, 1, 1, 1)$ and $\mathbf{g}^{(2)} = (1, 0, 1, 1)$. These are written in matrix form as

$$\mathbf{G} = \begin{pmatrix} g_0^{(1)} g_0^{(2)} & g_1^{(1)} g_1^{(2)} & g_2^{(1)} g_2^{(2)} & \ldots & g_m^{(1)} g_m^{(2)} & & & \\ & g_0^{(1)} g_0^{(2)} & g_1^{(1)} g_1^{(2)} & \ldots & g_{m-1}^{(1)} g_{m-1}^{(2)} & g_m^{(1)} g_m^{(2)} & & \\ & & g_0^{(1)} g_0^{(2)} & \ldots & g_{m-2}^{(1)} g_{m-2}^{(2)} & g_{m-1}^{(1)} g_{m-1}^{(2)} & g_m^{(1)} g_m^{(2)} & \\ & & & \ddots & & & & \ddots \end{pmatrix},$$

i.e. encoding the information sequence $\mathbf{u} = (1, 0, 1, 1, 1, 0, \ldots)$ yields (see example 8.4) the code sequence

$$\begin{aligned} \mathbf{v} &= \mathbf{u} \cdot \mathbf{G} \\ &= (1, 0, 1, 1, 1, 0, \ldots) \cdot \begin{pmatrix} 11 & 10 & 11 & 11 & & & & \\ & 11 & 10 & 11 & 11 & & & \\ & & 11 & 10 & 11 & 11 & & \\ & & & 11 & 10 & 11 & 11 & \\ & & & & 11 & 10 & 11 & 11 \\ & & & & & \ddots & & & \ddots \end{pmatrix} \\ &= (11, 10, 00, 10, 10, 10, 00, 11, \ldots). \end{aligned}$$

◇

Example 8.8 (Generator matrix of a $(3,2,[2])$ encoder) Consider the $(3,2,[2])$ convolutional encoder from figure 8.3. The information sequence $\mathbf{u}$ is shifted into the encoder in blocks with $k=2$ bits. These information blocks are then split into two parallel input sequences:

$$\mathbf{u}^{(1)} = (u_0^{(1)}, u_1^{(1)}, u_2^{(1)}, \ldots) \text{ and } \mathbf{u}^{(2)} = (u_0^{(2)}, u_1^{(2)}, u_2^{(2)}, \ldots).$$

The information sequence can be written as

$$\mathbf{u} = (u_0^{(1)}\, u_0^{(2)}, u_1^{(1)}\, u_1^{(2)}, \ldots) = (\mathbf{u}_0, \mathbf{u}_1, \mathbf{u}_2, \ldots), \text{ with } \mathbf{u}_i = (u_i^{(1)}\, u_i^{(2)}).$$

Both input sequences $\mathbf{u}^{(i)}$ are fed into a shift register of length $m=1$. Generator sequences can be defined similarly to the previous examples. Three generator sequences exist, one for each output sequence, i.e.

$$\mathbf{g}_i^{(j)} = (g_{i,0}^{(j)}, g_{i,1}^{(j)}, \ldots, g_{i,m}^{(j)}), \text{ for } i \in [1,k] \text{ and } j \in [1,n].$$

We know from example 8.5 that

$$\begin{aligned} \mathbf{g}_1^{(1)} &= (1\,1), & \mathbf{g}_1^{(2)} &= (0\,1), & \mathbf{g}_1^{(3)} &= (1\,1), \\ \mathbf{g}_2^{(1)} &= (0\,1), & \mathbf{g}_2^{(2)} &= (1\,0), & \mathbf{g}_2^{(3)} &= (1\,0). \end{aligned}$$

We want to find the form of the encoding matrix, as in the previous example. In general, the generator matrix of a $(3,2,[\nu])$ encoder is given by

$$\mathbf{G} = \begin{pmatrix} g_{1,0}^{(1)} g_{1,0}^{(2)} g_{1,0}^{(3)} & g_{1,1}^{(1)} g_{1,1}^{(2)} g_{1,1}^{(3)} & \cdots & g_{1,m}^{(1)} g_{1,m}^{(2)} g_{1,m}^{(3)} & & \\ g_{2,0}^{(1)} g_{2,0}^{(2)} g_{2,0}^{(3)} & g_{2,1}^{(1)} g_{2,1}^{(2)} g_{2,1}^{(3)} & \cdots & g_{2,m}^{(1)} g_{2,m}^{(2)} g_{2,m}^{(3)} & & \\ & g_{1,0}^{(1)} g_{1,0}^{(2)} g_{1,0}^{(3)} & \cdots & g_{1,m-1}^{(1)} g_{1,m-1}^{(2)} g_{1,m-1}^{(3)} & g_{1,m}^{(1)} g_{1,m}^{(2)} g_{1,m}^{(3)} & \\ & g_{2,0}^{(1)} g_{2,0}^{(2)} g_{2,0}^{(3)} & \cdots & g_{2,m-1}^{(1)} g_{2,m-1}^{(2)} g_{2,m-1}^{(3)} & g_{2,m}^{(1)} g_{2,m}^{(2)} g_{2,m}^{(3)} & \\ & & \ddots & & & \ddots \end{pmatrix}.$$

The first two rows are repeated periodically, always shifted by three positions to the right. For the considered encoder, we find that

$$\mathbf{G} = \begin{pmatrix} 101 & 111 & & & \\ 011 & 100 & & & \\ & 101 & 111 & & \\ & 011 & 100 & & \\ & & 101 & 111 & \\ & & 011 & 100 & \\ & & & \ddots & \ddots \end{pmatrix}$$

◇

In general, the procedure to mathematically describe convolutional encoders shown in figures 8.1–8.3 and 8.5 can be summarized by the following steps:

- Determination of the generator sequences $\mathbf{g}_i^{(j)}$ using the convolutional encoder.
- The generator sequences represent the $k \times n$ generator submatrices $\mathbf{G}_i$, $i \in [0,m]$. The submatrices determine the generator matrix.

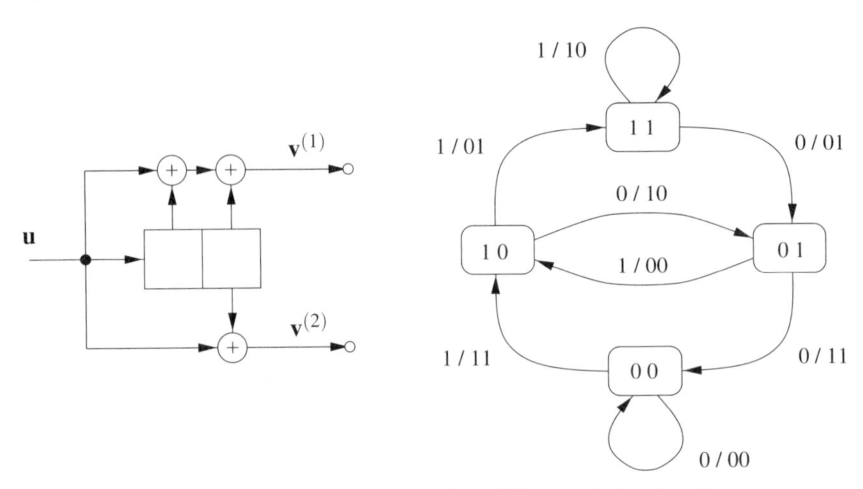

Figure 8.6 State diagram of a $(2, 1, [2])$ convolutional code.

8.1.5 State diagram, code tree and trellis

So far we have used an implementation-oriented representation to describe convolutional codes based on sequential circuits with linear logic. Alternatively, a convolutional encoder or a generated code sequences can be described with a *graph*. We shall discuss three of these graphs in this section.

State diagram A convolutional encoder has a finite number of memory elements ν and therefore a finite number 2^{ν} of memory states σ. Hence an encoder can also be thought of as *finite-state machine*: the output of a code block $\mathbf{v}_t$ at time t depends on the state of the memory σ_t and the information block $\mathbf{u}_t$. At the next time instance $t+1$, the encoder is in state σ_{t+1}. Each change of state $\sigma_t \rightarrow \sigma_{t+1}$ is associated with the input of an information block and the output of a code block.

The *state diagram* is obtained by drawing a graph. In this graph, nodes are the possible states of the convolutional encoder, and the state transitions are labeled with the appropriate inputs and outputs $(\mathbf{u}_t \,/\, \mathbf{v}_t)$ as illustrated in the following example:

Example 8.9 (State diagram) The $(2, 1, [2])$ convolutional encoder with $\nu = m = 2$ memory elements from examples 8.1 and 8.4 can be transformed to the state diagram shown in figure 8.6. The encoder has $2^{\nu} = 4$ memory states $\sigma \in \{(00), (10), (01), (11)\}$. As an example, consider the state transition $(10) \rightarrow (01)$, which is labeled with $0 \,/\, 10$, i.e. the associated information block is $\mathbf{u}_t = (0)$ and the output code block is $\mathbf{v}_t = (10)$. Not all states can be attained by the simple input of one information bit. The machine's memory limits which transitions are possible. For example, state (00) cannot be reached from state (11) with one transition. $\diamond$

Note Convolutional encoders exist with state diagrams that have parallel transitions. This is possible when $k > 1$ and one or more of the constraint lengths of the k input sequences are $\nu_i = 0$.

A relation between code sequence and state sequence exists, which can be expressed using equation 8.1:

$$\mathbf{v}_t = \text{fct}(\mathbf{u}_{t-m}, \ldots, \mathbf{u}_t) = \text{fct}(\sigma_t, \mathbf{u}_t). \tag{8.4}$$

Therefore the state of an encoder depends on the m previous information blocks. For each input sequence $\mathbf{u}^{(i)}$, $i \in 1, \ldots, k$, the encoder stores ν_i bits according to the particular constraint length. Therefore the state of the encoder can be expressed as ν-tuple of the memory values:

$$\sigma_t = \left((u_{t-1}^{(1)} \ldots u_{t-\nu_1}^{(1)}), (u_{t-1}^{(2)} \ldots u_{t-\nu_2}^{(2)}), \ldots, (u_{t-1}^{(k)} \ldots u_{t-\nu_k}^{(k)}) \right). \tag{8.5}$$

We interprete this vector of ν bits as a binary representation of a decimal number. This way, we can describe each state unambiguously with a value $\sigma \in [0, 2^\nu - 1]$. The sequence

$$S = (\sigma_0, \sigma_1, \sigma_2, \ldots, \sigma_t, \ldots),$$

is called a *state sequence*, where the initialization of the encoder is defined as $\sigma_0 = 0$. If no parallel transitions exist then there is a one-to-one correspondence between code sequences and state sequences.

Code tree A convolutional code is the set of all code sequences or codewords, which can be generated with a convolutional encoder. These codewords will now be represented in graphical form as a *tree diagram*, according to the following example. The tree diagram is then called a *code tree*.

Example 8.10 (Code tree) The $(2, 1, [2])$ convolutional encoder from figure 8.6 generates the code tree shown in figure 8.7. Branches pointing up represent an information bit 0, while branches pointing down represent 1. For example, the information sequence $\mathbf{u} = (1, 0, 1, 1, \ldots)$ produces the code sequence $\mathbf{v} = (11, 10, 00, 01, \ldots)$, which corresponds to the path marked in the code tree. ◇

In contrast to the state diagram, the code tree has a time-dependent structure. At each time instance t, also referred to as depth, the tree consists of 2^{kt} nodes. Since the complexity of the code tree grows exponentially with t, it is impractical to use this kind of representation.

When we have a closer look at the code tree, we see that it consists of periodically repeating sub trees. Each node in the tree can be associated with a state σ_t. This state is defined according to equation 8.5, and can be uniquely determined by the path to that node. When labeling the nodes this way, one can see that exactly the same branches leave equal states. These transitions from nodes with equal states can be merged for depth $t > m$, which results in the so-called code *trellis* representation.

Trellis The trellis also contains all code sequences of a convolutional code as paths (similar to the code tree). The code sequences share nodes with equal states. This way, one obtains (after initially growing exponentially) a graph with a constant number of nodes 2^ν at every level $t \geq m$. This kind of representation is especially useful for ML decoding of convolutional codes using the Viterbi algorithm (see section 8.4).

Example 8.11 (Trellis) The convolutional encoder in figure 8.6 has the trellis shown in figure 8.8. All nodes with the same state are placed at one level. It is common practice to put

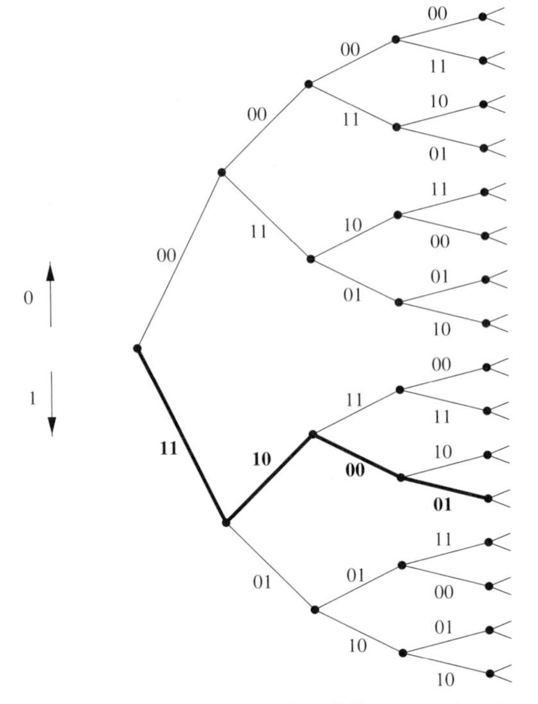

Figure 8.7 Code tree for the $\mathcal{C}(2,1,[2])$ convolutional code.

nodes in the zero state at the highest level. The states are defined by ν bits, according to equation 8.5. They can also be interpreted as binary numbers. Each state therefore corresponds to an integer $\sigma \in \{0, 1, 2, \ldots, 2^{\nu} - 1\}$.

Furthermore, each branch is labeled with a corresponding code block. The labels in figure 8.8 were placed at the right end of the branches, for clarity. Dotted branches correspond to an information bit 0 (solid branches correspond to an information bit 1). As for the code tree in figure 8.7, the code sequence $\mathbf{v} = (11, 10, 00, 01, \ldots)$ is drawn with thicker lines.

The subtrellis in figure 8.9 shows the structure of the trellis in steady state. When looking at equation 8.4 together with this subtrellis, we see that each node always contains information from the past, i.e. the state σ_t of the encoder, while the transitions leaving the node correspond to the current information bits. The codeword is obtained by the branch labels. ◇

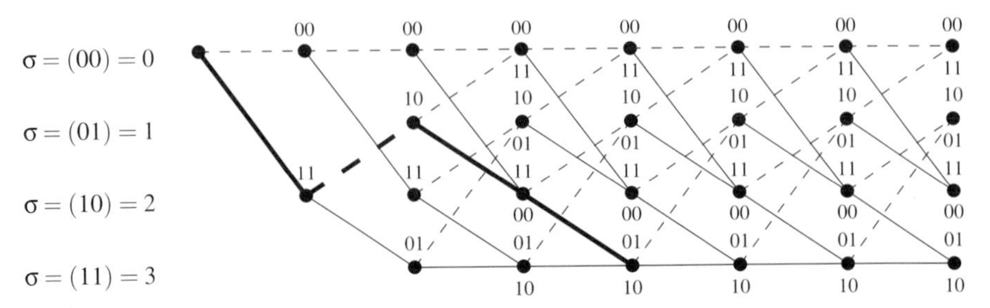

Figure 8.8 Trellis of a $(2,1,[2])$ convolutional code.

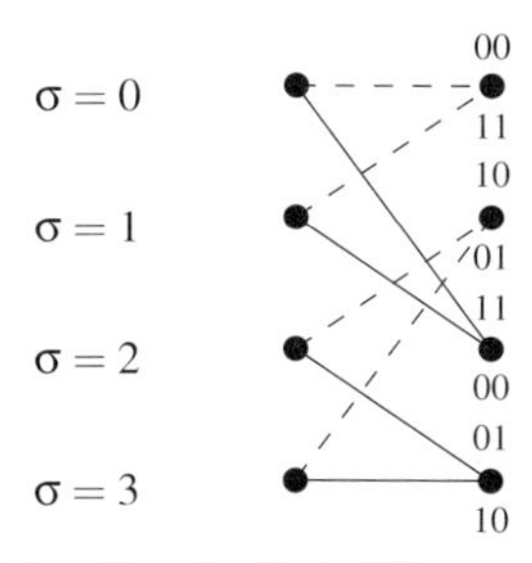

Figure 8.9 Subtrellis of a $(2, 1, [2])$ convolutional code.

All three representations of a convolutional encoder discussed in this section are graphs consisting of labeled branches and nodes. This representation provides an overview of the structure of a convolutional code, and also allows a variety of methods and procedures known from graph theory to be applied to convolutional codes. Some of these considerations are discussed in the next section.

8.1.6 Free distance and path enumerators

Describing the distance properties of convolutional codes is much more complex then for block codes. The problem is to define appropriate distance measures that account for the fact that a convolutional code consists of an infinite number of code sequences with infinite length. Therefore it is not possible (as for linear block codes) to completely describe the distance properties in terms of Euclidean geometry or as a distance or weight distribution. Instead, we shall define subsets from the set of all code sequences and describe the distance properties for these subsets.

Free distance The *free distance* of a convolutional code is the minimum Hamming distance between any two code sequences of the code:

$$d_f = \min_{\mathbf{v}' \neq \mathbf{v}''} \operatorname{dist}(\mathbf{v}', \mathbf{v}''),$$

with $\mathbf{v}', \mathbf{v}'' \in \mathcal{C}$. Because of the linearity of convolutional codes, we may also write

$$d_f = \min_{\mathbf{v} \neq \mathbf{0}} \operatorname{wt}(\mathbf{v}), \tag{8.6}$$

with $\mathbf{v} \in \mathcal{C}$. The free distance is a code property, and therefore does not depend on the mapping of the information sequence to the code sequence.

The calculation of the free distance according to equation 8.6 appears to be difficult because the number of code sequences is infinite. However, the free distance can be determined fairly easily when using the state diagram of the code. We demonstrate this using an example.

Example 8.12 (Free distance of a $\mathcal{C}(2, 1, [2])$ convolutional code) Figure 8.10 shows a modified version of the state diagram of the $(2, 1, [2])$ convolutional encoder already introduced in example 8.9. The transitions are labeled with an operator W. Its power corresponds to the Hamming weight of the code block associated with the transition. All possible code sequences of the code start by definition at time $t = 0$ and state $\sigma_0 = (00)$, and run through an arbitrary

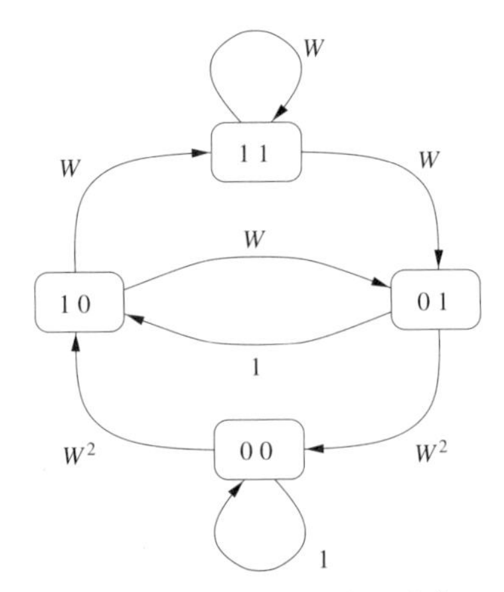

Figure 8.10 Modified state diagram of a $(2, 1, [2])$ convolutional code.

sequence $S = (\sigma_0, \sigma_1, \ldots)$. If the labels of all transitions made are multiplied then the power of W yields the Hamming weight of the corresponding code sequence.

The number of code sequences that are necessary for calculation can be limited in order to evaluate equation 8.6:

- Only code sequences that leave the zero state at time $t = 0$ have to be considered. All other code sequences are just delayed versions.

- Only code sequences with finite weight are considered. These code sequences have to end at zero state again. State diagrams with zero loops, except the one from zero to zero, will be covered in section 8.1.9, and are therefore excluded here.

- In addition, code sequences that leave the zero state more than once can also be excluded, since a code sequence with a smaller weight always exists.

- All code sequences that are not in zero state after $t > 2^\nu$ state transitions can also be excluded. This would require the path to pass through the zero state $\sigma \neq 0$ at least twice, and would form a loop in the state diagram. Therefore a code sequence with smaller weight always exists.

The free distance of the $C(2, 1, [2])$ code can be obtained by considering only those code sequences that leave the zero state at time $t = 0$ and return to it for the first time at the latest when $t = 4$. Looking at figure 8.10, we find that only the following two code sequences satisfy these requirements:

$$
\begin{aligned}
S_1 &= ((00), (10), (01), (00), (00), \ldots), \\
S_2 &= ((00), (10), (11), (01), (00), (00), \ldots).
\end{aligned}
$$

There are no parallel state transitions, and so only two code sequences with these state sequences exist. The multiplication of the labels of the transitions yield $W^5 = W^2 W^1 W^2$ and $W^6 = W^2 W^1 W^1 W^2$ for S_1 and S_2 respectively. Therefore the free distance of the investigated code is $d_f = 5$. ◇

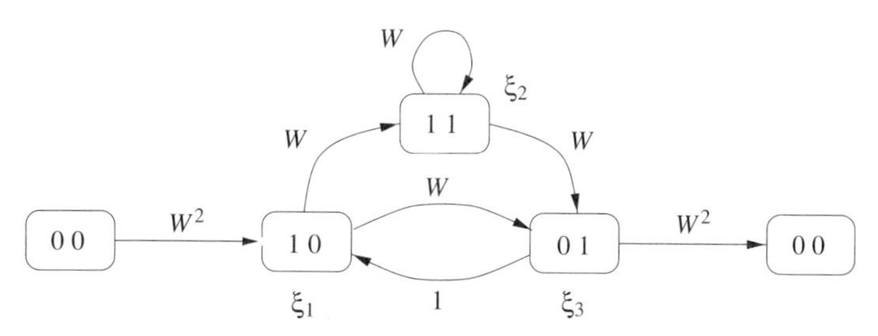

Figure 8.11 Signal flow chart of the $(2,1,[2])$ convolutional code.

The free distance is one of the most important distance measures for convolutional codes. Therefore convolutional codes are typically specified by their rate R, their overall constraint length ν and their free distance d_f (see section 8.8).

Path enumerator Other distance properties of convolutional codes exist, for example the *path enumerator* or the *extended path enumerator*. First we want to explain the calculation of these functions using two examples, and then discuss possible interpretations.

In analogy to example 8.12, the calculation of the path enumerators is based on a subset of the set of all possible code sequences. All sequences of this subset leave the zero state at time $t=0$ and remain there after reaching it again. Hence all sequences that have the history

$$S=(0,\sigma_1,\sigma_2,\ldots,\sigma_{l-1},0,0,\ldots), \tag{8.7}$$

with $\sigma_t \neq 0$ for $t \in [1,l-1]$, will be considered. l must be greater than m, because the zero state cannot be reached before $m+1$ transitions.

Example 8.13 (Distance function) A procedure from signal flow theory is used to calculate the path enumerator. The state diagram shown in example 8.12 (figure 8.10) is sliced through the zero state. The resulting signal flow diagram is shown in figure 8.11. The left node is considered as a source and the right one as a sink.

If the variables ξ_i for $i \in [1,3]$ are assigned to the nodes between source and sink then the following equations hold:

$$\begin{aligned} \xi_1 &= \xi_3 + W^2 \\ \xi_2 &= W\xi_1 + W\xi_2 \qquad \text{and} \qquad T(W) = W^2\xi_3 \\ \xi_3 &= W\xi_1 + W\xi_2 \end{aligned}$$

Each ξ_i is obtained by adding up all branches that point to the node. The weights of the branches are multiplied by the variable and are assigned to the original node. The corresponding system of equations in matrix form is

$$\begin{pmatrix} 1 & 0 & -1 \\ -W & 1-W & 0 \\ -W & -W & 1 \end{pmatrix} \cdot \begin{pmatrix} \xi_1 \\ \xi_2 \\ \xi_3 \end{pmatrix} = \begin{pmatrix} W^2 \\ 0 \\ 0 \end{pmatrix}.$$

The solution of this system of equations, substituted in $T(W)=W^2\xi_3$, yields the closed form of the path enumerator:

$$T(W) = \frac{W^5}{1-2W} = W^5 + 2W^6 + \ldots + 2^j W^{j+5} + \ldots.$$

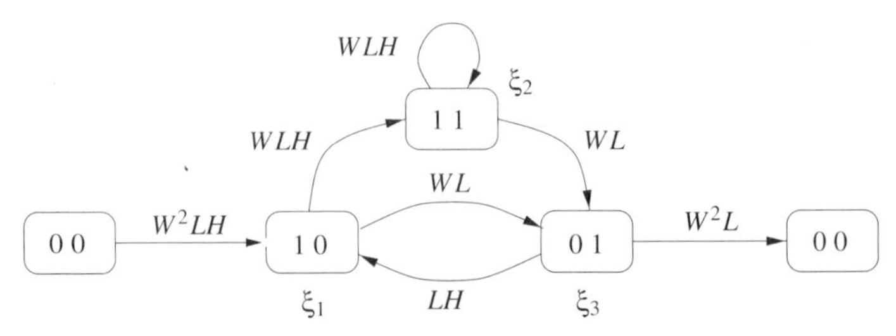

Figure 8.12 Extended signal flow diagram of a $(2, 1, [2])$ convolutional encoder.

which can be expanded in a series with positive powers of W. Their coefficients are equal to the number of code sequences with corresponding weight. For example, a code sequence with weight 5 exists, two with weight 6, and in general 2^j with weight $j+5$.

The power of the first member of this series corresponds to the free distance of the code. We obtain $d_f = 5$, as already calculated in example 8.12. ◇

Example 8.14 (Extended path enumerator) Often, additional information is desired, for example the distance distribution of code sequences of certain length, where length means the number of state transitions before the sequence returns to the zero state. The extended signal flow diagram in figure 8.12 corresponds to that in figure 8.11, but the labels on the transitions have changed. The additional operator L helps counting the state transitions, and the power of H represents the Hamming weight of the information block that is associated with the transition. If the linear system of equations is written out and solved as in the previous example then the closed form of the extended path enumerator is obtained as

$$T(W,H,L) = \frac{W^5HL^3}{1-WH(1+L)L}.$$

We can expand in different series, for example a series in W:

$$T(W,H,L) = L^3HW^5 + L^4(1+L)H^2W^6 + \ldots + L^{3+j}(1+L)^jH^{j+1}W^{5+j} + \ldots.$$

When considering the coefficient $H^2L^4 + H^2L^5$ of W^6, the extended path enumerator says that there are two code sequences with weight 6. Both have their origin in an information sequence with weight 2, where one has length 4 and the other length 5, and so on. ◇

In general, the extended path enumerator can be expressed as a series with three variables:

$$T(W,H,L) = \sum_w \sum_h \sum_l N(l,h,w)\, L^l H^h W^w. \tag{8.8}$$

The coefficients $N(l,h,w)$ of this series specify the numbers of code sequences with Hamming weight w, information weight h and length l. The length is thereby the number of state transitions before the code sequence returns to the zero state. Note that the above statements are valid for the subset according to equation 8.7.

Equation 8.8 can now be interpreted or evaluated differently:

- The path enumerator is obtained as the special case

$$T(W,1,1) = T(W) = \sum_w N(w)W^w, \tag{8.9}$$

where $N(w)$ is the number of code sequences with weight w. $N(w)$ is also called the *distance spectrum* of the convolutional code, which is a property of the code. The free distance of the code is obtained from the distance spectrum:

$$d_f = \arg\left(\min_{w, N(w)\neq 0} N(w)\right)$$

- The weight distribution of all paths with length l,

$$A_l(W) = \sum_w N(l,w)W^w, \qquad (8.10)$$

is obtained from

$$T(W,1,L) = \sum_l \sum_w N(l,w)\,W^w L^l = \sum_l A_l(W)\,L^l.$$

The path enumerator and the extended path enumerator of a convolutional code are of very limited use in practice, since the computational complexity grows exponentially with the memory length ν of the convolutional encoder. In principle, a $((2^\nu - 1) \times (2^\nu - 1))$-dimensional linear system of equations has to be solved. We shall derive other distance measures for convolutional codes in section 8.3.

8.1.7 Termination, truncation and tail-biting

The code sequences of a convolution code are of infinite length. In practical applications, code sequences with finite length are used almost exclusively. Therefore *termination*, *truncation* or *tail-biting* is necessary. These methods will be explained below.

Termination In order to terminate a convolutional code, some bits are appended onto the information sequence in a way that all storage elements in the encoder return to the zero state at the end of the input sequence. This way, the memory of the encoder is in the initial zero state. In the case of an encoder with memory m, which can be described as an FIR system, km zeros are appended to the information sequence in order to terminate it. The rate of the terminated convolutional code is diminished by the *fractional rate loss* $L/(L+m)$ compared with the original code, and one obtains

$$R_T = \frac{kL}{n(L+m)} = R\frac{L}{L+m},$$

where L is the number of encoded information blocks. If L is sufficiently large then it can be assumed that $R_T \approx R$.

The terminated code sequences are obtained from

$$\mathbf{v}_{[L+m]} = \mathbf{u}_{[L]} \cdot \mathbf{G}^r_{[L]},$$

where the generator matrix is clipped after the Lth row (r stands for *row*):

$$\mathbf{G}^r_{[L]} = \begin{pmatrix} \mathbf{G}_0 & \mathbf{G}_1 & \dots & \mathbf{G}_m & & & \\ & \mathbf{G}_0 & \mathbf{G}_1 & \dots & \mathbf{G}_m & & \\ & & \ddots & & & \ddots & \\ & & & \mathbf{G}_0 & \mathbf{G}_1 & \dots & \mathbf{G}_m \end{pmatrix}. \qquad (8.11)$$

Note Convolutional encoders that can be described as IIR systems (see section 8.1.3) are called *recursive* (see section 8.1.8). When recursive encoders have to be terminated, km termination bits can be added to truncate the sequence. A particular termination sequence depends on the state of the encoder, which is not necessarily the all-zeros sequence.

Example 8.15 (Termination) If the encoder from example 8.1 for the $\mathcal{C}(2,1,[2])$ convolutional code with rate $R = 1/2$ is terminated after $L = 8$ information blocks then we obtain from the semi-finite code matrix $\mathbf{G}$ the 8×20 generator matrix

$$\mathbf{G}^r_{[8]} = \begin{pmatrix} 11 & 10 & 11 & & & & & & & \\ & 11 & 10 & 11 & & & & & & \\ & & 11 & 10 & 11 & & & & & \\ & & & 11 & 10 & 11 & & & & \\ & & & & 11 & 10 & 11 & & & \\ & & & & & 11 & 10 & 11 & & \\ & & & & & & 11 & 10 & 11 & \\ & & & & & & & 11 & 10 & 11 \end{pmatrix}.$$

The fractional rate loss is $L/(L+m) = 0.8$, and the rate of the terminated convolutional code is $R_T = 0.4 < 0.5$. ◇

Truncation A second option for generating finite code sequences is to simply stop the output of the encoder for $t > L$, no matter what the present state is. Although code blocks $\mathbf{v}_t \neq \mathbf{0}$ for $t \geq L$ exist, the code sequence is clipped after L code blocks, i.e. for the last encoded information block $\mathbf{u}_{L-1}$, the corresponding code block $\mathbf{v}_{L-1}$ is generated, but subsequent blocks are not. Therefore the rate does not change, as was the case with termination. After discussing the decoding of convolutional codes (sections 8.4 to 8.6), it will become clear that with this procedure the information bits at the end of a block have substantially worse error protection.

The clipped code sequences are obtained from

$$\mathbf{v}_{[L]} = \mathbf{u}_{[L]} \cdot \mathbf{G}^c_{[L]},$$

where the generator matrix is clipped after the Lth column (c stands for *column*):

$$\mathbf{G}^c_{[L]} = \begin{pmatrix} \mathbf{G}_0 & \mathbf{G}_1 & \dots & \mathbf{G}_m & & & \\ & \mathbf{G}_0 & \mathbf{G}_1 & \dots & \mathbf{G}_m & & \\ & & \ddots & & & \ddots & \\ & & & \mathbf{G}_0 & & & \mathbf{G}_m \\ & & & & \ddots & & \vdots \\ & & & & & \mathbf{G}_0 & \mathbf{G}_1 \\ & & & & & & \mathbf{G}_0 \end{pmatrix}. \quad (8.12)$$

Example 8.16 (Truncation) The same $\mathcal{C}(2,1,[2])$ code is clipped after $L = 8$ information blocks. We obtain the 8×16 generator matrix

$$\mathbf{G}^c_{[8]} = \begin{pmatrix} 11 & 10 & 11 & & & & & \\ & 11 & 10 & 11 & & & & \\ & & 11 & 10 & 11 & & & \\ & & & 11 & 10 & 11 & & \\ & & & & 11 & 10 & 11 & \\ & & & & & 11 & 10 & 11 \\ & & & & & & 11 & 10 \\ & & & & & & & 11 \end{pmatrix}.$$

The last information bit u_7 has less error protection: it only affects the code block $\mathbf{v}_7$, whereas, for example, the information bit u_0 at the beginning of the information sequence affects $\mathbf{v}_0$, $\mathbf{v}_1$ and $\mathbf{v}_2$, according to the constraint length $\nu_1 = 2$ (section 8.1.3). If an error occurs in code block $\mathbf{v}_7$ then this can lead to an error when decoding u_7. As we demonstrated in example 8.12, the free distance of the unclipped code is $d_f = 5$. Therefore it can always correct $e = \lfloor (d_f - 1)/2 \rfloor = 2$ errors. For a truncated code, the minimum distance of the last block is decreased, which in turn decreases the error correcting capability ◇

Tail-biting A third possibility to generate finite code sequences using convolutional encoding is called tail-biting. The underlying idea is to start the convolutional encoder in the same state in which it will stop after the input of L information blocks. This requires knowledge of the last m blocks before encoding can be started. The final state of the encoder can be calculated and used for initialization. This way, an equal protection of all information bits of the entire block is possible. The generated code is termed a quasicyclic block code.

The encoding equations in matrix form are

$$\mathbf{v}_{[L]} = \mathbf{u}_{[L]} \cdot \tilde{\mathbf{G}}^c_{[L]}.$$

The generator matrix has to be clipped after the Lth column and manipulated as follows:

$$\tilde{\mathbf{G}}^c_{[L]} = \begin{pmatrix} \mathbf{G}_0 & \mathbf{G}_1 & \dots & \mathbf{G}_m & & & \\ & \mathbf{G}_0 & \mathbf{G}_1 & \dots & \mathbf{G}_m & & \\ & & \ddots & & & \ddots & \\ & & & \mathbf{G}_0 & & & \mathbf{G}_m \\ \mathbf{G}_m & & & & \ddots & & \vdots \\ \vdots & \ddots & & & & \mathbf{G}_0 & \mathbf{G}_1 \\ \mathbf{G}_1 & \dots & \mathbf{G}_m & & & & \mathbf{G}_0 \end{pmatrix}.$$

The submatrices that have been clipped in the last km rows are mapped to the appropriate columns in the lower left corner of the matrix.

Example 8.17 (Tail-biting) In analogy to the above example, the $C(2, 1, [2])$ code is clipped after $L = 8$. The corresponding encoding matrix is

$$\tilde{\mathbf{G}}^c_{[8]} = \begin{pmatrix} 11 & 10 & 11 & & & & & \\ & 11 & 10 & 11 & & & & \\ & & 11 & 10 & 11 & & & \\ & & & 11 & 10 & 11 & & \\ & & & & 11 & 10 & 11 & \\ & & & & & 11 & 10 & 11 \\ 11 & & & & & & 11 & 10 \\ 10 & 11 & & & & & & 11 \end{pmatrix}.$$

As can be seen, the last two information bits affect the first code bits. When looking at the convolutional encoder in figure 8.1, one can see that this encoding matrix can be implemented with a properly initialized convolutional encoder. The initial state is obtained from $\mathbf{u}_6$ and $\mathbf{u}_7$. ◇

Relation between block and convolutional codes There is a duality between block and convolutional codes. This can be seen through the different techniques for creating finite-length codes.

- A convolutional code maps information blocks of length k to code blocks of length n. This linear mapping contains memory, because the code block depends on m previous information blocks. In this sense, block codes are a special case of convolutional codes, i.e. convolutional codes without memory.
- On the other hand, in practical applications, convolutional codes have code sequences of finite length. When looking at the finite generator matrix of the created code in time domain, we find that it has a special structure. However, because the generator matrix of a block code with corresponding dimension generally does not have to have a special structure, convolutional codes with finite length can be considered as a special case of block codes.

8.1.8 Generator matrix in the $\mathcal{Z}$ domain

In the field of systems theory, time-discrete linear time-invariant (LTI) systems, and therefore convolutional codes, can be described by *difference equations*. As shown in equation 8.2, the jth output sequence $\mathbf{v}^{(j)} = (v_0^{(j)}, v_1^{(j)}, \ldots)$ is obtained from the convolution of the k input sequences $\mathbf{u}^{(i)}$ with the associated impulse responses $\mathbf{g}_i^{(j)}$ of the encoder:

$$\mathbf{v}^{(j)} = \mathbf{u}^{(1)} * \mathbf{g}_1^{(j)} + \mathbf{u}^{(2)} * \mathbf{g}_2^{(j)} + \ldots + \mathbf{u}^{(k)} * \mathbf{g}_k^{(j)} = \sum_{i=1}^{k} \mathbf{u}^{(i)} * \mathbf{g}_i^{(j)}. \tag{8.13}$$

The difference equations upon which the encoding equations are based upon are

$$\begin{aligned} v_t^{(j)} &= \sum_{i=1}^{k} \sum_{l=0}^{m} u_{t-l}^{(i)} g_{i,l}^{(j)} \\ &= \sum_{i=1}^{k} (u_t^{(i)} g_{i,0}^{(j)} + u_{t-1}^{(i)} g_{i,1}^{(j)} + \ldots + u_{t-m}^{(i)} g_{i,m}^{(j)}). \end{aligned} \tag{8.14}$$

We can use the $\mathcal{Z}$ transform [OpWi] to represent this equation in the $\mathcal{Z}$ *domain* (also termed the *frequency domain*). The $\mathcal{Z}$ transform $\mathcal{Z}\{\mathbf{x}\} = \sum_{t=0}^{+\infty} x_t z^{-t}$ of a sequence $\mathbf{x} = (x_0, x_1, x_2, \ldots)$ is defined as $X(z)$. To describe convolutional codes, it is common practice to introduce the *delay operator* $D = z^{-1}$ and we obtain $X(D) = \sum_{t=0}^{+\infty} x_t D^t$. The input and output sequences are therefore given in the $\mathcal{Z}$ domain by

$$\mathbf{u}^{(i)} \quad \circ\!\!-\!\!\bullet \quad U_i(D) = \sum_{t=0}^{+\infty} u_t^{(i)} D^t,$$

$$\mathbf{v}^{(j)} \quad \circ\!\!-\!\!\bullet \quad V_j(D) = \sum_{t=0}^{+\infty} v_t^{(j)} D^t.$$

System function According to systems theory, the transformed impulse responses are called *system functions*:

$$\mathbf{g}_i^{(j)} \quad \circ\!\!-\!\!\bullet \quad G_{ij}(D) = \sum_{t=0}^{+\infty} g_{i,t}^{(j)} D^t. \tag{8.15}$$

The convolutional relation of the $\mathcal{Z}$ transform $\mathcal{Z}\{\mathbf{x} * \mathbf{y}\} = X(D)Y(D)$ is used to transform the convolution in equation 8.13 to a multiplication in the $\mathcal{Z}$ domain.

$$V_j(D) = \sum_{i=1}^{k} U_i(D) \cdot G_{ij}(D). \tag{8.16}$$

This gives the frequency-domain solution of the difference equation in equation 8.14.

Example 8.18 (System function of a $(2, 1, [3])$ encoder) First we consider an encoder with $k = 1$, namely the $(2, 1, [3])$ encoder in figure 8.2. Its generator sequences are

$$\mathbf{g}_1^{(1)} = (1, 1, 1, 1) \quad \text{and} \quad \mathbf{g}_1^{(2)} = (1, 0, 1, 1).$$

The associated transformed equations are

$$G_{11}(D) = 1 + D + D^2 + D^3 \quad \text{and} \quad G_{12}(D) = 1 + D^2 + D^3.$$

The associated difference equations are

$$v_t^{(1)} = u_t + u_{t-1} + u_{t-2} + u_{t-3} \quad \text{and} \quad v_t^{(2)} = u_t + u_{t-2} + u_{t-3}.$$

We can derive the transformed equations directly from the difference equations:

$$\begin{aligned}
V_1(D) &= \sum_{t=0}^{+\infty} (u_t + u_{t-1} + u_{t-2} + u_{t-3}) D^t \\
&= \sum_{t=0}^{+\infty} u_t D^t + \sum_{t=0}^{+\infty} u_{t-1} D^t + \sum_{t=0}^{+\infty} u_{t-2} D^t + \sum_{t=0}^{+\infty} u_{t-3} D^t \\
&= \sum_{t=0}^{+\infty} u_t D^t + D \sum_{t=0}^{+\infty} u_t D^t + D^2 \sum_{t=0}^{+\infty} u_t D^t + D^3 \sum_{t=0}^{+\infty} u_t D^t \\
&= U_1(D)(1 + D + D^2 + D^3) \\
&= U_1(D) G_{11}(D),
\end{aligned}$$

where the information bits $u_t = 0$ for $t < 0$. ◇

Generator matrix If the k input and n output sequences are written as vectors in the transformed domain,

$$\begin{aligned}\mathbf{U}(D) &= (U_1(D),\ldots,U_k(D)),\\ \mathbf{V}(D) &= (V_1(D),V_2(D),\ldots,V_n(D)),\end{aligned}$$

then equation 8.16 can be expressed in matrix form:

$$\mathbf{V}(D) = \mathbf{U}(D)\cdot\mathbf{G}(D)$$

with the matrix elements $G_{ij}(D)$ from equation 8.15. The matrix $\mathbf{G}(D)$ is the generator matrix in the $\mathcal{Z}$ domain.

Example 8.19 (Generator matrix of a $(3,2,[2])$ encoder) The generator sequences $\mathbf{g}_i^{(j)}$ of the $(3,2,[2])$ encoder (example 8.5, figure 8.3) make up the generator polynomials $G_{ij}(D)$ in the $\mathcal{Z}$ domain. These are, as in the time domain, the system function of the ith input sequence $U_i(D)$ to the jth output $V_j(D)$. The transforms of the generator sequences in the $\mathcal{Z}$ domain are

$$\begin{array}{lll} G_{11}(D) = 1+D, & G_{12}(D) = D, & G_{13}(D) = 1+D,\\ G_{21}(D) = D, & G_{22}(D) = 1, & G_{23}(D) = 1.\end{array}$$

These generator polynomials can be expressed as the $k\times n$ generator matrix

$$\begin{aligned}\mathbf{G}(D) &= \begin{pmatrix} 1+D & D & 1+D\\ D & 1 & 1\end{pmatrix} = \begin{pmatrix} 1 & 0 & 1\\ 0 & 1 & 1\end{pmatrix} + \begin{pmatrix} 1 & 1 & 1\\ 1 & 0 & 0\end{pmatrix} D\\ &= \mathbf{G}_0 + \mathbf{G}_1 D,\end{aligned}$$

where the relation to the submatrices $\mathbf{G}_0$ and $\mathbf{G}_1$ (defined in the time domain) can be seen in the last equation (compare with example 8.8). ◇

All previous examples have dealt with FIR systems. We can also describe IIR (infinite impulse response) systems by using the representation in the $\mathcal{Z}$ domain. Let us first consider an example.

Example 8.20 (IIR system) An IIR system is shown in figure 8.13. If the encoder is described using its impulse response then one faces the problem of an infinite output sequence:

$$\begin{aligned}\mathbf{g}_1^{(1)} &= (1),\\ \mathbf{g}_1^{(2)} &= (1,1,1,0,1,1,0,1,1,0,\ldots).\end{aligned}$$

Although the encoder only has a finite number of memory elements $\nu_1 = 2$, the second generator sequence is of infinite length. This is because of the recursive structure of the encoder: a linear combination of the memory contents is fed back to its input.

To carry out the analysis in the $\mathcal{Z}$ domain, we first have to describe the encoder with difference equations:

$$\begin{aligned} v_t^{(1)} &= u_t^{(1)},\\ v_t^{(2)} + v_{t-1}^{(2)} + v_{t-2}^{(2)} &= u_t^{(1)} + u_{t-2}^{(1)}.\end{aligned}$$

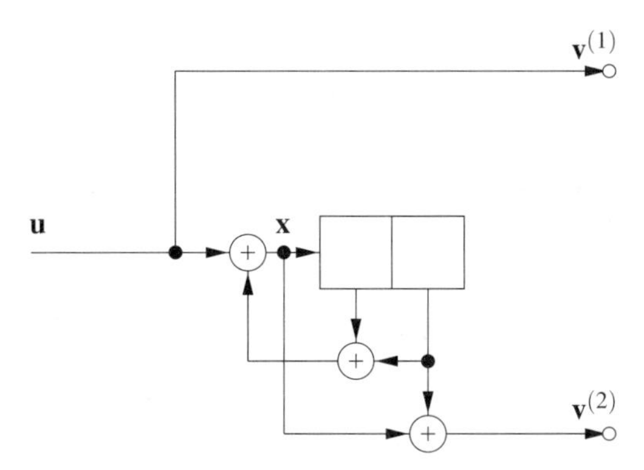

Figure 8.13 Recursive $(2, 1, [2])$ convolutional encoder.

The second equation is obtained by introducing the variable **x**, i.e. from $x_t = u_t + x_{t-1} + x_{t-2}$ and $v_t^{(2)} = x_t + x_{t-2}$. If we now apply the $\mathcal{Z}$ transform to these equations, we have

$$
\begin{aligned}
\sum_{t=0}^{+\infty} v_t^{(2)} D^t + \sum_{t=0}^{+\infty} v_{t-1}^{(2)} D^t + \sum_{t=0}^{+\infty} v_{t-2}^{(2)} D^t &= \sum_{t=0}^{+\infty} u_t^{(1)} D^t + \sum_{t=0}^{+\infty} u_{t-2}^{(1)} D^t, \\
V_2(D) + DV_2(D) + D^2 V_2(D) &= U_1(D) + D^2 U_1(D).
\end{aligned}
$$

Hence the difference equations, which could only be solved recursively in the time domain, become a polynomial equation in D, giving the system function:

$$
V_2(D) = \frac{1+D^2}{1+D+D^2} U_1(D) = G_{12}(D) U_1(D).
$$

The first output sequence corresponds to the input sequence that gives $V^{(1)}(D) = U_1(D)$, i.e. $G_{11}(D) = 1$. The generator matrix of the convolutional encoder is obtained:

$$
\mathbf{G}(D) = \begin{pmatrix} 1 & \frac{1+D^2}{1+D+D^2} \end{pmatrix}.
$$

To verify this equation, the impulse response can be calculated with $G_{12}(D) = \mathcal{Z}^{-1}(\mathbf{g}_1^{(2)})$, the inverse of the system function. This is done by formal expansion in a series of increasing powers of D,

$$
g_{1,l}^{(2)} = \left. \frac{\partial^l G_{12}(D)}{\partial D^l} \right|_{D=0},
$$

and one obtains the impulse response $g_1^{(2)} = (1\,110\,110\,110\ldots)$, as at the beginning of the example. The calculation of this series can be done as a simple polynomial division! ◇

Systems with feedback are very hard to describe in the time domain because of their infinite impulse response. If the difference equations of the system are written out and then transformed, one obtains a representation of the system function having rational functions of

D. The mapping from information sequence to code sequence can be carried out in the $\mathcal{Z}$ domain by a matrix multiplication for arbitrary linear systems, i.e. FIR and IIR systems:

$$\mathbf{v}(D) = \mathbf{u}(D) \cdot \mathbf{G}(D).$$

The generator matrix $\mathbf{G}(D)$ is then

$$\mathbf{G}(D) = \begin{pmatrix} G_{1,1}(D) & G_{1,2}(D) & \dots & G_{1,n}(D) \\ G_{2,1}(D) & G_{2,2}(D) & \dots & G_{2,n}(D) \\ \vdots & \vdots & & \vdots \\ G_{k,1}(D) & G_{k,2}(D) & \dots & G_{k,n}(D) \end{pmatrix},$$

where the matrix elements $G_{ij}(D)$ describe the system function of the ith input to the jth output. These are, in general, rational functions

$$G_{ij}(D) = \frac{p_0 + p_1 D + \ldots + p_m D^m}{1 + q_1 D + \ldots + q_m D^m}.$$

Table 8.1 shows a summary of the representations in the time and $\mathcal{Z}$ domains.

8.1.9 Systematic and catastrophic generator matrices

The generator matrix $\mathbf{G}(D)$ of a convolutional encoder describes the mapping from information to code sequence. We shall cover the algebraic properties of generator matrices in section 8.2; however, we want to introduce two important types of generator matrices now.

Systematic generator matrix As with block codes, the redundancy and information bits can be distinct. The encoder is then called *systematic*:

$$\begin{aligned} \mathbf{u}(D) &= (U_1(D), U_2(D), \ldots, U_k(D)) \rightarrow \\ \mathbf{v}(D) &= (U_1(D), U_2(D), \ldots, U_k(D), V_{k+1}(D) \ldots, V_n(D)). \end{aligned} \tag{8.17}$$

A generator matrix that incorporates a systematic mapping from the information sequence to the code sequence has the following structure:

$$\begin{aligned} \mathbf{G}(D) &= \begin{pmatrix} 1 & & & G_{1,k+1}(D) & \dots & G_{1,n} \\ & 1 & & \vdots & & \vdots \\ & & \ddots & & & \\ & & 1 & G_{k,k+1}(D) & & G_{k,n}(D) \end{pmatrix} \\ &= (\mathbf{I}_k \mid \mathbf{R}(D)), \end{aligned} \tag{8.18}$$

where $\mathbf{I}_k$ is the $k \times k$ unit matrix and $\mathbf{R}(D)$ is a $k \times (n-k)$ matrix with rational elements. Systematic encoders can be FIR as well as IIR systems. Of course, the input sequences do not have to be equal to the first k output sequences, as shown in equation 8.17. Any permutation is allowed. The generator matrix in equation 8.18 is then changed accordingly.

Example 8.21 (Systematic encoder) The convolutional encoder shown in figure 8.13 is systematic. ◇

Table 8.1 Summary: time and $\mathcal{Z}$ domains.

$\mathbf{u} \circ\!\!-\!\!\bullet\ \mathbf{u}(D)$	
ith input sequence $\mathbf{u}^{(i)} = (u_0^{(i)}, u_1^{(i)}, u_2^{(i)}, \ldots)$, for $i \in [1,k]$	ith input sequence $U_i(D) = u_0^{(i)} + u_1^{(i)}D + u_2^{(i)}D^2 + \ldots,$ for $i \in [1,k]$
information sequence $\mathbf{u} = ((u_0^{(1)}, \ldots, u_0^{(k)}), (u_1^{(1)}, \ldots, u_1^{(k)}), \ldots)$	information sequence $\mathbf{u}(D) = (U_1(D), U_2(D), \ldots, U_k(D))$

$\mathbf{v} \circ\!\!-\!\!\bullet\ \mathbf{v}(D)$	
jth output sequence $\mathbf{v}^{(j)} = (v_0^{(j)}, v_1^{(j)}, v_2^{(j)}, \ldots)$, for $j \in [1,n]$	jth output sequence $V_j(D) = v_0^{(j)} + v_1^{(j)}D + v_2^{(j)}D^2 + \ldots,$ for $j \in [1,n]$
code sequence $\mathbf{v} = ((v_0^{(1)}, \ldots, v_0^{(n)}), (v_1^{(1)}, \ldots, v_1^{(n)}), \ldots)$	code sequence $\mathbf{v}(D) = (V_1(D), V_2(D), \ldots, V_n(D))$

$\mathbf{G} \circ\!\!-\!\!\bullet\ \mathbf{G}(D)$	
generator matrix of FIR systems $\mathbf{G} = \begin{pmatrix} \mathbf{G}_0 & \ldots & \mathbf{G}_m & & \\ & \mathbf{G}_0 & \ldots & \mathbf{G}_m & \\ & & \ddots & & \ddots \end{pmatrix},$ with the $k \times n$ submatrices $\mathbf{G}_l$ for $l \in [0,m]$	generator matrix of FIR systems $\mathbf{G}(D) = \mathbf{G}_0 + \mathbf{G}_1 D + \ldots + \mathbf{G}_m D^m$

$\mathbf{G} \circ\!\!-\!\!\bullet\ \mathbf{G}(D)$	
generator matrix of IIR systems $\mathbf{G} = \begin{pmatrix} \mathbf{G}_0 & \mathbf{G}_1 & \ldots & \\ & \mathbf{G}_0 & \mathbf{G}_1 & \ldots \\ & & \ddots & \end{pmatrix},$ with the $k \times n$ submatrices $\mathbf{G}_l = (1/l!)[\partial^l/\partial D^l\ \mathbf{G}(D)]\big\rvert_{D=0}$ for $l \in [0,+\infty)$, where $\mathbf{G}_l$ is a Taylor series expansion.	generator matrix of IIR systems $\mathbf{G}(D) = \begin{pmatrix} G_{1,1}(D) & \ldots & G_{1,n}(D) \\ \vdots & & \vdots \\ G_{k,1}(D) & \ldots & G_{k,n}(D) \end{pmatrix},$ with $G_{ij}(D) = \frac{p_0 + p_1 D + \ldots + p_m D^m}{1 + q_1 D + \ldots + q_m D^m}$

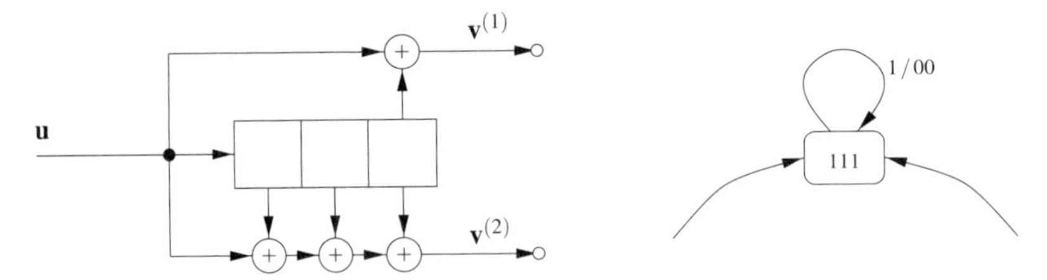

Figure 8.14 Catastrophic encoder.

Catastrophic generator matrix Another class of generator matrices comprises *catastrophic* generator matrices. These have to be avoided under all circumstances, because they perform a mapping from information sequences with infinite Hamming weight to code sequences with finite Hamming weight. Therefore a finite number of transmission errors can cause an infinite number of errors in the decoded information sequence.

We shall discuss methods to test whether or not a generator matrix is catastrophic in section 8.2.6. Furthermore, a number of special generator matrices exists that guarantee non-catastrophic behavior – for example, systematic generator matrices are never catastrophic.

A loop in the state diagram that produces a zero output for a non-zero input is a necessary and sufficient condition for a catastrophic matrix. The loop $0_t \xrightarrow{\mathbf{u}_t=0} 0_{t+1}$ in the zero state is an exception to this rule. A loop is generally defined as a closed path in the state diagram, possibly including many different states.

Example 8.22 (Catastrophic encoder) The convolutional encoder in figure 8.14 is catastrophic. The code block for the state transition $(111) \rightarrow (111)$ is (00). Hence the encoder has a zero loop and is catastrophic. ◇

Although the existence of a zero loop in the state diagram is a very clear condition for a catastrophic encoder, it is of less importance in practice, since other effective methods to test for a catastrophic behavior of an encoding matrix exist, which we shall describe later on.

8.1.10 Punctured convolutional codes

The set of *punctured convolutional codes* is a subset of all convolutional codes. Before we describe the applications and advantages of these codes, we shall explain their construction.

Mother code, deletion map and punctured code Punctured convolutional codes are derived from the mother code. This is done by periodically deleting (puncturing) certain code bits in the code sequence of the mother code:

$$\mathcal{C}_m(n_m, k_m, [\nu_m]) \xrightarrow{\mathcal{S}(p,w)} \mathcal{C}_p(n_p, k_p, [\nu_p]). \tag{8.19}$$

The pattern $\mathcal{S}$ that is used for deletion of symbols (*deletion map*) is characterized by the period p and the number $p - w$ of punctured bits per period. The parameter w corresponds to the number of remaining bits per period. Therefore the rate of the punctured code is

$$R_p = \frac{p}{w} R_m. \tag{8.20}$$

The punctured rate is always greater than the rate of the mother code but less than or equal to 1:

$$R_m \leq R_p \leq 1.$$

Various deletion maps are possible, depending on the rate of the mother code and the desired rate of the punctured code.

Mother codes with rate $R_m = 1/n_m$ and puncturing matrices When punctured convolutional codes with rate $R_p = k_p/n_p$ are derived from a mother code with rate $R_m = 1/n_m$, the deletion map $\mathcal{S}(p,w)$ can be represented as an $n_m \times k_p$ puncturing matrix $\mathbf{P}$ with matrix elements $p_{ij} \in \{0,1\}$. A 0 means the deletion of a bit and a 1 means that this position is maintained in the punctured sequence. The number of 1s in the puncturing matrix $w = \sum_{i,j} p_{ij} = n_p$ is obtained from equations 8.19 and 8.20, and the puncture period is $p = k_p n_m$.

Example 8.23 (Puncturing of convolutional codes) The encoder from figure 8.1 with the generator matrix $\mathbf{G}(D) = (1+D+D^2 \quad 1+D^2)$ is used for the encoding of a mother code $\mathcal{C}_m$ with rate $R_m = 1/2$ and memory $m_m = 2$. To generate a punctured code with rate $R_p = (n_p - 1)/n_p = 6/7$, the 2×6 puncturing matrix

$$\mathbf{P} = \begin{pmatrix} 1 & 0 & 0 & 0 & 0 & 1 \\ 1 & 1 & 1 & 1 & 1 & 0 \end{pmatrix}$$

is used. The convolutional encoder of the mother code generates at each point in time t a code block $\mathbf{v}_t = (v_t^{(1)} \, v_t^{(2)})$ of 2 code bits. The (t mod 6)th column of the periodically punctured matrix determines which of these two bits is deleted. So the punctured code sequence is obtained in the following manner:

$$\begin{aligned} \mathbf{v}_m &= ((v_0^{(1)} v_0^{(2)}), (v_1^{(1)} v_1^{(2)}), (v_2^{(1)} v_2^{(2)}), (v_3^{(1)} v_3^{(2)}), (v_4^{(1)} v_4^{(2)}), \ldots) \\ \mathbf{v}_p &= ((v_0^{(1)} v_0^{(2)} v_1^{(2)} v_2^{(2)} v_3^{(2)} v_4^{(2)} v_5^{(1)}), (v_6^{(1)} v_6^{(2)} v_7^{(2)} \ldots), \ldots). \end{aligned}$$

Hence the punctured code has, as can be seen above, code block length $n_p = 7$. Then the deletion map specified by the puncturing matrix repeats itself. ◇

A punctured code with a fixed rate can of course be derived from totally different mother codes. Furthermore, different permutations of the deletion map are possible, depending on the mother code.

Good punctured codes Good punctured codes are typically found using a computer-aided search. The terminology 'good' refers to those codes $\mathcal{C}_p(n_p, k_p, [\nu_p])$ that have the largest free distance d_f among all other punctured codes with the same rate $R_p = k_p/n_p$ and the same overall constraint length ν_p.

Because many codes with rate $R_p \rightarrow 1$ have the same free distance, we also consider the number of paths with the minimum free distance as an evaluation criteria. When evaluating the extended path enumerator in equation 8.8, the *information weight* $I(w)$ of the paths with code weight w is obtained from

$$\frac{\partial}{\partial H} T(W,H,L=1)\bigg|_{H=1} = \sum_w I(w) W^w.$$

The information weight of all the paths with Hamming weight equal to the free distance is therefore $I(d_f)$. The information weight refers to the Hamming weight of an information sequence associated with a code sequence. How the information weight can be used to calculate an upper bound for the bit error rate of a corresponding code will be shown in section 8.4.3 (equation 8.45).

Note For the practical calculation of the value $I(d_f)$, the extended path enumerator is not used, but instead the trellis of the mother code. To find a punctured code with the largest possible free distance d_f and smallest information weight $I(d_f)$, the best-known codes are chosen as mother codes. The reason behind this is that the distance properties of a punctured code can never be better than the mother code. All possible punctured codes with fixed rate obtained with a deletion map $S(p, w)$ having a certain period are investigated. Since the combination of generator matrix of the mother code and puncturing can lead to a catastrophic encoder, the test on catastrophic behavior has to be done first. If non-catastrophic behavior is ensured, the free distance and the associated information weight are determined.

Example 8.24 (Generator matrix of a punctured convolutional code) The test for catastrophic behavior of an encoder is carried out using the generator matrix of the encoder (see section 8.2.6). In this example, we shall calculate the generator matrix of the punctured code with rate $R_p = 6/7$, which we described in example 8.23. The generator matrix of the mother code $C(2, 1, [2])$ in the time domain is

$$\mathbf{G} = \begin{pmatrix} 11 & 10 & 11 & 11 & & \\ & 11 & 10 & 11 & 11 & \\ & & \ddots & & & \ddots \end{pmatrix}.$$

When considering the matrix multiplication $\mathbf{v} = \mathbf{u} \cdot \mathbf{G}$, the lth code bit of the code sequence $\mathbf{v}$ is obtained by multiplying the information sequence $\mathbf{u}$ with the lth column of the generator matrix. If the lth code bit is removed then the respective column of the generator matrix of the mother code must be deleted.

11	01	01	01	01	10	11	01	01	01	01	10	11	10	10	10
11	10	11													
	11	10	11												
		11	10	11											
			11	10	11										
				11	10	11									
					11	10	11								
						11	10	11							
							11	10	11						
								11	10	11					
									11	10	11				
										11	10	11			
											11	10	11		
												11	10	11	
													11	10	11
														⋱	⋱

The cyclically repeated puncturing matrix **P** is shown in the topmost row. The generator matrix of the mother code is shown below. Columns that are not deleted are printed in bold. This way the generator matrix of the punctured code is obtained:

$$\mathbf{G}_P = \begin{pmatrix} 1 & 1 & 0 & 1 & & & & & & & & \\ & & 1 & 0 & 1 & & & & & & & \\ & & & 1 & 0 & 1 & & & & & & \\ & & & & 1 & 0 & 1 & & & & & \\ & & & & & 1 & 1 & 1 & 1 & & & \\ & & & & & & 1 & 1 & 0 & 1 & & \\ & & & & & & & 1 & 1 & 0 & 1 & \\ & & & & & & & & & 1 & 0 & 1 \\ & & & & & & & & & & \ddots & & \ddots \end{pmatrix},$$

which consists of the two 6×7 submatrices

$$\mathbf{G}_0 = \begin{pmatrix} 1 & 1 & 0 & 1 & 0 & 0 & 0 \\ 0 & 0 & 1 & 0 & 1 & 0 & 0 \\ 0 & 0 & 0 & 1 & 0 & 1 & 0 \\ 0 & 0 & 0 & 0 & 1 & 0 & 1 \\ 0 & 0 & 0 & 0 & 0 & 1 & 1 \\ 0 & 0 & 0 & 0 & 0 & 0 & 1 \end{pmatrix} \quad \text{and} \quad \mathbf{G}_1 = \begin{pmatrix} 0 & 0 & 0 & 0 & 0 & 0 & 0 \\ 0 & 0 & 0 & 0 & 0 & 0 & 0 \\ 0 & 0 & 0 & 0 & 0 & 0 & 0 \\ 0 & 0 & 0 & 0 & 0 & 0 & 0 \\ 1 & 1 & 0 & 0 & 0 & 0 & 0 \\ 1 & 0 & 1 & 0 & 0 & 0 & 0 \end{pmatrix}.$$

The generator matrix in the frequency domain is then $\mathbf{G}(D) = \mathbf{G}_0 + D\mathbf{G}_1$. The punctured code has memory $m_p = 1$. It is smaller than the memory of the mother code. However, the overall constraint length $\nu_m = \nu_p = 2$ remains the same. ◇

In general, the relation $\nu_p \leq \nu_m$ holds true for all good punctured codes.
The reasons for the use of punctured codes are as follows:

- The possibility to decode high-rate punctured codes with the decoding complexity of the lower-rate mother code. As we shall see later in section 8.4, the decoding complexity of a Viterbi decoder for a convolutional code trellis[2] increases exponentially with km, for example for higher-rate codes where $k > 1$.
- The construction of new convolutional codes based on already-known codes with good distance properties. This is particularly important, because, in contrast to block codes, there are no procedures for constructing good convolutional codes.

Areas of application for punctured convolutional codes are, for example,

- the use of codes with an extremely high-rate in concatenated code schemes (see chapter 9);
- flexible error protection with the same decoder by using the class of RCPC codes (*rate-compatible punctured convolutional codes*).

Punctured convolutional codes are very important in practical applications, and are used in many areas. Tables and references for the best known punctured codes can be found in section 8.8.

Note At this point, the fundamentals of convolutional codes is complete. The reader may now skip to section 8.4 on Viterbi decoding of convolutional codes without considering the algebraic structure of codes or special distance measures.

[2]Convolutional code trellises classified by Forney [For73, For74], which are most frequently considered.

8.2 Algebraic description

We have introduced the basic terminology and properties of convolutional codes in the previous section. We shall mathematically derive and prove them in this section. We shall only consider codes with infinite code sequences, and shall use the notation from [JoZi]. At this point, we want to focus on aspects needed to introduce the algebraic description. It will be shown how equivalent generator matrices can be constructed. The most important forms, i.e. systematic and catastrophic generator matrices, basic generator matrices and minimal basic generator matrices will be discussed. The parity check matrix of a code and the dual code will also be described.

8.2.1 Code, generator matrix and encoder

We present three definitions now, which clearly distinguish between code, generator matrix and their realization as an encoder.

Definition 8.1 (Convolutional code) *The set of all code sequences that is obtained from a linear mapping is called a convolutional code.*

In this definition, we intentionally avoid the term 'information sequence'. Of course, information sequences are mapped to code sequences, but this mapping is not a code property. The code properties are derived based only on the generated set of codewords $\mathbf{V}(D) = (V_1(D), V_2(D), \dots, V_n(D))$. Hence a convolutional code consists of an infinite number of codewords.

Definition 8.2 (Generator matrix) *A generator matrix is a $k \times n$ matrix $\mathbf{G}(D)$ with rank k and matrix elements $G_{ij}(D)$ that are rational functions.*

The generator matrix $\mathbf{G}(D)$ determines the mapping between *information words* (information sequences) $\mathbf{U}(D) = (U_1(D), U_2(D), \dots, U_k(D))$ and the codewords $\mathbf{V}(D)$. It is a concrete linear mapping. The generator matrix $\mathbf{G}(D)$ must have full rank, since this mapping should be reversible.

Definition 8.3 (Convolutional encoder) *A convolutional encoder with rate $R = k/n$ is the implementation of a $k \times n$ generator matrix $\mathbf{G}(D)$ as sequential linear circuit.*

This definition does not make any constructive statements concerning the concrete implementation of the encoder. Arbitrary circuit designs with storage elements and linear circuit logic are conceivable. The same generator matrix can be implemented with different encoders.

From now on, we shall distinguish strictly between these three terms. Therefore every statement applies either to a code, generator matrix or encoder property.

8.2.2 Convolutional encoder in controller and observer canonical form

The elements $G_{ij}(D)$ of a generator matrix $\mathbf{G}(D)$ are generally rational system functions. We limit our investigation to realizable system functions, i.e. those that can be implemented. These functions can be defined as follows:

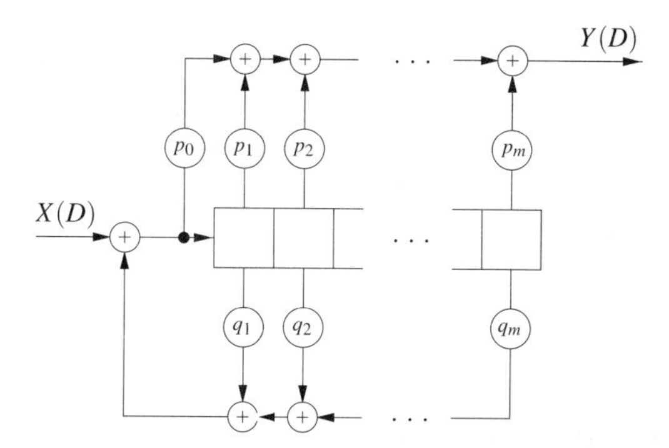

Figure 8.15 Controller canonical form of a realizable transmission function.

Realizable system functions A rational system function $H(D) = P(D)/Q(D)$ consists of a numerator polynomial $P(D) = \sum_{i=0}^{m} p_i D^i$ and a denominator polynomial $Q(D) = \sum_{i=0}^{m} q_i D^i$, with $p_i, q_i \in \{0,1\}$. If $q_0 = 1$ then $H(D)$ is termed *realizable*, and the denominator polynomial $Q(D)$ is called *not delaying*. The output is written $Y(D) = X(D)H(D)$, with

$$H(D) = \frac{p_0 + p_1 D + \ldots + p_m D^m}{1 + q_1 D + \ldots + q_m D^m}. \tag{8.21}$$

In general, two important implementations of a realizable system function are known from systems theory. These are the *controller canonical form* and the *observer canonical form*.

Controller canonical form The controller canonical form implementation of the system function from equation 8.21 is shown in figure 8.15. The input sequence $X(D)$ is fed into a shift register with length m. The circuitry around the shift register corresponds to the coefficients of the numerator and denominator polynomials of the system function $H(D)$.

Observer canonical form The observer canonical form implementation of the system function from equation 8.21 is shown in figure 8.16. The storage elements are here not grouped in a shift register. The computation of the system function $H(D)$ using this circuit is carried out in two steps. First we find $Y(D) = X(D)\,(p_0 + p_1 D + \ldots + p_m D^m) + Y(D)\,(q_1 D + \ldots + q_m D^m)$ directly from the circuit. The system function is obtained from equation 8.21 after some straightforward mathematical operations.

Ways to implement a generator matrix Similar to the concept of the controller or observer canonical form of a realizable system function, a generator matrix can be implemented with these two designs. Then $\mathbf{G}(D)$ must be realizable, i.e. all elements $G_{ij}(D) = P_{ij}(D)/Q_{ij}(D)$ are realizable system functions. The following concept can then be used for circuit design:

- *Generator matrix in controller canonical form* All k input sequences $U_i(D)$, for $i \in [1,k]$, are treated independently, i.e. each of these sequences is fed into its own associated shift register (see figure 8.15). The feedback of the ith shift register is obtained from the least common multiple (lcm) of the respective denominator polynomials:

$$Q_i(D) = \mathrm{lcm}(Q_{i1}(D), Q_{i2}(D), \ldots, Q_{in}(D)).$$

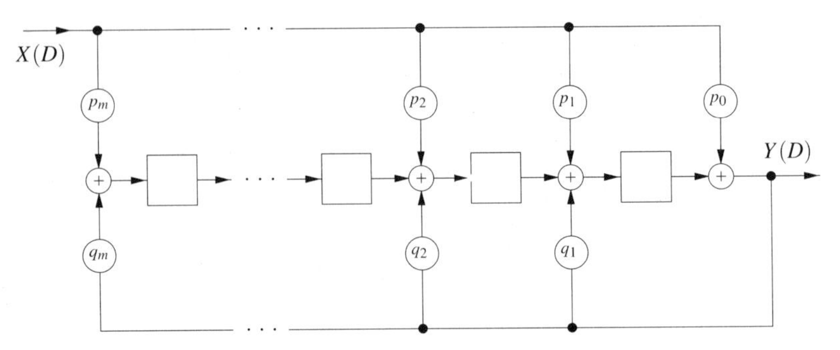

Figure 8.16 Observer implementation of a realizable transmission function.

When the k system functions $G_{ij}(D)$ of the ith row are extended to $\tilde{G}_{ij}(D) = \tilde{P}_{ij}(D)/Q_i(D)$, for $j \in [1,n]$, we obtain the circuitry for the shift registers in the forward branch. Hence each of the k shift registers has exactly one external circuit for each of the n output sequences.

- *Generator matrix in observer canonical form* Each one of the n output sequences $V_j(D)$, for $j \in [1,n]$, has one memory unit, which is independent of the other output sequences (see figure 8.16). The circuit elements around the memory elements of this storage unit in the 'backward branch' are obtained from

$$Q_j(D) = \mathrm{lcm}(Q_{1j}(D), Q_{2j}(D), \ldots, Q_{kj}(D)).$$

Similar to the controller canonical form, we extend the n system functions $G_{ij}(D)$ of the jth column to $\tilde{G}_{ij}(D) = \tilde{P}_{ij}(D)/Q_j(D)$, for $i \in [1,k]$, and obtain the circuit storage elements in the 'forward branch'. Hence every one of the n storage elements has exactly one external circuit for each of the k input sequences.

Of course, convolutional encoders can be designed in an arbitrary way. Among others, procedures exist to implement an encoder with the smallest possible number of storage elements for a given generator matrix. These minimization techniques are especially important when implementing very large circuits. However, we are interested in structural properties, and shall therefore limit our discussion to the controller and observer canonical form.

Example 8.25 (Controller and observer canonical form [JoZi]) The generator matrix with rational elements,

$$\mathbf{G}(D) = \begin{pmatrix} \frac{1}{1+D+D^2} & \frac{D}{1+D^3} & \frac{1}{1+D^3} \\ \frac{D^2}{1+D^3} & \frac{1}{1+D^3} & \frac{1}{1+D} \end{pmatrix},$$

is shown in figures 8.17 and 8.18 in controller and observer canonical forms, respectively. The controller canonical form consists of $k = 2$ shift registers with feedback $Q_1(D) = Q_2(D) = 1 + D^3$. The external circuitry of the two registers in the forward branch is combined at the level of the output sequences. The observer canonical form consists of $n = 3$ storage elements. Their feedback is also $Q_1(D) = Q_2(D) = Q_3(D) = 1 + D^3$. The logic in the forward branch is obtained as explained above. ◇

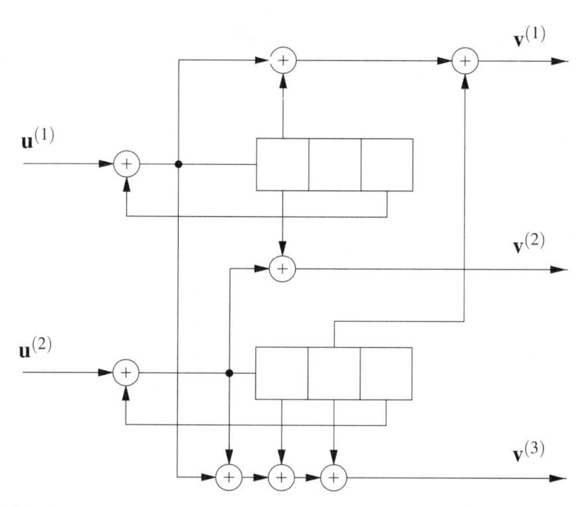

Figure 8.17 Controller implementation of a rate $R = 2/3$ convolutional code.

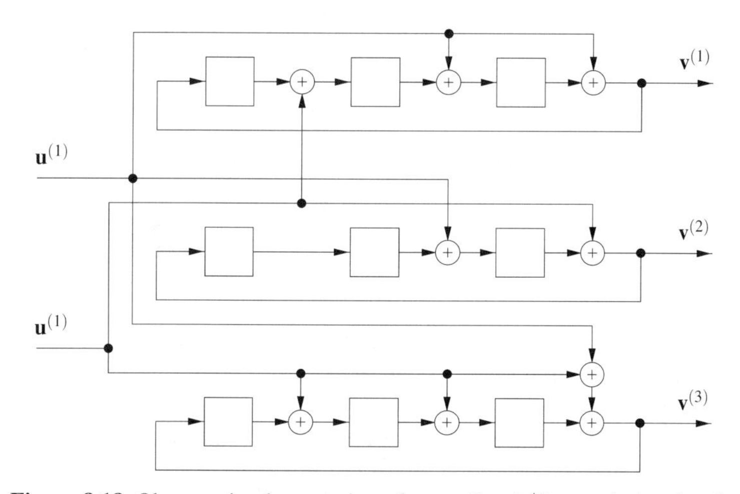

Figure 8.18 Observer implementation of a rate $R = 2/3$ convolutional code.

8.2.3 Equivalent generator matrices

A generator matrix can be realized with different convolutional encoders. Equivalently, a convolutional code can be generated with different generator matrices.

Definition 8.4 (Equivalent generator matrices) *Two generator matrices $\mathbf{G}(D)$ and $\mathbf{G}'(D)$ are said to be equivalent if they create the same code.*

Theorem 8.5 (Equivalent generator matrices) *Two generator matrices $\mathbf{G}(D)$ and $\mathbf{G}'(D)$ are equivalent if a $k \times k$ scrambler matrix $\mathbf{T}(D)$ exists such that*

$$\mathbf{G}'(D) = \mathbf{T}(D) \cdot \mathbf{G}(D),$$

where the determinant $\det(\mathbf{T}(D)) \neq 0$. Hence the equivalent generator matrix $\mathbf{G}'(D)$ has full rank.

Proof By definition 8.4, both generator matrices must generate the same code, i.e. create the same set of codewords. The same codeword

$$\begin{aligned}\mathbf{V}(D) &= \mathbf{U}(D)\cdot\mathbf{G}'(D) = \underbrace{\mathbf{U}(D)\cdot\mathbf{T}(D)}_{\mathbf{U}'(D)}\cdot\mathbf{G}(D)\\ &= \mathbf{U}'(D)\cdot\mathbf{G}(D)\end{aligned}$$

can be generated with $\mathbf{G}'(D)$ as well as with $\mathbf{G}(D)$. Since $\mathbf{T}(D)$ is non-singular, a pair of information words can always be found such that $\mathbf{U}'(D) = \mathbf{U}(D)\cdot\mathbf{T}(D)$. Both create the same codeword. □

The set of equivalent generator matrices of a convolutional code contains a subset of polynomial generator matrices, which can be implemented as FIR systems. These FIR implementations are of special interest.

Theorem 8.6 (Polynomial generator matrices) *Each rational generator matrix has an equivalent polynomial generator matrix.*

Proof Consider the matrix $\mathbf{G}(D) = \big(G_{ij}(D)\big)$, with $G_{ij}(D) = P_{ij}(D)/Q_{ij}(D)$ for $i \in [1,k]$ and $j \in [1,n]$. The elements are rational functions of D. The least common multiple of the denominator polynomials

$$T(D) = \mathrm{lcm}\big(Q_{11}(D),\dots,Q_{ij}(D),\dots,Q_{kn}(D)\big)$$

can be used to calculate an equivalent generator matrix:

$$\mathbf{G}'(D) = \mathbf{T}(D)\cdot\mathbf{G}(D).$$

The scrambler matrix is obtained from $\mathbf{T}(D) = T(D)\mathbf{I}_k$. The elements $G'_{ij}(D)$ can therefore be reduced so that they are all polynomials. □

Except for the section on systematic generator matrices, all further investigations will be with respect to polynomial generator matrices, .

The constraint lengths that we have introduced in section 8.1.3 for generator sequences of non- recursive convolutional encoders can now be generally defined as properties of the generator matrix.

Definition 8.7 (Memory) *Let $\mathbf{G}(D)$ be a $k \times n$ generator matrix of polynomials. Then*

$$\nu_i = \max_{1\le j\le n}\{\deg G_{ij}(D)\}, \tag{8.22}$$

is referred to as the i-th constraint length. The memory of the generator matrix is defined as

$$m = \max_{1\le i\le k}\{\nu_i\},$$

and the overall constraint length is given by

$$\nu = \sum_{i=1}^{k}\nu_i.$$

Note A definition of constraint lengths for rational generator matrices is also possible, but is not presented here due to the associated degree of complexity.

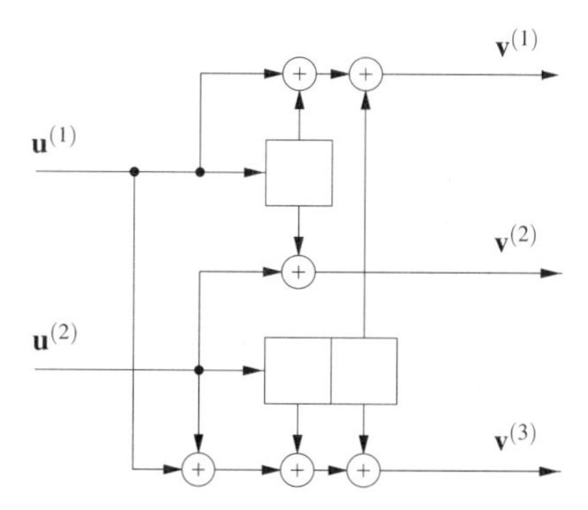

Figure 8.19 Convolutional encoder with rate $R = 2/3$.

Example 8.26 (Equivalent polynomial generator matrix) We want to describe the implementation of the convolutional encoder with rate $R = 2/3$, shown in figures 8.17 and 8.18. We derive an equivalent polynomial generator matrix and construct a controller implementation circuit.

The underlaying rational generator matrix is

$$\mathbf{G}(D) = \begin{pmatrix} \frac{1}{1+D+D^2} & \frac{D}{1+D^3} & \frac{1}{1+D^3} \\ \frac{D^2}{1+D^3} & \frac{1}{1+D^3} & \frac{1}{1+D} \end{pmatrix}.$$

An equivalent polynomial generator matrix is obtained by multiplying the least common multiple of all denominator polynomials,

$$T(D) = \operatorname{lcm}\left(1+D+D^2, 1+D^3, 1+D\right) = 1+D^3,$$

by the rational generator matrix

$$\mathbf{G}'(D) = (1+D^3)\begin{pmatrix} \frac{1}{1+D+D^2} & \frac{D}{1+D^3} & \frac{1}{1+D^3} \\ \frac{D^2}{1+D^3} & \frac{1}{1+D^3} & \frac{1}{1+D} \end{pmatrix} = \begin{pmatrix} 1+D & D & 1 \\ D^2 & 1 & 1+D+D^2 \end{pmatrix}.$$

A realization of this generator matrix in controller canonical form is shown in figure 8.19. $\diamond$

8.2.4 Generator matrix in Smith form

According to definition 8.2, a $k \times n$ generator matrix $\mathbf{G}(D)$ always has full rank. Therefore a right-inverse $\mathbf{G}^{-1}(D)$, which is an $n \times k$ matrix, always exists. The inverse operation of $\mathbf{V}(D) = \mathbf{U}(D) \cdot \mathbf{G}(D)$ is $\mathbf{U}(D) = \mathbf{V}(D) \cdot \mathbf{G}^{-1}(D)$. Hence

$$\mathbf{I}_k = \mathbf{G}(D) \cdot \mathbf{G}^{-1}(D),$$

where $\mathbf{I}_k$ is the $k \times k$ unit matrix. To calculate the right-inverse $\mathbf{G}^{-1}(D)$ we use a linear algebraic method, which represents the generator matrix in so-called *Smith form*.

Theorem 8.8 (Smith form) *Let* $\mathbf{G}(D)$ *be a* $k \times n$ *polynomial generator matrix, which can be written as*

$$\mathbf{G}(D) = \mathbf{A}(D) \cdot \Gamma(D) \cdot \mathbf{B}(D), \tag{8.23}$$

where $\mathbf{A}(D)$ *is a* $k \times k$ *matrix and* $\mathbf{B}(D)$ *is an* $n \times n$ *matrix, both having unit determinant, i.e. they are scrambler matrices. The* $k \times n$ *matrix* $\Gamma(D)$ *is referred to as the Smith form of the generator matrix* $\mathbf{G}(D)$*, and has diagonal form:*

$$\Gamma(D) = \begin{pmatrix} \gamma_1(D) & & & & & \\ & \ddots & & & & \\ & & \gamma_k(D) & 0 & \dots & 0 \end{pmatrix}. \tag{8.24}$$

The diagonal elements $\gamma_i(D)$, $1 \leq i \leq k$, *are called invariant factors of* $\mathbf{G}(D)$*. For them, the following relation holds:*

$$\gamma_i(D) \quad | \quad \gamma_{i+1}(D), \tag{8.25}$$

for $i = 1, 2, \dots, k-1$, *i.e. the first diagonal element divides all others with zero reminder, and so on. Furthermore, for the greatest common divisor* $\Delta_i(D)$ *of the determinants of all* $i \times i$ *submatrices of* $\mathbf{G}(D)$*, it holds that*

$$\Delta_i(D) = \gamma_1(D) \dots \gamma_{i-1}(D)\gamma_i(D). \tag{8.26}$$

Proof The proof of this theorem is a constructive one, i.e. an algorithm to calculate the Smith form will be described and with it the relations from equations 8.23–8.26 validated. However, since even a simple description of the algorithm is relatively complex, we refer the reader to [JoZi] or [McE98]. □

The Smith form plays a crucial role in the discussion of the structural properties of generator matrices. With equations 8.23 and 8.24, the right-inverse of the generator matrix becomes

$$\mathbf{G}^{-1}(D) = \mathbf{B}^{-1}(D) \cdot \Gamma^{-1}(D) \cdot \mathbf{A}^{-1}(D), \tag{8.27}$$

where the $n \times k$ matrix $\Gamma^{-1}(D)$ is defined as

$$\Gamma^{-1}(D) = \begin{pmatrix} \gamma_1^{-1}(D) & & \\ & \ddots & \\ & & \gamma_k^{-1}(D) \\ 0 & \dots & 0 \\ \vdots & & \vdots \\ 0 & \dots & 0 \end{pmatrix}. \tag{8.28}$$

The inverted quadratic scrambler matrices $\mathbf{A}^{-1}(D)$ and $\mathbf{B}^{-1}(D)$ exist, since $\mathbf{A}(D)$ and $\mathbf{B}(D)$ have determinant 1.

Example 8.27 (Smith form [JoZi]) The polynomial generator matrix

$$\mathbf{G}(D) = \begin{pmatrix} 1+D & D & 1 \\ D+D^2+D^3+D^4 & 1+D+D^2+D^3+D^4 & 1+D+D^2 \end{pmatrix}$$

has the Smith form $\mathbf{G}(D) = \mathbf{A}(D) \cdot \Gamma(D) \cdot \mathbf{B}(D)$, where

$$\begin{aligned}
\mathbf{A}(D) &= \begin{pmatrix} 1 & 0 \\ 1+D+D^2 & 1 \end{pmatrix}, \\
\Gamma(D) &= \begin{pmatrix} 1 & 0 & 0 \\ 0 & 1+D & 0 \end{pmatrix}, \\
\mathbf{B}(D) &= \begin{pmatrix} 1+D & D & 1 \\ 1+D^2+D^3 & 1+D+D^2+D^3 & 0 \\ D+D^2 & 1+D+D^2 & 0 \end{pmatrix}.
\end{aligned}$$

Hence the invariant factors are $\gamma_1 = 1$ and $\gamma_2 = 1+D$. For the right-inverse, we obtain, using equations 8.27 and 8.28,

$$\mathbf{G}^{-1}(D) = \frac{1}{1+D} \begin{pmatrix} 1+D^2+D^4 & 1+D+D^2 \\ D+D^4 & D+D^2 \\ D^3+D^4 & 1+D^2 \end{pmatrix}. \qquad \diamond$$

8.2.5 Basic generator matrix

The set of equivalent polynomial generator matrices of a convolutional code can be investigated with respect to the overall constraint length ν. The goal is to find generator matrices with the smallest possible overall constraint length. This will lead to the least complexity in the trellis, which is a measure of complexity for decoding (see sections 8.4 and 8.5).

First we consider polynomial generator matrices $\mathbf{G}(D)$, which have a polynomial right-inverses $\mathbf{G}^{-1}(D)$.

Definition 8.9 (Basic generator matrix) *A polynomial generator matrix where all invariant factors $\gamma_i(D)$, for $i \in [1,k]$, are equal to one is called a basic generator matrix $\mathbf{G}_b(D)$.*

Since the determinants of the scrambler matrices in Smith form are $\det(\mathbf{A}(D)) = 1$ and $\det(\mathbf{B}(D)) = 1$, and the k invariant factors of a basic generator matrix are $\gamma_i = 1$ for $i \in [1,k]$, all three inverses $\mathbf{B}^{-1}(D)$, $\Gamma^{-1}(D)$ and $\mathbf{A}^{-1}(D)$ in equation 8.27 are polynomial, and therefore so is their product. Hence every basic generator matrix $\mathbf{G}_b(D)$ has a polynomial right-inverse.

When considering the Smith form (equation 8.23) of an arbitrary polynomial generator matrix $\mathbf{G}(D)$, we obtain an equivalent basic generator matrix $\mathbf{G}_b(D)$ from the first k rows of the scrambler matrix, and $\mathbf{B}(D)$:

$$\begin{aligned}
\mathbf{G}(D) &= \mathbf{A}(D) \cdot \Gamma(D) \cdot \mathbf{B}(D) \\
&= \mathbf{A}(D) \begin{pmatrix} \gamma_1(D) & & \\ & \ddots & \\ & & \gamma_k(D) \end{pmatrix} \underbrace{\begin{pmatrix} 1 & & & & & \\ & \ddots & & & & \\ & & 1 & 0 & \cdots & 0 \end{pmatrix} \mathbf{B}(D)}_{\mathbf{G}_b(D)} \\
&= \mathbf{T}(D) \cdot \mathbf{G}_b(D).
\end{aligned}$$

The generator matrices with minimal overall constraint length are found in the set of all equivalent basic generator matrices.

Definition 8.10 (Minimal basic generator matrix) *A basic generator matrix $\mathbf{G}_b(D)$, whose overall constraint length is minimal among all equivalent basic generator matrices, is called a minimal basic generator matrix $\mathbf{G}_{mb}(D)$.*

Starting with a basic generator matrix $\mathbf{G}_b(D)$, we can calculate an equivalent minimal basic generator matrix $\mathbf{G}_{mb}(D)$. To do this, we first define the matrix $[\mathbf{G}(D)]_h$ with elements $[G_{ij}(D)]_h$ that are either 0 or 1. A 1 is placed at position ij if the degree of $G_{ij}(D)$ is equal to the ith constraint length (equation 8.22). If the degree is lower then a 0 is entered. A basic generator matrix is minimal when $[\mathbf{G}_b(D)]_h$ has full rank. If this is not the case then an equivalent minimal basic generator matrix $\mathbf{G}_b(D)$ can be calculated by a series of basic row operations, as shown in the following example.

Example 8.28 (Minimal basic generator matrix [JoZi]) The smith form of the polynomial generator matrix $\mathbf{G}(D)$ from example 8.27 gives

$$\mathbf{B}(D) = \begin{pmatrix} 1+D & D & 1 \\ 1+D^2+D^3 & 1+D+D^2+D^3 & 0 \\ D+D^2 & 1+D+D^2 & 0 \end{pmatrix}.$$

We obtain an equivalent basic generator matrix from the first k rows of this matrix:

$$\mathbf{G}_b(D) = \begin{pmatrix} 1+D & D & 1 \\ 1+D^2+D^3 & 1+D+D^2+D^3 & 0 \end{pmatrix}.$$

This way, the matrix

$$[\mathbf{G}_b(D)]_h = \begin{pmatrix} 1 & 1 & 0 \\ 1 & 1 & 0 \end{pmatrix}$$

is obtained, and its rows are linearly dependent. Therefore $[\mathbf{G}_b(D)]_h$ does not have full rank. However, if the first row is multiplied by $1+D+D^2$ and added to the second then an equivalent matrix is obtained:

$$\mathbf{G}_{mb}(D) = \begin{pmatrix} 1+D & D & 1 \\ D^2 & 1 & 1+D+D^2 \end{pmatrix},$$

with

$$[\mathbf{G}_b(D)]_h = \begin{pmatrix} 1 & 1 & 0 \\ 1 & 0 & 1 \end{pmatrix},$$

which has full rank ($\text{rank}([\mathbf{G}_{mb}(D)]_h) = 2$) and is minimal. An implementation of this generator matrix as controller canonical form is shown in figure 8.19. ◇

A realization of a minimal basic generator matrix in controller canonical form leads to a convolutional encoder with the smallest number of memory elements among all possible encoders of a certain convolutional code. An encoder with this property is also referred to as *minimal*.

Note Other minimal encoders exist; however, they have an arbitrary realization. Generator matrices that can be implemented with a minimal encoder are called *canonical*, and can generally be composed of rational functions. Hence, canonical encoders with feedback also exist (IIR system).

8.2.6 Catastrophic generator matrices

Catastrophic generator matrices are a subset of all equivalent polynomial generator matrices. This form must be avoided under all circumstances.

Definition 8.11 (Catastrophic generator matrices) *A generator matrix* $\mathbf{G}(D)$ *is said to be catastrophic if an information sequence with infinite weight is mapped to a code sequence with finite weight.*

Example 8.29 (Catastrophic generator matrix) The polynomial generator matrix $\mathbf{G}(D)$ from example 8.27 is catastrophic. To show this, we consider the inverse mapping $\mathbf{G}^{-1}(D)$ and demonstrate that a code sequence with finite weight is mapped to an information sequence with infinite weight.

When considering the minimal basic generator matrix $\mathbf{G}_{mb}(D)$ calculated in example 8.28, the information word $\mathbf{U}(D) = (0, 1)$ with Hamming weight $\text{wt}(\mathbf{U}(D)) = 1$ is mapped to the codeword $\mathbf{V}(D) = (D^2, 1, 1+D+D^2)$ with Hamming weight $\text{wt}(\mathbf{V}(D)) = 5$. On the other hand, this codeword could have also been generated by an equivalent generator matrix. Therefore we use as inverse mapping the right-inverse from example 8.27:

$$\begin{aligned}
\mathbf{U}(D) &= \mathbf{V}(D) \cdot \mathbf{G}^{-1}(D) \\
&= \frac{1}{1+D}(D+D^2+D^3 \;\; 1) \\
&= (D+D^3+D^4+D^5+D^6+\dots \\
&\qquad 1+D+D^2+D^3+D^4+D^5+D^6+\dots).
\end{aligned}$$

The information sequences are obtained from the closed form by formal series expansion with positive powers of D. Hence an information sequence with infinite weight is mapped to a code sequence with finite weight. A finite number of transmission errors can lead to an infinite number of errors in the estimated information sequence. ◇

The following theorem allows one to test for catastrophic generator matrices.

Theorem 8.12 (Test for catastrophic behavior from Massey–Sain) *A polynomial generator matrix* $\mathbf{G}(D)$ *is non-catastrophic if and only if* $\Delta_k(D) = D^s$, $s \geq 0$, *where* Δ_k *is the greatest common divisor of all determinants of all* $k \times k$ *submatrices of* $\mathbf{G}(D)$.

Proof According to theorem 8.8 (equation 8.26), the greatest common divisor Δ_k of all determinants of all $k \times k$ submatrices of a generator matrix $\mathbf{G}(D)$ corresponds to the product of the invariants $\gamma_1\gamma_2\dots\gamma_k$ of the generator matrix. Because of their series expansion, the denominator polynomials $1/\Delta_k$ of the inverse $\Gamma^{-1}(D)$ indicate a catastrophic generator matrix by producing a factor of the form $D^s(1+D^r+\dots)$. □

A consequence of this theorem is that all basic generator matrices and therefore all minimal basic generator matrices are non-catastrophic. Since the invariant factors γ_i for $i \in [1, k]$ are all unity, these matrices have a polynomial right-inverse, or equivalently $\Delta_k = 1$. In general, all generator matrices with polynomial right-inverse are non-catastrophic, because, through the inverse mapping, no finite codewords can be mapped to infinite information words.

Example 8.30 (Check for catastrophic behavior) As shown above, the polynomial generator matrix from example 8.27 is catastrophic. The test on catastrophic behavior according to theorem 8.12 for the three determinants d_1, d_2 and d_3 of the 2×2 submatrix of $\mathbf{G}(D)$ yields

$$d_1 = \begin{vmatrix} 1+D & D \\ D+D^2+D^3+D^4 & 1+D+D^2+D^3+D^4 \end{vmatrix}$$
$$= 1+D^2+D^3+D^4 = (1+D+D^3)(1+D),$$

$$d_2 = \begin{vmatrix} D & 1 \\ 1+D+D^2+D^3+D^4 & 1+D+D^2 \end{vmatrix}$$
$$= 1+D^4 = (1+D)^4,$$

$$d_3 = \begin{vmatrix} 1+D & 1 \\ D+D^2+D^3+D^4 & 1+D+D^2 \end{vmatrix}$$
$$= 1+D+D^2+D^4 = (1+D^2+D^3)(1+D).$$

The greatest common divisor is $\Delta_2 = \gcd(d_1, d_2, d_3) = 1+D \neq D^s$, and this shows the catastrophic behavior of the generator matrix. When considering the Smith form of this generator matrix in example 8.27, the product of the invariant factors is $\Delta_2 = \gamma_1\gamma_2 = 1+D$. This term emerges as a denominator polynomial in $\Gamma^{-1}(D)$ and again in the right-inverse of the generator matrix. Therefore codewords with finite weight are mapped to information words with infinite weight. ◇

Example 8.31 (Basic generator matrices and catastrophic behavior) In figure 8.14, the convolutional encoder with rate $R = 1/2$ and generator matrix

$$\mathbf{G}(D) = \begin{pmatrix} 1+D^3 & 1+D+D^2+D^3 \end{pmatrix}$$

is catastrophic, as has already been demonstrated in example 8.22 due to the zero loop in the state diagram of the encoder. Alternatively, we obtain with the above test

$$\begin{aligned} \Delta_1 &= \gcd(1+D^3, 1+D+D^2+D^3) \\ &= \gcd(\,(1+D+D^2)(1+D), (1+D)^3\,) = 1+D. \end{aligned}$$

If this term is factored out of the parenthesis then the equivalent generator matrix is obtained with $\mathbf{T}(D) = 1/(1+D)$:

$$\mathbf{G}'(D) = \begin{pmatrix} 1+D+D^2 & 1+D^2 \end{pmatrix}.$$

An implementation of $\mathbf{G}'(D)$ in controller canonical form is shown in figure 8.1. It is non-catastrophic because $\Delta_1 = 1$. Furthermore, $\gamma_1 = 1$ and $\mathrm{rank}([\mathbf{G}(D)]_h) = 1$. Consequently, $\mathbf{G}'(D)$ is a minimal basic generator matrix. ◇

Catastrophic behavior is a property of the generator matrix and not of the convolutional code. Every convolutional code has catastrophic and non-catastrophic generator matrices. A catastrophic convolutional code does not exist!

8.2.7 Systematic generator matrices

In this section, we shall show that every convolutional code can be encoded systematically and a systematic generator matrix can be calculated from a given generator matrix.

Definition 8.13 (Systematic generator matrices) *The generator matrix of a convolutional code with rate $R = k/n$ is said to be systematic if all k information sequences $U_i(D)$ are distinct in the n-element code sequence $V_j(D)$.*

Example 8.32 (Systematic generator matrix) When considering the minimal basic generator matrix

$$\mathbf{G}_{mb}(D) = \begin{pmatrix} 1+D+D^2 & 1+D^2 \end{pmatrix}$$

from example 8.31, an equivalent systematic generator matrix is

$$\mathbf{G}_{sys}(D) = \frac{1}{1+D+D^2} \quad \mathbf{G}_{mb}(D) = \begin{pmatrix} 1 & \frac{1+D^2}{1+D+D^2} \end{pmatrix}$$

The convolutional encoder for this generator matrix in controller canonical form is given in figure 8.13. ◇

In general, a basic generator matrix $\mathbf{G}_b(D)$ of the code is necessary for the calculation of a systematic generator matrix. It has a $k \times k$ invertible submatrix $\mathbf{T}^{-1}(D)$ with delay free polynomial determinants, i.e. $\det\left(\mathbf{T}^{-1}(D)\right)\big|_{D=0} = 1$. By swapping columns of an equivalent code, we can obtain

$$\mathbf{G}_{sys}(D) \quad = \quad \mathbf{T}(D) \cdot \mathbf{G}_b(D) = (\mathbf{I}_k \quad \mathbf{R}(D)).$$

Note The set of all equivalent convolutional codes is obtained by permutation of the n output sequences $V_j(D)$ in the code sequence $\mathbf{V} = (V_1(D), V_2(D), \dots, V_n(D))$.

Example 8.33 (Systematic generator matrix [JoZi]) The first two columns of the minimal basic generator matrix from example 8.28,

$$\mathbf{T}^{-1}(D) = \begin{pmatrix} 1+D & D \\ D^2 & 1 \end{pmatrix},$$

have the delay-free determinant $\det(\mathbf{T}^{-1}(D)) = 1 + D + D^3$. Consequently, the systematic generator matrix

$$\begin{aligned}
\mathbf{G}_{sys}(D) &= \mathbf{T}(D) \cdot \mathbf{G}_{mb}(D) \\
&= \frac{1}{1+D+D^3} \begin{pmatrix} 1 & D \\ D^2 & 1+D \end{pmatrix} \begin{pmatrix} 1+D & D & 1 \\ D^2 & 1 & 1+D+D^2 \end{pmatrix} \\
&= \begin{pmatrix} 1 & 0 & \frac{1+D+D^2+D^3}{1+D+D^3} \\ 0 & 1 & \frac{1+D^2+D^3}{1+D+D^3} \end{pmatrix}
\end{aligned}$$

is obtained.

The implementation of this systematic generator matrix in controller canonical form leads to a convolutional encoder with two shift registers that contain three storage elements each. In contrast, the implementation of the observer canonical form shown in figure 8.20 only requires three storage elements, as many as the implementation of the associated equivalent minimal basic generator matrix using the controller canonical form (figure 8.19). ◇

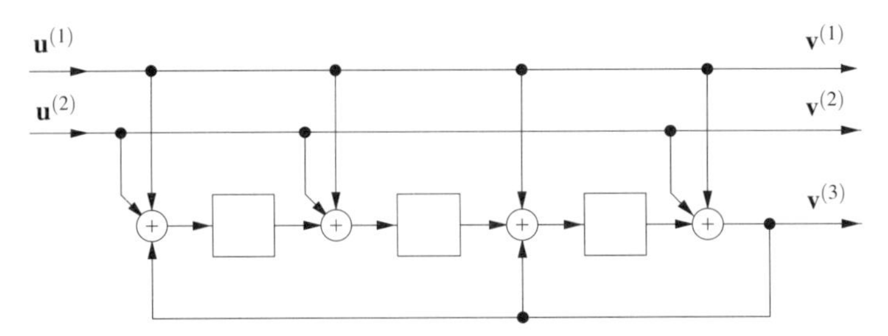

Figure 8.20 Systematic convolutional encoder with $R = 2/3$.

Regarding systematic generator matrices

- Every systematic generator matrix can be implemented with a minimal encoder, which is not limited to controller or observer canonical form.
- A systematic generator matrix is non-catastrophic, and always has a polynomial right-inverse.
- Every convolutional code has a systematic generator matrix, where the elements can be polynomial, but generally take the form of rational functions. Polynomial systematic generator matrices are very restrictive, so that generally only codes with worse distance properties are obtained.

8.2.8 Parity check matrix and dual code

In analogy to block codes, a *parity check matrix* and consequently a *dual code* can be defined for convolutional codes.

Definition 8.14 (Parity check matrix) *If* $\mathrm{rank}(\mathbf{H}(D)) = n - k$ *and*

$$\mathbf{V}(D) \cdot \mathbf{H}^T(D) = \mathbf{0}$$

is valid for every codeword $\mathbf{V}(D)$ *of a convolutional code* C *with rate* $R = k/n$ *then the* $n \times (n-k)$ *matrix* $\mathbf{H}^T(D)$ *is called a parity check matrix or syndrome former of the convolutional code. The elements* $H_{ij}^T(D)$ *for* $i \in [1,n]$ *and* $j \in [1, n-k]$ *are generally realizable rational system functions.*

Since the rows of the generator matrix are also codewords, it follows directly from this definition that

$$\mathbf{G}(D) \cdot \mathbf{H}^T(D) = \mathbf{0}.$$

The Smith form of a polynomial generator matrix of a code $\mathbf{G}(D) = \mathbf{A}(D) \cdot \Gamma(D) \cdot \mathbf{B}(D)$ can be used to calculate a parity check matrix. When considering the $n \times n$ right-inverse of the scrambler matrix

$$\mathbf{B}(D) = \begin{pmatrix} \mathbf{G}_b(D) \\ (\mathbf{H}^{-1}(D))^T \end{pmatrix},$$

we obtain

$$\mathbf{B}^{-1}(D) = \left(\begin{array}{cc} \mathbf{G}_b^{-1}(D) & \mathbf{H}^T(D) \end{array} \right).$$

The first k rows of $\mathbf{B}(D)$ are an equivalent basic generator matrix $\mathbf{G}(D) = \mathbf{T}(D) \cdot \mathbf{G}_b(D)$ to $\mathbf{G}(D)$, as already shown in section 8.2.5. When looking at

$$\begin{aligned} \mathbf{I}_n &= \mathbf{B}(D) \cdot \mathbf{B}^{-1}(D) \\ &= \left(\begin{array}{cc} \mathbf{G}_b(D) \cdot \mathbf{G}_b^{-1}(D) & \mathbf{G}_b(D) \cdot \mathbf{H}^T(D) \\ (\mathbf{H}^{-1}(D))^T \cdot \mathbf{G}_b^{-1}(D) & (\mathbf{H}^{-1}(D))^T \cdot \mathbf{H}^T(D) \end{array} \right) = \left(\begin{array}{cc} \mathbf{I}_k & \mathbf{0} \\ \mathbf{0} & \mathbf{I}_{n-k} \end{array} \right), \end{aligned}$$

we find that a parity check matrix for $\mathbf{G}_b(D)$ can be calculated from the last $n-k$ rows of the scrambler matrix $\mathbf{B}(D)$, which is also a parity check matrix for the original generator matrix: $\mathbf{G}(D) \cdot \mathbf{H}^T(D) = \mathbf{T}(D) \cdot \mathbf{G}_b(D) \cdot \mathbf{H}^T(D) = \mathbf{0}$.

When using the $(n-k) \times n$ parity check matrix $\mathbf{H}(D)$ for encoding a code with rate $R = (n-k)/n$, the transposed codewords $\mathbf{V}^T(D) = (\mathbf{U}(D) \cdot \mathbf{H}(D))^T$ of the convolutional code thus generated are orthogonal to the codewords $\mathbf{V}'(D) = \mathbf{U}'(D) \cdot \mathbf{G}(D)$ of the original code:

$$\begin{aligned} \mathbf{V}'(D) \cdot \mathbf{V}^T(D) &= \mathbf{U}'(D)\mathbf{G}(D) \cdot (\mathbf{U}(D)\mathbf{H}(D))^T \\ &= \mathbf{U}'(D) \cdot \mathbf{G}(D)\mathbf{H}^T(D) \cdot \mathbf{U}^T(D) = \mathbf{0}. \end{aligned}$$

Definition 8.15 (Dual code) *The dual code $C^\perp$ with rate $R^\perp = (n-k)/n$ of a convolutional code C with rate $R = k/n$ is the set of all code sequences that can be generated with the transposed parity check matrix $\mathbf{H}(D)$ of the code C.*

Example 8.34 (Parity check matrix) The generator matrix of the convolutional encoder shown in figure 8.17 was shown in Smith form in figure 8.27. The scrambler matrix is

$$\mathbf{B}(D) = \left(\begin{array}{ccc} 1+D & D & 1 \\ 1+D^2+D^3 & 1+D+D^2+D^3 & 0 \\ D+D^2 & 1+D+D^2 & 0 \end{array} \right).$$

If it is inverted then one obtains

$$\mathbf{B}^{-1}(D) = \left(\begin{array}{ccc} 0 & 1+D+D^2 & 1+D+D^2+D^3 \\ 0 & D+D^2 & 1+D^2+D^3 \\ 1 & 1+D^2 & 1+D+D^3 \end{array} \right),$$

and the parity check matrix is

$$\mathbf{H}^T(D) = \left(\begin{array}{c} 1+D+D^2+D^3 \\ 1+D^2+D^3 \\ 1+D+D^3 \end{array} \right).$$

◇

It can be shown that the overall constraint length of a minimal basic generator matrix $\mathbf{G}_{mb}(D)$ of a convolutional code C and the overall constraint length of a minimal basic generator matrix $\mathbf{H}_{mb}(D)$ of the dual code $C^\perp$ are identical. This relation will be used later on when estimating the ML decoding complexity of a convolutional code.

8.3 Distance measures

In general, two different kinds of distance measures exist, which are either properties of the code or the generator matrix, i.e. of the encoder. In section 8.1.6, we have the example where the path enumerator is a code property, whereas the extended path enumerator is a generator matrix property.

The definitions and calculations of distance measures introduced here have one thing in common: a subset of the set of all possible code sequences is defined first. This is done by restricting the information sequences used for encoding. The distances in this subset can then be calculated from the weight distribution of the respective code sequences or the weight distribution of segments of these code sequences, based on the linear behavior of convolutional codes.

8.3.1 Row and column distance

The *column distance* and *row distance* were introduced in [Cos69]. A number of other distance measures can be derived from the column distance: the *minimal distance*, the *distance profile* and the *free distance*.

Column distance The column distance is a property of the generator matrix and not of the code. It is defined as follows:

Definition 8.16 (Column distance) *The column distance d_j^c of order j of a generator matrix $\mathbf{G}(D)$ is the minimum Hamming distance of the first $j+1$ code blocks of two codewords $\mathbf{v}_{[j+1]} = (\mathbf{v}_0, \mathbf{v}_1, \dots, \mathbf{v}_j)$ and $\mathbf{v}'_{[j+1]} = (\mathbf{v}'_0, \mathbf{v}'_1, \dots, \mathbf{v}'_j)$, where the information sequences used for encoding $\mathbf{u}_{[j+1]} = (\mathbf{u}_0, \mathbf{u}_1, \dots, \mathbf{u}_j)$ and $\mathbf{u}'_{[j+1]} = (\mathbf{u}'_0, \mathbf{u}'_1, \dots, \mathbf{u}'_j)$ differ in the first information block, i.e. $\mathbf{u}_0 \neq \mathbf{u}'_0$.*

Because of the linearity of convolutional codes, the column distance is equal to the minimum Hamming weight of all sequences $\mathbf{v}_{[j+1]}$:

$$d_j^c = \min_{\mathbf{u}_0 \neq \mathbf{0}} \mathrm{wt}(\mathbf{v}_{[j+1]}),$$

where $\mathbf{v}_{[j+1]} = (\mathbf{v}_0, \mathbf{v}_1, \dots, \mathbf{v}_j)$ are the first $j+1$ code blocks of all possible code sequences with $\mathbf{u}_0 \neq \mathbf{0}$. Therefore we are only considering code sequence segments, and not the entire code sequence. When considering the semi-infinite generator matrix $\mathbf{G}$ of the code (see table 8.1), we obtain

$$d_j^c = \min_{\mathbf{u}_0 \neq \mathbf{0}} \mathrm{wt}(\mathbf{u}_{[j+1]} \cdot \mathbf{G}^c_{[j+1]}),$$

with the $k(j+1) \times n(j+1)$ generator matrix

$$\mathbf{G}^c_{[j+1]} = \begin{pmatrix} \mathbf{G}_0 & \mathbf{G}_1 & \dots & \mathbf{G}_j \\ & \mathbf{G}_0 & & \mathbf{G}_{j-1} \\ & & \ddots & \vdots \\ & & & \mathbf{G}_0 \end{pmatrix}.$$

If the generator matrix $\mathbf{G}(D)$ of the code is polynomial then $\mathbf{G}^c_{[j+1]}$ is the same as the generator matrix defined for truncation in equation 8.12 (section 8.1.7) and $\mathbf{G}_j = \mathbf{0}$ for $j > m$ is the termination sequence.

We shall now derive two more distance measures from the column distance:

Definition 8.17 (Distance profile of the generator matrix) *The first $m+1$ values of the column distance $\mathbf{d} = (d^c_0, d^c_1, \ldots, d^c_m)$ of a given generator matrix $\mathbf{G}(D)$ with memory m are referred to as the distance profile of the generator matrix.*

A steep rise in the distance profile is particularly important for sequential decoding (section 8.6).

Definition 8.18 (Minimum distance) *The column distance d^c_m of order m of a given generator matrix $\mathbf{G}(D)$ with memory m is called the minimum distance d_m.*

Note that the only thing in common between the minimum distance of the generator matrix of a convolutional code and the minimum distance of a block code is the name!

Consider the state diagram of the generator matrix $\mathbf{G}(D)$. The paths of length $j+1$ that are used for the calculation of d^c_j leave the zero state $\sigma_0 = 0$ at time $t = 0$ and are in an arbitrary state $\sigma_{j+1} \in (0, 1, \ldots, 2^\nu - 1)$ after $j+1$ state transitions. Consequently, the paths relevant for $j = 0$ are a part of the paths relevant for $j = 1$, and so on. This means that all paths, considered for order j, correspond to the extension of the column path enumerator of order $j-1$ by one transition. Therefore the column distance is a monotonically increasing function in j,

$$d^c_0 \le d^c_1 \le \ldots \le d^c_j \le \ldots \le d^c_\infty, \tag{8.29}$$

where a finite limit d^c_∞ for $j \longrightarrow \infty$ exists. We consider an example to illustrate this distance bound. There is some code sequence with Hamming weight d that returns to the zero state after at most $2^\nu + 1$ transitions. If we stay in the zero loop for any subsequent number of transitions, we shall not increase the Hamming weight of the sequence. Therefore the minimum weight of the sequence for any arbitrary number of additional transitions will not be larger than d.

Row distance The row distance is also a property of the generator matrix and not of the code. It is defined as follows:

Definition 8.19 (Row distance) *The row distance d^r_j of order j of a generator matrix $\mathbf{G}(D)$ with memory m is the minimum Hamming distance of the $j+m+1$ code blocks of two codewords $\mathbf{v}_{[j+m+1]} = (\mathbf{v}_0, \mathbf{v}_1, \ldots, \mathbf{v}_{j+m})$ and $\mathbf{v}'_{[j+m+1]} = (\mathbf{v}'_0, \mathbf{v}'_1, \ldots, \mathbf{v}'_{j+m})$, where the two information sequences used for encoding $\mathbf{u}_{[j+m+1]} = (\mathbf{u}_0, \mathbf{u}_1, \ldots, \mathbf{u}_{j+m})$ and $\mathbf{u}'_{[j+m+1]} = (\mathbf{u}'_0, \mathbf{u}'_1, \ldots, \mathbf{u}'_{j+m})$ with $(\mathbf{u}_{j+1}, \ldots, \mathbf{u}_{j+m}) = (\mathbf{u}'_{j+1}, \ldots, \mathbf{u}'_{j+m})$ are different in at least the first coordinate of the $j+1$ information blocks.*

Graphically, d^r_j is the minimum distance between all paths in the trellis of the generator matrix that diverge from the zero state and run through the same state (which does not necessarily have to be the zero state) at time $j+m+1$. When deriving code or encoder properties, we may compare any non-zero sequence with the zero sequence without loss of generality, based on the linearity of the convolutional code.

The row distance can also be calculated as the minimum of the Hamming weights of all sequences $\mathbf{v}_{[j+m+1]}$, in analogy with the column distance:

$$d_j^r = \min_{\mathbf{u}_0 \neq \mathbf{0}} \text{wt}(\mathbf{v}_{[j+m+1]}),$$

where $\mathbf{v}_{[j+m+1]} = (\mathbf{v}_0, \ldots, \mathbf{v}_j, \ldots, \mathbf{v}_{j+m})$ are terminated code sequences of length $j+m+1$ (section 8.1.7). We obtain

$$d_j^r = \min_{\mathbf{u}_0 \neq \mathbf{0}} \text{wt}(\mathbf{u}_{[j+m+1]} \mathbf{G}_{[j+m+1]}^r),$$

with the information sequence $\mathbf{u}_{[j+m+1]} = (\mathbf{u}_0, \ldots, \mathbf{u}_j, \mathbf{u}_{j+1}, \ldots, \mathbf{u}_{j+m})$, where $(\mathbf{u}_{j+1}, \ldots, \mathbf{u}_{j+m})$ is equal to the termination sequence. The $k(j+m+1) \times n(j+m+1)$ generator matrix $\mathbf{G}_{[j+m+1]}^r$ is given by

$$\mathbf{G}_{[j+m+1]}^r = \begin{pmatrix} \mathbf{G}_0 & \mathbf{G}_1 & \mathbf{G}_2 & \ldots & \mathbf{G}_j & \ldots & \mathbf{G}_{j+m} \\ & \mathbf{G}_0 & \mathbf{G}_1 & & & & \mathbf{G}_{j+m-1} \\ & & \ddots & & & & \vdots \\ & & & \mathbf{G}_0 & \mathbf{G}_1 & \ldots & \mathbf{G}_m \\ & & & & \mathbf{G}_0 & \ldots & \mathbf{G}_{m-1} \\ & & & & & \ddots & \vdots \\ & & & & & & \mathbf{G}_0 \end{pmatrix}.$$

If the generator matrix $\mathbf{G}(D)$ of the code is polynomial then $\mathbf{G}_j = \mathbf{0}$ for $j > m$ and $(\mathbf{u}_{j+1}, \ldots, \mathbf{u}_{j+m}) = \mathbf{0}$ is the termination sequence.

In order to develop a bound for the column distance, we consider the condition when $j = 0$. There is one sequence that diverges from the zero state at time $t = 0$ and returns to the zero state at time $t = m$ and that has a specific weight d_0^r. Increasing the sequence length to $j = 1$ corresponds to increasing the number of valid code sequences that must be considered in finding the minimum-weight sequence. If a code sequence is found for $j = 1$ that has a Hamming weight less than d_0^r then it is selected. Otherwise, we choose the original code sequence, and remain in the zero loop of the zero state for the required number of j transitions without increasing the Hamming weight d_0^r. This argument can be applied to all subsequent values of $j > 1$. Therefore the row distance is a monotonically decreasing function in j, and we can write

$$d_0^r \geq d_1^r \geq \ldots \geq d_j^r \geq \ldots \geq d_\infty^r, \tag{8.30}$$

where the limit $j \to \infty$ gives $d_\infty^r > 0$. Because every path of length $2m+1$ goes through at least one loop in the state diagram, the path with minimum Hamming weight is back in the zero state after at most $j = 2m+1$ transitions and remains there.

The paths in the state diagram that were used for calculating the column distance d_∞^c go from the zero state to the zero loop, which is, according to the Hamming weight of the necessary state transitions, the 'closest'. On the other hand, the row distance d_∞^r is the minimum Hamming weight of all paths in the state diagram from the zero state to the zero state and therefore in the zero loop. However, every other node can be reached as well with $m+1$ state transitions. Therefore $d_\infty^r \geq d_0^c$, and with equations 8.29 and 8.30 we obtain

$$d_0^c \leq d_1^c \leq \ldots \leq d_j^c \leq \ldots \leq d_\infty^c \leq d_\infty^r \leq \cdots \leq d_j^r \leq \cdots \leq d_0^r.$$

Theorem 8.20 (Limit of row and column distance) *If $\mathbf{G}(D)$ is a non-catastrophic generator matrix then the following relation holds true for the limits of the row and column distances:*

$$d_\infty^c = d_\infty^r.$$

Proof In the derivation of equation 8.29, it was shown that the path d_∞^c returns to the zero loop of the state diagram and remains there. Only one zero loop, namely from zero directly to zero, exists for non-catastrophic generator matrices. Therefore d_∞^c is the minimum Hamming weight of a path leaving from and returning to zero. It is equal to the row distance d_∞^r, as shown in the derivation of equation 8.30. □

Free distance The free distance is a property of the code, as already mentioned in section 8.1.6. The formal definition is as follows:

Definition 8.21 (Free distance) *The free distance d_f of a convolutional code C with rate $R = k/n$ is defined as the minimum distance between two arbitrary codewords of C:*

$$d_f = \min_{\mathbf{v} \neq \mathbf{v}'} \mathrm{dist}(\mathbf{v}, \mathbf{v}').$$

Due to linearity, the problem of calculating distances can be interpreted as the problem of calculating Hamming weights. Since the distance between code sequences is a code property, any generator matrix may be considered.

Theorem 8.22 (Row, column and free distance) *If $\mathbf{G}(D)$ is a non-catastrophic generator matrix then the following equation holds for the limits of the row and column distances:*

$$d_\infty^c = d_\infty^r = d_f. \tag{8.31}$$

Proof A possible way to calculate the free distance has been shown in example 8.12 by using the state diagram. By taking into account the above description of row and column distances in the state diagram, equation 8.31 can be proved. □

Example 8.35 (Row and column distance) Figure 8.21 shows the column and row distances of the generator matrix $\mathbf{G}(D) = \begin{pmatrix} 1+D+D^2+D^3 & 1+D^2+D^3 \end{pmatrix}$ of a convolutional code with rate $R = 1/2$ and $m = 3$. A controller design encoder for this generator matrix is shown in figure 8.2. ◇

8.3.2 Extended distance measures

In many cases, the traditional distance measures defined in the previous section do not allow for a sufficient description of the distance properties of convolutional codes or their generator matrices. For this reason, so-called extended distance measures that vary with the length of the considered code sequence will be introduced.

All distance measures discussed here are based on the same principle: segments of code sequences are investigated, where the set of considered code sequences is a subset of the code. The definition of this subset is carried out by restricting possible state sequences. Therefore the considered segments can be depicted in the trellis of the generator matrix and easily determined. The considered distance measure is then the minimum Hamming weight of these segments. Furthermore, all extended distances introduced here are properties of the generator matrix. We shall consider polynomial generator matrices exclusively.

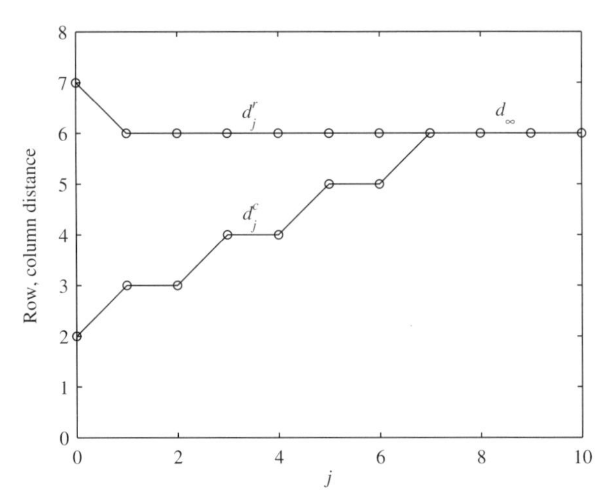

Figure 8.21 Row and column distance.

Extended column distance The *extended column distance* d_j^{ec} is the minimum Hamming weight of the segments $\mathbf{v}_{[j+1]} = (\mathbf{v}_0, \mathbf{v}_1, \ldots, \mathbf{v}_j)$ of length $j+1$, where the set of considered codewords $\mathbf{v}$ is defined using the permitted state sequences $S = (\sigma_0, \sigma_1, \ldots, \sigma_t, \ldots, \sigma_j, \sigma_{j+1})$ with $\sigma_0 = 0$ and $\sigma_t \neq 0$ for $1 \leq t \leq j$. This means only those code segments $\mathbf{v}_{[j+1]}$ are considered that leave the zero state in the trellis at time $t = 0$ and do not return to the zero state in the subsequent j state transitions. The state σ_{j+1} after the $(j+1)$th information block $\mathbf{v}_j$ is not relevant, and can be the zero state.

Definition 8.23 (Extended column distance) *Let the generator matrix* $\mathbf{G}$ *of a convolutional code* $\mathcal{C}$ *with rate* $R = k/n$ *be given. Then the extended column distance of order* j *is*

$$d_j^{ec} = \min_{\sigma_0 = 0, \sigma_t \neq 0, 0 < t \leq j} \left\{ \mathrm{wt}\left((\mathbf{u}_0, \mathbf{u}_1, \ldots, \mathbf{u}_j) \mathbf{G}_{[j+1]}^c \right) \right\},$$

where

$$\mathbf{G}_{[j+1]}^c = \begin{pmatrix} \mathbf{G}_0 & \mathbf{G}_1 & \cdots & \mathbf{G}_m & & & & \\ & \mathbf{G}_0 & \mathbf{G}_1 & \cdots & \mathbf{G}_m & & & \\ & & \ddots & \ddots & & & \ddots & \\ & & & \mathbf{G}_0 & \mathbf{G}_1 & & \cdots & \mathbf{G}_m \\ & & & & \mathbf{G}_0 & & \cdots & \mathbf{G}_{m-1} \\ & & & & & & \ddots & \vdots \\ & & & & & & & \mathbf{G}_0 \end{pmatrix}$$

is the $k(j+1) \times n(j+1)$ *truncated generator matrix.*

Just like the column distance d_j^c, the extended column distance d_j^{ec} is also a monotonically increasing function in j. For non-catastrophic generator matrices, the zero state and consequently the associated zero loop are only permitted after j transitions, and a limit for $j \to \infty$ does not exist for the extended column distance.

Extended row distance The *extended row distance* is an important distance measure of a generator matrix. This will become clear in section 8.4. The state sequence $S = (\sigma_0, \sigma_1, \ldots, \sigma_t, \ldots, \sigma_j, \sigma_{j+1}, \ldots, \sigma_{j+1+m})$ used to define the code segments starts and ends in the zero state, i.e. $\sigma_0 = 0$ and $\sigma_{j+1+m} = 0$. Additionally, the state transitions are $\sigma_t \neq 0$ for $1 \leq t \leq j$.

Definition 8.24 (Extended row distance) *Let the generator matrix* **G** *of a convolutional code* $\mathcal{C}$ *with rate* $R = k/n$ *be given. Then the extended row distance of order* j *is*

$$d_j^{er} = \min_{\substack{\mathbf{u}_j \neq 0 \\ \sigma_0 = 0, \sigma_t \neq 0, 0 < t \leq j}} \left\{ \mathrm{wt}\left((\mathbf{u}_0, \mathbf{u}_1, \ldots, \mathbf{u}_j)\, \mathbf{G}^r_{[j+m+1]} \right) \right\},$$

where

$$\mathbf{G}^r_{[j+m+1]} = \begin{pmatrix} \mathbf{G}_0 & \mathbf{G}_1 & \cdots & \mathbf{G}_m & & & \\ & \mathbf{G}_0 & \mathbf{G}_1 & \cdots & \mathbf{G}_m & & \\ & & \ddots & \ddots & & \ddots & \\ & & & \mathbf{G}_0 & \mathbf{G}_1 & \cdots & \mathbf{G}_m \end{pmatrix}$$

is a $k(j+m+1) \times n(j+1)$ *truncated generator matrix.*

Extended segment distance The *extended segment distance* is defined by the state sequences $S = (\sigma_0, \sigma_1, \ldots, \sigma_m, \sigma_{m+1}, \ldots, \sigma_{j+1+m})$ with $\sigma_0 = 0$. The states σ_m and σ_{j+m+1} can assume arbitrary values. The states in between are not equal to the zero state.

Definition 8.25 (Extended segment distance) *Let the generator matrix* **G** *of a convolutional code* $\mathcal{C}$ *with rate* $R = k/n$ *be given. The extended segment distance of order* j *is*

$$d_j^{es} = \min_{\sigma_t \neq 0, m < t \leq j+m} \left\{ \mathrm{wt}\left((\mathbf{u}_0, \mathbf{u}_1, \ldots, \mathbf{u}_{j+m})\, \mathbf{G}^s_{[j+1]} \right) \right\},$$

where

$$\mathbf{G}^s_{[j+1]} = \begin{pmatrix} \mathbf{G}_m & & & \\ \mathbf{G}_{m-1} & \mathbf{G}_m & & \\ \vdots & \mathbf{G}_{m-1} & \ddots & \\ \mathbf{G}_0 & \vdots & \ddots & \mathbf{G}_m \\ & \mathbf{G}_0 & \cdots & \mathbf{G}_{m-1} \\ & & \ddots & \vdots \\ & & & \mathbf{G}_0 \end{pmatrix}$$

is a $k(j+1+m) \times n(j+1)$ *truncated generator matrix.*

The generator matrices $\mathbf{G}^c_{[j+1]}$, $\mathbf{G}^r_{[j+1]}$ and $\mathbf{G}^s_{[j+1]}$ introduced in definitions 8.23–8.25 are truncated versions of the semi-infinite generator matrix **G** in the time domain. This is demonstrated graphically in figure 8.22.

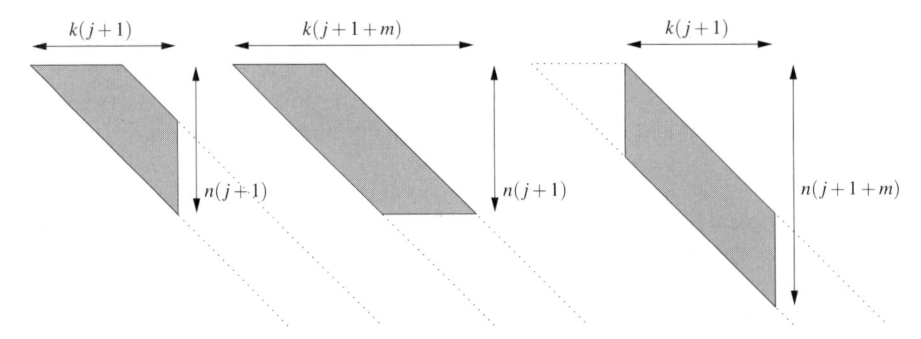

Figure 8.22 Graphical representation of the generator matrices corresponding to the extended distance properties of $\mathbf{G}^{ec}_{[j+1]}$, $\mathbf{G}^{r}_{[j+1]}$ and $\mathbf{G}^{s}_{[j+1]}$ (from left to right).

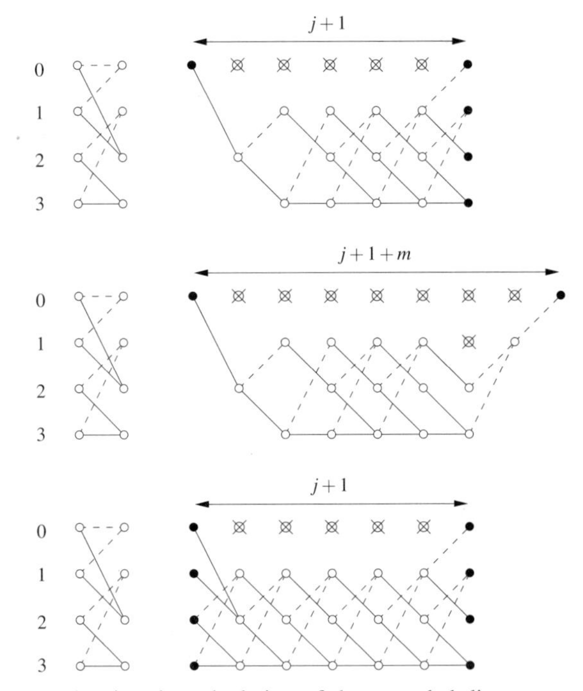

Figure 8.23 Diagrams showing the calculation of the extended distance measurements in trellis form: column, row and segment distances (from top to bottom).

Example 8.36 (Calculation of the extended distance measures) The equations given in the definitions can be used directly for calculating the extended distances. However, the computational complexity increases exponentially with j. The Viterbi algorithm, which we shall describe in the next section, is an appropriate method to calculate the extended distance measures. The trellis in figure 8.23 shows the code segments for the extended row, column and segment distance for $j = 5$. The generator matrix is $\mathbf{G}(D) = \begin{pmatrix} 1+D+D^2 & 1+D^2 \end{pmatrix}$. It is, as demonstrated in example 8.32, a minimal basic generator matrix with $m = 2$. A subtrellis, from which a complete code trellis can be built (example 8.11), is always shown left of the trellis. The states are drawn as circles: the initial and final states of the code segments are

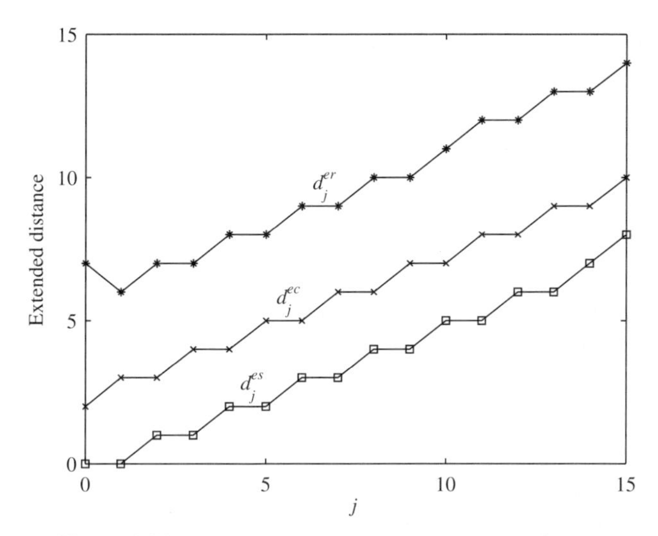

Figure 8.24 Extended row, column and segment distances.

black, the crossed states are prohibited. ◇

Example 8.37 (Extended row, column and segment distance) The extended distance measures of the generator matrix $\mathbf{G}(D)$ of the code $\mathcal{C}$ with rate $R = 1/2$ from example 8.35 are shown in figure 8.24. ◇

Active distances The extended distances can also be modified to give *active* distances. The difference is that the permitted state sequences can pass through the zero state; however they can not remain in the zero state. All practically relevant convolutional codes and their generator matrices show a steep increase in the distance profile, and for the parameters n, k and ν have the largest possible free distance. Active and free distances for these codes are equally large.

The major advantage of the definitions of active distances only becomes clear when calculating limits for these distance measures. For example, the Schwarz inequality holds for the active row distance, which substantially simplifies a number of proofs.

8.4 Maximum-likelihood (Viterbi) decoding

In this section, we discuss the decoding of convolutional codes based on sequence estimation. We briefly review the underlying transmission system (see chapter 7, figure 7.1). The binary information sequence $\mathbf{u}$ is encoded giving the code sequence $\mathbf{v}$ using a convolutional encoder. It is then mapped to the code symbol sequence $\mathbf{x} = 1 - 2\mathbf{v}$ with a BPSK modulator, i.e. $\{0,1\} \rightarrow \{-1,+1\}$, with $0 \rightarrow +1$ and $1 \rightarrow -1$. Channel models include BSC, AWGN and AWGN channels with fading. The sequence $\mathbf{y}$ is received, demodulated and decoded with the Viterbi decoder, i.e. there is no separation of these blocks. The sequences $\mathbf{x}$ and $\mathbf{y}$ consist of blocks of length n, just like $\mathbf{v}$.

The goal of the sequence estimation is to find the code sequence $\mathbf{x}_{opt}$ for which the a posteriori probability $P(\mathbf{x}\,|\,\mathbf{y})$ is maximized (see equation 7.5): $\mathbf{x}_{opt} = \arg\left(\max_{\mathbf{x}\in\mathcal{C}} P(\mathbf{x}\,|\,\mathbf{y})\right)$. If

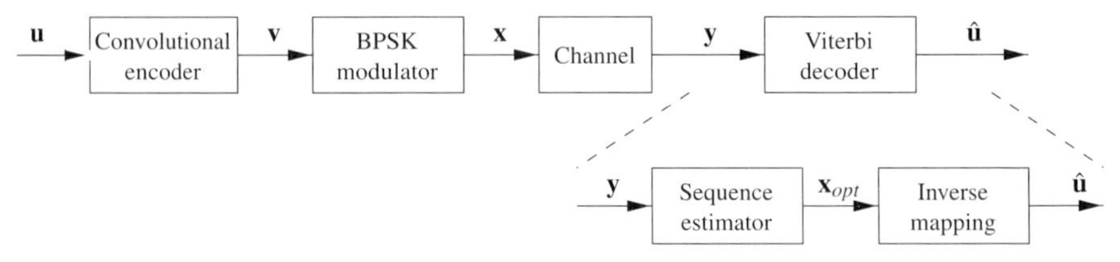

Figure 8.25 Viterbi decoding based on the transmission system.

this equation is applied to Bayes' law $P(\mathbf{x})P(\mathbf{y}\,|\,\mathbf{x}) = P(\mathbf{y})P(\mathbf{x}\,|\,\mathbf{y})$ and the constant factor $P(\mathbf{y})$ is neglected, one obtains

$$\mathbf{x}_{opt} = \arg\left(\max_{\mathbf{x}\in C}\left(P(\mathbf{x})P(\mathbf{y}\,|\,\mathbf{x})\right)\right), \tag{8.32}$$

where $P(\mathbf{x})$ is the a priori probability of the sequence $\mathbf{x}$. If all sequences $\mathbf{x} \in C$ are equally likely then $P(\mathbf{x})$ is constant and can be neglected:

$$\mathbf{x}_{opt} = \arg\left(\max_{\mathbf{x}\in C} P(\mathbf{y}\,|\,\mathbf{x})\right). \tag{8.33}$$

Note that this assumption is often used, even if $P(\mathbf{x})$ is unknown. These methods, based upon sequence estimation, are known as MAP decoding (equation 8.32) and ML decoding (equation 8.33). However, we shall refer to the first case also as ML decoding – under the assumption of a priori probabilities. Therefore the term MAP decoding is reserved for the symbol-by-symbol decoding procedures, described in the next section.

Conceptually, the ML decoding of convolutional codes can be divided into a module that estimates sequences and a module that performs the inverse mapping, as shown in figure 8.25. Thereby the inverse mapping is independent from the sequence estimation. Properties of the generator matrix and code can be distinguished at this point. Because all of the equivalent generator matrices of a convolutional code generate the same set of code sequences, and the sequence estimation of ML decoding always takes place with regard to this set, its result should be interpreted as a code property. In contrast, the inverse mapping is a property of the generator matrix. In this section, we shall deal with the problem of sequence estimation. The properties of the inverse mapping have already been discussed in section 8.2, and are important for the determination of the bit error rate. Depending on the mapping, the same error in the sequence estimation can lead to a different number of bit errors in the estimated information sequence.

8.4.1 Metrics

In order to describe a method to calculate equations 8.32 and 8.33, we first face the problem of having codewords $\mathbf{x} = (\mathbf{x}_0, \mathbf{x}_1, \ldots)$ with infinite length. Fortunately, this can be solved with a step-by-step calculation. Therefore a finite sequence $\mathbf{x}_{[t]} = (\mathbf{x}_0, \mathbf{x}_1, \ldots, \mathbf{x}_{t-1})$ of length t and the current code block $\mathbf{x}_t$ are considered. The sum of the metric at the previous time instance t and the submetric of the tth state transition is called the *metric* (see section 7.1.4) of a code sequence $\mathbf{x}$ at time $t+1$:

$$\Lambda_{t+1}^{\mathbf{x}} = \Lambda_t^{\mathbf{x}} + \lambda_t^{\mathbf{x}}, \tag{8.34}$$

with $\Lambda_0^{\mathbf{x}} = 0$. The *submetric* is generally given by

$$\lambda_t^{\mathbf{x}} = c_1 \left(\ln P(\mathbf{x}_t) + \ln P(\mathbf{y}_t \,|\, \mathbf{x}_t)\right) + c_2. \tag{8.35}$$

Hence the metric of a code sequence $\mathbf{x}$ is obtained as the sum of all submetrics $\Lambda^{\mathbf{x}} = \sum_t \lambda_t^{\mathbf{x}}$. Since taking the logarithm and adding or multiplying by a constant has no effect on the maximization in equation 8.32, we obtain

$$\mathbf{x}_{opt} = \arg\left(\max_{\mathbf{x}\in C} \Lambda^{\mathbf{x}}\right). \tag{8.36}$$

A memoryless information source and a memoryless channel are generally assumed for this calculation. Furthermore, we want to point out that the use of the term 'metric' is not correct in a strict mathematical sense (see appendix B). However, its usage in the above context is quite common.

The calculation of the submetric is done using the L values (see appendix C). Based on equation 8.35, we get

$$\lambda_t^{\mathbf{x}} = \sum_{j=1}^{n} L(y_t^{(j)} \,|\, x_t^{(j)}) x_t^{(j)} + \sum_{i=1}^{k} L(u_t^{(i)})(1 - 2u_t^{(i)}). \tag{8.37}$$

The L values of the a priori probabilities of the channel are

$$L(u_t^{(i)}) = \ln \frac{P(u_t^{(i)} = 0)}{P(u_t^{(i)} = 1)}, \tag{8.38}$$

$$L(y_t^{(j)} \,|\, x_t^{(j)}) = \ln \frac{P(y_t^{(j)} \,|\, x_t^{(j)} = +1)}{P(y_t^{(j)} \,|\, x_t^{(j)} = -1)} = L_{ch} y_t^{(j)}, \tag{8.39}$$

with

$$\begin{aligned} L_{ch} &= \ln \frac{1-p}{p} && \text{(BSC)}, \\ L_{ch} &= \frac{4E_s}{N_0} && \text{(AWGN channel)}, \\ L_{ch} &= a_t^{(j)} \frac{4E_s}{N_0} && \text{(AWGN channel with fading)}. \end{aligned} \tag{8.40}$$

p is the bit error rate of the BSC, E_s the signal energy, N_0 the power of the noise of the AWGN and $a_t^{(j)}$ the attenuation factor of the AWGN channel with fading. Note that the attenuation factor $a_t^{(j)}$ can vary during the transmission, and consequently it can be different for each code bit $x_t^{(j)}$. Using these numbers, the metric calculation for ML decoding can now be performed by taking the a priori probabilities into account. The derivation of equations 8.37–8.40 is complex and is not presented explicitly here.

Without taking the a priori probabilities into account, and for time-invariant channels (BSC and AWGN channels), equation 8.37 is reduced to

$$\lambda_t^{\mathbf{x}} = \sum_{j=1}^{n} y_t^{(j)} x_t^{(j)}. \tag{8.41}$$

In this case, constant factors can be neglected, since they do not affect the result of the maximization in equation 8.33.

Consider the special case of hard-decision decoding. The maximization of equation 8.36 is not affected by the addition of 1 and multiplication by $1/2$ of equation 8.41. Further, we assume a BSC channel. If the bit of the received sequence $y_t^{(j)}$ and the bit of the relevant code sequence $x_t^{(j)}$ are multiplied then we obtain

$$\frac{1}{2}\left(1-y_t^{(j)}x_t^{(j)}\right)=\begin{cases} +1 & \text{for } y_t^{(j)} \neq x_t^{(j)}, \\ 0 & \text{for } y_t^{(j)} = x_t^{(j)}. \end{cases}$$

The metric $\Lambda^{\mathbf{x}}$ of a code sequence $\mathbf{x}$ is therefore the number of positions where $\mathbf{x}$ and $\mathbf{y}$ are equal. Consequently the estimated code sequence $\mathbf{x}_{opt}$ is the sequence among all code sequences with the smallest Hamming distance $\operatorname{dist}(\mathbf{x}_{opt},\mathbf{y}) = \min_{\mathbf{x}\in C}\{\operatorname{dist}(\mathbf{x},\mathbf{y})\}$ to the received sequence $\mathbf{y}$. This metric agrees with the mathematically defined axioms of a metric (see appendix B). This example also shows that either a minimization or a maximization operation can be used to find the optimal code sequence, depending on the metric used.

8.4.2 Viterbi algorithm

The Viterbi algorithm has already been described in sections 7.3.4 and 7.4.2 for hard- and soft-decision ML decoding of block codes. We want to repeat it now for convolutional codes.

The calculation of $\mathbf{x}_{opt}$ with equation 8.32 is done by maximizing the metric $\Lambda^{\mathbf{x}}$. When considering the code tree of a convolutional encoder, a possible (but unrealistic) method is the step-by-step calculation of the metric $\Lambda_t^{\mathbf{x}}$ of all possible code sequences $\mathbf{x}_{[t]}$ at the 2^{kt} nodes with depth t in the code tree. This however leads to an exponentially increasing number of code sequences and metrics in t.

As has already been shown in section 8.1.5, when introducing the trellis, the structure of the code tree repeats itself, and nodes with equal states can be combined. This leads to a substantial simplification of the ML decoding: if multiple code sequences with depth $t > m$ go through the same node σ_t then only the code sequence with the largest metric need be considered for further calculations. It is called the *survivor*. All other sequences can be neglected when maximizing the metric for subsequent times. This can be explained as follows: if the path $\mathbf{x}_{opt}$ goes through a certain node σ_t in the trellis then only the path with the maximum metric Λ_t^{σ} will contribute to the maximum overall metric $\Lambda^{\mathbf{x}_{opt}}$ when maximizing over all alternative paths through this node. Because the trellis consists of 2^{ν} nodes at all times $t \geq m$, only one survivor path per node need be considered and stored.

Each of the 2^{ν} states $\sigma_t \in [0,\ldots,2^{\nu}-1]$ at time $t \geq m$ is associated with a metric Λ_t^{σ}. This and the submetric $\lambda_t^{\sigma'\sigma}$ of all 2^k possible transitions $\sigma_t' \to \sigma_{t+1}$ determine the metric Λ_{t+1}^{σ} of the states σ_{t+1}:

$$\Lambda_{t+1}^{\sigma} = \max_{\sigma'}\left(\Lambda_t^{\sigma'} + \lambda_t^{\sigma'\sigma}\right) \quad \text{for } \sigma \in [0, 2^{\nu}-1],$$

where ν is the overall constraint length and m is the memory of the generator matrix used to form the trellis. Therefore 2^k transitions end in every state σ_{t+1}. A submetric is obtained for every transition. The sum of this value and the old metric of the corresponding initial state σ_t' yields the new metric.

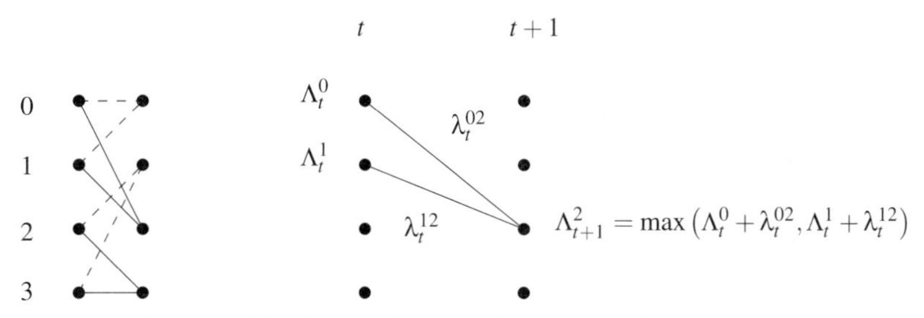

Figure 8.26 Metric calculation for Viterbi trellis decoding.

This method of decoding seems to result in a contradiction. The sequence estimation of the ML decoding of convolutional codes is independent of the selection of the generator matrix. On the other hand, the trellis in which decoding is performed is created using the generator matrix (see section 8.1.5). However, all generator matrices of the same code contain all possible codeword paths. Therefore any valid generator matrix can be used for ML decoding. The property that depends on the generator matrix is the decoding complexity (i.e. the number of states to be stored and compared). It is therefore recommended that a minimal basic trellis be used for Viterbi decoding.

Example 8.38 (Viterbi decoding) We want to describe decoding using the Viterbi algorithm on a corrupted received sequence. The terminated information sequence $\mathbf{u}_{[l]}$ is mapped with the generator matrix $\mathbf{G}(D) = \left(\begin{array}{cc} 1+D+D^2 & 1+D^2 \end{array}\right)$ of a code with rate $R = 1/2$ to a code sequence $\mathbf{v}_{[l+m]}$ and transmitted through a BSC. The decoding is performed in the trellis, which has four states σ_t for each transition $m \leq t \leq l$. Consequently, four metrics Λ_t^σ with $\sigma \in [0,3]$ are calculated at all times. As shown in figure 8.26, the metric Λ_{t+1}^2 of state $\sigma_{t+1} = 2$ at time $t+1$ is obtained from the metric Λ_t^0 of state $\sigma_t = 0$ and Λ_t^1 of state $\sigma_t = 1$ at time t and the respective submetric. For example, λ_t^{02} is the submetric of the state transition, $0_t \rightarrow 2_{t+1}$. The metric calculation is performed for each state at that time in the same manner.

The terminated information sequence $\mathbf{u}_{[l]}$ of length $l = 4$ is encoded and transmitted. Transmission errors have occurred, and they are marked in the received sequence $\mathbf{y}_{[4+2]}$:

$$\begin{array}{rcllllll} \mathbf{u}_{[4+2]} & = & (1 & 0 & 1 & 1 & 0 & 0) \\ \mathbf{v}_{[4+2]} & = & (11 & 10 & 00 & 01 & 01 & 11) \\ \mathbf{y}_{[4+2]} & = & (1\mathbf{0} & 10 & 00 & 0\mathbf{0} & 01 & 11) \end{array}$$

The Viterbi decoding is carried out in the six steps shown in figure 8.27. The metric is here the number of bits in which the received and the original code sequence correlate. The most likely path through the trellis is the one with the largest metric. We have kept the familiar binary representation of the sequences. The considered sequences consist of six blocks, i.e., all together, six steps are necessary to calculate the associated submetric, which are shown in figure 8.27.

Step 1 We begin by considering the 0th transition in the trellis. Because the trellis is still in its starting phase, only two transitions are possible, i.e. $0 \rightarrow 0$ and $0 \rightarrow 2$. These transitions are labeled with the respective code blocks (00) and (11). The metric $\Lambda_0^0 = 0$ (at the root of the trellis) is zero by definition. Additionally, the 0th transition is labeled with the received block (10). The two metrics $\Lambda_1^0 = 1$ and $\Lambda_1^2 = 1$ are obtained from the comparison between code blocks of the corresponding state transitions and the received block.

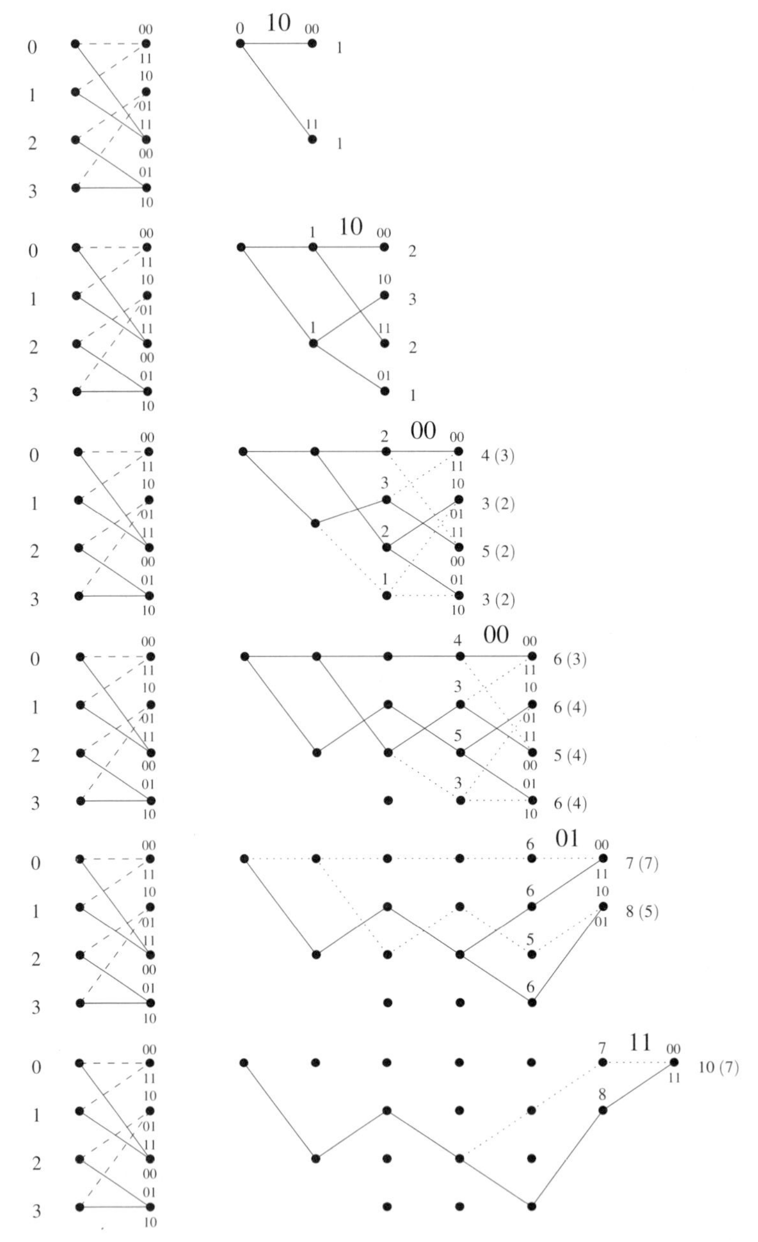

Figure 8.27 Steps in the Viterbi decoding of a terminated convolutional code.

Step 2 Also in the second step, the metrics $\Lambda_2^0 = 2$, $\Lambda_2^1 = 3$, $\Lambda_2^2 = 2$ and $\Lambda_2^3 = 1$ are calculated using the metric $\Lambda_1^0 = 1$ and $\Lambda_1^2 = 1$ and the corresponding submetric. The submetrics are again obtained from the matching bits of the received sequence (10) and the code blocks (00), (10), (11) and (01).

Step 3 In the third step, all states of the trellis are active. Two transitions always arrive at each state. As demonstrated in figure 8.26, a new metric Λ_3^σ is calculated for all of these states. Consider the state $\sigma_3 = 0$. The metric is obtained from $\Lambda_3^0 = \max(4,3) = 4$, where for the metric of the two paths arriving at this node, $\Lambda_2^0 + \lambda_2^{00} = 2+2 = 4$ and $\Lambda_2^1 + \lambda_2^{10} = 3+0 = 3$. Suboptimal paths are drawn with dotted lines, and the survivors with solid lines.

Step 4 The fourth step is performed in the same way as for step 3.

Step 5 In the fifth step in the state $\sigma_5 = 0$, two paths with the same metric meet. In this case, either path can be chosen arbitrarily as the survivor, since both are equally likely. Furthermore, the number of states decreases due to the requirement that all paths must terminate in the zero state.

Step 6 In the sixth step, only the zero state is possible in the trellis, which terminates the sequence. In this way, we only obtain one survivor - just the one in the zero state - and the decoding is finished. If we trace back the survivor then we obtain the original transmitted code sequence $\mathbf{v}_{[4+2]}$ as the estimated code sequence $\hat{\mathbf{v}}_{[4+2]}$. The transmission errors have been corrected. Also the inverse mapping from code to information sequence can be done in the trellis, and we obtain the transmitted information sequence $\hat{\mathbf{u}}_{[4]} = \mathbf{u}_{[4]}$. ◇

This example has demonstrated the functionality of the Viterbi algorithm for terminated code sequences.

No terminated final state in the trellis exists for code sequences with infinite length. Therefore all 2^ν survivors of the relevant states σ_t at time $t \geq m$ must be stored. A decoding decision is made when all survivors have a common starting path. However, a decision may be made before this condition is satisfied. This is due to the practical limitations of the decoder's memory. A rule of thumb for the necessary memory size of the survivor paths is $10m$ for 'good' channels and $100m$ for 'bad' channels. Generally, this situation should be avoided by appropriate termination and a choice of memory size equal to the block size. If a decision for an information block has to be made before the survivors of all states have merged then two strategies exist: either the information block associated with the survivor with the largest metric is chosen, or the one that is associated with the most survivors.

Example 8.39 (Bit error probability) Some good convolutional codes with rate $R = 1/2$ are given in section 8.8 (table 8.10). As shown in example 8.52, any convolutional code can be described with n octal numbers $(O_G^{(1)}, \ldots, O_G^{(n)})$.

The bit error rates over an AWGN channel of the three convolutional codes $(5,7)$, $(23,35)$ and $(133,171)$ are shown in figure 8.28. Because the free distance d_f increases with the overall constraint length ν of the convolutional code, an asymptotically better correction capability can be observed for codes with larger free distance. ◇

Note The Viterbi algorithm can be used for a variety of problems; however, it is necessary that the given problem can be represented in trellis form. The term 'trellis' is not limited to the representation of convolutional codes, but in general refers to any graph having a trellis structure. Consider for example the calculation of free distances (section 8.3.2). This problem (see example 8.36) can be presented in terms of a trellis and solved using the Viterbi algorithm.

8.4.3 Bounds for decoding performance

We shall now demonstrate the relation between the distance measures and the correction capability of a convolutional code. Therefore it is first necessary to find an appropriate measure to

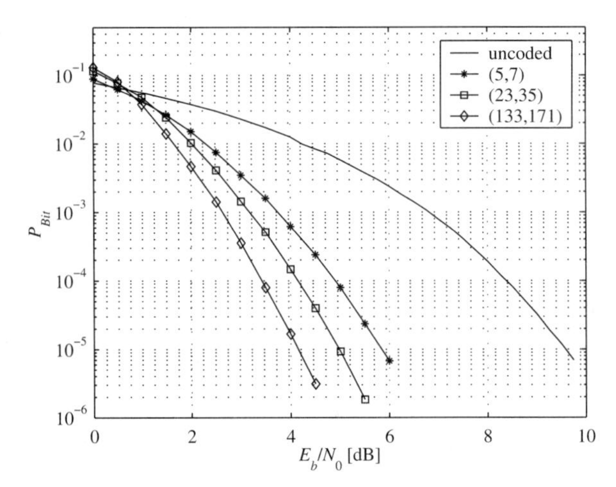

Figure 8.28 Error probability of three convolutional codes with rate $R = 1/2$ and total constraint length $\nu \in \{2,4,6\}$ over an AWGN channel.

describe possible decoding errors. The definition of a codeword error probability P_{Block}, as it is common for block codes, is impractical for convolutional codes. When considering the above described Viterbi decoding using a code trellis, the probability of a wrongly decoded code segment and consequently P_{Block} reaches unity for a code sequence with infinite length. However, a satisfactory protection of the information can still be accomplished. This reasoning leads to the definition of *error bursts*.

Burst errors An error burst is a segment of the estimated code sequence that differs from the correct path in the trellis, takes another path through the trellis, and finally merges with the correct path again. The event that an error burst starts at time t in the trellis and has length l is called $\mathcal{F}_t(l)$. Our goal is to give an upper limit for the probability $P(\mathcal{F}_t(l)) = P_{\mathcal{F}_t}(l)$ of such an error burst, independent of time t.

First, we can assume without restrictions that the zero sequence is a correct path in the trellis because of the linearity of convolutional codes. Therefore an error burst is a path that leaves the zero sequence at time t and returns after l state transitions at time $t+l$ to the zero state. Further, an error burst can only start at times t, at which the correct path and the estimated path are the same, i.e. previous error bursts have merged with the correct path. The stochastic process is in a state that is equivalent to the initial state, and the probability of an error burst $P_{\mathcal{F}}(l)$ at time $t = 0$ is an upper limit for $P_{\mathcal{F}_t}(l)$ with $t > 0$.

Example 8.40 (Error burst) Consider the trellis of the code with rate $R = 1/2$ with the generator matrix $\begin{pmatrix} 1+D+D^2 & 1+D^2 \end{pmatrix}$. A possibly wrong decoded segment $\mathbf{v}_{[6]} = (11,01,01,00,10,11)$ with length $l = 6$ is shown in figure 8.29. If the BSC is used as the channel model then the correct and the wrong path have Hamming distance $w = \mathrm{wt}\left(\mathbf{v}_{[6]}\right) = 7$ in this section. If $e > \lfloor (w-1)/2 \rfloor = 3$ bit errors happened at the respective 7 positions then a decoding error occurs. A possible error pattern that will lead to a decoding error is, for example, $\mathbf{e}_{[6]} = (01,00,01,00,10,11)$. It occurs with probability $P(\mathbf{e}_{[6]}) = (1-p)^7 p^5$, where p is the bit error rate of the channel. ◇

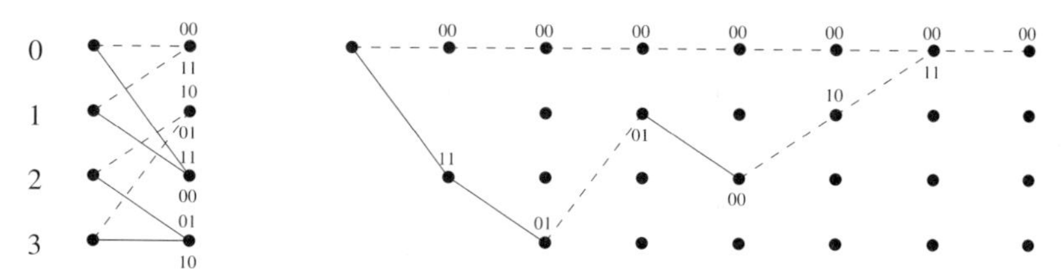

Figure 8.29 Decoding of a received vector with transmission errors.

Probability of an error burst Depending on the weight of a possibly false code segment $\mathbf{v}_{[l]}$ with Hamming weight $w = \mathrm{wt}(\mathbf{v}_{[l]})$, an upper limit for the probability of an associated decoding error can be given. Therefore

$$e > \lfloor (w-1)/2 \rfloor$$

bit errors have to occur at the respective positions of the sequence. When only considering these positions, the probability

$$p_w = \begin{cases} \sum\limits_{e=(w+1)/2}^{w} \binom{w}{e} p^e (1-p)^{w-e} & \text{for } w \text{ odd,} \\ \frac{1}{2}\binom{w}{w/2} p^{w/2}(1-p)^{w/2} + \sum\limits_{e=w/2+1}^{w} \binom{w}{e} p^e (1-p)^{w-e} & \text{for } w \text{ even} \end{cases} \tag{8.42}$$

is an upper limit that the considered code segment $\mathbf{v}_{[l]}$ will lead to a decoding error, i.e. it is ensured that the probability of decoding a false segment is p_w, but it will not necessarily be decoded to the considered segment.

The probability p_w, described in equation 8.42 can be substituted by the upper limit

$$p_w < \left(2\sqrt{p(1-p)}\right)^w. \tag{8.43}$$

This will allow a compact presentation of further results.

Union bound We can give an upper limit for the probability of false-decoding a certain code segment, which can be applied to every possible code segment, $\mathbf{v}_{[l]}$. Because the events $\mathcal{E}_i$ are not exclusive, we use the so-called *union bound* $P(\cup_i \mathcal{E}_i) \leq \sum_i P(\mathcal{E}_i)$ to derive an upper limit for the probability of an error burst with length l. With $P(\mathcal{E}_w) = p_w$, it holds that

$$P_{\mathcal{F}}(l) \leq \sum_w N(l,w) p_w,$$

where $N(l,w)$ are the coefficients of the weight distribution $A_l(W)$ defined in equation 8.10. The number $N(l,w)$ of all paths with weight w and length l that are subject to possible decoding errors can be calculated using the extended path enumerator. When summing over all possible lengths

$$P_{\mathcal{F}} \leq \sum_l P_{\mathcal{F}}(l),$$

an upper limit for the probability that an error occurs at an arbitrary state in the trellis can be given, independent of the length of the error. This upper limit is called the *first-event probability*. With this, we obtain

$$P_{\mathcal{F}} \leq \sum_{w}^{\infty} N(w) p_w = T(W)\Big|_{W=2\sqrt{p(1-p)}} \quad , \tag{8.44}$$

where $N(w)$ is the distance spectrum (see equation 8.9) of the code and $T(W)$ is the path enumerator of the code, defined in section 8.1.6. This very compact limit for the first-event probability is also referred to as *Viterbi bound*. Whereas with good channels, i.e. low bit error rate, this limit is almost equal to the real values, this bound becomes poor for bad channels.

An upper limit of the bit error probability P_{Bit} can also be calculated using the above-derived limit. Depending on the probability p_w, we obtain (without derivation)

$$P_{Bit} < \frac{1}{k} \sum_{w} I(w) p_w = \frac{1}{k} \frac{\partial T(W,H,L=1)}{\partial H}\Bigg|_{\substack{W=2\sqrt{p(1-p)} \\ H=1}} \quad , \tag{8.45}$$

where $T(W,H,L)$ is the extended path enumerator from equation 8.8, and the approximation from equation 8.43 was used for p_w. The term $I(w)$ is called the information weight, and is obtained from the extended path enumerator by $\partial T(W,H,L=1)/\partial H|_{H=1} = \sum_w I(w) W^w$.

All statements made here can be extended to AWGN channels. In this case, we obtain, in analogy to equations 8.42 and 8.43 (without derivation),

$$p_w = \operatorname{erfc}\left(\sqrt{2R\frac{E_b}{N_0}w}\right) < e^{-R\frac{E_b}{N_0}w} ,$$

with the complementary error function $\operatorname{erfc}(z) = 1/\sqrt{2\pi} \int_z^{+\infty} e^{-\eta^2/2}\, d\eta$. Using this the limits provided in equations 8.44 and 8.45 can be extended to AWGN channels.

Example 8.41 (Approximation for bit error probability) A good approximation for the bit error probability P_{Bit} is obtained if the series given in equation 8.45 is solved for the first terms only. The extended path enumerator need not be calculated in closed form. Only the first coefficients in the series expansion need be calculated. This computation can be done in a much more efficient manner. No upper limit is obtained when using this procedure, but it yields a sufficient approximation for practical purposes.

Figure 8.30 shows the approximations and the simulated bit error curves for the $(5,7)$ and $(133,171)$ codes. In both curves, the series stopped after four terms, i.e. for $w = d_f + 3$. ◇

8.4.4 Interleaving

The Viterbi algorithm described in the previous section and the discussion about error bursts show that convolutional codes can efficiently correct statistically independent single errors occurring in BSC or AWGN channels. Correct decoding becomes worse for the case of bursty channels, (e.g. the Gilbert–Elliot channel in section 7.1.3). *Interleaving* is a method of converting error bursts to single independent errors through the permutation of bit positions.

In principle, we can define an interleaver as a code with rate $R = 1$, which maps a stochastic process creating error bursts to a process creating independent single errors. Finding an appropriate interleaver for a given system is a non-trivial task. The stochastic process at the input of

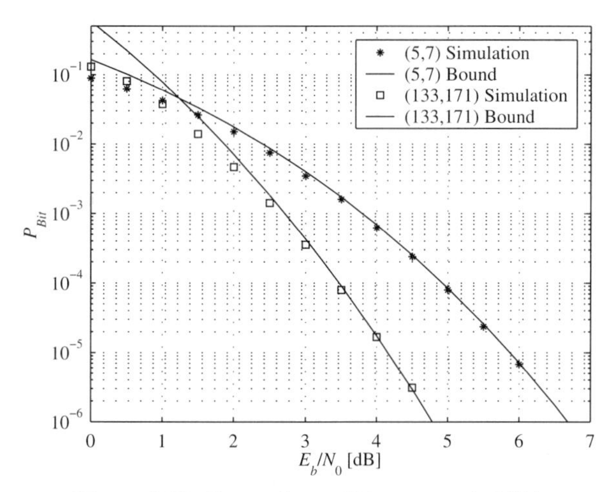

Figure 8.30 Comparison of bit error probabilities.

the interleaver has to be described. Therefore knowledge of the properties of the underlying channel model is necessary. This knowledge allows us to specify the size of the interleaving block. Thereby a *trade-off* has to take place between the delay and the spreading effect, both caused by the interleaving. Finally, an appropriate permutation has to be found.

A block interleaver is used in many cases. It writes the input sequence row by row into a matrix and reads the output sequence column by column. However, other interleaver constructions are conceivable!

8.4.5 Soft-output Viterbi algorithm (SOVA)

The Viterbi algorithm described in the previous section provides 'hard' decisions at the output. Often it is desirable to have information about the reliability of the decisions. This can be accomplished with a *soft-output Viterbi algorithm* (SOVA).

In order to give a reliability measure to a decision, we have to evaluate possible alternatives. As shown in equations 8.32 and 8.33, the probability $P(\mathbf{x})P(\mathbf{y}\,|\,\mathbf{x})$ of all code sequences $\mathbf{x}$ and hence all paths in the trellis is maximized depending on the received sequence $\mathbf{y}$. Consider the calculation of the metric or submetric (see equations 8.34–8.37) in the trellis. It holds that

$$P(\mathbf{x})P(\mathbf{y}\,|\,\mathbf{x}) \sim e^{\Lambda^{\mathbf{x}}/2},$$

and for a segment of the code sequence $\mathbf{x}_{[t]}$ of length t,

$$P(\mathbf{x}_{[t]})P(\mathbf{y}_{[t]}\,|\,\mathbf{x}_{[t]}) \sim e^{\Lambda_t^{\mathbf{x}}/2}.$$

The factor $1/2$ is caused by the choice of the multiplying constants $c_1 = 2$ in the calculation of the submetric $\lambda_t^{\mathbf{x}}$ in equation 8.35. The path $\mathbf{x}_{opt}$, based on ML decoding, is the most likely transmitted code sequence. Possible alternatives to this decision are deleted in the decoding process at all times $t > m$, since at each state σ_t only the survivor is stored in memory.

When using a convolutional code with rate $R = 1/n$, two paths are compared at each decision level (node). Depending on the metric $\Lambda_t^{\mathbf{x}}$ of the survivors $\mathbf{x}_{[t]}$ and $\Lambda_t^{\mathbf{x}'}$ of the terminated

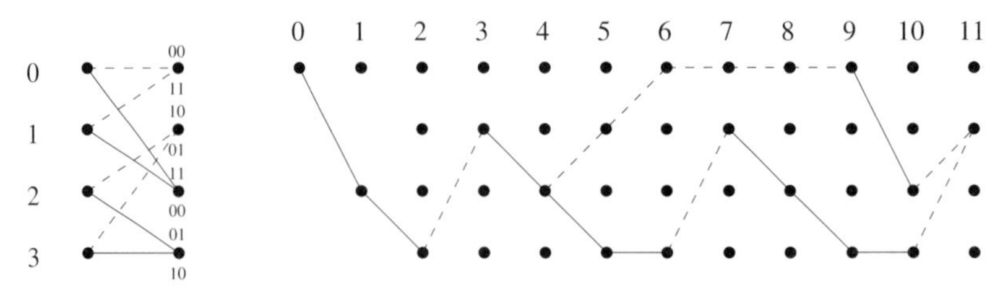

Figure 8.31 Iterative calculation of the reliability at the output of the Viterbi decoder.

path $\mathbf{x}'_{[t]}$, the probability of a false decision can be calculated:

$$\frac{P(\mathbf{x}'_{[t]})P(\mathbf{y}_{[t]}\,|\,\mathbf{x}'_{[t]})}{P(\mathbf{x}'_{[t]})P(\mathbf{y}_{[t]}\,|\,\mathbf{x}'_{[t]})+P(\mathbf{x}_{[t]})P(\mathbf{y}_{[t]}\,|\,\mathbf{x}_{[t]})}$$

$$= \frac{e^{\Lambda_t^{\mathbf{x}'}/2}}{e^{\Lambda_t^{\mathbf{x}'}/2}+e^{\Lambda_t^{\mathbf{x}}/2}} = \frac{1}{1+e^{(\Lambda_t^{\mathbf{x}}-\Lambda_t^{\mathbf{x}'})/2}} = \frac{1}{1+e^{\Delta_t^{\sigma}}},$$

where the difference between the metrics of the survivor and the terminated path at state σ_t at time t is

$$\Delta_t^{\sigma} = (\Lambda_t^{\mathbf{x}} - \Lambda_t^{\mathbf{x}'})/2.$$

The probability of an error associated with each of these decisions can be given as

$$p_f(\Delta_t^{\sigma}) = \frac{1}{1+e^{\Delta_t^{\sigma}}}.$$

Consequently, a false decision at state σ_t is made, and hence the correct path is left with this probability. The reliability of single bits of the estimated information sequence is calculated based upon the probability $p_f(\Delta_t^{\sigma})$. Consider the following example:

Example 8.42 (SOVA) Figure 8.31 shows a part of the Viterbi-decoding process. As already mentioned, at all times t, a survivor is determined for each state σ_t. Figure 8.31 shows only the calculation of the survivor at state $\sigma_{11} = 1$ and the two incoming paths $\mathbf{x}_{[11]}$ and $\mathbf{x}'_{[11]}$. Let Δ_{11}^1 be the metric difference of these two paths; then the probability of a false decision at this point is $p_f(\Delta_{11}^1)$. If we assume that the path $\mathbf{x}_{opt}$ continues beyond this state then only a part of the estimated information sequence is affected by this decision. When looking at the estimated information sequence (dotted transitions correspond to an information bit 0 and solid transitions to a 1), we obtain

$$\begin{aligned} \mathbf{u}_{[11]} &= (1\ \ 1\ \ 0\ \ 1\ \ 1\ \ 1\ \ 0\ \ 1\ \ 1\ \ 1\ \ 0), \\ \mathbf{u}'_{[11]} &= (1\ \ 1\ \ 0\ \ 1\ \ \mathbf{0}\ \ \mathbf{0}\ \ 0\ \ \mathbf{0}\ \ \mathbf{0}\ \ 1\ \ 0). \end{aligned}$$

Therefore only information bits u_t at times $4 \le t \le 8$ are affected by the decision at state 1_{11}. On the one hand, the first four transitions of the two paths are equal, and $u_i = u'_i$ for $i \in \{0,3\}$, on the other hand, the last $m = 2$ information bits $u_i = u'_i$ for $i \in \{9,10\}$ of the two code sequences $\mathbf{x}_{[11]}$ and $\mathbf{x}'_{[11]}$ are also equal. Otherwise, the two code sequences would not merge in the same state! The two information sequences do not necessarily have to differ in the false area; for example, $u_6 = u'_6$, but $u_i \neq u'_i$ for $i \in \{4,5,7,8\}$. ◇

As this example shows, the e information bits

$$u_i \neq u_i', \quad i \in \{i_1, \ldots, i_e\},$$

of the survivor are affected by the decision at state σ_{t+1}. The relevant bit positions have to be determined using the respective information sequences of the survivor and the terminated path. Using these, the probability $p_{t+1}^{\sigma}(i)$ of a false bit u_i can be calculated:

$$p_{t+1}^{\sigma}(i) = \begin{cases} p_t^{\sigma'}(i)(1 - p_f(\Delta_{(t+1)}^{\sigma})) + (1 - p_t^{\sigma'}(i))p_f(\Delta_{(t+1)}^{\sigma}), & i \in \{i_1, \ldots, i_e\}, \\ p_t^{\sigma'}(i) & \text{otherwise,} \end{cases} \tag{8.46}$$

where $p_t^{\sigma'}(i)$ is the probability of a false bit based on previous decisions, which led to the considered survivor. Note that σ_t' is the state of the survivor at the previous state. Equation 8.46 can be interpreted in the following way:

- If the information bits of the survivor and the terminated path do not differ then their error probability is not affected by the decision.
- If the respective information bits are affected by the decision then their error probability is affected, and we obtain two terms. The *first* term is the probability of a correct decision $(1 - p_f(\Delta_t^{\sigma}))$ multiplied by the probability $p_t^{\sigma'}(i)$ that errors have occurred in previous decisions leading to this survivor. The *second* term assumes a false decision, and the associated probability $p_f(\Delta_t^{\sigma})$ is multiplied by the probability $(1 - p_t^{\sigma'}(i))$ that the bits of previous decisions have been determined correctly.

This procedure has to be carried out at all times $t > m$ and for all states σ_t. If all survivors merge, or at the end of a block a terminated convolutional code is used, then the reliability of the estimated bit $\hat{u}_i$ can be calculated from the error probabilities $p(i)$ thus calculated, and we obtain

$$L(\hat{u}_i) = \ln \frac{1 - p(i)}{p(i)}.$$

This reliability can be calculated directly while performing the Viterbi algorithm. This is done by modifying equation 8.46, i.e., instead of the probability of a false decision $p_t^{\sigma}(i)$, we calculate the associated reliability using log likelihood algebra (see appendix C):

$$L_{t+1}^{\sigma}(i) = \begin{cases} \ln \dfrac{1 + e^{L_t^{\sigma'}(i) + \Delta_t^{\sigma}}}{e^{\Delta_t^{\sigma}} + e^{L_t^{\sigma'}(i)}} & \text{for } i \in \{i_1, \ldots, i_e\}, \\ L_t^{\sigma'}(i) & \text{otherwise.} \end{cases}$$

When using the approximation

$$L_{t+1}^{\sigma}(i) = \begin{cases} \min\{L_t^{\sigma'}(i), \Delta_t^{\sigma}\} & \text{for } i \in \{i_1, \ldots, i_e\}, \\ L_t^{\sigma'}(i) & \text{otherwise,} \end{cases}$$

which is frequently used in log lokelihood algebra calculations, we obtain a calculation procedure with low complexity, which is especially suited for a SOVA implementation.

List decoding Let us now consider the metric differences $\Delta_t^{\mathbf{x}_{opt}}$ at all state transitions of the decoded path $\mathbf{x}_{opt} = \mathbf{x}_{1st}$. The smallest of these differences $\Delta_{min} = \min_t \Delta_t^{\mathbf{x}_{opt}}$ is the most unreliable decision, upon which selection of the decoded sequence is based. Hence with the path that has been terminated at this point, we obtain the code sequence $\mathbf{x}_{2nd}$, which is the second most likely code sequence after the decoded sequence $\mathbf{x}_{1st}$ for a given received sequence $\mathbf{y}$.

If the third most likely path $\mathbf{x}_{3rd}$ is to be calculated then the second most likely path $\mathbf{x}_{2nd}$ has to be taken into account. Therefore we consider the minimum metric difference Δ_{min} that occurs in the code path $\mathbf{x}_{2nd}$. We construct $\mathbf{x}_{3rd}$ based on the path that was terminated at Δ_{min}. This calculation can be done recursively and we obtain a list of the L most likely code sequences. This list $(\mathbf{x}_{1st}, \mathbf{x}_{2nd}, \dots, \mathbf{x}_{Lth})$ can now be used in different decoding procedures (see [ZPS93]).

8.5 Maximum a posteriori decoding (MAP)

The combination of source and channel coding and the decoding of convolutional codes has brought soft-output decoding techniques to the forefront of research over the past several years. A code can incorporate modulation as well as equalization (see chapter 10). One technique uses reliability information to conceal errors. For example, unreliable speech blocks, which cause distortion are deleted. Furthermore, soft-output decoding of the inner decoder serves as soft input information of a second decoder.

In the previous section, we have explained a modification of the Viterbi algorithm (SOVA) that can be used for soft-output decoding. In this section, we shall discuss the MAP decoder, which is capable of processing soft input data as well as providing soft-output information. The algorithm minimizes the symbol error probability and also calculates the reliability of a decision for a symbol. Furthermore, a simplification termed the max log MAP algorithm will be presented.

8.5.1 BCJR algorithm

In contrast to the estimation of the entire sequence, as was done for the Viterbi algorithm, a method will be introduced that performs the estimation of a single symbol while taking the entire received sequence into consideration. The *symbol-by-symbol* MAP algorithm [BCJR74] for the trellis representation of a code (block or convolutional code) is used. This method has in principle already been introduced in section 7.4.1. Now we present the BCJR algorithm for binary convolutional codes. Generally, the description of the BCJR algorithm can be done based on probabilities or log-likelihood ratios. Whereas we have used L values for block codes, we want to use probabilities now. To keep things simple, we only consider terminated convolutional codes with rate $1/n$. Further, it is shown in [Rie98] how a convolutional code can be MAP-decoded using its dual code. We assume a received vector $\mathbf{y}$ with BPSK modulation and an AWGN channel according to section 7.1.2.

Let the state of the encoder at time t be $\sigma_t \in \{0, 1, \dots, 2^\nu - 1\}$. The information bit u_t corresponds to the state transition between $t-1$ and t. With the help of the BCJR algorithm, the *a posteriori probability* APP will be calculated for each information bit u_t. The logarithm of the APP can be used as information reliability, according to section 7.2.2. Thereby the APP of an information bit $u_t = 0$ is divided by the APP of $u_t = 1$, i.e. one obtains

$$L(\hat{u}_t) = \ln \frac{\sum\limits_{\sigma_{t-1},\sigma_t} \alpha_{t-1}(\sigma_{t-1})\gamma_0(y_t,\sigma_{t-1},\sigma_t)\beta_t(\sigma_t)}{\sum\limits_{\sigma_{t-1},\sigma_t} \alpha_{t-1}(\sigma_{t-1})\gamma_1(y_t,\sigma_{t-1},\sigma_t)\beta_t(\sigma_t)} . \tag{8.47}$$

The parameters are defined as follows. The forward recursion numbers are

$$\alpha_t(\sigma_t) = \sum_{\sigma_{t-1},i=0,1} \alpha_{t-1}(\sigma_{t-1})\gamma_i(y_t,\sigma_{t-1},\sigma_t) \tag{8.48}$$

$$\alpha_0(\sigma_0) = \begin{cases} 1 & \text{for } \sigma_0 = 0, \\ 0 & \text{otherwise}, \end{cases}$$

and the backward recursion numbers are

$$\beta_t(\sigma_t) = \sum_{\sigma_{t+1},i=0,1} \gamma_i(y_{t+1},\sigma_t,\sigma_{t+1})\beta_{t+1}(\sigma_{t+1}), \tag{8.49}$$

$$\beta_l(\sigma_l) = \begin{cases} 1 & \text{for } \sigma_l = 0, \\ 0 & \text{otherwise}. \end{cases}$$

The path transition probabilities are given by

$$\gamma_i(y_t,\sigma_{t-1},\sigma_t) = q(u_t = i|\sigma_t,\sigma_{t-1})p(y_t|u_t = i,\sigma_t,\sigma_{t-1})P_a(\sigma_t|\sigma_{t-1}) .$$

The value of $q(u_t = i|\sigma_t,\sigma_{t-1})$ is either 1 or 0, depending on whether a transition with information bit i from state σ_{t-1} to state σ_t exists. That means that the function $q(\cdot)$ selects a subset of branches from the full trellis (another possibility is to work directly in the code trellis, see section 7.4.1). The term $P_a(\sigma_t|\sigma_{t-1})$ is the *a priori* probability for the bit u_t (for the case of equally likely bits, it is 0.5).

This representation of the BCJR algorithm can only be used for terminated convolutional codes with block length l. The implementation is based on the trellis diagram, according to the following example.

Example 8.43 (BCJR algorithm) A code with parameters $k = 1$, $n = 3$ and overall constraint length $\nu = 2$ is used for this example. The three generator polynomials are $g_1^{(1)} = (1,0,0)$, $g_1^{(2)} = (1,0,1)$ and $g_2^{(3)} = (1,1,1)$, or, in octal representation, $(4,5,7)$. Three information bits are transmitted. To end up in the zero state, two termination bits are appended, so that the size of the information block $l = 5$. Because the code has rate $R = 1/3$, $l = 5$ code symbols generate 15 code bits. Figure 8.32 shows the associated trellis diagram. The path that corresponds to the assumed code sequence is drawn with a dashed line. The information and code bits are shown for each vertex.

A BPSK transmission through an AWGN channel with the mapping $0 \rightarrow 1$ and $1 \rightarrow -1$ is used. Table 8.2 contains the information bits $\mathbf{u}$, the associated code bits $\mathbf{v}$ and the received values $\mathbf{y}$, based on a signal-to-noise ratio of -4dB. For simplicity, we assume that zeros and ones appear equally often, i.e. no a priori information is available, or it is equal to 1/2 for each symbol.

To insure that the metric values do not become increasingly small when dealing with large block lengths, the accumulated probabilities $\alpha_t(\sigma_t)$ and $\beta_t(\sigma_t)$ are scaled at each time step t,

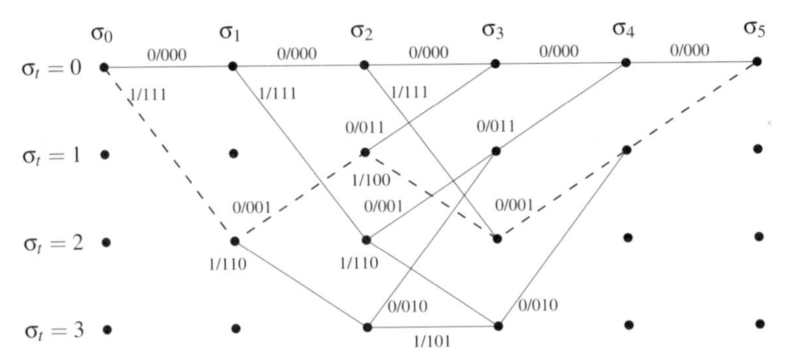

Figure 8.32 Trellis diagram of a code with rate $R = 1/3$, $\nu = 2$ and block length $l = 5$. The paths are labeled with the input and output bits.

Table 8.2 Information sequence **u**, code sequence **v** and received signal **y**.

	$t=0$	$t=1$	$t=2$	$t=3$	$t=4$
$\mathbf{u}_t$	1	0	1	0	0
$\mathbf{v}_t$	1 1 1	0 0 1	1 0 0	0 0 1	0 1 1
$\mathbf{y}_t^T$	−1.024 −0.169 +0.111	+0.241 −0.825 −1.044	−2.143 +0.796 +0.560	−0.260 +1.101 −0.006	+0.266 +0.244 −0.632

Table 8.3 Scaled transition probabilities of all state transitions $\gamma_i(y_t, \sigma_{t-1}, \sigma_t)$.

$\gamma_i(y_t, \sigma_{t-1}, \sigma_t)$	$t=1$	$t=2$	$t=3$	$t=4$	$t=5$
$\sigma_0 \to \sigma_0$	0.083	0.040	0.018	0.354	0.223
$\sigma_1 \to \sigma_0$	0.000	0.000	0.002	0.062	0.413
$\sigma_2 \to \sigma_1$	0.000	0.211	0.007	0.357	0.000
$\sigma_3 \to \sigma_1$	0.000	0.000	0.005	0.061	0.000
$\sigma_0 \to \sigma_2$	0.464	0.535	0.062	0.000	0.000
$\sigma_1 \to \sigma_2$	0.000	0.000	0.542	0.000	0.000
$\sigma_2 \to \sigma_3$	0.000	0.101	0.152	0.000	0.000
$\sigma_3 \to \sigma_3$	0.000	0.000	0.222	0.000	0.000

Table 8.4 Accumulated probabilities $\alpha_t(\sigma_t)$ of the forward recursion.

$\alpha_t(\sigma_t)$	$t=0$	$t=1$	$t=2$	$t=3$	$t=4$	$t=5$
$\sigma_t = 0$	1.000	0.152	0.017	0.004	0.006	1.000
$\sigma_t = 1$	0.000	0.000	0.508	0.008	0.994	0.000
$\sigma_t = 2$	0.000	0.848	0.230	0.747	0.000	0.000
$\sigma_t = 3$	0.000	0.000	0.244	0.241	0.000	0.000

so that $\sum_{\sigma_t} \alpha(\sigma_t) = 1$ and $\sum_{\sigma_t} \beta(\sigma_t) = 1$. This way, the factor in equation 7.1 can be neglected, and this leads in our example to the (scaled) transition probabilities listed in table 8.3. The scaled forward and backward probabilities according to equations 8.49 and 8.50 are shown in tables 8.4 and 8.5. Consequently, the reliabilities $L(\hat{u}_t)$ for the estimated bits $\hat{u}_t$ are $(-3.650, 2.877, -5.679, 1.0, 1.0)$, according to equation 8.47. ◇

Table 8.5 Accumulated probabilities $\beta_t(\sigma_t)$ of the backward recursion.

$\beta_t(\sigma_t)$	$t=0$	$t=1$	$t=2$	$t=3$	$t=4$	$t=5$
$\sigma_t=0$	1.000	0.127	0.106	0.297	0.351	1.000
$\sigma_t=1$	0.000	0.000	0.798	0.052	0.649	0.000
$\sigma_t=2$	0.000	0.873	0.039	0.556	0.000	0.000
$\sigma_t=3$	0.000	0.000	0.057	0.095	0.000	0.000

Note A version of the MAP algorithm for general convolutional concatenated codes is used in section 9.3.4. In section 10.3.4, a version for soft-output demodulation is used.

8.5.2 Max log MAP algorithm

A simplification of the MAP algorithm is described in [RVH95], which we shall explain now.

In contrast to the MAP algorithm, only the most 'likely' path that arrives at a state is accounted for in the calculation of the forward and backward recursion numbers in the max log MAP algorithm. To determine the reliabilities $L(\hat{u}_t)$, only the most 'likely' paths that correspond at time t to the to be estimated bit $\hat{u}_t=0$ and $\hat{u}_t=1$ are considered. This means that the summations in equations 8.47, 8.49 and 8.50 are substituted by finding the maximum. This way, the following equations are obtained:

$$L(\hat{u}_t) = \ln \frac{\max\limits_{\sigma_t,\sigma_{t-1}} \alpha_{t-1}(\sigma_{t-1})\gamma_0(y_t,\sigma_{t-1},\sigma_t)\beta_t(\sigma_t)}{\max\limits_{\sigma_t,\sigma_{t-1}} \alpha_{t-1}(\sigma_{t-1})\gamma_1(y_t,\sigma_{t-1},\sigma_t)\beta_t(\sigma_t)}, \tag{8.50}$$

forward recursion numbers

$$\begin{aligned} \alpha_t(\sigma_t) &= \max_{\sigma_{t-1},i=0,1} \alpha_{t-1}(\sigma_{t-1})\cdot\gamma_i(y_t,\sigma_{t-1},\sigma_t), \\ \alpha_0(\sigma_0) &= \begin{cases} 1 & \text{for } \sigma_0=0, \\ 0 & \text{otherwise}, \end{cases} \end{aligned} \tag{8.51}$$

backward recursion numbers

$$\begin{aligned} \beta_t(\sigma_t) &= \max_{\sigma_{t+1},i=0,1} \gamma_i(y_{t+1},\sigma_t,\sigma_{t+1})\cdot\beta_{t+1}(\sigma_{t+1}); \\ \beta_l(\sigma_l) &= \begin{cases} 1 & \text{for } \sigma_l=0, \\ 0 & \text{otherwise}. \end{cases} \end{aligned} \tag{8.52}$$

Example 8.44 (Max log MAP algorithm) In this example, we consider the information / code sequence as in example 8.43 and the received sequence $\mathbf{y}$ (see table 8.2), using an AWGN channel. The signal-to-noise ratio is again $-4\,$dB.

Since only paths with maximum probability are considered, maximization of the probabilities can be substituted by minimization of the Euclidean distances. Thereby the multiplication of the probabilities in equations 8.50, 8.51 and 8.52 can be reduced to an addition of the distances. Table 8.6 lists the quadratic Euclidean distances $\gamma_i^*(y_t,\sigma_{t-1},\sigma_t)$ of all trellis transitions. The forward and backward recursion numbers are presented in tables 8.7 and 8.8.

Using these, the reliabilities $L(\hat{u}_t)$ are $(-9.603, 8.509, -15.57, \infty, \infty)$. ◇

Table 8.6 Quadratic Euclidean distance of all state transitions.

$\gamma_i^*(y_t, \sigma_{t-1}, \sigma_t)$	$t=1$	$t=2$	$t=3$	$t=4$	$t=5$
$\sigma_0 \to \sigma_0$	6.252	8.084	10.11	2.609	3.772
$\sigma_1 \to \sigma_0$	∞	∞	15.54	6.989	2.223
$\sigma_2 \to \sigma_1$	∞	3.908	12.35	2.585	∞
$\sigma_3 \to \sigma_1$	∞	∞	13.30	7.013	∞
$\sigma_0 \to \sigma_2$	1.927	1.572	6.967	∞	∞
$\sigma_1 \to \sigma_2$	∞	∞	1.541	∞	∞
$\sigma_2 \to \sigma_3$	∞	5.749	4.726	∞	∞
$\sigma_3 \to \sigma_3$	∞	∞	3.781	∞	∞

Table 8.7 Accumulated forward recursion numbers $\alpha_t(\sigma_t)$ as a function of the quadratic Euclidean distance.

$\alpha_t(\sigma_t)$	$t=0$	$t=1$	$t=2$	$t=3$	$t=4$	$t=5$
$\sigma_t = 0$	0.000	4.325	8.502	14.00	14.02	0.000
$\sigma_t = 1$	∞	∞	0.000	12.80	0.000	∞
$\sigma_t = 2$	∞	0.000	1.990	0.000	∞	∞
$\sigma_t = 3$	∞	∞	1.841	4.081	∞	∞

Table 8.8 Accumulated backward recursion numbers $\beta_t(\sigma_t)$ as a function of the quadratic Euclidean distance.

$\beta_t(\sigma_t)$	$t=0$	$t=1$	$t=2$	$t=3$	$t=4$	$t=5$
$\sigma_t = 0$	0.000	5.278	5.426	1.573	1.549	0.000
$\sigma_t = 1$	∞	∞	0.000	5.953	0.000	∞
$\sigma_t = 2$	∞	0.000	7.613	0.000	∞	∞
$\sigma_t = 3$	∞	∞	6.668	4.428	∞	∞

Simulation results Two convolutional codes in series, as in [HH89], are referred to as outer and inner convolutional codes respectively (see chapter 9), and have been used in the following simulations. Both codes have rate $R = 1/2$ and identical generator polynomials $(15, 17)$ (in octal representation). The outer code is punctured according to table 8.15 (section 8.8) to a rate $R = 2/3$ and decoded in the mother code trellis with rate $1/2$ using the Viterbi algorithm. The inner code was decoded soft-in and soft-out. Thereby the BCJR algorithm, the max log MAP algorithm and the SOVA, introduced in section 8.4.5, have been used. Block interleaving between inner and outer stages has been incorporated. BPSK modulation over an AWGN channel was considered.

The simulated bit error rates are shown in figure 8.33. It can be seen that the suboptimal algorithms max log MAP and SOVA show an only slightly worse decoding capability compared with the optimal BCJR algorithm. The discrepancy becomes bigger when using iterative decoding (algorithm 7.5).

8.6 Sequential decoding

In contrast to Viterbi decoding, where the 2^ν (ν is the overall constraint length) survivors are considered, only one code sequence is processed in sequential decoding. This reduces the necessary memory requirements for decoding substantially. Sequential decoders are currently

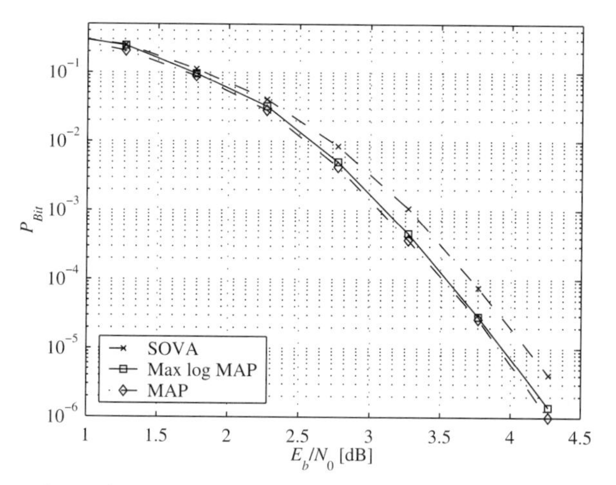

Figure 8.33 Comparison of the soft-output decoding algorithms MAP, max log MAP and SOVA over an AWGN channel using BPSK modulation.

feasible up to an overall constraint length of $\nu \approx 100$. The decoding time depends on the number of errors that have occurred, i.e. the processing time is a variable, in contrast to Viterbi decoding. Therefore the decoding complexity is low with relatively good channels but increases greatly with bad channels. Since a system is typically designed for the *worst case*, this behavior can become be a major disadvantage.

Sequential decoding can best be explained when the code is represented in a code tree (see section 8.1.5). However, we shall first derive an appropriate metric.

8.6.1 Fano metric

The metric $\Lambda^{\mathbf{x}}$, introduced in section 8.4.1 for Viterbi decoding, was defined to compare sequences of equal lengths. The sequential decoding method, however, is based on a different principle. It compares paths with different lengths. Therefore a metric has to be defined that takes this into account. Let $\mathbf{x}_{[t]}$ be the transmitted code sequence and $\mathbf{y}_{[t]}$ the received sequence. If we now consider all sequences

$$\mathbf{x}_{[q]} = (\mathbf{x}_0, \dots, \mathbf{x}_l, \dots, \mathbf{x}_{q-1}) \quad \text{with } \mathbf{x}_l = (x_l^{(1)}, x_l^{(2)}, \dots, x_l^{(n)}), \tag{8.53}$$

$$\mathbf{y}_{[q]} = (\mathbf{y}_0, \dots, \mathbf{y}_l, \dots, \mathbf{y}_{q-1}) \quad \text{with } \mathbf{y}_l = (y_l^{(1)}, y_l^{(2)}, \dots, y_l^{(n)}), \tag{8.54}$$

with lengths $q \leq t$, then we are interested in the most likely transmitted code sequence

$$\max_{\mathbf{x}_{[q]} \in C, q \in [1,t]} \frac{P(\mathbf{x}_{[q]}) P(\mathbf{y}_{[q]} \mid \mathbf{x}_{[q]})}{P(\mathbf{y}_{[q]})}. \tag{8.55}$$

Therefore the *Fano metric*

$$\Lambda^{\mathbf{x}_{[q]}} = \log_2 \frac{P(\mathbf{x}_{[q]}) P(\mathbf{y}_{[q]} \mid \mathbf{x}_{[q]})}{P(\mathbf{y}_{[q]})}, \tag{8.56}$$

is defined. If we assume a memoryless information source, i.e. $P(u_l^{(i)}=0)=P(u_l^{(i)}=1)=1/2$ for $l \in [1,t]$ and $j \in [1,k]$, we obtain

$$P(\mathbf{x}_{[q]}) = \prod_{l=0}^{q-1} P(\mathbf{x}_l) = \prod_{l=0}^{q-1}\prod_{i=1}^{k} P(u_l^{(i)}) = \left(\frac{1}{2}\right)^{qk} = 2^{-Rnq}, \tag{8.57}$$

for the a priori probability of a code sequence $\mathbf{x}_{[q]}$ with length q. Accordingly,

$$P(\mathbf{y}_{[q]} \mid \mathbf{x}_{[q]}) = \prod_{l=0}^{q-1} P(\mathbf{y}_l \mid \mathbf{x}_l) = \prod_{l=0}^{q-1}\prod_{j=1}^{n} P(y_l^{(j)} \mid x_l^{(j)}) \tag{8.58}$$

is obtained for a memoryless time invariant channel. If we use equations 8.58 and 8.57 in equation 8.56, we obtain

$$\Lambda^{\mathbf{x}_{[q]}} = \sum_{l=0}^{q-1}\sum_{j=1}^{n} \lambda_{l,j} \tag{8.59}$$

with the so-called bit metric

$$\lambda_{l,j} = \log_2 \frac{P(y_l^{(j)} \mid x_l^{(j)})}{P(y_l^{(j)})} - R. \tag{8.60}$$

This metric can be used for hard- and soft-decision decoding, as demonstrated in the following examples.

Example 8.45 (Bit metric for a binary symmetric channel) This is $P(y_l^{(j)}=+1)=P(y_l^{(j)}=-1)=1/2$ and

$$P(y_l^{(j)} \mid x_l^{(j)}) = \begin{cases} (1-p) & \text{for } y_l^{(j)} = x_l^{(j)}, \\ p & \text{for } y_l^{(j)} \neq x_l^{(j)}. \end{cases}$$

For a code with rate $R=1/2$ and bit error probability $p=2^{-5}$, the bit metric $\lambda_{l,j}$ is therefore

$$\lambda_{l,j} = \begin{cases} \log_2(2(1-p)) - R & \approx & 0.45 & \rightarrow & 1 & \text{for } y_l^{(j)} = x_l^{(j)}, \\ \log_2(2p) - R & \approx & -4.5 & \rightarrow & -10 & \text{for } y_l^{(j)} \neq x_l^{(j)}. \end{cases}$$

Note that in practical applications integer numbers are typically used for the scaled metric values. ◇

Example 8.46 (Probabilities for an AWGN channel) Let the y-axis be partitioned in intervals I_j of width Δr, according to figure 7.5:

$$r_i \in I_j: \quad j\Delta r < r_i < (j+1)\Delta r\,.$$

The likelihood that the received value r_i is in the interval I_j, under the condition that $x_i = \pm 1$ was transmitted, is

$$P(r_i \in I_j | x_i = \pm 1) = \tfrac{1}{2}\,\mathrm{erfc}\left(\frac{\mp 1 + j\Delta r}{\sqrt{2}\sigma}\right) - \tfrac{1}{2}\,\mathrm{erfc}\left(\frac{\mp 1 + (j+1)\Delta r}{\sqrt{2}\sigma}\right).$$

The probability that r_i is in the interval I_j is:

$$P(r_i \in I_j) = \tfrac{1}{2} P(r_i \in I_j | x_i = 1) + \tfrac{1}{2} P(r_i \in I_j | x_i = -1) \ .$$

This information can be used to calculate the bit metric according to example 8.45. ◇

8.6.2 Zigangirov–Jelinek (ZJ) decoder

A sequential decoding algorithm was proposed by Zigangirov [Zig66] and Jelinek [Jel69]. The decoder is therefore referred to as the ZJ decoder. It uses a *stack* in which the paths are listed according to their metric. The path with the largest metric is on top of the stack, giving the name 'stack decoder'. A decoding scheme is to take the uppermost path, and calculate all continuing branches from the last node of that path and then reorder the resulting paths into the stack.

Example 8.47 (ZJ-decoder) Figure 8.34 illustrates the decoding methodology. In the received sequence $\mathbf{y}_{[4+2]}$, transmission errors have occurred at the two marked positions:

$$\begin{aligned} \mathbf{v}_{[4+2]} &= (11 \quad 10 \quad 00 \quad 01 \quad 01 \quad 11), \\ \mathbf{y}_{[4+2]} &= (\mathbf{10} \quad 10 \quad \mathbf{01} \quad 01 \quad 01 \quad 11). \end{aligned}$$

The same bit metric as in example 8.45, derived for the BSC, is used. If two paths have the same metric then the one ending with a 0 is entered above the path ending with a 1. ◇

When implementing a ZJ decoder, the stack size as well as the path length have to be limited. The path memory for each stack element is implemented as shift register with fixed length. It can happen that the correct path is not accounted for because of limited memory. The sorting of the calculated paths in a very large stack involves a great deal of computation time. Therefore fields of paths with the same metrics are introduced to reduce the computational complexity. The ZJ decoder can also integrate available reliability information by modifying the metric (see example 8.46).

8.6.3 Fano decoder

Another sequential decoding algorithm is implemented using the Fano decoder. It only stores one path, and transverses back and forth on the paths of the code tree (figure 8.7). The metric of the current path is not allowed to fall below a threshold T. The threshold T has a variable magnitude, and it is constantly updated by the decoder. Table 8.9 shows the rules for calculating the threshold and for sideward (S), backward (R) or forward (V) translations. A path with depth t in the code tree is associated with metric Λ_t. The initial values for the metric are $\Lambda_{-1} = -\infty$, $\Lambda_0 = 0$. If needed, the Fano decoder increases or decreases the threshold by Δ. The change of the threshold Δ must be optimized through simulations, to minimize the number of translations in the code tree and hence the number of processing operations. In addition, the decoder stores how many side branches it has already tried for each node. An example of Fano decoding is given in problem 8.2.

The path memory in the Fano decoder can be made very large, since only one path has to be stored. However, it has to be limited to keep the decoding delay reasonable. This decoding is very effective for good channels, but the number of decoding steps per decoded symbol increases greatly for bad channels.

0	−9
1	−9

00 : −18
01 : −18

1	−9
00	−18
01	−18

10 : −7
11 : −29

10	−7
00	−18
01	−18
11	−29

100 : −16
101 : −16

100	−16
101	−16
00	−18
01	−18
11	−29

1000 : −25
1001 : −25

101	−16
00	−18
01	−18
1000	−25
1001	−25
11	−29

1010 : −36
1011 : −14

1011	−14
00	−18
01	−18
1000	−25
1001	−25
11	−29
1010	−36

10110 : −12
10111 : −34

10110	−12
00	−18
01	−18
1000	−25
1001	−25
11	−29
10111	−34
1010	−36

101100 : −10
101101 : −38

101100	−10
00	−18
01	−18
1000	−25
1001	−25
11	−29
10111	−34
1010	−36
101101	−38

...

Figure 8.34 Illustration of the ZJ decoder.

8.7 (Partial) unit memory codes, (P)UM codes

In this section, we present the representation of convolutional codes as unit memory codes (UM codes) or partial unit memory codes (PUM codes), introduced by Lee [Lee76]. Thereby we deal with convolutional codes $\mathcal{C}(k,n,[\nu])$ with memory $m = 1$. The overall constraint length satisfies $\nu \leq k$. Codes satisfying this condition with equality are termed UM codes; all others are termed PUM codes. We shall introduce a construction method for (P)UM codes where algebraic construction methods from the block code theory are used. These (P)UM codes are generally not binary and typically have a very large overall constraint length ν. Because of

Table 8.9 Rules for the Fano decoder.

last move	condition	threshold change	move
F or S	$\Lambda_{t-1} < T+\Delta, \Lambda_t \geq T$	$+\Delta$ if possible*	$V^\dagger$
F or S	$\Lambda_{t-1} \geq T+\Delta, \Lambda_t \geq T$	none	$V^\dagger$
F or S	$\Lambda_t < T$	none	S or $B^\ddagger$
B	$\Lambda_{t-1} < T$	$-\Delta$	$F^\dagger$
B	$\Lambda_{t-1} \geq T$	none	S or $B^\ddagger$

*Add $j\Delta$ to threshold, if j satisfies $\Lambda_t - \Delta < T + j\Delta \leq \Lambda_t$.
†Move forward to the first of the 2^k following nodes.
‡Move sidewards. If last of the 2^k nodes, move backwards.

the high state complexity of the Viterbi trellis (in the q-ary case the number of states is q^ν), ML decoding of these codes corresponds to a high degree of complexity. Therefore we shall describe an algorithm for BMD decoding of these (P)UM codes, where in principle all block code decoding algorithms described in chapter 7 can be used.

8.7.1 Definition of (P)UM codes

Consider the non-binary information sequence $\mathbf{i}$, in contrast to the binary sequence $\mathbf{u}$,

$$\mathbf{i} = (\mathbf{i}_0, \mathbf{i}_1, \ldots, \mathbf{i}_t, \ldots), \quad \text{with } \mathbf{i}_t = (i_t^{(1)}, i_t^{(2)}, \ldots, i_t^{(k_0)}), \tag{8.61}$$

where $i_t^{(i)} \in GF(q)$ for $i \in [1, k_0]$. Accordingly, we denote the code sequence $\mathbf{c}$ instead of $\mathbf{v}$:

$$\mathbf{c} = (\mathbf{c}_0, \mathbf{c}_1, \ldots, \mathbf{c}_t, \ldots), \quad \text{with } \mathbf{c}_t = (c_t^{(1)}, c_t^{(2)}, \ldots, c_t^{(n)}), \tag{8.62}$$

where $c_t^{(j)} \in GF(q)$ for $j \in [1, n]$. A UM code is a convolutional code where a code block of the code sequence only depends on the last and the current information blocks. The corresponding mapping can therefore be written as $\mathbf{c}_t = \text{fct}(\mathbf{i}_{t-1}, \mathbf{i}_t)$.

Definition 8.26 (Unit memory code) *Let $\mathbf{G}_a$ and $\mathbf{G}_b$ be two $k_0 \times n$ matricies with rank k_0 and matrix elements from $GF(q)$. An (n, k_0) UM code is defined by*

$$\mathbf{c}_t = \mathbf{i}_t \cdot \mathbf{G}_a + \mathbf{i}_{t-1} \cdot \mathbf{G}_b, \tag{8.63}$$

with $\mathbf{i}_{-1} = \mathbf{0}$, $\mathbf{i}_t \in GF(q)^{k_0}$ and $\mathbf{c}_t \in GF(q)^n$.

If the matrix $\mathbf{G}_b$ does not have full rank, then a partial unit memory code is obtained.

Definition 8.27 (Partial unit memory code) *Let $\mathbf{G}_a$ and $\mathbf{G}_b$ be two $k_0 \times n$ matrices with matrix elements from $GF(q)$ and ranks k_0 and $k_1 < k_0$ respectively. An $(n, k_0|k_1)$ PUM code is defined by*

$$\mathbf{c}_t = \mathbf{i}_t \cdot \mathbf{G}_a + \mathbf{i}_{t-1} \cdot \mathbf{G}_b,$$

where the two generator matrices are

$$\mathbf{G}_a = \begin{pmatrix} \mathbf{G}_{00} \\ \mathbf{G}_{01} \end{pmatrix} \quad \text{and} \quad \mathbf{G}_b = \begin{pmatrix} \mathbf{G}_{10} \\ 0 \end{pmatrix} \tag{8.64}$$

The submatrices $\mathbf{G}_{00}$ and $\mathbf{G}_{10}$ are $k_1 \times n$ and $\mathbf{G}_{01}$ is $(k_0 - k_1) \times n$. The ranks of the submatrices are $\text{rank}(\mathbf{G}_{00}) = \text{rank}(\mathbf{G}_{10}) = k_1$ *and* $\text{rank}(\mathbf{G}_{01}) = k_0 - k_1$.

With this definition, the calculation of a code block becomes

$$\mathbf{c}_t = \mathbf{i}_t \cdot \mathbf{G}_a + \mathbf{i}_{t-1} \cdot \mathbf{G}_b = (\mathbf{i}_t, \mathbf{i}_{t-1}) \cdot \begin{pmatrix} \mathbf{G}_a \\ \mathbf{G}_b \end{pmatrix} = (\mathbf{i}_{t,0}, \mathbf{i}_{t,1}, \mathbf{i}_{t-1,0}) \cdot \begin{pmatrix} \mathbf{G}_{00} \\ \mathbf{G}_{01} \\ \mathbf{G}_{10} \end{pmatrix}, \tag{8.65}$$

with $\mathbf{i}_t = (\mathbf{i}_{t,0}|\mathbf{i}_{t,1}) = (i_t^{(0)}, \ldots, i_t^{(k_1-1)}|i_t^{(k_1)}, \ldots, i_t^{(k_0-1)})$, i.e. $\mathbf{i}_{t,0} \in GF(q)^{k_1}$ and $\mathbf{i}_{t,1} \in GF(q)^{k_0-k_1}$. Before we make the connection to block codes, we want to demonstrate that every convolutional code can be represented as a (P)UM code.

Theorem 8.28 (Convolutional code as a UM code) *Every convolutional code can be represented as a PUM code or as a UM code.*

Proof When encoding a convolutional code with rate $\hat{k}/\hat{n}$ and memory m according to equation 8.3, a code sequence $\mathbf{v}$ is obtained by multiplication of the information sequence $\mathbf{u} = (\mathbf{u}_0, \mathbf{u}_1, \mathbf{u}_2, \ldots)$ by the semi-infinite generator matrix $\hat{\mathbf{G}}$, i.e. $\mathbf{v} = \mathbf{u} \cdot \hat{\mathbf{G}}$. The generator matrix looks like

$$\hat{\mathbf{G}} = \begin{pmatrix} \hat{\mathbf{G}}_0 & \hat{\mathbf{G}}_1 & \cdots & \hat{\mathbf{G}}_m & 0 & 0 & 0 & \cdots \\ 0 & \hat{\mathbf{G}}_0 & \hat{\mathbf{G}}_1 & \cdots & \hat{\mathbf{G}}_m & 0 & 0 & \cdots \\ 0 & 0 & \hat{\mathbf{G}}_0 & \hat{\mathbf{G}}_1 & \cdots & \hat{\mathbf{G}}_m & 0 & \cdots \\ \vdots & \vdots & \ddots & \vdots & & & & \vdots \end{pmatrix}.$$

The two matrices $\mathbf{G}_a$ and $\mathbf{G}_b$ of a (P)UM code can always be written as

$$\mathbf{G}_a = \begin{pmatrix} \hat{\mathbf{G}}_0 & \hat{\mathbf{G}}_1 & \cdots & \hat{\mathbf{G}}_{m-1} \\ 0 & \hat{\mathbf{G}}_0 & \cdots & \hat{\mathbf{G}}_{m-2} \\ \vdots & \vdots & \ddots & \vdots \\ 0 & 0 & 0 & \hat{\mathbf{G}}_0 \end{pmatrix} \quad \text{and} \quad \mathbf{G}_b = \begin{pmatrix} \hat{\mathbf{G}}_m & 0 & \cdots & 0 \\ \hat{\mathbf{G}}_{m-1} & \hat{\mathbf{G}}_m & \cdots & 0 \\ \vdots & \vdots & \ddots & 0 \\ \hat{\mathbf{G}}_1 & \hat{\mathbf{G}}_2 & \cdots & \hat{\mathbf{G}}_m \end{pmatrix}.$$

In this case, information and code blocks are

$$\mathbf{c}_t = (c_t^{(1)}, c_t^{(2)}, \ldots, c_t^{(m\hat{n})}) \quad \text{and} \quad \mathbf{i}_t = (i_t^{(1)}, i_t^{(2)}, \ldots, i_t^{(\hat{k}m)}).$$

Of course, numerous other configurations exist if $n > m\hat{n}$ is chosen. □

Block codes as (P)UM codes We interpret the code sequence $\mathbf{c}$ as an addition of a sequence of codewords $\mathbf{a}_t = \mathbf{i}_t \cdot \mathbf{G}_a$ of a block code $\mathcal{C}_a$ and a sequence of codewords $\mathbf{b}_t = \mathbf{i}_t \cdot \mathbf{G}_b$ of a block code $\mathcal{C}_b$:

$$\begin{aligned} \mathbf{c}_0 &= \mathbf{a}_0, \\ \mathbf{c}_1 &= \mathbf{a}_1 + \mathbf{b}_0, \\ &\vdots \\ \mathbf{c}_t &= \mathbf{a}_t + \mathbf{b}_{t-1}, \\ &\vdots \end{aligned}$$

The code block $\mathbf{c}_t$ can then be interpreted either as the codeword of the code $\mathcal{C}_\alpha$ with generator matrix $\mathbf{G}_\alpha = (\mathbf{G}_a^T \cdot \mathbf{G}_b^T)^T$, or as the result of an addition of two block codewords of the codes $\mathcal{C}_a$ and $\mathcal{C}_b$. Assuming that information blocks are known, we can state:

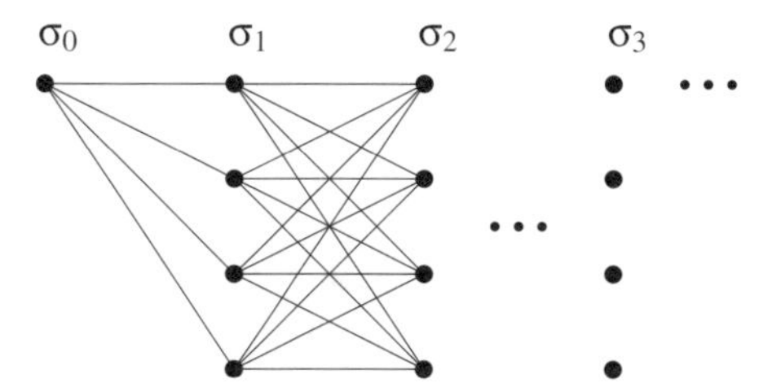

Figure 8.35 Trellis of the UM$(3,2)$ code.

(i) $\mathbf{i}_{t-1}$ known: The codeword $\mathbf{c}_t$ can be used to calculate the codeword $\mathbf{a}_t$:

$$\mathbf{c}_t - \mathbf{i}_{t-1} \cdot \mathbf{G}_b = \mathbf{a}_t \in \mathcal{C}_a \,, \tag{8.66}$$

and one could decode with respect to the code $\mathcal{C}_a$.

(ii) $\mathbf{i}_t$ known: The codeword $\mathbf{c}_t$ can be used to calculate the codeword $\mathbf{b}_t$:

$$\mathbf{c}_t - \mathbf{i}_t \cdot \mathbf{G}_a = \mathbf{b}_t \in \mathcal{C}_b \,, \tag{8.67}$$

and one could decode with respect to the code $\mathcal{C}_b$.

We shall take advantage of this fact when performing BMD decoding in section 8.7.5. In principle one could recursively decode a (P)UM code terminated after L blocks, since $\mathbf{i}_{-1} = \mathbf{0}$ and $\mathbf{i}_L = \mathbf{0}$. That means that when decoding in the forward direction (increasing t), $\mathbf{i}_{t-1}$, $t \in [-1, L-1]$, is always known. When decoding in the reverse direction (decreasing t), $\mathbf{i}_t$, $t \in [L, 0]$, is known. However, this decoding method has not been investigated.

8.7.2 Trellis of (P)UM codes

Another way to represent a (P)UM code is by using a trellis, where the state transitions correspond to the code blocks $\mathbf{c}_t$. Consequently, a trellis of a UM code consists of q^{k_0} states σ_t at all times $t > 0$. q^{k_0} paths leave each state, and exactly one of them is connected to each of the q^{k_0} subsequent states. In contrast, the trellis of a PUM code consists of q^{k_1} states at all times, and $q^{k_0-k_1}$ parallel transitions exist between any two consecutive states.

Example 8.48 (Trellis of the codes PUM$(3,2|1)$ and UM$(3,2)$) Figure 8.35 shows the structure of a trellis of the $(3,2)$ UM code. It has $2^{k_0} = 4$ states, and each state with depth t is connected with all states of adjacent depth.

A trellis for the PUM$(3,2|1)$ code is shown in figure 8.36. It consists of $2^{k_1} = 2$ states and $2^{k_0-k_1} = 2$ parallel transitions per state. The figure also includes the distance measures, which are explained in the next section. ◇

Obviously, this representation of (P)UM codes leads to a very complex trellis for $q > 2$ and large k_0 or $k_0 - k_1$. Consequently ML decoding with the Viterbi algorithm requires a great deal of computational effort. Therefore we shall provide an algorithm for BMD decoding in section 8.7.5.

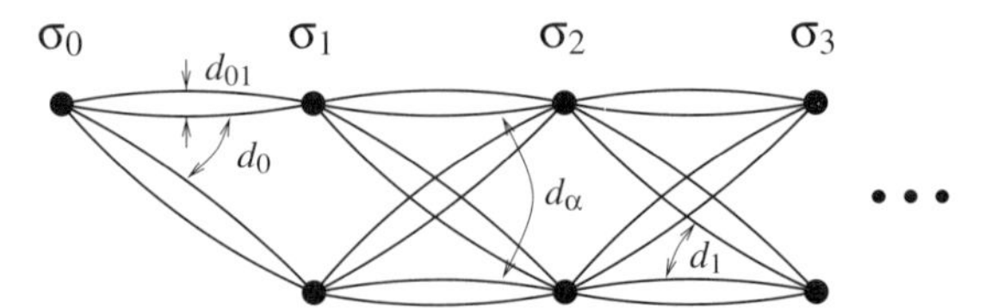

Figure 8.36 Trellis of the $(3,2|1)$ PUM code.

8.7.3 Distance measures for (P)UM codes

The extended distance measures for convolutional codes were originally introduced for (P)UM codes [TJ83]. We want to present the distance measures in a unified manner, and therefore adapt the definitions for convolutional codes from section 8.3 for (P)UM codes. Therefore, for PUM codes, we first define the following block codes using generator matrices and their minimum distances:

$$\begin{aligned}
d_0 &= \text{dist}(\mathcal{C}_a), && \text{with } \mathcal{C}_0 = \mathcal{C}_a : \mathbf{G}_0 = \mathbf{G}_a = \left(\mathbf{G}_{00}^T \cdot \mathbf{G}_{01}^T\right)^T, \\
d_1 &= \text{dist}(\mathcal{C}_1), && \text{with } \mathcal{C}_1 \quad : \mathbf{G}_1 = \left(\mathbf{G}_{01}^T \cdot \mathbf{G}_{10}^T\right)^T, \\
d_{01} &= \text{dist}(\mathcal{C}_{01}), && \text{with } \mathcal{C}_{01} \quad : \mathbf{G}_{01} = \mathbf{G}_{01}, \\
d_\alpha &= \text{dist}(\mathcal{C}_\alpha), && \text{with } \mathcal{C}_\alpha \quad : \mathbf{G}_\alpha = \left(\mathbf{G}_{00}^T \cdot \mathbf{G}_{01}^T \cdot \mathbf{G}_{10}^T\right)^T.
\end{aligned} \tag{8.68}$$

The definitions for UM codes are obtained accordingly. The distance measures can be interpreted as follows (see also figure 8.36):

(1) d_{01} is the minimum distance between the $q^{k_0-k_1}$ parallel branches;

(2) d_0 is the minimum distance between all code blocks that have the same initial state;

(3) d_1 is the minimum distance between all code blocks that merge in the same final state;

(4) d_α is the minimum distance between all code blocks that have neither the initial state nor the final state in common;

Extended row, column and segment distance The extended distances have been defined for convolutional codes in section 8.3. We now present a lower bound for them for (P)UM codes:

- estimated extended row distance d_j^{er}:

$$\begin{aligned}
d_1^{er} &= d_{01}, \\
d_j^{er} &\geq d_0 + (j-2)d_\alpha + d_1, \qquad j > 1;
\end{aligned} \tag{8.69}$$

- estimated extended column distance d_j^{ec}:

$$d_j^{ec} \geq d_0 + (j-1)d_\alpha, \qquad j \geq 1; \tag{8.70}$$

- estimated backwards extended column distance d_j^{rec}:

$$d_j^{rec} = d_1 + (j-1)d_\alpha, \qquad j \geq 1; \tag{8.71}$$

- estimated extended segment distance d_j^{es}:

$$d_j^{es} \geq j \cdot d_\alpha, \qquad j \geq 1. \tag{8.72}$$

The error correction capability of (P)UM codes can be assessed and the quality of designed codes can be analyzed with these lower bounds of the extended distances.

8.7.4 Construction of (P)UM codes

When designing (P)UM codes from block codes, one chooses a known block code $\mathcal{C}_\alpha$. It is then divided in three parts, so that the distances d_1, d_0 and d_{01} of the corresponding subcodes are maximized. In general, the determination of the optimum partitioning is not easy to calculate. The following designs are based on the fact that the distance measures of the underlying block codes and their subcodes are known.

(P)UM codes based on RS codes We shall refer to a (P)UM code with length n, dimension k_0 and overall constraint length k_1 that is based on RS codes as an RS PUM$(n, k_0 \mid k_1)$. It is constructed in the following manner:

1. Select for $\mathbf{G}_{00}$ the generator matrix of an RS code with length n, and make sure that the transformed codewords satisfy $C_j = 0$, for $j = k_1, k_1+1, \dots, n-1$.

2. Select for $\mathbf{G}_{10}$ the generator matrix of an RS code with length n and make sure that its transformed codewords satisfy $C_j = 0$, for $j = 0, \dots, k_0 - 1$ and $j = k_0 + k_1, k_0 + k_1 + 1, \dots, n-1$.

3. Select for $\mathbf{G}_{01}$ the generator matrix of an RS code with length n, and make sure that its transformed codewords satisfy $C_j = 0$, for $j = 0, \dots, k_1 - 1$ and $j = k_0, k_0 + 1, \dots, n-1$.

The distances of the subcodes are then

$$\begin{aligned} d_{01} &= n - (k_0 - k_1) + 1, \\ d_0 &= n - k_0 + 1, \\ d_1 &= n - k_0 + 1, \\ d_\alpha &= n - (k_0 + k_1) + 1. \end{aligned}$$

The following bound on the free distances of the (P)UM codes hold:

$$\text{PUM code:} \quad d_f \geq n - (k_0 - k_1) + 1; \qquad \text{UM code:} \quad d_f \geq 2(n - k_0 + 1)\,.$$

Example 8.49 ((P)UM codes based on RS codes) We use RS codes with length 7 in $GF(2^3)$. The generator polynomials of the PUM code are

$$\begin{aligned} g_{00}(x) &= (x-\alpha^2)(x-\alpha^3)(x-\alpha^4)(x-\alpha^5)(x-\alpha^6), \\ g_{01}(x) &= (x-\alpha^0)(x-\alpha^1)(x-\alpha^4)(x-\alpha^5)(x-\alpha^6), \\ g_{10}(x) &= (x-\alpha^0)(x-\alpha^1)(x-\alpha^2)(x-\alpha^3)(x-\alpha^6)\,. \end{aligned}$$

The information polynomial $i_t(x)$ is divided into $i_t^{(0)}(x)$ and $i_t^{(1)}(x)$:

$$\begin{aligned} i_t(x) &= i_{t,0} + i_{t,1}x + i_{t,2}x^2 + i_{t,3}x^3 \\ &= i_t^{(0)}(x) + i_t^{(1)}(x) \\ &= i_{t,0}^{(0)} + i_{t,1}^{(0)}x + i_{t,0}^{(1)}x^2 + i_{t,1}^{(1)}x^3 . \end{aligned}$$

Consequently, the encoding rule is

$$c_t(x) = i_t^{(0)}(x)\,g_{00}(x) + i_t^{(1)}(x)\,g_{01}(x) + i_{t-1}^{(0)}(x)\,g_{10}(x) = a_t^{(0)}(x) + a_t^{(1)}(x) + b_t(x)\ ,$$

where $i_t^{(0)}(x)$ and $i_t^{(1)}(x)$ have order 2. Note that, because the construction was based upon a given $c_t(x)$, all codewords ($a_t^{(0)}(x), a_t^{(1)}(x)$ and $b_t(x)$) can be calculated explicitly; see also example 9.37 and figure 9.27 in section 9.3.1. The PUM code has parameters $k=4$ and $k_1=2$ and distances

$$\begin{aligned} d_f &\geq n-(k-k_1)+1 = 7-(4-2)+1 = 6 \quad , \\ d_\alpha &\geq n-(k+k_1)+1 = 7-(4+2)+1 = 2 . \end{aligned}$$

◇

(P)UM codes based on BCH codes The transformed codewords of a BCH code contain a series of zeros corresponding to the designed minimum distance, as well as zeros in the coordinates corresponding to the other complex-conjugate roots in the cyclotomic coset (section 4.1.1). The challenge in the design of a (P)UM code from a BCH code is to determine the generator polynomials of the code $\mathcal{C}_\alpha$ and of the subcodes such that the corresponding transformed codewords contain as many consecutive zero factors as possible.

1. Form a BCH code of length n with distance d_α by setting $d_\alpha - 1$ consecutive positions and their complex-conjugate positions in the transformed domain to zero.

2. Associate the information positions of this code with the codes $\mathcal{C}_0, \mathcal{C}_{10}$ and $\mathcal{C}_1$, so that the free distance d_f is maximized.

Because of the interdependences of the complex-conjugate code symbols, it is not possible to find a closed-form expression for the free distance of the constructed code.

Example 8.50 ((P)UM codes based on BCH codes) We want to construct PUM codes based on BCH codes with length 15 (see [DS92]). According to section 4.1.1, example 4.1, the irreducible polynomials $m_i(x)$ of the associated cyclotomic coset K_i,

$$m_0(x),\ m_1(x),\ m_3(x),\ m_5(x),\ m_7(x)$$

are given. We choose the generator polynomials of the PUM codes to be

$$\begin{aligned} g_{00}(x) &= m_0(x)\,m_1(x)\,m_5(x)\,m_7(x) \\ g_{01}(x) &= m_0(x)\,m_3(x)\,m_7(x) \\ g_{10}(x) &= m_0(x)\,m_1(x)\,m_3(x)\,m_5(x) . \end{aligned}$$

The constructed PUM code is $\mathcal{C}(n=15, k=10|k_1=4)$, with free distance $d_f=6$. See also example 9.38 in section 9.3.1 and figure 9.28.

◇

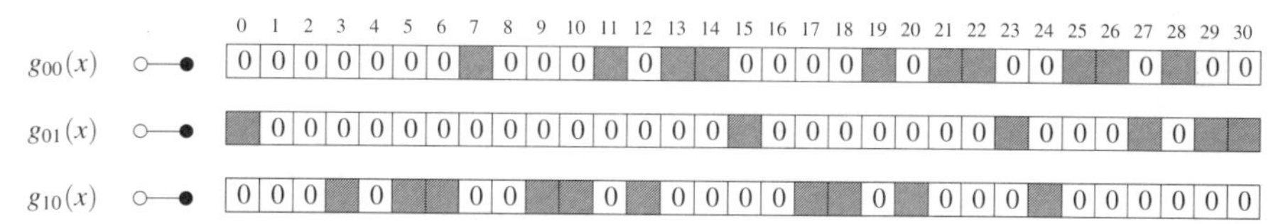

Figure 8.37 Transformation of the generator polynomial of the RM code from example 8.51.

(P)UM codes based on RM codes RM codes can be interpreted as extended BCH codes. Consequently, (P)UM codes based on RM codes are constructed by first determining BCH sub codes with the above specified method and then extending each of the subcodes by one code position. The distance of the resulting RM code is one larger than distance of the underlying BCH codes.

Example 8.51 ((P)UM codes based on RM codes) We first determine the generator polynomials of BCH codes with length 31 according to figure 8.37:

$$\begin{aligned} g_{00}(x) &= m_0(x)m_1(x)m_3(x)m_5(x)m_{15}(x) \\ g_{01}(x) &= m_1(x)m_3(x)m_5(x)m_7(x)m_{11}(x) \\ g_{10}(x) &= m_0(x)m_1(x)m_7(x)m_{11}(x)m_{15}(x). \end{aligned}$$

Then we append a parity bit to each of these subcodes. The resulting new codes are then specified by the generator matrices $\mathbf{G}_{00}$, $\mathbf{G}_{01}$ and $\mathbf{G}_{10}$. $\mathbf{G}_{01}$ defines an $\mathcal{R}(1,5)$ code, $\mathbf{G}_0 = \begin{pmatrix} \mathbf{G}_{00} \\ \mathbf{G}_{01} \end{pmatrix}$ defines an $\mathcal{R}(2,5)$ code, and $\mathbf{G}_\alpha = \begin{pmatrix} \mathbf{G}_{00} \\ \mathbf{G}_{01} \\ \mathbf{G}_{10} \end{pmatrix}$ defines an $\mathcal{R}(3,5)$ code. The resulting PUM code has parameters $n = 32$, $k = 16$, $k_1 = 10$, $d_\alpha = 4$ and $d_f = 16$. This construction method can be applied to arbitrary lengths $n = 2^m$ with odd m, according to [Jus93]. ◇

In [DS93] and [DS92], this construction method and others are presented. Constructions focusing on PUM codes based on RM codes are proposed in [ZP91] and [Mau95].

8.7.5 BMD decoding

ML decoding of convolutional codes was discussed in section 8.4. As already mentioned, it is in general not applicable to (P)UM codes, because of the exponentially increasing decoding complexity. In [Jus93], it was demonstrated that a hard-decision BMD decoder for (P)UM codes can be realized in three steps, shown below. We assume that a codeword $\mathbf{c}$ of a (P)UM code, terminated to L blocks, is transmitted and the vector $\mathbf{r}$ was received (*hard decision*).

Step 1 Decode each block $\mathbf{r}_t, t = [0, L-1]$ according to code $\mathcal{C}_\alpha$, where the blocks are treated independently from each other. Let the outcome of decoding be $\hat{\mathbf{c}}_t, t = [0, L-1]$. Note that without errors, the initial state of $\hat{\mathbf{c}}_t$ corresponds to the final state of $\hat{\mathbf{c}}_{t-1}$ and the final state of $\hat{\mathbf{c}}_t$ corresponds to the initial state of $\hat{\mathbf{c}}_{t+1}$. A possible decoding result containing errors is shown in figure 8.38. Three final states correspond to three initial states. If more than one state is obtained at depth t, one can conclude that some blocks have been incorrectly decoded, but these blocks are not identified.

Step 2 Starting from each final state $\hat{\mathbf{c}}_t$ that is not equal to the initial state of $\hat{\mathbf{c}}_{t+1}$, decode $\mathbf{r}_{t+i}$, $i = 1, 2, \ldots$, according to code $\mathcal{C}_0$ (equation 8.67). This is done until a state is reached that

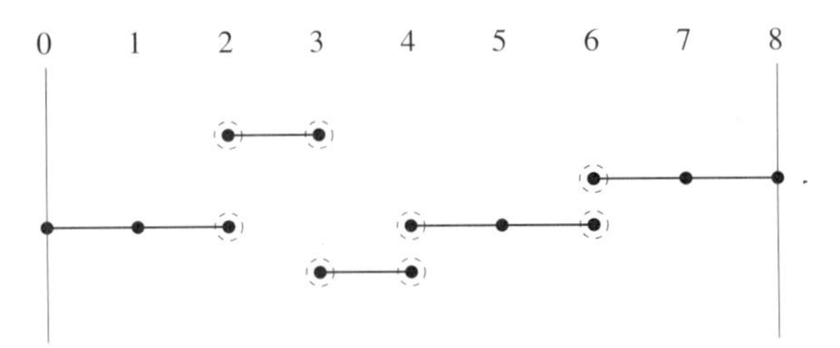

Figure 8.38 Step 1: decoding according to $\mathcal{C}_\alpha$.

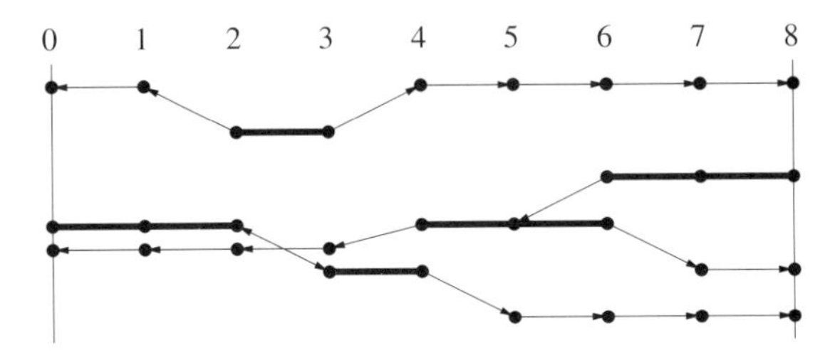

Figure 8.39 Step 2: forward ($\mathcal{C}_0$) and reverse ($\mathcal{C}_1$) BMD decoding of a PUM code.

was already decoded in step 1. Then backward decoding is performed accordingly. We start from each initial state of $\hat{\mathbf{c}}_t$ that is not equal to the final state of $\hat{\mathbf{c}}_{t+1}$ and decode $\mathbf{r}_{t-i}$, $i = 1, 2, \ldots$, according to code $\mathcal{C}_1$ (equation 8.66). This is also done until a state is reached that was already decoded in step 1. This means that forward decoding is carried out using the code $\mathcal{C}_0$ and backward decoding using $\mathcal{C}_1$, which have generally better distances than the code $\mathcal{C}_\alpha$. This is shown schematically in figure 8.39.

Step 3 First all nodes that were created in steps 1 and 2 are used to build a complete graph, i.e. all nodes are connected. Then the Viterbi algorithm is carried out in the constructed trellis and this way the best path of this reduced trellis is selected. This is shown in figure 8.40.

In [Jus93] it is proved that this hard-decision BMD decoding method can correct all errors $\mathbf{e} = (\mathbf{e}_0, e_1, \ldots \mathbf{e}_{L-1})$, under the condition that the error weight satisfies

$$\sum_{t=j}^{j+l-1} \mathrm{wt}(\mathbf{e}_t) < d^{er}(l)/2.$$

BMD decoding of PUM codes allows block codes to be decoded using the Viterbi algorithm through the construction of a simplified trellis. Normally the trellis representation of a block

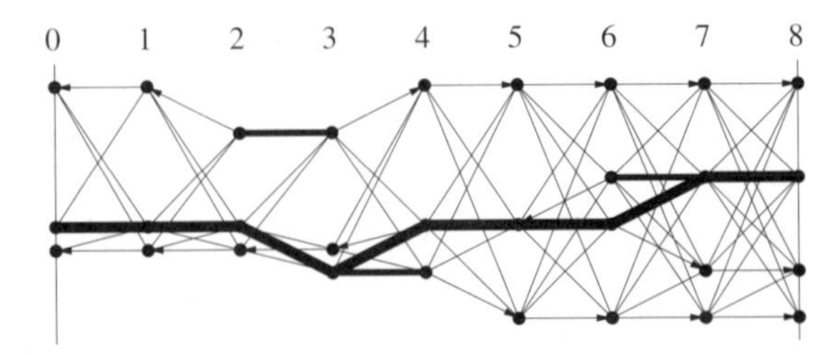

Figure 8.40 Step 3: ML decoding in the BCH reduced trellis.

code is too large for practical Viterbi decoding. This method is related to list decoding (section 7.4.2). If convolutional codes are considered as (P)UM codes and BMD decoding is applied, then the results are rather poor compared with ML decoding. This is due to the typically small distance d_α of convolutional codes.

Nothing has yet been published about the use of soft-decision decoding for block codes using the PUM decoding method. Therefore the question concerning the performance soft decision MAP decoding for $\mathcal{C}_\alpha, \mathcal{C}_0$ and $\mathcal{C}_1$ instead of hard-decision decoding is as yet unanswered.

8.8 Tables of good codes

In this section, we provide some tables of good convolutional codes. This is not a complete list of all known good convolutional codes, but only a very limited selection. Some of the most common codes are introduced, and are available for practical applications. The notation used is standard for such code tables. We shall refer to publications that contain complete code tables later in this section.

In contrast to block codes, no construction procedure to generate classes of good codes exists for convolutional codes. The only exception is (P)UM code construction introduced in section 8.7. However, (P)UM codes have little practical relevance at present. All other known convolutional codes have typically been found using a computer-aided code search.

Convolutional codes $\mathcal{C}(n,k,[\nu])$ in tables are sorted by their rate $R = k/n$ and overall constraint length ν and specified by a minimal basic generator matrix $\mathbf{G}_{mb}(D)$. This $k \times n$ matrix is represented with n octal numbers $O_G^{(j)}$ with $j \in [1,n]$, as in the following example.

Example 8.52 (Representation of convolutional codes with octal numbers) A possible representation of the minimal basic generator matrix $\mathbf{G}(D)$ is

$$\begin{aligned}\mathbf{G}(D) &= \begin{pmatrix} 1+D & D & 1 \\ D+D^2 & 1 & 1+D^2 \end{pmatrix} \\ &= \begin{pmatrix} 1 & 0 & 1 \\ 0 & 1 & 1 \end{pmatrix} + D\begin{pmatrix} 1 & 1 & 0 \\ 1 & 0 & 0 \end{pmatrix} + D^2\begin{pmatrix} 0 & 0 & 0 \\ 1 & 0 & 1 \end{pmatrix} \\ &= \mathbf{G}_0 + D\mathbf{G}_1 + D^2\mathbf{G}_2.\end{aligned}$$

If the matrix

$$\hat{\mathbf{G}} = \begin{pmatrix} \mathbf{G}_0 \\ \mathbf{G}_1 \\ \mathbf{G}_2 \end{pmatrix}^T = \begin{pmatrix} 1 & 0 & 1 & 1 & 0 & 1 \\ 0 & 1 & 1 & 0 & 0 & 0 \\ 1 & 1 & 0 & 0 & 0 & 1 \end{pmatrix} = \begin{pmatrix} (55)_8 \\ (30)_8 \\ (61)_8 \end{pmatrix}$$

is introduced then this $2 \times n$ generator matrix $\mathbf{G}(D)$ can be defined using the three octal numbers $(55, 30, 61)$, where leading zeros are added if necessary. ◇

Codes are said to be good codes if they have the best distance properties among all codes with the same rate $R = k/n$ and the same overall constraint length ν. Therefore we consider

Table 8.10 Some OFD convolutional codes with rate $R = 1/n$ with $n \in [2,4]$ (from [Pal95]).

R	m	$O_G^{(1)}$	$O_G^{(2)}$	$O_G^{(3)}$	$O_G^{(4)}$	d_f
1/2	2	5	7			5
	3	15	17			6
	4	23	35			7
	5	53	75			8
	6	133	171			10
	7	247	371			10
	8	561	753			12
1/3	2	5	7	7		8
	3	13	15	17		10
	4	25	33	37		12
	5	47	53	75		13
	6	133	145	175		15
	7	225	331	367		16
	8	557	663	711		18
1/4	2	5	7	7	7	10
	3	13	15	15	17	13
	4	25	27	33	37	16
	5	53	67	71	75	18
	6	135	135	147	163	20
	7	235	275	313	357	22
	8	463	535	733	745	24

the path enumerator

$$
\begin{aligned}
T(W) &= \sum_{w=d_f}^{+\infty} N(w)W^w \\
&= N(d_f)W^{d_f} + N(d_f+1)W^{d_f+1} + N(d_f+2)W^{d_f+2} + \ldots
\end{aligned}
$$

or the distance spectrum $N(w)$ with $w \in [d_f, +\infty]$ of the code (see equation 8.9 in section 8.1.6). Consequently, the best code has the maximum free distance d_f. If multiple codes with the same free distance exist then the one with the smallest number $N(d_f)$ is chosen. If the two codes are equal in this term too then higherorder terms of $N(w)$ are considered in increasing order. Codes with this property are called optimum free-distance (OFD) codes. These are the asymptotically best codes with a given rate and overall constraint length when using ML and MAP decoding.

As an alternative, the optimization can also be carried out with respect to the distance profile (see definition 8.17 in section 8.3.1) and one obtains the optimum distance profile (ODP) codes. These are the preferred choice over OFD codes when using sequential decoding procedures.

Some OFD codes with rate $R = 1/n$ (table 8.10) A more extensive list of OFD codes with rate $R = 1/n$ can be found in [Pal95]. If using ML and MAP decoding, then the exponentially increasing with ν state complexity of the trellis limits the code selection to codes with a small overall general constraint length, i.e. $\nu \in [2,8]$. However, larger values can be implemented

Table 8.11 Some OFD codes with rate $R = 2/3$ and $R = 3/4$ (from [LiCo]).

R	m	ν	$O_G^{(1)}$	$O_G^{(2)}$	$O_G^{(3)}$	$O_G^{(4)}$	d_f
2/3	1	2	13	6	16		3
	2	3	41	30	75		4
	2	4	56	23	65		5
	3	5	245	150	375		6
	3	6	266	171	367		7
3/4	1	3	400	630	521	701	4
	2	5	442	270	141	763	5
	2	6	472	215	113	764	6

Table 8.12 Low-rate OFD codes with memory $m = 4$ (from [Pal95]).

R	$O_G^{(1)}$	$O_G^{(2)}$	$O_G^{(3)}$	$O_G^{(4)}$	$O_G^{(5)}$	$O_G^{(6)}$	$O_G^{(7)}$	$O_G^{(8)}$	$O_G^{(9)}$	$O_G^{(10)}$	d_f
1/5	25	27	33	35	37						20
1/6	25	27	33	35	35	37					24
1/7	25	27	27	33	35	35	37				28
1/8	25	25	27	33	33	35	37	37			32
1/9	25	25	27	33	33	35	35	37	37		36
1/10	25	25	25	33	33	33	35	37	37	37	40

for specific constructions. At present, most applications incorporate codes with rate $R = 1/n$ with $n \in [2,4]$, shown in table 8.10.

Some OFD codes with rate $R = k/n$ (table 8.11) Typically, convolutional codes with rate $R = k/n$ for $k > 1$ are generated by puncturing the mother code with rate $R = 1/n$ (see section 8.1.10). This way, very good distance properties can be attained in almost all cases. Decoding in the trellis of the mother code involves less effort then decoding in the trellis of the high-rate code, especially when using very high-rate punctured codes. Nonetheless, we shall provide some codes with rate $R = k/n$ that have not been generated by puncturing mother codes in table 8.11. Detailed tables can be found in [LiCo].

Some low-rate OFD codes with $m = 4$ (table 8.12) Low-rate convolutional codes with rate $R = 1/n$ are generated with n generator polynomials. Good high-rate polynomials are usually used, and are then modified for the desired lower rate. Detailed tables can be found in [Pal95].

Rule to calculate OFD codes with rate $R = 1/n$ with memory $m = 2$ and $m = 3$ (table 8.13) A general calculation rule can be given for these two cases.

Some OFD codes with polynomial systematic generator matrices (table 8.14) Of course, all codes introduced so far can also be encoded systematically. However, the associated generator matrices are generally rational (see section 8.2.7). The codes in table 8.14 have worse distance properties than other good codes with the same rate and overall constraint length, but have a systematic polynomial generator matrix. Detailed tables of these codes can be found in [JoZi].

Table 8.13 General calculation rule for OFD codes with $m \in [2,3]$ (from [Pal95]).

Rate R	$O_G^{(1)}$	$O_G^{(2)}$	$O_G^{(3)}$	d_f
$1/3n$	5^n	7^{2n}		$8n$
$1/(3n+1)$	5^{n+1}	7^{2n}		$8n+2$
$1/(3n+2)$	5^{n+1}	7^{2n+1}		$8n+5$
$1/3n$	13^n	15^n	17^n	$10n$
$1/(3n+1)$	13^n	15^{n+1}	17^n	$10n+3$
$1/(3n+2)$	13^n	15^{n+1}	17^{n+1}	$10n+6$

Table 8.14 Systematic convolutional codes with rate $R = 1/2$ (from [JoZi]).

m	$O_G^{(2)}$	d_f
1	3	3
2	7	4
3	15	4
4	35	5
5	73	6
6	153	6
7	153	6
8	715	7

Table 8.15 Some punctured codes (from [Lee94]).

mother code			punctured code		
m	$O_G^{(1)}$	$O_G^{(2)}$	puncturing matrix	rate R	d_f
2	5	7	$\begin{pmatrix}1&0\\1&1\end{pmatrix}$	2/3	3
	5	7	$\begin{pmatrix}1&1&0\\1&0&1\end{pmatrix}$	3/4	3
	5	7	$\begin{pmatrix}1&0&0&1\\1&1&1&0\end{pmatrix}$	4/5	2
	5	7	$\begin{pmatrix}0&0&0&0&1\\1&1&1&1&1\end{pmatrix}$	5/6	2
3	15	17	$\begin{pmatrix}1&1\\0&1\end{pmatrix}$	2/3	4
	15	17	$\begin{pmatrix}0&1&1\\1&1&0\end{pmatrix}$	3/4	4
	15	17	$\begin{pmatrix}1&1&1&0\\1&0&0&1\end{pmatrix}$	4/5	3
	15	17	$\begin{pmatrix}1&0&0&1&0\\0&1&1&1&1\end{pmatrix}$	5/6	3
6	133	171	$\begin{pmatrix}1&1\\1&0\end{pmatrix}$	2/3	6
	133	171	$\begin{pmatrix}1&1&0\\1&0&1\end{pmatrix}$	3/4	5
	133	171	$\begin{pmatrix}1&1&1&1\\1&0&0&0\end{pmatrix}$	4/5	4
	133	171	$\begin{pmatrix}1&1&1&1&1\\1&0&0&0&0\end{pmatrix}$	5/6	3

Tables of punctured codes (table 8.15) Punctured convolutional codes and the class of RCPC codes (*rate-compatible punctured convolutional codes*) [Hag88] play an important role in practical applications. RCPC codes are a set of punctured codes with the same period of the puncturing matrices. Every higher rate code can be generated from a code with a lower rate by puncturing. Table 8.15 shows an example of how these codes can be presented in tabular form. More methods can be found in [Lee94].

8.9 Summary

Convolutional codes were first described by Elias [Eli55] in 1955. A sequential decoding procedure was then proposed in [WoRe] to decode convolutional codes efficiently. In 1963, Massey [Mas] introduced suboptimal *threshold decoding* for convolutional codes. The main advantage was the simplicity of the algorithm. Fano [Fano63] described a new variant of sequential decoding, which is now known as the Fano algorithm, in the same year. Another algorithm for sequential decoding was published in 1966 by Zigangirov [Zig66] and later, in 1969, by Jelinek [Jel69]. It is known as the stack algorithm, and we have referred to it as the ZJ algorithm. These breakthroughs made real-world applications of convolutional codes for digital data transmission possible.

Another decoding algorithm was published in 1967 by Viterbi [Vit67]. Forney ([For73], [For74]) demonstrated later that this algorithm performs a maximum-likelihood decoding of convolutional codes.

Sequential as well as Viterbi decoding yields a hard decision. This is a major problem when concatenated codes are considered. The BCJR algorithm proposed by Bahl, Cocke, Jelinek and Raviv in 1974 [BCJR74] does not have this restriction. It provides the a posteriori probability for each symbol. In 1989, Hagenauer ([Hag90]) proposed a modified Viterbi algorithm that is also capable of providing symbol decisions and a measure for their reliability with the SOVA.

Convolutional codes are used in many systems. For example, the two convolutional codes $(1+D^3+D^4, 1+D+D^3+D^4)$ with rate 1/2 and $(1+D^2+D^4, 1+D+D^3+D^4, 1+D+D^2+D^3+D^4)$ with rate 1/3 are used in the European mobile radio standard GSM. For these two codes, the overall constraint length is equal to the memory ($m = \nu = 4$).

In this chapter, we have first given the basic terms and definitions of convolutional codes. We have distinguished between code, generator matrix and encoder. We have described the representation as a LTI system and have explained the representation as FIR and IIR systems. An important classification was made when introducing systematic and catastrophic generator matrices. Further, we have discussed the trellis, code tree and state diagram representations, and have derived some associated distance measures.

In the second section, we have presented the algebraic description of convolutional codes. Some terms from the first section have been refined and formally derived. Additionally, the basic generator matrix and the minimal basic generator matrix have been defined. The minimal basic form can be thought of as the normalized representation of the a generator matrix, i.e. the one with the lowest state complexity. The parity check matrix and the dual convolutional code have also been introduced. This section is based mostly on the work of Johannesson and Wan [JW93], and on the book [JoZi].

After a section dealing with distance measures, we have described the Viterbi algorithm for ML decoding of convolutional codes. This is one of the most used algorithms, and can be applied to many situations, such as for the equalization of multiple path channels. Since the reliability of a decoded symbol is becoming increasingly important for applications, other algorithms are needed. Therefore we have described the SOVA and the BCJR algorithm, which are both capable of calculating not only a symbol decision but also a measure of its reliability. A good approximation of the BCJR algorithm is the max log algorithm, which is less complex and therefore very attractive for practical applications.

Sequential decoding has the property that the decoding complexity depends on the signal-to-noise ratio of the channel. The less a channel is distorted, the lower is the corresponding decoding complexity. Although this is an interesting aspect for practical applications, sequential decoding procedures are not very common, even though convolutional codes with substantially larger overall constraint length can be decoded with the Fano or ZJ algorithms.

In the field of (P)UM codes, techniques based on block codes can be used to design convolutional codes with specified distance properties. The best practical decoding method for PUM codes is a BMD technique. The fact that no ML decoding technique is known is a major limitation to the application of these codes.

In section 8.8, we have listed tables of good convolutional codes, which are meant for practical use. In some cases, codes with certain rates are required, which can be generated from good convolutional codes by means of puncturing. Some tables concerning this have also been provided, and further literature references can be found in [Lar73, Joh75].

8.10 Problems

Problem 8.1

You are given the following convolutional encoder:

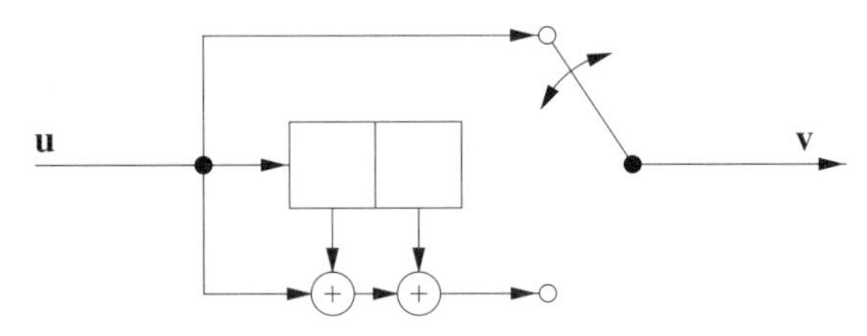

(a) Find the state diagram of the encoder.

(b) Determine the free distance d_f.

(c) Find the trellis diagram of the encoder.

(d) A received sequence is (11 00 10 01 01 00 00 ...). Determine the transmitted sequence using the Viterbi algorithm.

Problem 8.2

Consider the following convolutional encoder:

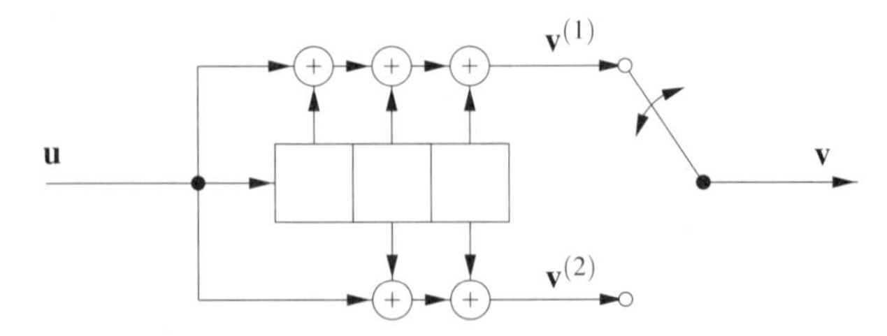

A received sequence is 10 10 11 11 ....

Specify the steps that a Fano decoder would carry out in order to determine the transmitted code sequence. Use a code tree and a table. Set $\Delta = 6$. The bit error probability of the channel is $p = 0.1$.

9

Generalized code concatenation

The concatenation of codes was first investigated in 1966 by Forney [For66a]. He defined an *inner code* together with the channel as a *superchannel*. A second code, called the *outer code*, encoded the data to be sent over the superchannel.

In his work, Forney showed three clear advantages of concatenated codes

- *Very long* codes can be constructed by concatenation based on short codes.
- Concatenated codes can correct burst errors and independent single errors at the same time
- There is a smaller decoding complexity compared with the decoding of a single code of the same length, because decoding of the shorter component codes can be done in separate stages.

Unfortunately, the problem of soft-decision decoding of concatenated codes has not been satisfactorily solved.

In 1974, Blokh and Zyablov published the concept of generalized concatenated codes [BZ74]. We shall focus on this representation of concatenated codes in this chapter. In contrast to Forney's model, a generalized concatenated code treats the inner and outer code as *one* code. The two models are shown in figures 9.1 and 9.2. The models have the same construction; however, by grouping the components differently, certain advantages can be recognized.

The Blokh–Zyablov model makes the concept of generalized concatenated codes possible. This concept has been employed in many communication transmission systems. Better concatenated code parameters can be achieved based on one inner code and several outer codes. In addition to some of the already mentioned advantages of GC codes (*generalized concatenated codes*) such as simultaneous single error and burst error correction, there are other properties. In particular, GC codes can be designed to produce *unequal error protection* codes (UEP codes) (see also section 8.1.10). Codes with unequal error protection are gaining increasing

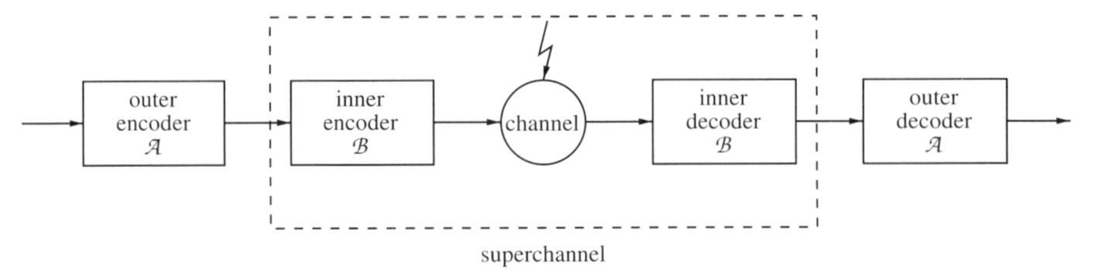

Figure 9.1 Concatenation according to Forney.

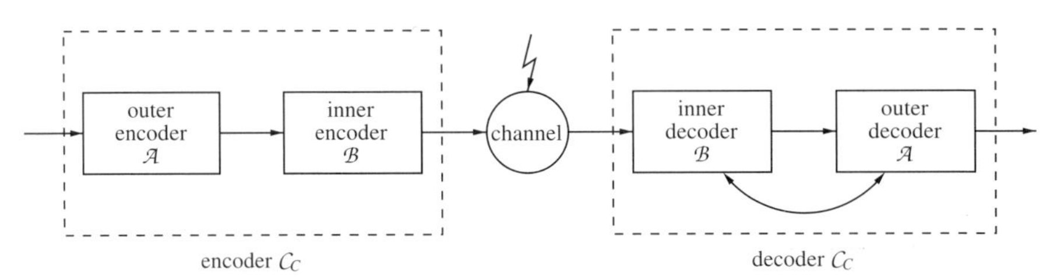

Figure 9.2 Concatenation according to Blokh and Zyablov.

importance in channel coding. For example, in the reconstruction of speech or music, the information bits have different levels of importance (e.g. *least significant* and *most significant* bits). It is important to mention that, for these codes, the correction properties are maintained in the case of UEP. Furthermore, the decoding method remains the same, meaning that relatively long codes can be decoded using several short codes.

It will be shown that generalized concatenated codes have the following advantages over the classical concatenation method:

- For the same minimum distance, the GC code has more codewords, therefore the GC code has a higher code rate.
- For the same number of codewords, i.e. for the same code rate, GC codes have a larger minimum distance.

We consider the principle of *coded modulation* as a special case of generalized concatenation whereby the modulation method is considered as the inner code. Using this concept, one has an elegant method of describing both coded modulation and the associated decoding methods.

The hard-decision decoding of concatenated codes was for a long time not satisfactorily solved. Traditionally, one decoded the inner code and then the outer code. With this technique, it can be shown that errors cannot be corrected up to half the minimum distance. We shall introduce a decoding algorithm for GC codes that can correct errors up to half the minimum distance. The same algorithms can be used for unequal error protection codes. Furthermore, single and burst errors can be simultaneously corrected.

In this chapter, we shall describe several techniques, methods and properties for the construction of GC codes for specific practical applications.

Strategy First, we present the concept of generalized concatenation using several examples, and contrast them with *classical* concatenated codes. We deal with the two methods separately because a general notation would be difficult and would cause more confusion than necessary for the understanding of this powerful concept. We then proceed by defining GC codes.

The classes of block and convolutional codes give four different possibilities for concatenation depending on which code is used as the inner or the outer code. The first GC codes used block codes for both the inner and outer encoding. This construction is the best understood, and will be presented in section 9.2. In section 9.3, we introduce GC codes that use convolutional codes for the inner and outer codes. Less research has been done for this construction method. At this point, we present results from [BDS96a] first published in 1996.

The concatenation of convolutional codes under the title 'turbo codes' was presented in 1993. This technique comes very close to achieving the channel capacity limit for the AWGN channel, and as a result prompted a great deal of research on this topic. We show that this turbo construction is based on the *classical* concatenation perspective, and that the coding gain is solely a result of the iterative decoding steps. Relatively few theoretical results have been proved in this area, although excellent decoding performance is observed. Appendix A focuses on turbo codes and the related problems of so-called parallel and serial concatenation together with iterative decoding. In conclusion, section 9.4 describes GC codes with mixed code classes. Although several system standards fall into this class of codes, there are still many open questions.

In section 9.2, we focus on the construction and variations of GC codes. In particular, section 9.2.3 describes several possibilities for modifying a code's parameters. A well-known principle that is used in the case of burst error correction codes is called interleaving [WHPH87]. In section 9.2.3, we apply this principle to GC codes. In section 9.2.5, the construction of codes with unequal error protection of information blocks is investigated. Cyclic codes are then presented as GC codes. The interesting concept of describing GC codes through the re-encoding of the syndrome is also introduced, which gives rise to so-called error locating codes (EL). These codes are effective for small redundancy, and posses special error correction properties. This concept will also be generalized to give so-called GEL codes.

Section 9.2.4 describes a decoding method for GC codes that can decode errors up to half the minimum distance. At this point, we explain the technique of Blokh and Zyablov [ZZ79a]. We then investigate a modification of this method that improves the decoding properties of the code while at the same time reducing the complexity of the decoding. The modification can be applied to any decoding method, for example *soft-decision decoding* or ML decoding methods.

The decoding of GC codes is a key problem for practical systems. Therefore this chapter includes decoding methods and the associated performance capabilities. Often, the performance capabilities of a method can be calculated by the use of simple bounds. However, for an exact performance measure, simulations are necessary. An important insight for the use of GC codes is stated in section 9.2.4, namely that there are two different construction criterion for GC codes. The most common criterion for code construction is the optimization of the minimum distance. Most publications to date assume that this is equivalent to the reduction of the error probability after decoding. This is not necessarily true. An optimization of the error probability after decoding almost always leads to a different code construction than by optimizing the minimum distance.

For the generalized concatenation of convolutional codes, the central problem is partitioning. In section 9.3.3, we present and investigate several GC code constructions.

In section 9.4 on generalized concatenation with block and convolutional codes, we investigate the natural extension of the concepts of GC codes, including multilevel concatenation in

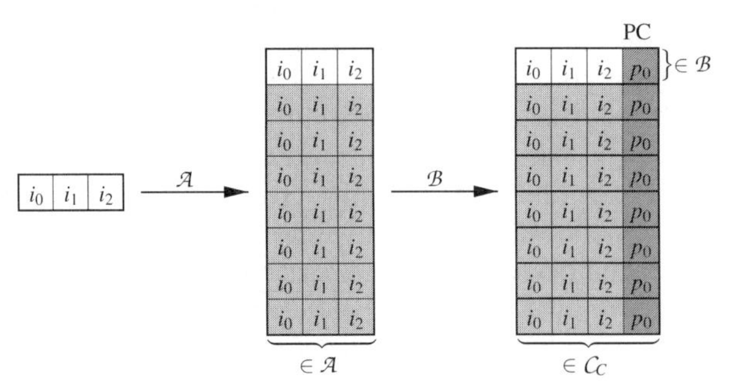

Figure 9.3 Example of classical concatenation (systematic).

section 9.5. This means that the inner code is already a GC code. An interesting example of this is the class of binary Reed–Muller codes, which can be described as generalized multiple concatenated codes (GMC).

Using the description of RM codes as generalized multiple concatenated codes, we also derive an associated decoding method (GMC algorithm). This technique has a very low complexity and a very simple use of reliability information. Furthermore, we introduce an extension of the GMC algorithm that is based on the list decoding concept.

9.1 Introductory examples

'Long is the way through teaching, short and successful through example.' Seneca

In this section, the concept of generalized concatenation is explained using simple examples. We first concatenate the codes $\mathcal{B}$ and $\mathcal{A}$ *classically* according to Forney (figure 9.1) and then we present the generalized concatenation perspective according to Blokh and Zyablov (figure 9.2).

Example 9.1 (Classic concatenation) We use a single parity check (SPC) code[1] $\mathcal{B}(2;4,3,2)$ with length 4 as the inner code. A repetition code $\mathcal{A}(2^3;8,1,8)$ in $GF(2^3)$ with length 8 is used as the outer code (see examples 1.2 and 1.3 respectively). Therefore three information bits (i_0,i_1,i_2) are coded with the outer code $\mathcal{A}$, giving $\mathbf{a}=(a_0,a_1,\dots,a_7)$. Each element a_j, $j=0,\dots,7$, of the outer code is considered in terms of the basic field and are encoded with the inner code $\mathcal{B}$ to produce 8 codewords. This technique is shown in figure 9.3.

The codeword of the concatenated code $\mathcal{C}_C$ consists of an 8×4 matrix where the rows are codewords of the the parity check code $\mathcal{B}$. The length of the code is $n=4\cdot 8=32$ and the dimension is $k=3$, which corresponds to the three encoded information bits. The minimum distance is $d=d_a d_b=8\cdot 2=16$. The concatenated code has parameters

$$\mathcal{C}_C(2;32,3,16).$$

◇

In the following example, we construct a generalized concatenated code based on the same codes as above.

[1]The usual notation for codes $\mathcal{B}(n,k,d)$ will be replaced by the notation $\mathcal{B}(q;n,k,d)$ in order to differentiate between binary and non-binary codes. The parameter q indicates the code alphabet.

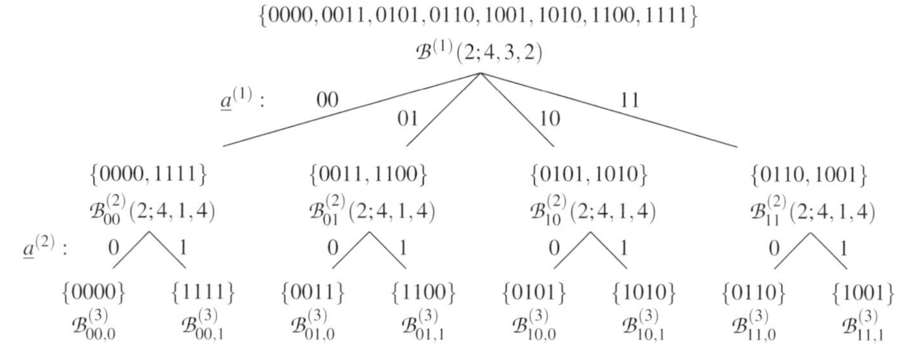

Figure 9.4 Partitioning of the code $\mathcal{B}^{(1)}(2;4,3,2)$.

Example 9.2 (Generalized concatenated code) The inner code is the binary single parity check (SPC) code with length $n = 4$, which we term $\mathcal{B}^{(1)}$. We group this code into four subcodes. Each subcode is made up of two codewords of the original code set. This division is called *partitioning*, and is carried out such that the minimum distance within a subcode is maximized. All 8 possible codewords of $\mathcal{B}^{(1)}(2;4,3,2)$ are listed below:

$$\begin{array}{llll} (0000), & (0011), & (0101), & (0110), \\ (1001), & (1010), & (1100), & (1111)\,. \end{array}$$

The four subcodes with best minimum distance are shown to be

$$\begin{array}{rcl} \mathcal{B}^{(2)}_0(2;4,1,4) & = & \{\,(0000),(1111)\,\}, \\ \mathcal{B}^{(2)}_1(2;4,1,4) & = & \{\,(0011),(1100)\,\}, \\ \mathcal{B}^{(2)}_2(2;4,1,4) & = & \{\,(0101),(1010)\,\}, \\ \mathcal{B}^{(2)}_3(2;4,1,4) & = & \{\,(0110),(1001)\,\}\,. \end{array}$$

We can uniquely determine a codeword of the single parity check (SPC) code $\mathcal{B}^{(1)}$ when we know the the subcode label $(0,1,2,3)$ and the label of the codeword in the subcode $(0,1)$. This means that, in order to identify a codeword based on this partitioning, we need two indices[2]:

$$\underline{a}^{(1)} \in GF(2)^2 \quad \text{and} \quad \underline{a}^{(2)} \in GF(2).$$

The two binary indices $\underline{a}^{(1)} = (10)$ and $\underline{a}^{(2)} = 1$ specify the codeword $\mathcal{B}^{(3)}_{10,1} = (1010)$. The partitioning is presented again in figure 9.4.

The idea is to protect the enumeration of the partitioning using the outer repetition code $\mathcal{A}^{(1)}(2^2,8,1,8)$ and an extended Hamming code (see section 1.10) $\mathcal{A}^{(2)}(2;8,4,4)$. The encoding of 6 information bits is carried out as follows. Two information bits (i_0,i_1) are encoded by $\mathcal{A}^{(1)}$, giving the codeword $\underline{\mathbf{a}}^{(1)} = (\underline{a}_0^{(1)},\underline{a}_1^{(1)},\underline{a}_2^{(1)},\underline{a}_3^{(1)},\underline{a}_4^{(1)},\underline{a}_5^{(1)},\underline{a}_6^{(1)},\underline{a}_7^{(1)})$, $\underline{a}_j^{(1)} \in GF(2)^2$. The remaining four information bits (i_2,i_3,i_4,i_5) are encoded by $\mathcal{A}^{(2)}$, giving $\underline{\mathbf{a}}^{(2)} = (\underline{a}_0^{(2)},\underline{a}_1^{(2)},$ $\underline{a}_2^{(2)},\underline{a}_3^{(2)},\underline{a}_4^{(2)},\underline{a}_5^{(2)},\underline{a}_6^{(2)},\underline{a}_7^{(2)})$, $\underline{a}_j \in GF(2)^1$. Every coordinate of the codeword $\underline{\mathbf{a}}^{(1)}$ selects a

[2]In order to differentiate vector elements from different fields, we require a special notation, which is given in section 9.2. This notation is used in these examples.

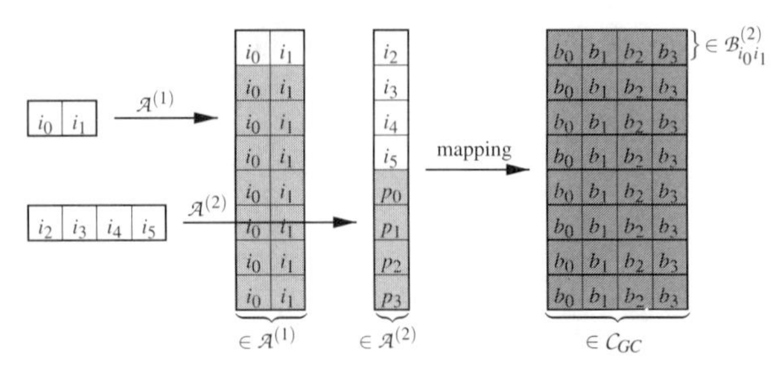

Figure 9.5 Example of generalized concatenated encoding.

subcode from $\mathcal{B}^{(1)}$, namely $\mathcal{B}^{(2)}_{\underline{a}_i^{(1)}}$, $i = 0, \ldots, 7$. Every coordinate of the codeword $\underline{\mathbf{a}}^{(2)}$ selects a codeword of the subcode $\mathcal{B}^{(2)}_{\underline{a}_i^{(1)}}$, namely $\mathcal{B}^{(3)}_{\underline{a}_i^{(1)},\underline{a}_i^{(2)}} \in \mathcal{B}^{(1)}$, $i = 0, \ldots, 7$:

$$\underline{a}_i^{(1)}, \underline{a}_i^{(2)} \implies \mathcal{B}^{(3)}_{\underline{a}_i^{(1)},\underline{a}_i^{(2)}} \in \mathcal{B}^{(1)}.$$

This results in an 8×4 matrix with each row corresponding to a codeword of the parity check code $\mathcal{B}^{(1)}$. In comparison with example 9.1, we have encoded 6 information bits instead of 3. Figure 9.5 shows a schematic representation of the encoding.

The general concatenation code C_{GC} has length $n = 4 \cdot 8 = 32$ and dimension $k = 6$. In considering the minimum distance, we first identify the subcode indicated by the received row. This corresponds to the minimum distance of $\mathcal{B}^{(1)}$, which is 2. This labeling is protected with the outer code which has a minimum distance of 8. Therefore we get $d = 2 \cdot 8 = 16$. For the second partition level, a subcode is selected such that the minimum distance between codewords is 4, and this labeling is protected with an outer code with distance 4. Therefore $d = 4 \cdot 4 = 16$. We show that (see theorem 9.2)

$$d \geq \min\{2 \cdot 8, 4 \cdot 4\} .$$

The GC code has parameters

$$C_{GC}(2; 32, 6, 16).$$

We have constructed a code that has 8 times more codewords and the same minimum distance as compared with the classic concatenated code.

Note This code is a Reed–Muller code of first order (section 5.1), which is written as a generalized concatenated code. The decoding of this code is particularly simple using generalized concatenation techniques.

In the next example, we consider a construction that uses an inner SPC code and an outer RS code.

Example 9.3 (Interleaving and GC codes) In general, interleaving is achieved by concatenation by writing the outer codewords columnwise in the outer code matrix, encoding by the appropriate inner code, and then transmitting the GC codeword matrix row-wise. Interleaving is also achieved by using more than one outer code.

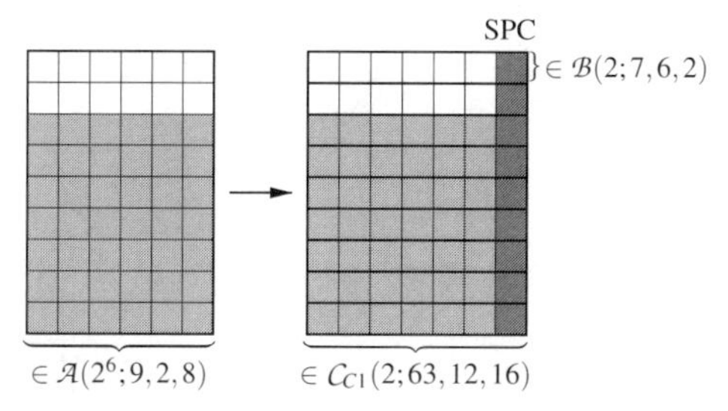

Figure 9.6 Classical concatenation with one outer code and a single parity check (SPC) inner code.

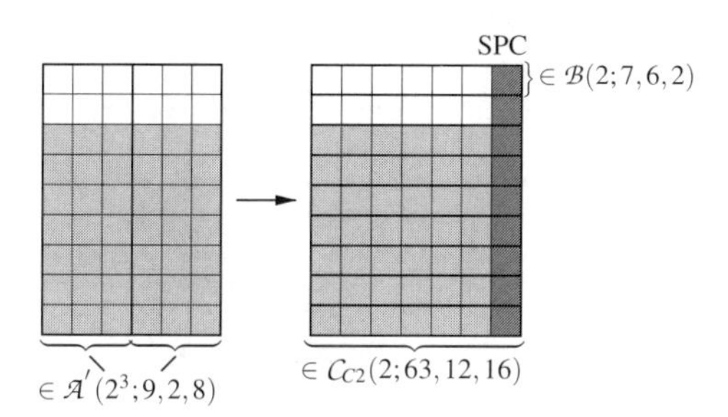

Figure 9.7 Classical concatenation with two outer codes.

Consider the binary parity check code $\mathcal{B}(2;7,6,2)$. A *classical* concatenation with the shortened RS code $\mathcal{A}(2^6;9,2,8)$ results in the code $\mathcal{C}_{CI}(2;63,12,16)$. The concatenation proceeds as follows. Twelve information bits are encoded as two information elements in $GF(2^6)$ to produce the the RS codeword. Next, each coordinate of the RS codeword is written in the basic field as 6 information bits of the inner code $\mathcal{B}$, which adds a parity bit. This is shown in figure 9.6.

In order to interleave the code through the use of multiple outer codes, we consider the outer RS code $\mathcal{A}'(2^3;9,2,8)$ to generate two codewords with elements from $GF(2^3)$. In total, 12 information bits are encoded. The 6 information coordinates of the inner code are filled by writing the elements of the outer code in terms of the base field $GF(2)$. The encoding is shown in figure 9.7. The concatenated code has parameters $\mathcal{C}_{C2}(2;63,12,16)$.

In order to construct a GC code, we use the fact that we can partition the SPC code $\mathcal{B}^{(1)}(2;7,6,2)$ into 8 codes $\mathcal{B}_i^{(2)}(2;7,3,4)$, $i=0,\ldots,7$ (see figure 9.8) This corresponds to cosets of even-weight Hamming codes. The enumeration of the partition is further protected with an outer code. The outer codes are two RS codes: $\mathcal{A}^{(1)}(2^3;9,2,8)$ and $\mathcal{A}^{(2)}(2^3;9,6,4)$.

Figure 9.9 shows the encoding of 24 information bits: 6 information bits are encoded by $\mathcal{A}^{(1)}$, giving the codeword $\underline{\mathbf{a}}^{(1)} = (\underline{a}_0^{(1)}, \underline{a}_1^{(1)}, \ldots, \underline{a}_8^{(1)})$, $\underline{a}_j^{(1)} \in GF(2)^3$. The remaining 18 information bits are encoded by $\mathcal{A}^{(2)}$, giving the codeword $\underline{\mathbf{a}}^{(2)} = (\underline{a}_0^{(2)}, \underline{a}_1^{(2)}, \ldots, \underline{a}_8^{(2)})$, $\underline{\mathbf{a}}_j^{(2)} \in GF(2)^3$. Every coordinate of the codeword $\underline{\mathbf{a}}^{(1)}$ selects a subcode from $\mathcal{B}^{(1)}$, namely $\mathcal{B}^{(2)}_{\underline{a}_j^{(1)}}$, and every coordinate from $\underline{a}_j^{(2)}$ selects a subcode of $\mathcal{B}^{(2)}_{\underline{a}_j^{(1)}}$, namely $\mathcal{B}^{(3)}_{\underline{a}_j^{(1)},\underline{a}_j^{(1)}}$:

$$\underline{a}_j^{(1)}, \underline{a}_j^{(2)} \implies \mathcal{B}_{\underline{a}_j^{(1)},\underline{a}^{(3)}_{j^{(2)}}} \in \mathcal{B}^{(1)}.$$

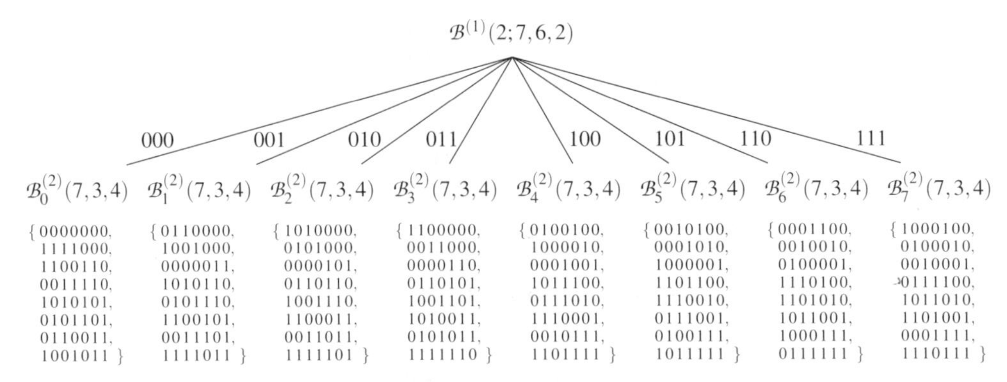

Figure 9.8 Partitioning of the inner codes.

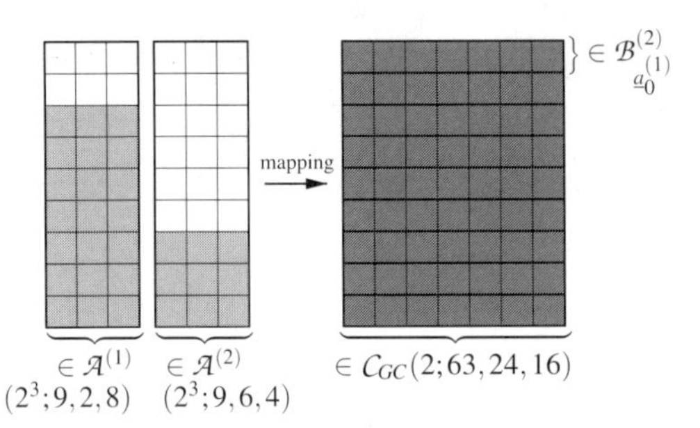

Figure 9.9 GC code.

A codeword is composed of $n = 7 \cdot 9 = 63$ bits. The dimension of the code is $k = 24$ and the minimum distance is $d = \min\{2 \cdot 8, 4 \cdot 4\} = 16$. The parameters of the GC code are $\mathcal{C}_{GC}(2;63,24,16)$.

A comparison of the parameters of the codes shows that the GC code has 4096 more codewords and the same minimum weight:

$$\mathcal{C}_{C1}(2;63,12,16), \quad \mathcal{C}_{C2}(2;63,12,16) \quad \text{and} \quad \mathcal{C}_{GC}(2;63,24,16).$$

One observes that the code $\mathcal{C}_{GC}(2;63,24,16)$ has better parameters that a similar BCH code (see table 4.1). $\diamond$

As a further example, a code with length 180 is constructed that belongs to the group of best known codes according to appendix A in [McWSl].

Example 9.4 (GC code) The inner code is a $\mathcal{B}(2;6,6,1)$ code with no redundancy. We concatenate this code with the RS code $\mathcal{A}(2^6;30,1,30)$, and thereby get the code $\mathcal{C}_C(2;180,6,30)$. The code $\mathcal{C}_C$ is the binary interpretation of a RS code (see section 5.6). It can be shown that more than 14 binary errors can be corrected if they are spread over ≤ 14 symbols.

In order to construct a GC code, we partition the inner code $\mathcal{B}^{(1)}(2;6,6,1)$ into two codes $\mathcal{B}_i^{(2)}(2;6,5,2)$, $i=0,1$. For the outer codes, we use the binary repetition code $\mathcal{A}^{(1)}(2;30,1,30)$ and the RS code $\mathcal{A}^{(2)}(2^5;30,16,15)$. A coordinate of $\mathcal{A}^{(1)}$ selects one of the subcodes, and the corresponding coordinate of the code $\mathcal{A}^{(2)}$ selects the codeword in the subcode. The GC code has parameters $\mathcal{C}_{GC}(2;180,81,30)$. $\diamond$

The examples have shown that GC codes have much better parameters compared with a similar *classical* concatenation. By using generalized concatenation instead of classical concatenation, we can achieve one of the following goals. For the same minimum distance,we can construct a code that has more codewords, and therefore a higher code rate. On the other hand, we may construct a GC code with a larger minimum distance while maintaining the same code rate. The examples also bring to light the basic principle behind generalized concatenation, namely the partitioning of the inner codes into subcodes with larger minimum distances through the appropriate design of the outer codes.

Note on the concept of generalized concatenation The GC coding concept is not restricted to linear codes, but can be generally applied to sets of vectors defined over any metric space. In addition, the inner metric does not necessarily have to be the same as the outer metric. It is critical that the partitioning of the sets of the inner vectors into appropriate subsets result in a larger minimum distance than the original set. The enumeration of the inner subsets is then protected by the outer code according to its minimum distance. The minimum distance of the overall code is calculated according to the product of the inner and outer codes. From the large number of possible cases, we limit our focus to code constructions achieving reliable data transmission based on the Hamming, Euclidean or Lee metric. In order to gain a deeper understanding of GC codes, we investigate several specific construction techniques. The first method uses block codes for inner and outer GC encoding. The second is the use of convolutional codes for both inner and outer GC encoding. The final two cases consider block codes as the outer code with an inner convolutional code, and vice versa. Furthermore, a separate chapter is dedicated to the important special case of coded modulation where the modulation technique is used as the inner code. Here, the Euclidian metric is used.

9.2 GC codes with block codes

This section has the following structure. First we give a general definition for GC codes and discuss the possibilities for the partitioning of the inner codes. Next, we discuss several practical points of view concerning the decoding of GC codes. Finally we describe the special case of UEP and cyclic GC codes. An improvement in decoding may be realized if cyclic codes are employed for GC code construction. Little research in this area has been published, and many of the associated problems are still open. We close this section by looking at the construction of GC codes through syndrome re-encoding, which are named GEL codes (generalized error locating codes).

Notation In order to describe generalized concatenation for linear and nonlinear component codes, we require a new notation. A code $\mathcal{B}$ with length n in the field $GF(q)$ with M codewords and minimum distance d is described as

$$\mathcal{B}(q;n,M,d) \subseteq GF(q)^n .$$

For linear codes, we have $M = q^k$. The notation $\mathcal{B}(q;n,k,d)$ and $\mathcal{B}(n,k,d)$ for $q = 2$ will be used in parallel in subsequent sections. Confusion over these parameters is not likely because usually $M > n$. If the notation is not clear, an explanation will be provided.

9.2.1 Definition of GC codes

The idea of generalized code concatenation is based on the partitioning of an inner code into subcodes. The labeling of this partitioning is protected by the outer code. In order to give a general definition of GC codes, we must first define partitioning.

Consider the partitioning of order s for the code $\mathcal{B}^{(1)}(q;n_b,M_b^{(1)},d_b^{(1)})$, where $M_b^{(j)}$ gives the code cardinality:

$$\begin{aligned}
\mathcal{B}^{(1)} &= \bigcup_{i_1=1}^{\nu_1} \mathcal{B}^{(2)}_{i_1}\,,\\
\mathcal{B}^{(2)}_{i_1} &= \bigcup_{i_2=1}^{\nu_2} \mathcal{B}^{(3)}_{i_1,i_2}\,,\\
&\vdots\\
\mathcal{B}^{(s)}_{i_1,i_2,\ldots,i_{s-1}} &= \bigcup_{i_s=1}^{\nu_s} \mathcal{B}^{(s+1)}_{i_1,i_2,\ldots,i_s}\,.
\end{aligned}$$

The parameter i_1 labels the subcode $\mathcal{B}^{(2)}_{i_1}$ of the code $\mathcal{B}^{(1)}$, and i_2 labels the subcodes $\mathcal{B}^{(3)}_{i_1,i_2}$ of the code $\mathcal{B}^{(2)}_{i_1}$, and so on. Finally, i_s labels the codeword $\mathcal{B}^{(s+1)}_{i_1,i_2,\ldots,i_s}$ in the subcode $\mathcal{B}^{(s)}_{i_1,i_2,\ldots,i_{s-1}}$. The codes $\mathcal{B}^{(j)}_{i_1,i_2,\ldots,i_{j-1}}(q;n_b,M_b^{(j)},d_b^{(j)})$, $j=1,\ldots,s$, are called inner codes. The minimum distance $d_b^{(j)}$, $j=1,\ldots,s$, is known. This means that a codeword $\mathbf{b}\in\mathcal{B}^{(1)}$ is uniquely determined by the label $i_1,i_2,\ldots,i_s$. Therefore $\mathcal{B}^{(s+1)}_{i_1,i_2,\ldots,i_s}=\mathbf{b}$ defines a single codeword. The basic idea from generalized code concatenation is to protect a set of symbols $a_1^{(i)},a_2^{(i)},\ldots,a_{k_i}^{(i)}$ using outer codes $\mathcal{A}^{(i)}(q^{\mu_i};n_a,M_a^{(i)},d_a^{(i)})$. We assume that the alphabet of the corresponding outer code labels all ν_i subcodes of the ith partitioning:

$$\nu_i = q^{\mu_i}, \quad \text{code symbol: } a_j^{(i)} \in GF(q^{\mu_i})\,.$$

For linear codes, the dimension corresponds to the number of information symbols $k_a^{(i)}$. The cardinality $M_a^{(i)}$ of a partition level i is calculated with respect to the field and dimension of the of the outer code, which gives $M_a^{(i)}=(q^{\mu_i})^{k_a^{(i)}}$.

Note on symbols Each symbol of the ith outer code $a_j^{(i)}\in GF(q^{\mu_i})$, $j=1,\ldots,n_a$, that is used to label the subcodes of partition step i can be interpreted as a vector of μ_i elements of the field $GF(q)$. The elements of the vector interpretation are given by the coefficients of the polynomial representation of an element of the extension field. A symbol $a_j^{(i)}\in GF(q^{\mu_i})$ written as a vector is denoted by $\underline{a}_j^{(i)}\in GF(q)^{\mu_i}$. Similarly, $\underline{\mathbf{a}}^{(i)}$ is a column vector with length n_a having all elements written as vectors in $GF(q)^{\mu_i}$. $\underline{\mathbf{a}}^{(i)}$ can be written as a matrix in $GF(q)$ with dimension $n_a^{(i)}\times\mu_i$.

We arrange the codewords $(a_1^{(i)},a_2^{(i)},\ldots,a_{n_a}^{(i)})$, $i=1,2,\ldots,s$, from s outer codes as columns of a matrix:

$$\begin{pmatrix} a_1^{(1)} & a_1^{(2)} & \dots & a_1^{(s)} \\ a_2^{(1)} & a_2^{(2)} & \dots & a_2^{(s)} \\ \vdots & \vdots & \ddots & \vdots \\ a_{n_a}^{(1)} & a_{n_a}^{(2)} & \dots & a_{n_a}^{(s)} \end{pmatrix} . \quad (9.1)$$

The matrix elements $a_j^{(i)} \in GF(q^{\mu_i})$, $j = 1,2,\dots,n_a$, $i = 1,2,\dots,s$, correspond to a matrix with n_a rows. Each row has the form

$$\left(a_j^{(1)}, a_j^{(2)}, a_j^{(3)}, \dots, a_j^{(s)}\right) .$$

Every row determines a codeword through a unique partitioning of $\mathcal{B}^{(1)}$. We encode the outer code symbols by the inner code to obtain a codeword of the GC code C_{GC}. This gives an $n_a \times n_b$ matrix with elements from $GF(q)$. Every row is a codeword from $\mathcal{B}^{(1)}$:

$$\begin{pmatrix} c_{11} & c_{12} & \dots & c_{1,n_b} \\ c_{21} & c_{22} & \dots & c_{2,n_b} \\ \vdots & \vdots & \ddots & \vdots \\ c_{n_a,1} & c_{n_a,2} & \dots & c_{n_a,n_b} \end{pmatrix} . \quad (9.2)$$

Now we may formally define GC codes.

Definition 9.1 (GC codes) *A generalized concatenated code, (GC code) $C_{GC}(q;n,M,d)$ of order s is composed of s outer codes $\mathcal{A}^{(i)}(q^{\mu_i};n_a,k_a^{(i)},d_a^{(i)})$ and a partitioning of order s of the inner code $\mathcal{B}^{(1)}(q;n_b,k_b^{(1)},d_b^{(1)}) \supset \dots \supset \mathcal{B}^{(s+1)}(q;n_b,k_b^{(s+1)} = 0, d_b^{(s+1)} = \infty) = \mathbf{b}$. The subcodes of level $s+1$ consist of single codewords. The symbol $a_j^{(i)} \in GF(q^{\mu_i})$ of the i-th outer code determines a subset of $\mathcal{B}^{(i)}_{a_j^{(1)},a_j^{(2)},\dots,a_j^{(i-1)}}$. Thus the symbols $a_j^{(i)}$, $i = 1,\dots,s$, of all outer codes together form the label $(a_j^{(1)},\dots,a_j^{(s)})$ of a unique codeword $\mathbf{b}_j$. The codeword of the GC code consists of all the codewords $\mathbf{b}_j$, $j = 1,\dots,n_a$.*

Note that the code $\mathcal{B}^{(s+1)}$ has minimum distance $d_b^{(s+1)} = \infty$, since it consists of a single codeword of $\mathcal{B}^{(1)}$.

Theorem 9.2 (GC code) *Let C_{GC} be a GC code according to definition 9.1 with partitioning order s, $\mathcal{A}^{(i)}(q^{\mu_i};n_a,M_a^{(i)},d_a^{(i)})$ and $\mathcal{B}^{(1)}(q;n_b,M_b^{(1)},d_b^{(1)})$ corresponding to the outer codes and the first inner code respectively. C_{GC} has length $n = n_a n_b$, cardinality $M = \prod_{i=1}^{s} M_a^{(i)}$ and minimum distance*

$$d \geq \min_{i=1,\dots,s} \{d_b^{(i)} d_a^{(i)}\} .$$

Proof The length $n = n_a n_b$ and cardinality M are clear. We shall prove only the minimum distance. Consider two information sequences $\mathbf{i} \neq \mathbf{i}'$ encoded by the s outer codes. We obtain a matrix with at least one different column i, having at least $d_a^{(i)}$ different positions. Each different column element is in turn encoded by the inner code $\mathcal{B}^{(i)}$, which generates a row vector having at least $d_{\mathbf{b}}^{(i)}$ different elements. Therefore there are at least $d_a^{(i)} d_b^{(i)}$ different elements between the GC codewords $\mathbf{c} \neq \mathbf{c}'$. The lower bound on the minimum distance is obtained by taking the minimum inner and outer code product for each partition level $i \in 1, \ldots, s$ (see also [Zin76]). □

In many cases, the true minimum distance of a GC code is larger than the above bound, although not known exactly.

Classical concatenation is a special case of a GC code construction where the partition order $s = 1$. Note that the definition imposes no restrictions on the parameters of inner and outer codes. For example, the code $\mathcal{B}^{(1)}(q; n_b, n_b, 1)$ can represent the set of all n_b-tuples. The outer code $\mathcal{A}(q^{\mu}; n_a, n_a, 1)$ can also be defined as to consist of a single codeword (in such a case we define $d = \infty$ and write $\mathcal{A}(q^{\mu}; n_a, 0, \infty)$). We shall use this degree of freedom in section 9.2.7 to describe the class of error locating codes as a special case of GC codes.

Encoding can be done by first encoding $k_a^{(i)}$ information symbols into a codeword $\mathbf{a}^{(i)}$, $i \in 1, \ldots, s$, and then mapping the label $(a_j^{(1)}, \ldots, a_j^{(s)})$ onto the codeword $\mathbf{b}_j$. This encoding scheme, is not necessarily systematic. A systematic encoding scheme that can be used also for GC codes will be presented in section 9.2.7. Encoding can also be done in a single step using a generator matrix for the concatenated code (see [ZMB99]).

Note The alphabet $GF(q^{\mu_i})$ of symbols $a_j^{(i)}$ defines the partitioning. This results in three partitioning cases. The number of elements of the alphabet is larger than, smaller than or equal to the number ν_i of subcodes:

$$q^{\mu_i} \begin{cases} > \nu_i\,, \\ < \nu_i\,, \\ = \nu_i\,. \end{cases}$$

If all symbols of the alphabet of an outer code $\mathcal{A}^{(i)}$ correspond to a subcode then $q^{\mu_i} \leq \nu_i = M_b^{(i)}/M_b^{(i-1)}$ must be valid. It is assumed that all $M_b^{(i)}$ have the same cardinality, although this is not required. If q^{μ_i} is smaller than ν_i then not all of the subcodes from the GC code construction can be labeled by the alphabet.

It should again be mentioned that the concept of generalized concatenation is not limited to inner and outer block codes. The only requirement on the partitioning scheme is that the minimum distance of the subcodes at level i must be greater than or equal to the subcodes at level $i-1$ (see section 7.1 and appendix B). In general, codes may consist of a set of points in a vector space defined by a metric, and need not be linear. This is demonstrated in chapter 10 on coded modulation. It is understandable that these different concepts cannot be encompassed by one short definition.

9.2.2 Partitioning of block codes

In this section, different block code partitioning methods are given, and properties are derived. A construction is necessary whereby long codes can be partitioned without the need for listing all codewords of the code.

We now examine partitioning in more detail and limit ourselves mostly to first-order partitioning ($s = 1$). Higher-order partitions can be constructed through a repeated application of the described methods.

Definition 9.3 (Partitioning) *Consider a code $\mathcal{B}^{(1)}$ that is a set of M vectors of an n_b-dimensional space. The minimum distance among the vectors $\mathbf{b} \in \mathcal{B}^{(1)}$ is $d^{(1)}$. An ordering of the set $\mathcal{B}^{(1)}$ into ν disjoint subsets (subcodes) $\mathcal{B}_i^{(2)}$ gives*

$$\mathcal{B}^{(1)} = \bigcup_{i=1}^{\nu} \mathcal{B}_i^{(2)},$$

and is called a partition of $\mathcal{B}^{(1)}$. Every subset corresponds again to a code $\mathcal{B}_i^{(2)}$ with a minimum distance $d_i^{(2)}$, $d_i^{(2)} \geq d^{(1)}$, $i = 1,2,\ldots,\nu$. The minimum distance of the set of subcodes is defined by the minimum of the minimum distance of the individual codes.

$$d^{(2)} = \min_{i=1,\ldots,\nu} \{d_i^{(2)}\}.$$

Clearly, there are many possible ways to partition a code. For the construction of GC codes it is important that the subcode with the smallest minimum distance be as large as possible.

A technique for the partitioning of a $\mathcal{B}^{(1)}(q;n_b,M^{(1)},d^{(1)}) \subseteq GF(q)^{n_b}$ code is to methodically divide the codewords into subsets. This means that definition 9.3 is applied directly. We call this technique method 1, and it is only useful for codes that have a relatively small number of codewords.

Method 1 (Arbitrary partitioning) Consider the code $\mathcal{B}^{(1)}(q;n,M^{(1)},d^{(1)}) \subseteq GF(q)^n$. We partition $\mathcal{B}^{(1)}$ into ν subcodes $\mathcal{B}_i^{(2)}(q;n,M_i^{(2)},d_i^{(2)})$, $i = 1,2,\ldots,\nu$:

$$\mathcal{B}^{(1)} = \bigcup_{i=1}^{\nu} \mathcal{B}_i^{(2)}.$$

Therefore, every $\mathbf{b} \in \mathcal{B}^{(1)}$ belongs to one of the subcodes $\mathcal{B}_i^{(2)}$, and can be uniquely determined. This can be written in terms of the cardinality of the inner codes as follows:

$$M^{(1)} = \sum_{i=1}^{\nu} M_i^{(2)}.$$

In order to label the selected partitioning, each codeword $\mathbf{b} \in \mathcal{B}^{(1)}$ is mapped to two symbols $\underline{a}^{(1)} \in GF(q)^{\mu_1}$ and $\underline{a}^{(2)} \in GF(q)^{\mu_2}$. Therefore $\underline{a}^{(1)}$ identifies the subcode $\mathcal{B}_i^{(2)}$ of the codeword $\mathbf{b}$, and $\underline{a}^{(2)}$ identifies the codeword within the subset $\mathcal{B}_i^{(2)}$.

There are ν subcodes with the basis q, which can be labeled with the indices $0,1,\ldots,\nu-1$. There are three possibilities.

- For the case when $q^{\mu_1} < \nu$, not every subcode can be given a label $\underline{a}^{(1)}$. There are more subcodes than indices. Also, if $q^{\mu_2} < M_i^{(2)}$, not every codeword can be identified by the index $\underline{a}^{(2)}$.

- If $q^{\mu_1} > \nu$ then not every element $\underline{a}^{(1)}$ corresponds to a subcode. There are fewer subcodes as indices. Also, $q^{\mu_2} > M_i^{(2)}$ does not require every element of $\underline{a}^{(2)}$ in order to identify a codeword of a subcode.

- For $q^{\mu_1} = \nu$, every subcode corresponds exactly to one label $\underline{a}^{(1)}$, and for $q^{\mu_2} = M_i^{(2)}$, every codeword in a subcode corresponds exactly to a label $\underline{a}_j^{(2)}$.

We require that each $\underline{a}^{(i)} \in GF(q)^{\mu_i}$. It follows that

$$q^{\mu_1} \leq \nu \qquad \text{and} \qquad q^{\mu_2} \leq \max_{i=1,\dots,q^{\mu_1}} \{M_i^{(2)}\}.$$

Therefore the partitioning is described by the following mapping:

$$\begin{aligned} \left(\underline{a}^{(1)}, \underline{a}^{(2)}\right) &\iff \mathbf{b} \in \mathcal{B}^{(1)}, \\ \underline{a}^{(1)} &\iff i: \mathcal{B}_i^{(2)}, \\ \underline{a}^{(2)} &\iff \mathbf{b} \in \mathcal{B}_i^{(2)}. \end{aligned} \tag{9.3}$$

In general, method 1 requires a partitioning table to define the mapping between $\underline{a}^{(1)}$, $\underline{a}^{(2)}$ and the codewords. $\mathcal{B}_i^{(2)}$ can be subdivided again in order to achieve a higher-order partitioning. For *longer* codes, it is necessary to consider other construction methods for the partitioning. This is possible for linear codes, and is presented next.

Linear codes

Consider a code $\mathcal{B}^{(1)}$ that is partitioned into subcodes of different levels. Our goal is to be able to calculate the subcode labels of a particular mapping. In general, this technique is not optimal with respect to maximizing the minimum distance of the subcodes; however, it can be applied to the partitioning of long codes.

Method 2 (Partitioning by cosets) Consider the two linear codes with parameters

$$\mathcal{B}^{(2)}(q; n_b, k_b^{(2)}, d_b^{(2)}) \subset \mathcal{B}^{(1)}(q; n_b, k_b^{(1)}, d_b^{(1)}).$$

Furthermore, $k_b^{(1)} = m + k_b^{(2)}$, $m > 0$ and $d_b^{(2)} > d_b^{(1)}$. The code $\mathcal{B}^{(1)}$ is partitioned into subcodes $\mathcal{B}_i^{(2)}$ as follows:

$$\mathcal{B}^{(1)} = \bigcup_{i=1}^{\nu = q^m} \mathcal{B}_i^{(2)},$$

where

$$\mathcal{B}_i^{(2)} = \{\mathbf{v}_i + \mathcal{B}^{(2)}\}, \qquad \mathbf{v}_i \in GF(q)^{n_b}, \qquad i = 1, \dots, \nu.$$

The set $\{\mathbf{v}_i + \mathcal{B}^{(2)}\}$ is called a *coset* of $\mathcal{B}^{(2)}$ (see section 1.3 and problem 7.1). The vector $\mathbf{v}_i$ is called the *coset representative*, and is used to identify the appropriate subcode.

We obtain the construction rule for the partitioning of the code $\mathcal{B}^{(1)}$ in the following way. Choose from the set of all cosets of code $\mathcal{B}^{(2)}$ the ν cosets that when joined make up the code $\mathcal{B}^{(1)}$.

Note Subcodes constructed using this method all have the same cardinality; however, the codes $\mathcal{B}_i^{(2)}$ are no longer linear, with the exception of the code containing $\mathbf{v}_i = \mathbf{0}$.

Elements from $\underline{\mathbf{a}}^{(1)}$ label the partition, while elements from $\underline{\mathbf{a}}^{(2)}$ label the codewords of the subcode:

$$\underline{a}^{(1)} \in GF(q)^m \quad \text{and} \quad \underline{a}^{(2)} \in GF(q)^{k^{(2)}} .$$

The calculation procedure for partitioning is as follows. There exists a codeword $\mathbf{b} \in \mathcal{B}^{(1)}$ which corresponds to the label of the subcode $\mathcal{B}_i^{(2)}$ containing the codeword $\mathbf{b}$. Therefore $\mathbf{b}$ is decoded according to $\mathcal{B}^{(2)}$. The vector $\mathbf{v}_i$ is the coset representative, and is given by the error vector found by decoding $\mathbf{b}$ according to the code $\mathcal{B}^{(2)}$. The coset representative has a direct mapping to the subcode label. In general it is only possible to find all coset representatives if a ML decoder exists for $\mathcal{B}^{(2)}$ (see section 1.3). Decoding with a BMD decoder is not always successful, because in this case, ν is not necessarily smaller than the number of identifiable coset representatives. This problem is illustrated in the following example.

Example 9.5 (Partitioning (subset identification) through decoding) We consider a method of identifying the coset representatives by decoding the subcode $\mathcal{B}^{(2)}$ defined as

$$\mathcal{B}^{(2)}(2;15,6,6) \subset \mathcal{B}^{(1)}(2;15,14,2).$$

Therefore we must distinguish among 2^8 different cosets of the code $\mathcal{B}^{(2)}$. For a BMD decoder, it follows that only coset leaders with a weight smaller than half the minimum distance can be decoded. Because the minimum distance of the code $\mathcal{B}^{(2)}(2;15,6,6)$ is equal to 6, only coset representatives with weight smaller than two can be decoded. This gives the inequality

$$\binom{15}{0} + \binom{15}{1} + \binom{15}{2} = 1 + 15 + 15 \cdot 7 = 16 + 105 = 121 < 256.$$

Therefore, in general, not all of the subcodes can be identified. ◇

Method 2 for cyclic codes in the transform domain The description in the transformation domain (see section 3.1.1) can also be used for cyclic codes. The inner codewords are denoted by the polynomials $b(x)$ and $B(x)$ corresponding to the original and transform domains respectively:

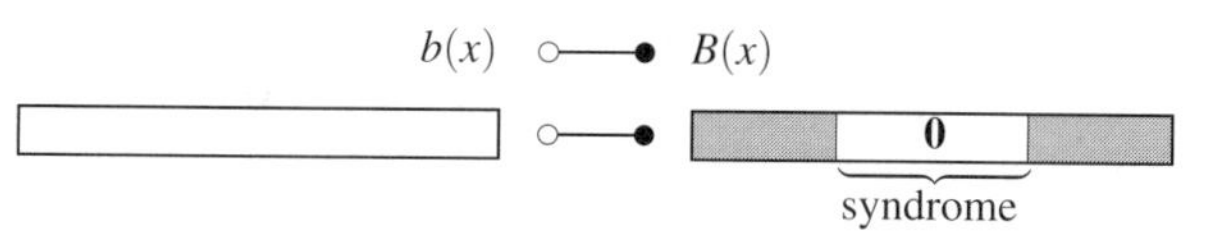

The syndrome is zero for all codewords and non-zero if $b(x)$ is not equal to a codeword. A subset of codewords from $b^{(1)}(x) \in \mathcal{B}^{(1)}$ can be written in terms of the original code and a subcode $b^{(2)}(x) \in \mathcal{B}^{(2)} \subset \mathcal{B}^{(1)}$. One recognizes in the following sketch that a codeword from $\mathcal{B}^{(1)}$ can give a non-zero syndrome if decoded in $\mathcal{B}^{(2)}$:

syndrome

		0	$\in \mathcal{B}^{(1)}$
	0	0	$\in \mathcal{B}^{(2)}$

Any codeword of the subcode $\mathcal{B}^{(2)}$ can be added to $\mathbf{b}$ without changing the syndrome. In other words, there is a mapping of the coset representative $\mathbf{v}_i$ of the coset $\{\mathbf{b}^{(2)} + \mathbf{v}_i \mid \mathbf{b}^{(2)} \in \mathcal{B}^{(2)}\}$ and the syndrome:

0	syndrome	0

$\bullet\!\!-\!\!\circ \quad \mathbf{v}_i : \{\mathbf{b}^{(2)} + \mathbf{v}_i \mid \mathbf{b}^{(2)} \in \mathcal{B}^{(2)}\}$

Description of coset partitioning by a linear code There are sets of coset representatives that form a linear code and can be described by a generator matrix. This follows immediately from the following consideration. Let $\mathbf{G}_{\mathcal{B}^{(i)}}$ be a generator matrix of the code $\mathcal{B}^{(i)}$. $\mathbf{G}_{\mathcal{B}^{(i)}}$ consists of $k_b^{(i)}$ rows of length n_b. A generator matrix $\mathbf{G}_{\mathcal{B}^{(i+1)}}$ of the subcode $\mathcal{B}^{(i+1)}$ has $k_b^{(i+1)} < k_b^{(i)}$ rows. It is obvious that $\mathbf{G}_{\mathcal{B}^{(i)}}$ can be transformed via elementary row operations into a form that contains $\mathbf{G}_{\mathcal{B}^{(i+1)}}$ as a submatrix. The remaining rows of $\mathbf{G}_{\mathcal{B}^{(i)}}$ form a $(k_b^{(i)} - k_b^{(i+1)}) \times n_b$ generator matrix of coset representatives, and this matrix will be denoted by $\mathbf{G}_{\mathcal{B}^{(i)}/\mathcal{B}^{(i+1)}}$. The coset representative of the linear subcode is the zero codeword $\underline{0}$ and the label is $\underline{a} = \underline{0}$. $\mathbf{G}_{\mathcal{B}^{(i)}/\mathcal{B}^{(i+1)}}$ describes the mapping of the label $\underline{a}^{(i)}$ to a coset representative of level i. Hence it is possible to calculate a codeword corresponding to the subcode labels $(\underline{a}^{(1)}, \underline{a}^{(2)}, \ldots, \underline{a}^{(s)})$:

$$\mathbf{v} = \mathbf{G}^T_{\mathcal{B}^{(1)}/\mathcal{B}^{(s+1)}} \cdot \begin{pmatrix} \underline{a}^{(s)} \\ \vdots \\ \underline{a}^{(1)} \end{pmatrix}, \quad \mathbf{G}_{\mathcal{B}^{(1)}/\mathcal{B}^{(s+1)}} = \begin{pmatrix} \mathbf{G}_{\mathcal{B}^{(s)}/\mathcal{B}^{(s+1)}} \\ \vdots \\ \mathbf{G}_{\mathcal{B}^{(1)}/\mathcal{B}^{(2)}} \end{pmatrix}, \quad \mathbf{v} \in \mathcal{B}^{(s)}_{\underline{a}^{(1)},\ldots,\underline{a}^{(s)}}.$$

$\mathbf{G}_{\mathcal{B}^{(1)}/\mathcal{B}^{(s+1)}}$ is the combination of all generator matrices of coset representatives. Note that $\mathbf{G}_{\mathcal{B}^{(1)}/\mathcal{B}^{(s+1)}}$ cannot be transformed to a systematic matrix without destroying the nested structure for the partitioning.

Because the mapping done by the generator matrices is unique, there exists an inverse mapping that produces the label of a partition at level i if a coset element is given. The matrix describing this mapping is a (partial) parity check matrix of the linear subcode $\mathcal{B}_0^{(i+1)}$, and will be denoted by $\mathbf{H}_{\mathcal{B}^{(i)}/\mathcal{B}^{(i+1)}}$. Therefore the label $\underline{a}^{(i)}$ may also be called the syndrome. Note that there exists a one-to-one correspondence between syndromes and coset representatives, and therefore the number of possible syndromes is equal to the number of coset representatives. $\mathbf{H}_{\mathcal{B}^{(i)}/\mathcal{B}^{(i+1)}}$ corresponds (similar to the generator matrix of coset representatives $\mathbf{G}_{\mathcal{B}^{(i)}/\mathcal{B}^{(i+1)}}$) to all the rows needed to extend the parity check matrix $\mathbf{H}_{\mathcal{B}^{(i)}}$ in order to produce a parity check matrix $\mathbf{H}_{\mathcal{B}^{(i+1)}}$ of $\mathcal{B}^{(i+1)}$. The combination of all the matrices $\mathbf{H}_{\mathcal{B}^{(i)}/\mathcal{B}^{(i+1)}}$, $i = 1, \ldots, s$, defines the linear mapping from a given coset to the label:

$$\mathbf{H}_{\mathcal{B}^{(1)}/\mathcal{B}^{(s+1)}} \cdot \mathbf{v} = \begin{pmatrix} \underline{a}^{(s)} \\ \vdots \\ \underline{a}^{(1)} \end{pmatrix}, \qquad \mathbf{H}_{\mathcal{B}^{(1)}/\mathcal{B}^{(s+1)}} = \begin{pmatrix} \mathbf{H}_{\mathcal{B}^{(s)}/\mathcal{B}^{(s+1)}} \\ \vdots \\ \mathbf{H}_{\mathcal{B}^{(1)}/\mathcal{B}^{(2)}} \end{pmatrix}.$$

We shall use this description in the definition of generalized error locating codes (see section 9.2.7).

The generator matrix of coset representatives $\mathbf{G}_{\mathcal{B}^{(1)}/\mathcal{B}^{(s+1)}}$ can be transformed into a representation with systematic submatrices $\mathbf{G}_{\mathcal{B}^{(i)}/\mathcal{B}^{(i+1)}}$ by row operations within $\mathbf{G}_{\mathcal{B}^{(i)}/\mathcal{B}^{(i+1)}}$ and column permutations of the whole matrix $\mathbf{G}_{\mathcal{B}^{(1)}/\mathcal{B}^{(s+1)}}$. This is explained in the following for a partitioning of order $s=1$. Note that $\mathbf{G}_{\mathcal{B}^{(1)}/\mathcal{B}^{(s+1)}}$ cannot be made systematic while preserving the partition structure. Therefore, in contrast to systematic encoding of classical concatenated codes, there is only a non-systematic mapping that combines outer and inner codes.

Method 3 Consider

$$\mathcal{B}^{(2)}(q;n_b,k^{(2)},d_b^{(2)}) \subset \mathcal{B}^{(1)}(q;n_b,k^{(1)},d_b^{(1)}),$$

with $k^{(2)}+m=k^{(1)}$, $m>0$, $d_b^{(2)}>d_b^{(1)}$. Let $\mathbf{G}_{\mathcal{B}^{(1)}}$ and $\mathbf{G}_{\mathcal{B}^{(2)}}$ be the generator matrices of the codes $\mathcal{B}^{(1)}$ and $\mathcal{B}^{(2)}$ in systematic form. We calculate

$$\begin{aligned}
\mathbf{b}^{(2)} &= \underline{a}^{(2)}\cdot\mathbf{G}_{\mathcal{B}^{(2)}} \in \mathcal{B}^{(2)}, &\quad \text{with} \quad \underline{a}^{(2)} \in GF(q)^{k^{(2)}},\\
\mathbf{b}^{(1)} &= (\mathbf{0}|\underline{a}^{(1)})\cdot\mathbf{G}_{\mathcal{B}^{(1)}} \in \mathcal{B}^{(1)}, &\quad \text{with} \quad \underline{a}^{(1)} \in GF(q)^{m}.
\end{aligned}$$

Therefore $(\mathbf{0}|\underline{a}^{(1)}) \in GF(q)^{k^{(1)}}$ is a vector constructed from $k^{(2)}$ zeros and the vector $\underline{a}^{(1)}$.

Because $\mathcal{B}^{(2)} \subset \mathcal{B}^{(1)}$, $\mathbf{b}=\mathbf{b}^{(2)}+\mathbf{b}^{(1)} \in \mathcal{B}^{(1)}$, and $\mathbf{b}$ is a codeword of $\mathcal{B}^{(1)}$. Furthermore, the vector $(\underline{a}^{(2)},\underline{a}^{(1)})$ has the same length as the information block of the code $\mathcal{B}^{(1)}$.

If a codeword $\mathbf{b} \in \mathcal{B}^{(1)}$ is given then the information block can be calculated. It has the information block $(\underline{a}^{(2)}|\tilde{\underline{a}}^{(1)})$. From the systematic form of the partition method, we can directly recover $\underline{a}^{(2)}$, and from this $\mathbf{b}-\underline{a}^{(2)}\cdot\mathbf{G}_{\mathcal{B}^{(2)}}$ is calculated. The difference is a codeword from $\mathcal{B}^{(1)}$ having the information block $(\mathbf{0}|\underline{a}^{(1)})$, also in systematic form (see section 3.1.4).

Partitioning by means of method 3 is preferred for linear codes because the partitioning is directly *calculable* through the systematic form, and therefore codes of arbitrary length can be used.

Example 9.6 (Partitioning (subset) identification through coding) For the case of cyclic codes that can be represented in systematic form, we can use the polynomial description. Consider the parity check code $\mathcal{B}^{(1)}(2;7,2^6,2)$ with the generator polynomial $g_1(x)$ and the simplex code $\mathcal{B}^{(2)}(2;7,2^3,4)$ with $g_2(x)$. We get $\mathcal{B}^{(2)} \subset \mathcal{B}^{(1)}$, and $g_2(x)=\Delta(x)\,g_1(x)$. We assume that the generator polynomials are in systematic form. At this point, we use method 3 in order to partition the parity check code.

The vectors $\underline{a}^{(1)} \in GF(2)^3$ and $\underline{a}^{(2)} \in GF(2)^3$ are represented by the polynomials $a_i(x)=a_0^{(i)}+a_1^{(i)}x+a_2^{(i)}x^2$, $i=1,2$. This gives a codeword from $\mathcal{B}^{(1)}$:

$$b(x)=a_2(x)\,g_2(x)+x^3\,a_1(x)\,g_1(x) \in \mathcal{B}^{(1)}.$$

For a given codeword $b(x)$ and the known generator polynomials, we can calculate the codeword labels $\underline{a}^{(1)}(x)$ and $\underline{a}^{(2)}(x)$ as follows:

$$\begin{aligned}
\frac{b(x)}{g_1(x)} &= \frac{\left(a_2(x)\,g_1(x)\,\Delta(x)+x^3\,a_1(x)\,g_1(x)\right)}{(g_1(x))}\\
&= a_2(x)\,\Delta(x)+x^3\,a_1(x)\\
&= w(x),
\end{aligned}$$

where $\Delta(x)$ is also known. Due to the systematic encoding, the coefficients of the codeword $b(x)$ correspond directly to $a^{(2)}(x)$. From this, we calculate $a^{(1)}(x)=w(x)-a_2(x)\,g_2(x)$, which gives the partition labels. The partition labels can now be mapped back to the information sequence. ◇

From the construction of GC codes, it is necessary to have flexibility in selection of the parameters μ_1 and μ_2, ($\underline{a}^{(1)} \in GF(q)^{\mu_1}$ and $\underline{a}^{(2)} \in GF(q)^{\mu_2}$). This way, the principle of code concatenation can be exploited. Before we prove theorem 9.4, we consider three examples.

Example 9.7 (Partitioning through inner code shortening) Assume that we are restricted to using the two outer code alphabets $\underline{a}^{(2)} \in GF(2)^3$ and $\underline{a}^{(1)} \in GF(2)^6$ for partitioning. We use the BCH codes $\mathcal{C}^{(2)}(2;31,2^{10},12)$ and $\mathcal{C}^{(1)}(2;31,2^5,16)$, where $\mathcal{C}^{(2)} \subset \mathcal{C}^{(1)}$. The generator polynomials $g_1(x)$ and $g_2(x)$ are used to calculate $\mathcal{C}^{(1)}$ and $\mathcal{C}^{(2)}$ respectively, where $g_2(x) = \Delta(x)\, g_1(x)$. To obtain the inner codes, we shorten both BCH codes such that we have 3 and 9 information bits instead of 5 and 10. The minimum distance of both codes remains the same:

$$\mathcal{B}^{(2)}(2;29,2^3,16) \quad \text{and} \quad \mathcal{B}^{(1)}(2;30,2^9,12)\ .$$

When we partition using method 3, we get

$$b(x) = a_2(x)\, g_2(x) + x^3 a_1(x)\, g_1(x) \in \mathcal{B}^{(1)}\ .$$

The polynomials $a_1(x)$ and $a_2(x)$ correspond to the vectors $\underline{a}^{(1)}$ and $\underline{a}^{(2)}$. For a codeword $b(x)$, we can calculate $\underline{a}^{(1)}$ and $\underline{a}^{(2)}$ corresponding to example 9.6 ◇

Example 9.8 (Partitioning through inner code dimension modification) In this example, we calculate a partitioning with the labels $\underline{a}^{(1)} \in GF(2)^8$ and $\underline{a}^{(2)} \in GF(2)^8$. We use the BCH code $\mathcal{C}^{(1)}(2;31,2^{16},7)$. A subcode of $\mathcal{C}^{(1)}$ is the BCH code $\mathcal{C}^{(2)}(2;31,2^{10},12)$. The corresponding generator polynomials are $g_1(x)$ and $g_2(x)$, with $g_2(x) = \Delta(x)\, g_1(x)$. According to method 3, the systematic form of the partitioning requires that the polynomials $a_1(x)$ and $a_2(x)$ do not overlap. This cannot be achieved directly with the original codes without increasing the overall length of the GC code. Therefore we shorten the second inner code by fixing the appropriate two information bits to zero. The minimum distance of $\mathcal{B}^{(2)}$ remains the same. This gives

$$\mathcal{B}^{(2)}(2;29,2^8,12) \quad \text{and} \quad \mathcal{B}^{(1)}(2;31,2^{16},7)\ .$$

The partitioning can now be calculated as in example 9.7. ◇

Example 9.9 (Partitioning of order 5) We use method 3 and the BCH code $\mathcal{B}^{(1)}(2;31,2^{25},4)$ to perform an order-5 partitioning. The information bits are encoded by outer codes of length n_a. We may represent all outer code elements as vectors in the basic field, where $\underline{a}_j^{(i)} \in GF(2)^5$. We partition the inner codes in the following manner:

$$\begin{aligned} \mathcal{B}^{(5)}(2;31,5,16) \ &\subset \mathcal{B}^{(4)}(2;31,10,12) \ \subset \mathcal{B}^{(3)}(2;31,15,8) \\ &\subset \mathcal{B}^{(2)}(2;31,20,6) \quad \subset \mathcal{B}^{(1)}(2;31,25,4)\ . \end{aligned}$$

We use method 3 to obtain a encoding of the GC code. Each row $\mathbf{c}_j$, $j \in 1 \ldots n_a$ of the GC codeword is a codeword of the first outer code $\mathcal{B}^{(1)}$, and is constructed as follows:

$\mathbf{b}_j^{(5)}$:	$\underline{a}^{(5)}$						$\in \mathcal{B}^{(5)}$
$\mathbf{b}_j^{(4)}$:	0	$\underline{a}^{(4)}$					$\in \mathcal{B}^{(4)}$
$\mathbf{b}_j^{(3)}$:	0	0	$\underline{a}^{(3)}$				$\in \mathcal{B}^{(3)}$
$\mathbf{b}_j^{(2)}$:	0	0	0	$\underline{a}^{(2)}$			$\in \mathcal{B}^{(2)}$
$\mathbf{b}_j^{(1)}$:	0	0	0	0	$\underline{a}^{(1)}$		$\in \mathcal{B}^{(1)}$

$$\mathbf{c}_j = \mathbf{b}_j^{(5)} + \mathbf{b}_j^{(4)} + \mathbf{b}_j^{(3)} + \mathbf{b}_j^{(2)} + \mathbf{b}_j^{(1)} \in \mathcal{B}^{(1)} .$$

Each vector component of $\mathbf{c}_j$ is obtained by the corresponding generator matrix of the subcode. The column outer codeword vector $\underline{\mathbf{a}}^{(5)}$ is unchanged in the GC codeword. For every given $\mathbf{c}_j$, every vector $\underline{a}^{(i)}$, $i = 1, 2, \ldots, 5$, can be calculated by taking advantage of the systematic encoding structure of the subcodes. ◇

The previous examples are generalized in the following theorem:

Theorem 9.4 (Partitioning condition for method 3) *Consider the code* $\mathcal{B}^{(2)}(q; n_b, k^{(2)}, d_b^{(2)}) \subset \mathcal{B}^{(1)}(q; n_b, k^{(1)}, d_b^{(1)})$, *with* $k^{(2)} + m = k^{(1)}$, $d_b^{(2)} > d_b^{(1)}$. *A partitioning using method 3 can be constructed for the alphabet* $\underline{a}^{(1)} \in GF(q)^{\mu_1}$ *and* $\underline{a}^{(2)} \in GF(q)^{\mu_2}$ *if*

$$\mu_2 \leq k^{(2)} \quad \text{and} \quad \mu_1 \leq m + (k^{(2)} - \mu_2) .$$

Proof $\mathbf{G}^{(2)}$ is the generator matrix of $\mathcal{B}^{(2)}$. Let $\underline{a}_j^{(2)} \in GF(q)^{\mu_2}$, $\underline{a}_j^{(1)} \in GF(q)^{\mu_1}$, where $\underline{\mathbf{a}}^{(2)} \cdot \mathbf{G}^{(2)} = \mathbf{b}^{(2)} \in \mathcal{B}^{(2)}$. If $\mu_2 < k^{(2)}$ then we may shorten $\mathcal{B}^{(2)}$ by removing information coordinates without changing the minimum distance (see examples 9.7 and 9.8). The GC codeword $\mathbf{b}^{(1)} \in \mathcal{B}^{(1)}$ is obtained by calculating $(\mathbf{0}|\underline{a}^{(1)}) \in GF(q)^{k^{(1)}}$, with $\mathbf{0} \in GF(q)^{\mu_2}$, and adding this vector to $\mathbf{b}^{(2)}$. From the construction method, it holds that $\underline{a}^{(1)} \in GF(q)^{k^{(1)}-\mu_2}$ and $k^{(1)} = k^{(2)} + m$. □

Note For the case that $\mu_1 + \mu_2 < k^{(1)}$, we can shorten both codes $\mathcal{B}^{(1)}$ and $\mathcal{B}^{(2)}$. In addition, the codes $\mathcal{B}^{(1)}$ and $\mathcal{B}^{(2)}$ may already be shortened codes.

Several properties of method 3 are presented that will be useful in later decoding descriptions. They state that two different codewords that are not in the same subcode must have a different index $\underline{a}^{(1)}$.

Theorem 9.5 (Uniqueness of partitioning) *Consider a partitioning of the code* $\mathcal{B}^{(1)}$ *using method 3:*

$$\mathcal{B}^{(2)}(q; n, k^{(2)}, d^{(2)}) \subset \mathcal{B}^{(1)}(q; n, k^{(1)}, d^{(1)}) .$$

The codeword $\mathbf{b}$ *is defined by* $(\underline{a}^{(1)}, \underline{a}^{(2)})$, *and* $\hat{\mathbf{b}}$ *is defined by* $(\underline{\hat{a}}^{(1)}, \underline{\hat{a}}^{(2)})$ *and* $\mathbf{b} \neq \hat{\mathbf{b}}$. *Based on the requirements of method 3, for each* $\mathbf{e} = \mathbf{b} - \hat{\mathbf{b}} \in \mathcal{B}^{(1)}$ *with* $\mathbf{e} \neq \mathbf{0}$, *and* $\mathbf{e} \notin \mathcal{B}^{(2)}$, *we have* $\underline{\hat{a}}^{(1)} \neq \underline{a}^{(1)}$.

Proof $\mathbf{G}_{\mathcal{B}^{(1)}}$ and $\mathbf{G}_{\mathcal{B}^{(2)}}$ are the generator matrixes for the two codes. Because both codes are linear, we can assume $\underline{a}^{(1)} = \underline{a}^{(2)} = \mathbf{0}$ without loss of generality. Therefore

$$\mathbf{e} = \mathbf{0} - (\underline{\hat{a}}^{(2)} \cdot \mathbf{G}_{\mathcal{B}^{(2)}} - (\mathbf{0}|\underline{\hat{a}}^{(1)}) \cdot \mathbf{G}_{\mathcal{B}^{(1)}}) .$$

From the requirements of the theorem, $\mathbf{e} \notin \mathcal{B}^{(2)}$. It follows that $\underline{\hat{a}}^{(1)} \neq \mathbf{0}$ □

9.2.3 Code construction

In the following, we use examples of partitioning from the preceding section in order to describe several constructions for GC codes. For comparison, we shall consider a classical code concatenation first. A comparison with *classical* code concatenation is also considered (see example 9.20 from section 9.2.4).

Example 9.10 (Concatenation of QR and RS codes) We construct a classical concatenated code using the QR code$\mathcal{B}(2;17,8,6)$ as the inner code and the RS code $\mathcal{A}(2^8;255,223,33)$ as the outer code. (This code is the standard code used in satellite communications see [WHPH87]. The label $\underline{a} \in GF(2)^8$ identifies the codewords of code $\mathcal{B}$ (one can view these as information blocks). The concatenated code has length $n_a n_b = 255 \cdot 17 = 4335$. The number of information symbols from $GF(2)$ is $223 \cdot 8 = 1784$. The minimum distance is calculated as $d \geq 6 \cdot 33 = 198$. We construct a GC code as follows. The first inner code is a parity check code $\mathcal{B}^{(1)}(2;17,16,2)$, and the second inner code $\mathcal{B}^{(2)} = \mathcal{B}$ is the QR code defined above. The first outer code is the repetition code $\mathcal{A}^{(1)}(2^8;255,1,255)$, and the second outer code $\mathcal{A}^{(2)} = \mathcal{A}$ is the RS code defined above. The partitioning is carried out using method 3. We compare the codes C_C and C_{GC}:

$$C_C(2;4335,1784,198) \quad \text{and} \quad C_{GC}(2;4335,1792,198).$$

The GC code has 64 times as many codewords for the same minimum distance. ◇

In the following, we construct a GC code with a partitioning of higher order.

Example 9.11 (Concatenation of BCH and RS codes) In example 9.9, we gave a partitioning of order 5 for the BCH code $\mathcal{B}^{(1)}(2;31,2^{25},4)$. We used the labeling $\underline{a}^{(i)} \in GF(2)^5$, $i = 1,2,\ldots,5$. Therefore 5 outer codes from the alphabet $GF(2)^5$ are required.

According to section 9.2.2, the BCH codes have the following minimum distances:

$$\mathcal{B}^{(i)}(2;31,2^{5(6-i)},d_b^{(i)}), \quad i = 1,2,\ldots,5\,,$$

$$d_b^{(1)} = 4, \quad d_b^{(2)} = 6, \quad d_b^{(3)} = 8, \quad d_b^{(4)} = 12, \quad d_b^{(5)} = 16\,.$$

We want to construct a GC code with the minimum distance $d \geq 96$. Based on the inequality

$$d \geq \min_{i=1,\ldots,5} \{d_b^{(i)} d_a^{(i)}\}\,,$$

we require outer codes with the following minimum distances:

$$d_a^{(1)} = 24, \quad d_a^{(2)} = 16, \quad d_a^{(3)} = 12, \quad d_a^{(4)} = 8, \quad d_a^{(5)} = 6\,.$$

We select the following outer RS codes:

$$\mathcal{A}^{(i)}(2^5;31,k_a^{(i)} = 31 - d_a^{(i)} + 1, d_a^{(i)}), \quad i = 1,2,\ldots,5\,.$$

The GC code has parameters

length:	$n = n_a n_b = 31^2 = 961\,,$
cardinality:	$M = \prod_{i=1}^{5} M_a^{(i)} = 2^{470}\,,$
minimum distance:	$d \geq 96\,.$

$$C_{GC}(2;961,2^{470},96)\,.$$

◇

Modification of rate, length and minimum distance of GC codes

As already mentioned in section 8.1.10, there are channel coding methods that modify the rate of a code such that they are appropriate for a particular application. In the following, we investigate modifications used to create GC codes with different rates and minimum distances. Only methods using the extension and shortening of codes (see section 4.3) are considered.

We denote s as the partitioning order of the code $\mathcal{B}^{(1)}(q; n_b, M_b^{(1)}, d_b^{(1)})$. Each interdependent subset of the code partitioning can be organized to give a different GC code partitioning. In particular, we focus on how the code length and distance are affected through the deletion of a set of partition levels. Consider the encoding of the information bits by the outer codes. This produces a matrix of dimension $n_a \times s$ with elements in the extension field. Each row of this matrix can be written as

$$\left(a_l^{(j_0)},\, a_l^{(j_0+1)},\, \ldots,\, a_l^{(s-j)}\right), \quad j_0 = 1, 2, \ldots, s-j, \quad j = 0, \ldots, s-1\,, \tag{9.4}$$

where $l \in 1, \ldots, n_a$. The variables j_0 and j can be viewed as indices corresponding to the start and end of the row vector with index l respectively.

The minimum distances of the inner and outer codes used to construct a GC code with order s partitioning are given:

$$\begin{aligned} &\text{for the inner codes,} \quad d_b^{(1)},\, d_b^{(2)},\, \ldots,\, d_b^{(s)}\,; \\ &\text{for the outer codes,} \quad d_a^{(1)},\, d_a^{(2)},\, \ldots,\, d_a^{(s)}\,. \end{aligned}$$

The selection of a subset $j_0, j_0+1, \ldots, s-j$ of the GC code defines the minimum distance, code rate and cardinality.

Our first example considers the shortening of a GC code C_{GC} constructed using method 3. The indices from equation 9.4 are set to $j_0 = 1$ and $j > 0$. This choice of parameters is equivalent to removing the outer codes $s-j+1, \ldots, s$. If we consider the elements $a_l^{(i)} \in GF(q^{\mu_i})$ in terms of the basic field, we may write them as vectors $\underline{a}_l^{(i)}$ of length μ_i:

$$\underline{a}_l^{(1)},\, \underline{a}_l^{(2)},\, \ldots,\, \underline{a}_l^{(s-j)},$$

The inner codes $\mathcal{B}^{(i)}, i = 1, \ldots, s$, are all shortened by $\mu_{s-j+1} + \ldots + \mu_s$ information coordinates, which does not affect the minimum distance. The GC codeword is constructed from row vectors $\mathbf{c}_l$ that come from the shortened code $\mathcal{B}^{(1)}$. The corresponding outer and inner codes are

$$\mathcal{A}^{(1)},\, \mathcal{A}^{(2)},\, \ldots,\, \mathcal{A}^{(s-j)} \quad \text{and} \quad \mathcal{B}^{(1)},\, \mathcal{B}^{(2)},\, \ldots,\, \mathcal{B}^{(s-j)}$$

respectively. The shortened GC code $C_{GC'}$ has length $n_{GC'} = n_a\left(n_b - \sum_{i=s-j+1}^{s} \mu_{(i)}\right)$, cardinality $M_{GC'} = \prod_{i=1}^{s-j} M_a^{(i)}$ and minimum distance $d_{GC'} \geq d_{GC}$. The rate depends on the which outer codes were deleted.

It is worth mentioning that shortening of the outer coordinates through the deletion of information coordinates allows the same encoder and decoder to be used.

In the next example, we again modify a GC code C_{GC} partitioned according to method 3, without changing the original codes. Our goal is to increase the minimum distance of the code.

We do this by eliminating the weaker inner and outer codes as follows. From equation 9.4, we select $j_0 = 1$ and $j > 0$, which corresponds to the outer code with the smallest minimum distance according to method 3. These outer code elements are written in the basic field as in the above method. We add the requirement that $\mu = \mu_i$ for all $i \in 1, \ldots, s$. Therefore

$$\underline{a}_l^{(1)}, \underline{a}_l^{(2)}, \ldots, \underline{a}_l^{(s-j)}, \quad j > 1,$$

is the set of vectors in the basic field where $l \in 1, \ldots, n_a$. Similar to the above example, this is equivalent to removing μj rows of the GC code; however, in this case, we use these rows to increase the redundancy of the inner codes. This allows us to select the more powerful inner codes $\mathcal{B}^{(i)}$, $i \in j+1, \ldots, s$ for encoding. The final codes used for encoding are $\mathcal{A}^{(i)}$ and $\mathcal{B}^{(i+j)}$ where $i \in 1, \ldots, s-j$. The modified GC code $C_{GC'}$ has length $n_{GC'} = n_{GC}$ and cardinality $M_{GC'} = \prod_{i=1}^{s-j} M_a^{(i)}$. The minimum distance becomes

$$d_{GC'} = \min_{i=1,\ldots,s-j} \{d_a^i d_b^{j+i}\} \geq d_{GC}$$

The resulting code rate is smaller than the original code.

A GC code can also be modified by shortening or lengthening all of the inner or outer codes by the same number of coordinates. Through the extension by one coordinate, we add redundancy and obtain a code with the same cardinality, but larger minimum distance and length.

There are many more variations possible if we consider the use of different outer codes, increasing the set of codes $\mathcal{A}^{(i)}, i = 1, 2, \ldots, s$, and for different partitions. When using outer RS codes, the resulting modified code is very similar to the original code. This is due to the MDS property, which allows the cardinality to be easily changed while the decoding method is only slightly modified.

It is clear that we cannot present all possible modification methods. The necessary changes to the inner and outer codes are determined by the particular application. We next give an examples illustrating the above two techniques.

Example 9.12 (Modification of a GC code) In example 9.9, we performed an order-5 partition of the BCH code $\mathcal{B}^{(1)}(2; 31, 2^{25}, 4)$. In example 9.11, the GC code

$$C_{GC}(2; 961, 2^{470}, 96)$$

was constructed. This code is now modified.

In this example, $s = 5$, and $\mu_i = 5, i = 1, \ldots, 5$. The partitioning is determined according to the parameters j_0 and j as defined above. We set $j_0 = 1$ and $j = 2$ from equation 9.4, which is equivalent to shortening removing partitions 4 and 5. This shortens the inner code by 10 information coordinates. We use the labels $\underline{a}^{(i)}$, $i = 1, 2, 3$, which corresponds to the first three outer codes:

$$\mathcal{A}^{(i)}(2^5; 31, k_a^{(i)} = 31 - d_a^{(i)} + 1, d_a^{(i)}), \quad i = 1, 2, 3.$$

These codes have minimum distances

$$d_a^{(1)} = 24, \quad d_a^{(2)} = 16, \quad d_a^{(3)} = 12\,.$$

Now we use method 3 from section 9.2.2 to partition the inner code. We obtain the GC code $C_{GC'}$:

$$C_{GC'}(2;651,2^{220},96)\ .$$

Compared with the original code we have a shorter code with the same minimum distance; however, the rate is lower.

We next modify the same GC code in order to improve the minimum distance of the GC code. We use the inner codes $\mathcal{B}^{(l)} \in GF(2)^{31}$, $l = 3,4,5$, and the same outer codes:

$$\mathcal{A}^{(l)}(2^5;31,k_a^{(i)} = 31 - d_a^{(l)} + 1, d_a^{(l)}),\quad i = 1,2,3\ .$$

This equivalent to setting $j_0 = 1$ and $j = 2$ from equation 9.4. Instead of shortening the inner codes by two partition levels (or 10 binary coordinates) as in the above example, we use these coordinates to increase the redundancy of the inner codes. The fact that all $\mu_i = 5$ allows us to use the three most powerful inner codes. We get the GC code with parameters

$$C_{GC'}(2;961,2^{220},192)\ .$$

The length and cardinality of $C_{GC'}$ are easily calculated. The inner codes have minimum distances $d_b^{(3)} = 8$, $d_b^{(4)} = 12$ and $d_b^{(5)} = 16$, which give the minimum distance

$$d_{GC'} \geq \min\{8\cdot 24, 12\cdot 16, 16\cdot 12\} = 192\ . \qquad \diamond$$

Comparison of classical concatenated and GC interleaved codes

Burst errors often occur in telecommunication systems. Therefore *interleaving* is used in order to transform burst errors into single errors (see section 8.4.4). Interleaving can be thought of as the permutation of the coordinates of a codeword in time. This is done implicitly for concatenated codes through their construction. The outer codes are encoded and decoded as column vectors in $GF(q^{\mu_i})$, $i \in 1,\ldots,s$, but transmitted over the channel as row vectors in $GF(q)$. Permutation or scrambler matrices (corresponding to a rate-1 code) can also be used for interleaving (see e.g. section 9.3.2). For convolutional codes, interleaving is the only possibility for improving the correction properties in the case of burst errors. For block codes, the problem is different. The codeword of a block code can be decoded correctly if a *sufficient number* of error-free coordinates can be ensured. Based on these coordinates, the *corrupted* coordinates can be corrected. This property motivates the use of long block codes. On the other hand, concatenated codes are constructed such that if a burst error completely corrupts some of the inner codewords, the remaining codewords are error-free. Therefore block GC codes have a particular advantage in burst error channels (see section 9.2.4).

We consider two methods of interleaving classical concatenated codes in order to protect against burst errors. The first method has been presented in [DC87], [HEK87, KL87] and [KTL86]. We consider linear inner and outer codes with parameters $\mathcal{B}^{(1)}(q;n_b,k_b,d_b)$ and $\mathcal{A}^{(1)}(q^{\mu};n_a,k_a,d_a)$ respectively. The $k_a\mu$ information symbols are first encoded by the outer code, and the codeword is divided columnwise to form a $u \times k_b$ matrix, where u is some positive integer. This is shown in equation 9.5, where each matrix element is a row vector in

$GF(q)^{\mu}$:

$$\underbrace{\begin{pmatrix} \underline{a}_1 & \underline{a}_{u+1} & \cdots & \underline{a}_{n_a-u+1} \\ \underline{a}_2 & \underline{a}_{u+2} & \cdots & \underline{a}_{n_a-u+2} \\ \vdots & \vdots & \ddots & \vdots \\ \underline{a}_u & \underline{a}_{2\cdot u} & \cdots & \underline{a}_{n_a} \end{pmatrix}}_{k_b} . \tag{9.5}$$

The outer code parameters must fulfill the requirement that $\mu u = k_b$ and that $n_a = 0 \bmod u$. Each row of this matrix is encoded by the outer code $\mathcal{B}^{(1)}$ to give the final interleaved classical concatenated codeword, which is written as a $u \times n_b$ matrix.

A second method of interleaving of classical concatenated codes is presented in [Bla, p. 116]. Again we consider inner and outer codes with parameters $\mathcal{B}^{(1)}(q;n_b,k_b,d_b)$ and $\mathcal{A}^{(1)}(q^{\mu};n_a,k_a,d_a)$ respectively. A total of uk_a information symbols from $GF(q)$ are encoded to produce u column vectors of the code $\mathcal{A}^{(1)}$. The chosen parameters must satisfy the condition $u\mu = k_b$, where u is usually defined as the interleaving of depth. The outer code matrix is shown in equation 9.6, with elements from $GF(q)^{\mu}$:

$$\underbrace{\begin{pmatrix} \underline{a}_1^{(1)} & \underline{a}_1^{(2)} & \cdots & \underline{a}_1^{(u)} \\ \underline{a}_2^{(1)} & \underline{a}_2^{(2)} & \cdots & \underline{a}_2^{(u)} \\ \vdots & \vdots & \ddots & \vdots \\ \underline{a}_{n_a}^{(1)} & \underline{a}_{n_a}^{(2)} & \cdots & \underline{a}_{n_a}^{(u)} \end{pmatrix}}_{k_b} . \tag{9.6}$$

We have already seen this form of matrix in equation 9.1 for GC codes.

In contrast, a GC code has a partitioning of the inner code. This allows us the flexibility of increasing the code rate while keeping the minimum distance constant, or vice versa.

What is the gain that we can achieve with a GC code over interleaving of a classical concatenated code? Let C_u be a classically interleaved code according to the second method presented. This code has the length $n_u = n_a n_b$, the cardinality $M_u = (q^{\mu})^{uk_a}$ and minimum distance $d_u = d_a d_b$ (corresponding to theorem 9.1 on GC codes). The following theorem gives a bound on the gain that can be achieved by using GC codes instead of C_u under the condition that the minimum distance remains constant:

Theorem 9.6 (Improvement of the code rate) *Let R_a and R_b be the code rates of the inner code $\mathcal{B}$ and the outer RS code $\mathcal{A}$ that make up the code C_u. ΔR is the difference of the code rates of the code C_u and a GC code with the same length n and minimum distance. The code rate gain of the GC code is given by*

$$\frac{d_a}{n}\left(u-2+\frac{1}{2^{u-1}}\right) \leq \Delta R < (1-R_a)R_b .$$

Proof The left inequality comes from the assumption that the minimum distance is increased by at least 2 for each inner code partition. If the product of the minimum distances of the inner and outer code is kept constant then we may increase the number of information bits of the outer code (i.e. reduce the outer code redundancy). This allows additional outer code symbols to be used for information for each partition level. The increase in the code rate is at least

$$\frac{1}{n}\sum_{i=2}^{u}\left(d_a - \frac{d_a}{2^{i-1}}\right) = \frac{d_a}{n}\left(u-2+\frac{1}{2^{u-1}}\right).$$

The right inequality is obtained based on the relationship that $R_{GC} < R_b$. The upper bound on the gain is obtained by subtracting $R_a R_b$ from both sides of the inequality. □

Theorem 9.6 shows that the attained gain with the GC code is small if the interleaved code is constructed with a low rate inner code and a high rate outer code We clarify this fact using an example.

Example 9.13 (Improvement of the interleaved code rate) We choose an example from [HEK87]. Here we investigate the concatenation of RS and BCH codes for satellite channels with crosstalk. In this publication, the conclusion is reached that, due to the large amount of corruption caused by burst errors, only concatenated codes are applicable for the channel. A $\mathcal{B}(2;31,21,5)$ BCH code and an $\mathcal{A}(2^5;31,23,9)$ RS code are used for the inner and outer codes respectively. This gives a classical concatenated code C_u with code rate $R_a R_b = 0.5$ and minimum distance 45.

We proceed by comparing these parameters with other possible GC codes. First, we hold the minimum distance constant. The partitioning of the BCH code $\mathcal{B}^{(1)}(2;31,20,6)$ gives (see example 9.9)

$$\mathcal{B}^{(4)}(2;31,5,16) \subset \mathcal{B}^{(3)}(2;31,10,12) \subset \mathcal{B}^{(2)}(2;31,15,8) \subset \mathcal{B}^{(1)}(2;31,20,6),$$

with the labels

$$\underline{a}^{(i)} \in GF(2)^5, \quad i = 1,2,3,4.$$

Only even-weight codewords are used, in order for the dimensions to match. Compare this with method 3 from section 9.2.2. For the outer code, we select the RS code with length 31 and minimum distances

$$d_a^{(1)} = 8, \quad d_a^{(2)} = 6, \quad d_a^{(3)} = 4, \quad d_a^{(4)} = 3.$$

We have constructed a GC code with minimum distance 48 and code rate 0.56. The code rate difference is $\Delta R = 0.06$. We verify this calculation based on theorem 9.6:

$$0.02 \le 0.06 < 0.17.$$

The number 0.06 appears small, but one observes that the number of additional information bits per codeword is $0.06 \cdot 961 \approx 57$, or 2^{57} times as many codewords.

Another possible GC code construction proceeds as follows. We partition the BCH code $\mathcal{B}^{(1)}(2;31,30,2)$ into 5 subcodes. We use the same outer RS codes as above, having length 31. This produces a GC code with minimum distance 48 and rate 0.666.

In example 9.9, we constructed a GC code with minimum distance 96 and rate 0.49. ◇

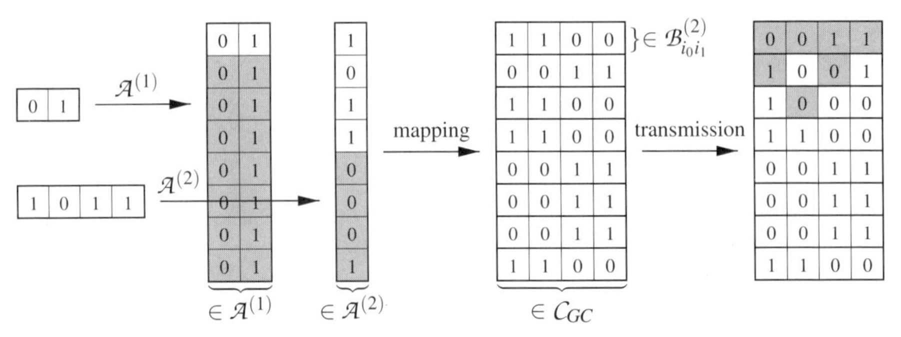

Figure 9.10 Encoding and transmission of the GC code from example 9.2.

9.2.4 Decoding of GC codes

The decoding of generalized concatenated codes is carried out in several steps, where the step number corresponds to index of the inner code and outer codes. In each decoding level i, the received word is row-wise decoded with respect to the inner subcode $\mathcal{B}^{(i)}$. Next, the decoding is carried out with respect to the outer code $\mathcal{A}^{(i)}$, whereby the coset of $\mathcal{B}^{(i+1)}$ is determined for the next coding step. This principle is carried out in the following example:

Example 9.14 (Decoding of generalized concatenated codes) A GC codeword from example 9.2 is corrupted during transmission. Seven errors occur as indicated in figure 9.10 in the last matrix

The decoding is carried out in two steps according to figure 9.11. First, all rows of the received GC codeword are decoded with respect to $\mathcal{B}^{(1)}$, and remapping according to the partitioning scheme is performed. If an error is identified in a row, the entire row is defined as an erasure for the decoding of the first outer code $\mathcal{A}^{(1)}$ (e.g. row 3 is identified as an error by the inner PC code $\mathcal{B}^{(1)}$). The outer code $\mathcal{A}^{(1)}$ decodes the corresponding columns with the help of the erasure information from the code $\mathcal{B}^{(1)}$. The decoding result from $\mathcal{A}^{(1)}$ determines the subcode $\mathcal{B}^{(2)}_{01}$ (see figure 9.4). The subcode is the basis for decoding of the received word at the next level; for example, all inner codewords must be in the subcode defined by the label $\mathcal{A}^{(1)}$. The codeword is then decoded according to the selected subcode $\mathcal{B}^{(2)}_{01}$, which has a larger minimum distance than $\mathcal{B}^{(1)}$ and can therefore correct more errors. In this example, the first outer code gives the partition label for the subcode (01). The second inner code $\mathcal{B}^{(2)}$ can correct the single error in row 3, and identifies the double error in row 2 of the GC codeword. This row is defined as an erasure for the decoding step $\mathcal{A}^{(2)}$. In the last step, column 3 is correctly decoded with $\mathcal{A}^{(2)}$. All seven transmission errors are corrected, which is equal to the half the minimum distance of the GC code:

$$\left\lfloor \frac{16-1}{2} \right\rfloor = 7.$$

◇

In this section, a modification of the decoding algorithm of Blokh and Zyablov is given for GC codes. The decoding is composed of two different algorithms. The algorithm for the first decoding step is called GCD-1, and the second algorithm is called GCD-i and is used in all subsequent decoding steps. GCD-1 is a generalization of the algorithm from [Zin81], where it is used for all decoding levels s. The change is necessary in order to allow the use of decoding

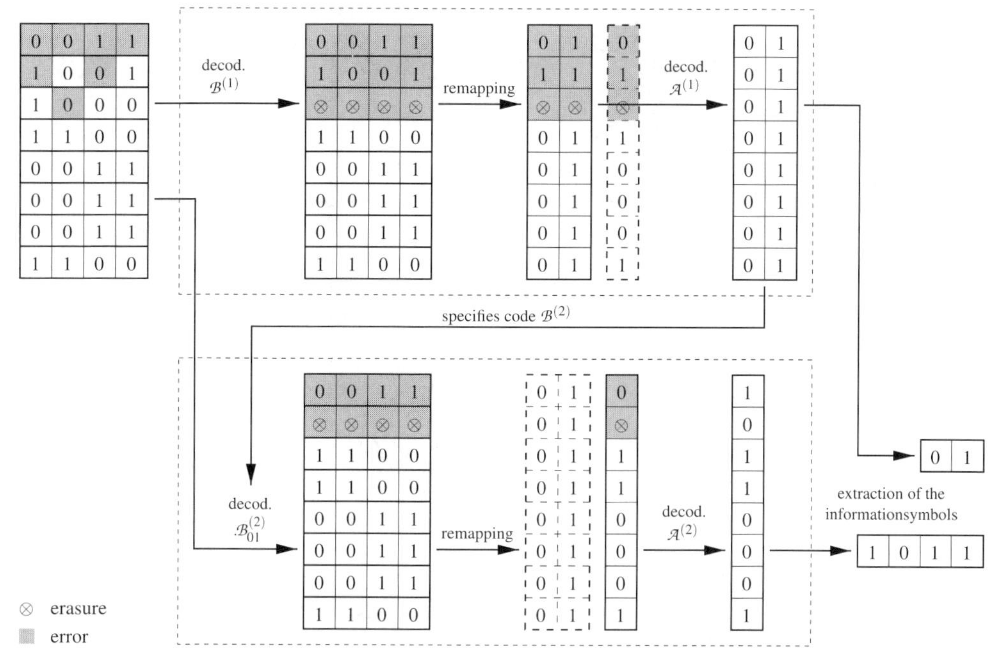

Figure 9.11 Decoding with the GCD algorithm.

algorithms that can decode errors with weight larger than half the minimum distance (e.g. ML decoding, which can use reliability information). The use of GCD-i in decoding levels > 1 incorporates results from the preceding step. We shall show that by using GCD-i, instead of the previously described algorithm, an improvement in the decoding capacity is achieved, while at the same time the complexity is reduced.

In this chapter, we limit ourselves to the investigation of linear inner and outer codes over a finite field. Modifications for nonlinear codes, or codes over spaces that use metrics other than the Hamming metric, are very difficult to work with, although they are in principle straightforward extensions.

Next, we define the notation that we shall be using in this section.

As described in section 9.2.1, the code $C(q;n,M,d)$ is a GC code with partitioning of order s of the inner code $\mathcal{B}^{(1)}(q;n_b,k_b^{(1)},d_b^{(1)})$ and the outer codes $\mathcal{A}^{(i)}(q^{\mu_i};n_a,k_a^{(i)},d_a^{(i)})$, $i=1,2,\ldots,s$. A GC codeword $\mathbf{c} \in C$ is composed of n_a codewords of the inner code $\mathbf{b}_j^{(1)} \in \mathcal{B}^{(1)}$, $j=1,2,\ldots,n_a$ with length n_b.

A codeword $\mathbf{c}$ is transmitted over a q-ary symmetric channel (see figure 7.2). Therefore every row $\mathbf{b}_j^{(1)}$ of the matrix $\mathbf{c}$ is corrupted by the channel error $\mathbf{e}_j \in GF(q)^{n_b}$. The received rows are $\mathbf{r}_j = \mathbf{b}_j^{(1)} + \mathbf{e}_j$, $j=1,2,\ldots,n_a$. The error vector $\mathbf{e}_j$ is defined as the channel error. Erasures are denoted by the symbol $\otimes$.

We now describe the decoding algorithm for the first step.

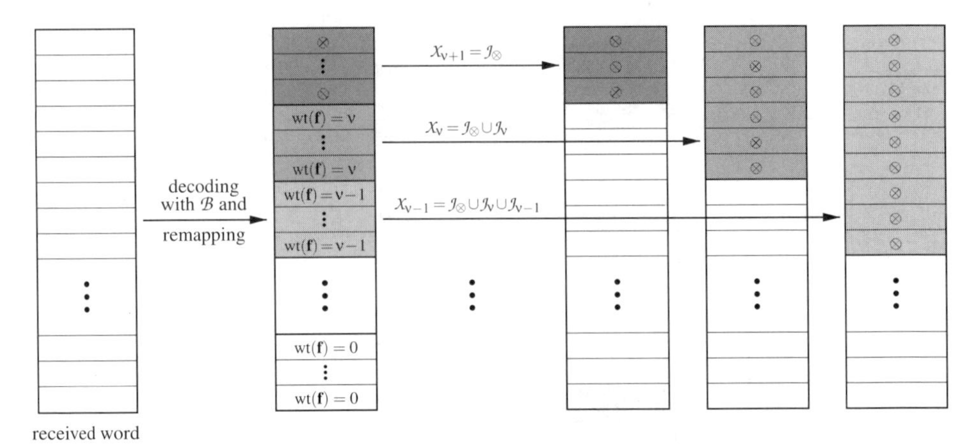

Figure 9.12 Erasure correction in the first GCD decoding step.

GCD-1: Decoding for the first step

Error correction decoding of the inner codes $\mathcal{B}^{(i)}, i = 1, \ldots, s$, is carried out based on any decoder that can correct up to half the minimum distance. Rows with the greatest number of identified errors are the most unreliable. In the case of a decoding failure, the entire row is defined as an erasure for the corresponding outer code decoding. This is done in much the same way as in section 7.4.3 for the GMD method. The outer code is repeatedly decoded, with each time more unreliable symbols being replaced by erasures (see figure 9.12). We stop when the number of erasures is less than the correction capability of the code. If a HD decoder is used for the inner code then the number of corrected inner code symbols of a codeword is given by the information reliability of the corresponding symbol of the outer code. The decoded labels of the outer code divides the inner codes into subsets with increasing minimum distance. It is clear that the concept of MAP decoding can be incorporated as well in order to compute the associated reliability information.

Each received row $\mathbf{r}_j^{(1)} = \mathbf{b}_j^{(1)} + \mathbf{e}_j$, $j = 1, 2, \ldots, n_a$, is decoded by the inner code $\mathcal{B}^{(1)}$ in the first step. The result of this decoding is

$$\hat{\mathbf{b}}_j^{(1)} = \begin{cases} \mathbf{b}_j^{(1)} + \mathbf{e}_j - \mathbf{f}_j, \\ \otimes. \end{cases} \tag{9.7}$$

An erasure is indicated if the decoder cannot find a solution.

From every $\hat{\mathbf{b}}_j^{(1)} \neq \otimes$, a unique symbol $\underline{\hat{a}}_j^{(1)} \in GF(q)^{\mu_1}$ of the first outer code $\mathcal{A}^{(1)}$ can be determined. Assuming that the symbol $\underline{a}_j^{(1)}$ was encoded at the transmitter, the following decoding results are possible for $\underline{\hat{a}}_j^{(1)}$:

For the case that $f_j = e_j$, $\underline{\hat{a}}_j^{(1)}$ is correctly decoded. If $\hat{b}_j = \otimes$, then $\underline{\hat{a}}_j^{(1)} = \otimes$ and we have a decoding failure. If $f_j = e_j$ then a decoding error in $\hat{a}_j^{(1)}$ is dependent on the error vector $\mathbf{f}$.

We divide the decoding result into different sets $\mathcal{J}_w$, depending on the number of the coordinates corrected by the inner code. w is the number of coordinates corrected.

Definition 9.7 (Corrected coordinates) *The n_a rows of the GC code with inner decoder $\mathcal{B}^{(1)}$ where w coordinates have been corrected is associated with the set $\mathcal{J}_w$. The set $\mathcal{J}_{\otimes}$ contains all* *rows that contain a decoding failure.*

$$\mathcal{J}_w := \{j \mid \mathrm{wt}(\mathbf{f}_j) = w, \quad j = 1,2,\ldots,n_a\}, \quad w = 0,1,2,\ldots,\nu;$$
$$\mathcal{J}_\otimes := \{j \mid \underline{\hat{a}}_j = \otimes, \quad j = 1,2,\ldots,n_a\} .$$

An important difference compared with the definition in [Zin81] is that the number of correctable errors ν is not bounded by $\lfloor \frac{d-1}{2} \rfloor$. The number ν depends on the decoder, and for the case of a ML decoder, it is larger than half the minimum distance.

We next determine the set $\mathcal{E}$, defined as the set of erroneously decoded rows by the inner code:

$$\mathcal{E} := \{j \mid \mathbf{f}_j \neq \mathbf{e}_j, \underline{\hat{a}}_j^{(1)} \neq \otimes, \quad j = 1,2,\ldots,n_a\} . \tag{9.8}$$

The set $\mathcal{J}_w$ can contain rows in which the number of corrected coordinates corresponds to the number of channel errors (correct decoding). However, it can also contain rows that are not correct. We want to indicate the number of erroneously decoded rows. This is the number of rows t_w that are erroneous in the set $\mathcal{J}_w$. The number of correctly decoded rows is s_w. This is formally written as

$$\begin{aligned} t_w &:= |\mathcal{J}_w \cap \mathcal{E}| , \\ s_w &:= |\mathcal{J}_w| - t_w , \end{aligned} \qquad w = 0,1,\ldots,\nu . \tag{9.9}$$

We define the union of the sets $\mathcal{J}_i$ by

$$\mathcal{X}_j := \mathcal{J}_\otimes \cup \left(\bigcup_{i=j}^{\nu} \mathcal{J}_i \right), \quad j = \nu, \nu-1, \ldots, 0 . \tag{9.10}$$

Finally, we define the integer v as the cardinality of the set $\mathcal{X}_v$, which is less than or equal to the minimum distance of the outer code:

$$v: \ |\mathcal{X}_v| < d_a^{(1)}, \ |\mathcal{X}_{v-1}| \geq d_a^{(1)} . \tag{9.11}$$

We can now calculate the weight of the channel error associated with the row elements of a set $\mathcal{J}_w$. The weight is exactly equal to w if the element is not included in the set $\mathcal{E}$; otherwise it is greater than or equal to $d_b^{(1)} - w$:

$$\forall\, j \in \mathcal{J}_w : \quad \begin{cases} \mathrm{wt}(\mathbf{e}_j) \geq \frac{d_b^{(1)}}{2} , & j \in \mathcal{J}_\otimes \\ \mathrm{wt}(\mathbf{e}_j) \geq \max\left\{ d_b^{(1)} - w, \frac{d_b^{(1)}}{2} \right\}, & j \in \mathcal{E} \\ \mathrm{wt}(\mathbf{e}_j) = w , & j \notin \mathcal{E} . \end{cases} \tag{9.12}$$

Algorithm 9.1 lists the necessary steps for GCD-1. It remains to be proved that this algorithm can correct codes up to half the minimum distance. To do this, we need an estimation of the number of channel errors.

Note In principle, the subsequent decoding steps can be carried out with this algorithm. This means that the reader who is only interested in understanding the principles of GC decoding need not focus on the following proof and modification for further steps (algorithm GCD-i). We return to the general concepts of concatenated codes after example 9.2.4.

Algorithm 9.1 The GCD-1 decoding algorithm.

Received: $\mathbf{r}_j,\ j = 1,2,\dots,n_a$.

Result of the decoding of $\mathbf{r}_j$ is $\hat{b}_j = \begin{cases} b_j^{(1)} + \mathbf{e}_j - \mathbf{f}_j\,, \\ \otimes\,. \end{cases}$

Determine: $\hat{\mathbf{b}}_j \neq \otimes \longrightarrow \underline{\hat{a}}_j,\ j = 1,2,\dots,n_a,$
$v\,,$ (equation 9.11)
$\mathcal{J}_w,\ w = 0,1,2,\dots,\nu$ and $\mathcal{J}_\otimes\,,$ (definition 9.7)
$X_j,\ j = \nu,\nu-1,\dots,v\,,$ (equation 9.10)

Initialization: $j = \nu + 1,\ X = \infty\,.$

Step 1: If $j < v$ then step 5 .

Step 2: $\mathbf{z}_i = \begin{cases} \underline{\hat{a}}_l^{(1)}, & l \notin X_j\,, \\ \otimes, & l \in X_j\,, \end{cases} \qquad l = 1,2,\dots,n_a\,.$

Step 3: decode the rows $(\mathbf{z}_1,\mathbf{z}_2,\dots,\mathbf{z}_{n_a})^T$ with the decoder for code $\mathcal{A}^{(1)}$ to $(\underline{\tilde{a}}_1,\underline{\tilde{a}}_2,\dots,\underline{\tilde{a}}_{n_a})^T$.

Step 4: The set of rows $\mathcal{T}_j$ corrected by the outer code, is $\mathcal{T}_j := \{l \mid \underline{\hat{a}}_l \neq \underline{\tilde{a}}_l, \underline{\hat{a}}_l \neq \otimes, \quad l = 1,2,\dots,n_a\}$.
For a decoding failure of the outer code, then: $j = j-1$, step 1.

Calculation of the number of errors Γ_j: $\quad \Gamma_j := \sum_{i=1}^{n_a} \gamma_i,$

with: $\gamma_l = \begin{cases} \frac{d_b^{(1)}}{2}, & l \in \mathcal{J}_\otimes\,, \\ w, & l \in \mathcal{J}_w \backslash \mathcal{T}_j\,, \quad w = 0,1,\dots,\nu\,, \\ \max\left\{d_b^{(1)} - w \frac{d_b^{(1)}}{2}\right\}, & l \in \mathcal{J}_w \cap \mathcal{T}_j\,, \end{cases}$

If $\Gamma_j < X$ then $X = \Gamma_j$, $\underline{a}^* = \underline{\tilde{a}}$, $\mathcal{L} = \mathcal{T}_j$.

If $X < \frac{1}{2} d_b^{(1)} \left(d_a^{(1)} - |\mathcal{J}_\otimes|\right)$ then step 5 .

$j := j-1$, step 1 .

Step 5: The decoding decision is $\mathbf{a}^*$ (X and $\mathcal{L}$ are necessary for further decoding steps). Decoding errors can occur.

We derive two estimations for the number of channel errors from the set $\mathcal{J}_w$ using definition 9.7. The first estimation in theorem 9.8 is a lower bound that estimates the number of channel errors from two elements of the same or different sets $\mathcal{J}_i$. The second estimation in theorem 9.9 likewise describe a lower bound that estimates the number of channel errors within a set $\mathcal{J}_i$.

Theorem 9.8 (Channel errors in two rows) *Let $m \in \mathcal{J}_w$, $j \in \mathcal{J}_l$ and $m \in \mathcal{E}$, $j \notin \mathcal{E}$; then the weight of the channel error is*

$$\forall l \geq w: \qquad \mathrm{wt}(\mathbf{e}_m) + \mathrm{wt}(\mathbf{e}_j) \geq d_b^{(1)}\,.$$

Proof From equation 9.12, we know that $\mathrm{wt}(\mathbf{e}_m) \geq d_b^{(1)} - w$, $\mathrm{wt}(\mathbf{e}_j) = l$. This gives directly $\mathrm{wt}(\mathbf{e}_m) + \mathrm{wt}(\mathbf{e}_j) \geq d_b^{(1)} - w + l \geq d_b^{(1)}$ for $l \geq w$. □

Theorem 9.9 (Channel errors in set $\mathcal{J}_w$) *The number of channel errors in the set $\mathcal{J}_w$ can be* *estimated by means of the parameters s_w and t_w:*

$$\forall w \in \{0,\dots,\nu\}: \quad \sum_{j\in\mathcal{J}_w} \mathrm{wt}(\mathbf{e}_j) \geq \begin{cases} |\mathcal{J}_w|\frac{d_b^{(1)}}{2} & \textit{for} \quad s_w \leq t_w, \\ t_w d_b^{(1)} + (s_w - t_w)w & \textit{for} \quad s_w > t_w \,. \end{cases}$$

Proof First we estimate the channel error corresponding to one element. We must assume that a decoder is used that can decode errors with a weight larger than half the minimum distance. We obtain

$$\forall j \in \mathcal{J}_w: \quad \begin{cases} \mathrm{wt}(\mathbf{e}_j) \geq \max\left\{\frac{d_b^{(1)}}{2}, d_b^{(1)} - w\right\} & \text{for} \quad j \in \mathcal{E} \\ \mathrm{wt}(\mathbf{e}_j) = w & \text{for} \quad j \notin \mathcal{E}\,. \end{cases}$$

From theorem 9.8, we have the relation

$$\mathrm{wt}(\mathbf{e}_j) + \mathrm{wt}(\mathbf{e}_l) \geq d_b^{(1)} \qquad \text{for} \quad l,j \in \mathcal{J}_w, l \in \mathcal{E}, j \notin \mathcal{E}\,. \tag{9.13}$$

If $s_w \leq t_w$ then there are s_w pairs for which equation 9.13 can be used. The remaining $t_w - s_w$ elements correspond to decoding errors, and the weight of the channel error is at least half the minimum distance of the inner code. If, however, $s_w > t_w$ then there are t_w pairs for which equation 9.13 can be used. In this case, the remaining $t_w - s_w$ elements are correctly decoded, and the weight of the channel error is therefore equal to w:

$$s_w \leq t_w: \qquad \sum_{j\in\mathcal{J}_w} \mathrm{wt}(\mathbf{e}_j) \geq s_w d_b^{(1)} + (t_w - s_w)\frac{d_b^{(1)}}{2} = |\mathcal{J}_w|\frac{d_b^{(1)}}{2}$$

$$s_w > t_w: \qquad \sum_{j\in\mathcal{J}_w} \mathrm{wt}(\mathbf{e}_j) \geq t_w d_b^{(1)} + (s_w - t_w)w\,. \qquad \square$$

To prove that the algorithm GCD-1 from figure 9.1 can correct all errors with weight smaller than half the minimum distance, we use nearly the same method as presented in [Zin81]. We first prove that each correctable error can be correctly decoded, and then that no two values for Γ_j (i.e. different error weights for one decoding result) can exist that are both smaller then $\frac{1}{2}d_a^{(1)}d_b^{(1)}$.

Theorem 9.10 (Correction capabilities of GCD-1) *If fewer than half the minimum distance number of errors occur then the outer code can correct the error in some step of GCD-1. As long as*

$$\sum_{i=1}^{n_a} \mathrm{wt}(\mathbf{e}_i) < \frac{d_a^{(1)}d_b^{(1)}}{2},$$

there is one $j \in \{v, v+1, \dots, \nu+1\}$, giving

$$2\left(|\mathcal{E}| - \sum_{i=j}^{\nu} t_i\right) + |X_j| < d_a^{(1)}.$$

The set X_j, $\mathcal{E}$ and t_i are defined in equations 9.8, 9.9 and 9.10 respectively. $|\mathcal{E}| - \sum_{i=j}^{\nu} t_i$ is the number of errors that have not been declared as erasures.

Proof We assume that the contrary holds, namely that, for all j,

$$2\left(|\mathcal{E}| - \sum_{i=j}^{\nu} t_i\right) + |\mathcal{X}_j| \quad \geq \quad d_a^{(1)}. \tag{9.14}$$

With the relation $|\mathcal{X}_j| = \sum_{i=j}^{\nu}(s_i + t_i)$, we obtain

$$2|\mathcal{E}| + \sum_{i=j}^{\nu}(s_i - t_i) \geq d_a^{(1)} - |\mathcal{I}_{\otimes}|,$$

which is equivalent to

$$2\sum_{i=0}^{j-1} t_i + \sum_{i=j}^{\nu}(t_i + s_i) \geq d_a^{(1)} - |\mathcal{I}_{\otimes}| = d_a.$$

The estimation for the channel error gives

$$\sum_{i=0}^{\nu}\sum_{j\in\mathcal{I}_i} \mathrm{wt}(\mathbf{e}_j) \geq \sum_{i=0}^{\nu} t_i\,(d_b^{(1)} - i) + \sum_{i=1}^{\nu} s_i\, i,$$

where the errors that are not correctable are not included in the sum. The index of the right sum starts at 1 because $i = 0$ corresponds to a multiplication by zero. We want to represent the numbers s_i and t_i in a different way:

$$t_i = \sum_{l=i}^{\nu} t_l - \sum_{l=i+1}^{\nu} t_l \qquad \text{and} \qquad s_i = \sum_{l=i}^{\nu} s_l - \sum_{l=i+1}^{\nu} s_l.$$

We apply this representation, and obtain

$$\sum_{i=0}^{\nu}\sum_{j\in\mathcal{I}_i} \mathrm{wt}(\mathbf{e}_j) \geq \sum_{i=0}^{\nu}\left(\sum_{l=i}^{\nu} t_l - \sum_{l=i+1}^{\nu} t_l\right)(d_b^{(1)} - i) + \sum_{i=1}^{\nu}\left(\sum_{l=i}^{\nu} s_l - \sum_{l=i+1}^{\nu} s_l\right) i.$$

After some calculation, we get

$$\sum_{i=0}^{\nu}\sum_{j\in\mathcal{I}_i} \mathrm{wt}(\mathbf{e}_j) \geq \sum_{i=0}^{\nu}\left(2\sum_{l=0}^{j-1} t_l + \sum_{l=j}^{\nu}(t_l + s_l)\right).$$

For the right side, we use the assumption 9.14:

$$\sum_{i=0}^{\nu}\sum_{j\in\mathcal{I}_i} \mathrm{wt}(\mathbf{e}_j) \geq \frac{d_a d_b^{(1)}}{2},$$

or

$$\sum_{i=1}^{n_a} \mathrm{wt}(\mathbf{e}_i) \geq \sum_{i=0}^{\nu}\sum_{j\in\mathcal{I}_i} \mathrm{wt}(\mathbf{e}_j) + |\mathcal{I}_{\otimes}|\frac{d_b^{(1)}}{2} \geq \frac{d_a d_b^{(1)}}{2} + |\mathcal{I}_{\otimes}|\frac{d_b^{(1)}}{2} = \frac{d_a^{(1)} d_b^{(1)}}{2}.$$

This is a contradiction to the assumption. □

Theorem 9.11 (Uniqueness of the solution of GCD-1) *If* $\Gamma_j \neq \Gamma_{j'}$ *for* $j \neq j'$ *and* $\Gamma_j < \frac{1}{2}d_a^{(1)}d_b^{(1)}$ *then* $\Gamma_{j'} > \frac{1}{2}d_a^{(1)}d_b^{(1)}$.

Proof We consider the following cases:

(1) $l \in \mathcal{J}_\otimes$: then $\gamma_l = \gamma'_l = d_b^{(1)}/2$;

(2) $l \in \mathcal{T}_j \cap \mathcal{J}_w$ and $l \notin \mathcal{T}_{j'} \cap \mathcal{J}_w$: then $\gamma_l = d_b^{(1)} - w, \gamma'_l = w$;

(3) $l \notin \mathcal{T}_j \cap \mathcal{J}_w$ and $l \in \mathcal{T}_{j'} \cap \mathcal{J}_w$: then $\gamma_l = w, \gamma'_l = d_b^{(1)} - w$;

(4) $l \notin \mathcal{T}_j \cap \mathcal{J}_w$ and $l \notin \mathcal{T}_{j'} \cap \mathcal{J}_w$: then $\gamma_l = \gamma'_l = w$.

For cases 1 - 3, $\gamma_l + \gamma'_l \geq d_b^{(1)}$ holds. With the relation $|\mathcal{T}_j \cap \mathcal{T}_{j'}| \geq d_a^{(1)} - |\mathcal{J}_\otimes|$, we obtain

$$\Gamma_j + \Gamma_{j'} = \sum_{l=1}^{n_a} \gamma_l + \gamma_{l'} \geq \left(d_a^{(1)} - |\mathcal{J}_\otimes|\right) d_b^{(1)} + |\mathcal{J}_\otimes| d_b^{(1)} = d_a^{(1)} d_b^{(1)} .$$

□

With theorems 9.10 and 9.11, we have proved that GCD-1 can decode all channel errors with weight smaller than half the minimum distance. However, GCD-1 can also correct many error patterns with a larger weight (e.g. in the case of burst errors).

In the algorithm from [Zin81] and [Eri86], $\nu = d_b^{(1)}/2$ is fixed. This means that if the inner code receives a vector with an error weight larger than half the minimum distance, then this row is considered as a decoding failure. Clearly, this is a limitation of the decoding capability. The algorithm in [Zin81] works only for a subset of cases that are corrected by the GCD-1 algorithm. In the GCD-1 algorithm, ν is greater than or equal to $d_b^{(1)}/2$ (see definition 9.7). Therefore, from step $j = d_b^{(1)}/2$ on, the GCD-1 algorithm is equivalent to the algorithm in [Zin81].

GCD-i: decoding for subsequent steps

From a previous decoding step, we know the set $\mathcal{J}_\otimes$, $\mathcal{J}_w$, $w = 0, 1, \ldots, \nu$, and the set $\mathcal{L}$ consists of the indices of all rows in the outer decoder that are identified as having decoding errors by the outer code. Before the next step, we determine the corresponding subcode of the inner code according to the received outer code row elements $\underline{a}_j^{(i)}$. We now decode the received row elements $\mathbf{r}_j^{(i)}$ with respect to the corresponding subcode. This method has been described previously in section 9.2.2. However, it is not necessary to make a new decoding attempt for every row. Each row that is an element of the set $\mathcal{L}$ corresponds to a decoding error of the previous inner decoder. There is a case when a decoding error of the inner code can be predicted:

Theorem 9.12 (Decoding capability of $\mathcal{B}^{(i)}$) *$d_b^{(i)} = d_b^{(i-1)} + \Delta$ is the minimum distance of the code $\mathcal{B}^{(i)}$. The decoding of row $\mathbf{r}_j^{(i)}$ can lead to a decoding error if the following situation occurs:*

$$j \in \mathcal{J}_w \cap \mathcal{L},\ w < d_b^{(i)} - \Delta - \left\lfloor \frac{d_b^{(i)} - 1}{2} \right\rfloor .$$

Proof From the decoding of the previous section, we know

$$\forall j \in \mathcal{J}_w \cap \mathcal{L} : \quad \mathrm{wt}(\mathbf{e}_j) \geq \max\left\{ d_b^{(i-1)} - w, \frac{d_b^{(i-1)}}{2} \right\} .$$

Therefore, in this step a correct decoding result can only occur if $d_b^{(i-1)} - w \leq \left\lfloor \frac{d_b^{(i)}-1}{2} \right\rfloor$. From this, it follows directly that

$$w \geq d_b^{(i)} - \Delta - \left\lfloor \frac{d_b^{(i)} - 1}{2} \right\rfloor .$$

This is a contradiction to the assumption. □

Theorem 9.12 has two consequences for decoding:

1. Rows should be defined as an erasure if the possibility of a decoding error of the inner decoder can be predicted according to theorem 9.12.

2. The decoding result of the preceeding step should satisfy the prediction according to theorem 9.12; otherwise the corresponding row should be declared as an erasure.

One row can be incorrectly decoded but the symbol of the corresponding outer code can nevertheless be correct; therefore a decoding error of the inner code would not be recognized by the outer decoder. One can assume [Zin81] that each row in this step must be decoded again using the decoder of the corresponding subcode. In the following theorem, we shall show that only rows that have decoding errors of the inner decoder that are identified by the previous outer code $i-1$ need be decoded in step i.

Theorem 9.13 (Identical solution according to $\mathcal{B}^{(i)}$ and $\mathcal{B}^{(i-1)}$) *$\mathcal{L}$ is the set of errors of the inner decoder that have been discovered by the outer decoder. The results of the inner decoding $\mathcal{B}^{(i-1)}$ of the previous step are $\mathbf{f}_j$, $j = 1, 2, \ldots, n_a$. The decoder of the inner code $\mathcal{B}^{(i)}$ delivers the decoding results $\mathbf{h}_j$, $j = 1, 2, \ldots, n_a$:*

$$j \notin \mathcal{L} \quad \Longrightarrow \quad \mathbf{h}_j = \mathbf{f}_j \,, \qquad \textit{and} \qquad j \in \mathcal{L} \quad \Longrightarrow \quad \mathbf{h}_j \neq \mathbf{f}_j \,.$$

Proof According to theorem 9.5, if the channel error $\mathbf{e}_j$ was decoded by $\mathbf{f}_j$ and is not detected by the outer decoder then $\mathbf{e}_j - \mathbf{f}_j \in \mathcal{B}^{(i)}$, where $\mathcal{B}^{(i)} \subset \mathcal{B}^{(i-1)}$.

For $j \in \mathcal{L}$: $\mathbf{e}_j - \mathbf{f}_j \in \mathcal{B}^{(i-1)}$, but $\mathbf{e}_j - \mathbf{f}_j \notin \mathcal{B}^{(i)}$ and $\mathbf{e}_j - \mathbf{h}_j \in \mathcal{B}^{(i)}$; therefore $\mathbf{h}_j \neq \mathbf{f}_j$.

For $j \notin \mathcal{L}$: either $\mathbf{e}_j = \mathbf{f}_j = \mathbf{h}_j$ or $\mathbf{e}_j - \mathbf{f}_j \in \mathcal{B}^{(i-1)}$.

For the case $j \notin \mathcal{L}$, the decoding is either correct or the the decoder of the code $\mathcal{B}^{(i-1)}$ produces the same decoding result as code $\mathcal{B}^{(i)}$. Therefore $\mathbf{h}_j = \mathbf{f}_j$. □

Theorem 9.13 shows that in the ith step, only the rows from the set $\mathcal{L}$ need again be decoded at the next level, as well as the decoding failures from the preceeding step. This means that at most $d_a^{(i-1)}$ rows in step i need be decoded.

If the inner decoder can correct over half the minimum distance then we must modify theorem 9.12. Corresponding to the decoder of $\mathcal{B}^{(i-1)}$ and the subcode $\mathcal{B}^{(i)}$, we define

$$\lambda = d_b^{(i-1)} - \left\lfloor \frac{d_b^{(i)}}{2} \right\rfloor - \text{constant} .$$

For a BMD decoding method, the constant is equal to zero.

Algorithm 9.2 shows the required steps for GCD-i. Therefore, the results of the previous steps for $\mathcal{J}'_{\otimes}$, $\mathcal{J}'_w$, $w = 0, 1, \ldots, \nu'$, and $\mathcal{L}'$ are assumed to be known. The determination of the

Algorithm 9.2 The GCD-i decoding algorithm.

Decode: $\mathbf{r}_j,\ j \in \{\mathcal{I}'_{\otimes} \cup \{\mathcal{I}'_w \cap \mathcal{L}'\} \mid w \geq \lambda\} = \mathcal{M}$;
the result is $\mathbf{f}_j$ or $\otimes$ and ν .

Determine: $\mathcal{I}_w = \mathcal{I}'_w \setminus \{\mathcal{I}'_w \cap \mathcal{M}\}$,
$\mathcal{I}_{\otimes} = \{j \in \{\mathcal{I}'_w \cap \mathcal{L}',\ w < \lambda\}\}$.

From decoding: for $j \in \mathcal{M}$,
$\mathcal{I}_w = \mathcal{I}_w \cup \{j,\ \text{wt}(\mathbf{f}_j) = w \geq d_b^{(i-1)} - v,\ v : j \in \mathcal{I}'_v\}$,
$\mathcal{I}_{\otimes} = \mathcal{I}_{\otimes} \cup \{j,\ \mathbf{r}_j \to \otimes,\ \text{wt}(\mathbf{f}_j) = w < d_b^{(i-1)} - v,\ v : j \in \mathcal{I}'_v\}$.

Initialization: $j = \nu + 1,\ X = n_a d_b^{(i)},\ X_{\nu+1} = \{\ \}$.

Step 1: if $j < v$ then step 5.

Step 2: $\mathbf{z}_j = \begin{cases} \hat{\underline{a}}_l^{(i)}, & l \notin X_j, \\ \otimes, & l \in X_j, \end{cases} \quad l = 1, 2, \ldots, n_a.$

Step 3: decode rows $(\mathbf{z}_1, \mathbf{z}_2, \ldots, \mathbf{z}_{n_a})$ with the decoder for the code $\mathcal{A}^{(i)}$ as $(\tilde{\underline{a}}_1, \tilde{\underline{a}}_2, \ldots, \tilde{\underline{a}}_{n_a})$.

Step 4: $\mathcal{T}_j := \{l \mid \hat{\underline{a}}_l \neq \tilde{\underline{a}}_l, \hat{\underline{a}}_l \neq \otimes, l = 1, 2, \ldots, n_a\}$.
Estimation of the number of errors Γ_j: $\Gamma_j := \sum_{i=1}^{n_a} \gamma_i$,
where $\gamma_l = \begin{cases} w, & l \notin \mathcal{I}_w \cap \mathcal{T}_j, \\ \max\left\{d_b^{(i)} - w, \frac{d_b^{(i)}}{2}\right\} & l \in \mathcal{I}_w \cap \mathcal{T}_j, \end{cases} \quad w = 0, 1, \ldots, \nu.$
If $\Gamma_j < X$ then $X = \Gamma_j$, $\underline{a}^* = \tilde{\underline{a}}$, $\mathcal{L} = \mathcal{T}_j$.
If $X < \frac{d_b^{(i)}}{2}\left(d_a^{(i)} - |\mathcal{I}_{\otimes}|\right)$ then step 5, otherwise $j := j - 1$, step 1.

Step 5: The decoding decision is $\underline{a}^*$, X, $\mathcal{L}$.

corresponding subcode is not given explicitly. However, we stress that not all rows need be decoded for the inner code.

The major difference between the GCD-i algorithm and the one in [Zin81] is that in the ith step, less than $d_a^{(i-1)}$ rows are decoded, which greatly reduces the complexity. In addition, the decoding errors can be predicted and immediately identified as erasures, which increases the reliability of the result.

Theorem 9.14 (Decoding over half the minimum distance) *Channel errors with weight larger than half the minimum distance can be decoded with the GCD-i algorithm in step i if and only if the set*

$$\{\mathcal{L}' \cap \mathcal{I}'_w, w < \lambda\}, \quad \lambda \leq d_b^{(i-1)} - \left\lfloor \frac{d_b^{(i)} - 1}{2} \right\rfloor,$$

is not empty.

Proof Let $\xi = |\{\mathcal{L}' \cap \mathcal{I}'_w, w < \lambda\}|$ be the cardinality of the subset of the discovered decoding errors and the set of the corrected errors of weight w. The distances of the inner and outer codes are $d_b^{(i)}$ and $d_a^{(i)}$. All channel errors with weight smaller than $\frac{1}{2} d_b^{(i)} d_a^{(i)}$ can be decoded. Based on the number of erasures ξ, the channel error is at least

$$\sum_{j \in \mathcal{I}'_w \cap \mathcal{L}', w < \lambda} \text{wt}(e_j) \geq \sum_{j \in \mathcal{I}'_w \cap \mathcal{L}', w < \lambda} \max\{d_b^{(i)} - w, d_b^{(i)}/2\}\ .$$

Because, however, $\lambda \leq d_b^{(i-1)} - \left\lfloor \frac{d_b^{(i)}-1}{2} \right\rfloor$, we obtain:

$$\frac{d_a^{(i)} d_b^{(i)}}{2} - 1 + \sum_{j \in \mathcal{I}'_w \cap \mathcal{L}', w < \lambda} d_b^{(i)} - w \geq \frac{d_a^{(i)} d_b^{(i)}}{2} .$$

□

Example 9.15 (Decoding over half the minimum distance) A simple example should clarify when an error pattern can be decoded with GCD-i, even though the error weight is larger than half the minimum distance. We assume that the first step produces an accurate decoding result, and that the following channel errors have occurred: $d_a^{(2)} - 1$ errors with weight $d_b^{(1)} - w > \left\lfloor \frac{1}{2}(d_b^{(2)} - 1) \right\rfloor$. All errors have been falsely decoded when an error vector of weight w occurs. Furthermore, an additional $d_a^{(1)} - d_a^{(2)}$ errors with weight $w + d_b^{(2)} - d_b^{(1)}$ occurred. The first step will be correctly decoded, because there are less than $d_a^{(1)}$ erroneous rows. GCD-i is also correct, because all corrected errors from the first step are identified as erasures in the second step. In contrast, the algorithm from [Zin81] cannot perform corrections if the weight $w + d_b^{(2)} - d_b^{(1)}$ of one of the $d_a^{(1)} - d_a^{(2)}$ errors is greater than or equal to the minimum of the $d_a^{(2)} - 1$ error weights $d_b^{(1)} - w$ of all decoding attempts. All of these $d_a^{(2)} - 1$ errors are falsely decoded, since a BMD decoder is assumed. ◇

Burst errors and single errors

We know that, for GC codes, burst errors and single errors can often be corrected even when the channel error weight is larger than half the minimum distance of the GC code. An exact analysis of the error correction capabilities of burst and single errors based on the parameters of a GC code is difficult. In the following, we consider some of the best known theorems and results on the topic of single and burst error correction; however, simulations are unfortunately the best means for determining the performance of GC codes.

We proceed by citing the most important theorems from [ZZ79b] without proof.

The first theorem estimates the length of the longest correctable burst error under the assumption that no other errors have occurred.

Theorem 9.15 (Correction of a burst error) *GCD can correct every burst error with length*

$$\beta^{(i)} = \begin{cases} \frac{n_b}{2}(d_a^{(i)} - 3) + d_b^{(i)} + \left\lfloor \frac{d_b^{(i)}-1}{2} \right\rfloor, & d_a^{(i)} \text{ odd,} \\ \frac{n_b}{2}(d_a^{(i)} - 4) + 2d_b^{(i)} - 1, & d_a^{(i)} \text{ even,} \end{cases}$$

in the i-th step. Therefore the correction capability of a single error burst for the case of a GC code with a partitioning of order s of the inner code is

$$\beta = \min_{i=1,\dots,s} \{\beta^{(i)}\} .$$

The correction capability of a single error burst depends on the smallest minimum distance $d_a^{(s)}$ of the outer code. If this distance is increased then the correction capabilities of the code is also increased.

For the case of error bursts and independent errors, the following theorem is proved in [ZZ79b].

Theorem 9.16 (Correction of multiple bundle errors) *With the GCD, ν error bursts with length smaller than β_j, $j = 1,2,\ldots,\nu$, and additional t single errors can be corrected if*

$$\beta_j \leq (z_j - 1)n_b + d_b^{(1)} \quad \text{and} \quad 2t < \min_{i=1,2,\ldots,s} \left\{ (d_a^{(i)} - 2\kappa) d_b^{(i)} \right\},$$

where

$$\sum_{j=1}^{\nu} z_j = \kappa \leq \min_{i=1,2,\ldots,s} \left\{ \frac{d_a^{(i)} - 1}{2} \right\}.$$

It should again be stated that the algorithm need not be modified for the correction of burst errors and single errors.

Example 9.16 (Burst error correction of $C_{GC}(2;1024,455,128)$) In [ZZ79b], the following impressive result is given:

The inner code is a binary parity check code $\mathcal{B}^{(1)}(2;8,7,2)$, and the outer codes is an extended RS code $\mathcal{A}^{(1)}(2^7;128,65,64)$. The GC code has the parameters $C(2;1024,455,128)$. This code has the following correction properties:

- all single error bursts up to length 243;
- two error bursts with lengths $\beta_1 \leq 90$ and $\beta_2 \leq 50$ and all single errors with weight ≤ 25;
- four error bursts with lengths ≤ 50 and all single errors with weight ≤ 7.

There are also many other possible combinations. ◇

Decoding complexity

The decoding complexity of a block code depends on the length and the minimum distance. Consider a GC code $C(q;n,M,d)$ based on an order-s inner code partitioning of $\mathcal{B}^{(1)}$ with length n_b The s outer codes $\mathcal{A}^{(i)}, i = 1,2,\ldots,s$ have length n_a. The overall code length $n = n_a n_b$ and the minimum distance $d \geq \min_{i=1,\ldots,s}\{d_a^{(i)} d_b^{(i)}\}$. The decoding complexity of the GC code depends on the decoding complexity of the inner and outer codes. Although s outer codes must be decoded, their length and minimum distance are smaller by factors n_b and $d_b^{(i)}$ respectively. For the s inner codes, the length and minimum distance are smaller by factors n_a and $d_a^{(i)}$ respectively. Furthermore, smaller fields can be used for the calculations, because, in general, the length of the code also determines the field. Section 9.2.4 shows that in the ith step, at most $d_a^{(i-1)}$ codewords of the inner code $\mathcal{B}^{(i)}_{i_1,\ldots,i_{i-1}}$ must be decoded. This results in a major decrease in the decoding complexity, because the decoding complexity increases with length and minimum distance.

Unfortunately, no generalized complexity analysis for GC codes exists. Such an analysis would depend on the specific decoding method and code employed.

Further examples of good GC codes are presented in [Zin76]. Very simple inner codes are used for the GC construction with a maximum length of 16. The overall GC code length is up to 200. These codes belong to the list of the best known codes shown in tabular form [McWSl, p. 675]. The inner codes can be decoded using tables. This guarantees that an ML decoding algorithm can be used for decoding that has a low complexity and simple implementation.

However, note that the ML decoding of the inner and outer codes is not equivalent to the ML decoding of the entire GC code.

In order to get a feeling for the computational complexity, we compare a GC code from [Zin76] with a block code of the same length.

Example 9.17 (Decoding complexity) We compare the extended QR code $\mathrm{QR}(2;72,36,12)$ with the GC code $\mathcal{C}(2;72,41,12)$. The GC code is constructed from the inner codes

$$\mathcal{B}^{(1)}(2;8,7,2),\ \mathcal{B}^{(2)}(2;8,4,4) \quad \text{and} \quad \mathcal{B}^{(3)}(2;8,1,8)\ .$$

These codes are a single parity check code, a Hamming code and a repetition code. The outer codes are

$$\mathcal{A}^{(1)}(2^3;9,4,6),\ \mathcal{A}^{(2)}(2^3;9,7,3) \quad \text{and} \quad \mathcal{A}^{(3)}(2;9,8,2)\ .$$

These are two extended RS codes and a parity check code. The parameters of the GC codes are $n = 9 \cdot 8 = 72$, $d \geq \min\{12,12,16\}$ and $k = 3 \cdot 4 + 3 \cdot 7 + 8 = 41$.

GC decoding is relatively simple and flexible because all component codes are small. For the QR code, there are few practical known decoding methods. A possible algorithm is described in section 7.3.3 (see also [Boss87]). Furthermore, the QR code has a factor $2^5 = 32$ fewer codewords. ◇

Performance of the GCD algorithm

The error probability for block codes can be calculated when BMD decoding is used, as was described in section 1.4. For GC codes, this is not the case due to the fact that many error patterns with weight larger than half the minimum distance can be corrected (see section 9.2.4). The exact decoding error probability can only be determined by simulations. In this section, we make several observations for the estimation of the error probability of GC codes.

First, we define several block error probabilities:

- P_{GCD}: the block error probability after the decoding of a GC code with the GCD algorithm.

- P_{ND}: the block error probability after the decoding of the inner and outer codes with individual BMD decoders.

- P_{SD}: the block error probability after the decoding of the inner code with a soft-decision decoding method and of the outer code by a BMD decoder. P_{SD} corresponds to P_{ND} with the inclusion of reliability information for inner code decoding.

- P_{BMD}: the block error probability after decoding with a fictional BMD decoder for the entire code (i.e. not component-wise decoding of levels). One observes that P_{ND} can be smaller than P_{BMD}.

Note When decoding the inner and outer codes with a BMD decoder, the inner code is considered together with the channel, giving a superchannel model. Therefore the block error probability after decoding of the inner code corresponds to the symbol error probability of the superchannel. The decoding of errors up to a quarter of the minimum distance is guaranteed. For GC codes with multiple partitions, we can calculate the block error probability $P_{ND}^{(i)}$ for each pair of inner and outer codes $\mathcal{B}^{(i)}$ and $\mathcal{A}^{(i)}$. Finally, the block error probability for a GC code can be estimated with the relationship $P_{ND} \leq \sum P_{ND}^{(i)}$.

The GCD algorithm guarantees decoding up to at least half the minimum distance of a GC code, and therefore $P_{GCD} \leq P_{BMD}$. Furthermore, the method where the inner and outer codes are decoded with a BMD decoder is a subset of the GCD algorithm, and therefore it is also insured that $P_{GCD} \leq P_{ND}$. This is also valid if information reliability is used, i.e. $P_{GCD} \leq P_{SD}$. We have proved the following theorem.

Theorem 9.17 (Calculation of the block error probability) *The block error probability P_{GCD} after decoding of a GC code with the GCD algorithm is*

$$P_{GCD} \leq \min\{P_{BMD}, P_{ND}, P_{SD}\} .$$

It is interesting to mention that the bound for the block error probability according to theorem 9.17 can be calculated based on the given channel.

For the construction of GC codes, we can use two different criteria. One method optimizes the minimum distance of the GC code, and minimizes the error probability after decoding. Both approaches do not necessarily generate the same GC code.

Optimization of the minimum distance The construction of the GC code with the largest possible minimum distance d is sought. The product $d_a^{(i)} d_b^{(i)}$ is selected to give minimum distances that are as close as possible at each partition level. This case is implicitly assumed in all publications on GC codes to date.

Optimization of the error probability after decoding In this section, it will be shown how to construct a GC code that has the smallest possible error probability after decoding. In almost all cases, this corresponds to a UEP code (see section 9.2.5). Therefore the product $d_a^{(i)} d_b^{(i)}$ must decrease as i increases. This is plausible, because for the case of partitioning of higher order, the minimum distance $d_b^{(i)}$ increases. After a particular subcode $\mathcal{B}^{(i_0)}$ is used for decoding, its minimum distance is large enough to protect the information from channel distortion. There is a level i_0 after which the decoding properties (i.e. the minimum distances $d_b^{(i)}$ for $i > i_0$) delivers a sufficient measure of the error probability, even though the product $d_a^{(i)} d_b^{(i)}$ for $i > i_0$ can be smaller than for $i \leq i_0$.

In practice, it is preferable to optimize the error probability after decoding (see example 10.9).

In the next section, we investigate the decoding of the inner code with and without reliability information, and present several examples.

Decoding without information reliability

In this section, we give several examples of decoding of the inner code based on theorem 9.17 without reliability information. Transmission is over a BSC (binary symmetric channel). The bit error probability is described in section 1.4.

Example 9.18 (Error probability calculation 1) In this example, we construct a GC code based on an order-2 partition. For the outer code, we use two RS codes $\mathcal{A}^{(1)}(2^8;255,223,33)$ (satellite transmission code standard [WHPH87]) and $\mathcal{A}^{(2)}(2^8;255,236,20)$. We need two inner codes with dimension 8 and dimension $2 \cdot 8 = 16$. This requirement is satisfied by the shortened BCH code $(2;29,8,12)$, which is a subcode of $BCH(2;31,16,7)$ and satisfies the required dimensions. We partition according to method 3. To reiterate, the two codewords $\underline{\mathbf{a}}^{(1)} = (\underline{a}_1^{(1)}, \underline{a}_2^{(1)}, \ldots, \underline{a}_{n_a}^{(1)}) \in \mathcal{A}^{(1)}$ and $\underline{\mathbf{a}}^{(2)} = (\underline{a}_1^{(2)}, \underline{a}_2^{(2)}, \ldots, \underline{a}_{n_a}^{(2)}) \in \mathcal{A}^{(2)}$ of the outer code determine the n_a codewords from $\mathcal{B}^{(1)}$ as follows: $\underline{a}_j^{(2)} \in GF(2)^8$ labels the codewords $\mathbf{b}_j^{(2)} \in \mathcal{B}^{(2)}$, and the direct sum $(\mathbf{0} \mid \underline{a}_j^{(1)})$ determines the codeword $\mathbf{b}_j^{(1)} \in \mathcal{B}^{(1)}$ that identifies the subcode set. The sum of the codewords gives $\mathbf{b}_j = \mathbf{b}_j^{(1)} + \mathbf{b}_j^{(2)} \in \mathcal{B}^{(1)}$. The resulting GC codeword has rate $R = 0.46$ and minimum distance 231, where $d \geq \min\{231, 240\}$. The GC code parameters are $C(2;7905,3672,231)$.

We can now calculate the different block error probabilities using theorem 9.17, where P_{sym} denotes the symbol error probability, i.e. the error probability of the superchannel. For a BSC with a symbol error probability $p = 10^{-2}$ given by simulations, the block error rate can be calculated:

$$
\begin{array}{rrcrcl}
 & P_{BMD} & = & 5 \cdot 10^{-5}, & & \\
P_{ND}: & P_{sym}^{(1)} & = & 2.5 \cdot 10^{-4} & \Longrightarrow & P_{Block}^{(1)} < 10^{-34}, \\
 & P_{sym}^{(2)} & = & 6 \cdot 10^{-7} & \Longrightarrow & P_{Block}^{(2)} \approx 0\,.
\end{array}
$$

It follows from this that $P_{GCD} < 10^{-34}$ for $p = 10^{-2}$. Because $P_{Block}^{(2)} \approx 0$, an outer code with lower minimum distance can be used.

We replace the code $\mathcal{A}^{(2)}(2^8;255,236,20)$ by the code $\mathcal{A}^{(2)}(2^8;255,243,13)$. This does not change the bound. The code rate is increased and the minimum distance is $d \geq \min\{231, 156\}$. The resulting GC code with rate $R = 0.47$ has the parameters

$$
C(2;7905,3728,156)\,. \qquad \diamond
$$

Example 9.19 (Error probability calculation 2) In this example, we construct a GC code with a partition of order 5. We use method 3 (section 9.2.2 and example 9.11).

The inner codes are BCH codes with length 31:

$$
\begin{aligned}
\mathcal{B}^{(1)}(2;31,25,4) \subset \mathcal{B}^{(2)}(2;31,20,6) \subset \mathcal{B}^{(3)}(2;31,15,8) \\
\subset \mathcal{B}^{(4)}(2;31,10,12) \subset \mathcal{B}^{(5)}(2;31,5,16)\,.
\end{aligned}
$$

The outer RS codes are

$$
\begin{aligned}
\mathcal{A}^{(1)}(2^5;31,8,24),\ \mathcal{A}^{(2)}(2^5;31,16,16),\ \mathcal{A}^{(3)}(2^5;31,20,12), \\
\mathcal{A}^{(4)}(2^5;31,24,8),\ \mathcal{A}^{(5)}(2^5;31,26,6)\,.
\end{aligned}
$$

We obtain a GC code $C(2;961,470,96)$.

We calculate the block error probability P_{BMD} and P_{ND}, which gives $P_{BMD} < P_{ND}$. Several values for the block error probability P_{BMD} are

$$P_{BMD} = 4.6 \cdot 10^{-19} \quad \text{for } p = 1 \cdot 10^{-2},$$
$$P_{BMD} = 1.5 \cdot 10^{-8} \quad \text{for } p = 2 \cdot 10^{-2},$$
$$P_{BMD} = 5.5 \cdot 10^{-4} \quad \text{for } p = 3 \cdot 10^{-2}.$$

◇

In this example, the GC code is constructed such that the minimum distance is optimized; therefore $d_a^{(i)} d_b^{(i)} = 96$, $i = 1, 2, \ldots, 5$. It is easy to see that this is not optimal in terms of the block error probability after decoding. Based only on the inner code $\mathcal{B}^{(5)}(2; 31, 5, 16)$, a block error probability of $P_{Block} = 1.3 \cdot 10^{-7}$ occurs for a channel error probability of $p = 2 \cdot 10^{-2}$. It follows that a single error correcting outer RS code is sufficient, although the minimum distance at level 5 would only be 48.

Example 9.20 (Error probability calculation 3) Consider the code $\mathcal{C}(2; 4335, 1792, 198)$ from example 9.10. The inner codes are $\mathcal{B}^{(1)}(2; 17, 16, 2)$ and $\mathcal{B}^{(2)}(2; 17, 8, 6)$, which are a parity check code and a QR code respectively (see [Boss87]). The outer code is a repetition code $\mathcal{A}^{(1)}(2^8; 255, 1, 255)$ and a Reed–Solomon code $\mathcal{A}^{(2)}(2^8; 255, 223, 33)$.

A theoretical BMD decoding algorithm gives the following error rates:

$$P_{BMD} = 10^{-1} \quad \text{for } p = 2 \cdot 10^{-2},$$
$$P_{BMD} = 2.2 \cdot 10^{-13} \quad \text{for } p = 10^{-2},$$

while

$$P_{ND} = 10^{-15} \quad \text{for } p = 2 \cdot 10^{-2},$$
$$P_{ND} = 10^{-29} \quad \text{for } p = 10^{-2}.$$

◇

A small improvement in the symbol error probability of the superchannel results in a large improvement in the error probability P_{ND} and reduces the upper error bound in the GCD algorithm. The symbol error probability of the superchannel can be further improved if one uses a decoder that can correct beyond half the minimum distance (e.g. methods from section 7.3.3 or an ML decoder). The following two examples illustrate this technique:

Example 9.21 (Error probability calculation 4) We consider the GC code from example 9.18. In [BH86], the block error probability for both inner codes is given. A BMD decoding method is used based on the algorithm from example 7.3.3. The possible values are as follows. The symbol error probability of the superchannel is $5 \cdot 10^{-2}$ when using BMD decoding, and $3 \cdot 10^{-2}$ when the other algorithm is employed. This results in an improvement by a factor of 50 for the P_{ND} error probability as compared with example 9.18. The improvement for the second inner code is even larger. ◇

Example 9.22 (Error probability calculation 5) We assume a BCH inner code $(2; 63, 30, 13)$. The block error probability after decoding with the algorithm from section 7.3.3 over a BSC channel with channel error probability of $p = 3 \cdot 10^{-2}$ is equal to $P_{Block} = 3.5 \cdot 10^{-3}$, as compared with $P_{Block} = 6 \cdot 10^{-4}$ for a BMD decoding method. If we use an outer RS code $\mathcal{A}(2^7; 127, 115, 11)$ then the P_{ND} error probability is improved by a factor of 10^5, namely from $P_{ND} = 3 \cdot 10^{-7}$ in the case of a BMD decoder to $P_{ND} = 2.5 \cdot 10^{-12}$. ◇

In [BH86] and [Boss87], some examples are given for hard-decision decoding of BCH, QR and RM codes using algorithms from section 7.3.3. In the majority of cases, a clear improvement in the block error probability after decoding is achieved, which corresponds to an improvement in the associated GC code. On including soft-decision information, the performance again increases.

Decoding with information reliability

It is generally known that the use of reliability improves the performance of a decoder. In previous sections, we have shown that a small improvement in the block error probability of the inner code corresponds to a large improvement in the overall GC code error probability. For GC codes, many different inner codes can be used that allow for a variety of different decoding methods. For block codes, there are different possibilities for incorporating information reliability. This is explained in chapter 7. Furthermore, there are relatively short codes with smaller minimum distances that allow for ML decoding using reliability information.

Another possibility for the use of reliability information can be achieved by the decoding algorithm from section 7.3.3. We present examples for soft-decision decoding of QR, RM and BCH codes using the algorithms from [LBB96].

Instead of investigating simulation examples, we use theorem 9.17 in order to calculate the error probability bound P_{GCD} for the decoding of the inner codes with information reliability.

Example 9.23 (Calculation using soft-decision 1) We use $\mathcal{B}^{(i)}(2;31,15,7)$ as the inner and $\mathcal{A}^{(i)}(2^5;31,16,16)$ as the outer code. For an AWGN channel, the algorithm offers an improvement in the block error probability of a factor > 10 (see [LBB96]) within the standard signal-to-noise ratio range. If, for example, the bit error probability of the channel is $p = 2\cdot 10^{-2}$ (corresponding to a AWGN channel with a signal-to-noise ratio of 1.86 dB) then

$$p = 4\cdot 10^{-2} \stackrel{\text{hard}}{\Longrightarrow} P_{sym} \approx 3.4\cdot 10^{-2} \stackrel{\text{outer}}{\Longrightarrow} P_{Block} \approx 7\cdot 10^{-6},$$
$$p = 4\cdot 10^{-2} \stackrel{\text{soft}}{\Longrightarrow} P_{sym} < 2\cdot 10^{-3} \stackrel{\text{outer}}{\Longrightarrow} P_{Block} < 2\cdot 10^{-15}\,. \qquad \diamond$$

Example 9.24 (Calculation using soft-decision 2) We consider the BCH code $\mathcal{B}^{(i)}(2;63,24,15)$ as the inner code and as outer code the RS code $\mathcal{A}^{(i)}(2^8;255,223,33)$. If the algorithm from section 7.3.3 is used for decoding with information reliability, then the error probability of the inner code is decreased by a factor of about 100 (see [LBB96]). Furthermore, the bit error probability of the chanel $p = 4\cdot 10^{-2}$ gives the following results:

$$p = 4\cdot 10^{-2} \stackrel{\text{hard}}{\Longrightarrow} P_{sym} \leq 4\cdot 10^{-3} \stackrel{\text{outer}}{\Longrightarrow} P_{Block} < 1\cdot 10^{-15}$$
$$p = 4\cdot 10^{-2} \stackrel{\text{soft}}{\Longrightarrow} P_{sym} \leq 5\cdot 10^{-5} \stackrel{\text{outer}}{\Longrightarrow} P_{Block} \approx 1\cdot 10^{-47}\,. \qquad \diamond$$

9.2.5 Unequal error protection (UEP) codes

Codes with unequal error protection (*unequal error protection codes*, UEP codes) are increasing in importance because they can be designed to encode information depending on the relative importance of different bits. Source coding and channel coding are unified, which allows for optimization of the entire system. This is achieved when the importance of the source coding is incorporated in the channel coding scheme [Hag95]. Unequal error protection is achieved for convolutional codes through puncturing (see section 8.1.10).

The channel coding considers the source coding as well as the modulation method for the case of coded modulation (presented in chapter 10).

UEP codes allow different information coordinates to be protected by different minimum distances. If t errors occur, shich we can design codes which guarantee that certain information bits are protected for t or more errors. This is independent of what happens in the other information coordinates.

Definition 9.18 (UEP codes) *A GC code is constructed such that the product of the minimum distance increases as i becomes smaller:*

$$d_a^{(i)} d_b^{(i)} \text{ becomes smaller as } i \text{ increases.}$$

The information that is coded with the outer code $\mathcal{A}^{(i)}$ has a minimum distance $d_a^{(i)} d_b^{(i)}$.

We do not need to go into special details for the decoding of UEP codes, because GC decoding methods also apply (see section 9.2.4). It is obvious why the product of the minimum distances must decrease as the level index increases. The first level of decoding can then correct more errors than the second, and so on. It is worth mentioning that the decoding of a *normal* GC code always behaves as though one is correcting a UEP code (see section 9.2.4).

9.2.6 Cyclic codes as GC codes

All cyclic codes can be described as GC codes. This interesting property has unfortunately not been exploited for the decoding of cyclic codes.

Introductory example

The following concatenated cyclic code examples are from [Jen96]:

Example 9.25 (Concatenated cyclic code) Consider a double extended cyclic RS code $\mathcal{A}(2^3;9,6,4)$ (see section 3.1.6) as the outer code and a cyclic simplex code $\mathcal{B}(2;7,3,4)$ as the inner code. A code having the irreducible parity check polynomial $h(y) = (y^n - 1)/g_b(y)$ is called minimal, and the code $\mathcal{B}$ is minimal.

Let $\beta \in GF(2^3)$ be a zero of the irreducible polynomial $h(y) = (y^7 - 1)/g(y)$. For a minimal cyclic code, we have the following isomorphism:

$$\gamma = b(\beta),\ b(y) \in \mathcal{B} \iff \psi(\gamma) : b(y) = \sum_{j=0}^{6} b_j y^j,\ b_j = \mathrm{tr}(\beta^{-j} \cdot \gamma) .$$

Each of the 2^3 codewords $b(y)$ from $\mathcal{B}$ is uniquely constructed from $b(\beta^{-j})$ and an element $\gamma \in GF(2)^3$. $\mathrm{tr}(\beta^{-j} \cdot \gamma)$ is the trace function from definition 2.26. With the primitive polynomial $y^3 + y + 1$, we obtain

$$\beta^0 = 001,\ \beta^1 = 010,\ \beta^2 = 100,\ \beta^3 = 011,\ \beta^4 = 110,\ \beta^5 = 111,\ \beta^6 = 101.$$

The generator polynomial is given by (where the cyclotomic coset $K_3 = \{3, 6, 12 = 5 \bmod 7\}$):

$$g_b(y) = (y-1)(y-\beta^3)(y-\beta^6)(y-\beta^5) = (y-1)(y^3+y^2+1) = y^4+y^2+y+1 .$$

The codeword for the information polynomial $i(y) = y^2 + 1$ is $b(y) = i(y)\,g_b(y) = y^6 + y^3 + y + 1$. This gives

$$\gamma = b(\beta) = \beta^6 + \beta^3 + \beta^1 + \beta^0 = 101 + 011 + 010 + 001 = 101 = \beta^6 .$$

Using the trace function, we can again calculate the code coordinates (i.e. coordinates 2 and 3):

$$\begin{aligned} b_2 &= \mathrm{tr}(\gamma\cdot\beta^{-2}) = \mathrm{tr}(\beta^6\cdot\beta^{-2}) = \mathrm{tr}(\beta^4) = \beta^4 + (\beta^4)^2 + (\beta^4)^4 = \beta^4 + \beta + \beta^2 = 0 , \\ b_3 &= \mathrm{tr}(\gamma\cdot\beta^{-3}) = \mathrm{tr}(\beta^6\cdot\beta^{-3}) = \mathrm{tr}(\beta^3) = \beta^3 + (\beta^3)^2 + (\beta^3)^4 = \beta^3 + \beta^6 + \beta^5 = 1 . \end{aligned}$$

A cyclic shift $v(y) = yb(y)$ of $b(y)$ by multiplying γ by β gives $v_i = \mathrm{tr}(\beta^{-i}\cdot\gamma\cdot\beta)$. The generator polynomial of the outer code $\mathcal{A}(2^3;9,6,4)$ has primitive elements from $\eta \in GF(2^6)$. The primitive element $\alpha = \eta^7$ has order 9 from the cyclotomic cosets $K_0 = \{0\}$ and $K_1 = \{1,8\}$:

$$g_a(x) = \prod_{i\in\{0,1,8\}} (x - \alpha^i) = \prod_{i\in\{0,1,8\}} (x - \eta^{7i}) = x^3 + \eta^9 x^2 + \eta^9 x + 1 .$$

The concatenated code C_C with length 63 and dimension 18 is described by a binary 9×7 matrix where every row is a codeword from $\mathcal{B}$. By considering the elements of the outer code $a_i \in GF(2)^3$ in the basic field, we can describe a codeword from C_C using another method. Let $a(x) = a_0 + a_1x + a_2x^2 + \ldots + a_8x^8 \in \mathcal{A}$ and let each coordinate a_i be coded with the code $\mathcal{B}$:

$$b_i(y) = \psi(a_i) = \sum_{j=0}^{6} b_{i,j}y^j, \quad b_{i,j} = \mathrm{tr}(\beta^{-j}\cdot a_i) ,$$

or

$$b_i(y) = y^k g_b(y), \quad a_i = \beta^k .$$

From this, we can describe every codeword of C_C as a polynomial using the variables x and y:

$$c(x,y) = \sum_{i=0}^{8} \psi(a_i)x^i = \psi(a_0) + \psi(a_1)x + \ldots + \psi(a_8)x^8 = \sum_{i=0}^{8}\sum_{j=0}^{6} b_{i,j}y^jx^i .$$

The lengths of the inner and outer codes are relatively prime. By using the Chinese remainder theorem [McWSl, chapter 10, theorem 5] a value ℓ can be found for every pair (i,j), $0 \le i \le 8$, $0 \le j \le 6$ where $\ell = j \bmod 7$ (see figure 9.13). It is now possible to transform a codeword of the concatenated code having the polynomial description in variables x and y as $z = xy$:

$$c(x,y) \iff c(z) = \sum_{\ell=0}^{9\cdot 7-1} c_\ell z^\ell = (c_0, c_1, \ldots c_{62}) .$$

The coefficients c_ℓ are are found in the matrix representation of figure 9.13 in the increasing rows of the extended main diagonal. A single shift of a codeword of a cyclic code defined by x and y (see figure 9.13) gives

$$\begin{aligned} zc(z) \bmod (z^{63}-1) &= \big((xyc(x,y) \bmod (x^9-1)\big) \bmod (y^7-1) \\ &= \psi(\beta a_8) + \psi(\beta a_0)x + \ldots + \psi(\beta a_7)x^8 . \end{aligned}$$

$i \downarrow$ / $j \rightarrow$	0	1	2	3	4	5	6
0	c_0	c_{36}	c_9	c_{45}	c_{18}	c_{54}	c_{27}
1	c_{28}	c_1	c_{37}	c_{10}	c_{46}	c_{19}	c_{55}
2	c_{56}	c_{29}	c_2	c_{38}	c_{11}	c_{47}	c_{20}
3	c_{21}	c_{57}	c_{30}	c_3	c_{39}	c_{12}	c_{48}
4	c_{49}	c_{22}	c_{58}	c_{31}	c_4	c_{40}	c_{13}
5	c_{14}	c_{50}	c_{23}	c_{59}	c_{32}	c_5	c_{41}
6	c_{42}	c_{15}	c_{51}	c_{24}	c_{60}	c_{33}	c_6
7	c_7	c_{43}	c_{16}	c_{52}	c_{25}	c_{61}	c_{34}
8	c_{35}	c_8	c_{44}	c_{17}	c_{53}	c_{26}	c_{62}

$i \downarrow$ / $j \rightarrow$	0	1	2	3	4	5	6
0	c_{62}	c_{35}	c_8	c_{44}	c_{17}	c_{53}	c_{26}
1	c_{27}	c_0	c_{36}	c_9	c_{45}	c_{18}	c_{54}
2	c_{55}	c_{28}	c_1	c_{37}	c_{10}	c_{46}	c_{19}
3	c_{20}	c_{56}	c_{29}	c_2	c_{38}	c_{11}	c_{47}
4	c_{48}	c_{21}	c_{57}	c_{30}	c_3	c_{39}	c_{12}
5	c_{13}	c_{49}	c_{22}	c_{58}	c_{31}	c_4	c_{40}
6	c_{41}	c_{14}	c_{50}	c_{23}	c_{59}	c_{32}	c_5
7	c_6	c_{42}	c_{15}	c_{51}	c_{24}	c_{60}	c_{33}
8	c_{34}	c_7	c_{43}	c_{16}	c_{52}	c_{25}	c_{61}

Figure 9.13 Matrix representation of a codeword of a concatenated code and the corresponding cyclic shift.

This corresponds to the polynomial $\beta(a_8 + a_0x + \ldots + a_7x^8)$, and because the outer code is cyclic and linear, it again produces a codeword from $\mathcal{A}$. With this result, it is plausible that the concatenated code C_C is also a cyclic code, and we can describe the entire code with one generator polynomial. Consider the primitive element $\eta \in GF(2)^6$, $\beta = \eta^9$ an element of order 7, and $\alpha = \eta^7$ an element of order 9. The zero coordinates of the parity check polynomial of the concatenated code are given by the product of the zero coordinates of the parity check polynomial of the inner and outer component codes. Here, every element of the cyclotomic coset of the component code must be considered in order to determine the cyclotomic coset of the concatenated code:

$$\beta \cdot \alpha^2 = \eta^9 \cdot \eta^{14} = \eta^{23}, \quad 23 \in K_{23} = \{23, 29, 43, 46, 53, 58\},$$
$$\beta \cdot \alpha^3 = \eta^9 \cdot \eta^{21} = \eta^{30}, \quad 30 \in K_{15} = \{15, 30, 39, 51, 57, 60\},$$
$$\beta \cdot \alpha^4 = \eta^9 \cdot \eta^{28} = \eta^{37}, \quad 37 \in K_{11} = \{11, 22, 25, 37, 44, 50\}.$$

The generator polynomial of the concatenated code C_C can be determined using the cyclotomic coset of $GF(2^6)$ (primitive polynomial: $z^6 + z + 1$):

$$h_C(z) = \prod_{\ell \in K_{11} \cup K_{15} \cup K_{23}} (z - \eta^\ell), \qquad g_C(z) = \frac{z^{63} - 1}{h(z)},$$

$$g(z) = z^{45} + z^{44} + z^{42} + z^{40} + z^{38} + z^{37} + z^{35} + z^{33} + z^{31} + z^{29} + z^{28}$$
$$+ z^{26} + z^{24} + z^{22} + z^{20} + z^{19} + z^{18} + z^9 + z^8 + z^6 + z^4 + z^2 + z + 1\,.$$

This code has a designed minimum distance (BCH bound) of $d = 14$. A calculation of the minimum distance based on the concatenated component codes gives $d \geq d_a \cdot d_b = 4 \cdot 4 = 16$, which is also the minimum distance. In conclusion, we mention that there is a BCH$(63, 18, 21)$ code with a better minimum distance. ◇

In the preceding example, we have shown the construction of a cyclic concatenated code. Using such codes, cyclic GC codes with good parameters can be constructed, as shown in the following example:

Example 9.26 (Cyclic GC code) This example illustrates the construction of a GC code using cyclic component codes with a partition level of 2. Let $\mathcal{A}^{(1)}(2^3; 9, 2, 8)$ and $\mathcal{A}^{(2)}(2^3; 9, 6, 4)$

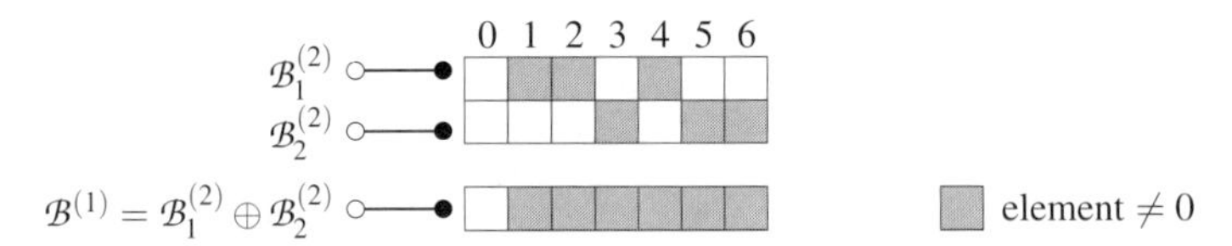

Figure 9.14 Direct addition of the codes $\mathcal{B}_1^{(2)}$ and $\mathcal{B}_2^{(2)}$ and their representation in the transform domain.

be two outer RS codes. $\mathcal{B}_1^{(2)}(2;7,3,4)$ and $\mathcal{B}_2^{(2)}(2;7,3,4)$ are two simplex codes of the second partition level having generator polynomials $g_1(x) = (x-1)(x-\beta^3)(x-\beta^6)(x-\beta^5)$ and $g_2(x) = (x-1)(x-\beta^1)(x-\beta^2)(x-\beta^4)$. From definition 6.19, we can add both level-two inner codes together to get the inner code $\mathcal{B}^{(1)}(2,7,6,2) = \mathcal{B}_1^{(2)} \oplus \mathcal{B}_2^{(2)}$. This principle is illustrated in figure 9.14 in the transform domain.

We encode 6 information bits by the outer code $\mathcal{A}^{(1)}$ to produce a column vector $\underline{\mathbf{a}}^{(1)}$. Each element $\underline{a}_j^{(1)} \in GF(2)^3$, $j \in 1,\ldots,9$, is encoded row-wise by the inner code $\mathcal{B}_1^{(2)}$, which results in a codeword matrix with parameters $C_1(2;63,6,32)$. Next we encode 18 information bits by the outer code $\mathcal{A}^{(2)}$ to produce a column vector $\underline{\mathbf{a}}^{(2)}$. Each element $\underline{a}_j^{(2)} \in GF(2)^3$ $j \in 1,\ldots,9$ is encoded row-wise by the inner code $\mathcal{B}_2^{(2)}$, which results in a codeword matrix with parameters $C_2(2;63,18,16)$. The final GC codeword from $C_1 \oplus C_2$ is $C_{GC}(2;63,24,16)$. The GC code length is found by the matrix dimensions $n_a \times n_b = 9 \times 7 = 63$, and the GC code dimension is $k_a^{(1)} + k_a^{(2)} = 6 + 18 = 24$. The minimum distance is calculated by considering the decoding method.

We decode the 9 rows of the received GC codeword $\hat{C}_{GC}$ with respect to the SPC inner code $\mathcal{B}^{(1)}$. This vector is mapped to the transform domain. The associated information coordinates from $\mathcal{B}_1^{(2)}$ are selected, and the inverse transformation is performed to give $\hat{C}_1$. $\mathcal{A}^{(1)}$ is then used to recover the information bits, which should now be error-free. The information bits are re-encoded to give C_1'. It should be mentioned that, although C_1 has minimum distance 32, the inner code $\mathcal{B}^{(1)}$ with minimum distance 2 was used to recover the row codewords and the according outer code has distance 8. The second estimated codeword is obtained by $\hat{C}_2 = \hat{C}_{GC} \oplus \hat{C}_1'$. $\hat{C}_2$ is then decoded by $\mathcal{B}^{(2)}$ and $\mathcal{A}^{(2)}$ to recover the associated information bits. The minimum distance of the GC code is therefore $d = \min\{d(\mathcal{A}^{(1)})\, d(\mathcal{B}^{(1)} \oplus \mathcal{B}^{(2)}), d(\mathcal{A}^{(2)}) \cdot d(\mathcal{B}^{(2)})\} = \min\{8 \cdot 2, 4 \cdot 4\} = 16$. The code is cyclic because the addition of two cyclic codes is cyclic. The parameters of a BCH code with the same length are BCH$(63,24,15)$. ◇

This example from [Jen96] is the basis for many of the best known code parameters (see [BJ74], [Jen85] and [Jen92]).

Definition of cyclic GC codes

Example 9.25 shows that the concatenation of a binary minimal code and a cyclic code produces a cyclic code. From this result, the following theorem has been derived:

Theorem 9.19 (Cyclic concatenated codes [BJ74]) *Consider $\mathcal{B}(2;n_b,k_b,d_b)$, n_a odd, a minimal binary cyclic code and $\mathcal{A}(2^{k_b};n_a,k_a,d_a)$ a cyclic code. The concatenated code*

$$C_C(2;n_a n_b, k_a k_b, \geq d_a d_b)$$

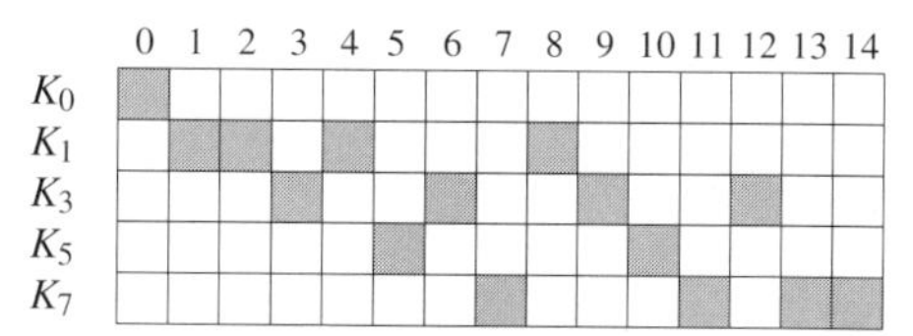

Figure 9.15 The cyclotomic coset for the construction of all minimal binary codes with length 15.

is cyclic if $\gcd(n_a, n_b) = 1$. *The zeros of the parity check polynomial of* C_C *are* $(\beta\alpha_j)^{2^i}$, *where the non-zero coordinates of the inner and outer code are and* $\beta, \beta^2, \ldots, \beta^{2^{k-1}}$ *and* $\alpha_1, \alpha_2, \ldots, \alpha_{k_a}$ *respectively.*

Using this theorem, we can construct simple GC codes as shown in example 9.26.

Theorem 9.20 (Cyclic GC code [Jen96]) *Let* $\gcd(n_a, n_b) = 1$. *Every cyclic code constructed from an inner minimal cyclic binary code and outer cyclic code with length* $n = n_a n_b$ *is cyclic. The code is termed a direct sum binary cyclic concatenated code.*

Example 9.27 (Cyclic GC code of length 255**)** We construct a cyclic code of length 255 based on BCH codes of length 15. Figure 9.15 shows the associated cyclotomic cosets. We use the cyclotomic cosets K_3, K_5 and K_7 to construct the component codes $\mathcal{B}_{K_3}(2;15,4,6), \mathcal{B}_{K_2}(2;15,2,10)$ and $\mathcal{B}_{K_7}(2;15,4,8)$. The partitioning of the inner code is $\mathcal{B}^{(1)}(2;15,10,4) = \mathcal{B}_{K_3} \oplus \mathcal{B}_{K_2} \oplus \mathcal{B}_{K_7}$, $\mathcal{B}^{(2)}(2;15,6,6) = \mathcal{B}_{K_2} \oplus \mathcal{B}_{K_7}$ and $\mathcal{B}^{(3)}(2;15,4,8) = \mathcal{B}_{K_7}$. The outer codes are $\mathcal{A}^{(1)}(2^4;17,2,16)$, $\mathcal{A}^{(2)}(2^2;17,4,12)$, and $\mathcal{A}^{(3)}(2^4;17,10,8)$, where $\mathcal{A}^{(1)}$ and $\mathcal{A}^{(3)}$ are double extended RS codes. In contrast, $\mathcal{A}^{(2)}$ is a non-binary BCH code with symbols from $GF(2^2)$ (see section 4.4). The elements $\alpha \in GF(2^8-1)$ have order 15, and result in a code with length $n = \frac{2^8-1}{15} = 17$. The remaining code parameters can be determined using the cyclotomic cosets ($q = 2^2 = 4$, $m = 8$):

$$K_i = \{iq^j \bmod n,\ j = 0,1,\ldots,m-1\} = \{i4^j \bmod 17,\ j = 0,1,\ldots,7\}\ .$$

We select the cyclotomic coset $K_6 = \{6,7,10,11\}$ for the zero coordinates of the parity check polynomial. This gives $k = 4$ and the designed minimum distance of $d = 12$.

The GC codeword is found by encoding 8, 8 and 40 information bits by the outer codes $\mathcal{A}^{(1)}, \mathcal{A}^{(2)}$ and $\mathcal{A}^{(3)}$ respectively. These outer codewords are encoded by the component codes $\mathcal{B}_{K_3}, \mathcal{B}_{K_5}$ and $\mathcal{B}_{K_7}$ to produce the sub-codewords C_1, C_2 and C_3. The final GC codeword is $C_{GC} = C_1 \oplus C_2 \oplus C_3$. The length and dimension are $n = n_a n_b = 255$ and $k_a^{(1)} + k_a^{(2)} + k_a^{(3)} = 8+8+40 = 56$. The minimum distance can be calculated by considering the decoding method.

Each row of the received codeword $\hat{C}_{GC}$ is decoded with respect to the inner code $\mathcal{B}^{(1)}$ and mapped to the transform domain. The non-zero coordinates of K_3 are selected, and their inverse transformation produces the estimated codeword $\hat{C}_{GC}^{(1)}$. $\mathcal{A}^{(1)}$ is used to recover the information bits, which should now be error-free. These bits are re-encoded to give the best possible estimate C_1'. We proceed by calculating $\hat{C}_{GC} \ominus C_1 = \hat{C}_2 \oplus \hat{C}_3$ and decoding row-wise according to $\mathcal{B}^{(2)}$. We map each row vector to the transform domain, select the non-zero coordinates corresponding to K_5, and perform the inverse transformation to obtain $\hat{C}_2$. $\mathcal{A}^{(2)}$ is used to correct errors and recover the information bits, which are again encoded to give the estimate C_2'. In the final step, $\hat{C}_{GC} \ominus C_1' \ominus C_2' = C_3'$, which is decoded with respect to $\mathcal{B}^{(3)}$. The

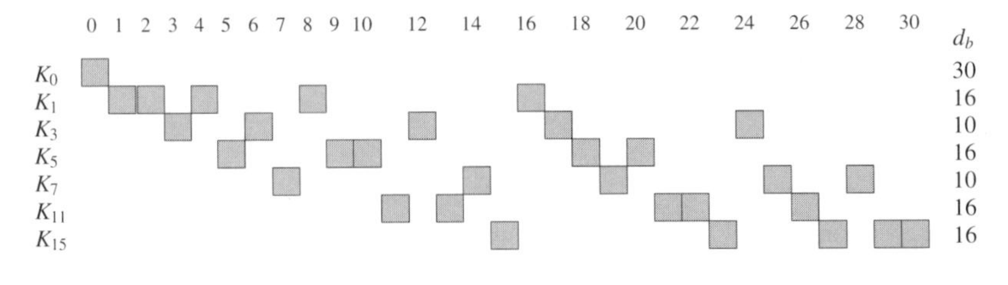

Non-zero coordinates of the transformed codeword, and zero coordinates of the parity check polynomial

Figure 9.16 Minimal binary code of length 31 derived from the cyclotomic coset for $GF(2^5)$.

final information block is recovered by decoding according to $\mathcal{A}^{(3)}$. The minimum distance is therefore

$$\begin{aligned}\min\{d(\mathcal{A}^{(1)})d(\mathcal{B}^{(1)}), d(\mathcal{A}^{(2)})d(\mathcal{B}^{(2)}), d(\mathcal{A}^{(3)})d(\mathcal{B}^{(3)})\} \\ = \min\{d(\mathcal{A}^{(1)})d(\mathcal{B}_{K_3}\oplus\mathcal{B}_{K_2}\oplus\mathcal{B}_{K_7}), d(\mathcal{A}^{(2)})d(\mathcal{B}_{K_2}\oplus\mathcal{B}_{K_7}), d(\mathcal{A}^{(2)})d(\mathcal{B}_{K_7})\} \\ = \min\{16\cdot 4, 12\cdot 6, 8\cdot 8\} = 64\end{aligned}$$

The final cyclic GC code parameters are $C_{GC}(255,56,64)$, which is better than the corresponding $(255,55,63)$ BCH code. ◇

Example 9.28 (Cyclic code with length 1023) We construct a cyclic code of length 1024 based on BCH codes of length 31. Figure 9.16 shows the associated cyclotomic cosets. We use the cyclotomic cosets K_7, K_{11}, K_{15} to construct the component codes $\mathcal{B}_{K_7}(2;31,5,10)$, $\mathcal{B}_{K_{11}}(2;31,5,16)$ and $\mathcal{B}_{K_{15}}(2;31,5,16)$. The partitioning of the inner code is $\mathcal{B}^{(1)}(2;31,15,8) = \mathcal{B}_{K_7}\oplus\mathcal{B}_{K_{11}}\oplus\mathcal{B}_{K_{15}}$, $\mathcal{B}^{(2)}(2;31,10,12) = \mathcal{B}_{K_{11}}\oplus\mathcal{B}_{K_{15}}$ and $\mathcal{B}^{(3)}(2;31,5,16) = \mathcal{B}_{K_{15}}$. The outer RS codes are $\mathcal{A}^{(1)}(2^5;31,2,32)$, $\mathcal{A}^{(2)}(2^5;31,12,22)$ and $\mathcal{A}^{(3)}(2^5;31,18,16)$. Based on the same encoding and decoding method as in example 9.27, we obtain a cyclic GC code $C_{GC}(2;1023,160,256)$, which is better than the same length $BCH(1023,153,251)$ code. ◇

This final example shows clearly that for low rate codes, the calculation of the minimum distance using the generalized concatenated structure of a cyclic code results in better codes than for single BCH codes. Furthermore, these examples demonstrate that long codes can be constructed through the use of much shorter codes to form generalized concatenated codes. As a consequence, it becomes feasible to use soft-decision decoding methods.

Jensen [Jen92] showed that each cyclic code can be represented as a GC code. Instead of limiting the application to only cyclic codes, non-cyclic codes can also be considered consisting of the same inner and outer codes, and therefore the decoding properties also do not change.

9.2.7 Error locating codes

The principle of error locating codes (EL codes) was first described by Wolf [WE63, Wolf65]. This concatenation concept is suitable for the detection of errors in sub-blocks of the code, and therefore has been given the name ‘error locating’. Zyablov [Zya72] investigated this class of

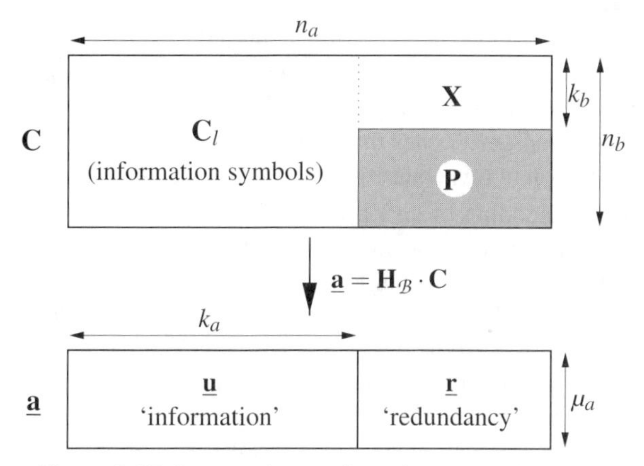

Figure 9.17 Systematic encoding of error locating codes.

codes and generalized the construction in a similar way as was done for concatenated schemes. This led to a so-called generalized error locating code (GEL code).

It has been shown in [ZMB98, ZMB99] that the code classes of generalized concatenated codes and generalized error locating codes are equivalent. Any GC code can be represented by a GEL code, and vice versa. We shall illustrate these results through some examples.

Definition 9.21 (EL code) *Let $\mathbf{H}_{\mathcal{B}}$ be a parity check matrix of an inner code $\mathcal{B}(q; n_b, k_b, d_b)$ and $\underline{\mathbf{a}}$ a codeword of an outer code $\mathcal{A}(q^{\mu}; n_a, k_a, d_a)$. Each codeword $\mathbf{C}$ of an error locating code (EL code) in matrix form fulfills*

$$\underline{\mathbf{a}} = \mathbf{H}_{\mathcal{B}} \cdot \mathbf{C}, \ \underline{\mathbf{a}} \in \mathcal{A}. \tag{9.15}$$

Notice that the columns of the codeword matrix $\mathbf{C}$ are in general not codewords of $\mathcal{B}$. $r_a = n_a - k_a$ and $r_b = n_b - k_b$ are the redundancies of the outer and inner codes respectively. For a matching concatenation, we assume that $\mu_a = r_b$. The EL code has length $n_c = n_a n_b$ and redundancy $r_c = r_a r_b$, or equivalently dimension $k_c = n_c - r_c$.

We consider systematic encoding, which is illustrated in figure 9.17. It will be assumed that both outer and inner encoders are systematic. The parity check matrix of the inner code can be represented as $\mathbf{H}_{\mathcal{B}} = (\mathbf{Q}_b | \mathbf{I}_{r_b})$, where $\mathbf{I}_{r_b}$ is the $r_b \times r_b$ identity matrix. A codeword of the EL code can be encoded as follows. First the information symbols are filled into the codeword matrix $\mathbf{C}$ corresponding to regions $\mathbf{C}_l$ and $\mathbf{X}$. The area $\mathbf{P}$ marked by hatching is left blank. Next, the 'information' part of the code vector $\underline{\mathbf{a}}$ is calculated:

$$\underline{\mathbf{a}} = (\underline{\mathbf{u}} | \underline{\mathbf{r}}) \ \Rightarrow \ \underline{\mathbf{u}} = \mathbf{H}_{\mathcal{B}} \cdot \mathbf{C}_l,$$

where $\mathbf{C}_l$ is the leftmost $n_b \times k_a$ submatrix of $\mathbf{C}$, which is known completely. Then $\underline{\mathbf{u}}$ is encoded with code $\mathcal{A}$ to give a codeword $\underline{\mathbf{a}}$. The last step is the calculation of the submatrix $\mathbf{P}$, which corresponds to the redundancy symbols of the codeword $\mathbf{C}$:

$$\underline{\mathbf{a}} = (\underline{\mathbf{u}} | \underline{\mathbf{r}}) = \mathbf{H}_{\mathcal{B}} \cdot \mathbf{C} = (\mathbf{Q}_b | \mathbf{I}_{r_b}) \cdot \left(\mathbf{C}_l \middle| \begin{matrix} \mathbf{X} \\ \hline \mathbf{P} \end{matrix} \right)$$
$$\Rightarrow \underline{\mathbf{r}} = \mathbf{Q}_b \cdot \mathbf{X} + \mathbf{I}_{r_b} \cdot \mathbf{P}$$
$$\Rightarrow \mathbf{P} = \underline{\mathbf{r}} - \mathbf{Q}_b \cdot \mathbf{X}\,.$$

The defining equation 9.15 of an EL code is equivalent to the syndrome calculation. Therefore the code vector $\underline{\mathbf{a}}$ is often referred to as the syndrome vector of the code $\mathbf{B}$. It is possible to apply this coding concept to columns or columns and rows simultaneously. This has been used in [BBZS98, BBZS99] to construct codes that can correct a two-dimensional rectangular burst of errors. For an introduction to this code construction, see section 9.2.8.

We next describe the decoding of an EL code. Consider the case when the EL code can correct all transmitted errors of the received codeword $\mathbf{C}' = \mathbf{C} + \mathbf{E}$, where $\mathbf{E}$ is the transmission error matrix, and $\mathbf{C}$ and $\mathbf{C}'$ are the transmitted and received codewords respectively. We first calculate the syndrome vector $\mathbf{H}_{\mathcal{B}} \cdot \mathbf{C}' = \underline{\mathbf{a}}'$. The code $\mathcal{A}$ corrects any errors that have occurred, producing the vector $\underline{\mathbf{a}}$. The syndrome error vector is calculated as $\underline{\mathbf{e}} = \underline{\mathbf{a}}' - \underline{\mathbf{a}}$. The non-zero elements of $\underline{e}_j, j \in 1, \ldots, n_a$, identify which columns of $\mathbf{C}'$ are corrupted, and provide the syndrome leader of the error pattern. This property gives rise to the name 'error locating' code. The syndrome leader for each $\underline{e}_j$ is mapped to the ML error pattern $\mathbf{b}_j$, which is then added to column j of $\mathbf{C}'$. The information can now be recovered directly from the corrected codeword matrix C due to the systematic encoding method:

$$\underline{a}'_j = \mathbf{H}_{\mathcal{B}} \mathbf{c}'_j, \quad \underline{a}'_j \overset{\text{decode}}{\Rightarrow} \underline{a}_j, \quad \underline{e}_j = \underline{a}'_j - \underline{a}_j, \quad \mathbf{b}_j = \underline{e}_j^T \cdot \mathbf{H}_{\mathcal{B}}^T, \quad \mathbf{c}_j = \mathbf{c}'_j + \mathbf{b}_j$$

Example 9.29 (EL code) An EL code combining the inner code $\mathcal{B}(7,4,3)$ and the outer code $\mathcal{A}(2^3;7,5,3)$ will be constructed. The resulting code has length $n_c = 49$ and dimension $k_c = n_c - (n_a - k_a)(n_b - k_b) = 43$. ◇

Definition 9.22 (GEL code) *An s-th-order generalized error locating code (GEL code) consists of s outer codes $\mathcal{A}^{(i)}(q^{\mu_i}; n_a, k_a^{(i)}, d_a^{(i)})$, $i = 0, \ldots, s-1$, and s inner codes $\mathcal{B}^{(i)}(q; n_b, k_b^{(i)}, d_b^{(i)})$, $i = 1, \ldots, s$. Let the inner code matrix $\mathcal{B}^{(s)}$ be*

$$\begin{pmatrix} \mathbf{H}_{\mathcal{B}^{(i-1)}/\mathcal{B}^{(i)}} \\ \vdots \\ \mathbf{H}_{\mathcal{B}^{(1)}}, \end{pmatrix},$$

where $\mathbf{H}_{\mathcal{B}^{(i)}}$ is the parity check matrix of the inner code $\mathcal{B}^{(i)}$ and $\underline{a}^{(i)}$ is a codeword of the i-th outer code $\mathcal{A}^{(i)}$. Each codeword $\mathbf{C}$ *of a GEL code in matrix form fulfills*

$$\underline{\mathbf{a}}^{(i)} = \mathbf{H}_{\mathcal{B}^{(i)}/\mathcal{B}^{(i+1)}} \cdot \mathbf{C}, \quad \underline{\mathbf{a}}^{(i)} \in \mathcal{A}^{(i)}, \tag{9.16}$$

for all $i = 0, \ldots, s-1$.

The outer and inner code redundancies are given by $r_a^{(i)} = n_a - k_a^{(i)}$ and $r_b^{(i)} = n_b - k_b^{(i)}$ respectively. We require that $\mu_i = k_b^{(i)} - k_b^{(i+1)}$, $k_b^{(0)} = n_b$, for all $i = 0, \ldots, s-1$. The GEL code has length $n_c = n_a n_b$, redundancy $r_c = \sum_{i=0}^{s-1} r_a^{(i)} \mu_i$ and dimension $k_c = n_a k_b^{(s)} + \sum_{i=0}^{s-1} k_a^{(i)} \mu_i = n_c - r_c$.

For $s = 1$, the GEL code reduces to an ordinary EL coding scheme. Systematic encoding can be done quite similarly as for the EL code. The encoders for the outer codes should be

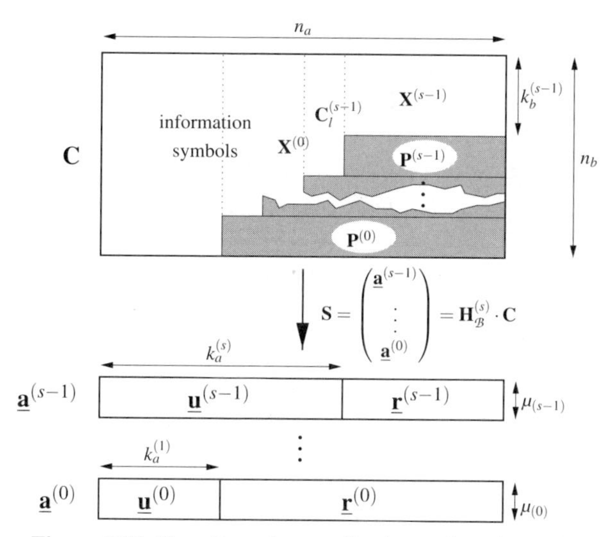

Figure 9.18 Encoding of generalized error locating codes.

systematic. Further, we assume that the matrices $\mathbf{H}_{\mathcal{B}^{(i-1)}/\mathcal{B}^{(i)}}$ are given by

$$\mathbf{H}_{\mathcal{B}^{(s)}} = \begin{pmatrix} \mathbf{H}_{\mathcal{B}^{(s-1)}/\mathcal{B}^{(s)}} \\ \mathbf{H}_{\mathcal{B}^{(s-2)}/\mathcal{B}^{(s-1)}} \\ \vdots \\ \mathbf{H}_{\mathcal{B}^{(1)}/\mathcal{B}^{(2)}} \\ \mathbf{H}_{\mathcal{B}^{(1)}} \end{pmatrix} = \begin{pmatrix} \mathbf{Q}_b^{(s-1)} & \mathbf{I}_{\mu_{(s-1)}} & \mathbf{0} & \mathbf{0} & \dots & \mathbf{0} \\ \mathbf{Q}_b^{(s-2)} & & \mathbf{I}_{\mu_{(s-2)}} & \mathbf{0} & \dots & \mathbf{0} \\ \vdots & & & \ddots & \ddots & \vdots \\ \mathbf{Q}_b^{(1)} & & & & \mathbf{I}_{\mu_1} & \mathbf{0} \\ \mathbf{Q}_b^{(0)} & & & & & \mathbf{I}_{\mu_0} \end{pmatrix},$$

where $\mathbf{I}_{\mu_i}$ is the $\mu_i \times \mu_i$ identity matrix. Note that any $\mathbf{H}_{\mathcal{B}^{(s)}}$ describing the partitioning and its labeling of the inner code can be transformed to such a representation without destroying the nested partition (compare partitioning method 3 after example 9.2.2). This is possible if row operations within $\mathbf{H}_{\mathcal{B}^{(i-1)}/\mathcal{B}^{(i)}}$ and column permutations of $\mathbf{H}_{\mathcal{B}^{(s)}}$ are allowed. Permuting the columns of $\mathbf{H}_{\mathcal{B}^{(s)}}$ will lead to an equivalent code, without changing the properties of the concatenated code.

Figure 9.18 illustrates the encoding process. After filling the white part of the codeword matrix $\mathbf{C}$ with information symbols, the information part $\underline{\mathbf{u}}^{(i)}$ of every outer codeword can be calculated using equation 9.16. Note that this can be done without knowing any of the redundancy matrices in $\mathbf{C}$, which are denoted by $\mathbf{P}^{(i)}$. The redundancy part $\underline{\mathbf{r}}^{(i)}$ of the outer codewords, $i = 0, \dots, s-1$, results from ordinary systematic encoding. In the last step, the matrices $\mathbf{P}^{(i)}$ are calculated. If we denote the upper right submatrix by $\mathbf{X}^{(i)}$ then the following equation holds for all $i = 0, \dots, s-1$ (see figure 9.18):

$$\underline{\mathbf{r}}^{(i)} = \left(\mathbf{Q}_b^{(i)} \middle| \mathbf{I}_{\mu_a^{(i)}}\right) \cdot \left(\frac{\mathbf{X}^{(i)}}{\mathbf{P}^{(i)}}\right) = \mathbf{Q}_b^{(i)} \cdot \mathbf{X}^{(i)} + \mathbf{P}^{(i)}$$

$$\Rightarrow \mathbf{P}^{(i)} = \underline{\mathbf{r}}^{(i)} - \mathbf{Q}_b^{(i)} \cdot \mathbf{X}^{(i)}.$$

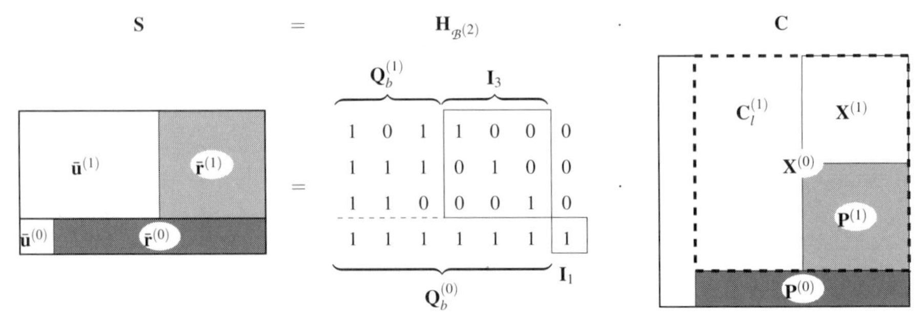

Figure 9.19 Encoding of a sample generalized error locating code (example 9.30).

The matrix $\mathbf{X}^{(i)}$ is calculated as

$$\mathbf{X}^{(i)} = \left(\mathbf{C}_l^{(i+1)} \left| \frac{\mathbf{X}^{(i+1)}}{\mathbf{P}^{(i+1)}} \right. \right)$$

where $\mathbf{C}_l^{(i)}$ is an information submatrix with dimension $(k_b^{(i+1)} + \mu_a^{(i+1)}) \times (r_a^{(i)} - r_a^{(i+1)})$ (we assume that $\mu_a^{(i)} = r_b^{(i)}$). $\mathbf{X}^{(s-1)}$ is defined as the information submatrix of the original EL codeword with dimension $k_b^{(s-1)} \times (n_a - k_a^{(s-1)})$.

Because $\mathbf{P}^{(i+1)}$ is a submatrix of $\mathbf{X}^{(i)}$ that is needed for the calculation of $\mathbf{P}^{(i)}$, the matrices $\mathbf{P}^{(i)}$ are calculated recursively, beginning with $i = s-1$ and finishing with $i = 0$.

Since an sth order GEL code is only a special case of an $(s+1)$th-order GC code, decoding algorithms for GC codes can be employed for GEL codes. Thus the Blokh–Zyablov algorithm [BZ74] can be used to decode all error patterns whose weight does not exceed $\frac{d_c-1}{2}$. However, this algorithm can correct many error patterns of higher weight.

Construction and encoding for GEL codes are illustrated in the following example of a second-order GEL code.

Example 9.30 (GEL code) A second-order GEL code will be constructed using the inner codes $\mathcal{B}^{(1)}(7,6,2)$ and $\mathcal{B}^{(2)}(7,3,4)$ and the outer codes $\mathcal{A}^{(0)}(2;7,1,7)$ and $\mathcal{A}^{(1)}(2^3;7,4,4)$. A parity check matrix $\mathbf{H}_{\mathcal{B}^{(2)}}$ for the code $\mathcal{B}^{(2)}$ is

$$\mathbf{H}_{\mathcal{B}^{(2)}} = \begin{pmatrix} \mathbf{H}_{\mathcal{B}^{(1)}/\mathcal{B}^{(2)}} \\ \mathbf{H}_{\mathcal{B}^{(1)}} \end{pmatrix} = \begin{pmatrix} 1 & 0 & 1 & 1 & 0 & 0 & 0 \\ 1 & 1 & 1 & 0 & 1 & 0 & 0 \\ 1 & 1 & 0 & 0 & 0 & 1 & 0 \\ 1 & 1 & 1 & 1 & 1 & 1 & 1 \end{pmatrix}.$$

The resulting concatenated code has length $n_c = 49$ and dimension $k_c = n_a k_b^{(2)} + k_a^{(0)} \mu_0 + k_a^{(1)} \mu_1 = 34$.

The equation for calculation of the syndrome and the corresponding matrices for systematic encoding are illustrated in figure 9.19. For encoding, the white area in the codeword matrix $\mathbf{C}$ is filled with information symbols; then $\underline{\mathbf{u}}^{(0)}$ and $\underline{\mathbf{u}}^{(1)}$ are calculated. After encoding both codewords of the outer code $\underline{\mathbf{a}}^{(0)}$ and $\underline{\mathbf{a}}^{(1)}$, the redundancy symbols of the codeword matrix are given by $\mathbf{P}^{(1)} = \underline{\mathbf{r}}^{(1)} - \mathbf{Q}^{(1)}\mathbf{X}^{(1)}$ and $\mathbf{P}^{(0)} = \underline{\mathbf{r}}^{(0)} - \mathbf{Q}^{(0)}\mathbf{X}^{(0)}$.

A simple algorithm for the decoding of the received code matrix $\mathbf{C}'$, $\mathbf{C}' = \mathbf{C} + \mathbf{F}$, is as follows. First the syndrome matrix $\mathbf{S}' = \mathbf{H}_{\mathcal{B}^{(2)}} \cdot \mathbf{C}'$ is calculated. Each symbol of the last row of $\mathbf{S}'$ is the syndrome of a column of $\mathbf{C}'$ with respect to the code $\mathcal{B}^{(1)}$, which is a parity check code for our example. The PC code has minimum distance $d = 2$; therefore it can identify single errors in each column. If one column of the code matrix has exactly *one* error then the last row of the syndrome matrix will be false. This row corresponds to the code $\mathcal{A}^0$, which is a repetition code of length 7 that can correct 3 errors. This allows the EL code to correct any pattern of 3 errors that occur in different columns of the EL codeword matrix. The corrupted columns found by $A^{(0)}$, which correspond to the redundancy coordinates of $\mathcal{A}^{(1)}$, are defined as erasures (see 3.2.6) for the decoding of the syndrome vector $\underline{\mathbf{a}}^1 = \mathbf{H}_{\mathcal{B}^1/\mathcal{B}^2} \cdot \mathbf{C}'$.

If no more than three errors occur in different columns of the code matrix during transmission then the corrected syndrome matrix is error-free. Each column of the syndrome matrix corresponds to the syndrome of each column of the code matrix with respect to the code $\mathcal{B}^{(2)}(7,3,4)$. The minimum distance of 4 insures that one error can be corrected. $\diamond$

Equivalence of GC and GEL codes It has been shown in [ZMB98, ZMB99] that the code classes of GC codes and GEL codes are equivalent. An sth-order GC code can be interpreted as a special case of a GEL code of order $s-1$, s or $s+1$, depending on the specific codes selected for the inner and outer codes. This property is also true of GEL codes, as shown in the following examples.

Both code concatenation descriptions are based on the same partition principle of the inner code. The main difference is the labeling of the partition. For GEL codes, the parity check matrix $\mathbf{H}_{\mathcal{B}^{(s)}}$ is specified, whereas for GC codes, the labeling is not defined. Usually, however, labeling of the subcodes is done by a linear mapping, and can therefore be described via a matrix $\mathbf{G}_{\mathcal{B}^{(1)}}$. Then both descriptions are identical:

$$\mathbf{G}^T_{\mathcal{B}^{(1)}} \cdot \mathbf{S} = \mathbf{C} \iff \mathbf{S} = \mathbf{H}_{\mathcal{B}^{(s)}} \cdot \mathbf{C},$$

provided that $\mathbf{G}^T_{\mathcal{B}^{(1)}} = \mathbf{H}^{-1}_{\mathcal{B}^{(s)}}$. Since $\mathbf{H}_{\mathcal{B}^{(s)}}$ has full rank, it is always possible to find such a $\mathbf{G}_{\mathcal{B}^{(1)}}$.

EL $\longrightarrow$ second-order GC An EL code with outer code $\mathcal{A}$ and inner code $\mathcal{B}$ can be interpreted as a second-order GC code. If we set $\mathcal{A} = \mathcal{A}^{(1)}$ and $\mathcal{B} = \mathcal{B}^{(2)}$ then a GC code consisting of the partitioning $\mathcal{B}^{(1)}(q;n_b,n_b,1) \supset \mathcal{B}^{(2)}(q;n_b,k_b,d_b) \supset \mathcal{B}^{(3)}(q;n_b,0,\infty)$ and the outer codes $\mathcal{A}^{(1)}$ and $\mathcal{A}^{(2)}(q^{k_b};n_a,n_a,1)$ has parameters equal to the EL code. This description immediately leads to an expression for the minimum distance of the EL code: $d_c \geq \min\{d_a^{(1)}d_b^{(1)}, d_a^{(2)}d_b^{(2)}\} = \min\{d_a, d_b\}$. An encoding matrix

$$\mathbf{G}_{\mathcal{B}^{(1)}} = \begin{pmatrix} \mathbf{G}_{\mathcal{B}^{(2)}} \\ \mathbf{G}_{\mathcal{B}^{(1)}/\mathcal{B}^{(2)}} \end{pmatrix}$$

for the GC code is the inverse of the mapping defined by $\mathbf{H}_{\mathcal{B}}$.

Example 9.31 The interpretation of the EL code from example 9.29 with $\mathcal{B}(7,4,3)$ and $\mathcal{A}(2^3;7,5,3)$ as a GC code requires the partitioning $\mathcal{B}^{(1)}(7,7,1) \supset \mathcal{B}^{(2)}(7,4,3) \supset \mathcal{B}^{(3)}(7,0,\infty)$ and the outer codes $\mathcal{A}^{(1)}(2^3;7,5,3)$ and $\mathcal{A}^{(2)}(2;7,7,1)$. The minimum distance of the code is given by $d_c \geq \min\{d_a, d_b\} = 3$. $\diamond$

CC $\longrightarrow$ second-order GEL A classical concatenated code (CC code) with outer code $\mathcal{A}$ and inner code $\mathcal{B}$ can be seen as a two-level GEL code consisting of the component codes $\mathcal{A}^{(0)}(q^{r_b};n_a,0,\infty)$, $\mathcal{A}^{(1)}=\mathcal{A}$, $\mathcal{B}^{(1)}=\mathcal{B}$ and $\mathcal{B}^{(2)}(n_b,0,\infty)$.

Example 9.32 We consider the CC code with the component codes $\mathcal{A}(2^3;7,4,4)$ and $\mathcal{B}(7,3,4)$. A GEL code of order two with $\mathcal{A}^{(0)}(2^4;7,0,\infty)$, $\mathcal{A}^{(1)}=\mathcal{A}(2^3;7,4,4)$, $\mathcal{B}^{(1)}=\mathcal{B}(7,3,4)$ and $\mathcal{B}^{(2)}(7,0,\infty)$ leads to an equivalent concatenated code. ◇

***s*th-order GEL $\longrightarrow$ $(s+1)$th-order GC** A sth-order GEL code with inner component codes $\mathcal{B}^{(i)}(q;n_b,n_b^{(i)},d_b^{(i)})$, $i=1,\ldots,s$, and outer component codes $\mathcal{A}^{(i)}(q^{\mu_i};n_a,n_a^{(i)},d_a^{(i)})$, $i=0,\ldots,s-1$, is a special case of an $(s+1)$th-order GC code with the additional codes $\mathcal{B}^{(0)}(q;n_b,n_b,1)$ and $\mathcal{A}^{(s)}(q^{k_b^{(s)}};n_a,n_a,1)$. Note that indexing of the component codes has only been shifted by 1 compared with the original definition of GC codes.

This description reveals the minimum distance of the code: $d_c\geq\min_{i=0,\ldots,s}\{d_a^{(i)}d_b^{(i)}\}=\min_{i=1,\ldots,s-1}\{d_a^{(0)},d_a^{(i)}d_b^{(i)},d_b^{(s)}\}$, taking into account that $d_b^{(0)}=d_a^{(s)}=1$.

Example 9.33 The second-order GEL code from example 9.30 will be interpreted as a third-order GC code. The parameters of the GEL component codes are $\mathcal{B}^{(1)}(7,6,2)$, $\mathcal{B}^{(2)}(7,3,4)$, $\mathcal{A}^{(0)}(2;7,1,7)$ and $\mathcal{A}^{(1)}(2^3;7,4,4)$. For the GC description, we need two more component codes: $\mathcal{B}^{(0)}(7,7,1)$ and $\mathcal{A}^{(2)}(2^3;7,7,1)$. The minimum distance of the code is $d_c\geq\min\{d_a^{(0)}1,d_a^{(1)}d_b^{(1)},1d_b^{(2)}\}=4$. This minimum distance allows the correction of one arbitrary error only, but nevertheless special error patterns up to weight 3 can be corrected (see example 9.30). ◇

***s*th-order GC $\longrightarrow$ $(s+1)$th-order GEL** An sth-order GC code with inner and outer component codes $\mathcal{B}^{(l)}(q;n_b,n_b^{(l)},d_b^{(l)})$ and $\mathcal{A}^{(l)}(q^{m_a^{(l)}};n_a,n_a^{(l)},d_a^{(l)})$, $l=1,\ldots,s$, is a special case of an $(s+1)$th-level GEL code with the additional codes $\mathcal{B}^{(s+1)}(q;n_b,0,\infty)$ and $\mathcal{A}^{(0)}(q^{r_b^{(1)}};n_a,0,\infty)$.

Example 9.34 An interpretation of the two-level GC code with $\mathcal{B}^{(1)}(7,4,3)$, $\mathcal{B}^{(2)}(7,3,4)$, $\mathcal{A}^{(1)}(2;7,1,7)$ and $\mathcal{A}^{(2)}(2^3;7,4,4)$ as a three-level GEL code requires the additional component codes $\mathcal{B}^{(3)}(7,0,\infty)$ and $\mathcal{A}^{(0)}(2^3;7,0,\infty)$. ◇

Note that the error correcting capability of GEL codes is the same as for GC codes.

9.2.8 Error locating codes in two dimensions

In this section, we illustrate how the concept of error locating codes (EL codes) can be applied in two dimensions, leading to a code that is able to correct a two-dimensional burst of errors.

We assume that an error occurs during transmission that corrupts the codeword matrix only in a rectangular area of size $b_1\times b_2$ bits. This can be considered a two-dimensional burst error. According to Gilbert [Gil60], the capacity of a burst error channel is larger than that of a channel with independent errors. Therefore it must be possible to construct a code with a higher rate. In this section, we present a method of encoding for transmission channels that contain burst errors. We present this technique for codes with a basis in $GF(2)$, although the technique can be extended to any field. Figure 9.20 shows a transmitted matrix protected by an interleaved single parity check code. Coordinates with the same shade are summed to give

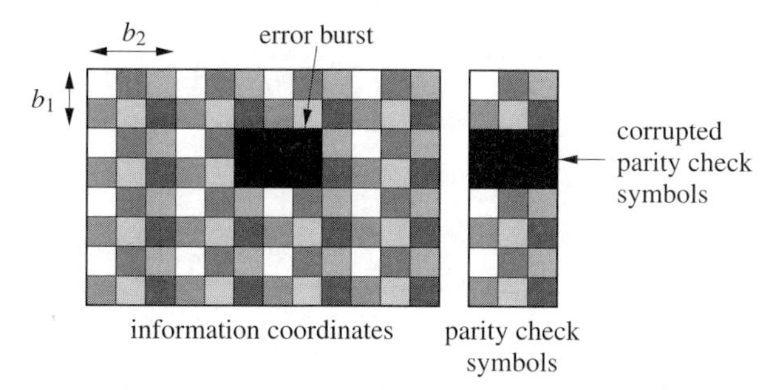

Figure 9.20 Parity check symbols on the receiver side.

a parity check bit. The black region corresponds to a burst error. We assume that the parity check symbols (figure 9.20) are not transmitted. Instead, they are calculated by the receiver. It is observed that most of the parity symbols can be calculated correctly based on the received information coordinates. This give rise to the question: Why should these bits be transmitted? In the following, we describe how a smaller number of parity check bits can be transmitted, which greatly reduces the number of two-dimensional burst errors, and increases the overall code rate.

The concept of a two-dimensional error locating code We describe a possible code construction from [BBZS98], which can correct a two-dimensional burst error of size $b_1 \times b_2$, which requires $4b_1 b_2$ redundancy symbols (a burst error with $b_1 = 1$ corresponds to the one dimensional case). We write the information bits in a binary $n_1 \times n_2$ matrix where sections in the upper right and lower left corners are set to zero (see figure 9.21). These coordinates are left empty for the initial encoding steps. The two zero-coordinate regions have the dimensions $2b_1$ rows and b_2 columns and b_1 rows and $2b_2$ columns for the upper right and lower left regions respectively. In an $n_1 \times n_2$ matrix, there are $n_1 n_2 - 2b_1 b_2 - b_1 2b_2 = n_1 n_2 - 4b_1 b_2$ coordinates for information bits. We calculate the redundancy symbols for the upper right zero coordinates as follows:

Step 1: Each row $i = 2b_1, \ldots, n_1 - 1$ is encoded with b_2 interleaved single parity check codes. Therefore

$$p_{im} = \sum_{\substack{j=m,m+b_2,m+2\cdot b_2 \ldots \\ j<n_2}} c_{ij}, \quad i = 2b_1, \ldots, n_1 - 1\,, \quad m = 0, \ldots, b_2 - 1, \tag{9.17}$$

where m is the column index of the single parity check code. The parity check bits are written in an $(n_1 - 2b_1) \times b_2$ matrix (see figure 9.21).

Step 2: Each row of the $(n_1 - 2b_1) \times b_2$ matrix is written as a single element of $GF(2^{b_2})$. These symbols are now considered as the information elements of a systematic RS code with parameters $(2^{b_2}; n_1, n_1 - 2b_1, 2b_1 + 1)$. The RS codeword is written next to the codeword matrix as shown in 9.22, with the redundancy coordinates labeled as r_{ij}, with $i = 0, \ldots, 2b_1 - 1$ and $j = n_2 - b_2, \ldots, n_2$.

Step 3: In order to calculate the elements of the empty $b_2 \times 2b_1$ region, we calculate the single parity check symbol of each row i, $i = 0, \ldots, 2b_1 - 1$, as in equation 9.17. All

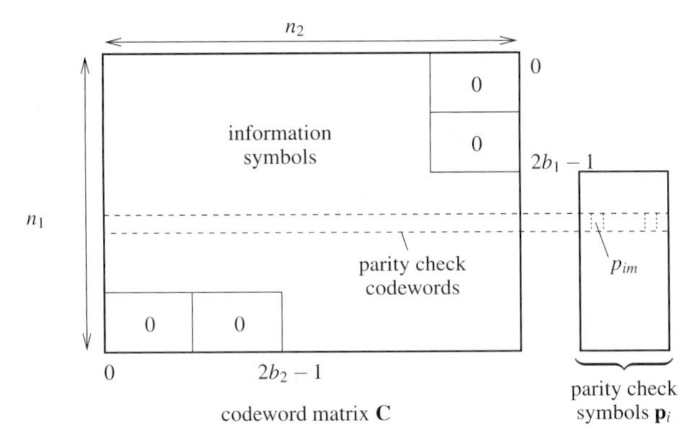

Figure 9.21 First encoding step in the row direction.

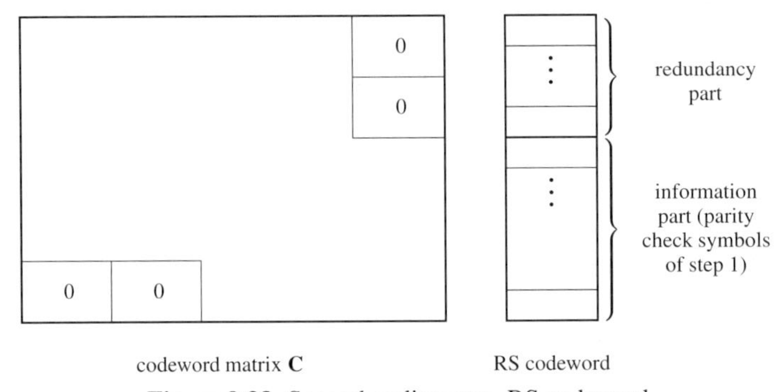

Figure 9.22 Second coding step, RS codeword.

elements c_{ij} for $j \in n_2 - b_2, \dots, n_2$ (i.e. the zero coordinates) are substituted with the corresponding redundancy symbols r_{ij} calculated for the RS codeword from step 2. The resulting $b_2 \times 2b_1$ single parity check symbols are written in the upper right zero coordinates.

The zero coordinates of the lower left corner of the original matrix (figure 9.21) are calculated following the same procedure as for steps 1–3 based on the columns. The bound in this case is $n_2 \leq 2^{b_2} + 1$.

After encoding, we have an $n_1 \times n_2$ matrix with a row RS codeword $(2^{b_2}; n_1, n_1 - 2b_1, 2b_1 + 1)$ and a column RS codeword $(2^{b_1}; n_2, n_2 - 2b_2, 2b_2 + 1)$. We transmit only the $n_1 \times n_2$ matrix as a codeword of the code $C(n_1 n_2, n_1 n_2 - 4b_1 b_2, d_T = 3)$. The minimum distance $d_T = 3$ is the correction capability of a burst error with dimension $b_1 \times b_2$ using the combinatoric metric (see appendix B).

Theorem 9.23 (Correction capability of the code C) *Each two-dimensional burst error with size $b_1 \times b_2$ can be corrected with the code C.*

Proof We assume that a burst error with dimensions $b_1 \times b_2$ occurs during the transmission of an $n_1 \times n_2$ matrix. We first calculate the parity bits of the b_2 row interleaved SPC codes and the b_1 column interleaved SPC codes (see figure 9.23). The matrices to the right and beneath the code matrix (figure 9.23) must give valid RS codewords $(2^{b_1}; n_1, n_1 - 2b_1, 2b_1 + 1)$ and $(2^{b_2}; n_2, n_2 - 2b_2, 2b_2 + 1)$ for the error-free

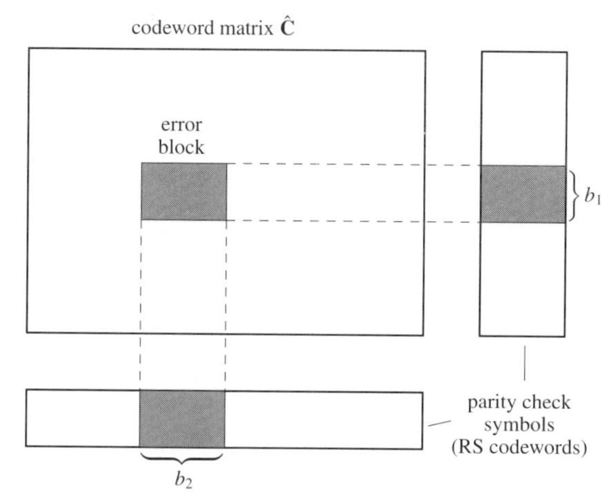

Figure 9.23 Parity check symbols corrupted by a two-dimensional burst error.

Table 9.1 Logarithm table for the Galois field $GF(2^4)$.

$\alpha^{-\infty}$	0000	α^3	0001	α^7	1101	α^{11}	0111
α^0	1000	α^4	1100	α^8	1010	α^{12}	1111
α^1	0100	α^5	0110	α^9	0101	α^{13}	1011
α^2	0010	α^6	0011	α^{10}	1110	α^{14}	1001

case. Figure 9.23 shows that a $b_1 \times b_2$ error block produces at most b_1 row errors and b_2 column errors in the respective RS words. Based on the minimum distance of both codes, the error can be corrected. Therefore error positions and values are known, and can be corrected using the $n_1 \times n_2$ matrix. Burst errors occurring in the upper right or lower left coordinates need not be corrected. □

The following example demonstrates the correction of two-dimensional burst errors:

Example 9.35 (Code correction of a 4×4 burst error) We construct a $(2; 150, 96, d_T = 3)$ code that can correct a 4×4 burst error. We write the information in a 10×15 matrix according to figure 9.24. The gray shaded areas of the lower left and upper right corners contain no information, and are initially set to zero. We first code in the row direction. One begins with the calculation of each $b_2 = 4$ parity check symbol for the rows $2b_1$ to $n_1 - 1 = 9$ according to equation 9.17. The result is entered to the right of the codeword. One of these parity check equations is marked with framed boxes (figure 9.24). The RS parity check symbols of each row can be calculated in the next step with help of the logarithm table 9.1 through the binary representation of the symbols in $GF(2^4)$ (the component representation is the same as for example 2.12).

The symbols in rows 8 and 9, α^2 and α^{10}, are information symbols of a shortened RS code. The information is encoded systematically using the generator polynomial $1 + \alpha^6 x + \alpha x^2 + \alpha^{10} x^3 + x^4 + \alpha^3 x^5 + \alpha^2 x^6 + x^7 + \alpha^2 x^8$, giving the codeword $\alpha^{13} + \alpha^5 x + \alpha^{-\infty} x^2 + \alpha^{12} x^3 + \alpha^8 x^4 + \alpha^{10} x^5 + \alpha^{12} x^6 + \alpha^9 x^7 + \alpha^2 x^8 + \alpha^{10} x^9$. A binary interpretation of the redundancy symbols of the RS code is taken. The bits of the upper right corner of the codeword (gray) are selected such that they satisfy the parity check equations of rows $0 \ldots 7$ (see figure 9.25).

The parity check symbols for the rows are calculated in exactly the same way. It must

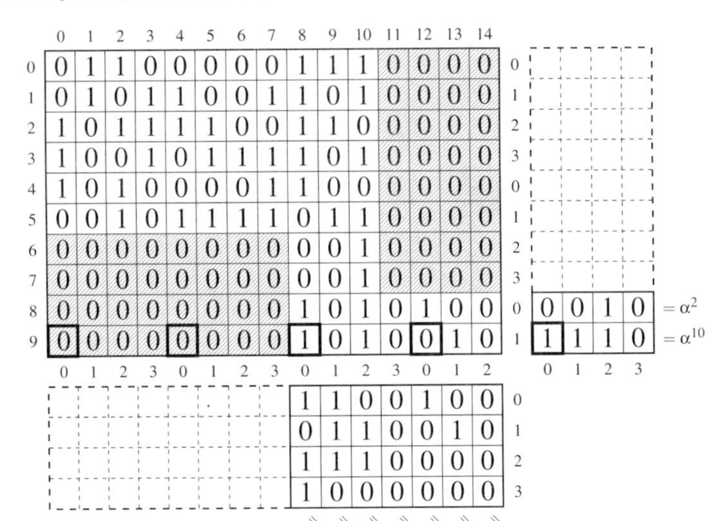

Figure 9.24 Calculation of the row and column parity check symbols.

be noted that the calculated redundancy symbols of the row redundancy quadrant (zero coordinates in the upper right corner of figure 9.21) are not considered when determining the column redundancy symbols, and vice versa. The symbols that are constructed from columns $8\ldots14$ correspond to the information symbols of a RS code. The completed code is shown in figure 9.25.

During the transmission of the codeword, a burst error occurs. The corrupted bits are shown in gray in figure 9.26. The calculation of the parity check symbols on the receiver side produces errors in both binary representations of the RS codewords. These errors can in all cases be identified by the vertical RS codeword. With this knowledge, it is possible to correct all errors of the corrupted codeword. For example, consider row 5. Instead of α^{10}, we decode α^8. A single bit error has occurred in this row, and, based on the interleaving scheme of the parity check equation, it follows that this error corresponds to an error in column 1, 5, 9 or 13. Because the vertical component code only detected a single error in column 5, it is clear that the error has occurred in coordinate $(5,5)$. The errors in positions $(6,8)$ and $(8,8)$ can be similarly corrected. The remaining errors in coordinates $(6,5)$, $(7,6)$ and $(8,7)$ are in the redundant positions of the codeword, and not detected by a horizontal parity check code. Because the information symbols of the code matrix are error-free, they can be used to recalculate the redundant bits, if needed. ◇

9.3 GC codes with convolutional codes

In order to construct GC codes using convolutional codes, the code partitioning must first be determined. Next a GC code construction is carried out based on a single inner and outer convolutional code, and a decoding technique is also presented. The partitioning of convolutional codes is again used in chapter 10 for a trellis representation of coded modulation with memory. This results in codes with very good distance properties.

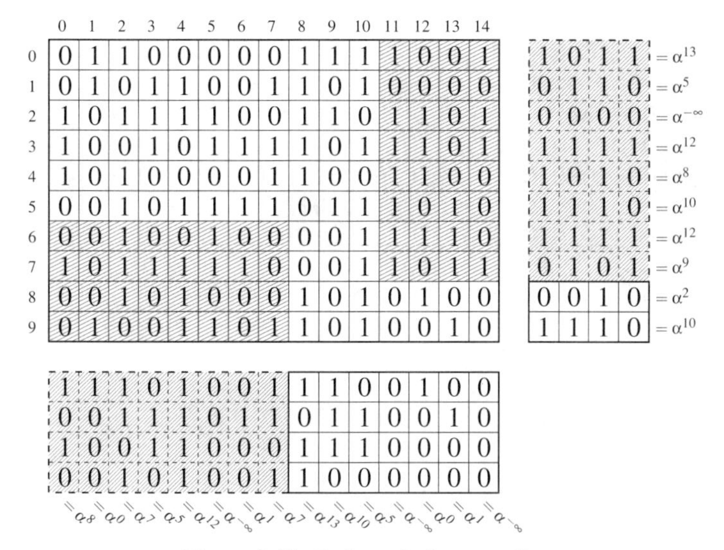

Figure 9.25 Codeword after encoding.

9.3.1 Partitioning of (P)UM codes

Through the representation of convolutional codes by (P)UM codes, we are able to exploit block code partitioning methods. This was the starting point for the construction of GC codes with inner convolutional codes (see [ZySha]).

In section 8.7, we introduced (P)UM codes, while section 8.7.4 described various construction methods on the basis of BCH, RS and RM codes. Each convolutional code can be represented as a (P)UM code. Encoding proceeds as follows:

$$\mathbf{c}_t = \mathbf{i}_t \cdot \mathbf{G}_a + \mathbf{i}_{t-1} \cdot \mathbf{G}_b, \quad \mathbf{i}_{-1} = \mathbf{0}\,. \tag{9.18}$$

Each $\mathbf{c}_t$ is a codeword of a block code $\mathcal{C}_\alpha$ (see section 8.7.1). A subcode of $\mathcal{C}_\alpha$ must be found to carry out the partitioning. This procedure is demonstrated in the following example:

Example 9.36 (Partitioning using the generator matrix of (P)UM codes) This example is taken from [ZySha]. Let the generator matrices $\mathbf{G}_a$ and $\mathbf{G}_b$ represent a $(n = 6, k = 4)$ UM code,

$$\mathbf{G}_a = \begin{pmatrix} 1 & 0 & 1 & 1 & 1 & 0 \\ 0 & 1 & 1 & 1 & 0 & 1 \\ 1 & 0 & 1 & 0 & 1 & 1 \\ 0 & 1 & 1 & 1 & 1 & 0 \end{pmatrix}, \quad \mathbf{G}_b = \begin{pmatrix} 1 & 1 & 1 & 1 & 1 & 1 \\ 0 & 0 & 0 & 1 & 1 & 1 \\ 1 & 0 & 0 & 1 & 1 & 0 \\ 0 & 1 & 1 & 1 & 1 & 0 \end{pmatrix}.$$

with free distance $d_f = 5$.

A PUM code $(n = 6, k_0 = 4|k_1 = 2)$ with free distance $d_f = 4$ has the following generator matrices:

$$\mathbf{G}_a = \begin{pmatrix} 1 & 0 & 1 & 1 & 1 & 0 \\ 0 & 1 & 1 & 1 & 0 & 1 \\ 1 & 0 & 1 & 0 & 1 & 1 \\ 0 & 1 & 1 & 1 & 1 & 0 \end{pmatrix}, \quad \mathbf{G}_b = \begin{pmatrix} 1 & 1 & 1 & 1 & 1 & 1 \\ 0 & 0 & 0 & 1 & 1 & 1 \\ 0 & 0 & 0 & 0 & 0 & 0 \\ 0 & 0 & 0 & 0 & 0 & 0 \end{pmatrix}.$$

	0	1	2	3	4	5	6	7	8	9	10	11	12	13	14
0	0	1	1	0	0	0	0	0	1	1	1	1	0	0	1
1	0	1	0	1	1	0	0	1	1	0	1	0	0	0	0
2	1	0	1	1	1	1	0	0	1	1	0	1	1	0	1
3	1	0	0	1	0	1	1	1	1	0	1	1	1	0	1
4	1	0	1	0	0	0	0	1	1	0	0	1	1	0	0
5	0	0	1	0	1	0	1	1	0	1	1	1	0	1	0
6	0	0	1	0	0	0	0	0	1	0	1	1	1	1	0
7	1	0	1	1	1	1	0	0	0	0	1	1	0	1	1
8	0	0	1	0	1	0	0	1	0	0	1	0	1	0	0
9	0	1	0	0	1	1	0	1	1	0	1	0	0	1	0

1	0	1	1	$=\alpha^{13}$
0	1	1	0	$=\alpha^{5}$
0	0	0	0	$=\alpha^{-\infty}$
1	1	1	1	$=\alpha^{12}$
1	0	1	0	$=\alpha^{8}$
1	0	1	0	$=\alpha^{8}$
0	1	1	1	$=\alpha^{11}$
0	1	0	1	$=\alpha^{9}$
1	0	1	0	$=\alpha^{8}$
1	1	1	0	$=\alpha^{10}$

1	1	1	0	1	0	0	0	0	1	0	0	1	0	0
0	0	1	1	1	1	1	1	0	1	1	0	0	1	0
1	0	0	1	1	1	0	0	0	1	1	0	0	0	0
0	0	1	0	1	0	1	1	1	0	0	0	0	0	0
$=\alpha^8$	$=\alpha^0$	$=\alpha^7$	$=\alpha^5$	$=\alpha^{12}$	$=\alpha^5$	$=\alpha^9$	$=\alpha^9$	$=\alpha^3$	$=\alpha^{10}$	$=\alpha^5$	$=\alpha^{-\infty}$	$=\alpha^0$	$=\alpha^1$	$=\alpha^{-\infty}$

Figure 9.26 Received matrix with a burst error and corrupted parity check symbols.

Both UM and PUM codes have the UM subcode $\mathcal{B}^{(2)}(n=6, k=2)$, with generator matrices

$$\mathbf{G}_a^{(2)} = \begin{pmatrix} 1 & 0 & 1 & 1 & 1 & 0 \\ 0 & 1 & 1 & 1 & 0 & 1 \end{pmatrix}, \quad \mathbf{G}_b^{(2)} = \begin{pmatrix} 1 & 1 & 1 & 1 & 1 & 1 \\ 0 & 0 & 0 & 1 & 1 & 1 \end{pmatrix}.$$

The subcode has free distance $d_f^{(2)} = 7$. For the corresponding GC construction, we use the partitioning of the UM and PUM codes into 2^2 subcodes with an increasing free distance. In section 9.4.1, we return to this example for the concatenation of codes having inner convolutional and outer block component codes. ◇

If the (P)UM codes are constructed from RS or BCH codes, a partitioning can be achieved where suitable subcodes of the RS or BCH codes can be chosen.

Example 9.37 (Partitioning of (P)UM codes with RS codes) We have already seen the following codes in example 8.49. We use the RS codes with length 7 over $GF(2^3)$. The corresponding generator polynomials of the PUM codes are

$$\begin{aligned} g_{00}(x) &= (x-\alpha^2)(x-\alpha^3)(x-\alpha^4)(x-\alpha^5)(x-\alpha^6), \\ g_{01}(x) &= (x-\alpha^0)(x-\alpha^1)(x-\alpha^4)(x-\alpha^5)(x-\alpha^6), \\ g_{10}(x) &= (x-\alpha^0)(x-\alpha^1)(x-\alpha^2)(x-\alpha^3)(x-\alpha^6). \end{aligned}$$

The information polynomial $i_t(x)$ is divided into polynomials $i_t^{(0)}(x)$ and $i_t^{(1)}(x)$, giving $i_t(x) = i_{t,0} + i_{t,1}x + i_{t,2}x^2 + i_{t,3}x^3 = i_t^{(0)}(x) + i_t^{(1)}(x) = i_{t,0}^{(0)} + i_{t,1}^{(0)}x + i_{t,0}^{(1)}x^2 + i_{t,1}^{(1)}x^3$. The codeword is found by

$$c_t(x) = i_t^{(0)}(x)\,g_{00}(x) + i_t^{(1)}(x)\,g_{01}(x) + i_{t-1}^{(0)}(x)\,g_{10}(x) = a_t^{(0)}(x) + a_t^{(1)}(x) + b_t(x) ,$$

where $i_t^{(0)}(x)$ and $i_t^{(1)}(x)$ have degree 2. One observes that, based on the construction of a given $c_t(x)$, all codewords ($a_t^{(0)}(x), a_t^{(1)}(x)$ and $b_t(x)$) can be calculated explicitly (see figure 9.27 and

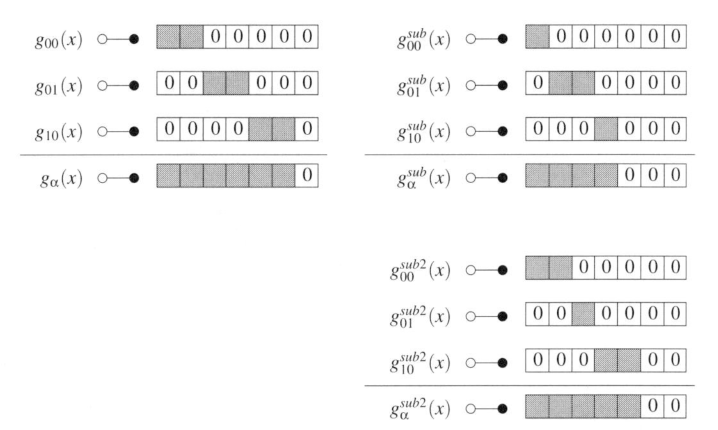

Figure 9.27 Transformation of the generator polynomial of the RS code.

section 8.7.4). The PUM code has parameters $k_0 = 4$ and $k_1 = 2$. The distances based on the RS code construction is

$$d_f \geq n-(k-k_1)+1 = 7-(4-2)+1 = 6,$$
$$d_\alpha \geq n-(k+k_1)+1 = 7-(4+2)+1 = 2.$$

where d_f and d_α are defined in section 8.7.3.

A possible subcode is given by

$$\begin{aligned} g_{00}^{sub}(x) &= (x-\alpha^1)(x-\alpha^2)(x-\alpha^3)(x-\alpha^4)(x-\alpha^5)(x-\alpha^6), \\ g_{01}^{sub}(x) &= (x-\alpha^0)(x-\alpha^3)(x-\alpha^4)(x-\alpha^5)(x-\alpha^6), \\ g_{10}^{sub}(x) &= (x-\alpha^0)(x-\alpha^1)(x-\alpha^2)(x-\alpha^4)(x-\alpha^5)(x-\alpha^6). \end{aligned}$$

This subcode has parameters $k_0 = 3$ and $k_1 = 1$, with distances

$$d_f \geq n-(k-k_1)+1 = 7-(3-1)+1 = 6,$$
$$d_\alpha \geq n-(k+k_1)+1 = 7-(3+1)+1 = 4.$$

The PUM code $\mathcal{B}^{(1)}(2^3; n=7, k_0=4|k_1=2)$ is divided into 2^3 subcodes $\mathcal{B}_i^{(2)}(2^3; n=7, k_0=3|k_1=1)$, $i \in GF(2^3)$. The subcodes $\mathcal{B}_i^{(2)}$ have the same free distance as the code $\mathcal{B}^{(1)}$; therefore one has no gain in the free distance due to the partitioning. However, the code $\mathcal{C}_\alpha^{sub}$ with generator polynomial g_α^{sub} (see figure 9.27) has a larger minimum distance. This partitioning is undesirable for the construction of GC codes, because the subcode free distance does not increase.

Another possible subcode is given by

$$\begin{aligned} g_{00}^{sub2}(x) &= (x-\alpha^2)(x-\alpha^3)(x-\alpha^4)(x-\alpha^5)(x-\alpha^6), \\ g_{01}^{sub2}(x) &= (x-\alpha^0)(x-\alpha^1)(x-\alpha^3)(x-\alpha^4)(x-\alpha^5)(x-\alpha^6), \\ g_{10}^{sub2}(x) &= (x-\alpha^0)(x-\alpha^1)(x-\alpha^2)(x-\alpha^5)(x-\alpha^6). \end{aligned}$$

This subcode has parameters $k_0 = 3$ and $k_1 = 2$, with distances

$$d_f \geq n-(k_0-k_1)+1 = 7-(3-2)+1 = 7,$$
$$d_\alpha \geq n-(k_0+k_1)+1 = 7-(3+2)+1 = 3.$$

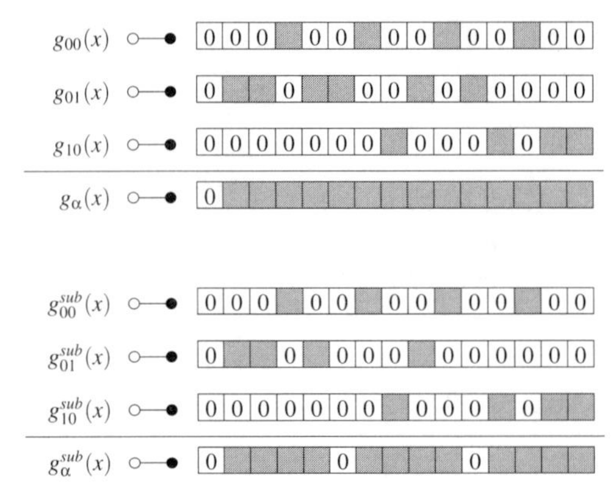

Figure 9.28 Transformation of the generator polynomial of the BCH code.

In this case, one has a gain in the free distance, and the partitioning is suited for a GC code, because the subcode free distance increases. ◇

Example 9.38 (Partitioning of (P)UM codes with BCH codes) The PUM code from example 8.50 was $\mathcal{B}^{(1)}(n=15, k_0=10|k_1=4)$, having free distance $d_f=6$ and generator polynomials

$$\begin{aligned} g_{00}(x) &= m_0(x)m_1(x)m_5(x)m_7(x), \\ g_{01}(x) &= m_0(x)m_3(x)m_7(x), \\ g_{10}(x) &= m_0(x)m_1(x)m_3(x)m_5(x). \end{aligned}$$

A possible subcode has generator polynomials

$$\begin{aligned} g_{00}(x) &= m_0(x)m_1(x)m_5(x)m_7(x), \\ g_{01}(x) &= m_0(x)m_1(x)m_3(x)m_5(x), \\ g_{10}(x) &= m_0(x)m_1(x)m_3(x)m_5(x). \end{aligned}$$

$\mathcal{B}^{(1)}$ is therefore partitioned into 2^2 subcodes $\mathcal{B}_i^{(2)}(n=15, k_0=8|k_1=4)$, having a free distance $d_f=8$. In figure 9.28 the transformation is shown for each generator polynomial of the constructed codes. ◇

Example 9.39 (Partitioning of (P)UM codes with RM codes) We consider the codes from example 8.51, and determine first the generator polynomial of BCH codes of length 31 using the same principle as in example 9.38:

$$\begin{aligned} g_{00}(x) &= m_0(x)m_1(x)m_3(x)m_5(x)m_{15}(x), \\ g_{01}(x) &= m_1(x)m_3(x)m_5(x)m_7(x)m_{11}(x), \\ g_{10}(x) &= m_0(x)m_1(x)m_7(x)m_{11}(x)m_{15}(x). \end{aligned}$$

We can now construct the subcode where we replace $g_{01}(x)$ with $g_{01}^{sub}(x) = m_1(x)m_3(x)m_5(x)m_7(x)m_{11}(x)m_{15}(x)$. The parameters of the subcode are $n=32$, $k=11$, $k_1=10$, $d_\alpha=4$ and $d_f=16$. The free distance d_f of the subcode is the same as the original code. The partitioning of parallel branches in the trellis has increased the distance from $d_{01}=16$ to $d_{01}^{sub}=32$. ◇

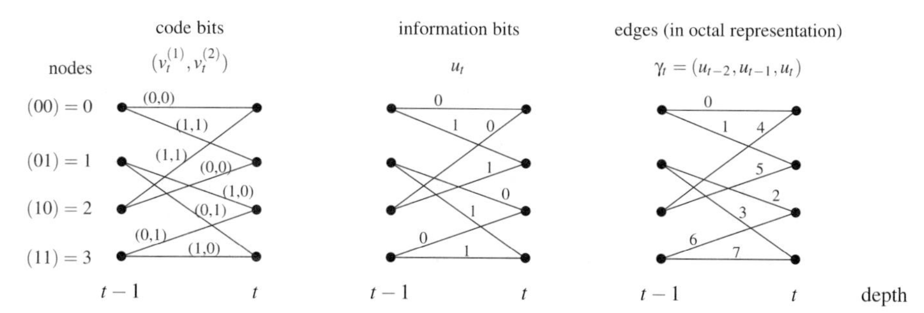

Figure 9.29 Construction of the code/input bits in trellis form.

Based on the above examples, we develop a general theory for the partitioning of PUM codes. In section 9.4.1, we again describe methods required for the construction of GC codes with inner (P)UM and outer RS codes.

9.3.2 Introductory example of trellis partitioning

We have stated above that (P)UM codes can be partitioned based on block codes. In the following, we consider an example of possible partitioning of the well-known rate-1/2 convolutional code from [Vit71] having the generator sequence (octal) $g^{(1)} = 7$ and $g^{(2)} = 5$ (see example 8.1). The partitioning method can be found in [BDS97]. The basis of this trellis construction is shown in figure 9.29. The trellis can be defined based on the input bits. Therefore a node is labeled $\sigma_t = (u_{t-1}, u_t)$ and an edge by $\gamma_t = (u_{t-2}, u_{t-1}, u_t)$. Thereby, two bits are necessary to define a node. One bit indicates the state of the system, and the other is an input bit. At a particular depth (level) there are $2^\nu = 4$ nodes. Two paths enter and leave each node.

Example 9.40 (Order-2 partitioning without a scrambler) Our example code can be considered as a $\mathcal{B}^{(1)}(n=4, k^{(1)}=2, d^{(1)}=5)$ UM code[3] having generator matrices

$$\mathbf{G}_0 = \begin{pmatrix} 1 & 1 & 1 & 0 \\ 0 & 0 & 1 & 1 \end{pmatrix} \quad \text{and} \quad \mathbf{G}_1 = \begin{pmatrix} 1 & 1 & 0 & 0 \\ 1 & 0 & 1 & 1 \end{pmatrix}.$$

In order to achieve a partitioning for the UM representation, the input bits u_t are divided into two groups. First we introduce the bit numbering $(z_\tau^{(1)}, z_\tau^{(2)})$ for the partition:

$$u_{t-1} = z_\tau^{(1)} \quad \text{and} \quad u_t = z_\tau^{(2)}.$$

This notation results in a change in how the trellis depth is labeled.

Figure 9.30 shows the partitioned trellis that results if all bits $z_\tau^{(1)}$, $\tau = 0, 1, \ldots,$ are set to 0. In the left section of the figure, all possible paths are shown. In the right section, the zero path and the path with the free distance are plotted. The nodes encircled by the dashed lines correspond to the nodes of the UM code trellis. Paths $\mathbf{p}_1$ and $\mathbf{p}_2$ have weight 5, and the free distance of the subcode $\mathcal{B}^{(2)}(n=4, k^{(2)}=1, d^{(2)}=5)$ is equal to the code $\mathcal{B}^{(1)}$. Therefore $d^{(1)} = d^{(2)} = 5$, due to the fact that $\mathbf{p}_2$ remains a codeword of the subcode. The partitioning does not increase the free distance of the subcodes. This partition is not suited for GC code construction. ◇

[3] For simplification we omit the indices of the subcodes as long as the meaning is clear.

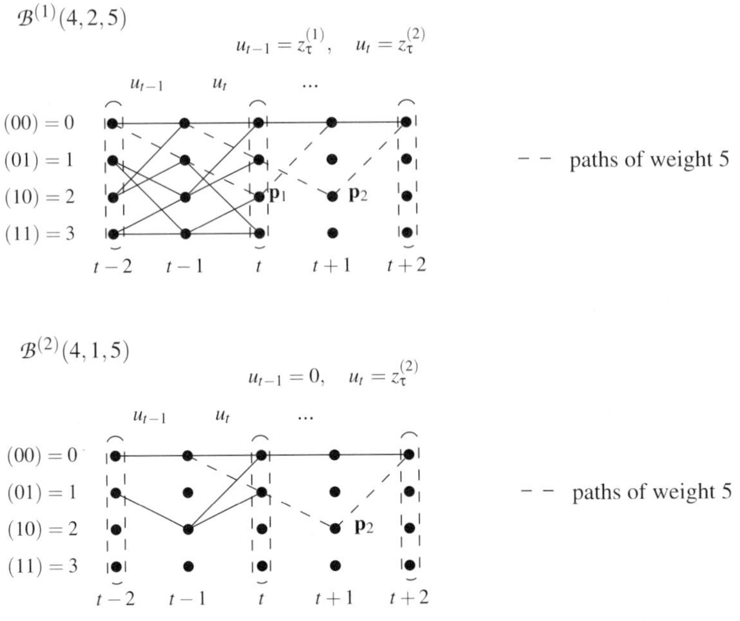

Figure 9.30 Trellis partitioning of order 2 without scrambler.

This example shows that the direct partitioning of codes viewed as UM codes does not necessarily lead to a better subcode. In the following section, we develop an equivalent convolutional encoder construction (the code sequences are identical). The difference lies in the mapping of the information sequence to the code sequence (see section 8.2.3 on equivalent code matrices).

Example 9.41 (Order-2 partitioning with a block scrambler) In figure 9.31, we consider the upper trellis as a partitioning of the code $\mathcal{B}^{(1)}(n=4, k^{(1)}=2, d^{(1)}=5)$. The partitioning is advantageous when the subcode $\mathcal{B}^{(2)}$ does not contain the path with the smallest weight. Our goal is to select the vertices from the $\mathcal{B}^{(1)}$ trellis in such a way that the minimum- weight paths $\mathbf{p}_1$ and $\mathbf{p}_2$ are punctured. This is achieved by removing states $\sigma_t \in \{1,2\}$ at depth $t = 2\tau$, $\tau \in 1,2,\ldots$ (see figure 9.31).

This requirement can be achieved by insuring the following condition for $t = 2\tau$, $\tau > 0$:

$$\mathcal{V}_t = \left\{\sigma_t \,:\, \sigma_t = \left(u_{t-1}, u_t\right) = \left(0, z_\tau^{(2)}\right) \cdot \mathbf{M};\quad z_\tau^{(2)} \in \{0,1\}\right\} = \{0,3\}\,.$$

where

$$\mathbf{M} = \mathbf{M}^{-1} = \begin{pmatrix} 1 & 0 \\ 1 & 1 \end{pmatrix}$$

are the scrambler and inverse scrambler matrices respectively. The subcode $\mathcal{B}^{(2)}(n=4, k^{(2)} = 1, d^{(2)} = 6)$ has free distance $d^{(2)} = 6$. We partition the code as in example 9.40 using the bit numbering $(z_\tau^{(1)}, z_\tau^{(2)})$. The scrambler maps the bit $z_\tau^{(1)} = 0$ to states $\sigma_t \in \{0,3\}$ and the bit $z_\tau^{(1)} = 1$ to states $\sigma_t \in \{1,2\}$.

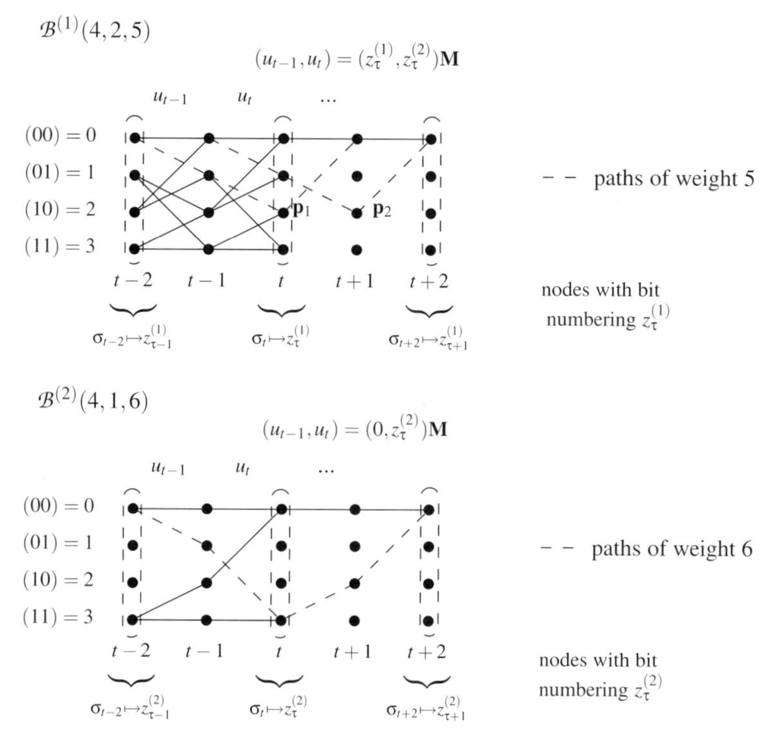

Figure 9.31 Trellis based on partitioning 2 with a block scrambler.

This can also be written in terms of the code matrices

$$\mathbf{G}_0=\begin{pmatrix}1&1&1&0\\0&0&1&1\end{pmatrix}\quad\text{and}\quad\mathbf{G}_1=\begin{pmatrix}1&1&0&0\\1&0&1&1\end{pmatrix}.$$

The scrambler gives the equivalent code matrices

$$\mathbf{G}_0'=\mathbf{M}\cdot\mathbf{G}_0=\begin{pmatrix}1&1&1&0\\1&1&0&1\end{pmatrix}\quad\text{and}\quad\mathbf{G}_1'=\mathbf{M}\cdot\mathbf{G}_1=\begin{pmatrix}1&1&0&0\\0&1&1&1\end{pmatrix}$$

Note (see section 8.2.3) For each code matrix $\mathbf{G}_j$, there is a set of equivalent generator matrices $\mathbf{G}_j'$ that produces the same code and that can be calculated with the aid of the scrambler matrix $\mathbf{M}$ ($|\mathbf{M}|=1$):

$$\mathbf{G}_j'=\mathbf{M}\cdot\mathbf{G}_j\,,\quad j=0,1.$$

The $k\times k$ matrix $\mathbf{M}$ is invertible, and is written as $\mathbf{M}^{-1}$. The difference between the generator matrices $\mathbf{G}'$ and $\mathbf{G}$ is the mapping from the information sequence to the code sequence. We use the same partitioning as in example 9.40, with the bit numbering $(z_\tau^{(1)},z_\tau^{(2)})$. Thereby we achieve a subcode with a larger free distance ($d^{(1)}=5$ and $d^{(2)}=6$), which is good for GC code construction.

This example shows that we must find a generator matrix that can produce an optional partitioning of the inner convolutional code. In the following example, we improve the partitioning by using a scrambler matrix with memory.

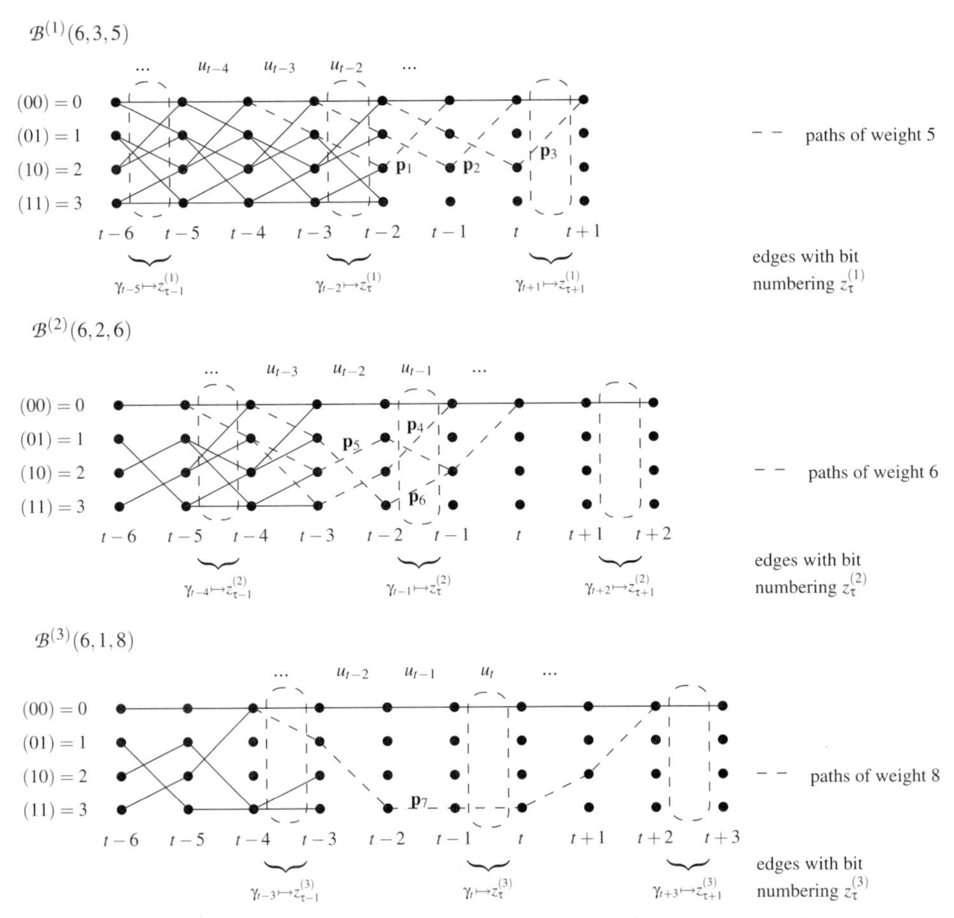

Figure 9.32 Trellises of the subcodes for the partitioning of order 3.

Example 9.42 (Order-3 partitioning with a convolutional scrambler) The basic goal of partitioning is to remove the trellis paths with the smallest weight, such that the subcode has the largest possible minimum distance. In contrast to the previous example, we consider the paths in the trellis instead of the nodes (see figure 9.32). We consider an order-three partitioning constructed from the enumeration bits $(z_\tau^{(1)}, z_\tau^{(2)}, z_\tau^{(3)})$ as shown in figure 9.32. At depth $t = 3\tau$, we make the following association between bits and edges: bits $z_\tau^{(1)}$ to edges γ_{t-2}, bits $z_\tau^{(2)}$ to edges γ_{t-1}, and bits $z_\tau^{(3)}$ to edges γ_t, as shown in the figure.

We proceed by identifying which path, when removed, will result in a partition with the largest possible free distance of the subcode. Based on the code trellis $\mathcal{B}^{(1)}$, we know that the edges $\gamma_{t-2} \in \{1,2,4\}$ should be punctured in order to remove the minimum-weight paths $\mathbf{p}_1$, $\mathbf{p}_2$ and $\mathbf{p}_3$. We set bits $z_\tau^{(1)} = 0$ that correspond to edges $\gamma_{t-2} \in \{0,3,5,6\}$ and $z_\tau^{(1)} = 1$ that correspond to edges $\gamma_{t-2} \in \{1,2,4,7\}$. This defines the first partition level, and gives the inner subcode $\mathcal{B}^{(2)}$.

In order to partition the code $\mathcal{B}^{(2)}$, the minimum-weight paths $\mathbf{p}_4$, $\mathbf{p}_5$ and $\mathbf{p}_6$ with weight 6 should be punctured. This means that edges $\gamma_{t-1} \in \{1,6\}$ should be removed. This can be achieved if $z_\tau^{(2)} = 0$ corresponds to edges $\gamma_{t-1} \in \{0,3,4,7\}$, and consequently $z_\tau^{(2)} = 1$ corresponds to edges $\gamma_{t-1} \in \{1,2,5,6\}$.

The resulting code $\mathcal{B}^{(3)}$ has a free distance $d^{(3)} = 8$, because the path $\mathbf{p}_7$ is not removed. The third enumeration bit $z_\tau^{(3)}$ selects the code sequence corresponding to $z_\tau^{(3)} = 0$, giving edges $\gamma_t \in \{0,3,4,7\}$, and $z_\tau^{(3)} = 1$, giving edges $\gamma_t \in \{1,2,5,6\}$.

At this point, we again formally describe a scrambler that achieves the desired subcode construction. Let $\mathbf{x}^{(j)} = (x_1^{(j)}, x_2^{(j)}, x_3^{(j)})^T$, $j = 1,2,3$ be a column 3-vector. Based on the above considerations, the numbering of the information bits (u_{t-2}, u_{t-1}, u_t) (corresponding to one edge in the trellis) is transformed to the three-bit enumeration $(z_\tau^{(1)}, z_\tau^{(2)}, z_\tau^{(3)})$:

$$(z_\tau^{(1)}, z_\tau^{(2)}, z_\tau^{(3)}) \cdot \mathbf{M} = \gamma \Longrightarrow (z_\tau^{(1)}, z_\tau^{(2)}, z_\tau^{(3)}) = \gamma \cdot \mathbf{M}^{-1} = \gamma \cdot \mathbf{X}$$

The inverse scrambler matrix $\mathbf{X}$ is found using the following system of equations:

$$\begin{aligned}
&\{\mathbf{x}^{(1)} \mid \forall \gamma_{t-2} \in \{1,2,4\} \;\; \gamma_{t-2}\mathbf{x}^{(1)} = z_\tau^{(1)} \neq 0\} && \Rightarrow && \mathbf{x}^{(1)} = (111)^T, \\
&\{\mathbf{x}^{(2)} \mid \forall \gamma_{t-1} \in \{1,6\} \;\; \gamma_{t-1}\mathbf{x}^{(2)} = z_\tau^{(2)} \neq 0\} && \Rightarrow && \mathbf{x}^{(2)} = (011)^T, \\
&\{\mathbf{x}^{(3)} \mid \forall \gamma_{t} \in \{1,2,5,6\} \;\; \gamma_{t}\mathbf{x}^{(3)} = z_\tau^{(3)} \neq 0\} && \Rightarrow && \mathbf{x}^{(3)} = (011)^T.
\end{aligned}$$

The mapping of the enumeration bits to edges is calculated as

$$z_\tau^{(j)} = \gamma_{t-3+j} \cdot \mathbf{x}^{(j)}, \quad t = 3\tau, \quad \tau = 1,2,\ldots .$$

This can be expanded to give

$$\begin{aligned}
z_\tau^{(1)} &= (u_{t-4}, u_{t-3}, u_{t-2}) \cdot (111)^T, \\
z_\tau^{(2)} &= (u_{t-3}, u_{t-2}, u_{t-1}) \cdot (011)^T, \\
z_\tau^{(3)} &= \;\; (u_{t-2}, u_{t-1}, u_t) \cdot (011)^T.
\end{aligned}$$

The inverse scrambler is now known, and it can be described in terms of the following matrices:

$$\mathbf{X}_0 = \begin{pmatrix} x_3^{(1)} & x_2^{(2)} & x_1^{(3)} \\ 0 & x_3^{(2)} & x_2^{(3)} \\ 0 & 0 & x_3^{(3)} \end{pmatrix} \quad \text{and} \quad \mathbf{X}_1 = \begin{pmatrix} 0 & 0 & 0 \\ x_1^{(1)} & 0 & 0 \\ x_2^{(1)} & x_1^{(2)} & 0 \end{pmatrix}.$$

Based on section 8.1.8, this can be written in the transformation domain as

$$\mathbf{M}^{-1}(D) = \mathbf{X}_0 + D\mathbf{X}_1 = \begin{pmatrix} 1 & 1 & 0 \\ D & 1 & 1 \\ D & 0 & 1 \end{pmatrix}.$$

One observes that this inverse scrambler has memory, and is therefore termed a convolutional scrambler. A matrix inversion produces the scrambler

$$\mathbf{M}(D) = \begin{pmatrix} 1 & 1 & 1 \\ 0 & 1 & 1 \\ D & D & 1+D \end{pmatrix}.$$

This matrix can calculate an equivalent encoder if the code matrix is first represented in the transform domain,

$$\mathbf{G}(D) = \mathbf{G}_0 + D\mathbf{G}_1 = \begin{pmatrix} 1 & 1 & 1 & 0 & 1 & 1 \\ D & D & 1 & 1 & 1 & 0 \\ D & 0 & D & D & 1 & 1 \end{pmatrix},$$

and then multiplied by the scrambler:

$$\mathbf{G}'(D) = \mathbf{M}(D) \cdot \mathbf{G}(D) = \begin{pmatrix} 1 & 1+D & D & 1+D & 1 & 0 \\ 0 & D & 1+D & 1+D & 0 & 1 \\ 0 & 1+D^2 & D+D^2 & D^2 & 1+D & 1 \end{pmatrix}.$$

We have derived an order-3 partitioning of $\mathcal{B}^{(1)}$ having free distances $d^{(1)} = 5$, $d^{(2)} = 6$ and $d^{(3)} = 8$. ◇

Is it possible to improve an order-2 partitioning using a convolutional scrambler? The next example confirms this idea.

Example 9.43 (Order-2 partitioning with a convolutional scrambler) We consider the scrambler

$$\mathbf{M}(D) = \begin{pmatrix} 0 & 1 \\ 1 & D \end{pmatrix} \quad \text{and} \quad \mathbf{M}^{-1}(D) = \begin{pmatrix} D & 1 \\ 1 & 0 \end{pmatrix}.$$

From the previous example, we can write the inverse scrambler using two column vectors $\mathbf{x}^{(1)}$ and $\mathbf{x}^{(2)}$:

$$\mathbf{M}^{-1}(D) = \mathbf{X}_0 + \mathbf{X}_1 D, \quad \text{with} \quad \begin{pmatrix} \mathbf{X}_0 \\ \mathbf{X}_1 \end{pmatrix} = \left(\mathbf{x}^{(1)}, \mathbf{x}^{(2)}\right) = \begin{pmatrix} 1 & 0 \\ 0 & 0 \\ 0 & 1 \\ 1 & 0 \end{pmatrix}.$$

According to the left trellis in figure 9.33, we require 4 input bits to define an edge:

$$\gamma_t = (u_{t-3}, u_{t-2}, u_{t-1}, u_t) \quad \text{and} \quad z_\tau^{(1)} = \gamma_t \mathbf{x}^{(1)},\ z_\tau^{(2)} = \gamma_t \mathbf{x}^{(2)},\ t = 2\tau,\ \tau > 0\,.$$

This is achieved because, instead of 4 nodes, we use 8. Therefore a node in depth $t-1$ is defined by $\sigma_{t-1} = (u_{t-3}, u_{t-2}, u_{t-1})$. This produces the edge $\gamma_t = (\sigma_{t-1}, u_t) = (u_{t-3}, u_{t-2}, u_{t-1}, u_t)$.

There are two further possible trellis representations that can achieve this construction. These graphs are shown in the center and right sections of figure 9.33. The middle trellis shows a time-variant trellis that has 8 nodes (instead of 4) for the odd levels. The right trellis uses the UM construction, and considers two successive edges at depth $t-1$ and t as one extended edge that is defined by $\gamma_{t,2} = (u_{t-3}, u_{t-2}, u_{t-1}, u_t)$. This concept of extended edges is used in the next example.

Independent of the trellis interpretation, we have the following equivalent generator matrix:

$$\mathbf{G}'(D) = \mathbf{M}(D) \cdot \mathbf{G}(D) = \begin{pmatrix} D & 0 & 1+D & 1+D \\ 1+D+D^2 & 1+D & 1+D+D^2 & D+D^2 \end{pmatrix}.$$

The free distance of the subcode $\mathcal{B}^{(2)}$ is $d^{(2)} = 8$. ◇

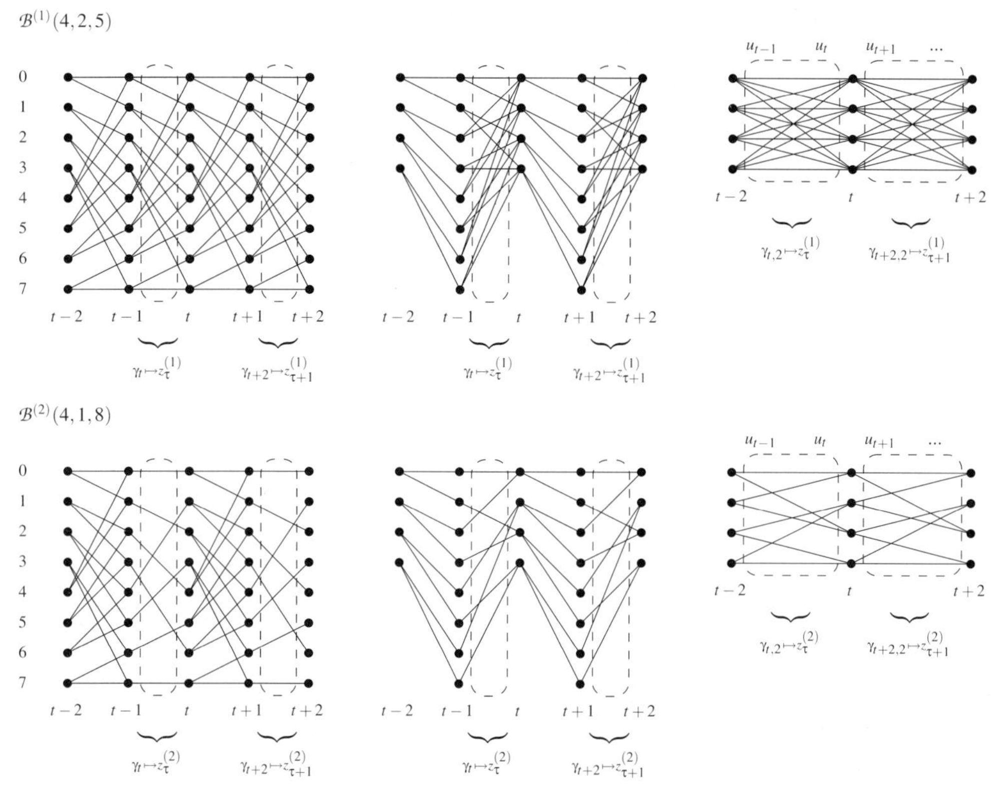

Figure 9.33 Subcode trellises for three construction methods.

In the following example, we derive a partitioning of order-4:

Example 9.44 (Order-4 partitioning using extended edges) Consider two successive edges γ_{t-1} and γ_t as one extended edge $\gamma_{t,2} = (u_{t-3}, u_{t-2}, u_{t-1}, u_t)$. We can use these 4 bits as enumeration bits in the same way as in the previous example. Figure 9.34 shows the extended edges. We remove all edges $\gamma_{t,2}, \gamma_{t+4,2}, \gamma_{t+8,2}, \ldots$ corresponding to the smallest-weight paths in the 4th step of the partitioning. The partitioning is

$$\mathcal{B}^{(4)}(8,1,10) \subset \mathcal{B}^{(3)}(8,2,6) \subset \mathcal{B}^{(2)}(8,3,6) \subset \mathcal{B}^{(1)}(8,4,5).$$

The corresponding scrambler matrix is calculated as

$$\mathbf{M} = \begin{pmatrix} 0&1&0&0\\ 1&1&0&0\\ 0&1&1&0\\ 1&0&0&1 \end{pmatrix}, \quad \text{with} \quad \mathbf{M}^{-1} = \begin{pmatrix} 1&1&0&0\\ 1&0&0&0\\ 1&0&1&0\\ 1&1&0&1 \end{pmatrix}.$$

At this point, we consider the code $\mathcal{B}^{(1)}$ as an $(n = 8, k^{(1)} = 4, d^{(1)} = 5)$ PUM code. We produce the 4×8 code matrices:

$$\mathbf{G}_0 = \begin{pmatrix} 1&1&1&0&1&1&0&0\\ 0&0&1&1&1&0&1&1\\ 0&0&0&0&1&1&1&0\\ 0&0&0&0&0&0&1&1 \end{pmatrix} \quad \text{and} \quad \mathbf{G}_1 = \begin{pmatrix} 0&0&0&0&0&0&0&0\\ 0&0&0&0&0&0&0&0\\ 1&1&0&0&0&0&0&0\\ 1&0&1&1&0&0&0&0 \end{pmatrix}.$$

Table 9.2 Error weight of subcode enumeration for codewords with weight w.

	Error weight c_w for distance w								
	Example 9.40		Example 9.41		Example 9.43		Example 9.42		
Weight w	$\mathcal{B}^{(1)}$	$\mathcal{B}^{(2)}$	$\mathcal{B}^{(1)}$	$\mathcal{B}^{(2)}$	$\mathcal{B}^{(1)}$	$\mathcal{B}^{(2)}$	$\mathcal{B}^{(1)}$	$\mathcal{B}^{(2)}$	$\mathcal{B}^{(3)}$
5	1	1	2	-	2	-	3	-	-
6	4	2	6	1	8	-	6	4	-
7	12	3	16	0	20	-	14	0	-
8	32	4	40	2	46	2	32	14	1
9	80	5	96	0	104	0	72	0	0
10	192	6	224	3	234	4	160	48	2
11	448	7	512	0	528	0	352	0	0
12	1024	8	1152	4	1184	8	768	156	3
$\vdots$									

We multiply the code matrix by the scrambler to get the equivalent code:

$$\mathbf{G}_0' = \begin{pmatrix} 0&0&1&1&1&0&1&0 \\ 1&1&0&1&0&1&1&1 \\ 0&0&1&1&0&1&0&1 \\ 1&1&1&0&1&1&1&1 \end{pmatrix} \quad \text{and} \quad \mathbf{G}_1' = \begin{pmatrix} 0&0&0&0&0&0&0&0 \\ 0&0&0&0&0&0&0&0 \\ 1&1&0&0&0&0&0&0 \\ 1&0&1&1&0&0&0&0 \end{pmatrix}.$$

Note In [BDS97], it is shown that this construction results in a higher decoding complexity and proceeds by presenting an order-4 partitioning that does not result in a decoding complexity increase.

In table 9.2, the results from the first four examples are presented. The error weight c_w is defined as the number of paths with weight w that deviate from the zero path at a particular point in time, multiplied by the associated number of information bits equal to 1. In our case, this corresponds to the information bits with the numbering $\mathbf{z}^{(i)}$ for the subcode $\mathcal{B}^{(i)}_{\mathbf{z}^{(1)},\dots,\mathbf{z}^{(i-1)}}$.

The partitioning of the same convolutional code can be carried out using different methods. The free distance is used as a measure of the quality of the code. It has not been shown whether any one of the distances defined in section 8.3 is consistently better than another. The concept of extended paths can be applied to derive a partitioning of higher order independent of the distance measure used.

9.3.3 Partitioning of convolutional codes

We have shown in the previous section that the performance of a code partitioning depends on the encoder. The principal partitioning possibilities of convolutional codes were demonstrated using examples. We generalize these techniques for rate $1/n$ (an extension to rate k/n is possible).

We illustrate the following important points from [BDS97]. Of special interest is a scrambler that can achieve a partitioning of order s for a given convolutional code. Our goal is to find an optimal equivalent code for partitioning based on the construction of an appropriate scrambler according to section 9.3.3.

Consider a convolutional code C of rate $1/n$ with overall constraint lenth ν and free distance d.

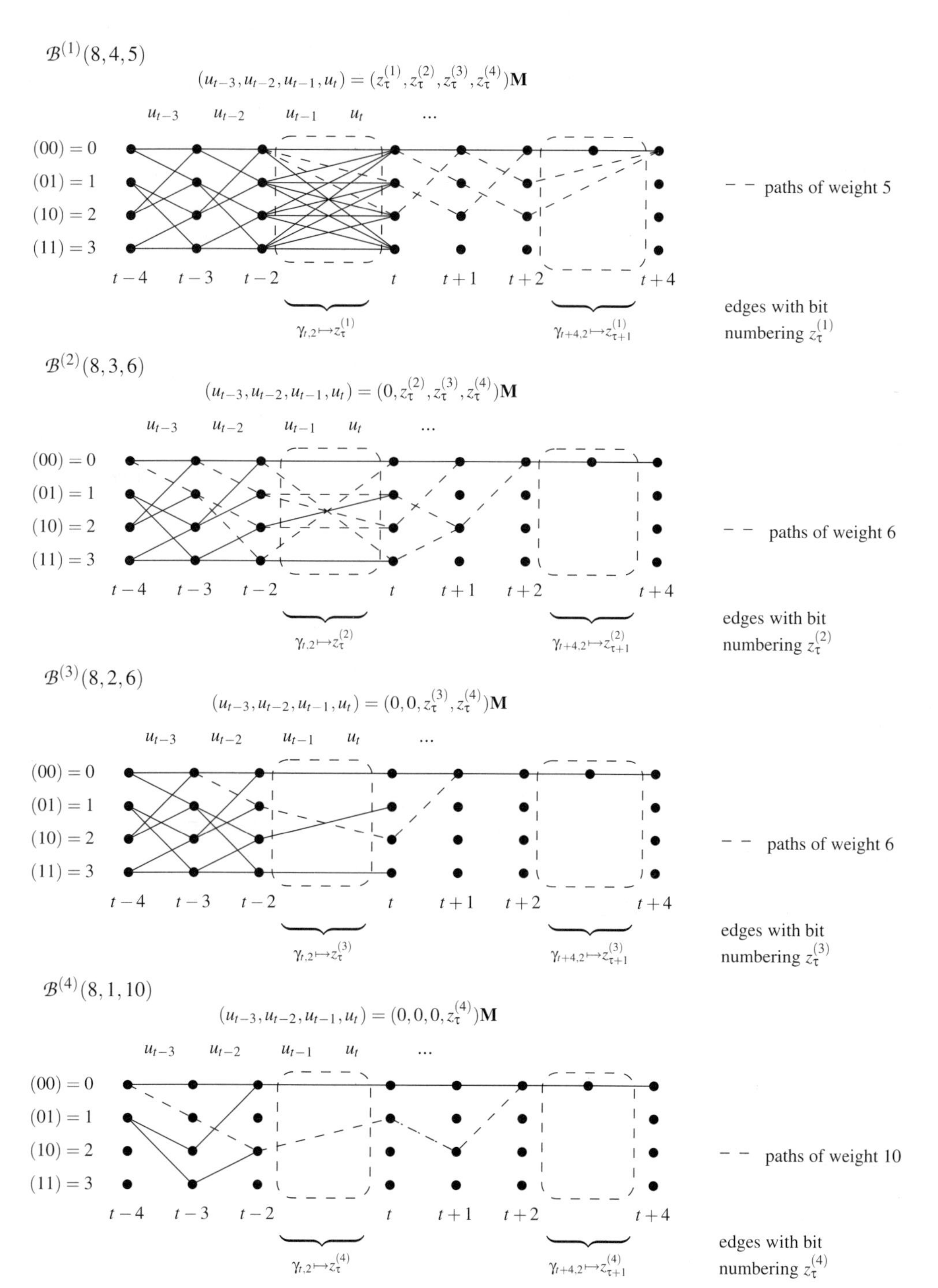

Figure 9.34 Subcode trellises for partitioning of order 4.

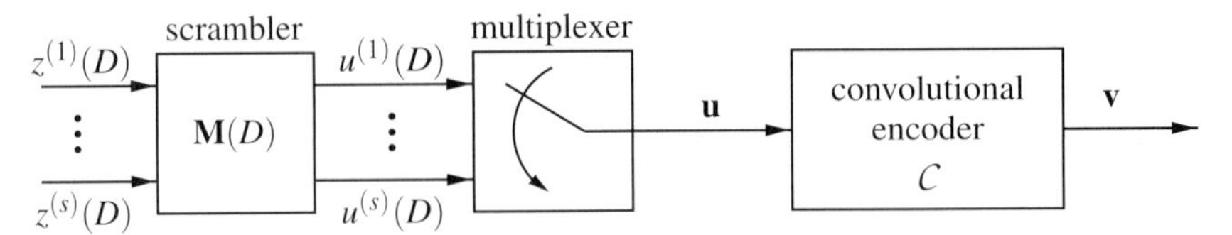

Figure 9.35 Equivalent encoder for the convolutional code C with a partitioning of order s.

The binary information sequence $\mathbf{i} = (i_1, i_2, \ldots, i_t, \ldots)$ is partitioned into s enumeration sequences $\mathbf{z}^{(j)} = (z_1^{(j)}, z_2^{(j)}, \ldots, z_\tau^{(j)}, \ldots)$, $j = 1, 2, \ldots, s$:

$$z_\tau^{(j)} = i_{s(\tau-1)+j}, \quad \tau = 1, 2, \ldots \quad .$$

The partitioning of order s is carried out with respect to this numbering as shown in figure 9.35.

For the encoding, the enumeration sequence is passed through the scrambler, which is an $s \times s$ matrix, and is then multiplexed to give the output sequence $\mathbf{u} = (u_1, u_2, \ldots, u_t, \ldots)$. This sequence is then encoded by the generator matrix of the convolutional code to produce the codeword $\mathbf{v}$. We achieve the partitioning

$$\mathcal{B}^{(s)}_{\mathbf{z}^{(1)}, \mathbf{z}^{(2)}, \ldots, \mathbf{z}^{(s-1)}} \subset \ldots \subset \mathcal{B}^{(2)}_{\mathbf{z}^{(1)}} \subset \mathcal{B}^{(1)} .$$

The subcode $B^{(j)}_{\mathbf{z}^{(1)}, \mathbf{z}^{(2)}, \ldots, \mathbf{z}^{(j-1)}}$ is defined as the codeword set of C, where

- $\mathbf{z}^{(1)}, \mathbf{z}^{(2)}, \ldots, \mathbf{z}^{(j-1)}$ correspond to the given numbering, and
- $\mathbf{z}^{(j)}, \mathbf{z}^{(j+1)}, \ldots, \mathbf{z}^{(s)}$ have any arbitrary value.

The code $\mathcal{B}^{(1)}$ is the original code C. The free distance of the subcode $d = d^{(1)} \leq d^{(2)} \leq \ldots \leq d^{(j)} \leq \ldots \leq d^{(s)}$, where d is the free distance of C.

For the enumeration, the following description is obtained using the transformation description:

$$\mathbf{z}(D) = (z^{(1)}(D), z^{(2)}(D), \ldots, z^{(s)}(D)) , \quad \text{with } z^{(j)}(D) = z_1^{(j)} + z_2^{(j)} D + z_3^{(j)} D^2 + \ldots \quad .$$

The output of the scrambler is written as

$$\mathbf{u}(D) = (u^{(1)}(D), u^{(2)}(D), \ldots, u^{(s)}(D)) , \quad \text{with } u^{(j)}(D) = u_j + u_{s+j} D + u_{2s+j} D^2 + \ldots \quad .$$

This gives

$$\mathbf{u}(D) = \mathbf{z}(D) \cdot \mathbf{M}(D), \quad \text{and correspondingly,} \quad \mathbf{z}(D) = \mathbf{u}(D) \cdot \mathbf{M}^{-1}(D) . \tag{9.19}$$

We require that a right-invertible matrix $\mathbf{M}^{-1}(D)$ exist and that $\mathbf{M}(D)$ be realizable. For the case when $\mathbf{M}(D) = \mathbf{M}$, it is termed a *block scrambler*, otherwise it is termed a *convolutional scrambler*. The encoding matrix of the equivalent convolutional generator matrix is

$$\mathbf{G}'(D) = \mathbf{M}(D) \cdot \mathbf{G}(D) ,$$

where $\mathbf{G}(D)$ is the $s \times sn$ generator matrix of the convolutional code C having a smaller memory (see section 8.7.4 concerning UM code representation). In [BDS97], it is proved that if the original encoder is non-catastrophic then the equivalent encoder is also non-catastrophic if the scrambler is polynomial.

In the previous examples, we have seen that, under certain conditions, subcodes can be represented through node or edge puncturing in the trellis of the original code. Algorithm 9.3 calculates the inverse scrambler $\mathbf{M}^{-1}(D)$.

Codes with rate $1/n$ can be described by a trellis where 2^{ν} paths enter and leave every node. The node at depth t

$$\sigma_t = (u_{t-\nu+1}, \dots, u_t)$$

can be described using ν input bits. The corresponding edge is

$$\gamma_t = (\sigma_{t-1}, u_t) = (u_{t-\nu}, \dots, u_t).$$

Each codeword is a possible path $\mathbf{p} = (\mathbf{p}_1, \dots, \mathbf{p}_t \dots,)$ in the trellis of the corresponding code, and can equivalently be written as a series of nodes $(\sigma_1, \dots, \sigma_t, \dots)$ or edges $(\gamma_1, \dots, \gamma_t, \dots)$

For node puncturing, we describe a path by nodes, while for edge puncturing, we use the edge description. We use the following inverse scrambler notation. An inverse scrambler $\mathbf{M}^{-1}(D)$ is constructed from s columns $\mathbf{m}_j^{-1}(D)$, $j = 1, 2, \dots, s$. Furthermore,

$$\mathbf{M}^{-1}(D) = \sum_{l=0}^{q} \mathbf{X}_l D^l$$

is a binary $s \times s$ submatrix $\mathbf{X}_l$, where q is the maximal degree of the polynomials of the inverse scrambler matrix. We write the matrix as

$$\mathbf{X} = \begin{pmatrix} \mathbf{X}_q \\ \vdots \\ \mathbf{X}_1 \\ \mathbf{X}_0 \end{pmatrix} = \left(\hat{\mathbf{x}}^{(1)}, \dots, \hat{\mathbf{x}}^{(j)}, \dots, \hat{\mathbf{x}}^{(s)}\right),$$

with s columns of length sq. Each column must satisfy the following condition:

$$\hat{\mathbf{x}}^{(j)} = (\mathbf{0}|\mathbf{x}^{(j)}|\overbrace{\mathbf{0}}^{\Delta^{(j)}\text{zeros}})^T, \quad 0 \le \Delta^{(j)} < s.$$

$\Delta^{(j)}$ is defined as a shift of the jth column. The subvector $\mathbf{x}^{(j)}$ contains all coordinates of the column that are not equal to zero. For node puncturing, this subvector has length ν, while for edge puncturing, it has length $\nu + 1$, which gives $q = \lceil \nu/s \rceil$. We obtain the mapping

$$(\mathbf{x}^{(j)}, \Delta^{(j)}) \Longrightarrow \hat{\mathbf{x}}^{(j)} \Longleftrightarrow \mathbf{m}_j^{(-1)}(D).$$

Note The first and/or last coordinate of $\mathbf{x}^{(j)}$ can contain zeros. A mapping between the enumerated bit $z_\tau^{(j)}$ and a node or a edge is given by

$$z_\tau^{(j)} = \mathbf{p}_{s\tau-\Delta^{(j)}} \mathbf{x}^{(j)} = \begin{cases} \sigma_{s\tau-\Delta^{(j)}} \mathbf{x}^{(j)} & \text{for node puncturing,} \\ \gamma_{s\tau-\Delta^{(j)}} \mathbf{x}^{(j)} & \text{for edge puncturing.} \end{cases} \tag{9.20}$$

Algorithm 9.3 Algorithm for the calculation of a scrambler matrix $\mathbf{M}(D)$.

Input: Order s of the partitioning, code trellis of C.

Initialization: $j = 1$.

Step 1: Let $\mathcal{E}_{d_H}$ correspond to the set of all paths corresponding to the subcode $\mathcal{B}^{(j)}$ with Hamming weight $d_H = d^{(j)}$.
Set $\eta = 1$.

Step 2: Given the columns $\mathbf{m}_i^{-1}(D)$, $i = 1, 2, \ldots, j-1$. Determine all solutions for η additional columns $\mathbf{m}_{j+l}^{-1}(D)$, $0 \leq l < \eta$, according to theorem 9.24, such that all paths in $\mathcal{E}_{d_h}$ are deleted.
If no solution exists then $\eta := \eta + 1$. $\Rightarrow$ Step 2.

Step 3: If many solutions exist, choose the one that maximizes the free distance of the subcode $\mathcal{B}^{(j+\eta)}$.

Step 4: If more than one solution exists, choose the one that minimizes the error weight $c_{d^{(j+l)}}^{(j+l)}$.
Therefore

$$c_{d^{(j)}}^{(j)} \geq \ldots \geq c_{d^{(j+l)}}^{(j+l)} \geq \ldots \geq c_{d^{(j+\eta-1)}}^{(j+\eta-1)}, \quad 0 \leq l < \eta.$$

Step 5: $j := j + \eta$.
If $(j \leq s) \Rightarrow$ Step 1.

Step 6: When possible, choose an inverse scrambler matrix with a determinant of the form $|\mathbf{M}^{-1}(D)| = 1 + \cdots$.
Calculate $\mathbf{M}(D) = \left(\mathbf{M}^{-1}(D)\right)^{-1}$.

Output: $\mathbf{M}(D)$, $\mathbf{M}^{-1}(D)$, $d^{(j)}$ and $c_{d^{(j)}}^{(j)}$, $j = 1, 2, \ldots, s$.

Theorem 9.24 (Path puncturing) *A codeword of C for a given column $\mathbf{m}^{(-1)}(x)$ (which means $\mathbf{x}^{(i)}$ and $\Delta^{(i)}$, $i < j$) is not contained in the subcode $\mathcal{B}^{(j)}_{\mathbf{z}^{(1)},\mathbf{z}^{(2)},\ldots,\mathbf{z}^{(j-1)}}$ if for the corresponding codeword paths it is true that*

$$\exists z_\tau^{(i)}: \quad z_\tau^{(i)} \neq \mathbf{p}_{s\tau-\Delta^{(i)}}\mathbf{x}^{(i)} \quad i < j \text{ and } \tau > 0.$$

Proof According to equation 9.20, we get the mapping of the codeword path $\mathbf{p}$ and the corresponding enumeration series $\mathbf{z}^{(i)}$, $i \in 1, 2, \ldots, s$. According to the definition of a subcode, $z_\tau^{(i)} = \mathbf{p}_{s\tau-\Delta^{(i)}}\mathbf{x}^{(i)}$ for $i < j$ and $\tau > 0$. The theorem follows directly. □

If one calculates the columns of the inverse scrambler (see example 9.42), it can be seen that there are different strategies for selecting columns. This in turn results in different scrambler matrices. Therefore there is no unique choice. We use two criteria for constructing the following scrambler matrix algorithm:

1. The free distance of the subcodes should be as large as possible.

2. The number of bit errors resulting from false decoding with the Viterbi algorithm must be minimized. For paths with the same weight, we puncture the one that causes the larger number of bit errors.

Algorithm 9.3 is one method for the construction of a scrambler matrix under these criteria.

Note For $|\mathbf{M}^{-1}(D)| = 1$, a non-recursive realizable encoder is obtained. Otherwise the encoder is realizable and recursive.

Any convolutional code can be partitioned with algorithm 9.3. A GC code can then be constructed after selection of appropriate outer codes. The result can also be generalized to convolutional codes with any rate, or for non-binary alphabets. [BDS97] contains methods for dealing with trellises with longer edge extensions.

9.3.4 Construction and decoding of a GC code

In this section, we construct a GC code [BDS96a] on the basis of a single convolutional code, namely the ESA/NASA convolutional code for satellite communications, which was recommended by the *Consultative Committee for Space and Data Systems* (CCSDS)[WHPH87]. This code C has rate $1/2$, constraint length $\nu = 6$, and the generator sequence (octal) $g^{(0)} = 133$ and $g^{(1)} = 171$. We use a partitioning of order 6 for this code, which is based on a block scrambler and node puncturing. The outer codes are chosen to be punctured versions of the same code (see section 8.1.10). This results in a gain of about 1 dB compared with the ESA/NASA standard. The ESA/NASA standard is a classical concatenation of this convolutional code as the inner code and a RS code as the outer code.

Construction of the inner code

First we derive the partitioning of the inner code corresponding to the preceeding section. Then we follow with the description of the code as a UM code:

$$\mathcal{B}^{(1)}(n = 12, k^{(1)} = 6, d^{(1)} = 10) \; ,$$

where $d^{(1)}$ is the free distance. The code matrices are

$$\mathbf{G}_0 = \begin{pmatrix} 1&1&0&1&1&1&1&1&0&0&1&0 \\ 0&0&1&1&0&1&1&1&1&1&0&0 \\ 0&0&0&0&1&1&0&1&1&1&1&1 \\ 0&0&0&0&0&0&1&1&0&1&1&1 \\ 0&0&0&0&0&0&0&0&1&1&0&1 \\ 0&0&0&0&0&0&0&0&0&0&1&1 \end{pmatrix} \quad \text{and} \quad \mathbf{G}_1 = \begin{pmatrix} 1&1&0&0&0&0&0&0&0&0&0&0 \\ 1&0&1&1&0&0&0&0&0&0&0&0 \\ 0&0&1&0&1&1&0&0&0&0&0&0 \\ 1&1&0&0&1&0&1&1&0&0&0&0 \\ 1&1&1&1&0&0&1&0&1&1&0&0 \\ 0&1&1&1&1&1&0&0&1&0&1&1 \end{pmatrix} .$$

Based on the algorithm in section 9.3.3, the columns of the inverse block scrambler are calculated:

$$\begin{aligned} \mathbf{m}_1^{-1} &= (001000)^T, & \mathbf{m}_2^{-1} &= (100010)^T, & \mathbf{m}_3^{-1} &= (010101)^T, \\ \mathbf{m}_4^{-1} &= (000101)^T, & \mathbf{m}_5^{-1} &= (000001)^T, & \mathbf{m}_6^{-1} &= (000010)^T. \end{aligned}$$

One observes that the first three columns $\mathbf{m}_1^{-1}, \mathbf{m}_2^{-1}$ and $\mathbf{m}_3^{-1}$ are necessary for the puncturing of all paths with weight 10. Columns $\mathbf{m}_4^{-1}$ and $\mathbf{m}_5^{-1}$ puncture paths with weight 12 and 14 respectively. The inversion gives the block scrambler

$$\mathbf{M} = \begin{pmatrix} 0&0&1&0&0&0 \\ 1&0&0&0&0&0 \\ 0&1&0&0&0&0 \\ 0&1&0&1&0&0 \\ 0&0&0&1&0&1 \\ 1&0&0&0&1&0 \end{pmatrix} ,$$

Table 9.3 Free distance and number of minimum weight paths.

	Subcode based on original code matrix						Subcodes based on equivalent code matrix					
	$\mathcal{B}^{(1)}$	$\mathcal{B}^{(2)}$	$\mathcal{B}^{(3)}$	$\mathcal{B}^{(4)}$	$\mathcal{B}^{(5)}$	$\mathcal{B}^{(6)}$	$\mathcal{B}'^{(1)}$	$\mathcal{B}'^{(2)}$	$\mathcal{B}'^{(3)}$	$\mathcal{B}'^{(4)}$	$\mathcal{B}'^{(5)}$	$\mathcal{B}'^{(6)}$
Free distance $d^{(i)}$	10	10	10	10	10	10	10	10	10	12	12	16
Number $a_{d^{(i)}}$	34	17	9	3	2	1	34	23	9	6	2	1

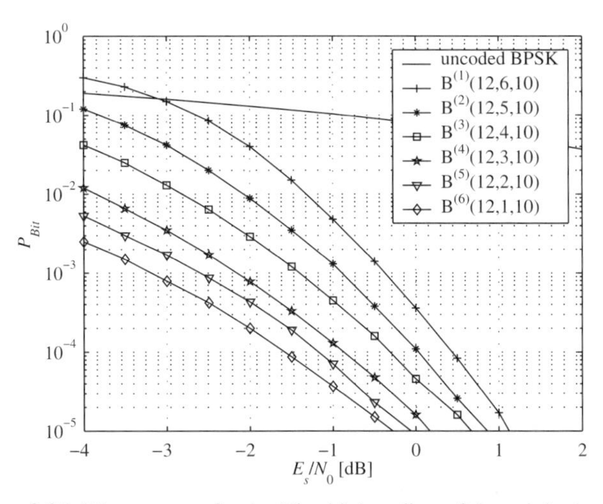

Figure 9.36 Bit error rate for the Viterbi decoding of the original encoder.

and with this we obtain the equivalent code matrices

$$\mathbf{G}_0' = \begin{pmatrix} 0&0&0&0&1&1&0&1&1&1&1&1 \\ 1&1&0&1&1&1&1&1&0&0&1&0 \\ 0&0&1&1&0&1&1&1&1&1&0&0 \\ 0&0&1&1&0&1&0&0&1&0&1&1 \\ 0&0&0&0&0&0&1&1&0&1&0&0 \\ 1&1&0&1&1&1&1&1&1&1&1&1 \end{pmatrix} \quad \text{and} \quad \mathbf{G}_1' = \begin{pmatrix} 0&0&1&0&1&1&0&0&0&0&0&0 \\ 1&1&0&0&0&0&0&0&0&0&0&0 \\ 1&0&1&1&0&0&0&0&0&0&0&0 \\ 0&1&1&1&1&0&1&1&0&0&0&0 \\ 1&0&1&1&0&1&1&1&1&0&1&1 \\ 0&0&1&1&0&0&1&0&1&1&0&0 \end{pmatrix}.$$

The code matrix of the inner subcode $\mathcal{B}'^{(i)}$ is obtained by deleting the first $i-1$ rows of the equivalent code matrices $\mathbf{G}_0'$ and $\mathbf{G}_1'$. The following partitioning is achieved:

$$\begin{aligned} &\mathcal{B}'^{(6)}(12,1,16) \subset \mathcal{B}'^{(5)}(12,2,12) \subset \mathcal{B}'^{(4)}(12,3,12) \\ \subset\ &\mathcal{B}'^{(3)}(12,4,10) \subset \mathcal{B}'^{(2)}(12,5,10) \subset \mathcal{B}'^{(1)}(12,6,10). \end{aligned}$$

If we partition the original code ($\mathbf{G}_0$ and $\mathbf{G}_1$) using the same method then no distance gain is achieved in the subcode. This fact is shown in table 9.3. Here, the additional number of paths with the corresponding free distance of each subcode is shown.

Figures 9.36 and 9.37 clearly show the gain achieved in the bit error rate for the corresponding subcodes over a Gaussian channel using BPSK modulation.

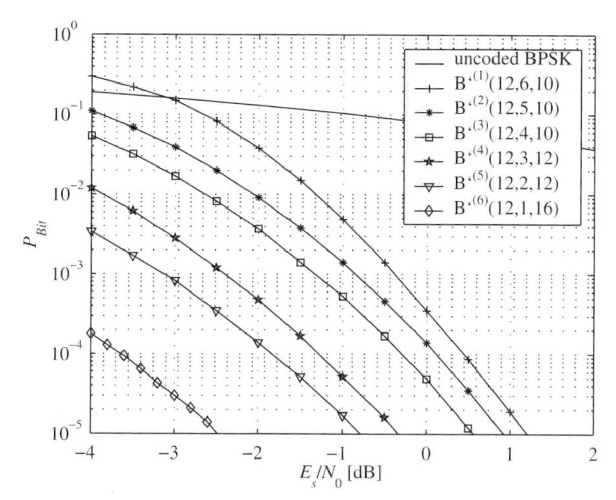

Figure 9.37 Bit error rate for the Viterbi decoding of the equivalent code.

Table 9.4 Punctured outer codes constructed from the ESA/NASA convolutional code.

Code	Rate R	Spectrum: Free distance d_A	Spectrum: Error weight c_{d_A}	Puncturing matrix (octal)
$\mathcal{A}^{(1)}$	1/2	10	36	1 / 1
$\mathcal{A}^{(2)}$	3/4	5	42	6 / 5
$\mathcal{A}^{(3)}$	6/7	3	5	72 / 45
$\mathcal{A}^{(4)}$	16/17	3	393	152205 / 125572
$\mathcal{A}^{(5)}$	19/20	3	991	1773425 / 1004352
$\mathcal{A}^{(6)}$	1/1 (uncoded)	–	–	–

Outer codes and interleaving

For the outer code, we consider various punctured versions of the ESA/NASA standard convolutional code. These codes have rates between 1/2 and 16/17 [WHPH87]. Table 9.4 shows the outer codes used. The code $\mathcal{A}^{(6)}$ is a code with no redundancy (i.e $(n,n,1)$), because the code $\mathcal{B}'^{(6)}$ already has a low enough bit error rate. The outer code $\mathcal{A}^{(5)}$ is the punctured code with rate $19/20$ constructed from the code $\mathcal{B}'^{(1)}$ using a computer search.

Note In [Hole88], punctured convolutional codes with 64 states are listed that have better parameters. However, the idea here is that a single code can be the basis for the entire GC code construction. The GC code $\mathcal{H}$ constructed on the basis of the ESA/NASA standard has the rate $R_{\mathcal{H}} \approx 0.416$.

In order to avoid error propagation (see [WH93a]) from one step to the other, we use the interleavers $I^{(i)}$ corresponding to figure 9.38, which behave as rectangular block interleavers having different row and column numbers. The outer convolutional codes are terminated to length n_a. We define the quasirectangular interleaver (figure 9.39). The interleaver $I^{(i)}$ has $I_r^{(i)}$

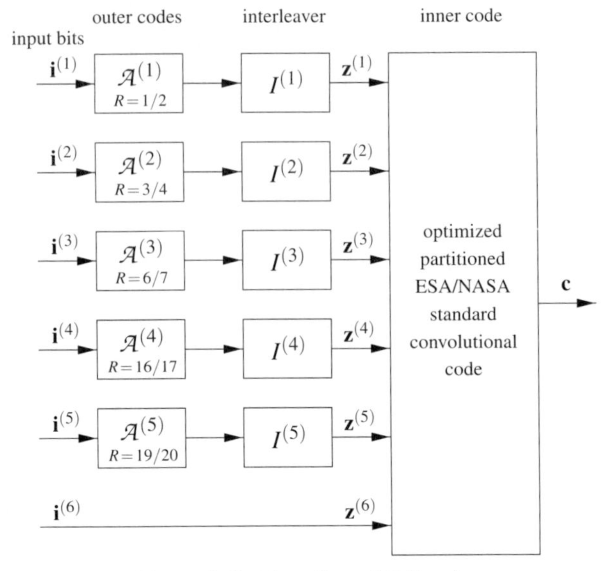

Figure 9.38 Encoding of GC codes.

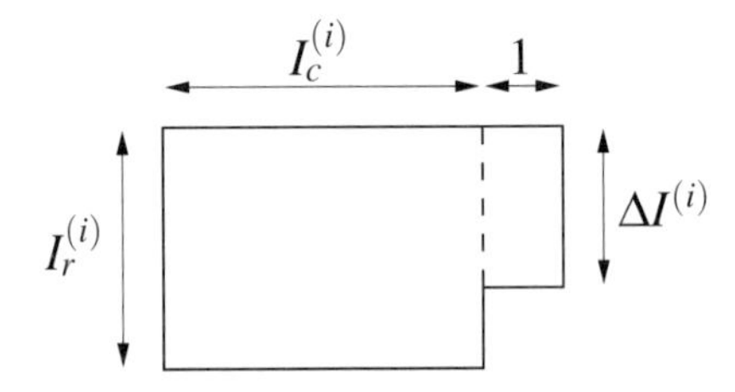

Figure 9.39 Quasirectangular block interleaver $I^{(i)}$.

rows and $I_c^{(i)}$ columns and a partially filled final column:

$$n_A = I_c^{(i)} I_r^{(i)} + \Delta I^{(i)}.$$

The bits are written in the interleaver columnwise and read row-wise. Six tail bits are used to terminate the inner code, i.e. one bit per interleaver. The parameters of this interleaver are presented in table 9.5. The entire code length $n_{\mathcal{H}} = n_B\,(n_A + 1) = 49164$. The GC code $\mathcal{H}$ has free distance $d_{\mathcal{H}} \geq 16$, which is determined by the subcode $\mathcal{B}'^{(6)}$ having free distance 16.

Table 9.5 Interleaver parameters.

	$I^{(1)}$	$I^{(2)}$	$I^{(3)}$	$I^{(4)}$	$I^{(5)}$
I_r	64	65	66	67	68
I_c	64	63	62	61	60
ΔI	0	1	4	9	16

Decoding and simulation results

The decoding of a GC code is carried out based on section 9.2.4 (this decoding principle is often termed *multistage decoding*). In the first step, the enumeration sequence $\hat{\mathbf{z}}^{(1)}$ is determined based on the decoder for the code $\mathcal{B}'^{(1)}$. This sequence is decoded with respect to the code $\mathcal{A}^{(1)}$, which gives $\tilde{\mathbf{z}}^{(1)}$ and the information sequence $\tilde{\mathbf{i}}^{(1)}$. In the second step, the sequence $\tilde{\mathbf{i}}^{(1)}$ is used to decode the received vector according to $\mathcal{B}'^{(2)}$. The result is the sequence $\tilde{\mathbf{z}}^{(2)}$ is decoded with the second outer code $\mathcal{A}^{(2)}$. The subsequent steps follow correspondingly.

Note Iterative decoding (section 7.4.1) between inner and outer codes can be used here (see [WH93a]). The optimal decoding of a GC code is to date an unsolved problem. Therefore we limit our investigation to the case when the inner decoder generates information reliability (i.e. *soft-output decoding*). The outer code produces binary output (*hard-output decoding*).

We consider the inner code decoding $\mathcal{B}'^{(i)}_{\tilde{\mathbf{z}}^{(1)},\dots,\tilde{\mathbf{z}}^{(i-1)}}$ of the ith step. The calculated enumeration sequence $\hat{\mathbf{z}}^{(i)} = (\hat{z}_1^{(i)}, \dots, \hat{z}_\tau^{(i)}, \dots)$ based on equation 9.20 is determined by

$$\hat{z}_\tau^{(i)} = \hat{\sigma}_{t_0} \mathbf{m}_i^{-1}, \tag{9.21}$$

where $\hat{\sigma}_{t_0}$ corresponds to the calculated nodes of depth $t_0 = s\tau$. The set of nodes to be considered for $i \in [1, s]$ is given by the relationship

$$\Omega_{t_0}^{(i)} = \left\{ \sigma : \sigma \mathbf{M}^{-1} = (\tilde{z}_\tau^{(1)}, \dots, \tilde{z}_\tau^{(i-1)}, \xi^{(i)}, \dots, \xi^{(s)}) \right\},$$

where $\xi^{(i)}, \dots, \xi^{(s)}$ can take any value $\{0, 1\}$.

The Viterbi algorithm can be used for the case of hard decision output decoding of the inner code. Only the nodes $\Omega_{t_0}^{(i)}$ of depth i are considered in this case. For soft-decision output decoding, the L values are used from section 7.2.2:

$$\lambda_\tau^{(i)} = \ln \frac{P(\hat{z}_\tau^{(i)} = 0)}{P(\hat{z}_\tau^{(i)} = 1)}, \tag{9.22}$$

where $P(\hat{z}_\tau^{(i)} = 1)$ and $P(\hat{z}_\tau^{(i)} = 0)$ represent the probability that a bit $\hat{z}_\tau^{(i)}$ is equal to 0 or 1. As described in sections 8.4.3 and 8.5, there are several algorithms for this decoding , for example SOVA [HH89] and the s/s-MAP algorithm [BCJR74]. Here we use a modified version of the s/s-MAP algorithm [BCJR74]. Consider

$$\Omega_{t_0}^{(i)}(0) = \left\{ \sigma : \ \sigma \mathbf{M}^{-1} = (\tilde{z}_\tau^{(1)}, \dots, \tilde{z}_\tau^{(i-1)}, 0, \xi^{(i+1)}, \dots, \xi^{(s)}) \right\},$$
$$\Omega_{t_0}^{(i)}(1) = \left\{ \sigma : \ \sigma \mathbf{M}^{-1} = (\tilde{z}_\tau^{(1)}, \dots, \tilde{z}_\tau^{(i-1)}, 1, \xi^{(i+1)}, \dots, \xi^{(s)}) \right\}, \quad 1 \le i \le s,$$

with any $\xi^{(i+1)}, \dots, \xi^{(s)} \in \{0, 1\}$. Therefore $\mathbf{p}_t i(\sigma)$ is the probability that the transmitted path passes through the node σ at depth t based on the received sequence. We can then write the L value according to equation 9.22 as

$$\lambda_\tau^{(i)} = \ln \left(\frac{\sum_{\sigma \in \Omega_{t_0}^{(i)}(0)} \mathbf{p}_{t_0}(\sigma)}{\sum_{\sigma \in \Omega_{t_0}^{(i)}(1)} \mathbf{p}_{t_0}(\sigma)} \right), \quad \text{with} \quad t_0 = s\tau. \tag{9.23}$$

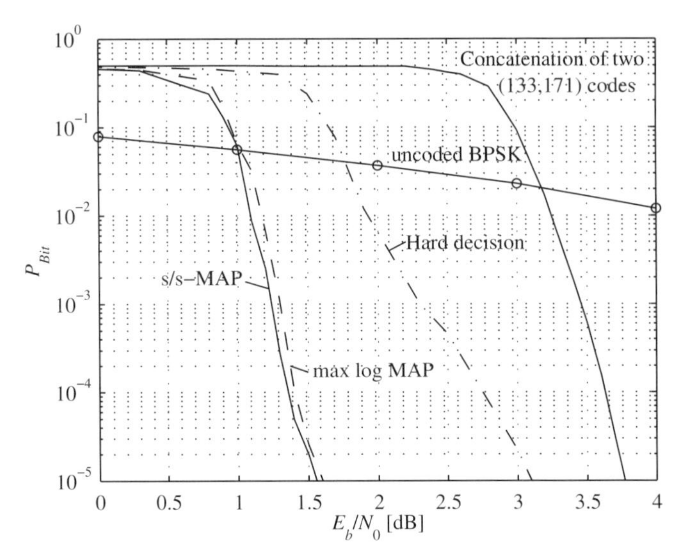

Figure 9.40 Bit error rate of generalized concatenated codes.

The codewords (paths) begin and depth 0 and end at t_e in the zero state. Corresponding to the algorithm from [BCJR74], the description of the probability $\mathbf{p}_t(\sigma)$ is carried out as follows:

$$\mathbf{p}_t(\sigma) = \alpha_t(\sigma) \cdot \beta_t(\sigma) \ . \tag{9.24}$$

$\alpha_t(\sigma)$ and $\beta_t(\sigma)$ indicate the probability that the transmitted path passes through node σ at depth t, for the forward (depth 0 to t) and reverse (depth t_e to t) directions. We next need the transition probability $\gamma_t(\sigma', \sigma)$ from node σ'_{t-1} to σ_t. The probability $\alpha_t(\sigma)$ and $\beta_t(\sigma)$ can be calculated recursively by

$$\alpha_t(\sigma) = \begin{cases} 0 & \text{for } t = s\tau,\ \sigma \notin \Omega^{(i)}_{t_0=s\tau}\ , \\ \sum_{\sigma'} \alpha_{t-1}(\sigma') \cdot \gamma_t(\sigma', \sigma) & \text{otherwise,} \end{cases} \tag{9.25}$$

with the values $\alpha_0(\mathbf{0}) = 1$ and $\alpha_0(\sigma) = 0$ for $\sigma \neq \mathbf{0}$. We obtain

$$\beta_t(\sigma) = \begin{cases} 0 & \text{for } t = s\tau,\ \sigma \notin \Omega^{(i)}_{t_0=s\tau}\ , \\ \sum_{\sigma'} \beta_{t+1}(\sigma') \cdot \gamma_{t+1}(\sigma', \sigma) & \text{otherwise,} \end{cases} \tag{9.26}$$

with $\beta_{t_e}(\mathbf{0}) = 1$ and $\beta_{t_e}(\sigma) = 0$ for $\sigma \neq \mathbf{0}$.

Up to this point, we have carried out all calculations by summing the probabilities. In equations 9.23, 9.25 and 9.26, we replace the summation by a maximization. In this way, the complexity becomes smaller, and equations 9.25 and 9.26 are like the Viterbi algorithm. With this modification, the metric values of the corresponding trellis can be calculated using the Viterbi algorithm. This computation makes two passes through the trellis, once from the start node 0 and once from the end node t_e. One observes that the path history must not be stored in memory. This technique corresponds to the max log MAP algorithm from section 8.5.2.

For the decoding of the GC code $\mathcal{H}$, the inner code is decoded based on two different methods: first using the s/s-MAP algorithm, and then with the max log MAP algorithm. As shown in figure 9.40, s/s-MAP produces a bit error rate of 10^{-5} for the value $E_b/N_0 = 1.55\,\text{dB}$, whereas max log MAP requires $E_b/N_0 = 1.6\,\text{dB}$ for the same BER. There is a difference of 0.05 dB between the two methods. A hard-decision decoding of the inner code is also carried out using a classical concatenation of two convolutional codes C with a 64×64 block interleaver. Compared with the classical concatenation, the GC code has a gain of 2.2 dB.

9.4 GC codes with block and convolutional codes

Until now, we have described GC codes where the inner and outer codes were either block or convolutional codes. The next logical step is to introduce GC codes where the inner code is a block or a convolutional code and the outer code corresponds to the other code class. Unfortunately, there are very few publications in this area. The number of possibilities for such GC code construction is large, and they cannot be treated in entirety here. Therefore we limit ourselves to a few special cases that illustrate the principles of this type of code construction, and that provide a basis for considering new GC codes.

9.4.1 Inner convolutional and outer block codes

The ESA/NASA classical code concatenation for satellite communication [WHPH87] is one example of this type of code. The inner code is the convolutional code $\mathcal{B}^{(1)}$ described in section 9.3.4. The outer code is a $(255, 223, 33)$ RS code over $GF(2^8)$. Using techniques from the preceeding section on code partitioning, a GC code can be constructed based on the convolutional code $\mathcal{B}^{(1)}$ and an outer RS code that achieves either a better minimum distance for the same code rate or a better code rate for the same minimum distance. Unfortunately, to date this code construction has not yet been analyzed.

In [ZySha], the construction of numerous codes with inner (P)UM codes (i.e. convolutional codes) and outer RS codes is presented. Several of these are discussed in this section. We compare the performance of these codes with a classically concatenated code $\mathcal{C}_C$ and a GC constructed code $\mathcal{C}_{GC}$ with the same rate. The GC code is expected to have a correspondingly larger minimum distance.

Classically concatenated code $\mathcal{C}_C$ A binary information sequence $\mathbf{i}$ is written as a matrix $\mathbf{M_i}$ according to figure 9.41 having $I_C\,\kappa$ rows and k_a columns. We define the constant $\kappa = \mu$ as the power of the extension field of the outer code. This definition becomes more meaningful with respect to GC code construction. Each $I_C\,\kappa \times k_a$ submatrix corresponds to the information symbols of the outer RS code $\mathcal{A}(n_a = 2^\kappa - 1, k_a, d_a)$ over the alphabet $GF(2^\kappa)$. We encode the corresponding matrices from I_C to the codewords $(\mathbf{a}_1, \mathbf{a}_2, \ldots, \mathbf{a}_{I_C})$ of the code $\mathcal{A}$ that produces the matrix $\mathbf{M_a}$.

As the inner code, we use a (P)UM code $\mathcal{B}(n_b, k_b | k_b^1, d_b)$ (d_b is the free distance), where $\mu_C = \kappa / k_b$ is an integer that defines the grouping of information bits of the outer code. The first $k_b \times n_a$ submatrix of $\mathbf{M_a}$ is encoded by $\mathcal{B}$, then the second submatrix, and so on up to μ_C. This produces $I_C\,\mu_C$ codeword submatrices $n_b \times n_a$, each with dimension $k_b\,k_a$.

The code $\mathcal{C}_C$ has the property that the convolutional code has a memory that influences successive blocks. In practice, this influence can be eliminated by introducing a periodic termination consisting of sequence of 0 symbols that decreases the code rate. For the code $\mathcal{C}_C$

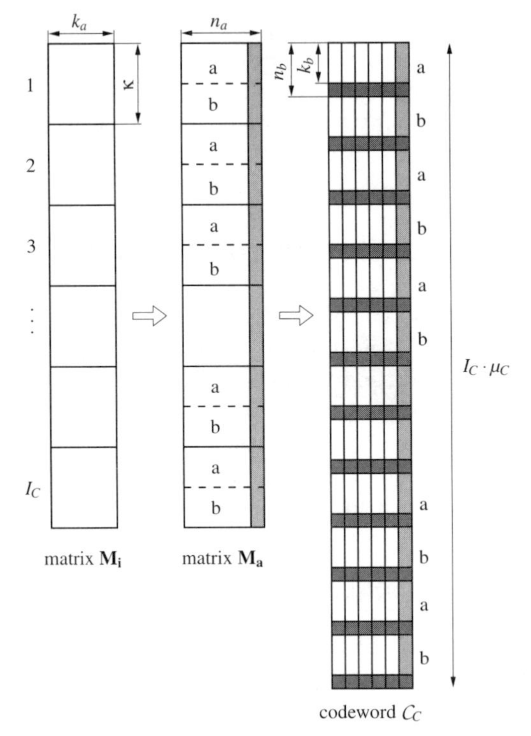

Figure 9.41 Simple concatenation of RS codes and PUM codes.

having code rate RC, length n_C and dimension k_C, we obtain

$$n_C = I_C \mu_C n_a n_b, \quad k_C = I_C \mu_C k_a k_b, \quad R_C = \frac{k_C}{n_C} = \frac{k_a k_b}{n_a n_b} = R_a R_b \ .$$

In [ZySha] and [ZS87], it is proved that the minimum distance can be written as:

$$d_C \geq \min\{d_a d_b, \omega \mu_C I_C d_a\}, \quad \mu_C = \kappa / k_b, \quad I_C \text{ interleaving depth.}$$

The parameter ω is the average weight of a loop in the state diagram loop of the inner (P)UM code, as described in section 9.3.1 (see also [TJ83]). The parameters of classical concatenated codes are

$$\mathcal{C}_C(2; n_C = I_C \mu_C n_a n_b, \ k_C = I_C \mu_C k_a k_b, \ d_C \geq \min\{d_a d_b, \ \omega \mu_C I_C d_a\}) \ .$$

Generalized concatenated code $\mathcal{C}_{GC}$ The binary information sequence $\mathbf{i}$ according to figure 9.42 is written in a matrix $\mathbf{M_i}$, which is subdivided into s $(\kappa \times k_a^{(i)})$ matrices $i = 1, 2, \ldots, s$. For this construction, we introduce the requirement that all outer code symbols belong to the same Galois field $\kappa = \mu_i$ and $a_j^{(i)} \in GF(q^\kappa), i \in 1, \ldots, s, j \in 1, \ldots, n_a$. All together, the matrix has $I_{GC}\kappa$ binary rows and a variable number of columns of value $k_a^{(i)}$. This is shown in figure 9.42. As a consequence, the order s must divide the interleaver depth I_{GC}. Each of the I_{GC} $(\kappa \times k_a^{(i)})$ submatrices, $i = 1, 2, \ldots, s$, corresponds to an information symbol

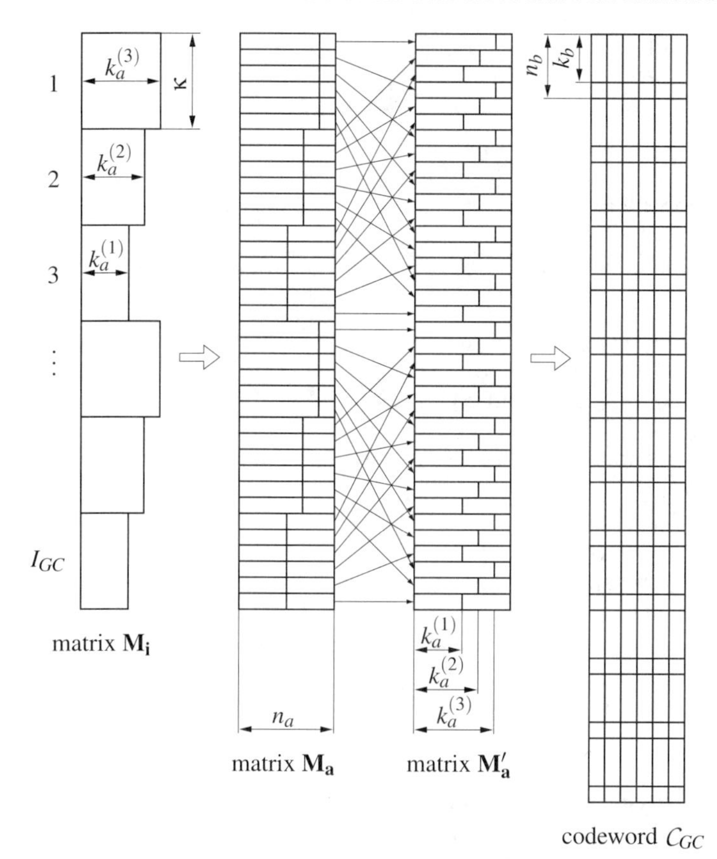

Figure 9.42 Generalized concatenation of RS codes and PUM codes.

of the outer RS code $\mathcal{A}^{(i)}(n_a = 2^\kappa - 1, k_a^{(i)}, d_a^{(i)})$ over the alphabet $GF(2^\kappa)$. One observes that this gives I_{GC}/s outer codes $\mathcal{A}^{(i)}$. We encode these I_{GC} matrices to I_{GC}/s codewords $(\mathbf{a}_1^{(i)}, \mathbf{a}_2^{(i)}, \ldots \mathbf{a}_{I_{GC}/s}^{(i)})$, $i = 1, 2, \ldots, s$, of the code $\mathcal{A}^{(i)}$, and obtain the matrix $\mathbf{M_a}$. This is a binary $(\kappa I_{GC} \times n_a)$ matrix. The rows are interchanged according to figure 9.42. We use the (P)UM code $\mathcal{B}^{(1)}(n_b, k_b^{(1)} | k_b^{(1),1}, d_b^{(1)})$ as the inner code, where $\mu_{GC} = \kappa / k_b^{(1)}$ is an integer. Furthermore, the code $\mathcal{B}^{(1)}$ can be partitioned into subcodes (see section 9.3.3), giving

$$\mathcal{B}^{(i)}(n_b, k_b^{(i)} | k_b^{(i),1}, d_b^{(i)}), \quad i = 1, 2, \ldots, s,$$

with

$$k_b^{(j)} = \frac{k_b^{(1)}(s - j + 1)}{s}, \quad j = 2, 3, \ldots, s.$$

We obtain the matrix $\mathbf{M'_a}$. A total of κ / k_b submatrices $(k_b \times n_a)$ of the matrix $\mathbf{M'_a}$ are encoded by the outer code $\mathcal{B}^{(1)}$. There are a total of $I_{GC}(\kappa / k_b)$ submatrices encoded by the inner code. The binary GC code C_{GC} thus constructed has the following parameters:

length: $n_{GC} = I_{GC}\mu_{GC}n_a n_b,$

dimension: $k_{GC} = \frac{I_{GC}}{s}\mu_{GC}k_b^{(1)}\sum_{i=1}^{s}k_a^{(i)},$

code rate: $R_{GC} = \frac{k_{GC}}{n_{GC}} = \frac{k_b^{(1)}}{n_b}\frac{1}{n_a}\sum_{i=1}^{s}k_a^{(i)},$

minimum distance: $d_{GC} \geq \min_{i=1,\ldots,s}\{d_a^{(i)}d_b^{(i)}, \omega^{(i)}\mu_{GC}I_{GC}d_a^{(i)}\}.$

The proof of the minimum distance follows directly from the proof of d_C and by considering the pair $\mathcal{A}^{(i)}$ and $\mathcal{B}^{(i)}$ [ZS87]. The parameter $\omega^{(i)}$ is the average value of the weight of a loop in the state diagram of the inner (P)UM code $\mathcal{B}^{(i)}$.

Note By using n_a identical encoders, this construction can theoretically employ infinitely long inner (P)UM codes.

We now give four examples of codes using the above construction principles, namely two classical and two generalized concatenated codes. As a basis, we consider the $(n = 6, k = 4, d = 5)$ UM and the $(n = 6, k = 4|k^{(1)} = 2, d = 4)$ PUM code from example 9.36 and the associated partitioning methods. The upper bounds on the free distance for these two codes are 6 and 4 for the UM and PUM codes respectively. For the common $(n = 6, k = 2, d = 7)$ UM subcode the bound has a value of 8 (see [HS73]). Furthermore, these codes have parameters $\omega_{UM}^{(1)} = 3/4$, $\omega_{PUM}^{(1)} = 1$ and $\omega_{PUM}^{(2)} = 2$.

Example 9.45 (Classical and GC codes of length 360**)** We use the RS code of length 15 over $GF(2^4)$ for the outer code. We consider the classical concatenation with the $(n = 6, k = 4, d = 5)$ UM code and with an RS code $\mathcal{A}(2^4; 15, 11, 5)$. We select the values $I_C = 4$ and $\mu_C = \kappa/k_b = 4/4 = 1$, which produces the 16×11 binary matrix $\mathbf{M_i}$. Every 4 rows build the 11 information symbols of the outer code $\mathcal{A}$. After this encoding, we obtain a binary 16×15 matrix which we encode columnwise with the inner UM code. The resulting 24×15 matrix is one codeword of the code $\mathcal{C}_C$, having parameters

$$n_C = I_C\mu_C n_a n_b = 4\,1\,15\,6 = 360\,,$$
$$k_C = I_C\mu_C k_a k_b = 4\,1\,11\,4 = 176\,,$$
$$d_C \geq \min\{d_a d_b, \omega\mu_C I_C d_a\} = \min\{5\,5, (3/4)\,1\,4\,5\} = 15.$$

If we change the inner code to an $(n = 6, k = 4|k^{(1)} = 2, d = 4)$ PUM code then we obtain

$$\mathcal{C}'_C(2; n_C = 360, k_C = 176, d_C \geq \min\{d_a d_b, \omega\mu_C I_C d_a\} = \min\{5\,4, 1\,1\,4\,5\} = 20)\,.$$

Next, we construct the corresponding GC codes based on the partitioning of the UM and PUM codes.

First, the $(n = 6, k = 4, d = 5)$ UM code is partitioned into 2^2 $(n = 6, k = 2, d = 7)$ UM subcodes. The interleaving field is defined as $\mu_{GC} = \kappa/k_b = 4/4 = 1$. As described in figure 9.42, we write the 176 information bits in a $\kappa \times k_a^{(2)} = 4 \times 13$ matrix, then in a $\kappa \times k_a^{(1)} = 4 \times 9$ matrix, followed by a $\kappa \times k_a^{(2)} = 4 \times 13$ matrix, and finally a $\kappa \times k_a^{(1)} = 4 \times 9$ matrix. The submatrices are then encoded with the outer codes $\mathcal{A}^{(1)}(2^4; 15, 9, 7)$ and $\mathcal{A}^{(2)}(2^4; 15, 13, 3)$. The 16×15

Table 9.6 Parameters for codes from example 9.45 (classical concatenation).

n_C	R_C	d_C	μ_C	I_C	$\mathcal{B}$	$\mathcal{A}(n_a, k_a, d_a)$
360	0.4889	15	1	4	UM	$(2^4; 15, 11, 5)$
360	0.4889	20	1	4	PUM	$(2^4; 15, 11, 5)$
2016	0.5	66	2	2	UM	$(2^8; 84, 63, 22)$
2016	0.5	88	2	2	PUM	$(2^8; 84, 63, 22)$
6120	0.4994	195	2	2	UM	$(2^8; 255, 191, 65)$
6120	0.4994	260	2	2	PUM	$(2^8; 255, 191, 65)$

matrix $\mathbf{M_a}$ is permuted such that the first two rows are copied, then rows 5 and 6 followed by rows 3 and 4, and so on, giving the matrix $\mathbf{M'_a}$. The inner code encoding proceeds as follows: bits 3 and 4 of the first column of the permuted matrix determine the inner subcode, and bits 1 and 2 determine the codeword of the inner subcode, and so forth up to column n_a. This step maps $k_b \times n_a$ bits of the matrix $\mathbf{M'_a}$ onto an $n_b \times n_a$ submatrix of the C_{GC} codeword. This process is repeated $I_C \mu_{GC}$ times. The GC code has the following parameters:

$$
\begin{aligned}
n_{GC} &= I_{GC}\mu_{GC} n_a n_b = 4\,1\,15\,6 = 360\ , \\
k_{GC} &= (I_{GC}/s)\,\mu_{GC} k_b^{(1)} \sum_{i=1}^{s} k_a^{(i)} = (4/2)\,1\,4\,(9+13) = 176\ , \\
d_{GC} &\geq \min_{i=1,2} \{d_a^{(i)}\, d_b^{(i)}, \omega^{(i)} \mu_{GC} I_{GC} d_a^{(i)}\} \\
&= \min\{7 \cdot 5, 3 \cdot 7, (3/4)\cdot, 1 \cdot 4 \cdot 7, 2 \cdot 1 \cdot 4 \cdot 3\} = 21\ .
\end{aligned}
$$

If we replace the UM code by the $(n=6, k=4|k^{(1)}=2, d=4)$ PUM code, which also contains the $(n=6, k=2, d=7)$ UM subcode, and use the outer codes $\mathcal{A}^{(1)}(2^4; 15, 10, 6)$ and $\mathcal{A}^{(2)}(2^4; 15, 12, 4)$, then we obtain the code C'_{GC} with parameters

$$
\begin{aligned}
n_{GC} &= 4 \cdot 1 \cdot 15 \cdot 6 = 360\ , \\
k_{GC} &= (4/2) \cdot 1 \cdot 4 \cdot (9+13) = 176, \\
d_{GC} &\geq \min_{i=1,2} \{d_a^{(i)}\, d_b^{(i)}, \omega^{(i)} \mu_{GC} I_{GC} d_a^{(i)}\} \\
&= \min\{6 \cdot 4, 4 \cdot 7, 1 \cdot 1 \cdot 4 \cdot 6, 2 \cdot 1 \cdot 4 \cdot 4\} = 24\ .
\end{aligned}
$$

◇

The results from this example are summarized in tables 9.6 and 9.7, along with other codes of length 2016 and 6120 and the associated parameters. The construction follows the exact steps as for the previous example, which uses the same inner codes. One observes a significant increase in the minimum distance between the classical and generalized concatenated codes for the same rate.

As an exercise, the inner $(n=6, k=4|k^{(1)}=2, d=4)$ PUM code with the partitioning from example 9.36 can be used to construct a $(2; 600, 240, 44)$ GC code in which two shortened RS codes with length 255 are used.

9.4.2 Inner block and outer convolutional codes

For this scenario, there are very few published results, especially for GC codes. We consider two examples, including one GC code construction.

Table 9.7 Parameters for the codes from example 9.45 (generalized concatenation).

n_{GC}	R_{GC}	d_{GC}	μ_{GC}	s	I_{GC}	$\mathcal{B}$	$\mathcal{A}^{(1)}(n_a,k_a^{(1)},d_a^{(1)})$	$\mathcal{A}^{(2)}(n_a,k_a^{(2)},d_a^{(2)})$
360	0.4889	21	1	2	4	UM	$(2^4;15,9,7)$	$(2^4;15,13,3)$
360	0.4889	24	1	2	4	PUM	$(2^4;15,10,6)$	$(2^4;15,12,4)$
2016	0.5	91	2	2	2	UM	$(2^8;84,54,31)$	$(2^8;84,72,13)$
2016	0.5	112	2	2	2	PUM	$(2^8;84,57,28)$	$(2^8;84,69,16)$
6120	0.4994	273	2	2	2	UM	$(2^8;255,165,91)$	$(2^8;255,217,39)$
6120	0.4994	329	2	2	2	PUM	$(2^8;255,173,83)$	$(2^8;255,209,47)$

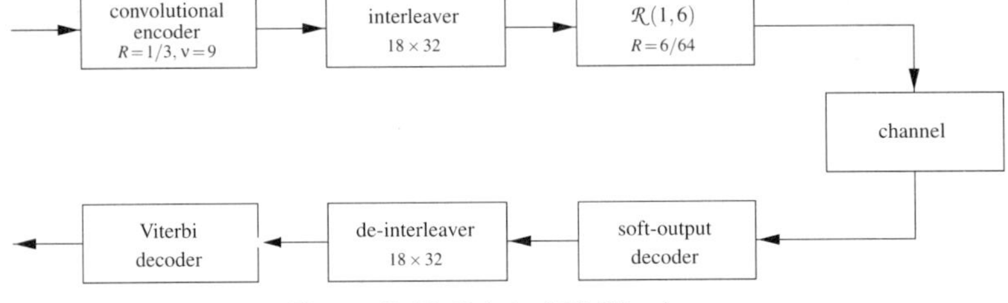

Figure 9.43 Original IS-95 scheme.

A good example of classical concatenation is used in a *direct sequence code division multiple access* (DS-CDMA) system (see [Pro]). We consider the spreading sequence as the inner code (usually a first-order RM code corresponding to section 5.1), and an outer convolutional code. In general, a block code of k bits is spread to n bits. A matched filter bank with the correlation for the corresponding sequence is performed. This can be interpreted as ML decoding, where all possible code sequences are compared with the received sequence.

We consider in the following example a modification of the code construction for the US mobile radio standard IS-95, which results in an improvement in the bit error rate. This example was first presented in [FB96].

Example 9.46 (IS-95 mobile radio standard) Here we compare the original IS-95 standard for mobile to base station transmission (figure 9.43) with a slightly modified concatenated version of the same standard (figure 9.44). In the modified code, the inner code rate is decreased, and the outer code rate is increased. The overall rate remains the same. A UM code based on an RM code (see section 8.7.4) is selected as the inner code. The IS-95 code has a minimum distance $d_{IS} = 14 \cdot 32 = 448$, while the modified code has a minimum distance $d_{UM} = 16 \cdot 32 = 512$. Figure 9.45 compares coherent detection simulation results for the two codes. A gain of 1.4dB is achieved for the modified code, at a bit error rate of 10^{-3}. In [FB96], the results for incoherent detection are given. In this case, the gain is reduced to 1 dB. Both decoders have approximately the same decoding complexity. ◇

The second example considers the RM construction (section 9.5) with convolutional codes. This method of GC code construction was published in [Che96].

Example 9.47 ($|\mathbf{u}|\mathbf{u}+\mathbf{v}|$ construction with convolutional codes) Consider the binary block code $\mathcal{B}^{(1)}(2,2,1)$ as the inner code. A partitioning gives the two subcodes $\mathcal{B}_i^{(2)}(2,1,2)$, $i=0,1$, where the minimum distance is twice as large as for $\mathcal{B}^{(1)}$. Next, two outer convolutional

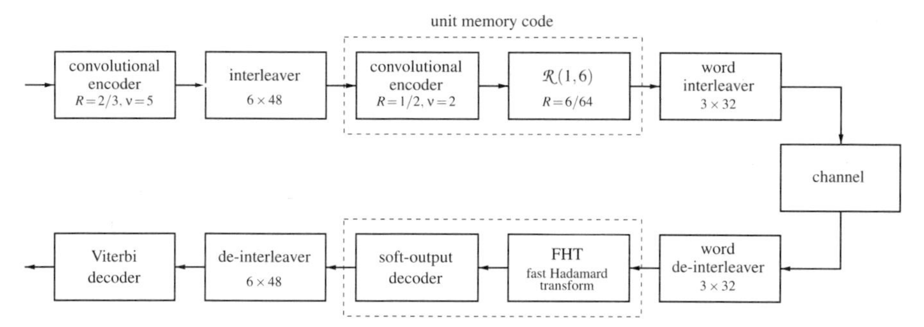

Figure 9.44 Modified IS-95 schematic.

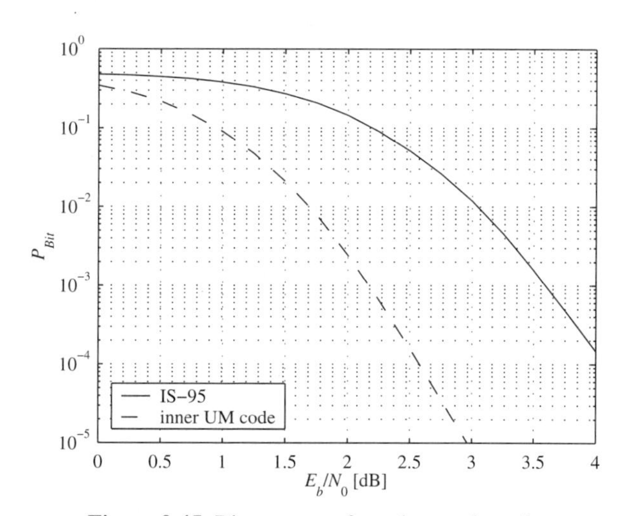

Figure 9.45 Bit error rate for coherent detection.

codes are used having a free distance difference of a factor of two. We assume that the first convolutional code $\mathcal{A}^{(1)}$ has free distance d_f, and the second $d_f/2$. The GC code free distance is d_f. In [Che96], several constructions are given. ◇

The case of inner block encoding and outer convolutional encoding occurs more often for the case of coded modulation, and is dealt with in chapter 10.

9.5 Multiple concatenation and Reed–Muller codes

We shall describe Reed–Muller codes that were defined in section 5.1 as generalized multiple concatenated codes.

An RM code $\mathcal{R}(r,m)$ is a binary code of length $n = 2^m$, dimension $k = 1 + \binom{m}{1} + \binom{m}{2} + \ldots + \binom{m}{r}$ and minimum distance $d = 2^{m-r}$. The corresponding dual code is the RM code $\mathcal{R}(m-r-1,m)$.

In order to construct RM codes as generalized concatenated codes, we consider the inner binary code of length 2, dimension 2 and minimum distance 1, i.e. the inner code consists of all binary vectors of length 2:

$$\mathcal{B}^{(1)}(n_b = 2, k_b^{(1)} = 2, d_b^{(1)} = 1) = \{(0,0),(1,1),(1,0),(0,1)\} .$$

The code $\mathcal{B}^{(1)}$ can be partitioned into subcodes $\mathcal{B}^{(2)}_{a^{(1)}}$ of minimum distance 2:

$$\mathcal{B}^{(1)} = \bigcup_{a^{(1)}=0}^{1} \mathcal{B}^{(2)}_{a^{(1)}}(n_b = 2, k_b^{(2)} = 1, d_b^{(2)} = 2) .$$

We have $\mathcal{B}^{(2)}_0 = \{(0,0),(1,1)\}$ and $\mathcal{B}^{(2)}_1 = \{(1,0),(0,1)\}$. The codewords are

$$\mathbf{b}_{00} = (0,0), \quad \mathbf{b}_{01} = (1,1), \quad \mathbf{b}_{10} = (1,0), \quad \mathbf{b}_{11} = (0,1) . \tag{9.27}$$

The enumerations of the partitions are binary:

$$(a^{(1)}, a^{(2)}) \iff \mathbf{b} \in \mathcal{B}^{(1)} . \tag{9.28}$$

Consider a $n_a \times 2$ matrix

$$\begin{pmatrix} a_1^{(1)} & a_1^{(2)} \\ a_2^{(1)} & a_2^{(2)} \\ \vdots & \vdots \\ a_{n_a}^{(1)} & a_{n_a}^{(2)} \end{pmatrix}$$

such that the first column is a codeword of the first outer code $\mathcal{A}^{(1)}(n_a, k_a^{(1)}, d_a^{(1)})$, and the second column is a codeword of the second outer code $\mathcal{A}^{(2)}(n_a, k_a^{(2)}, d_a^{(2)})$. Furthermore, each row $(a_j^{(1)}, a_j^{(2)})$ enumerates a codeword of the inner code $\mathcal{B}^{(1)}$. The corresponding mapping yields a codeword of the GC code $\mathcal{C}(2; n_c, k_c, d_c)$, i.e. a matrix of n_a codewords of the inner code:

$$\begin{pmatrix} \mathbf{b}_1 \\ \mathbf{b}_2 \\ \vdots \\ \mathbf{b}_{n_a} \end{pmatrix} = \begin{pmatrix} b_{11} & b_{12} \\ b_{21} & b_{22} \\ \vdots & \vdots \\ b_{n_a 1} & b_{n_a 2} \end{pmatrix} . \tag{9.29}$$

The length of the GC code is $n_c = 2n_a$, the dimension is $k_c = k_a^{(1)} + k_a^{(2)}$, and the minimum distance is $d_c \geq \min\{d_a^{(1)} d_b^{(1)}, d_a^{(2)} d_b^{(2)}\}$.

Theorem 9.25 (RM code as GC code) *Given the inner code $\mathcal{B}^{(1)}(n_b = 2, k_b^{(1)}, d_b^{(1)})$ and the outer RM codes $\mathcal{A}^{(1)} = \mathcal{R}(r,m)$, $\mathcal{A}^{(2)} = \mathcal{R}(r+1,m)$ of length $n_a = 2^m$, the GC code $\mathcal{C}(2; n_c, k_c, d_c)$ constructed as described above is an RM code $\mathcal{R}(r+1, m+1)$.*

Proof The mapping of a subcode according to equations 9.27 and 9.28 can be written as

$$\mathbf{b}_{a^{(1)}a^{(2)}} = (\mathbf{a}^{(1)} \oplus \mathbf{a}^{(2)}, \mathbf{a}^{(2)}), \tag{9.30}$$

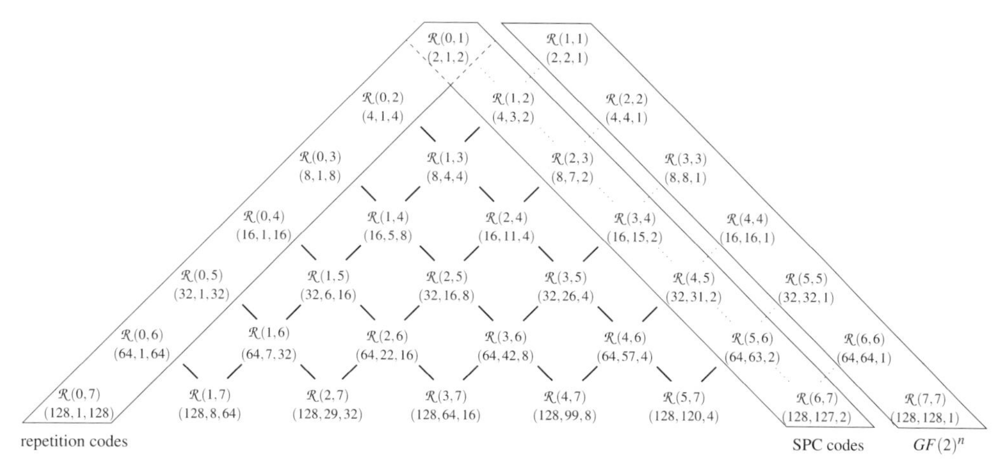

Figure 9.46 Structure of RM codes.

where $\oplus$ is addition modulo 2. We permute the elements of the codeword of equation 9.29 such that

$$\mathbf{c} = (b_{12}, b_{22}, \ldots, b_{n_a 2}, b_{11}, b_{21}, \ldots, b_{n_a 1}) \,. \tag{9.31}$$

Note that the vector $\mathbf{c}$ is of length $2n_a$. Substituting equation 9.30 into equation 9.31 yields

$$\mathbf{c} = \left(a_1^{(2)}, a_2^{(2)}, \ldots, a_{n_a}^{(2)}, a_1^{(1)} \oplus a_1^{(2)}, a_2^{(1)} \oplus a_2^{(2)}, \ldots, a_{n_a}^{(1)} \oplus a_{n_a}^{(2)}\right) .$$

We now find that the above notation for the vector $\mathbf{c}$ is identical to theorem 2, chapter 13, of [McWSl], which gives the $|\mathbf{u}|\mathbf{u} \oplus \mathbf{v}|$ construction of RM codes:

$$\mathcal{R}(r+1, m+1) = \{|\mathbf{u}|\mathbf{u} \oplus \mathbf{v}| : \mathbf{u} \in \mathcal{R}(r+1, m), \mathbf{v} \in \mathcal{R}(r, m)\}.$$

The corresponding GC code is the RM code $\mathcal{R}(r+1, m+1)$. □

Using this method, the RM code $\mathcal{R}(r+1, m+1)$ can be constructed as follows:

$$\mathbf{c} = (\mathbf{b}_1, \mathbf{b}_2, \ldots, \mathbf{b}_{n_a}) = \left(a_1^{(1)} \oplus a_1^{(2)}, a_1^{(2)}, a_2^{(1)} \oplus a_2^{(2)}, a_2^{(2)}, \ldots, a_{n_a}^{(1)} \oplus a_{n_a}^{(2)}, a_{n_a}^{(2)}\right) .$$

The outer codes $\mathcal{A}^{(1)}$ and $\mathcal{A}^{(2)}$ can again be interpreted as GC codes. Based on these considerations, the above code construction can be applied successively. Given the RM code $\mathcal{R}(0, \tilde{m}), 1 \leq \tilde{m} < m$, a repetition code, and the RM code $\mathcal{R}(\tilde{m}-1, \tilde{m}), 1 \leq \tilde{m} < m$, a single parity check code, as possible outer codes, and given the inner code $\mathcal{B}^{(1)}(2;2,2,1)$, every RM code $\mathcal{R}(r, m)$ can be constructed. Figure 9.46 shows that any RM code can be constructed based on the concatenation of repetition codes and single parity check codes. This fact will be used for an efficient decoding method presented in the next section.

9.5.1 GMC decoding algorithm for RM codes

We shall present a decoding procedure [SB95] using the property that RM codes can be viewed as multiple generalized concatenated codes. Moreover, a simple extension to list decoding is described that improves the performance significantly.

Algorithm 9.4 Decoding algorithm for Reed–Muller codes $\mathcal{R}(r,m)$.

Input: $\mathbf{y} = (y_1, \ldots, y_n)$ received vector.

Step 1: Decoding of a repetition code or SPC code
If $r = 0$ or $r = m-1$, SDML decoding of $\mathbf{y}$ according to repetition code or SPC code as $\hat{\mathbf{c}}$ then step 3.

Step 2: Decoding of the GC code

Step 2a: Determination of the metric $\mathbf{y}^{(1)} = \left(y_1^{(1)}, \ldots, y_{n/2}^{(1)}\right)$ of the first outer codeword:

$$y_j^{(1)} = \text{sign}(y_{2j-1}\,y_{2j}),\ \min\{|y_{2j-1}|, |y_{2j}|\}, \quad j = 1, \ldots, n/2\ .$$

First outer decoding:
Decode $\mathbf{y}^{(1)}$ according to $\mathcal{R}(r-1, m-1)$ as $\hat{\mathbf{a}}^{(1)}$ using the GMC algorithm.

Step 2b: Determination of the metric $\mathbf{y}^{(2)} = \left(y_1^{(2)}, \ldots, y_{n/2}^{(2)}\right)$ of the second outer codeword:

$$y_j^{(2)} = \tfrac{1}{2}\left(\hat{a}_j^{(1)}\,y_{2j-1} + y_{2j}\right), \quad j = 1, \ldots, n/2\ .$$

Second outer decoding:
Decode $\mathbf{y}^{(2)}$ according to $\mathcal{R}(r, m-1)$ as $\hat{\mathbf{a}}^{(2)}$ using the GMC algorithm.

Step 2c: Determination of the GC codeword $\hat{\mathbf{c}}$ from $\hat{\mathbf{a}}^{(1)}$ and $\hat{\mathbf{a}}^{(2)}$:

$$\hat{\mathbf{c}} = \left(\hat{a}_1^{(1)}\,\hat{a}_1^{(2)}, \hat{a}_1^{(2)}, \ldots, \hat{a}_{n/2}^{(1)}\,\hat{a}_{n/2}^{(2)}, \hat{a}_{n/2}^{(2)}\right).$$

Step 3: Decode $\hat{\mathbf{c}} = (\hat{c}_1, \ldots, \hat{c}_n)$.

Output: $\hat{\mathbf{c}} = (\hat{c}_1, \ldots, \hat{c}_n)$ decoded codeword.

The idea of the decoding algorithm is similar to the decoding of GC codes as presented in section 9.2.4. However, we shall use decoding algorithms for the inner and outer codes that employ soft-input information according to section 7.2.2. The decoding of RM codes can be carried out based exclusively on the decoding of repetition codes and single parity check codes. ML decoding of these codes can be achieved easily.

According to section 7.2.2, we assume BPSK modulation of the code symbols and an AWGN channel, and we use the following notation ($0 \to +1$ and $1 \to -1$):

$$\begin{aligned}
&\mathbf{c} = (c_1, \ldots, c_n) \in \mathcal{R}(r,m),\ n = 2^m, \quad \text{transmitted codeword,}\\
&\mathbf{y} = (y_1, \ldots, y_n) \quad \text{received vector.}
\end{aligned}$$

ML decoding:

$$\Lambda_{\hat{c}} = \sum_{j=1}^{n} \ln\left(\frac{p\{y_j \,|\, c_j = \hat{c}_j\}}{p\{y_j \,|\, c_j = \overline{\hat{c}_j}\}}\right).$$

The GMC decoding algorithm is described in algorithm 9.4. It consists of three steps. If the RM code is a repetition code ($\mathcal{R}(0,m)$) or an SPC code ($\mathcal{R}(m-1,m)$) then it will be decoded in step 1. Any other RM code is decoded recursively by the application of steps 2a and 2b. In step 3, the estimated codeword is the output of the GMC decoding algorithm.

For both the repetition code and the SPC code, SDML decoding algorithms are used that are defined in algorithm 9.5.

Example 9.48 (Decoding of $\mathcal{R}(1,3)$ using the GMC algorithm) Let us consider an example in order to illustrate the steps of the GMC decoding algorithm:

Algorithm 9.5 SDML decoding of repetition codes and SPC codes.

SDML decoding of a repetition code of length n:

$$\hat{c}_i = \text{sign}\left(\sum_{j=1}^{n} y_j\right), \quad i = 1, \dots, n.$$

SDML decoding of a SPC code of length n:

$$\hat{c}_j = \text{sign}(y_j), \quad j = 1, \dots, n.$$

If $p = \prod_{j=1}^{n} \hat{c}_j = -1$, find $i : |y_i| = \min_j\{|y_j|\}$ and set $\hat{c}_i := -\hat{c}_i$.

transmitted codeword: $\mathbf{c} = (-1.0, +1.0, -1.0, +1.0, +1.0, -1.0, +1.0, -1.0)$.
received vector: $\mathbf{y} = (-0.9, +1.2, -0.1, +0.6, -0.1, -1.3, +0.2, +0.3)$.
step 2a: $\mathbf{y}^{(1)} = (-0.9, -0.1, +0.1, +0.2)$.
SDML decoding of $\mathbf{y}^{(1)}$ according to the repetition code $\mathcal{R}(r = 0, m = 2)$ as
$\hat{\mathbf{a}}^{(1)} = (-1.0, -1.0, -1.0, -1.0)$.
step 2b: $\mathbf{y}^{(2)} = (+2.1, +0.7, -1.2, +0.1)$.
SDML decoding of $\mathbf{y}^{(2)}$ according to the SPC code $\mathcal{R}(r = 1, m = 2)$ as
$\hat{\mathbf{a}}^{(2)} = (+1.0, +1.0, -1.0, -1.0)$.
step 2c: $\hat{\mathbf{c}} = (-1.0, +1.0, -1.0, +1.0, +1.0, -1.0, +1.0, -1.0)$.

The received vector $\mathbf{y}$ is decoded correctly, i.e. $\hat{\mathbf{c}} = \mathbf{c}$. ◇

Metric derivation

In this section, we derive the metric values $y_j^{(1)}$ and $y_j^{(2)}$ used in steps 2a and 2b. We assume that the components of the received vector are statistically independent from each other. This assumption is only an approximation, since the received vector consists of the addition of the transmitted codeword and an error vector. For a memoryless channel, the errors in different components of the error vector are statistically independent; however, different components of a codeword are in general not independent.

The input of the decoding algorithm is the received vector $\mathbf{y} = (y_1, \dots, y_n)$. According to the considerations in section 7.2.2, y_j can be viewed as reliability information, and for a received value $\tilde{y}$, we have

$$y_j \sim \ln\left(\frac{p\{\tilde{y}\,|\,c_j = +1\}}{p\{\tilde{y}\,|\,c_j = -1\}}\right), \quad j = 1, \dots, n. \tag{9.32}$$

For the first outer decoding, we need the metric

$$y_j^{(1)} \sim \ln\left(\frac{p\{y\,|\,a_j^{(1)} = +1\}}{p\{y\,|\,a_j^{(1)} = -1\}}\right), \quad j = 1, \dots, n/2.$$

Using equation 9.30, the conditional probabilities are given by

$$\begin{aligned} p\{y|a_j^{(1)} = c_{2j-1}c_{2j} = +1\} \\ &= p\{y|c_{2j-1} = +1\}\,p\{y|c_{2j} = +1\} + p\{y|c_{2j-1} = -1\}\,p\{y|c_{2j} = -1\}\,, \\ p\{y|a_j^{(1)} = c_{2j-1}c_{2j} = -1\} \\ &= p\{y|c_{2j-1} = +1\}\,p\{y|c_{2j} = -1\} + p\{y|c_{2j-1} = -1\}\,p\{y|c_{2j} = +1\}\,. \end{aligned}$$

For the probability ratio, we obtain

$$\frac{p\{y|a_j^{(1)} = +1\}}{p\{y|a_j^{(1)} = -1\}} = \frac{1 + \frac{p\{y|c_{2j-1}=-1\}}{p\{y|c_{2j-1}=+1\}}\,\frac{p\{y|c_{2j}=-1\}}{p\{y|c_{2j}=+1\}}}{\frac{p\{y|c_{2j-1}=-1\}}{p\{y|c_{2j-1}=+1\}} + \frac{p\{y|c_{2j}=-1\}}{p\{y|c_{2j}=+1\}}}\,.$$

An approximation of the logarithm is (see appendix C on log likelihood algebra):

$$\begin{aligned} \ln\left(\frac{p\{y|a_j^{(1)} = +1\}}{p\{y|a_j^{(1)} = -1\}}\right) &\approx \operatorname{sign}\left(\ln\left(\frac{p\{y|c_{2j-1} = +1\}}{p\{y|c_{2j-1} = -1\}}\right)\ln\left(\frac{p\{y|c_{2j} = +1\}}{p\{y|c_{2j} = -1\}}\right)\right) \\ &\quad \cdot \min\left\{\left|\ln\frac{p\{y|c_{2j-1} = +1\}}{p\{y|c_{2j-1} = -1\}}\right|, \left|\ln\frac{p\{y|c_{2j} = +1\}}{p\{y|c_{2j} = -1\}}\right|\right\}. \end{aligned}$$

Thus we obtain for $y_j^{(1)}$

$$y_j^{(1)} \approx \operatorname{sign}(y_{2j-1}y_{2j})\min\{|y_{2j-1}|, |y_{2j}|\}\,, \quad j = 1, \ldots, n/2\,. \tag{9.33}$$

Simulations show that exact knowledge of the proportionality factor in equation 9.32 yields no significant improvement.

For the second outer decoding, we need the metric

$$y_j^{(2)} \sim \ln\left(\frac{p\{y|a_j^{(1)} = \hat{a}_j^{(1)}, a_j^{(2)} = +1\}}{p\{y|a_j^{(1)} = \hat{a}_j^{(1)}, a_j^{(2)} = -1\}}\right), \quad j = 1, \ldots, n/2\,.$$

With equation 9.30, the conditional probabilities are given by

$$\begin{aligned} p\{y|a_j^{(1)} = \hat{a}_j^{(1)}, a_j^{(2)} = +1\} &= p\{y|c_{2j-1} = \hat{a}^{(1)}(+1), c_{2j} = +1\} \\ &= p\{y|c_{2j-1} = \hat{a}^{(1)}\}\,p\{y|c_{2j} = +1\} \\ p\{y|a_j^{(1)} = \hat{a}_j^{(1)}, a_j^{(2)} = -1\} &= p\{y|c_{2j-1} = \hat{a}^{(1)}(-1)\}\,p\{y|c_{2j} = -1\}\,. \end{aligned}$$

The probability ratio is

$$\frac{p\{y|a_j^{(1)} = \hat{a}_j^{(1)}, a_j^{(2)} = +1\}}{p\{y|a_j^{(1)} = \hat{a}_j^{(1)}, a_j^{(2)} = -1\}} = \frac{p\{y|c_{2j-1} = +\hat{a}^{(1)}\}\;p\{y|c_{2j} = +1\}}{p\{y|c_{2j-1} = -\hat{a}^{(1)}\}\;p\{y|c_{2j} = -1\}}\,.$$

Thus we obtain $y_j^{(2)}$ as

$$y_j^{(2)} = \frac{1}{2}(\hat{a}_j^{(1)}y_{2j-1} + y_{2j})\,,\; j = 1, \ldots, n/2\,, \tag{9.34}$$

where a normalization factor $\frac{1}{2}$ has been introduced for convenience.

Two approximations, namely the assumption that the components of the received vector are statistically independent and the approximation in equation 9.33, prevent the decoding algorithm from being a SDML decoding algorithm.

Since the GMC decoding algorithm is similar to the steps of the algorithms GCD-1 and GCD-i from section 9.2.4, it is guaranteed that the GMC decoding algorithm can correct errors up to half the minimum distance.

Again let $\mathbf{c} = (c_1, \ldots, c_n) \in \mathcal{R}(r,m)$, $c_i \in \{-1, 1\}$, be the transmitted codeword, and let $\mathbf{y}$ be the received vector. The error introduced by the AWGN channel is $e = y - c$. The Hamming distance of the code $d = 2^{m-r}$, and the squared Euclidean distance $\delta = 4d = 4 \cdot 2^{m-r} = 2^{m-r+2}$.

Theorem 9.26 (Error correction capability of GMC) *The GMC decoding algorithm corrects all errors* $\mathbf{e} = (e_1, \ldots, e_n)$*, provided that*

$$\sum_{i=1}^{n} e_i^2 = \sum_{i=1}^{n} (y_i c_i)^2 < \left(\frac{\sqrt{\delta}}{2} \right)^2 = d = 2^{m-r} . \tag{9.35}$$

Proof Without loss of generality, we assume that $\mathbf{c} = (1, 1, \ldots, 1)$ was transmitted. Let us examine the different steps of the algorithm.

Step 1: If $r = 0$ or $r = m - 1$ then $\mathbf{y}$ is SDML-decoded. The set of SDBD decoding results is contained in the set of SDML results.

Step 2a: We shall show that if $\mathbf{y}$ is SDBD-decodable with respect to $\mathcal{R}(r,m)$ then $\mathbf{y}^{(1)}$ is SDBD-decodable with respect to $\mathcal{R}(r-1, m-1)$. We have

$$\mathbf{y}^{(1)} = (y_1^{(1)}, \ldots, y_{n/2}^{(1)}),$$

with

$$y_j^{(1)} = \text{sign}\left((1 + e_{2j-1})(1 + e_{2j})\right) \min\{|1 + e_{2j-1}|, |1 + e_{2j}|\}, \quad j = 1, \ldots, n/2 . \tag{9.36}$$

Without loss of generality, we assume that $|1 + e_{2j-1}| \leq |1 + e_{2j}|$. Consequently, we obtain for equation 9.36

$$y_j^{(1)} = \text{sign}(1 + e_{2j})\,(1 + e_{2j-1}) .$$

Consider the four possible cases:

(i) Both components are in error: $1 + e_{2j-1} \leq 0$ and $1 + e_{2j} \leq 0$
$\Longrightarrow (y_j^{(1)} - 1)^2 = (|1 + e_{2j-1}| - 1)^2 \leq (|1 + e_{2j-1}| + 1)^2 = e_{2j-1}^2.$

(ii) The first component is in error: $1 + e_{2j-1} \leq 0$ and $1 + e_{2j} \geq 0$
$\Longrightarrow (y_j^{(1)} - 1)^2 = ((1 + e_{2j-1}) - 1)^2 = e_{2j-1}^2.$

(iii) The second component is in error: $1 + e_{2j-1} \geq 0$ and $1 + e_{2j} \leq 0$
$\Longrightarrow (y_j^{(1)} - 1)^2 = (|1 + e_{2j-1}| + 1)^2 \leq (|1 + e_{2j}| + 1)^2 = e_{2j}^2.$

(iv) Both components are correct: $1 + e_{2j-1} \geq 0$ and $1 + e_{2j} \geq 0$
$\Longrightarrow (y_j^{(1)} - 1)^2 = ((1 + e_{2j-1}) - 1)^2 = e_{2j-1}^2 .$

From equation 9.35, we obtain:

$$\sum_{j=1}^{n/2} (y_j^{(1)} - 1)^2 \leq \sum_{j=1}^{n/2} \max\{e_{2j-1}^2, e_{2j}^2\} \leq \sum_{j=1}^{n/2} e_{2j-1}^2 + e_{2j}^2 = \sum_{i=1}^{n} e_i^2 < d .$$

In general, it follows that $\mathbf{y}^{(1)}$ can be decoded according to $\mathcal{R}(r-1,m-1)$.

Step 2b: We shall show that if $\mathbf{y}$ is SDBD decodable with respect to $\mathcal{R}(r,m)$, and if $a^{(1)}$ has been determined correctly, then $\mathbf{y}^{(2)}$ is SDBD decodable with respect to $\mathcal{R}(r,m-1)$. Determine

$$\mathbf{y}^{(2)} = \left(y_1^{(2)},\ldots,y_{n/2}^{(2)}\right) \quad \text{with} \quad y_j^{(2)} = 1 + \frac{1}{2}\left(e_{2j-1}+e_{2j}\right), \quad j=1,\ldots,n/2\,.$$

We have

$$\left(y_j^{(2)}-1\right)^2 = \frac{1}{4}(e_{2j-1}+e_{2j})^2 \leq \frac{1}{4}\left(e_{2j-1}^2 + 2\left|e_{2j-1}\cdot e_{2j}\right| + e_{2j}^2\right) \leq \frac{1}{4}\left(e_{2j-1}^2+e_{2j}^2\right)\,.$$

This yields an approximation for the sum:

$$\sum_{j=1}^{n/2}\left(y_j^{(2)}-1\right)^2 \leq \frac{1}{2}\sum_{j=1}^{n/2}\left(e_{2j-1}^2+e_{2j}^2\right) = \frac{1}{2}\sum_{i=1}^{n} e_i^2 < \frac{d}{2}\,.$$

The distance of $\mathcal{R}(r,m-1)$ is $\frac{d}{2} = 2^{m-r-1}$, and $\mathbf{y}^{(2)}$ can be decoded.

By applying the algorithm recursively, we eventually obtain the codes having $r=0$ or $r=m-1$, which can be decoded correctly, and thus the theorem is proved. □

According to the considerations on ordered statistics (algorithm 7.11, page 191), we can improve the error correction capability estimation. Let Ω be a subset with cardinality $d=2^{m-r}$ of the n codeword components for which the sum

$$\sum_{i\in\Omega}(y_i+c_i)$$

gives the maximum value. Then the GMC decoding algorithm can correct all errors, provided that [SB94]

$$\sum_{i\in\Omega}(y_i+c_i) < d = 2^{m-r}\,.$$

Correspondingly, we have $\sum_{i\in\Omega} c_i\,(y_i+c_i) < d = 2^{m-r}$.

9.5.2 L-GMC, list decoding of RM codes

We shall describe the extended GMC algorithm followed by the bit error rate and block error rate for L-GMC decoding. Finally, we will investigate the decoding complexity of L-GMC.

The main drawback of the GMC decoding algorithms described in the previous section is that in every decoding stage a *hard decision* is made that limits the use of information reliability. One attempt to overcome this drawback is to use the concept of *list decoding* according to section 7.4. This concept is incorporated into GMC decoding by passing a list of L codeword candidates. The codewords passed to the next higher level are those with smallest Euclidean distance to the corresponding soft input found by L-GMC decoding. The probability that the SDML codeword is an element of the list is higher than by considering only one codeword. An appropriate modification of the GMC algorithm for soft-decision list decoding of an RM code $\mathcal{R}(r,m)$ is shown in algorithm 9.6 (see also [LBD98b]).

Note that for $L=1$, we obtain the GMC algorithm. In step 2e, we reduce the list of length $L_1 L$ to the L codewords with the smallest Euclidean distance to $\mathbf{y}$. These are not necessarily the L codewords with the absolute smallest Euclidean distance to $\mathbf{y}$. Otherwise, L-GMC would be an SDML decoding procedure.

Algorithm 9.6 L-GMC algorithm.

Input:	Code length:	m: $n=2^m$, $m \geq 2$.
	Order of the code:	r.
	Received vector:	$\mathbf{y}=(y_1, y_2, \ldots, y_n)$.
	Depth of the list:	$1 \leq L \leq 6$ $(1 \leq L \leq 2$ for $r=0)$.
Output:	Decoded codewords:	$\hat{\mathbf{c}}[i]$, $i=1,\ldots,L$.

1. **(a)** If $(r=0)$: RC list decoding of $\mathbf{y}$ with depth of list L.
   (b) If $(r=m-1)$: SPC list decoding of $\mathbf{y}$ with depth of list L.
2. **(a)** If $(r=1)$: $L_1 := 2$, otherwise: $L_1 := L$.
   (b) If $(L=1 \wedge r=1)$: $L_1 := 1$.
   (c) Metric computation for the first codeword:
   $y_j^{(1)} \approx \operatorname{sign}(y_{2j-1} \cdot y_{2j}) \min\{|y_{2j-1}|, |y_{2j}|\}$, $j=1,\ldots,2^{m-1}$.
   (d) Decoding of $\mathbf{y}^{(1)}$ according to $\mathcal{R}(r-1, m-1)$:
   $\hat{\mathbf{a}}^{(1)}[i]$, $i=1,\ldots,L_1$, with L-GMC (depth of list L_1).
   (e) For $i=1,\ldots,L_1$:
   (i) Metric computation for the second outer codeword:
   $y_j^{(2)} = \hat{a}_j^{(1)}[i]\, y_{2j-1} + y_{2j}$, $j=1,\ldots,2^{m-1}$.
   (ii) Decoding of $\mathbf{y}^{(2)}$ according to $\mathcal{R}(r, m-1)$:
   $\hat{\mathbf{a}}^{(2)}[\ell]$, $\ell=1,\ldots,L$, with L-GMC (depth of list L).
   (iii) For $\ell=1,\ldots,L$, compute the GC codewords
   $\hat{\mathbf{c}}[(i-1)L+\ell] = (\hat{a}_1^{(1)}[i]\,\hat{a}_1^{(2)}[\ell], \hat{a}_1^{(2)}[\ell], \ldots, \hat{a}_{2^{m-1}}^{(1)}[i]\,\hat{a}_{2^{m-1}}^{(2)}[\ell], \hat{a}_{2^{m-1}}^{(2)}[\ell])$.
   (f) Ordering of the codewords $\hat{\mathbf{c}}[i]$, $i=1,\ldots,L_1 L$, according to their Euclidean distance $\mathbf{y}$ in increasing order.
3. Output of the (ordered) codewords $\hat{\mathbf{c}}[i]$, $i=1,\ldots,L$.

Example 9.49 (Comparison of GMC and L-GMC: decoding of the $R(2,4)$ code) We assume that the all-zero codeword of the RM code $\mathcal{R}(2,4)$, i.e. $(2;16,11,4)$, was transmitted over an AWGN channel and that the vector $\mathbf{y}$ was received. Figure 9.47 depicts the steps of the GMC decoding. The gray marked values correspond to hard-decision errors. We observe that the GMC algorithm gives a false decoding result. Now, we decode the same received vector using the L-GMC algorithm ($L=2$).This is shown in figure 9.48. Again, the gray-marked values correspond to hard-decision errors. The output of the algorithm is the 4 codewords $L_1\, L=4$ and their inner products with $\mathbf{y}$. We observe that the transmitted codeword is an element of the L-GMC list and has the largest inner product (12.4), i.e. the decoding was correct. ◇

The L-GMC decoding algorithm is in general not an SDML decoding algorithm, and in general it is not possible to compute the discrepancy to SDML decoding. Therefore we depend upon simulations (see section 9.5.3). For first-order RM codes, we can prove the following theorem:

Theorem 9.27 (L-GMC is equivalent to SDML decoding of $\mathcal{R}(1,m)$) *The L-GMC algorithm with $L=2$ is an SDML decoding algorithm for first order RM codes $\mathcal{R}(1,m)$, i.e. $(2;2^m, m+1, 2^{m-1})$.*

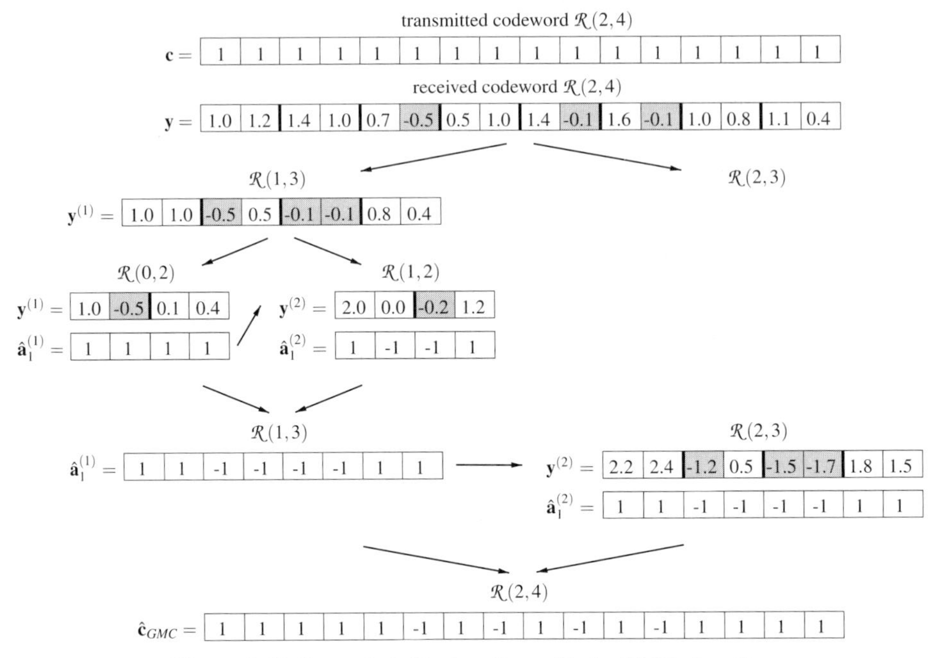

Figure 9.47 Example 9.49: decoding with the GMC algorithm.

Proof We first prove that if the code $\mathcal{R}(1,m-1)$ is SDML decoded with L-GMC, then $\mathcal{R}(1,m)$ is also SDML decoded. For simplicity, we use the following notation:

$$\begin{array}{|c|c|} \hline \multicolumn{2}{|c|}{\mathcal{R}(1,m)} \\ \hline \multicolumn{2}{c}{=} \\ \hline \mathcal{R}(1,m-1) & \mathcal{R}(1,m-1) \\ \hline \multicolumn{1}{c}{} & \multicolumn{1}{c}{\oplus} \\ \cline{2-2} \multicolumn{1}{c|}{} & \mathcal{R}(0,m-1) \\ \cline{2-2} \end{array}$$

A codeword of $\mathcal{R}(1,m)$ is then

$$\hat{\mathbf{c}} = \left(\hat{a}_1^{(2)}, \ldots, \hat{a}_{n/2}^{(2)}, \hat{a}_1^{(1)}\hat{a}_1^{(2)}, \ldots, \hat{a}_{n/2}^{(1)}\hat{a}_{n/2}^{(2)}\right), \tag{9.37}$$

where $\hat{\mathbf{a}}^{(2)} \in \mathcal{R}(1,m-1)$ and $\hat{\mathbf{a}}^{(1)} \in \mathcal{R}(0,m-1)$. For the inner product, we have

$$\begin{aligned} \langle \hat{\mathbf{c}}, \mathbf{y} \rangle &= \sum_{i=1}^{n} \hat{c}_i y_i = \sum_{i=1}^{n/2} \hat{c}_i y_i + \sum_{i=n/2+1}^{n} \hat{c}_i y_i = \sum_{i=1}^{n/2} \hat{a}_i^{(2)} y_i + \sum_{i=1}^{n/2} \hat{a}_i^{(1)} \hat{a}_i^{(2)} y_{i+n/2} \\ &= \sum_{i=1}^{n/2} \left(\hat{a}_i^{(2)} y_i + \hat{a}_i^{(1)} \hat{a}_i^{(2)} y_{i+n/2}\right) = \sum_{i=1}^{n/2} \hat{a}_i^{(2)} \cdot \underbrace{\left(y_i + \hat{a}_i^{(1)} y_{i+n/2}\right)}_{y_i^{(2)}} . \end{aligned} \tag{9.38}$$

The repetition code $\mathcal{R}(0,m-1)$ consists of the two codewords $\hat{\mathbf{a}}^{(1)} = +\mathbf{1}$ and $\hat{\mathbf{a}}^{(1)} = -\mathbf{1}$ only. Thus we obtain

$$\begin{aligned} &\{(\mathbf{u}|\mathbf{u}) \mid \mathbf{u} \in \mathcal{R}(1,m-1)\}, \quad \text{for } (\hat{\mathbf{a}}^{(1)} = +\mathbf{1}), \\ &\{(\mathbf{u}|-\mathbf{u}) \mid \mathbf{u} \in \mathcal{R}(1,m-1)\}; \quad \text{for } (\hat{\mathbf{a}}^{(1)} = -\mathbf{1}) . \end{aligned}$$

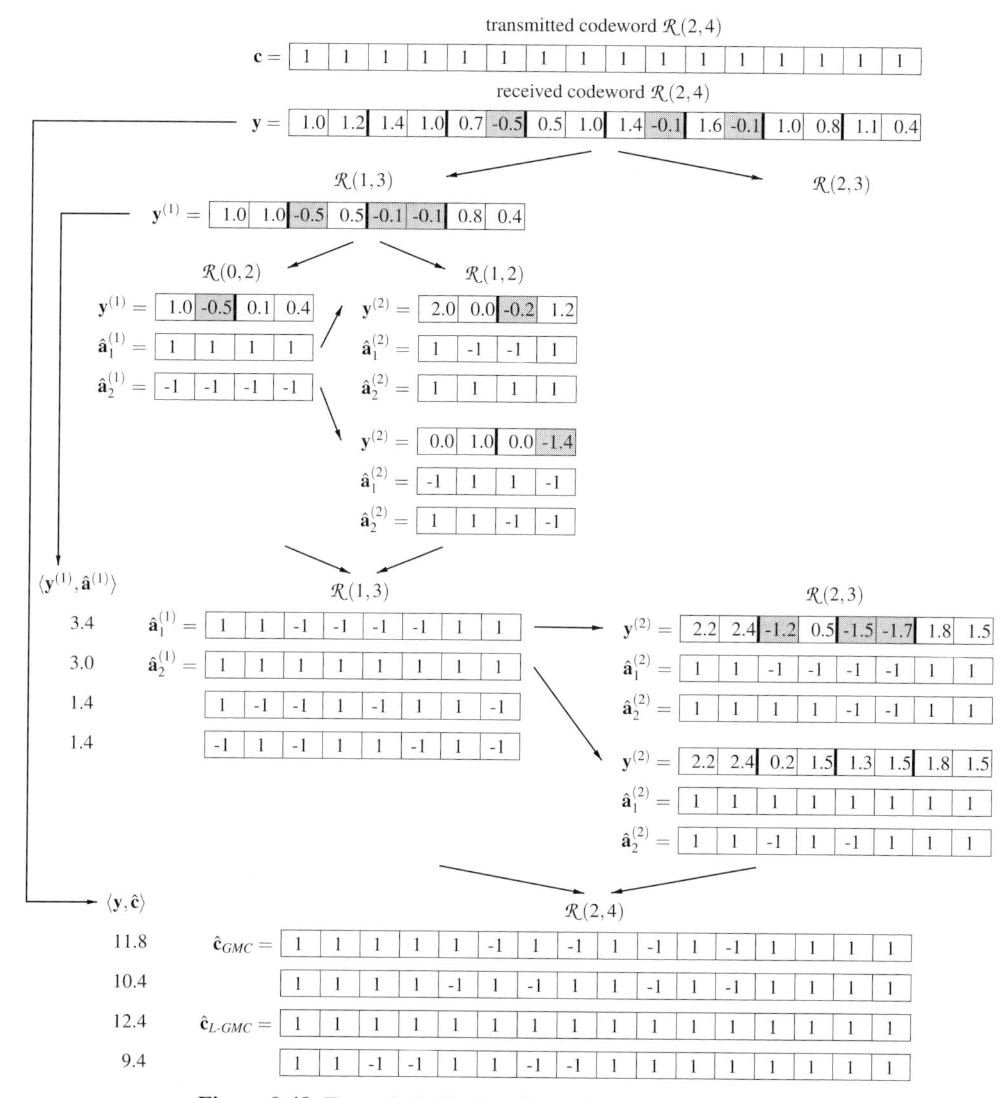

Figure 9.48 Example 9.49: decoding with the L-GMC algorithm.

According to the assumptions, L-GMC finds the SDML codeword of $\mathcal{R}(1, m-1)$. Consider the inner product $\langle \hat{\mathbf{c}}, \mathbf{y} \rangle$; then L-GMC yields for each of the subsets ($\hat{\mathbf{a}}^{(1)} = +\mathbf{1}$ and $\hat{\mathbf{a}}^{(1)} = -\mathbf{1}$) the codeword with the largest inner product (SDML codewords). One of these coset SDML codewords is the SDML codeword of the code $\mathcal{R}(1, m)$. Consequently, we have an SDML decoding procedure for $\mathcal{R}(1, m)$.

To complete the induction, we must show that L-GMC is an SDML decoding algorithm for $\mathcal{R}(1, 2)$. Obviously, this is fulfilled. □

9.5.3 Simulation results and computational complexity

We shall discuss some simulation results for bit error rates and block error rates for different RM codes. The GMC algorithm is viewed as a special case of the L-GMC algorithm with

$L = 1$. We again assume an AWGN channel and BPSK modulation. The simulation results are compared to SDML decoding. Since SDML decoding is possible for short codes only, we need an upper bound like the union bound (see [Pro]) and a lower bound that can be determined as follows:

Theorem 9.28 (Lower bound for block error probability) *(without proof) An SDML decoder cannot achieve a smaller block error probability than a decoder that makes the decoding decision* $\hat{\mathbf{a}}$ *based on the knowledge of the transmitted codeword* $\mathbf{c}$ *(*$\mathbf{y}$ *received) as follows:*

- $\hat{\mathbf{c}} = \mathbf{c} \Longrightarrow \hat{\mathbf{a}} = \mathbf{c} = \hat{\mathbf{c}}$;
- $\hat{\mathbf{c}} \neq \mathbf{c}$:

$$d_E(\mathbf{y}, \mathbf{c}) \leq d_E(\mathbf{y}, \hat{\mathbf{c}}) \Longrightarrow \hat{\mathbf{a}} = \mathbf{c},$$
$$d_E(\mathbf{y}, \mathbf{c}) > d_E(\mathbf{y}, \hat{\mathbf{c}}) \Longrightarrow \hat{\mathbf{a}} = \hat{\mathbf{c}};$$

- *decoding failure:* $\hat{\mathbf{a}} = \mathbf{c}$

This procedure can be used for simulation of a lower bound on the SDML block error probability. Simulations show that the closer the performance of a decoding algorithm to SDML decoding the tighter is the bound. Obviously, for an SDML decoding algorithm, both the bound and the SDML decoder are identical.

Furthermore, in the case of systematic encoding, we can use the approximation

$$P_{Bit} \approx \frac{d}{n} P_{Block}$$

for the bit error rate.

Example 9.50 (Simulation of the $\mathcal{R}(1,5)$ codes) First, we investigate the short RM code $\mathcal{R}(1,5)$, which is a $(32,6,16)$ code. In addition to GMC decoding, we show for comparison BMD, SDML and HDML decoding. The simulation results are depicted in figure 9.49. The block error probability is plotted versus the signal-to-noise ratio. In theorem 9.27, we showed that L-GMC with $L = 2$ is equivalent to SDML decoding. We observe at $E_b/N_0 = 6$ dB that the decoding with GMC yields a block error rate that is a factor 80 smaller than BMD decoding. Note that SDML decoding is identical to L-GMC decoding with $L = 2$. ◇

Example 9.51 (Simulation of $\mathcal{R}(r,6)$ codes) In this example, we investigate the codes $\mathcal{R}(2,6)$ and $\mathcal{R}(4,6)$. Figures 9.50 and 9.51 depict the block error rates for soft-decision decoding with L-GMC. The list depth was varied from $L = 1$ to $L = 4$. Decoding with $L = 1$ is GMC decoding. The RM codes are encoded systematically. In order to compare the simulation results with SDML decoding, we determined a lower bound on the SDML block error rate using theorem 9.28. The lower bound was obtained using list-4 decoders.

The block error rates together with the lower bounds show that we achieve approximately SDML decoding performance. Moreover, we observe that the decoding performance using GMC compared with SDML decoding performance decreases with increasing order of the RM codes. This can be explained as follows. For the recursive decoding of RM codes of higher order, the GMC algorithm uses mostly high-rate subcodes. On the other hand, RM codes of lower order contain mostly low-rate subcodes. For the computation of the soft input metrics, it was assumed that the coordinates are statistically independent. Code symbols of low-rate codes contain more dependences than high-rate codes. Therefore this approximation is rougher for RM codes of lower order than for RM codes of higher order. ◇

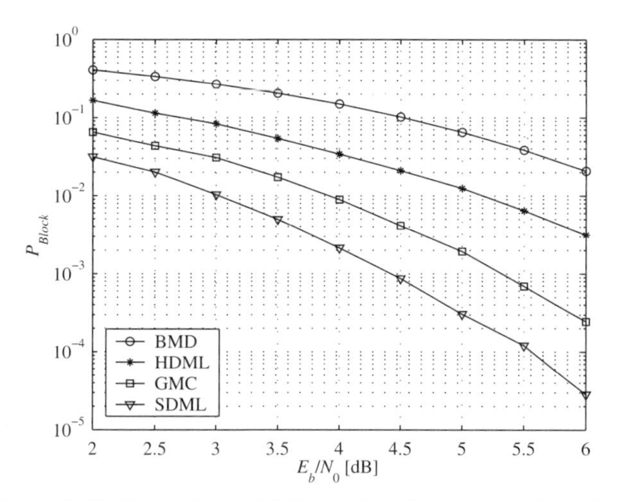

Figure 9.49 Comparison of different decoding concepts for $\mathcal{R}(1,5)$.

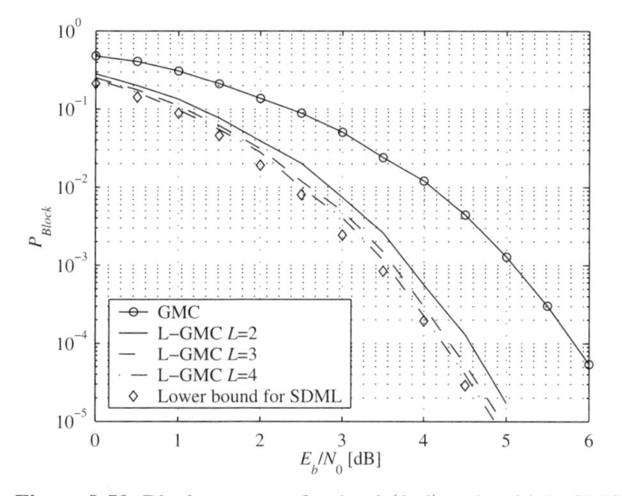

Figure 9.50 Block error rate for the $\mathcal{R}(2,6)$ code with L-GMC.

Since the L-GMC algorithm is an efficient soft-decision decoding procedure of low complexity, we are now able to decode the long RM codes $\mathcal{R}(4,9) = (512,256,32)$ and $\mathcal{R}(5,10) = (1024,638,32)$.

Example 9.52 (Simulation of the $\mathcal{R}(4,9)$ and the $\mathcal{R}(5,10)$ codes) For both the $\mathcal{R}(4,9) = (512,256,32)$ code and the $\mathcal{R}(5,10) = (1024,638,32)$ code, no SDML decoding procedure is known.

Figures 9.52 and 9.53 depict the bit error probabilities of the $\mathcal{R}(4,9)$ code and the $\mathcal{R}(5,10)$ code using L-GMC ($L = 1,\ldots,4$). At a bit error rate of 10^{-4}, the gains of list-4 decoding compared with a list-1 decoding are 1.1 dB and 0.8 dB respectively.

It was not possible to determine a lower bound on the bit error probability using theorem 9.28. This means that it is not possible to achieve an approximation to SDML decoding

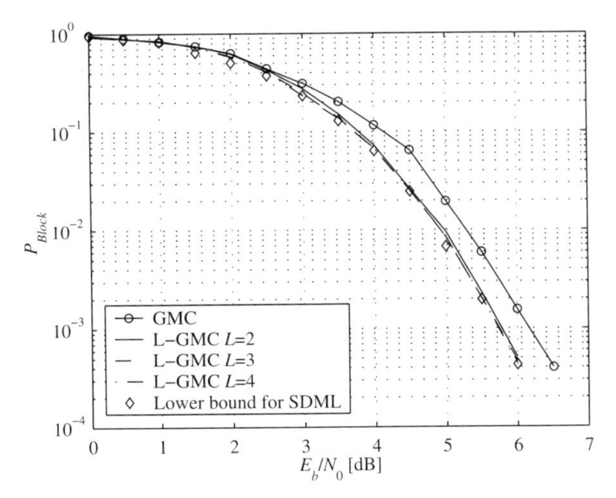

Figure 9.51 Block error rate for the $\mathcal{R}(4,6)$ code with L-GMC.

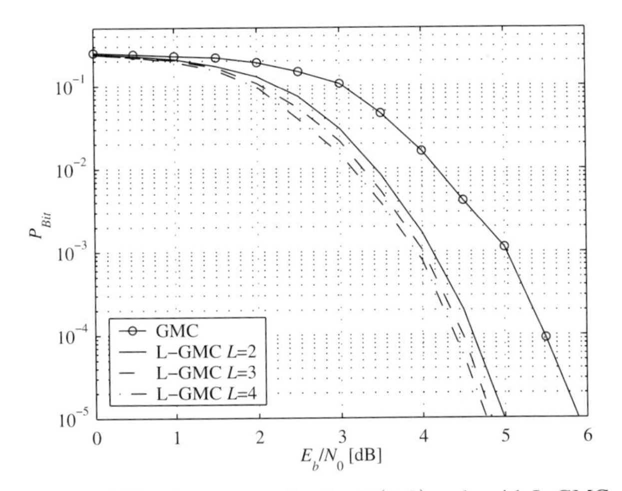

Figure 9.52 Bit error rate for the $\mathcal{R}(4,9)$ code with L-GMC.

performance with L-GMC. One reason for this is that, compared with shorter codes, equations 9.33 and 9.34 for the soft-input metric computation have to be applied more often. The assumption of statistical independence between the coordinates is therefore rougher and violated more often. Consequently, the metric becomes poorer, and with it the decoding performance, as the number of recursion steps increases. ◇

Computational complexity of the L-GMC algorithm Decoding with L-GMC has a lower computational complexity in comparison with all other known decoding algorithms for RM codes that use reliability information; for example, Viterbi decoding of the RM code $\mathcal{R}(4,9) = (512,256,32)$ using the minimum trellis needs more than 10^{30} times more floating point operations per codeword than the GMC algorithm.

The computational complexity is estimated by the number of floating point operations re-

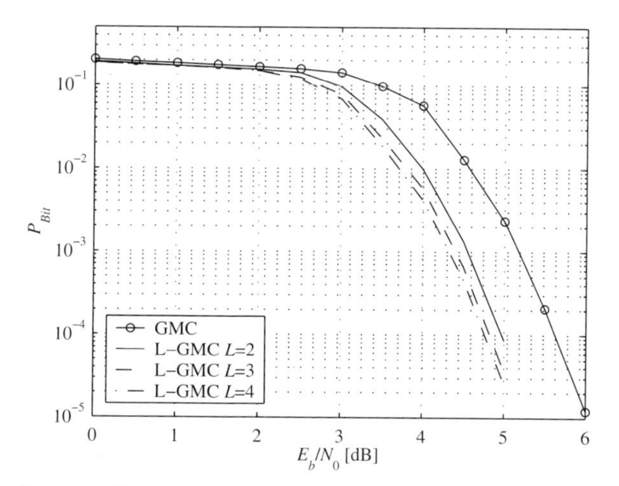

Figure 9.53 Bit error rate for the $\mathcal{R}(5,10)$ code with L-GMC.

Table 9.8 Computational complexity of the L-GMC algorithm.

Code	GMC	L-GMC		
		L=2	L=3	L=4
$\mathcal{R}(2,6)$	260	476	1,352	1,065
$\mathcal{R}(3,6)$	258	1,743	4,271	8,658
$\mathcal{R}(4,6)$	180	787	1,523	2,538
$\mathcal{R}(4,9)$	3,453	55,055	250,573	834,062
$\mathcal{R}(5,10)$	7,650	132,735	643,127	2,202,498

quired for decoding a codeword. Table 9.8 shows the number of floating point operations needed for decoding a single codeword for some of the above examples. We observe that, depending on the code length and order, the complexity of list-2 decoding is a factor of 2–17 more complex than list-1 decoding, and list-2 decoding achieves a large performance improvement compared to list-1 decoding. Moreover, the increase in complexity is slower as the list depth L increases. In [LBD98a], it is shown that the interpretation of RM codes of order $r \geq 2$ over the inner code $(4,4,1)$ leads to an even lower decoding complexity at the same performance.

9.6 Summary

Code concatenation was first introduced in 1954 by Elias using product codes. In 1966, Forney [For66a] investigated concatenated codes using the *superchannel* model. Several years later Blokh and Zyablov published their concept of generalized concatenation [BZ74] and hard-decision decoding of GC codes up to half the minimum distance (see [Zin81] and [Eri86]). In 1976, Zinoviev [Zin76] developed longer codes from relatively trivial length-16 codes. Codes with some of the best known parameters belong to this class (see the table in [McWSl, p. 675]).

The basic concept of generalized concatenation offers many possibilities, which we have divided into several different areas. In addition to the topics covered in this chapter, there are many other research areas related to GC codes which are constantly being introduced but that could not be mentioned here.

GC codes with block codes GC codes with the ability to correct burst errors and single errors were investigated by Zinoviev and Zyablov [ZZ79b]. Ahlin, Ericson and Zyablov ([AEZ85]) have analyzed the use of GC codes in fading channels. *Unequal error protection* (UEP) codes were analyzed in [ZZ79a] and [LL87]. Cyclic concatenated codes were introduced by Berelkamp and Justesen [BJ74] and Jensen [Jen85], [Jen92] and [Jen96] that each cyclic code can be viewed as a GC code. Therefore GC codes with the same inner and outer codes can be described as cyclic or non-cyclic. The GEL codes introduced by Wolf in 1962 are suited for the construction of high-rate codes with good decoding properties.

Concatenated codes designed for channel error patterns, termed blot correcting codes, were derived in [Bre97]. The concept is based on re-encoding the redundancy, and by transmitting only the re-encoded redundancy. The Singleton bound was proved in [BS96], and shows that in order to correct two-dimensional burst errors with size $b_1 b_2$, at least $2b_1b_2$ redundant bits are required. It is interesting that for the case where $b_1 = 1$, this corresponds to the Reiger bound (see section 6.6).

In 1980, Bos [Bos80] showed that GC codes with inner codes could be constructed based on any metric. This fact is exploited in chapter 10 for coded modulation. Furthermore, in [For88a] and [For88b], special cases of GC codes, called coset codes, are investigated in detail.

Numerous publications based on these topics have been written since 1985. These include [AEZ85, DC87, HEK87, KL87, KTL86, Eri86] and many more (see also the summary of chapter 10). The standard code for satellite transmission is a concatenated code (see [WHPH87]).

Decoding of GC codes The decoding of concatenated codes was for a long time not satisfactorily solved. By decoding the inner and outer codes, one can show that correction up to half the minimum distance cannot be guaranteed. In 1974, Blokh and Zyablov proved that one can correct concatenated codes up to half the minimum distance (see [Zin81]). Further work on decoding of GC codes is presented in [BS90], [BKS92] and [SB90] (see also the summary of chapter 10).

GC codes with convolutional codes The first construction of GC codes with inner convolutional codes based on (P)UM codes was described in [ZySha] and [ZSJ95]. Furthermore, in [JTZ88], [ZJTS96] and [ZS87], bounds for GC codes with inner UM codes were given. The partitioning can be carried out for any convolutional code, as shown in [BDS96a]. Block [BDS96a] and convolutional [BDS96b] scrambler matrices are used to transform the original convolutional code into an equivalent code. This equivalent code is designed to have better partitioning properties. An overview and description of all techniques can be found in [BDS97]. This work also gives an algorithm for the calculation of block and convolutional scramblers that can be applied to any code from section 9.3.3. The described method can be viewed as a generalization of the method from [ZySha], which has been presented in section 9.3.1. The basis for the partitioning comes from the algebraic description of convolutional codes in [For70] and [JW93]. Here it is shown that many different code matrices exist for one code. The view that a convolutional code is a specific case of a (P)UM code was presented by Lee [Lee76].

RM codes By applying the principle of multilevel error protection to generalized concatenation, we can describe a class of Reed–Muller codes (see the summary of chapter 5). This class gives a decoding algorithm which can use information reliability and at the same time have a low complexity. The $|u|u \oplus v|$ construction (also called the Plotkin construction [Plo60]) has been known since 1960 (1951 in Russian). A recursive description of RM codes was also given in [Gore70]. In 1988, Forney [For88b] published a recursive coset code construction that is a special case of generalized concatenation. Different decoding methods are known for the decoding of RM codes without reliability information. RM codes can be decoded using a multilevel decoding method (see section 7.3.2) i.e. using the Reed algorithm (see [McWSl, chapter 13]) or the method from [TSKN82]. Furthermore, an ML decoding method for punctured $\mathcal{R}(m-3,m)$ RM codes was described by Seroussi et al. [SL83]. A further ML decoding method for RM codes of first order $\mathcal{R}(1,m)$ was described in [McWSl, chapter 14] and in Karyakin [Kar87].

There are different ways in which reliability information can be used. Be'ery et al. [BS86] described an ML decoding method based on the Hadamard transformation. Litsyn et al. [LS83] and [LNSM85] described a suboptimal algorithm for RM codes of first order. Forney [For88b] presented an ML decoding method for RM codes that is based on the trellis representation of codes (section 6.8) and Viterbi decoding (sections 7.4.2 and 8.4.2).

What has been covered in this chapter We have defined and investigated several very powerful principles of generalized code concatenation. Starting with an inner code defined over a code space having a particular metric and distance measure, the code is partitioned into subcodes with larger minimum distances. The subcodes can again be partitioned, under the condition that the union of all subcodes contain all codewords of the original code. Each partitioning is labeled, and the labels are protected by an outer code. We have derived construction methods for partitioning, and have given decoding methods that can correct GC codes up to half the minimum distance.

After defining GC codes, the construction of UEP codes has been presented, and several results have been given. We have described the correction of burst errors of GC codes, and have compared the principle of generalized code concatenation with classical code concatenation with and without interleaving.

Cyclic GC codes have been defined and described, and the concept of re-encoding the redundancy bits has been introduced. The partitioning of convolutional codes allows them to be used as the inner code of a GC code.

All four possibilities for the construction of GC codes with block and convolutional codes have been dealt with to some degree. In many places, there are still unsolved problems and open questions. Further research in this area promises to produce more powerful coding techniques with many practical applications.

Note For the decoding of GC codes, there are still unsolved problems. In particular, the theory of ML decoding of the subcodes that will approach the ML decoding of the entire code is open. In many cases, this coding loss is not known. Another principal goal is to construct codes with low complexity and efficient decoding algorithms. Low-density parity check codes and turbo codes have resulted from research in this direction.

The topic of coded modulation also belongs to the area treated in this chapter. However, in order to emphasize its importance, we describe the special case of coded modulation separately in chapter 10.

10

Coded modulation

In this chapter, we use the concepts developed in chapter 9 to describe the fundamental concepts of coded modulation. Because this is an active area of research, we cannot discuss all the latest results in detail. Rather, our goal is to explain how generalized concatenation using modulation can increase the Euclidean distance.

Forney et al. [FGL$^+$84] asserted the following:

> *At the moment there is an explosive increase in interest in coded modulation, in research as well as in application.*

This statement is still valid, and thus much effort is being spent to obtain new results in this field.

We first consider two introductory examples to illustrate the basic principle. Then we define codes in Euclidean space (coded modulation) as a special case of generalized concatenation. As the outer code, we can use a block code as well as a convolutional code. As in chapter 9, we distinguish two cases: the inner code can be a block code (i.e. modulation without memory) or a convolutional code (i.e. modulation with memory). The first case corresponds to PSK (*phase shift keying*) and QAM (*quadrature amplitude modulation*) modulation schemes [Pro]. The second case belongs to the class of CPM (*continuous phase modulation*) modulation, for example CPFSK (*continuous phase frequency shift keying*) and MSK (*minimum shift keying*). These modulation methods may be described by means of a trellis.

Although the decoding process is identical to the decoding of generalized concatenated codes, we present coded modulation decoding separately to illustrate the use of different metrics.

Furthermore, we consider multidimensional spaces and lattices. These offer more diverse possibilities for partitioning of signals as compared with two-dimensional signal space, and allow for the concept of multilevel generalized concatenation of codes. The partitioning of modulation schemes with memory also allows for many diverse modifications.

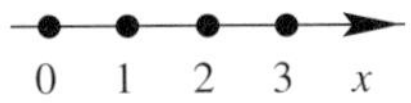

0 1 2 3 x

Figure 10.1 Modulation alphabet of 4-ASK constellation (one-dimensional).

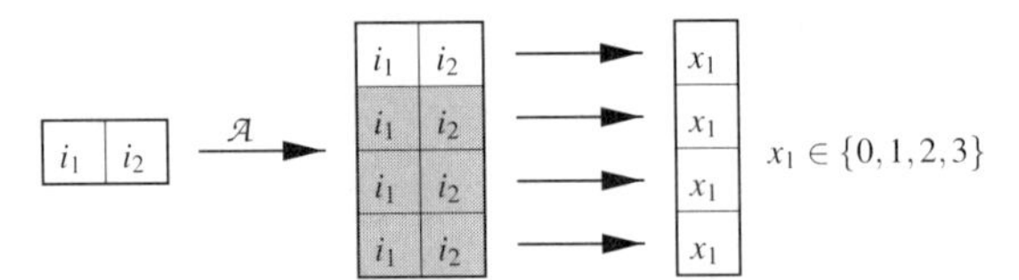

Figure 10.2 Classical concatenation.

10.1 Introductory examples

We first consider a simple example of a modulation scheme whose signals are represented in the one-dimensional signal space of real numbers $\mathbb{R}$ and thus can be visualized as points on a straight line. For example this could be a one-dimensional ASK (*amplitude shift keying*) modulation [Pro, p. 174]. As metric of the inner codes, we use the quadratic (squared) Euclidean distance according to section 7.1.4. In the following example, we compare classical concatenation with coded modulation.

Example 10.1 (One-dimensional inner modulation) Assume that a straight line with the points $\{0,1,2,3\}$ is given according to figure 10.1. Each point is unambiguously numbered by two bits. The minimum quadratic distance among these points is $\delta = 1$. As notation for the inner code, we use $\mathcal{X}(\mathbb{R}, M = 4, \delta = 1)$, where M labels the cardinality of the modulation alphabet. First we consider the classical concatenation of channel coding and modulation. Then we investigate the concept of coded modulation and compare the two methods. Two information bits are coded with the outer repetition code $\mathcal{A}(2^2;4,1,4)$ according to figure 10.2. Each of the four symbols chooses one point from $\mathcal{X}$. Therefore a codeword consist of four points of $\mathbb{R}$, which means that the codeword is a vector from $\mathbb{R}^4$, and the corresponding concatenated code is

$$C(\mathbb{R}^4, M = 2^2, \delta = 4).$$

We have to consider the minimum distance of the concatenated code using the same metric as for the inner code.

To obtain coded modulation, we now partition the code $\mathcal{X}^{(1)} = \mathcal{X}$ according to figure 10.3 into 2 times 2 points, that is in $\mathcal{X}_0^{(2)}$ and $\mathcal{X}_1^{(2)}$. As outer code, we use $\mathcal{A}^{(1)}(2;4,1,4)$ and

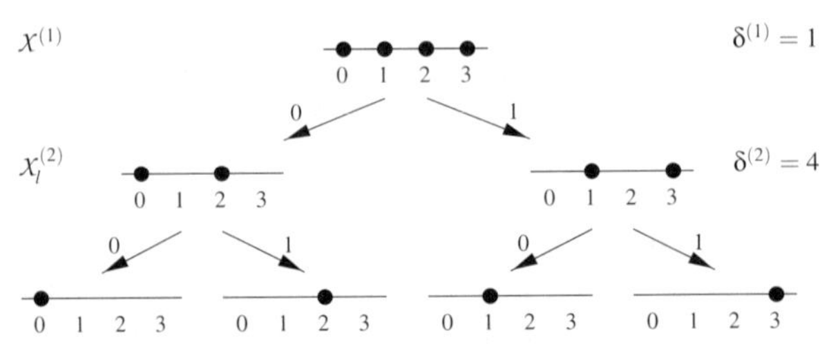

Figure 10.3 Partitioning of 4-ASK constellation.

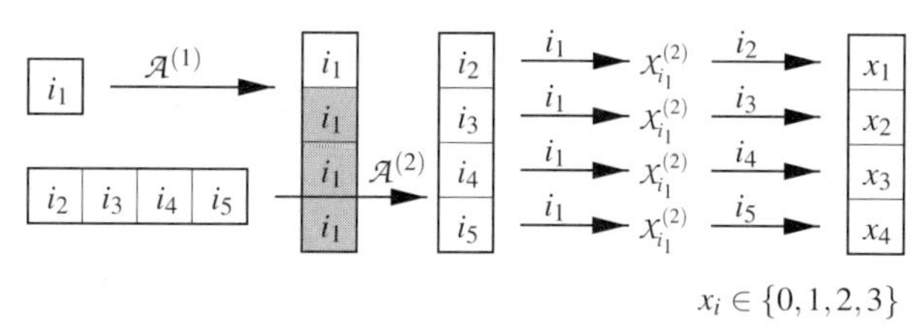

Figure 10.4 Coded modulation, coding of the GC code.

$\mathcal{A}^{(2)}(2;4,4,1)$ and obtain

$$C_{GC}(\mathbb{R}^4, M = 2^5, \delta = \min\{1 \cdot 4, 4 \cdot 1\}).$$

The GC code (with coded modulation) has the same minimum distance but 8 times more codewords, compared with classical concatenation. Figure 10.4 illustrates the coding procedure. A codeword is composed of 4 real numbers. ◇

The concepts of coded modulation and GC codes are identical. We merely use different metrics for the inner code. We now present a second example, where we focus on the common case of a two-dimensional modulation.

Example 10.2 (Two-dimensional inner modulation) Figure 10.5 illustrates a circular constellation that corresponds to 8-PSK. The energy of the signal points is normalized to 1. From the second step on, this can be considered as a partitioning of QPSK. Each of the 4 points is determined by 2 bits. We next construct a classic concatenated code and a GC code based on QPSK modulation and use the quadratic Euclidean distance as the metric. If we use two shortened BCH outer codes $(2;30,14,8)$ in parallel, we obtain the classic concatenated code C. Using the partitioning shown in figure 10.5 and two outer BCH codes $\mathcal{A}^{(1)}(2;30,14,8)$ and $\mathcal{A}^{(2)}(2;30,24,4)$, we create a generalized concatenated code C_{GC}. The parameters for the two codes are

$$C(\mathbb{R}^{60}, M = 2^{28}, \delta = 8 \cdot 2 = 16) \quad \text{and} \quad C_{GC}(\mathbb{R}^{60}, M = 2^{38}, \delta = \min\{2 \cdot 8, 4 \cdot 4\}) .$$

The code rate for C is $14/30$ and the rate for C_{GC} is $19/30$; thus C_{GC} has a higher code rate.

If we construct a GC code having a code rate of $14/30$ and use the two outer BCH codes $\mathcal{A}^{(1)}(2;30,9,12)$ and $\mathcal{A}^{(2)}(2;30,19,6)$, we obtain $C_{GC}(\mathbb{R}^{60}, M = 2^{28}, \delta = \min\{2 \cdot 12, 4 \cdot 6\})$, a code with a higher minimum distance ($24 > 16$). ◇

These examples demonstrate the fundamental concept of coded modulation. Now we describe concatenation with block coded modulation as the inner code (modulation without memory).

10.2 GC with block modulation

We use the term 'block modulation' as an expression for all modulation schemes without memory (see e.g. [Pro]), such as ASK (*amplitude shift keying*), FSK (*frequency shift keying*), PSK (*phase shift keying*), QAM (*quadrature amplitude modulation*), etc. For digital transmission, the most commonly used modulation schemes are PSK and QAM. For the moment, we focus only on the partitioning of these modulation schemes.

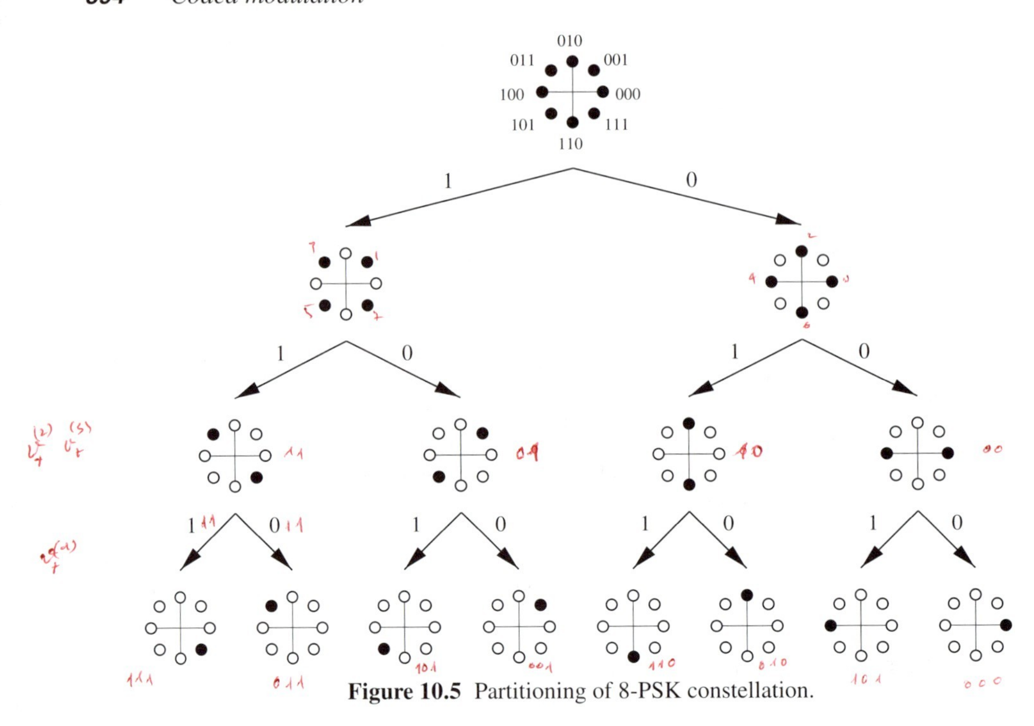

Figure 10.5 Partitioning of 8-PSK constellation.

10.2.1 Partitioning of signals

In example 10.1, we first partition the signals in the one-dimensional real space $\mathbb{R}$ (ASK). The partitioning of codes over arbitrary spaces with the Euclidean metric is basically the same as for codes with the Hamming metric. Here we are restricted to the Euclidean space $\mathbb{R}^2$, which contains modulated signals with respect to amplitude and phase, since signal theory describes modulated signals in this space (see [Pro, p. 106]).

A set of signals $s_i \in \mathbb{R}^2$, $i = 1, 2, \ldots, M^{(1)}$ is in this case a set of $M^{(1)}$ points that define a code $\mathcal{X}^{(1)}(\mathbb{R}^2, M^{(1)}, \delta^{(1)})$ over the Euclidean space. We define the metric, i.e. the distance δ between two points $s_1 = (x_1, y_1)$ and $s_2 = (x_2, y_2)$, as the *quadratic* Euclidean distance according to section 7.1.4:

$$\delta = \mathrm{dist}_E(s_1, s_2) = (x_1 - x_2)^2 + (y_1 - y_2)^2 \,. \tag{10.1}$$

So the minimum distance $\delta^{(1)}$ of a code is equal to the minimum of all distances between arbitrary points:

$$\delta^{(1)} = \min_{\substack{s_i, s_j \in \mathcal{X} \\ s_i \neq s_j}} \left\{ \mathrm{dist}_E(s_i, s_j) \right\} \,.$$

Next we partition the code $\mathcal{X}^{(1)} \subset \mathbb{R}^2$ into L subcodes such that

$$\mathcal{X}^{(1)} = \bigcup_{i=1}^{L} \mathcal{X}_i^{(2)} \,,$$

with $\mathcal{X}_i^{(2)}(2, M_i^{(2)}, \delta_i^{(2)})$ and $M^{(1)} = \sum_{i=1}^{L} M_i^{(2)}$. The minimum distance of the subcode $\mathcal{X}_i^{(2)}$ is defined as the minimum of the minimum distances of all subcodes:

$$\delta^{(2)} = \min_{i=1,\dots,L} \left\{ \delta_i^{(2)} \right\} .$$

As for codes over finite fields, each subcode $\mathcal{X}_i^{(2)}$ can again be partitioned into subcodes $\mathcal{X}_{i,j}^{(3)}$, and so forth.

The partitioning introduced in examples 10.1 and 10.2 is equivalent to the so-called *set partitioning* that was described by Ungerböck in his famous paper [Ung82] for 8-PSK.

Below we want to explain the partitioning of the modulation schemes QAM and M-PSK for binary numbering, that is $a^{(i)} \in GF(2)$.

2^s-PSK The signals s_i, $i = 1, 2, \dots, M = 2^s$, are points on the unit circle in $\mathbb{R}^2$ with phases

$$i \cdot \Delta\phi, \qquad i = 0, 1, \dots, 2^s - 1, \quad \text{where} \quad \Delta\phi = \frac{2\pi}{2^s} .$$

Therefore the minimum distance of $\mathcal{X}^{(1)}(\mathbb{R}^2, 2^s, \delta^{(1)})$ is given as

$$\delta^{(1)} = 4 \sin^2 \frac{\Delta\phi}{2} = 2(1 - \cos\Delta\phi) .$$

A partitioning of the code $\mathcal{X}^{(1)}$ of order s as binary numbering is constructed as follows. We first partition the code $\mathcal{X}^{(1)}$ into two subcodes $\mathcal{X}_0^{(2)}$ and $\mathcal{X}_1^{(2)}$, with

$$\mathcal{X}_i^{(2)}(\mathbb{R}^2, 2^{s-1}, \delta^{(2)}), \quad i = 0, 1, \qquad \text{with } \delta^{(2)} = 2(1 - \cos 2\Delta\phi) .$$

We have in general

$$\mathcal{X}_{i_1, i_2, \dots, i_{l-1}}^{(l)}(\mathbb{R}^2, 2^{s-(l-1)}, \delta^{(l)}) \quad \text{for} \quad l = 1, 2, 3, \dots, s ,$$

and, for the minimum distances,

$$\delta^{(l)} = 2\left(1 - \cos(2^{l-1} \cdot \Delta\phi)\right), \qquad l = 1, 2, 3, \dots, s .$$

The numbers $i_1, \dots, i_s$ are expressed as binary vectors $(\underline{a}^{(1)}, \underline{a}^{(2)}, \dots, \underline{a}^{(s)})$, $\underline{a}^{(i)} \in GF(2)$. We recall that $\underline{a}^{(s)}$ chooses the codeword in the subcode $\mathcal{X}_{i_1, i_2, \dots, i_{s-1}}^{(s)}$. A partitioning of the 8-PSK constellation is described in figure 10.5. The expansion to PSK schemes of higher order is self-explanatory.

M-QAM The partitioning of M-QAM signals (see e.g. [Pro, p. 189]) is similar to the partitioning of PSK constellations. For the partitioning with binary numbering, we only describe the special case $M = 2^s$. QAM is therefore a code $\mathcal{X}^{(1)}(\mathbb{R}^2, M^{(1)} = 2^s, \delta^{(1)})$ with minimum distance

$$\delta^{(1)} = \frac{6}{M-1} ,$$

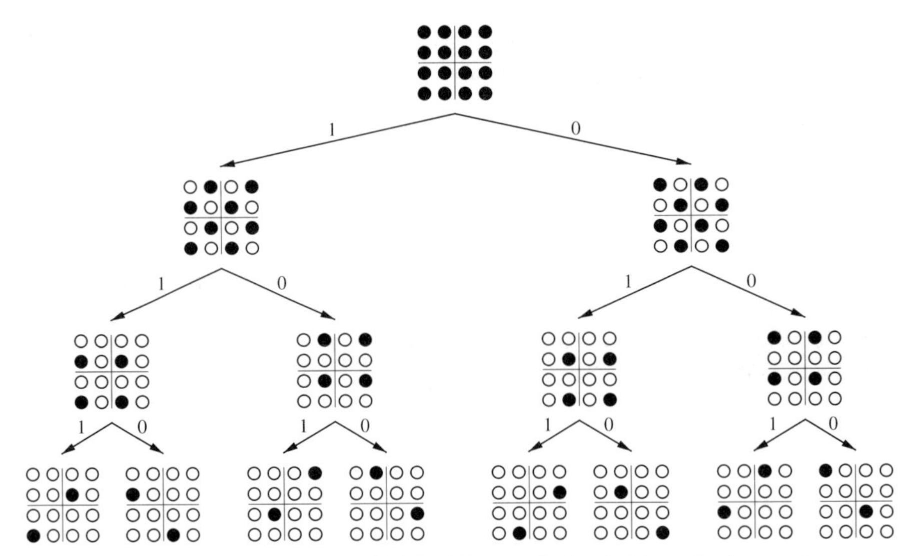

Figure 10.6 Partitioning of 16-QAM constellation (without the last step).

under the condition that the mean signal energy[1] $\bar{E}_s$ is normalized to one. For the minimum distances of the subcodes, the following expression holds:

$$\delta^{(j+1)} = 2\delta^{(j)}, \qquad j = 1, 2, \ldots, s-1 .$$

Figure 10.6 illustrates the partitioning of a constellation corresponding to 16-QAM. The last step of the partitioning is not shown here.

It is obvious that different modulation schemes without memory can be partitioned in the same manner. Non-binary numbering can also be used to label the partitions.

Note Orthogonal and biorthogonal signals correspond to simplex codes and first-order RM codes respectively (compare with section 5.1). For the simplex code, all codewords have the same weight. As a result, we cannot construct a subcode with a greater distance through partitioning.

10.2.2 Definition of coded modulation

We consider modulated signals as a code over the Euclidean space $\mathbb{R}^2$. Thus each modulation scheme that consists of $M_b^{(1)}$ signals can be regarded as an inner code $\mathcal{X}^{(1)}(\mathbb{R}^2, M_b^{(1)}, \delta^{(1)})$ of a GC code according to a particular partition. Usually (but not necessarily), binary numbering

[1] The mean signal energy is the sum over all individual signal energies multiplied by their event probability: $\bar{E}_s = \sum_i |s_i|^2 P(s_i)$

is used. For a partitioning, we have

$$\begin{aligned} \mathcal{X}^{(1)} &= \bigcup_{i_1=1}^{2} \mathcal{X}_{i_1}^{(2)}, \\ \mathcal{X}_{i_1}^{(2)} &= \bigcup_{i_2=1}^{2} \mathcal{X}_{i_1,i_2}^{(3)}, \\ &\vdots \\ \mathcal{X}_{i_1,\ldots,i_{s-1}}^{(s)} &= \bigcup_{i_s=1}^{2} x_{i_1,i_2,\ldots,i_s}. \end{aligned}$$

The binary numbers $i_1, i_2, \ldots, i_s$ with $i_j \in GF(2)$ label the partitioning, and $x_{i_1,i_2,\ldots,i_s}$ is a codeword (signal) of the code $\mathcal{X}^{(1)}$. With this partitioning, we have also defined the minimum distances $\delta^{(i)}$, $i = 1, 2, \ldots, s$. The s outer codes are binary codes $\mathcal{A}^{(i)}(2; n_a, M_a^{(i)}, d_a^{(i)})$. Hence the parameters for the GC code are

$$C_{GC}\left(\mathbb{R}^{2n_a}, M = \prod_{i=1}^{s} M_a^{(i)}, \delta \geq \min_{i=1,\ldots,s} \{d_a^{(i)} \delta^{(i)}\}\right),$$

over the Euclidean space $\mathbb{R}^{2n_a}$. The minimum distance δ is the minimum quadratic Euclidean distance between two arbitrary points (codewords) of the construction (of the GC code). This construction is frequently called *multilevel coding*.

Calculation of code rate for concatenated modulation Coded modulation is not a block concatenation technique; therefore the simple multiplication of subcode rates does not correspond to the overall code rate. The concatenated modulation code rate is calculated by

$$R = \left(\frac{1}{s}\right) \sum_{i=1}^{s} R^i, \tag{10.2}$$

$$R_{Info} = \text{ld}(M_{sym}) R, \tag{10.3}$$

where R is the overall code rate and R_{Info} is the information code rate. R^i is the code rate of the outer component codes, M is the cardinality of the transmission symbol and s is the number of partitions.

Coded modulation with outer convolutional codes If we use outer convolutional codes, we get coded modulation systems that correspond to the GC codes in section 9.4. As a rule, the free distance d_f (section 8.1.6) is used to calculate the code properties, giving

$$\delta \geq \min_{i=1,\ldots,s} \{d_f^{(i)} \delta^{(i)}\}.$$

Asymptotic coding gain For a constant data rate, the asymptotic coding gain for coded modulation is defined as

$$\Delta = 10 \log_{10}(\delta / \delta_{mod}). \tag{10.4}$$

Asymptotic corresponds to the coding gain that can be achieved with a very good additive white noise channel. We use δ for the minimum distance of the GC code and δ_{mod} for the minimum distance of the uncoded modulation. Specifically, δ_{mod} is the minimum distance of an uncoded modulation method that has the same information data rate as the GC code. For example, a GC code with rate $1/2$ and minimum distance δ based on an inner QPSK modulation is compared with BPSK modulation having the same information data rate. Therefore δ_{mod} corresponds to the minimum distance of BPSK (see also section 7.2.4 on coding gain).

10.2.3 Lattices and generalized multilevel concatenation

The possibilities are limited for the partitioning of a code (set of signal points) in $\mathbb{R}^2$ as compared with a multidimensional code. However, a GC code having an inner code signal space $\mathbb{R}^2$ and binary outer codes $\mathcal{A}_i$ with length n_a can be regarded as a code in $\mathbb{R}^{2n_a}$. A partitioning of such a GC code and the use of outer codes in order to protect the subcode labels produces a multilevel generalized concatenated code.

Multilevel spaces can also be constructed using spherical codes [EZ95] and lattices [CoSl]. These two methods are closely related. In the following, we describe some known lattices from [CoSl] as GC codes.

Lattices

Let $\mathbf{x}$ be points in the n-dimensional space $\mathbb{R}^n$:

$$\mathbf{x} = (x_1, x_2, \ldots, x_n), \quad x_i \in \mathbb{R} .$$

An n-dimensional sphere in $\mathbb{R}^n$ with radius ρ and center $\mathbf{u}$ is defined as

$$\left\{ \mathbf{x} : \ \sum_{i=1}^{n} (x_i - u_i)^2 = \rho^2 \right\} .$$

The volume of a unit sphere (see e.g. [CoSl]) is given as

$$V_n = \begin{cases} \dfrac{\pi^{\frac{n}{2}}}{\left(\frac{n}{2}\right)!} & \text{for } n \text{ even,} \\ \dfrac{2^n \pi^{\frac{n-1}{2}} \left(\frac{n-1}{2}\right)!}{n!} & \text{for } n \text{ odd} \end{cases} . \quad (10.5)$$

A lattice is now defined as all the centers of n-dimensional spheres in $\mathbb{R}^n$ that are in contact with each other with the following conditions:

(i) $\mathbf{0}$ is a center.

(ii) If $\mathbf{u}$ and $\mathbf{v}$ are centers than $\mathbf{u}+\mathbf{v}$ and $\mathbf{u}-\mathbf{v}$ have to be centers as well.

This is the classical sphere packing problem, which aims to maximize the ratio of sphere volume to the volume of the space:

$$\text{density:} \quad \Delta = \frac{\text{volume inside the spheres}}{\text{volume of the space}}$$

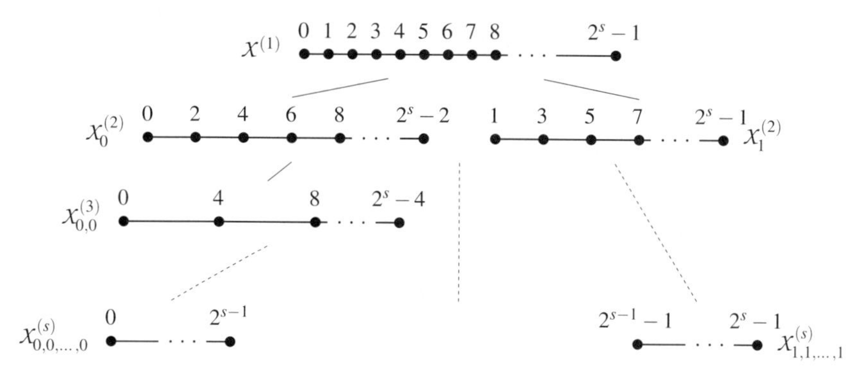

Figure 10.7 Partitioning of 2^s points on a straight line.

(see section 1.1.2 on the Hamming bound, which solves the same problem for the Hamming metric). Unfortunately the search for the greatest possible density has only been solved for $n \leq 8$ and $n = 24$ [CoSl]. The known solution for $n = 24$ is based on the perfect Golay codes (section 4.2).

GC lattice code construction We next construct a GC code with the s outer codes

$$\mathcal{A}^{(i)}(2; n_a, M_a^{(i)}, d_a^{(i)}), \quad i = 1, 2, \ldots, s\,.$$

We therefore need a binary partitioning of order s of the inner code defined by 2^s points on a straight line:

$$\mathcal{X}^{(1)}(\mathbb{R}, 2^s, \delta^{(1)} = 1) = \{0, 1, 2, 3, \ldots, 2^s - 1\}\,.$$

This partitioning is illustrated in figure 10.7. Hence the quadratic Euclidean minimum distances of the subcodes $\mathcal{X}^{(i)}, i = 1, \ldots, s$ is given by

$$\delta^{(i)} = (2^{i-1})^2, \quad i = 1, 2, \ldots, s\,,$$

and the GC code parameters are

$$C_{GC}\left(\mathbb{R}^{n_a}, M = \prod_{i=1}^{s} M_a^{(i)}, \delta \geq \min_{i=1,\ldots,s}\left\{d_a^{(i)}\,(2^{i-1})^2\right\}\right)$$

Now we can build spheres with radius $\frac{\sqrt{\delta}}{2}$ around each codeword that touch each other according to the lattice definition. From equation 10.5, the volume of a sphere is given by [CoSl]

$$V_n(\text{radius})^n = V_n\left(\frac{\sqrt{\delta}}{2}\right)^n, \quad V_n: \text{ volume of the unit sphere.}$$

The volume of the entire space in which all M codewords (spheres) are situated is equal to $(2^s)^n$. The density of a GC code yields

$$\Delta_{GC} \geq \frac{M}{(2^s)^n} V_n \left(\frac{\sqrt{\delta}}{2}\right)^n.$$

Next we consider some examples for this construction method.

Example 10.3 (Construction A [CoSl]) We choose as outer codes

$$\mathcal{A}^{(1)}(2;n,M,d) \quad \text{and} \quad \mathcal{A}^{(2)}(2;n,2^n,1)$$

and $s=2$; thus we obtain for the minimum distance of the GC codes $\delta \geq \min\{1\cdot d, 4\cdot 1\}$. The density in $\mathbb{R}^n$ is obtained as

$$\Delta_{GC} \geq \frac{M2^n}{(2^2)^n} V_n \left(\frac{\sqrt{\min\{d\cdot 1, 1\cdot 4\}}}{2}\right)^n = \frac{M}{2^n} V_n \min\left\{\left(\frac{\sqrt{d}}{2}\right)^n, 1\right\}.$$

This density is identical to the density of construction A [CoSl, p. 137]. An outer code distance $d > 4$ would decrease Δ_{GC} because as the distance of $\mathcal{A}^{(1)}$ increases, the dimension M decreases. ◇

Example 10.4 (Construction B [CoSl]) We again choose $s=2$ and outer codes

$$\mathcal{A}^{(1)}(2;n,M,d) \quad \text{and} \quad \mathcal{A}^{(2)}(2;n,2^{n-1},2)\,.$$

The GC code minimum distance is $\delta \geq \min\{1\cdot d, 4\cdot 2\}$. The sphere density in $\mathbb{R}^n$ can be calculated as

$$\Delta_{GC} \geq \frac{M2^{n-1}}{(2^2)^n} V_n \left(\frac{\sqrt{\min\{d\cdot 1, 2\cdot 4\}}}{2}\right)^n = \frac{M}{2^{n+1}} V_n \min\left\{\left(\frac{\sqrt{d}}{2}\right)^n, 2^{n/2}\right\}.$$

For a GC code that is optimized with regard to the minimum distance, we choose an outer code with $d=8$. The density is identical to construction B [CoSl, p. 141]. ◇

From the point of view of GC codes, we can use all known binary codes (even non-linear codes) as outer codes. The minimum distance is equivalent to the sphere radius, and is easily calculated. The first two examples were based on simple outer codes, whereas the following two examples use more complicated outer codes.

Example 10.5 (Construction C [CoSl]) Let us assume that s outer binary codes $\mathcal{A}^{(i)}(2;n_a,M_a^{(i)},d_a^{(i)})$, $i=1,2,\ldots,s$ are given that have minimum distances

$$d_a^{(i)} = \gamma 4^{s-i}, \quad i=1,2,\ldots,s\,, \quad \text{with parameter } \gamma = 1,2\,,$$

Thus we obtain for the minimum distance of the GC codes

$$\delta \geq \min_{i=1,\ldots,s}\left\{d_a^{(i)}(2^{i-1})^2\right\} = \min_{i=1,\ldots,s}\left\{\gamma 4^{s-i}4^{i-1}\right\} = \gamma 4^{s-1},$$

while the density is

$$\Delta_{GC} \geq \frac{M}{4^n} V_n \gamma^{n/2},$$

with $M = \prod_{i=1}^{s} M_a^{(i)}$. This density is identical to the density of construction C (see [CoSl, p. 150]). If we use the extended binary outer codes $\mathcal{A}^{(1)}(2^6,2^1,64)$, $\mathcal{A}^{(2)}(2^6,2^{28},16)$, $\mathcal{A}^{(3)}(2^6,2^{57},4)$ and $\mathcal{A}^{(4)}(2^6,2^{64},1)$, we obtain a GC code $C_{GC}(\mathbb{R}^{64};2^{150},\geq 64)$. The density is calculated as $(s=4, \gamma=1)$ $\Delta = 4^{11}V_n = 2^{22}V_n$. ◇

Example 10.6 (Construction D [CoSl]) As in example 10.5, we consider s outer codes $\mathcal{A}^{(i)}(2; n_a, M_a^{(i)}, d_a^{(i)})$, $i = 1, 2, \ldots, s$. We add the requirement that the outer codes be written as subcodes:

$$\mathcal{A}^{(1)}(2; n_a, M_a^{(1)}, d_a^{(1)}) \subset \mathcal{A}^{(2)}(2; n_a, M_a^{(2)}, d_a^{(2)}) \subset \ldots \subset \mathcal{A}^{(s)}(2; n_a, M_a^{(s)}, d_a^{(s)}).$$

The minimum distances are

$$d_a^{(i)} = \frac{4^{s-i}}{\gamma}, \quad i = 1, 2, \ldots, s, \quad \gamma = 1, 2 .$$

Thus the minimum distance of the GC code is obtained as

$$\delta \geq \min_{i=1,\ldots,s} \left\{ d_a^{(i)} (2^{i-1})^2 \right\} = \min_{i=1,\ldots,s} \left\{ \frac{1}{\gamma} 4^{s-i} 4^{i-1} \right\} = \frac{4^{s-1}}{\gamma},$$

and the density yields

$$\Delta_{GC} \geq \frac{M}{4^n} V_n \frac{1}{\gamma^{n/2}},$$

with $M = \prod_{i=1}^{s} M_a^{(i)}$. This density is identical to the density of construction D [CoSl, p. 232]. If we use RM codes as outer codes then we obtain the Barnes–Wall lattice. In general, GC codes do not require that the outer codes be partitioned into subcodes or that they be linear. ◇

These examples show that it is possible to achieve the same parameters with generalized concatenation as with lattice constructions. Therefore we have a variety of techniques for partitioning codes in multidimensional spaces. Investigations in this area were also carried out by Wei [Wei87] on 4D coded modulation. GC codes and their partitions can be used for coded modulation instead of lattices. This not only allows us to modify codes through partitioning, but also lets us use more powerful component codes, as described in the next section.

Multiple concatenation

The principle of generalized concatenation of codes can be extended further. We partition the GC code in $\mathbb{R}^{2n_a}$ and then protect this partitioning with an outer code. This defines a generalized, multiple concatenated code (see the construction of Reed–Muller codes in section 9.5). This gives a method of constructing codes (signal points) in multidimensional spaces. The partitioning of these GC codes depends directly on the outer codes $\mathcal{A}^{(i)}$. In other words, there exists a construction rule for finding the partitioning. It is obvious that signal constellations in multidimensional spaces can also be partitioned by hand; however, such constructions are in general nonlinear, and therefore are not easily calculated.

Next we give an example of a generalized multiple concatenated code, without considering the decoding dependence on the channel.

Example 10.7 (Multiple concatenated coded modulation) We first construct a GC code over a Euclidean space. We use as inner code QPSK modulation with labels according to figure 10.5 (from step 2 on). We use $\mathcal{A}^{(1)}(2; 8, 2^4, 4)$ and $\mathcal{A}^{(2)}(2; 8, 2^7, 2)$ as outer codes, which are an extended Hamming code and a parity check code. This gives the GC code $C_{GC}^{(1)}(\mathbb{R}^{16}, 2^{11}, 8)$ over the Euclidean space $\mathbb{R}^{16}$.

Another GC code over $\mathbb{R}^{16}$ is $C_{GC}^{(2)}(\mathbb{R}^{16}, 2^5, 16)$. The inner code is again a QPSK modulation, but the two binary outer codes are a repetition code and the extended Hamming code: $\mathcal{A}^{(1)}(2;8,2^1,8)$ and $\mathcal{A}^{(2)}(2;8,2^4,4)$. We see that $C_{GC}^{(2)} \subset C_{GC}^{(1)}$, since the inner codes are the same, and the outer codes of $C_{GC}^{(2)}$ are subcodes of those of $C_{GC}^{(1)}$. From this observation, we can use partitioning method 3 from section 9.2.2 and obtain a partitioning of the code $C_{GC}^{(1)}$ with labels $\mathbf{a}_i^{(1)} \in GF(2)^6$ and $\mathbf{a}_i^{(2)} \in GF(2)^5$.

We use the shortened RS code $\mathcal{A}^{(1)}(2^6;32,2^{150},8)$ and the extended RS code $\mathcal{A}^{(2)}(2^5;32,2^{145},4)$ to label the partitions. We then obtain the GC code $C_{GC}(\mathbb{R}^{512},2^{295},64)$ over the Euclidean space $\mathbb{R}^{512}$. The vector dimension for $C_{GC}^{(2)}$ is calculated by choosing 32 elements corresponding to $\mathcal{A}^{(2)}$, where each element comes from the code $C_{GC}^{(1)}$ having dimension $\mathbb{R}^{(16)}$. ◇

This method of generalized multiple concatenation reveals various possibilities. The example given above shows us that a code that is constructed with this method is a relatively long code composed of short codes.

10.2.4 Decoding

In principle, the GCD algorithm described in section 8.5 can be used for coded modulation decoding. For efficient decoding, a symbolwise MAP decoder (see section 8.5) should be used for the inner code, in order to provide reliability information. This allows us to decode *step by step*, where the inner code (the modulation) is decoded several times with respect to different subcodes, which gives rise to the term *multistage decoding*. Unfortunately, separate ML decoding of the outer codes does not guarantee ML decoding of the entire GC code. The ML decoding of GC codes is to date unsolved (see example 10.8).

Reliability information for M-PSK soft output decoding Let s_i be the signal that is sent in the ith symbol interval according to figure 10.5. The received signal is assumed to be

$$y_i = s_i + n_i,$$

corresponding to section 7.1.2, where n is a realization of an additive white Gaussian noise process (AWGN) with variance σ^2. With that the conditional probability density function of the received signals can be written as

$$p(y_i|s_i) = \frac{1}{\sqrt{2\pi}\sigma} \exp\left(-\frac{(y_i - s_i)^2}{2\sigma^2}\right). \tag{10.6}$$

We obtain the a posteriori probability $P(s_i|y_i)$ using Bayes' law:

$$P(s_i|y_i) = \frac{P(y_i|s_i)P(s_i)}{P(y_i)}. \tag{10.7}$$

With

$$P(y_i|s_i) = \lim_{\varepsilon \to 0} \int_{y_i}^{y_i+\varepsilon} p(r_i'|s_i)\, dr_i'$$

and

$$\frac{P(y_i|s_i)}{P(y_i)} = \lim_{\varepsilon \to 0} \frac{\int_{y_i}^{y_i+\varepsilon} p(r_i'|s_i)\,dr_i'}{\int_{y_i}^{y_i+\varepsilon} p(r_i')\,dr_i'} = \frac{p(y_i|s_i)}{p(y_i)},$$

we can express equation 10.7 as follows:

$$P(s_i|y_i) = \frac{p(y_i|s_i)P(s_i)}{p(y_i)}. \tag{10.8}$$

The combination of equations 10.6 and 10.8 leads directly to the calculation of the conditional probabilities $P(s_i|y_i)$.

We assume that all symbols M_{sym} have the same a priori probability, that is $P(s_i) = 1/M_{sym}$, and we consider the case of *binary* outer codes. Thus the reliability (probability) must be calculated for each bit. The vector $\mathbf{c}$ is the binary representation of cardinality of the transmitted symbol having length $m = \mathrm{ld}(M_{sym})$.

$$\mathbf{c}_i = (c_i^{(0)}, c_i^{(1)}, \ldots, c_i^{(l)}, \ldots, c_i^{m-1}), \quad \text{with} \quad c_i^{(l)} \in \{0,1\} \quad \text{and} \quad m = \mathrm{ld}(M).$$

Furthermore, let $X_i^{(l),1}$ be the subset of symbols for which the s_i have a 1 in the lth component of $c_i^{(l)}$. Analogously $X_i^{(l),0}$ labels the subset of symbols for which the s_i have a 0 in the lth component of $c_i^{(l)}$. So the probability that component l of c_i is equal to 1 is given as

$$P(c_i^{(l)} = 1) = \sum_{s_k \in X_i^{(l),1}} P(s_k|y_i) = \sum_{s_k \in X_i^{(l),1}} \frac{p(y_i|s_k)P(s_k)}{p(y_i)},$$

and the probability that the lth component of c_i is 0 is given as

$$P(c_i^{(l)} = 0) = \sum_{s_k \in X_i^{(l),0}} P(s_k|y_i) = \sum_{s_k \in X_i^{(l),0}} \frac{p(y_i|s_k)P(s_k)}{p(y_i)}.$$

The information reliability $L_i^{(l)}$ for the lth component of $\mathbf{c}_i$ can be written as

$$L_i^{(l)} = \ln \frac{P(c_i^{(l)} = 0)}{P(c_i^{(l)} = 1)}.$$

Simulation results

We want to use coded 8-PSK modulation over an AWGN channel (see section 7.1.2). We use the information reliability calculations described above.

Example 10.8 (ML decoding of individual codes as compared with the entire code) With this example, we show that the ML decoding of individual codes is not the same as ML decoding of the entire code. We use an 8-PSK constellation as inner code, which is partitioned as illustrated in figure 10.5. Additionally, the outer codes $\mathcal{A}^{(1)}(2;8,1,8)$, $\mathcal{A}^{(2)}(2;8,7,2)$ and

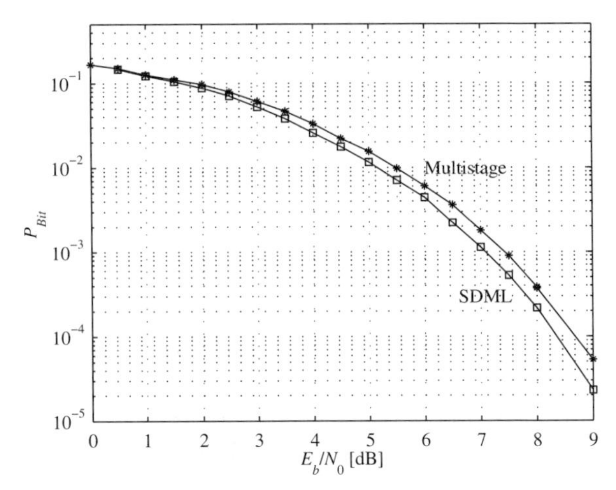

Figure 10.8 ML decoding of the entire code and the individual codes.

$\mathcal{A}^{(3)}(2;8,8,1)$ are used, corresponding to a repetition code and a parity check code, which can both be ML decoded. The GC code yields

$$\mathcal{C}_{GC}(\mathbb{R}^{16}, M = 2^16, \delta \geq \min\{2\cdot 8, 2\cdot 4, 1\cdot 8\}).$$

Figure 10.8 illustrates the additional 0.5 dB gain resulting from the ML decoding of the entire code. This gain can be increased by using larger codes and more levels. ◇

Example 10.9 (Minimum distance and bit error rate) This example illustrates that a GC code optimized with respect to the minimum distance can have a worse BER compared with codes designed to optimize the BER (see UEP codes in section 9.2.5). According to UEP construction, the minimum distance should therefore have its largest value at the first level, with decreasing distance in subsequent levels.

We construct two GC codes with identical rate $R = 0.5$, but with different minimum distances, having the following parameters (8-PSK):

$$\begin{aligned} &\mathcal{C}_{GC1}(\mathbb{R}^{64}, M = 2^{48}, \delta \geq 9.37), && \mathcal{C}_{GC2}(\mathbb{R}^{64}, M = 2^{48}, \delta \geq 8), \\ &\delta \geq \min\{9.37, 16, 16\}, && \delta \geq \min\{18.75, 16, 8\}, \\ &\mathcal{A}^{(1)}(2;32,6,16), && \mathcal{A}^{(1)}(2;32,1,32), \\ &\mathcal{A}^{(2)}(2;32,16,8), && \mathcal{A}^{(2)}(2;32,16,8), \\ &\mathcal{A}^{(3)}(2;32,26,4), && \mathcal{A}^{(3)}(2;32,31,2). \end{aligned}$$

Figure 10.9 clearly illustrates that $\mathcal{C}_{GC2}$ achieves (in the considered range) a much smaller bit error rate than $\mathcal{C}_{GC1}$, although the asymptotic behavior of $\mathcal{C}_{GC1}$ is 0.72 dB better. ◇

Example 10.10 (Longer codes are better) We construct two GC codes having rate ≈ 0.5 with

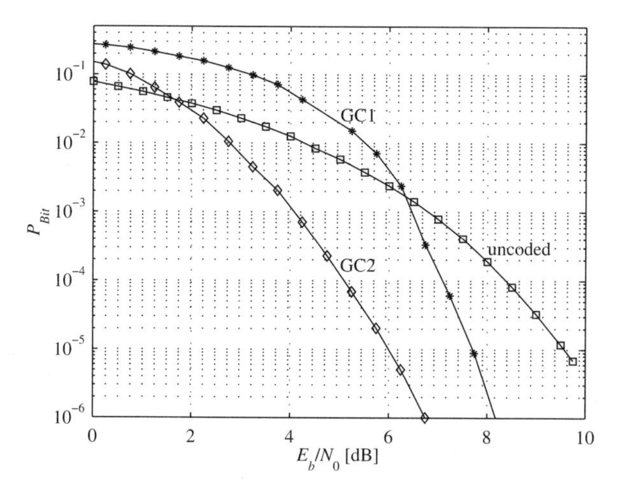

Figure 10.9 Optimization of a GC code with respect to the minimum distance is not the same as optimizing with respect to the bit error rate.

outer RM codes of length 8 (GC1) and 64 (GC2). Again, an 8-PSK modulation is used:

$$\begin{aligned} &C_{GC1}(\mathbb{R}^{16}, 2^{12}, \delta \geq 4.69), && C_{GC2}(\mathbb{R}^{128}, 2^{98}, \delta \geq 16), \\ &\delta \geq \min\{4.69, 8, 8\}), && \delta \geq \min\{37.50, 32, 16\}), \\ &\mathcal{A}^{(1)}(2; 8, 1, 8), && \mathcal{A}^{(1)}(2; 64, 1, 64), \\ &\mathcal{A}^{(2)}(2; 8, 4, 4), && \mathcal{A}^{(2)}(2; 64, 22, 16), \\ &\mathcal{A}^{(3)}(2; 8, 7, 2), && \mathcal{A}^{(3)}(2; 64, 75, 4). \end{aligned}$$

As in example 10.9, the codes are not maximized with respect to the minimum distance. Figure 10.10 illustrates that longer codes are better. ◇

10.2.5 Trellis-coded modulation systems

As already mentioned, we can use block codes as well as convolutional codes as outer code for a GC code with inner modulation. Thereby, it is less important whether the codes are linear or nonlinear. It is our goal to maximize the minimum quadratic Euclidean distance of the GC code (the signal sequence). To proceed, we describe the principle of Ungerböck coding [Ung82], which is a special case of GC codes for the construction of coded modulation schemes. Here convolutional codes or, more generally, trellis codes are used.

We begin with a trellis as shown in figure 10.11. Each state transition determines a code sequence consisting of n bits. These code sequences are mapped to the signal points in such a way as to maximize the Euclidean distance of the constellation. We choose the one mapping that generates the best distances.

The signals of trellis-coded modulation as described in [Ung82] are generated as follows. If k information bits are transmitted per time interval, a convolutional coder with rate $\tilde{k}/(\tilde{k}+1)$ generates $\tilde{k}+1$ code bits from $\tilde{k} < k$ information bits. Therefore a total of $k+1$ coded bits must

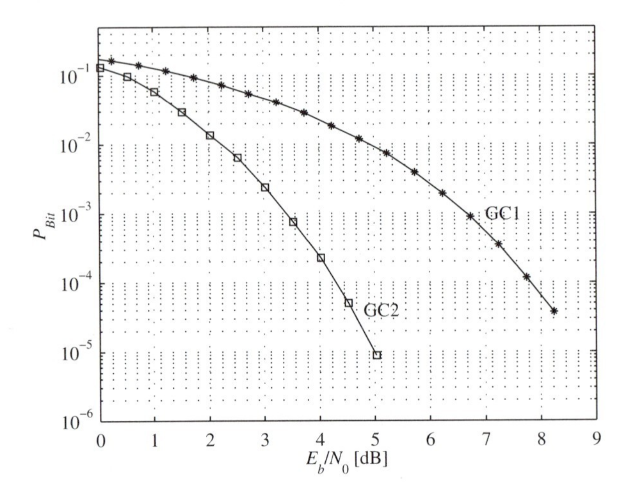

Figure 10.10 Generalized concatenated codes with different length.

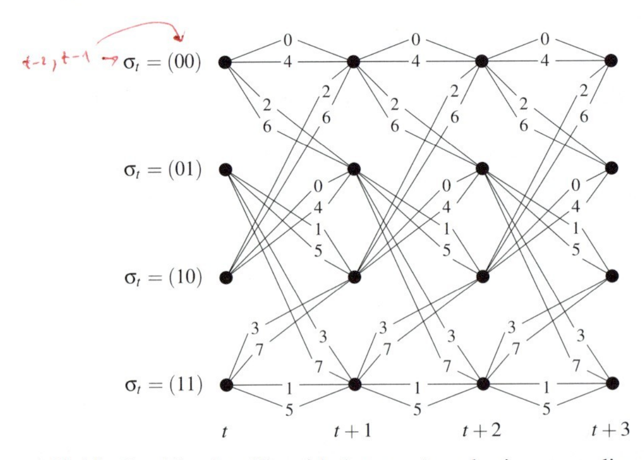

Figure 10.11 Combined trellis with 4 states (numbering according to figure 10.5).

be transmitted. The bits consist of $\tilde{k}+1$ and $k-\tilde{k}$ coded and uncoded bits respectively. For the transmission of $k+1$ bits, we need a modulation scheme consisting of 2^{k+1} signals. A binary partitioning of these signals is of order $k+1$. where the $\tilde{k}+1$ code bits select the first subsets of the partitioning, while $k-\tilde{k}$ uncoded information bits determine the following subsets. The last partitioning steps are therefore not protected by an outer code. To illustrate this concept we give an example from [Ung82], where a convolutional coder with 4 states is used.

Example 10.11 (Ungerböck coding) We use a systematic convolutional coder with rate $R = 1/2$ ($\tilde{k} = 1$) as illustrated in figure 10.12. The parallel state transitions of the encoder trellis in figure 10.5 are due to the fact that the constraint lengths of outputs $\nu^{(1)}$ and $\nu^{(2)}$ are 0.

The trellis shows parallel state transitions. This can be explained by the fact that the constraint lengths ν_1 and ν_2 are 0. The systematic encoder maps the information block $\mathbf{u}_t = (u_t^{(1)}, u_t^{(2)})$ to the code sequence $\mathbf{v}_t = (v_t^{(1)}, v_t^{(2)}, v_t^{(3)})$. The encoding is structured such that the code bits $v_t^{(2)}, v_t^{(3)}$ protect the selection of a subset of 8-PSK and $v_t^{(1)}$ selects the signal in the corresponding subset. The partitioning of the 8-PSK constellation can be seen in

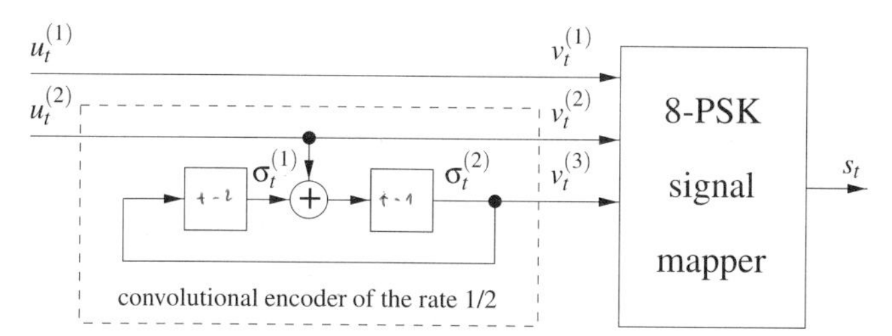

Figure 10.12 Trellis encoder/modulator for 8-PSK constellation.

figure 10.5. ◇

In practice, we achieve excellent results by using trellis-coded modulation. Perhaps the biggest advantage is that ML decoding can be implemented for the entire system using the Viterbi algorithm (see section 8.4). Feasible trellis-coded modulation systems are therefore under the same complexity constraints as convolutional codes with respect to ML decoding using the Viterbi algorithm.

Several examples of coded block modulation are given in [Cus84, Say86]. These examples are special cases of GC codes. The bit error performance of these methods are compared with the construction method from [Ung82], resulting in an asymptotic coding gain for the block coded modulation method. Obviously the GC codes used here are optimized based on the minimum distance that produces optimal results for the case of very good channels. For practical channels, it seems to be more advantageous to optimize the GC codes according to the error probability (see example 10.9). Although the coding gains are higher for GC codes than for trellis-coded modulation, the decoding capability of trellis-coded modulation is often better when typical channels are considered, due to possible ML decoding of trellis-coded modulation with the Viterbi algorithm. Some more special cases for GC codes and their decoding are examined in [HBS93]. The concept of trellis-coded modulation is a special case of generalized code concatenation, with the advantage that the ML decoding with the Viterbi algorithm is possible.

Wei [Wei87] describes an extension of the concept of trellis-coded modulation schemes. First he constructs a multidimensional Euclidean space by using a GC code (see section 10.2.3). The GC code is then partitioned to obtain a trellis-coded modulation system.

10.3 GC with convolutional modulation

In section 10.2 we focused on memoryless modulation schemes where the output signal at time t depends only on some combination of the input signals at time t. This is comparable to block codes, where a codeword corresponds to the information block, and vice versa. In contrast, convolutional codes construct the output signal based on some combination of present and past input signals. This is also true for convolutional modulation. The transmitted signal is determined by the input symbol and the present state of the modulation memory. CPM schemes (*continuous phase modulation*) found in [Sun86], [AS81a, AS81b], [AAS] and [Pro] are examples of convolutional encoded modulation. For this important class of modulation schemes, the memory is used to generate a signal with a continuous phase; that is, we avoid phase

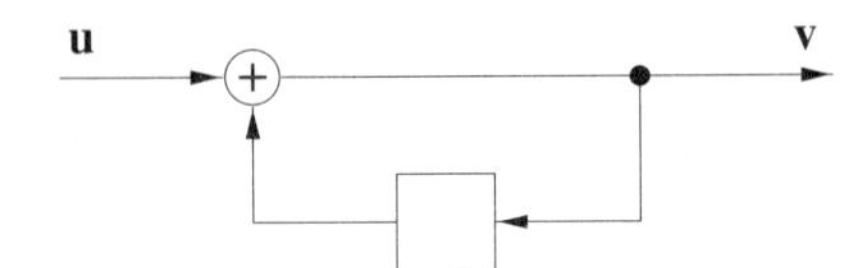

Figure 10.13 Differential encoder (precoder) for DBPSK.

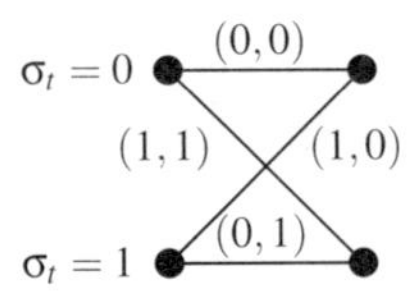

Figure 10.14 State transitions of the DBPSK trellis diagram, labeled by (u_t, v_t).

discontinuities, which are characteristic of PSK encoding. The motivation behind continuous phase signals is to decrease the signal bandwidth.

10.3.1 Introductory example

The differential precoding $\mathbf{G}(D) = 1/(1-D)$ of a differential BPSK (DBPSK) signal results in an output signal with memory.

Example 10.12 (DBPSK as convolutional code) The differential precoding $v_t = v_{t-1} + u_t$ is a recursive convolutional coder, as illustrated in figure 10.13. Figure 10.14 shows the DBPSK modulation as a trellis. A state is defined as the content of the memory element, that is $\sigma_t = v_{t-1}$. It follows that we can apply the partitioning of convolutional codes (see section 9.3.3) to this trellis diagram to obtain a partitioning of second order. The scrambling matrix yields

$$\mathbf{M}(D) = \frac{1}{1-D}\begin{pmatrix} 1 & 1 \\ D & 1 \end{pmatrix}.$$

With $\mathbf{M}(D)$, we create an equivalent coder according to figure 10.15 that is suitable for partitioning. The selection of the subcode $\mathcal{X}_{\mathbf{z}^{(1)}}$ is based on the numbering sequence $\mathbf{z}^{(1)}$. A codeword of the subcode $\mathcal{X}_{\mathbf{z}^{(1)}}$ is obtained based on the label sequence $\mathbf{z}^{(2)}$. The differential characteristics of DBPSK are reflected in the equivalent precoder. The labels $\mathbf{z}^{(1)}$ and $\mathbf{z}^{(2)}$ can be obtained directly from $\mathbf{u}$:

$$z_\tau^{(1)} = u_{2\tau}, \qquad z_\tau^{(2)} = u_{2\tau+1}.$$

The corresponding partitioning is shown in the trellis representation in figure 10.16. While

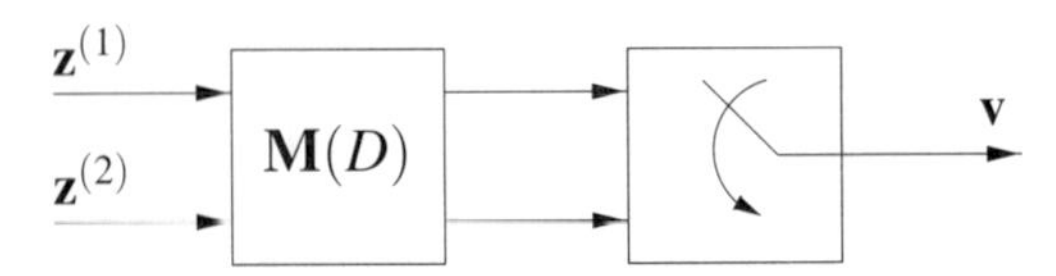

Figure 10.15 Equivalent precoder for DBPSK.

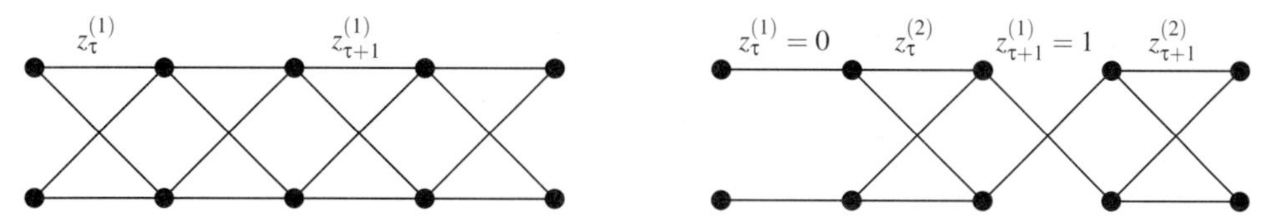

Figure 10.16 Initial trellis $\mathcal{X}^{(1)}$ (left) and subtrellis $\mathcal{X}^{(2)}_{\mathbf{z}^{(1)}}$ (right) for DBPSK.

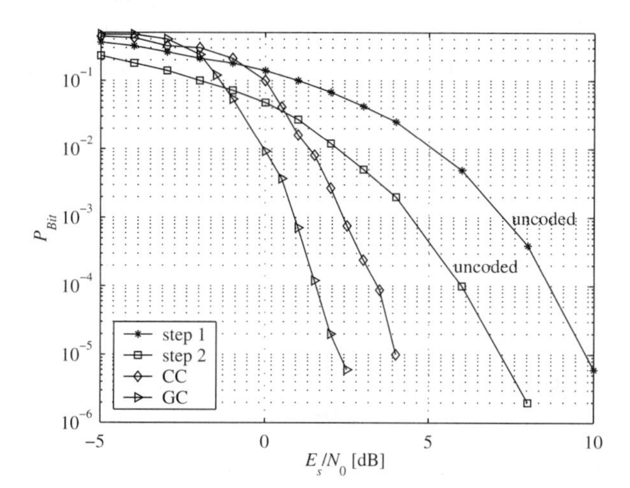

Figure 10.17 Simulation results for DBPSK.

the free Euclidean distance for DBPSK is $\delta^{(1)} = 4$, the partitioning produces an increase in distance of the second subcode trellis, giving $\delta^{(2)} = 8$.

We can now construct a GC code with two outer convolutional codes of rates $1/3$ and $2/3$ respectively. The resulting GC code has rate $1/2$. Non-partitioned DBPSK using a convolutional code of rate $1/2$ can be regarded as a classical concatenated code (CC). In both cases, convolutional codes with 16 states according to table 8.10 for $k = 1$ and [Lee94] for $k = 2$ were chosen. Interleaving was included between the convolutional encoding and modulation stages.

Figure 10.17 shows different simulation results with an AWGN channel. We clearly recognize that, for a bit error rate of 10^{-5}, the GC code is approximately 2 dB better than the classical concatenated code CC. ◇

This procedure can be generalized to all modulation schemes with memory represented as a trellis diagram. This motivates an algebraic description of the modulation schemes.

10.3.2 Algebraic description of convolutional modulation

The transmitted signal of M-valued CPM (*continuous phase modulation*) [AAS] can be written as

$$s(t, \mathbf{v}) = \sqrt{\frac{2E_s}{T_s}} \cos\left(2\pi f_0 t + \varphi(t, \mathbf{v}) + \varphi_0\right), \qquad (10.9)$$

with the modified (*tilted*) phase information [Rim88]

$$\varphi(t,\mathbf{v}) = 4\pi h \sum_{l=1}^{\infty} v_l g_\varphi(t), \qquad t \geq 0,$$

where $\mathbf{v} = (v_1, v_2, v_3, \ldots)$ is an M-valued information sequence $v_l \in \{0,1,\ldots,M-1\}$, E_s is the symbol energy, T_s is the symbol duration and h is the modulation index. Instead of the true carrier frequency f_T, the asymmetric carrier frequency $f_0 = f_T - (M-1)h/(2T_s)$ is considered. This produces a time-invariant trellis diagram. Furthermore, $g_\varphi(t)$ is the phase shaping pulse (phase pulse) and φ_0 is a constant phase offset. In the following, we assume that $\varphi_0 = 0$.

For *full-response* CPFSK modulation (*continuous phase frequency shift keying*), we have

$$g_\varphi(t) = \begin{cases} 0 & \text{for } t < 0\,, \\ t/2T_s & \text{for } 0 \leq t < T_s\,, \\ 1/2 & \text{for } t \geq T_s\,. \end{cases}$$

The phase values for the time coordinates $t = jT_s$, $j = 0,1,2,\ldots$, $\varphi(t,\mathbf{v})$, are

$$\varphi(t = jT_s, \mathbf{v}) = 2\pi h \left(\sum_{l=1}^{j} v_l \right) - j\pi h (M-1). \tag{10.10}$$

It is assumed that the modulation index $h = p/q$ is a rational number and that the positive integers p and q are relatively prime. Thus a finite number of phase states is obtained for $t = jT_s$. We consider the case $M \leq q$.

A trellis diagram is used to represent the set of all phase transitions (equation 10.10), generated with all possible information sequences $\mathbf{v}$ [AAS]. The trellis is completely determined by the expression $2\pi h \sum_{l=1}^{j} v_l$. A modified trellis can be calculated through the modification of this term according to the following equation:

$$\Phi(t = jT_s, \mathbf{v}) = 2\pi h \sum_{l=1}^{j} v_l\,.$$

The modified time-invariant trellis diagram consists of q states and is based on

$$\left[\Phi(t = jT_s, \mathbf{v})\right]_{\mathrm{mod}\, 2\pi} = \left[2\pi h \sum_{l=1}^{j} v_l \right]_{\mathrm{mod}\, 2\pi}. \tag{10.11}$$

Thus we obtain

$$\left[\Phi(t = jT_s, \mathbf{v})\right]_{\mathrm{mod}\, 2\pi} = \frac{2\pi}{q} \left[\sum_{l=1}^{j} (p v_l)_{\mathrm{mod}\, q} \right]_{\mathrm{mod}\, q}.$$

With

$$\sigma_j = \left[\sum_{l=1}^{j} (p v_l)_{\mathrm{mod}\, q} \right]_{\mathrm{mod}\, q}, \tag{10.12}$$

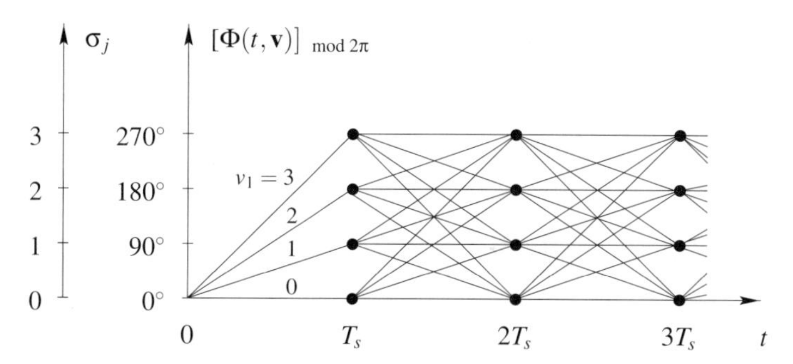

Figure 10.18 Time-invariant trellis diagram for $M = 4, h = 1/4$.

it follows that

$$\frac{q}{2\pi}\left[\Phi(t = jT_s, \mathbf{v})\right]_{\text{mod } 2\pi} = \sigma_j. \tag{10.13}$$

The phase states $[\Phi(\cdot)]_{\text{mod } 2\pi}$ in the modified phase trellis can be replaced by their isomorphic numbers (σ_j), and we can regard them as symbols of the code C (without redundancy) over the ring $\mathbb{Z}_q$ (Additions and multiplications are carried out modulo q).

Example 10.13 (Trellis diagram for 4-CPFSK with $h = 1/4$) Figure 10.18 shows an example of a modified phase trellis for $M = 4$, $h = 1/4$. ◇

Equation 10.12 yields a coding method that is illustrated in figure 10.19 (FSM A3). We represent the serial coder as a *finite state machine*(FSM). The pre-coding scheme (FSM A2) corresponding to equation 10.15 is also shown in figure 10.19. Figure 10.20 shows the parallel realization of the encoder (FSM A3) and the precoder (FSM A2). There are n input and output symbols.

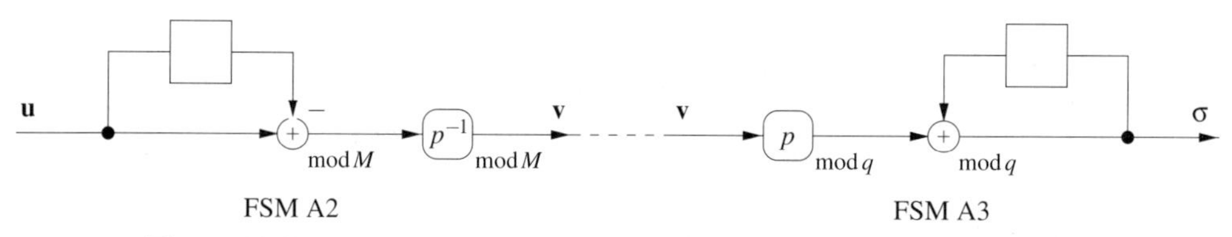

Figure 10.19 Serial realization of precoder (FSM A2) and encoder (FSM A3).

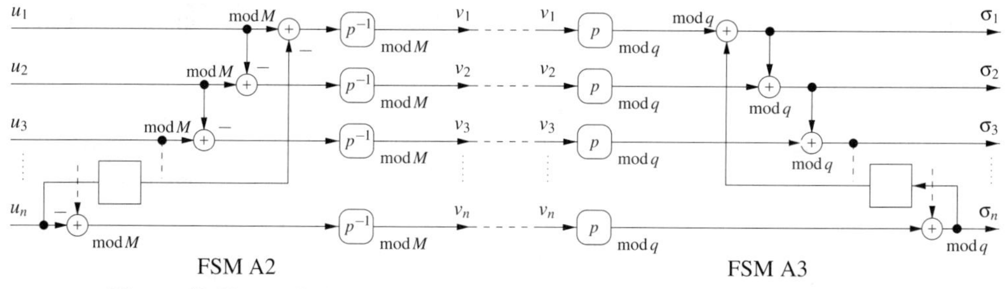

Figure 10.20 Parallel realization of precoder (FSM A2) and encoder (FSM A3).

If we use the description of the sequences $\mathbf{v}$ and σ and using the delay operator D (corresponding to section 8.1.8), the generator matrix of the code C yields

$$\mathbf{G}_n(D) = \frac{p}{1-D}\begin{pmatrix} 1 & 1 & 1 & \dots & 1 \\ D & 1 & 1 & \dots & 1 \\ D & D & 1 & \dots & 1 \\ \vdots & \vdots & \ddots & \ddots & \vdots \\ D & D & D & \dots & 1 \end{pmatrix}. \tag{10.14}$$

This is the algebraic description of CPFSK modulation, which in principle was published in [BU88], [BSU88], [Rim88], [Rim89] and [MMP88]. It is therefore possible to represent the signals as a code without redundancy (i.e. with rate 1) over an integer ring. This simplifies the design of coded CPFSK systems.

If the input symbols of CPFSK are differentially encoded, we get DCPFSK (*differential CPFSK*) modulation, which is known for $M = 2$ and $q = 2$ as MSK (*minimum shift keying*) and DMSK (*differential MSK*) respectively. The generator matrix of the precoder for CPFSK based on equation 10.14 gives

$$\mathbf{G}_n^*(D) = p^{-1}\begin{pmatrix} 1 & -1 & 0 & \dots & 0 \\ 0 & 1 & -1 & \dots & 0 \\ 0 & 0 & 1 & \dots & 0 \\ \vdots & \vdots & \ddots & \ddots & \vdots \\ -D & 0 & 0 & \dots & 1 \end{pmatrix}, \tag{10.15}$$

where the inverse element p^{-1} of p (if it exists) is defined by $p^{-1}p = 1 \bmod M$.

The memory description of the considered modulation scheme can be represented as a state diagram, or a trellis diagram if the time dimension is included (see chapter 8).

Note In principle, additional CPM scheme such as GMSK [MH81], TFM [JD78], MSK, etc., can be described algebraically (see [BSHD98, HBSD98]). CPFSK was chosen here in order to illustrate the principle.

10.3.3 Partitioning of convolutional modulation

According to section 9.3.3, we can partition all modulation schemes that can be described with a trellis. Therefore the partitioning leads to embedded subtrellis diagrams with higher distance that correspond to subcodes.

Example 10.14 (Partitioning for 8-DCPFSK) Figure 10.21 illustrates a partitioning trellis for 8-DCPFSK with binary labels. The original trellis has $h = 1/8$. It is partitioned into subsets corresponding to 4-DCPFSK modulation with $h = 1/4$ and having the corresponding improved distance properties. The subsequent partitioning step constructs a 2-DCPFSK scheme with $h = 1/2$ based on the 4-DCPFSK scheme with $h = 1/4$. This can be regarded as MSK modulation. These partitions can be represented with trellis diagrams as well. Figure 10.22 illustrates the original trellis and the corresponding subtrellis diagrams. The terminated convolutional

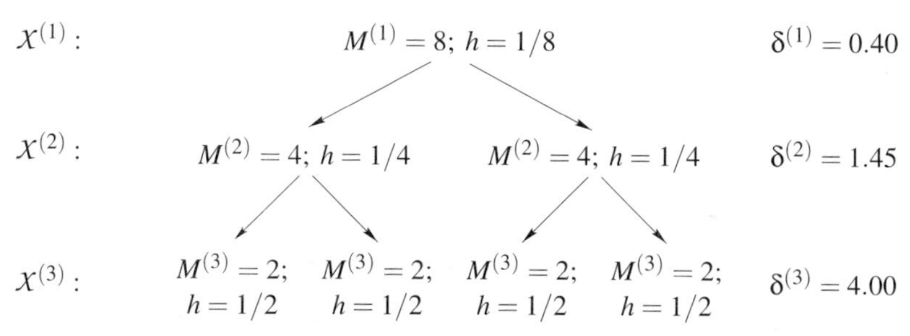

Figure 10.21 Partitioning tree for 8-valued DCPFSK with $h = 1/8$. Each edge is labeled by an 8-ary input symbol ($\mathbf{u}_n = (u_n^{(1)}, u_n^{(2)}, u_n^{(3)})$) of the precoder.

codeword consists of a block of L transmission symbols composed of $L-1$ information blocks and one termination block.

Figure 10.23 shows the simulated bit error rates of the individual partitioning levels without outer encoding. The curves illustrate the gain corresponding to the minimum distance increase of the subcodes. Based on the decoding of GC codes (section 9.2.4), the first level performs the demodulation in the original trellis (the upper trellis in figure 10.22). Here demodulation is performed with respect to the complete signal set with $\delta^{(1)} = 0.4$. Based on this decoding result, the next subtrellis is decoded, which has $\delta^{(2)} = 1.45$. This results in an asymptotic gain of $\Delta_1 = 10\log_{10}(1.45/0.4) = 5.6\,\mathrm{dB}$ based on equation 10.4. In the third and last level, we have $\delta^{(3)} = 4.0$. The distance properties of these subtrellis diagrams are equivalent to the original DMSK trellis. Compared with level 2, the last trellis has a coding gain of $\Delta_2 = 10\log_{10}(4.0/1.45) = 4.4\,\mathrm{dB}$. Therefore the asymptotic coding gain from the first to the third level is 10 dB. ◇

This example of DCPFSK modulation has shown a way of extending the principle of partitioning to higher-order modulation schemes. The same can be done for a two-valued modulation scheme with memory, such as GMSK, TFM, MSK, etc. Here, the theory of partitioning of binary convolutional codes based on scrambling matrices (see section 9.3.3) can be used to obtain an optimal partitioning of the original system (code) into subsystems (subcodes) with improved distance properties. This is shown in [BSHD98].

10.3.4 Outer convolutional codes

Before we apply the partitioning of CPM to the construction of GC codes or coded modulation, we want to outline the methods that are used in literature. Then we want to compare them with GC codes.

Trellis-coded modulation (matched convolutional codes) The example of MSK modulation in [MMP88] shows that 'matched' convolutional codes are an efficient way to construct powerful codes for MSK. In this way, a supertrellis is generated where modulation and a convolutional code are serially concatenated and afterwards ML decoded. This was one of many constructions applied to CPFSK modulation.

There are different possibilities of combining CPFSK signals with error-correcting codes. It turns out that *convolutional codes* over integer rings $\mathbb{Z}_M$ are the best candidates for coded CPFSK [US94, YT94, RL95].

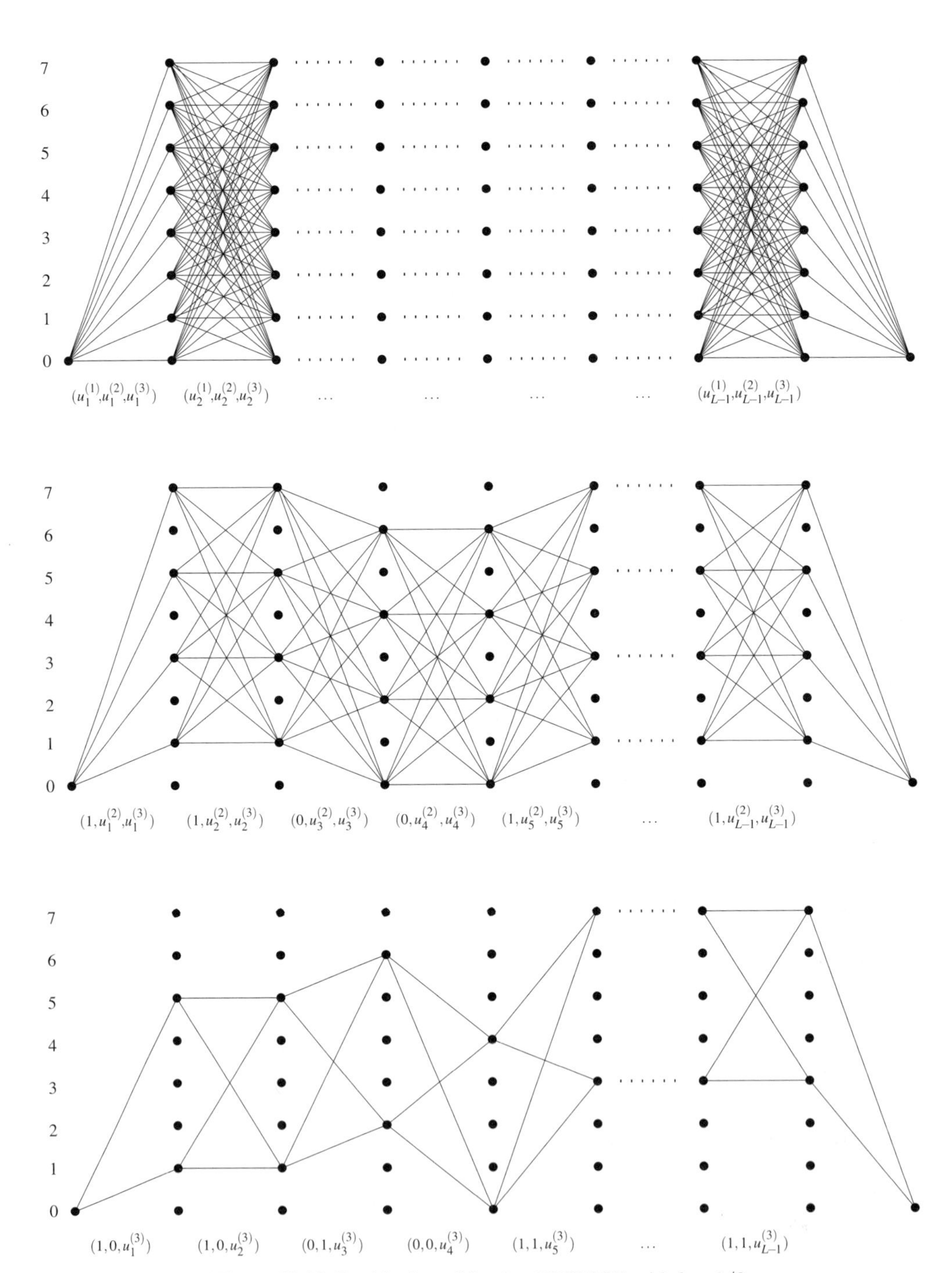

Figure 10.22 Partitioning of 8-valued DCPFSK with $h = 1/8$.

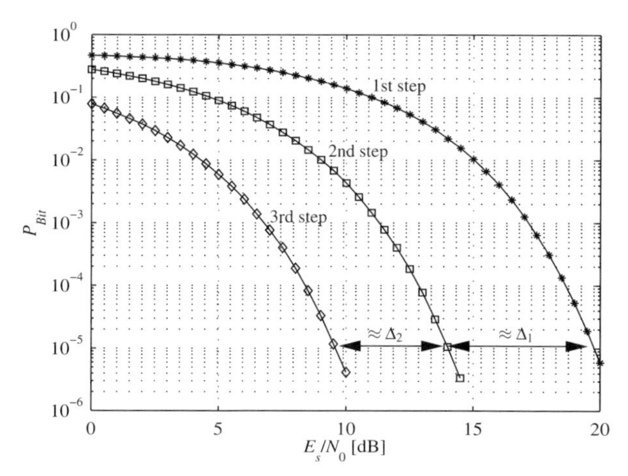

Figure 10.23 Simulated bit error rates of the uncoded partitioning levels for 8-DCFPSK with $h = 1/8$.

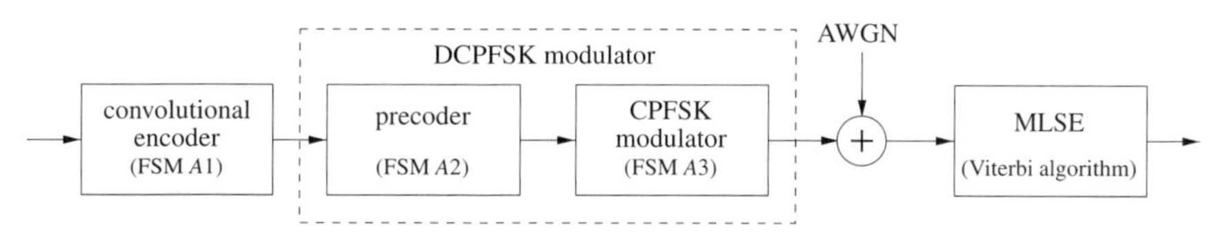

Figure 10.24 Trellis-coded CPFSK.

The DCPFSK modulator (FSM A2 and FSM A3) can be combined directly with an M-valued convolutional encoder (FSM A1). The entire transmission scheme of coded CPFSK is illustrated in figure 10.24.

The state complexity w of the MLSE receiver (Viterbi algorithm) is calculated based on the number of states in the combined trellis consisting of convolutional encoders (FSM $A1$) and DCPFSK modulator (FSM $A1 \cup A2$). In the literature, this trellis is often called a supertrellis. The aim is to find codes that have maximum free quadratic Euclidean distance for a certain state complexity. These codes are also called matched codes, since they are optimized, according to the corresponding modulations with memory. Here the quadratic Euclidean distance should be maximized with respect to a fixed state complexity.

For coded CPFSK, we can find (see e.g. [US94]) maximized convolutional codes, which have been found by a complete code search.

This method of 'matched' codes has also been applied to other CPM schemes, such as TFM/TMSK [MMHP94], MSK [MMP88] and GMSK [TH97]. Next we want to compare this concept of trellis-coded DCPFSK modulation (TC) and decoding in the supertrellis with GC codes based on inner DCPFSK modulation.

Example 10.15 (Comparison of *matched* codes with GC codes) We compare the best trellis codes known (TC), according to [YT94], with a GC code. The GC code is based on the partitioning of 8-DCPFSK (see example 10.14). As outer convolutional codes we use binary codes with different rate but equal state complexity ($w = 16$). For the first level, the rate-$1/6$ code according to [Pal95] was used. For the second level, the punctured rate-$5/6$ code from [YKH84] was chosen. The last level is uncoded. The overall code rate of the GC code

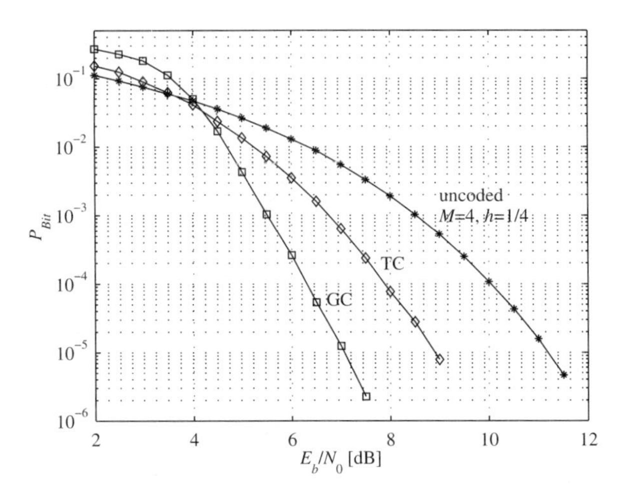

Figure 10.25 Coded 8-DCPFSK modulation with code rate $2/3$, based on outer *convolutional* codes.

is $R = (1/6 + 5/6 + 1)/3 = 2/3$. This GC code is compared with a trellis code with the same code rate and the same state complexity ($w = 16$).

Figure 10.25 illustrates different bit error simulations for coded 8-ary DCPFSK modulation schemes with modulation index $h = 1/8$ and code rate $R = 2/3$, as well as uncoded 4-ary CPFSK with $h = 1/4$ for comparison. We see that at 10^{-5}, the GC code is about 2 dB better than the trellis code.

In order to compare the decoding complexity of the two code constructions, we use the definition of the branch complexity proposed in [For88b]. The branch complexity for the trellis code is $170.67 = 16 \cdot 8 \cdot 8/(2 \cdot 3)$ branches per decoded information bit, whereas the branch complexity for the GC code is only $100 = 16 + 84$. The complexity is calculated as $16 = 6 \cdot 2 \cdot 16/12$ for the two decoding levels, where the last level is uncoded, and $84 = 6 \cdot (2 \cdot 8 \cdot 8 + 2 \cdot 4 \cdot 4 + 2 \cdot 2 \cdot 2)/12$ for the three MAP demodulation steps. For a bit error rate of 10^{-5}, the example shows a gain of about 2.0 dB, even though the decoding complexity is lower (average number of branches per decoded information bit).

If we use SOVA (soft-output Viterbi algorithm) for the soft-output demodulation, the branch complexity of the inner DCPFSK system can be further reduced by a factor of two, whereas we expect only a slight decrease in the bit error rate. ◇

10.3.5 Outer block codes

As mentioned in the last section, we can replace the outer convolutional codes with block codes and use soft-decision decoding according to section 7.4.

Example 10.16 (Block-coded 8-DCPFSK) As outer codes we use a repetition code with length 64 ($\mathcal{A}_1 = (2; 64, 1, 64)$), an extended Hamming code with length 64 ($\mathcal{A}_2 = (2; 64, 57, 4)$) and an uncoded level ($\mathcal{A}_3 = (64, 64, 1)$). All codes have been soft-decision decoded. The overall code rate is $\approx 2/3$ (as in example 10.15). Figure 10.26 illustrates the corresponding bit error rate. For a bit error rate of 10^{-5}, we see a gain of about 3.2 dB compared with 4-DCPFSK.

◇

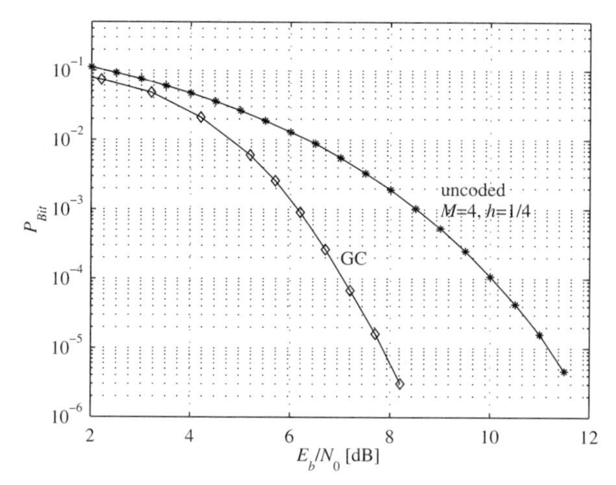

Figure 10.26 Coded 8-DCPFSK modulation of code rate 2/3, based on outer *block* codes.

10.4 Summary

Work done by Imai and Hirakawa [IH77] in 1977 and results from Ungerböck [Ung82] in 1982 have described coded modulation and have resulted in much research activity.

We have described coded modulation as a special case of generalized code concatenation, and have shown that most of the known constructions can be interpreted as special cases of GC codes. In [Bos80], it is shown that GC codes with inner modulation can be constructed using arbitrary metrics. In 1988, Zyablov, Portnoy and Shavgulidze [ZPS88] investigated GC codes over Euclidean spaces in detail. This work was based on [Por85], where it was shown that the decoding algorithm GCD in section 9.2.4 can also decode GC codes over the Euclidean space up to half the minimum distance. This resulted in an elegant description of coded modulation as well as decoding methods with respect to GC codes.

In [Cus84] and [Say86] coded modulation using memoryless modulation schemes with outer block codes, which can be interpreted as GC codes, are described. Cusack describes in his work a procedure where RM codes are used to encode QAM modulation. He incorporates results from Ungerböck [Ung82], and achieves considerable asymptotic coding gains for coded QAM. Sayegh's work contains calculations of the coding gain as well, but is not limited to RM codes and QAM. Furthermore, [SB90] and [BS90] give some results of coded modulation with multiple concatenated codes. In [HBS93], an asymptotic gain of more than 7 dB is achieved using QAM and PSK as modulation and RS outer codes. Additional works that investigate coding and decoding are those of Wörz and Hagenauer [WH93a, WH93b] and Calderbank [Cal89]. Forney [For88b] investigated coded modulation with lattice construction.

For practical use, there are tables for coded modulation with memoryless modulation and the best known convolutional codes based on certain parameters (see e.g. [Ung87a, Ung87b] and [Wei84a, Wei84b, Wei87]). Some aspects of decoding, including iterative decoding, can be found in [WH93b, WH93a]. Signal constellations in multidimensional spaces are investigated in [Wei87].

The partitioning of modulation with memory is based on [BDS96a, BDS96b, BDS97, BDS98]. Various results on concatenated systems with convolution as outer codes were investigated for CPFSK in [BHSD99] and for TFM in [HBSD98] and [BSHD98]. Results on trellis-coded modulation, based on matched convolutional codes and CPFSK, can be found in [Rim88] and [US94]. Trellis codes for MSK were published in [MMP88] and [USA94], for TFM in [MMHP94, MMHP95] and for GMSK in [TH95a, TH95b, TH97].

To date, there are no publications combining modulation with memory with block codes.

Appendix A

Serial and parallel concatenated codes and their iterative decoding: turbo codes

Turbo codes are a class of codes related to concatenated codes, which were presented in chapter 9. A great deal of research into serial and parallel coding schemes has been made since the introduction of turbo codes in [BGT93]. Since these codes are in almost all cases terminated (see section 8.1.7), they are block codes with a special structure. In fact, these codes can be considered as a particular construction of low-density parity check codes, introduced by Gallager [Gal62]. Nevertheless, the concept of iterative decoding introduced in section 7.4.1 gives excellent performance results for these codes as shown in numerous publications (e.g. [BM96, BM98, RS95, RW95, Svi95, WH95]).

Our presentation of turbo codes introduces little in the way of new theory. We proceed by illustrating how the code works by using examples. Our major goal is to show that the use of an equivalent encoder can increase the overall code distance, as does the use of an interleaver. In fact, the choice of the encoders and the interleaver determines the overall code properties. Furthermore, the memory of the convolutional code used influences the decoding performance. The length of the overall code also determines the decoding performance. However, the error exponent [Gal] of these codes is unknown.

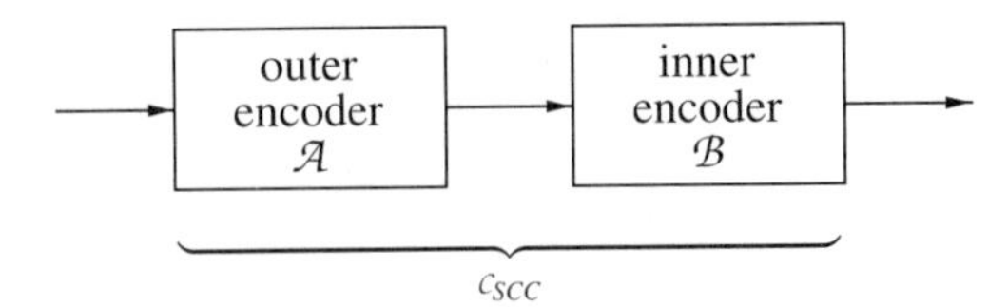

Figure A.1 Serial concatenation encoder.

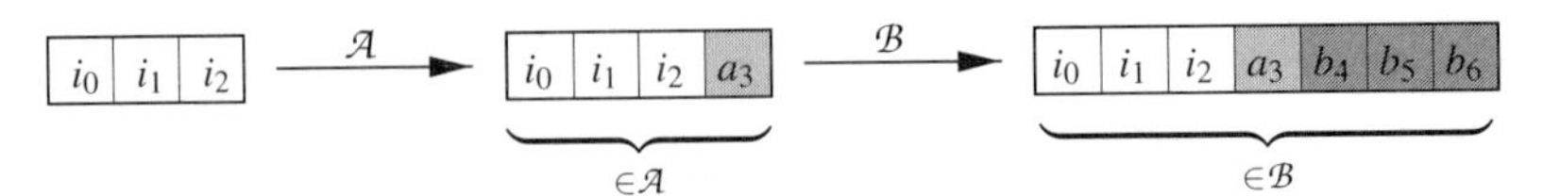

Figure A.2 Example of a serial concatenation.

We shall first describe serial concatenation, and afterwards consider parallel concatenation. Then we show the concept of iterative decoding for both cases. Finally, we illustrate some properties with the help of simulation results.

Note All examples considered in this section are performed in terms of the base field $GF(2)$.

A.1 Serial code concatenation

The serial concatenation of an outer and an inner encoder is shown in figure A.1. Our first example demonstrates that the distance properties of the resulting serial concatenated codes are determined by the distance properties of the component codes. In fact, the mapping of the inner encoder influences the overall code properties.

Example A.1 (Serial concatenation) We concatenate two codes $\mathcal{A}$ and $\mathcal{B}$ such that the generator matrix encoding the resulting code is the product of the generator matrices of the component codes. For this concatenation, we use a single parity check code with dimension $k_{SP} = 3$ and length $n_{SP} = 4$ as the outer code $\mathcal{A}(4,3,2)$. A Hamming code with dimension $k_H = 4$ and length $n_H = 7$ is used as the inner code $\mathcal{B}(7,4,3)$. Therefore three information bits (i_0, i_1, i_2) are coded by the outer code $\mathcal{A}$, giving $\mathbf{a} = (a_0, a_1, a_2, a_3)$. Next, these code bits are encoded by the inner code $\mathcal{B}$. Both codes are systematically encoded, and therefore the information bits are coordinates of the resulting codeword (see figure A.2).

The generator matrices of the component codes are

$$\mathbf{G}_{SP} = \begin{pmatrix} 1 & 0 & 0 & 1 \\ 0 & 1 & 0 & 1 \\ 0 & 0 & 1 & 1 \end{pmatrix} \quad \text{and} \quad \mathbf{G}_H = \begin{pmatrix} 1 & 0 & 0 & 0 & 0 & 1 & 1 \\ 0 & 1 & 0 & 0 & 1 & 0 & 1 \\ 0 & 0 & 1 & 0 & 1 & 1 & 0 \\ 0 & 0 & 0 & 1 & 1 & 1 & 1 \end{pmatrix}.$$

The generator matrix encoding the resulting serial concatenated code is

$$\mathbf{G}_{SCC} = \mathbf{G}_{SP} \cdot \mathbf{G}_H = \begin{pmatrix} 1 & 0 & 0 & 1 & 1 & 0 & 0 \\ 0 & 1 & 0 & 1 & 0 & 1 & 0 \\ 0 & 0 & 1 & 1 & 0 & 0 & 1 \end{pmatrix}.$$

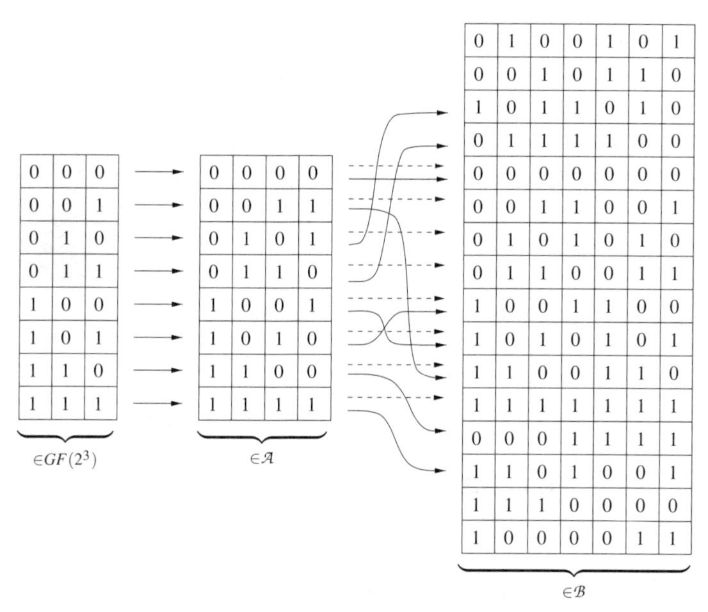

Figure A.3 Concatenated codes resulting from different inner mappings.

For the minimum distance of the overall code, we obtain $d_{SCC} = 3$, which is equal to the minimum distance of the inner Hamming code. Clearly, the distance properties of the resulting code depend on the properties of the inner and outer code. The codewords of the serial concatenated code $\mathcal{C}_{SCC}(7,3,3)$ can be grouped into subsets of the Hamming code. The particular subset is determined by the outer single parity check code and the mapping from information to code bits of the inner encoder.

In order to stress the fact that the mapping of the inner encoder influences the overall distance properties, we consider a second concatenation. Again, we use a single parity check code as outer code and a Hamming code as inner code. For the outer code, we employ the same generator matrix as above, while for the inner code, we use an equivalent generator matrix $\mathbf{G}'_H = \mathbf{M} \cdot \mathbf{G}_H$, with $\mathbf{M}$ a non-singular $k_H \times k_H$ scrambler matrix,

$$\mathbf{M} = \begin{pmatrix} 0 & 0 & 1 & 0 \\ 0 & 0 & 1 & 1 \\ 0 & 1 & 0 & 0 \\ 1 & 0 & 0 & 0 \end{pmatrix} .$$

The generator matrix $\mathbf{G}'_H$ produces the same code as $\mathbf{G}_H$, but with a different mapping from information to code bits. The serial concatenation results now in another code $\mathcal{C}'_{SCC}(7,3,4)$ with minimum distance $d_{SCC} = 4$ and generator matrix

$$\mathbf{G}'_{SCC} = \mathbf{G}_{SP} \cdot \mathbf{G}'_H = \mathbf{G}_{SP} \cdot \mathbf{M} \cdot \mathbf{G}_H = \begin{pmatrix} 1 & 0 & 1 & 0 & 1 & 0 & 1 \\ 1 & 0 & 1 & 1 & 0 & 1 & 0 \\ 1 & 1 & 0 & 0 & 1 & 1 & 0 \end{pmatrix} .$$

The influence of the inner encoder's mapping on the overall distance properties is illustrated

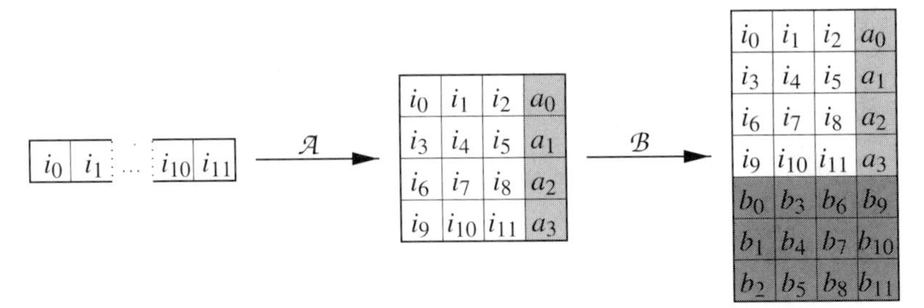

Figure A.4 Example of a product code.

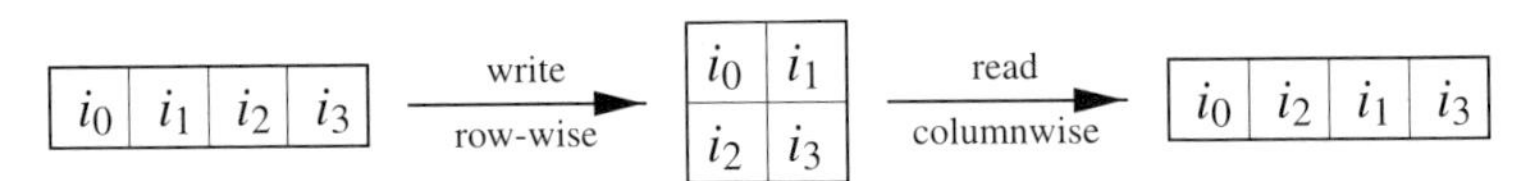

Figure A.5 Example of a block interleaver.

in figure A.3. Scrambling of the outer code does not change the resulting concatenated code, but $\mathbf{G}'_H$ (solid lines) produces an overall distance gain through the way in which it partitions the codewords of the outer code into subsets. ◇

In our next example, we show that by employing interleaving and more codes, we are able to construct long codes by serial concatenation. With an appropriate interleaver design, we can achieve overall minimum distances that are at least the products of the minimum distances of the component codes.

Example A.2 (Product code) We consider again the serial concatenation of an outer single parity check code and an inner Hamming code. The generator matrices $\mathbf{G}_{SP}$ and $\mathbf{G}_H$ from example A.1 are employed.

Now, four blocks of three information bits are encoded with the outer code $\mathcal{A}(4,3,2)$. The codewords of the outer code are written row-wise into a $k_H \times n_{SP}$ matrix. Then we encode columnwise with the inner code $\mathcal{B}(7,4,3)$ (see figure A.4).

The minimum distance of the serial concatenated code is equal to the product of the minimum distances of the component codes: $d_{SCC} = d_{SP}d_H = 6$. The resulting code has length $n_{SCC} = n_{SP}n_H = 28$ and dimension $k_{SCC} = k_{SP}k_H = 12$. For the overall rate, we obtain $R_{SCC} = R_{SP}R_H = 3/7$.

Before encoding the outer code bits with the inner code, we perform a permutation by writing them row-wise into a buffer and reading them columnwise. Such a permutation may be denoted by a function $\pi(\cdot)$ or by a multiplication of a vector with a permutation matrix $\mathbf{P}$. For example we may specify the permutation performed by a block interleaver as illustrated in figure A.5 by

$$\pi(\mathbf{i}) = \begin{pmatrix} i_0 & i_1 & i_2 & i_3 \\ i_0 & i_2 & i_1 & i_3 \end{pmatrix}$$

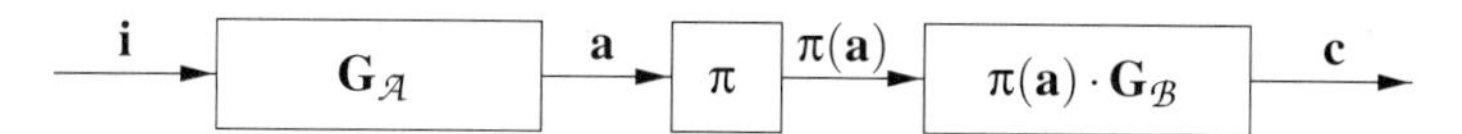

Figure A.6 Serial concatenation encoder with interleaver.

or

$$\mathbf{i}\cdot\mathbf{P} = (i_0, i_1, i_2, i_3)\cdot\begin{pmatrix} 1&0&0&0\\ 0&0&1&0\\ 0&1&0&0\\ 0&0&0&1 \end{pmatrix} = (i_0, i_2, i_1, i_3) \ .$$

Now let $\mathbf{P}$ denote a 16×16 matrix describing the block interleaving used in figure A.4. Then we are able to express the generator matrix of the resulting product code as follows:

$$\mathbf{G}_{SCC} = \begin{pmatrix} \mathbf{G}_{SP} & \underline{\mathbf{0}} & \underline{\mathbf{0}} & \underline{\mathbf{0}} \\ \underline{\mathbf{0}} & \mathbf{G}_{SP} & \underline{\mathbf{0}} & \underline{\mathbf{0}} \\ \underline{\mathbf{0}} & \underline{\mathbf{0}} & \mathbf{G}_{SP} & \underline{\mathbf{0}} \\ \underline{\mathbf{0}} & \underline{\mathbf{0}} & \underline{\mathbf{0}} & \mathbf{G}_{SP} \end{pmatrix} \cdot\mathbf{P}\cdot \begin{pmatrix} \mathbf{G}_{H} & \underline{\mathbf{0}} & \underline{\mathbf{0}} & \underline{\mathbf{0}} \\ \underline{\mathbf{0}} & \mathbf{G}_{H} & \underline{\mathbf{0}} & \underline{\mathbf{0}} \\ \underline{\mathbf{0}} & \underline{\mathbf{0}} & \mathbf{G}_{H} & \underline{\mathbf{0}} \\ \underline{\mathbf{0}} & \underline{\mathbf{0}} & \underline{\mathbf{0}} & \mathbf{G}_{H} \end{pmatrix} ,$$

where $\underline{\mathbf{0}}$ denotes 3×4 and 4×7 all-zero matrices respectively. Clearly, the particular permutation influences the overall distance properties. If we choose an identity matrix instead of $\mathbf{P}$, we obtain a serial concatenated code with minimum distance $d_{SCC} = 3$. ◇

Note that the theory presented in chapter 9 could be interpreted as serial concatenation with interleaving, due to the fact that the information bits of the outer code are written columnwise and then transmitted row-wise after encoding by the outer code. Now we shall give a description of serial concatenated codes for block and convolutional component codes.

Encoding of serial concatenated block codes The encoder scheme of a serial concatenation is shown in figure A.6. The outer encoder maps several information sequences of length k_{SCC} to codewords $\mathbf{a}$ of the outer code $\mathcal{A}$. Then we encode the interleaved codewords $\pi(\mathbf{a})$ with the inner code $\mathcal{B}$. Clearly, the encoder scheme does not depend on whether the inner encoder is systematic or non-systematic. Note that the particular mapping of the inner encoder influences the distance properties of the resulting concatenated code (see example A.1). Figure A.7 depicts the codeword of a serial concatenation of two systematically encoded block codes, where $\mathbf{A}_{\mathcal{A}}$ and $\mathbf{A}_{\mathcal{B}}$ denote the encoding of the codeword redundancy. The overall rate of a serial concatenated code is

$$R = R_{\mathcal{A}} R_{\mathcal{B}} \ . \tag{A.1}$$

Encoding of serial concatenated convolutional codes For the concatenation of convolutional codes, we terminate the code sequences in order to obtain finite codewords. Note that a terminated convolutional code is a block code. Thus all codes presented in the following are in fact block codes.

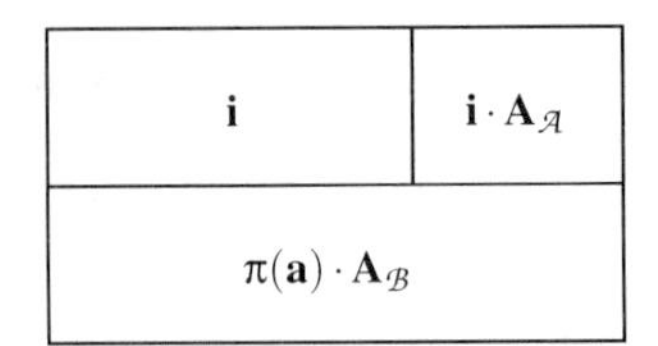

Figure A.7 Codeword of serial concatenated code.

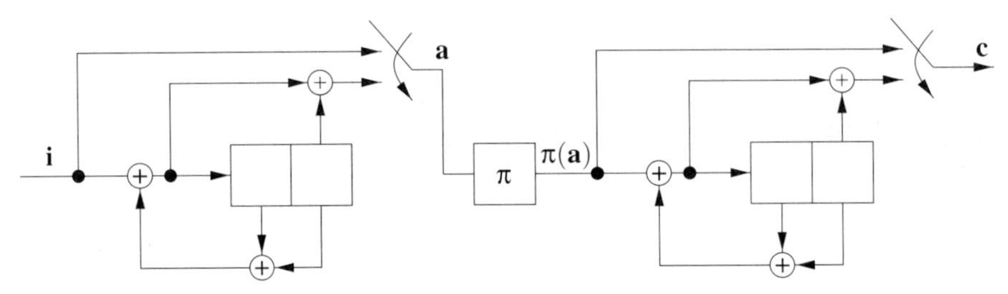

Figure A.8 Serial concatenated convolutional encoder.

Example A.3 (Serial concatenated convolutional encoders) An example of a serial concatenation of two identical convolutional encoders is shown in figure A.8. Both encoders are realizations of the systematic generator matrix in example A.5.

Both component codes have rate $R_c = 1/2$. Due to termination, the resulting rate of the serial concatenated convolutional code is approximately $R_{SCC} \lesssim R_c^2 = 1/4$. ◇

In order to obtain higher overall rates we may use punctured convolutional codes. Puncturing schemes that achieve the highest possible free distance for a given convolutional mother codes are presented in [YKH84]; however, it is not known whether optimizing the free distance of the component codes improves the distance properties of serial concatenated codes.

With proper interleaver design, the serial concatenation can obtain the same minimum distance as a product code, $d_{SCC} = d_{\mathcal{A}} d_{\mathcal{B}}$, where $d_{\mathcal{A}}$ and $d_{\mathcal{B}}$ are the minimum distances of $\mathcal{A}$ and $\mathcal{B}$ respectively. We have already discussed in example A.2 that the particular interleaving influences the distance properties of the concatenated code. It is important to note that, for a randomly chosen permutation, we can only guarantee the minimum distance

$$d_{SCC} \geq \max(d_{\mathcal{A}}, d_{\mathcal{B}}) \ . \tag{A.2}$$

A.2 Parallel code concatenation

In parallel concatenation, two codes are used. The component encoders are usually systematic. In general, each component code independently encodes the information or permutations of the information (see figure A.9). We get two codewords that contain the same information part. Therefore the redundancy parts of both codes are transmitted, and only one copy of the information part.

Note that in the case of serial concatenation, the information is included only once in a concatenated codeword, while in parallel concatenation, we have two codewords that encode identical information.

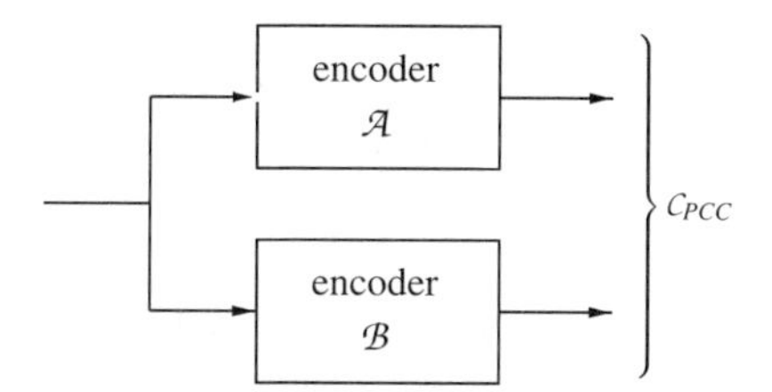

Figure A.9 Parallel concatenation encoder.

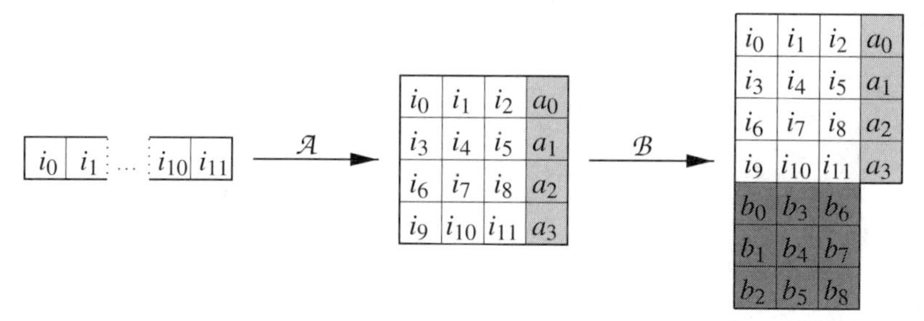

Figure A.10 Example of a parallel concatenated code.

First we give an example where we use several codes, which results in an implicit form of interleaving.

Example A.4 (Parallel concatenated code) We consider the parallel concatenation of a single parity check code and a Hamming code, where we again use the generator matrices $\mathbf{G}_{SP}$ and $\mathbf{G}_H$ from example A.1.

Consider the construction in figure A.10, which is similar to the product code presented in example A.2. As for the serial concatenation, $k_H = 4$ blocks of $k_{SP} = 3$ information bits are encoded by the component code $\mathcal{A}(4,3,2)$. The codewords of the first component code are written row-wise into a $k_H \times n_{SP}$ matrix. But now, only the information bits are encoded columnwise with the second component code $\mathcal{B}(7,4,3)$.

For the overall rate, we obtain

$$R_{SPC} = \frac{k_{SPC}}{n_{SPC}} = \frac{k_{SP}k_H}{k_{SP}k_H + k_{SP}(n_H - k_H) + k_H(n_{SP} - k_{SP})} = \frac{12}{25} \; .$$

The minimum distance of the parallel concatenated code is $d_{PCC} = d_{SP} + d_H - 1 = 4$. Comparing the parallel concatenation with the product code from example A.2, we notice that $C_{PCC}(25,12,4)$ has a higher code rate, but a lower minimum distance.

In order to show that the choice of the interleaver influences the overall distance properties of a parallel concatenated code, we shall compute the generator matrix for the case that both component codes are systematically encoded. Let $\mathbf{I}_{\mathbf{k}_{SP}}$ and $\mathbf{I}_{\mathbf{k}_H}$ denote the $k_{SP} \times k_{SP}$ and $k_H \times k_H$ identity matrices; then the systematic generator matrices can be expressed as

$$\mathbf{G}_{SP} = \begin{pmatrix} \mathbf{I}_{\mathbf{k}_{SP}} & \mathbf{A}_{SP} \end{pmatrix} \quad \text{and} \quad \mathbf{G}_H = \begin{pmatrix} \mathbf{I}_{\mathbf{k}_H} & \mathbf{A}_H \end{pmatrix} \; ,$$

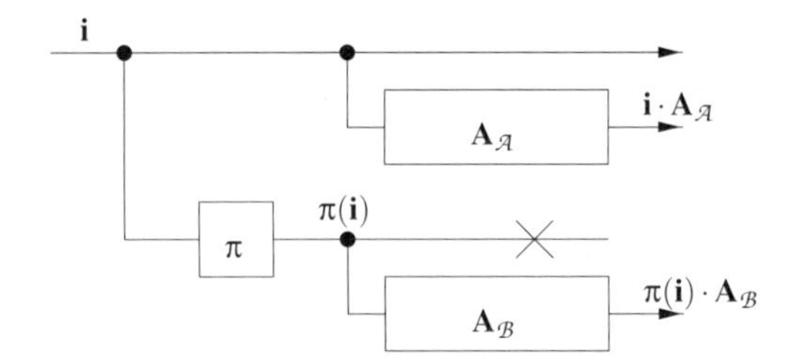

Figure A.11 Encoder scheme of a parallel concatenation.

with

$$\mathbf{A}_{SP} = \begin{pmatrix} 1 \\ 1 \\ 1 \end{pmatrix} \quad \text{and} \quad \mathbf{A}_H = \begin{pmatrix} 0 & 1 & 1 \\ 1 & 0 & 1 \\ 1 & 1 & 0 \\ 1 & 1 & 1 \end{pmatrix} \quad \text{respectively} \quad .$$

The generator matrix of the resulting concatenated code consists of three submatrices

$$\mathbf{G}_{PCC} = \left(I_{k_{SP}k_H}, \begin{pmatrix} \mathbf{A}_{SP} & \underline{\mathbf{0}} & \underline{\mathbf{0}} & \underline{\mathbf{0}} \\ \underline{\mathbf{0}} & \mathbf{A}_{SP} & \underline{\mathbf{0}} & \underline{\mathbf{0}} \\ \underline{\mathbf{0}} & \underline{\mathbf{0}} & \mathbf{A}_{SP} & \underline{\mathbf{0}} \\ \underline{\mathbf{0}} & \underline{\mathbf{0}} & \underline{\mathbf{0}} & \mathbf{A}_{SP} \end{pmatrix}, \mathbf{P} \cdot \begin{pmatrix} \mathbf{A}_H & \underline{\mathbf{0}} & \underline{\mathbf{0}} \\ \underline{\mathbf{0}} & \mathbf{A}_H & \underline{\mathbf{0}} \\ \underline{\mathbf{0}} & \underline{\mathbf{0}} & \mathbf{A}_H \end{pmatrix} \right) ,$$

where the permutation matrix **P** describes the block interleaving of the information bits. Other permutations may lead to codes with $d_{PCC} = \max\{d_{SP}, d_H\} = 3$. ◇

Despite the fact that random permutations may lead to poor minimum distances, they are often used in concatenated systems. There exist $l!$ different interleavers with interleaver length l. Investigating the complete set of possible interleavers soon becomes exceedingly complicated as the interleaver size increases. Therefore methods for interleaver design are often validated through simulation.

Encoding of parallel concatenated block codes In figure A.11, we depict the encoder scheme of a parallel concatenation with two component encoders. The two component codes $\mathcal{A}$ and $\mathcal{B}$ are encoded by systematic generator matrices

$$\mathbf{G}_{\mathcal{A}} = \begin{pmatrix} \mathbf{I}_{k_{\mathcal{A}}} & \mathbf{A}_{\mathcal{A}} \end{pmatrix} \text{ and } \mathbf{G}_{\mathcal{B}} = \begin{pmatrix} \mathbf{I}_{k_{\mathcal{B}}} & \mathbf{A}_{\mathcal{B}} \end{pmatrix} \quad . \tag{A.3}$$

We first encode the information with the code $\mathcal{A}$, and generate the first parity part of the resulting codeword (see figure A.12). In the next encoding step, the component encoder has the permuted information bits as an input, and generates the second parity part. The interleaver of size k_{PCC} performs a permutation of the information bits that is not necessarily a block interleaving. This parallel encoding scheme can be generalized to more than two component encoders.

The interleaver size k_{PCC} determines the code dimension of the resulting code. The code length is $n_{PCC} = k_{PCC} + k_{PCC}(n_{\mathcal{B}} - k_{\mathcal{B}}) + k_{PCC}(n_{\mathcal{A}} - k_{\mathcal{A}})$, and we obtain the following rate of

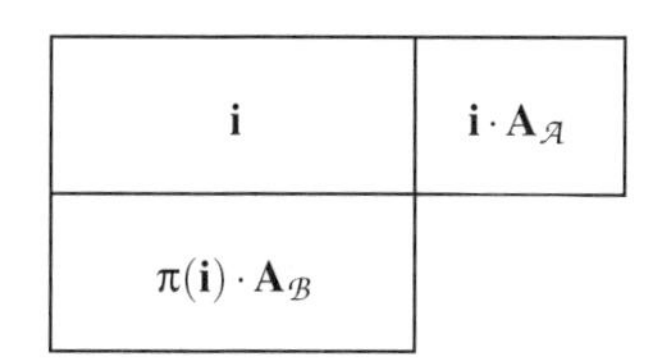

Figure A.12 Codeword of a parallel concatenated code.

the overall code:

$$R_{PCC} = \frac{k_{PCC}}{n_{PCC}} = \frac{k_{PCC}}{k_{PCC} + k_{PCC}(n_{\mathcal{B}} - k_{\mathcal{B}}) + k_{PCC}(n_{\mathcal{A}} - k_{\mathcal{A}})} = \frac{1}{\frac{1}{R_{\mathcal{A}}} + \frac{1}{R_{\mathcal{B}}} - 1} \quad . \tag{A.4}$$

If we use two identical component encoders, each with rate R_c, then it follows that

$$R_{PCC} = \frac{R_c}{2 - R_c} \quad . \tag{A.5}$$

Note that we usually employ systematically encoded component codes. If we use two identical, non-systematic encoders, we obtain $R_{PCC} = R_c/2$. There has been little investigation into non-systematic parallel encoders.

Encoding of parallel concatenated convolutional codes The case of parallel encoding for convolutional codes is straightforward. Let us give an example first.

Example A.5 (Parallel concatenated convolutional encoders) A parallel concatenation of two identical convolutional encoders with systematic generator matrix

$$\mathbf{G}(D) = (\mathbf{I}_k \ \mathbf{A}(D)) = \begin{pmatrix} 1 & \frac{1+D^2}{1+D+D^2} \end{pmatrix}$$

is presented in figure A.13. Again, we encode the information word with the first encoder and the permuted information bits with the second encoder. Transmitting the information part only once may be viewed as a puncturing of the information bits of the second encoder.

Both component codes have rate $R_c = 1/2$, but, because of termination, the resulting rate of the parallel concatenated code is approximately $R_{PCC} \lesssim 1/3$ ◇

We usually terminate the convolutional component codes after k_{PCC} information bits. Due to the additional tail bits, the overall rate is reduced, where this fractional rate loss depends on the overall code length and the memory of the component codes. Because usually relatively long codes are constructed, it is common in the literature to neglect this fractional rate loss.

In order to obtain higher overall rates, we may use component codes with higher rates; however, it is often more convenient to employ rate $R_c = 1/2$ convolutional codes and puncture those codes. In contrast to convolutional codes, no 'optimal' puncturing schemes for PCC are known.

The parallel encoder in figure A.13 generates two parity bits per information bit. Alternately deleting each second parity bit is the most common approach to obtain an overall rate $R_{PCC} = 1/2$ with $R_c = 1/2$ component encoders. For example, at even times we delete the parity bit generated by the first encoder, and at odd times we delete the parity bit generated by the second encoder. Most puncturing patterns presented in the literature do not puncture information bits; thus the resulting encoding remains systematic.

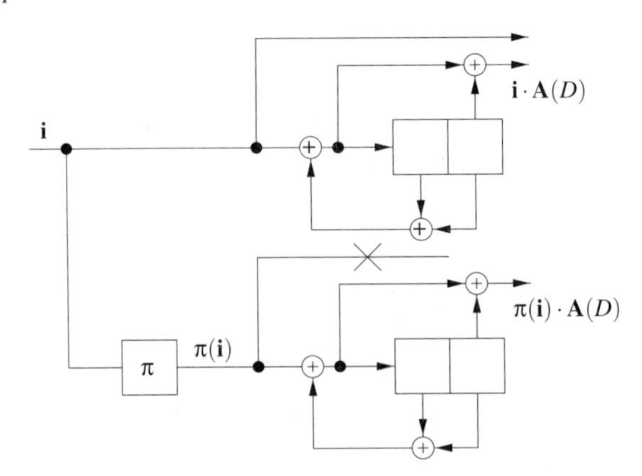

Figure A.13 Parallel concatenated convolutional encoder.

Figure A.14 Soft-input/soft-output decoder and L values.

A.3 Iterative decoding

We have already introduced iterative decoding in section 7.4.1 (see example 7.5). For serial and parallel concatenated codes, this method of decoding has to be adapted to reflect the code construction.

We need log likelihood ratios (L values), as presented in appendix C. Let $\mathbf{y}$ denote the codeword received after transmission over the Gaussian channel. A soft-input/soft-output (SISO) decoder as shown in figure A.14 maximizes the a posteriori probabilities and provides the following soft output for the jth information symbol:

$$L(\hat{i}_j) = L_{ch}y_j + L(i_j) + L_e(\hat{i}_j) \ ,$$

where we assume that the encoding is systematic. This soft information consists of three parts: reliability information from the channel, the a priori knowledge, and the so-called *extrinsic* part from code correlation. For the AWGN channel, we obtain the channel L values $L_{ch}y_j$, with $L_{ch} = 4E_s/N_0$. Possible decoding algorithms are, for example, the suboptimal soft-output Viterbi algorithm (SOVA) or the optimal BCJR algorithm, which evaluates symbol-by-symbol aposteriori probabilities (see section 8.5.1).

A SISO decoder requires channel L values and a priori information as an input. The output consists of reliabilities for the estimated information symbols $L(\hat{\mathbf{i}})$, and extrinsic L values for the redundancy bits $L_e(\hat{\mathbf{c}})$ and for the information bits $L_e(\hat{\mathbf{i}})$ (see figure A.14).

Example A.6 (Iterative decoding) Let us consider the parallel concatenation of two equal, systematically encoded single parity check codes $\mathcal{A}(3,2,2)$:

i_0	i_1	a_0^-
i_2	i_3	a_1^-
$a_0^{\vert}$	$a_1^{\vert}$	

With symbols from $\{+1,-1\}$, a possible codeword is

-1	$+1$	-1
$+1$	$+1$	$+1$
-1	$+1$	

For this particular codeword, we may obtain the following channel L values after transmission over a Gaussian channel:

$L_{ch} \cdot \mathbf{y} =$

-2	1	-5
-0.5	5	1
-5	1	

We assume that no a priori information is available; thus $L(i_j) = 0$ for all information bits. Let us start with decoding the first row. The hard decision for the information bit $\hat{i}_0$ should be equal to the result of the modulo addition $\hat{i}_1 \oplus \hat{a}_0^-$; thus for the corresponding extrinsic log likelihood ratio, we have to evaluate $L(y_1 \oplus y_0^-)$, denoted by $L(y_1) \boxplus L(y_0^-)$. Using log likelihood algebra, this can be approximated by (see appendix C)

$$\begin{aligned} L_e^-(\hat{i}_0) &= L(y_1) \boxplus L(y_0^-) \\ &\approx \operatorname{sign}(L(y_1)L(y_0^-)) \min |L(y_1)|, |L(y_0^-)| \\ &\approx \operatorname{sign}(1 \cdot (-5)) \min(|1|, |-5|) \\ &\approx -1\,. \end{aligned}$$

Similarly, we can evaluate the remaining extrinsic values of $L_e^-(\mathbf{i})$:

-1	2
1	-0.5

Now, we use this extrinsic information as a priori knowledge for the columnwise decoding. As we use log likelihood values and systematic encoded component codes, we can simply sum the channel and extrinsic values:

$L_{ch} \cdot \mathbf{y} + L_e^-(\mathbf{i}) =$

-3	3	-5
0.5	4.5	1
-5	1	

After decoding according to the second component code, we obtain

$L_e^{|}(\hat{\mathbf{i}}) =$

-0.5	1
3	1

$L(\hat{\mathbf{i}}) =$

-3.5	4
3.5	5.5

with $L(\hat{\mathbf{i}}) = L_{ch} \cdot \mathbf{y} + L_e^-(\mathbf{i}) + L_e^{|}(\mathbf{i})$. Now $L(\hat{\mathbf{i}})$ can be used as soft output of the iterative decoder or the iterations can be continued by using $L_e^{|}(\hat{\mathbf{i}})$ as a priori information for a new iteration. $\diamond$

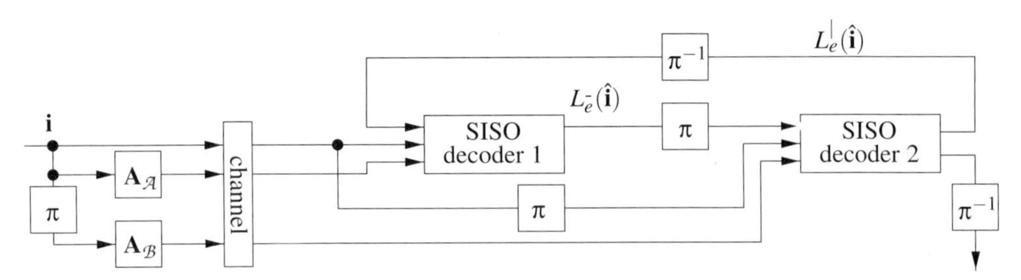

Figure A.15 Encoding and decoding of a PCC with interleaving.

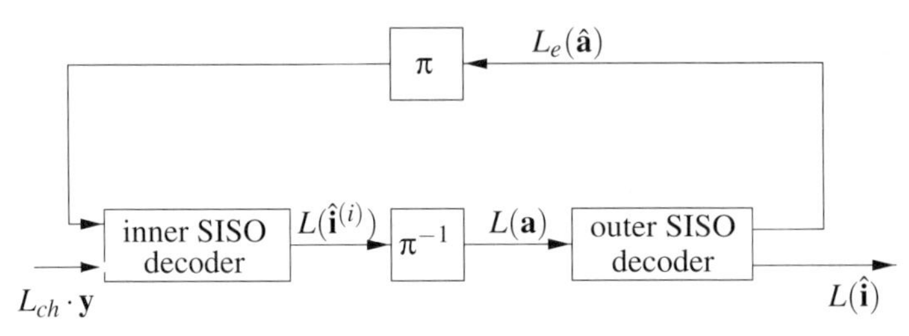

Figure A.16 Decoding scheme for serial concatenated convolutional codes with interleaving.

Decoding of parallel concatenated codes To apply the iterative decoding scheme for decoding of parallel concatenated codes with interleaving, we have to adapt it according to figure A.15. Here π and π^{-1} denote the interleaver and de-interleaver respectively.

Each SISO decoder accepts a priori and channel L values and delivers soft-output L values:

$$L(\hat{\mathbf{i}}) = L_{ch} \cdot \mathbf{y} + L(\mathbf{i}) + L_e(\hat{\mathbf{i}}) \ , \tag{A.6}$$

where we assume systematic encoded component codes. The soft output information of each information symbol consists of three parts: reliability information from the channel, the a priori knowledge, and the so-called *extrinsic* part from the other coordinates of the code. Only the extrinsic values should be used to gain the new a priori values for the next iteration. At least during the first iteration, these values are statistically independent.

Usually the a priori probabilities of information symbols are not known, and it is presumed they are equal probable information bits; thus $L(\mathbf{i}) = 0$ for the first decoding step. The first decoder calculates $L^-(\hat{\mathbf{i}})$ and $L_e^-(\hat{\mathbf{i}})$ from the soft-input channel information, and passes the extrinsic values as a priori information to decoder two. The second decoder feeds $L_e^|(\hat{\mathbf{i}})$ back to decoder one as a priori information for the next iteration, and so on. After the final iteration, the reliabilities $L^|(\hat{\mathbf{i}})$ are the soft-output values of the *iterative soft-input/soft-output decoder.*

Decoding serial concatenated codes The decoding scheme presented in figure A.16 is similar to the iterative decoding scheme for parallel concatenation. The decoder for the inner code generates reliability information for information bits $L(\hat{\mathbf{i}}^{(i)})$ that represent the permuted channel values of the outer decoder. The outer decoder calculates L values for the information word $L(\hat{\mathbf{i}})$ needed for the final decision, and additionally provides extrinsic values $L_e(\hat{\mathbf{a}})$ for outer code bits. These extrinsic values are interleaved and treated as a priori information for the next decoding step.

A.4 Properties and performance aspects of serial and parallel concatenated codes

Despite the great number of publications on turbo codes, the theoretical explanation of the code performance is still unsolved. Therefore we shall discuss the properties mainly based on simulation results.

Analysis In section 8.4.3, we have introduced the union bound as an upper limit on the probability of erroneously decoding a certain code segment of a convolutional code. We may also use the union bound to upper-bound the word error probability P_W for a given block code $C(n,k,d)$:

$$P_W \leq \sum_{w=d}^{n} A(w)P(w) \ , \tag{A.7}$$

where $A(w)$ denotes the number of codewords of weight w and $P(w)$ denotes the probability of an error pattern of weight w.

Although the union bound is based on the hypothesis of maximum-likelihood decoding, it is often used to explain the performance of turbo codes [BM96, BM98, Svi95]. ML decoding of parallel and serial concatenated codes would in general be far too complex, and therefore we use the suboptimal iterative decoding algorithm. Note that in general it is not known how close iterative decoding is to ML decoding; however, there is no bound known that considers suboptimal decoding.

If we consider an AWGN channel and calculate the sum of the union bounds for a parallel or serial concatenated code, we observe that some product terms $A(w)P(w)$ dominate the sum. With increasing signal-to-noise ratio, the lower-weight terms become more important, and asymptotically the minimum distance determines the bit error rate. In coding theory, codes with large minimum distances are desired in order to improve the code performance. In this sense, codes that fulfill, for example, the Varshamov bound are considered to be 'good'. On the other hand, for bad channel conditions, other regions of the weight distribution determine the word error probability. To obtain high coding gains for signal-to-noise ratios near the theoretical limit, the number of low-weight codewords is more important than the particular minimum distance.

Example A.7 (Weight distribution of a turbo code) Figure A.17 shows the number of low weight codewords of two turbo codes with equal dimension $k_{PCC} = 16$ and having different interleavers. Furthermore, we present the distribution for the terminated memory $m = 2$ convolutional mother code. In order to obtain a turbo code with a code rate equal to the rate of the mother code, we have to puncture the component codes. Therefore the free distance of the component codes is usually smaller than d_{free} of the mother code; in the example, $d_1 = d_2 = 2$.

While the overall distributions of the three presented codes hardly differ we notice a significant difference for the numbers of codewords with weight $w \leq 10$. Both turbo codes have a lower minimum distance than the convolutional mother code, but the number of codewords with small weights is much lower. Recall that the weight distribution and therefore the performance of a turbo code depends on the interleaving. ◇

The evaluation of the union bound requires the complete weight distribution. In general, the calculational complexity needed to obtain the complete weight distribution for a particular

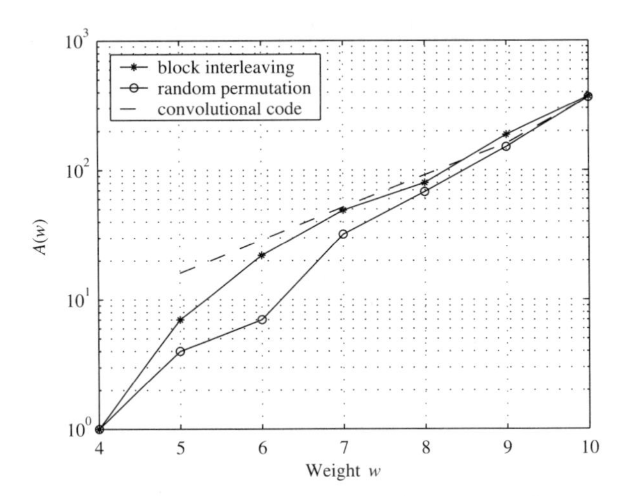

Figure A.17 Number of low-weight codewords.

code increases exponentially with the code dimension. To overcome this problem, Benedetto introduced the concept of *uniform interleaving*. In [BM96, BM98], he proved that one can obtain expectations of weight distributions for codes resulting from parallel or serial concatenation. As a prerequisite, certain weight enumeration functions of the component codes must be known. These expectations are often used to approximate the code performance for high signal-to-noise ratios. The average weight distribution of a class of codes with the same component codes is calculated. Then an upper limit on the code performance of the average code by employing the union bound is computed. Since it is not known if the union bound is valid for iterative decoding, one cannot assume that at least one code exists with identical or better performance. Thus these considerations do not lead to a bound.

Another approach to describe the code performance was introduced in [Svi95]. Assuming *fully optimal interleaving*, Svirid obtained an 'optimistic' weight distribution indicating the highest possible minimum distance for a given interleaver size. In comparison with fully optimal interleaving, Benedetto's approach is more practical, since we can estimate uniform interleaving with simulations. This can be done by randomly choosing a new interleaver for each transmitted block. However, neither concept leads to a construction method that can achieve the expected performance.

Simulation results We shall now present some simulation results for serial and parallel concatenated codes. Despite the restriction to constructions employing two convolutional component codes, there is still a great degree of freedom for the design of different concatenations. For example, we may choose component codes with different memory, different rates or puncturing schemes. Moreover, we may vary the overall code rate and dimension. The results presented in the following should demonstrate the effect of some parameters, namely

- memory of the component codes,
- code length,
- different interleaver types,

Table A.1 Encoding matrices of OFD component codes.

memory	2	3	4	5	6
$\mathbf{G}(D)$ polynomial	(7, 5)	(17, 15)	(35, 23)	(75, 53)	(171, 133)
$\mathbf{G}(D)$ systematic	(1, 5/7)	(1, 15/17)	(1, 23/35)	(1, 53/75)	(1, 133/171)

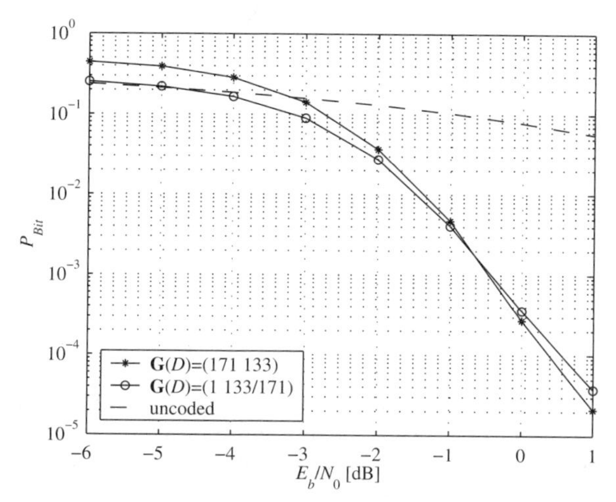

Figure A.18 s/s-MAP decoding of systematic and non-systematic encoded code.

- number of iterations.

We use the max log AAP version (see section 8.5.1) of the BCJR algorithm[2] for decoding the component codes and the number of iterations is ten. For all simulations, an additive white Gaussian noise channel and BPSK modulation are assumed.

In general, we restrict ourselves to constructions employing optimum free distance (OFD) and punctured OFD convolutional component codes. The encoding matrices of the OFD mother codes are presented in Table A.1, where we assume systematic encoding if not mentioned otherwise.

Bit error rate (BER) and systematic encoding An advantage of systematic encoding is that it gives better performance for low signal-to-noise ratios. In figure A.18, we present some simulation results for a memory $m = 6$, rate $R = 1/2$ convolutional code. We compare two different encoding schemes: recursive systematic and polynomial non-systematic encoding. Although both encoding schemes encode the same code, we observe that for high signal-to-noise ratios, non-systematic encoding yields a better performance. However, for low signal-to-noise ratios, we observe the opposite result. For the decoding of concatenated codes, the latter case is more important, since the component codes have larger code rate than the concatenated code and thus operate in relatively poor conditions.

Example A.8 (Reference codes) For our simulations we take rate $R = 1/2$ codes of dimension $k = 1000$ as a reference. Based on these codes, we investigate the influence of some code

[2]Usually the achievable coding gain for optimal decoding are 0.2 to 0.5 dB higher than for sub-optimal decoding.

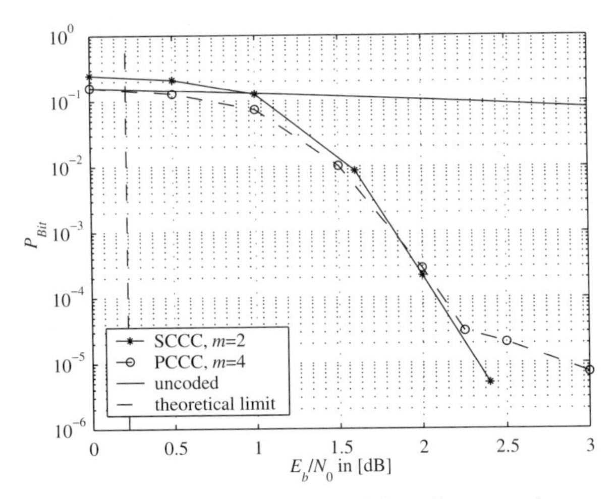

Figure A.19 Bit error rate of the reference code.

parameters. For the serial concatenated reference code, we employ the memory $m = 2$ mother code and deletion maps from [YKH84] for puncturing. For the parallel concatenation, we use $m = 4$ mother codes and randomly generated puncturing patterns. This choice of the memories showed the best performance in all our simulations.

The bit error rates of the two reference codes are depicted in figure A.19. We observe a typical behavior for turbo codes. The curve shows different behavior for low and high E_b/N_0. At 2.25 dB there is a 'kink'. At this point, a bit error rate of approximately 10^{-5} at an E_b/N_0 ratio of just 2 dB above the theoretical limit is achieved. After that, the bit error rate decreases with a much lower slope.

Note that the two constructions achieve comparable coding gains, while the decoding complexity of the serial concatenation is much lower, since we employ four-state component codes.

◇

It is known that the bit error rate for high E_b/N_0 ratios is determined by the minimum distance of a code. The progression of the bit error rate after the 'kink' is often explained as a consequence of the small minimum distance of turbo codes [BM96]. For serial concatenated codes, we do not observe this type of behavior for bit error rates greater than 10^{-5}, which indicates a larger minimum distance.

One might expect that increasing memory of the component codes may lead to more favorable weight distributions and therefore better code performance. However, simulation results show that increasing memory does not necessarily improve the code performance for low E_b/N_0 ratios. With regard to decoding near the coding limit, certain component codes seem to be optimal. For our rate $R = 1/2$ reference codes, we obtain the highest coding gain with $m = 4$ component codes for turbo codes. For serial concatenations, $m = 2$ codes are found to be best.

In the literature, several different generator matrices for memory $m = 4$ convolutional component codes are suggested, e.g. $\mathbf{G}(D) = (1 \;\; 31/33)$ and $\mathbf{G}(D) = (1 \;\; 27/31)$. In comparison with the $m = 4$ OFD code, we did not achieve higher coding gains, but the typical 'kink' appears at lower bit error rates.

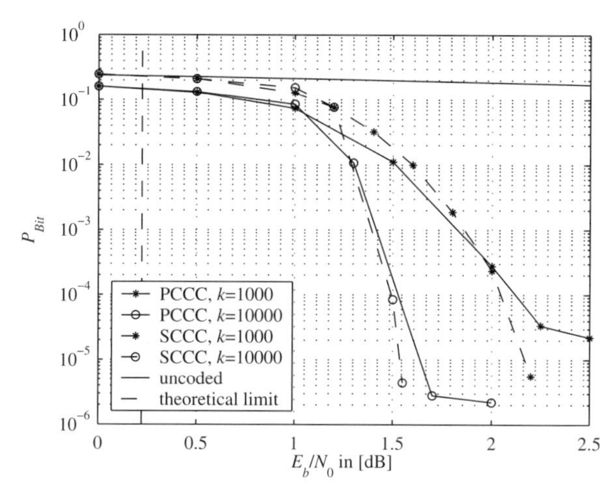

Figure A.20 Bit error rate for different code dimensions.

From information theory, we know that long codes should perform better. For random codes, we should expect an exponential decrease of the word and therefore also of the bit error rate. The aim of code concatenation is the construction of long codes with only a small increase in decoding complexity. The great success of turbo codes is mainly based on the fact that we are able to decode arbitrary long codes with the iterative decoding algorithm. This effect will be illustrated by the next example.

Example A.9 (Code length) In figure A.20, the bit error rate is plotted for different dimensions. We realize that, in fact, the probability of a bit error decreases as the code dimension increases. In the case of turbo codes, we recognize the typical 'kink', but at different bit error rates for different code dimensions. ◇

In examples A.2 and A.4, we have used block interleaving in order to guarantee a certain minimum distance for the resulting serial and parallel concatenated codes respectively. In example A.7, we have presented the weight distribution for two parallel concatenated codes employing block interleaving and a random permutation. There we have noticed that both types of interleaving lead to the same minimum distance, but the numbers of low-weight codewords are different. In the original turbo code proposal [BGT93], a random permutation of the information bits was employed. An examination of the literature shows that this is a widely used approach for interleaver design. But there is also evidence that we can do better [DZ97].

Example A.10 (Interleaving) In figure A.21, simulation results for two parallel concatenated codes with different types of interleaving are shown. We notice that the construction with random permutation achieves a higher coding gain. The slope after the 'kink' indicates an even higher minimum distance. ◇

Beside the impact of code parameters, the behavior of the iterative decoder influences the code performance. Concerning the iterative decoding, there are many open questions. How close is the proposed suboptimal iterative decoding algorithm to maximum-likelihood decoding? Does the iterative algorithm converge, and, if so, after how many iterations?

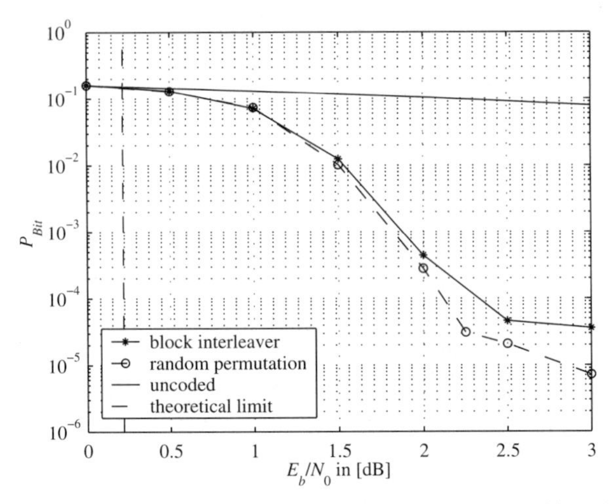

Figure A.21 Bit error rate for PCCC with different types of interleaving.

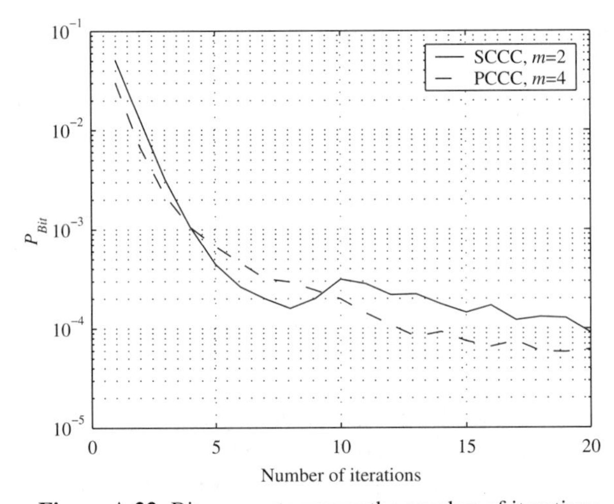

Figure A.22 Bit error rate versus the number of iterations.

Example A.11 (Iterative decoding) Figure A.22 shows the BER as a function of the number of iterations for $E_b/N_0 = 2\,\text{dB}$. Although there is no proof that the iterative decoding algorithm converges, we notice that the BER decreases rapidly with increasing number of iterations. For both concatenated codes, the BER decreases until it approaches a certain level. ◇

Appendix B
Metrics

We formally define several metrics and give several important representations for codes. Each different metric can define the minimum distance of a code.

Definition B.1 (Metric) *A function $d : \mathcal{A} \times \mathcal{A} \rightarrow \mathbb{R}$ that maps two elements a, b from a set $\mathcal{A}$ to a real number $d(a,b)$ is called a metric of $\mathcal{A}$ if the following axioms are satisfied:*

$$\begin{array}{lll} 1. \quad d(a,b) \geq 0, \quad d(a,b) = 0 \Leftrightarrow a = b & \textit{positive-definiteness,} & \\ 2. \quad d(a,b) = d(b,a) & \textit{symmetry,} & \quad (B.1) \\ 3. \quad d(a,b) \leq d(a,z) + d(z,b) & \textit{triangle inequality.} & \end{array}$$

A *metric space* $(\mathcal{A}, d)$ is a set $\mathcal{A}$ in which d is defined.

A vector $\mathbf{a}$ of a vector space $\mathcal{A}$ over $\mathbb{R}$, $\mathbb{C}$ or a similar scalar field, a *norm* can be defined. Every vector $\mathbf{a} \in \mathcal{A}$ is mapped to a real number, the norm $||\mathbf{a}||$, which must be positive-definite and homogenous. For two vectors, the norm of their sum must be smaller than the sum of their norms.

Definition B.2 (Normalized space) *A vector space $\mathcal{A}$ over $\mathbb{R}$ or $\mathbb{C}$ is defined as a normalized space if each $\mathbf{a} \in \mathcal{A}$ corresponds to a real number $||\mathbf{a}||$, such that:*

$$\begin{array}{lll} 1. \quad ||\mathbf{a}|| \geq 0, ||\mathbf{a}|| = 0 \Leftrightarrow \mathbf{a} = \mathbf{0} & \textit{positive-definiteness} & \\ 2. \quad ||\alpha \mathbf{a}|| = |\alpha| \, ||\mathbf{a}|| & \textit{homogeneity,} & \quad (B.2) \\ 3. \quad ||\mathbf{a} + \mathbf{b}|| \leq ||\mathbf{a}|| + ||\mathbf{b}|| & \textit{triangle inequality.} & \end{array}$$

In every space where a norm exists, a metric exists, since $d(\mathbf{a}, \mathbf{b}) = ||\mathbf{a} - \mathbf{b}||$ satisfies the properties of a metric (the reverse argument is not valid).

B.1 Lee metric

The Lee metric is not commonly discussed, and so is considered in detail in this section. The properties of the Lee metric are especially useful for non-binary codes in connection with M-PSK modulation [Pro]. The Lee metric is also well suited for the construction of negacyclic codes (section 5.5).

First, we define the representation of an element $\mathbf{x} \in GF(p)$ of a prime field as elements with the smallest absolute magnitude. In other words, we select either x or $x-p$, depending on which magnitude is smaller, i.e. $\min\{|x|, |x-p|\}$. This representation is indicated by $GF^A(p)$. The elements are $\{-(p-1)/2, \ldots, -1, 0, 1, \ldots, (p-1)/2\}$. The modulo calculation can be similarly modified. We therefore denote $\bmod\ ^a p$ as the calculation $\bmod\ p$ giving the result x or $x-p$, depending on which magnitude is smaller.

Example B.1 ($GF^A(p)$) The prime number field $GF(7)$ can be represented in the usual way as

$$GF(7) = \{0, 1, 2, 3, 4, 5, 6\},$$

or as a set of elements with the smallest magnitude $GF^A(7)$,

$$GF^A(7) = \{0, 1, 2, 3, -3, -2, -1\} = \{-3, -2, -1, 0, 1, 2, 3\}.$$

(Figure B.2 below illustrates this field.) For the modulo calculation, we get

$$5+6 = 4 \mod 7 \quad \text{or} \quad 5+6 = -3 \mod ^A 7\,. \qquad \diamond$$

Therefore the Lee metric for $x, y \in GF^A(p)$ is defined as follows:

$$d_L(x,y) = \left|(x-y) \bmod ^A p\right|\,. \tag{B.3}$$

It is unimportant whether x and y represent elements from $GF(p)$ or the corresponding isomorphic field $GF^A(p)$. This is also clear in example B.3 below. We proceed to show that the above definition is a metric by proving all the requirements from definition B.1.

Proof

1. Positive definiteness:

$$d_L(x,y) = \left|(x-y) \bmod ^A p\right| \geq 0, \qquad \text{because} \quad |z| \geq 0 \quad \forall\, z \in GF^A(p),$$
$$d_L(x,y) = 0 \Leftrightarrow x = y, \qquad \text{because} \quad 0 \bmod ^A p = 0.$$

2. Symmetry:

$$d_L(x,y) = d_L(y,x), \qquad \text{because} \quad (y-x) = -(x-y) \text{ and } |-z| = |z|.$$

3. Triangle inequality: $d_L(x,y) \leq d_L(x,z) + d_L(z,y)$.
Without loss of generality, we can assume $x < y$. Imagine that the coordinates are located on a wheel. The Lee metric $d_L(x,y)i$ is the shortest path from x to y; therefore either $x, x+1, \ldots, y-1, y$ or $y, y+1, \ldots, p-1, 0, 1, \ldots, x$. The element z can lie on the shortest path that results in equality, otherwise the path including z is longer. Compare with example B.2. □

The Lee metric is especially adapted for multilevel PSK modulation as shown in the following example.

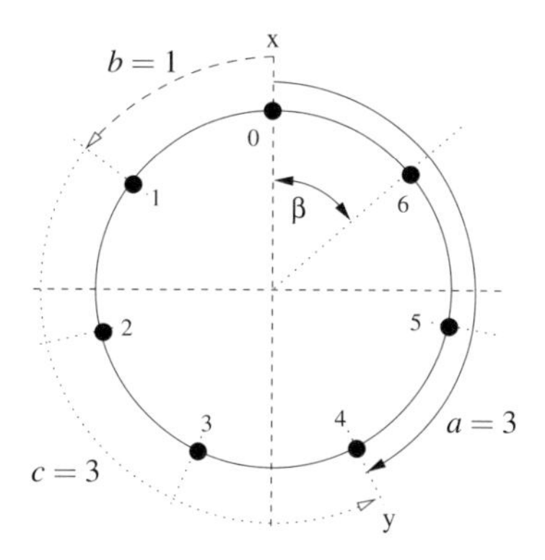

Figure B.1 7-PSK modulation.

Example B.2 (7-PSK modulation) The elements from $GF^A(p)$ correspond to figure B.1, which represents multilevel PSK transmission symbols. The distance to the next neighboring symbol is calculated as the minimum distance. The neighboring elements of the field must also correspond to neighboring transmission symbols.

The distance to the closest neighboring symbol is calculated. This symbol should also correspond to the neighbouring element in the field.

The Lee distance $d_L(x,y)$ between the transmitted symbols x and y corresponds to the number a of circular arcs with angle $\beta = 2\pi/7$. These are the arcs that must be considered in finding the shortest path on the unity circle from symbol x to y. Therefore multilevel PSK modulation of the transmitted symbols gives the elements of $GF(7)$. $d(x,y)$ is assumed to have a maximum value of $\lfloor 7/2 \rfloor$. The righthand side of the triangle inequality can be considered as a unified path from x to y. The number of necessary arcs for this path can be either equal to the number of arcs a that make up the shortest path, or larger than this number (e.g. equal to $b+c$) for the case that another path is selected. This gives $a \leq b+c$. $\diamond$

For n-dimensional vectors with components from $GF(p)$ (or $GF^A(p)$), we now construct a measure similar to the Hamming distance between vectors called the *Lee distance:*

$$d_L(\mathbf{x},\mathbf{y}) = \sum_{j=0}^{n-1} d_L(x_j,y_j) \,. \tag{B.4}$$

This can be extended to define the *Lee weight* (also known as the *Lee norm*) of a vector, which is calculated with respect to the zero vector:

$$\begin{aligned} w_L(\mathbf{x}) &= \sum_{j=0}^{n-1} d_L(x_j,0) && \text{generally} \\ &= \sum_{j=0}^{n-1} |x_j| && \text{for } x_j \in GF^A(p) \,. \end{aligned}$$

Example B.3 (Lee distance, Lee weight) In this example, we calculate the Lee metric, distance and weight with elements from $GF(7)$ and $GF^A(7)$ according to figure B.2.

First, the indicated distances a, b and c with respect to a_A, b_A, and c_A are calculated:

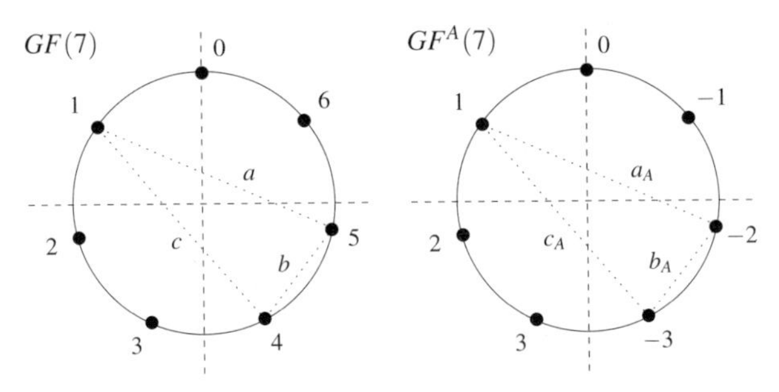

Figure B.2 Example B.3: coding of the 7-PSK transmission symbols using elements from $GF(7)$ or $GF^A(7)$.

$$
\begin{aligned}
a &= d(1,5) &&= |(1-5) \bmod {}^A 7|] &&= |(-4) \bmod {}^A 7| &&= |3| = 3,\\
a_A &= d(1,-2) &&= |(1-(-2)) \bmod {}^A 7| &&= |(3) \bmod {}^A 7| &&= |3| = 3,\\
b &= d(5,4) &&= |(5-4) \bmod {}^A 7| &&= |(1) \bmod {}^A 7| &&= |1| = 1,\\
b_A &= d(-2,-3) &&= |(-2-(-3)) \bmod {}^A 7| &&= |(1) \bmod {}^A 7| &&= |1| = 1,\\
c &= d(4,1) &&= |(4-1) \bmod {}^A 7| &&= |(3) \bmod {}^A 7| &&= |3| = 3,\\
c_A &= d(-3,1) &&= |(-3-1) \bmod {}^A 7| &&= |(-4) \bmod {}^A 7| &&= |3| = 3.
\end{aligned}
$$

The calculations produce the same distance, independent of whether $GF(7)$ or $GF^A(7)$ is selected. We get $a_A = a$, $b_A = b$ and $c_A = c$.

The Lee distance and the Lee weight of the vectors $\mathbf{x} = (1,3,4,0,3)$ and $\mathbf{y} = (5,3,2,6,0)$ with elements from $GF(7)$ are calculated by

$$
\begin{aligned}
d_L(\mathbf{x},\mathbf{y}) &= \sum_{j=0}^{4} d_L(x_j, y_j) = 3+0+2+1+3 = 9,\\
w_L(\mathbf{x}) &= \sum_{j=0}^{4} d_L(x_j, 0) = 1+3+3+0+3 = 10,\\
w_L(\mathbf{y}) &= \sum_{j=0}^{4} d_L(y_j, 0) = 2+3+2+1+0 = 8\,.
\end{aligned}
$$

◇

B.2 Manhattan and Mannheim metrics

The **Manhattan metric** is used [Ulr57] in order to describe the distances between symbols in a QAM alphabet [Pro]. Figure B.3 shows a 16 QAM signal constellation. The Manhattan distance between two points of a QAM alphabet is equal to the minimum number of vertical and horizontal steps needed to move from one point to another. A move directly to a horizontal or vertical neighboring point defines one step. In figure B.3, the Manhattan distance between **a** and **b** is calculated as $3+1=4$.

For vectors with components from the QAM alphabet, the distance is calculated by the addition of the distances of the individual components.

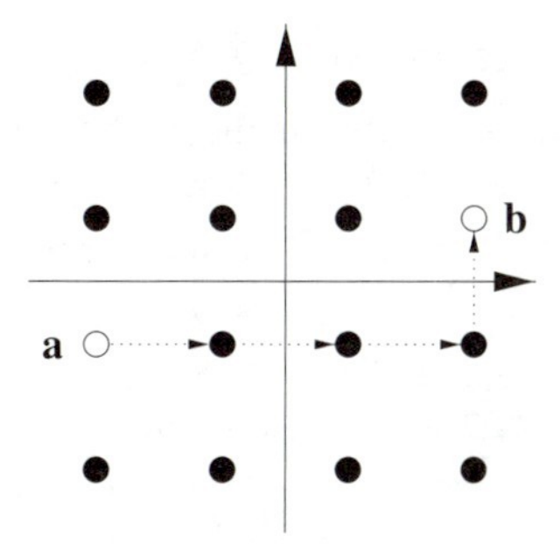

Figure B.3 Manhattan metric.

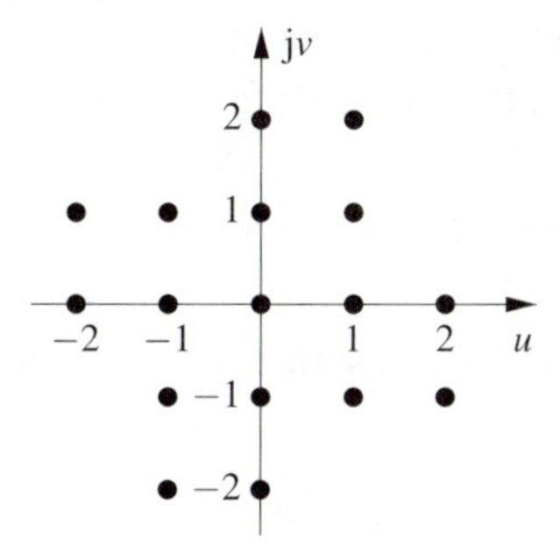

Figure B.4 Gaussian field $\mathbb{G}_{4+\mathrm{j}}$ as a modulation alphabet.

Mannheim metric For algebraic decoding, it is advantageous to have a metric whose values come from the same field as the code symbol. This is not the case for the Manhattan metric because $d_{MT} \in \mathbb{N}$ while the elements are paired numbers. The Mannheim metric $d_M(x,y)$ presented by Huber [Hub94] satisfies this criterion. It defines the value between two points x and y as $d_M(x,y) = d_{MT}(x,y) \bmod \Pi$, where Π is a Gaussian number as defined in section 2.3.3. If the Gaussian field is selected as the points for the modulation alphabet then the requirements for algebraic decoding are satisfied.

B.3 Combinational metrics

The combinatorial metric is defined for matrices [Gab71]. The subsequent description follows primarily from [BS96].

Consider $\mathcal{V} = \{(i,j) \mid 0 \leq i < n_1, 0 \leq j < n_2\}$, which is the set of all indices of an $n_1 \times n_2$ matrix. We define a set $\widehat{\mathcal{T}} = \{\mathcal{T}_1, \mathcal{T}_2, \ldots\}$ as a subset with $\mathcal{T}_i \in \mathcal{V}$. This means that each subset $\mathcal{T}$ can represent a bundle error with respect to the matrix. This bundle error may make the matrix impossible to decode. The set $\widehat{\mathcal{T}}$ represents the possible bundle errors. Consider $\mathcal{Q} \subseteq \mathcal{V}$, any subset of the set $\mathcal{V}$. It is now possible to construct a union of l elements $\mathcal{T}_{i_j}$ that belong to the set $\mathcal{Q}$. We write $\widehat{\mathcal{T}}$: $\mathcal{Q} \subseteq \bigcup_{j=1}^{l} \mathcal{T}_{i_j}$. We say that $\mathcal{Q}$ is 'covered' by the sets $\mathcal{T}_{i_j}$, $j = 1, \ldots, l$.

The distance between two matrices $\mathbf{X}$ and $\mathbf{Y}$ of the $\widehat{\mathcal{T}}$ metric can be defined using the set $\mathcal{Q}$ as follows. Consider $\mathcal{Q}(\mathbf{X},\mathbf{Y}) = \{(i,j) \mid x_{ij} \neq y_{ij}\}$, which have different sets of indices between $\mathbf{X}$ and $\mathbf{Y}$. The distance $d_{\widehat{\mathcal{T}}}(\mathbf{X},\mathbf{Y})$ is equal to the minimum number l for which the set

$\mathcal{Q}$ can be covered by $\mathcal{T}_{i_j}$, $j = 1,\ldots,l$. We write $\mathcal{Q}(\mathbf{X},\mathbf{Y}) \subseteq \bigcup_{j=1}^{l} \mathcal{T}_{i_j}$.

The definition of the $\widehat{\mathcal{T}}$ distance is a metric in the mathematical sense, because it is positive-definite ($d_{\widehat{\mathcal{T}}(\mathbf{X},\mathbf{Y})} \geq 0$), symmetric ($d_{\widehat{\mathcal{T}}}(\mathbf{X},\mathbf{Y}) = d_{\widehat{\mathcal{T}}}(\mathbf{Y},\mathbf{X})$) and satisfies the triangle inequality. The first two properties are apparent, therefore we shall only prove the triangle inequality.

Proof Consider $\mathcal{Q}(\mathbf{X},\mathbf{Z}) = \{(i,j) \mid x_{ij} \neq z_{ij}\}$ and $\mathcal{Q}(\mathbf{Z},\mathbf{Y}) = \{(i,j) \mid z_{ij} \neq y_{ij}\}$. Then $\mathcal{Q}(\mathbf{X},\mathbf{Y}) \subseteq \mathcal{Q}(\mathbf{X},\mathbf{Z}) \cup \mathcal{Q}(\mathbf{Z},\mathbf{Y})$, because no element $(i,j) \in \mathcal{Q}(\mathbf{X},\mathbf{Y})$ can satisfy $z_{ij} = x_{ij}$ and $z_{ij} = y_{ij}$ simultaneously. Therefore the element $(i,j) \in \mathcal{Q}(\mathbf{Z},\mathbf{Y})$ must be covered by either the set $\mathcal{Q}(\mathbf{X},\mathbf{Z})$ or $\mathcal{Q}(\mathbf{Z},\mathbf{Y})$. In order to cover $\mathcal{Q}(\mathbf{X},\mathbf{Z}) \cup \mathcal{Q}(\mathbf{Z},\mathbf{Y})$, we need as many elements in the set $\widehat{\mathcal{T}}$ as needed to cover $\mathcal{Q}(\mathbf{X},\mathbf{Z})$ and $\mathcal{Q}(\mathbf{Z},\mathbf{Y})$. This proves the triangle inequality $d_{\widehat{\mathcal{T}}}(\mathbf{X},\mathbf{Y}) \leq d_{\widehat{\mathcal{T}}}(\mathbf{X},\mathbf{Z}) + d_{\widehat{\mathcal{T}}}(\mathbf{Z},\mathbf{Y})$. □

The minimum distance of a code $\mathcal{C}$ is the distance between any two different codewords $\mathbf{C}_i$ and $\mathbf{C}_j$, $d_{\widehat{\mathcal{T}}}(\mathcal{C}) = \min_{i \neq j}\{\mathbf{C}_i, \mathbf{C}_j\}$, $\mathbf{C}_i, \mathbf{C}_j \in \mathcal{C}$. In addition, a maximum distance $D_{\widehat{\mathcal{T}}}$ can be given for this metric (in the case of the Hamming metric, $D_{\widehat{\mathcal{T}}}$ is the length of the vectors n). One considers the equality $D_{\widehat{\mathcal{T}}} = d_{\widehat{\mathcal{T}}}(\mathbf{X},\mathbf{Y})$, if $\mathcal{Q}(\mathbf{X},\mathbf{Y}) = \mathcal{V}$.

Similar to the Hamming metric, the number of correctable errors can be determined for a code with a particular minimum distance.

Theorem B.3 (Error correction ability with the $\widehat{\mathcal{T}}$ metric) *A code $\mathcal{C}$ with the minimum distance $d = d_{\widehat{\mathcal{T}}}(\mathcal{C}) = 2t + 1$ with respect to the $\widehat{\mathcal{T}}$ metric can correct up to t bundle errors.*

Proof Suppose that the matrix $\mathbf{C} \in \mathcal{C}$ was transmitted and $\mathbf{X}$ was received. If $t \leq \left\lfloor \frac{d-1}{2} \right\rfloor$ bundle errors occur during the transmission then the set of error coordinates is bounded by $\mathcal{Q}(\mathbf{C},\mathbf{X}) \subseteq \bigcup_{j=1}^{t} \mathcal{T}_j$. This means that the distance between $\mathbf{C}$ and $\mathbf{X}$ is equal to $d_{\widehat{\mathcal{T}}}(\mathbf{C},\mathbf{X}) \leq t < \frac{d}{2}$. Because $d_{\widehat{\mathcal{T}}}(\mathbf{C},\mathbf{C}') \geq d_{\widehat{\mathcal{T}}}(\mathcal{C}) = d$ is true for any other codeword $\mathbf{C}' \in \mathcal{C}$, $\mathbf{C} \neq \mathbf{C}'$, it follows from the triangle inequality that $d_{\widehat{\mathcal{T}}}(\mathbf{X},\mathbf{C}') \geq d_{\widehat{\mathcal{T}}}(\mathbf{C},\mathbf{C}') - d_{\widehat{\mathcal{T}}}(\mathbf{C},\mathbf{X}) > \frac{d}{2} > d_{\widehat{\mathcal{T}}}(\mathbf{C},\mathbf{X})$. Therefore if $\mathbf{X}$ is received, $\mathbf{C}$ can be uniquely decoded. □

Translational and cyclic combinatorial metrics Up to this point, the definitions are very general, and can be simplified if we consider only error bundles that have the same form. This means that the subsets of the set $\widehat{\mathcal{T}}$ all show the same form, and this can be represented as a template $T = \{\mathbf{t}_1,\ldots,\mathbf{t}_w\}$. They consist of the set of indices $\mathbf{t}_k = (i,j) \in \mathcal{V}$, with $\mathbf{t}_k \neq \mathbf{t}_l$ for $k \neq l$. The position of the bundle errors is shifted according to a vector $\mathbf{z}$. Therefore we define $T(\mathbf{z}) = T + \mathbf{z} = \{\mathbf{t}_1 + \mathbf{z},\ldots,\mathbf{t}_w + \mathbf{z}\}$ as a bundle error of the form T at the position $\mathbf{z}$. There are two ways in which to calculate $\mathbf{t}_i + \mathbf{z}$. The cyclic combinatorial method is carried out by calculating the translated burst error indicies based on the modulo of the row and column matrix dimensions n_1 and n_2. The translational combinatorial method is performed by considering the translated error burst indicies in the field of integer numbers. Both are special cases of the general combinatorial metric and therefore the statements concerning the correction properties also apply.

Examples Three examples are presented to illustrate the combinatorial metric. These include Hamming, burst and rectangle metrics, which are presented as special cases of the combinatorial metric.

Example B.4 (Hamming metric) Consider $T = \{(0,0)\}$, $T(\mathbf{z}) = \{\mathbf{z}\}$ and $\widehat{\mathcal{T}} = \{T(\mathbf{z}) \mid \mathbf{z} \in \mathcal{V}\}$. This results in the Hamming metric if each $T(\mathbf{z})$ represents one coordinate of a codeword. Therefore the T distance of two matrices $\mathbf{X}$ and $\mathbf{Y}$ is equal to the number of different coordinates. ◇

Example B.5 (Burst metric) An error burst with length b is usually [McWSl, Rei60] defined such that a series of b successive code symbols may contain errors. Therefore we can use the template $T = \{0, \ldots, b-1\}$. If the code length is n then, for $z = 0, \ldots, n-b-1$ we have the translational combinatorial metric and for $z = 0, \ldots, n-1$, the cyclic combinatorial metric. In both cases, the maximum possible distance is calculated as $D_T = \lceil \frac{n}{b} \rceil$. ◇

Example B.6 (Rectangle metric) The template $T = \{(i,j) \mid 0 \leq i < b_1, 0 \leq j < b_2\}$ describes a rectangular bundle error with dimensions $b_1 \times b_2$. This metric is used in connection with *array codes* see [Far92]. The maximum distance that can be achieved is $D_T = \lceil \frac{n_1}{b_1} \rceil \lceil \frac{n_2}{b_2} \rceil$. ◇

The last example demonstrates that the combinatorial metric unifies not only well-known metrics, but also new metrics.

Example B.7 (Cross-metric) The cross-metric [BSS90] is defined by the template

$$T = \{(0,1),(1,0),(1,1),(1,2),(2,1)\} = \left\{ \begin{array}{ccc} & (0,1) & \\ (1,0) & (1,1) & (1,2) \\ & (2,1) & \end{array} \right\}.$$

For the case of an $n \times n$ matrix, and for $5 \mid n$, the maximum possible distance is calculated as $5 \mid n$. ◇

The definitions of the combinatorial metric are used in [GB97] and [GB98] to construct codes with special phase correction abilities.

Appendix C
Log likelihood algebra

In the following, the log likelihood algebra developed by Hagenauer [HOP96] is presented. In many cases, this representation has proved helpful for the calculation of reliabilities.

Consider $x \in X = \{+1, -1\}$, a binary random variable with probability $P(x)$. $+1$ is the neutral element with respect to addition $\oplus$ in x. The log likelihood ratio $L(x)$ (L value) is defined by

$$L(x) := \ln\left(\frac{P(x=+1)}{P(x=-1)}\right). \tag{C.1}$$

If the random variable x is conditioned by another random variable y then the conditional log likelihood ratio $L(x\,|\,y)$ is given by

$$\begin{aligned} L(\hat{x}) \;:=\; L(x\,|\,y) \;&=\; \ln\frac{P(x=+1\,|\,y)}{P(x=-1\,|\,y)} = \ln\frac{P(y\,|\,x=+1)}{P(y\,|\,x=-1)} + \ln\frac{P(x=+1)}{P(x=-1)} \\ &= L(y\,|\,x) + L(x)\,. \end{aligned} \tag{C.2}$$

We observe that $L(x\,|\,y) = L(x,y)$, because the term $P(y)$ can be canceled. The sign of $L(y\,|\,x)$ is the hard-decision information and the magnitude of $L(y\,|\,x)$ is the reliability of the decision.

Consider two statistically independent binary symbols x_1 and x_2 under the operation $\oplus$. We obtain

$$P(x_1 \oplus x_2 = +1) = P(x_1 = +1)\,P(x_2 = +1) + (1 - P(x_1 = +1))\,(1 - P(x_2 = +1))\;.$$

With

$$P(x = +1) = \frac{e^{L(x)}}{1 + e^{L(x)}}, \tag{C.3}$$

the log likelihood ratio of this combination gives

$$\begin{aligned} L(x_1 \oplus x_2) &= \ln\left(\frac{1 + e^{L(x_1)} e^{L(x_2)}}{e^{L(x_1)} + e^{L(x_2)}}\right) \\ &\approx \operatorname{sign}(L(x_1)) \operatorname{sign}(L(x_2)) \min(|L(x_1)|, |L(x_2)|). \end{aligned} \tag{C.4}$$

With the above definitions, we present the log likelihood algebra for the calculation of logarithmic probability. The symbol $\boxplus$ defines the addition,

$$L(x_1) \boxplus L(x_2) := L(x_1 \oplus x_2), \tag{C.5}$$

with the following rules:

$$\begin{aligned} L(x) \boxplus \infty &= L(x), \\ L(x) \boxplus -\infty &= -L(x), \\ L(x) \boxplus 0 &= 0. \end{aligned} \tag{C.6}$$

Consider J statistically independent random variables $x_i, i \in [1, J]$. This gives $\tanh(\frac{x}{2}) = \frac{e^x - 1}{e^x + 1}$, and through induction we obtain

$$\begin{aligned} \sum_{i=1}^{J}{}^{\boxplus} L(x_i) := L\left(\sum_{i=1}^{J}{}^{\oplus} L(x_i)\right) &= \ln\left(\frac{1 + \prod_{i=1}^{J} \tanh\left(\frac{L(x_i)}{2}\right)}{1 - \prod_{i=1}^{J} \tanh\left(\frac{L(x_i)}{2}\right)}\right) \\ &\approx \left(\prod_{i=1}^{J} \operatorname{sign}(L(x_i)) \min_{i \in [1,J]} (|L(x_i)|)\right). \end{aligned} \tag{C.7}$$

The reliability of the combination of more statistically independent variables is an approximation method based on the smallest value of the reliability of the individual terms (compare this result with the calculation of the reliability according to section 9.5.1).

Appendix D
Solutions

Solution of Problem 1.1

$a = 9\mu_1 + \nu_1,\ b = 9\mu_2 + \nu_2$

$$\begin{aligned}
(a+b) \mod 9 &= (\underbrace{9\mu_1}_{=0} + \nu_1 + \underbrace{9\mu_2}_{=0} + \nu_2) \mod 9 \\
&= (\nu_1 + \nu_2) \mod 9 = \big((a \bmod 9) + (b \bmod 9)\big) \mod 9 \\
(ab) \mod 9 &= \big((9\mu_1 + \nu_1)(9\mu_2 + \nu_2)\big) \mod 9 \\
&= (\underbrace{9\mu_1}_{=0}\ \underbrace{9\mu_2}_{=0} + \nu_2\ \underbrace{9\mu_1}_{=0} + \nu_1\ \underbrace{9\mu_2}_{=0} + \nu_1\nu_2) \mod 9 \\
&= (\nu_1\nu_2) \mod 9 = \big((a \bmod 9)(b \bmod 9)\big) \mod 9
\end{aligned}$$

An integer Z can be represented as $Z = \sum_{k=0}^{\infty} z_k 10^k, \quad z_k \in \{0, \dots, 9\}$

It is $10^k \bmod 9 = \underbrace{10 \cdot 10 \cdots\cdots 10}_{\text{k times}} \bmod 9 = (10 \bmod 9)^k = 1^k = 1$, and therefore

$$Z \bmod 9 = \left(\sum_{k=0}^{\infty} z_k 10^k\right) \bmod 9 = \left(\sum_{k=0}^{\infty} z_k\right) \bmod 9.$$

Solution of Problem 1.2

(a) We check whether $\mathbf{c}_1$ satisfies the condition $\mathbf{H} \cdot \mathbf{c}^T = 0$ for codewords $\mathbf{c}$. Therefore we multiply $\mathbf{c}_1$ by the parity check matrix $\mathbf{H}$:

$$\mathbf{H} \cdot \mathbf{c}_1^T = \begin{pmatrix} 0 & 0 & 0 & 1 & 1 & 1 & 1 \\ 0 & 1 & 1 & 0 & 0 & 1 & 1 \\ 1 & 0 & 1 & 0 & 1 & 0 & 1 \end{pmatrix} (0\ 1\ 0\ 1\ 0\ 1\ 0)^T = \begin{pmatrix} 0 \\ 1 \\ 0 \end{pmatrix} + \begin{pmatrix} 1 \\ 0 \\ 0 \end{pmatrix} + \begin{pmatrix} 1 \\ 1 \\ 0 \end{pmatrix} = \begin{pmatrix} 0 \\ 0 \\ 0 \end{pmatrix}.$$

The vector $\mathbf{c}_1$ is a codeword.

(b) The length of the code is equal to the number of columns of the parity check matrix $\mathbf{H}$, i.e. the length is $n = 7$. The number of check bits is equal to the rank of the matrix $\mathbf{H}$, which is certainly ≤ 3. One calculates $n - k = 3$. Therefore the number k of information bits is $k = n - 3 = 4$, i.e. $2^k = 2^4 = 16$ codewords exist.

The minimum distance is the smallest number of linearly dependent columns of $\mathbf{H}$ (any two columns are linearly independent, three columns exist that are linearly dependent, e.g. the columns 1,2 and 3). Consequently the minimum distance is $d = 3$. The code rate R is calculated by $R = \frac{k}{n} = \frac{4}{7}$.

(c) We calculate the syndrome of the vector $\mathbf{c}_2$ by multiplication of $\mathbf{c}_2$ by the parity check matrix $\mathbf{H}$:

$$\mathbf{H} \cdot \mathbf{c}_2^T = \begin{pmatrix} 0 \\ 0 \\ 1 \end{pmatrix} + \begin{pmatrix} 0 \\ 1 \\ 0 \end{pmatrix} + \begin{pmatrix} 1 \\ 0 \\ 0 \end{pmatrix} + \begin{pmatrix} 1 \\ 0 \\ 1 \end{pmatrix} + \begin{pmatrix} 1 \\ 1 \\ 1 \end{pmatrix} = \begin{pmatrix} 1 \\ 0 \\ 1 \end{pmatrix}.$$

The syndrome is not equal to zero, and for a possible error $\mathbf{f}$ it has to hold that $\mathbf{H} \cdot \mathbf{f}^T = (1\ 0\ 1)^T$. Therefore the following possible error vectors are obtained:

$$\begin{aligned}
\mathbf{f}_1 &= 0\,0\,0\,0\,1\,0\,0 \quad \text{5th column}, \\
\mathbf{f}_2 &= 0\,1\,0\,0\,0\,0\,1 \quad \text{2nd column} + \text{7th column}, \\
\mathbf{f}_3 &= 0\,0\,1\,0\,0\,1\,0 \quad \text{3rd column} + \text{6th column}, \\
\mathbf{f}_4 &= 0\,1\,1\,1\,0\,0\,0 \quad \text{2nd column} + \text{3rd column} + \text{4th column}, \\
&\ \ \vdots
\end{aligned}$$

All vectors of the coset with $\mathbf{f}_1$ as coset leader generate the same syndrome. The coset, i.e. all possible error vectors, can be generated by adding all 16 codewords of the code to $\mathbf{f}_1$.

(d) We calculate the syndrome of the vector $\mathbf{c}_3$ by multiplication of $\mathbf{c}_3$ by the parity check matrix $\mathbf{H}$:

$$\mathbf{H} \cdot \mathbf{c}_3^T = \begin{pmatrix} 0 \\ 1 \\ 0 \end{pmatrix} + \begin{pmatrix} 0 \\ 1 \\ 1 \end{pmatrix} + \begin{pmatrix} 1 \\ 0 \\ 0 \end{pmatrix} + \begin{pmatrix} 1 \\ 1 \\ 0 \end{pmatrix} = \begin{pmatrix} 0 \\ 1 \\ 1 \end{pmatrix}.$$

The syndrome corresponds to the 3rd column of the parity check matrix $\mathbf{H}$. The columns are sorted in a way such that the column number corresponds to the binary representation of the integers $1, \ldots, 7$. Therefore the syndrome is the binary representation of the error positions. Because we have to assume that fewer errors are more likely than many errors, the third position is false. $\hat{\mathbf{c}}_3 = \mathbf{c}_3 - \mathbf{f} = (0,1,0,1,0,1,0) \in C$.

Solution of Problem 1.3

(a) 2^m binary vectors $\mathbf{h}$ of length m exist.

(b) There are $n = 2^m - 1$ vectors $\mathbf{h}$, without the zero vector. The number of parity check coordinates is m and the dimension k is $k = 2^m - m - 1$.

The code rate is therefore $R = \frac{2^m - m - 1}{2^m - 1}$.

(c) A possible parity check matrix for $m = 4$ is

$$\mathbf{H} = \begin{pmatrix}
0&0&0&0&0&0&0&1&1&1&1&1&1&1&1 \\
0&0&0&1&1&1&1&0&0&0&0&1&1&1&1 \\
0&1&1&0&0&1&1&0&0&1&1&0&0&1&1 \\
1&0&1&0&1&0&1&0&1&0&1&0&1&0&1
\end{pmatrix}.$$

The minimum distance of the code is the smallest number of linearly dependent columns. The minimum distance for Hamming codes is 3 (any two column vectors are linearly independent and there are three columns that are linearly dependent). Consequently the code can correct one error.

Solution of Problem 1.4

(a) The probability that in a block of length n, e arbitrary symbols are false is $P(e) = \binom{n}{e} p^e (1-p)^{n-e}$.

(b) With the result from (a), we obtain

$$\binom{n}{e+1} p^{e+1} (1-p)^{n-e-1} < \binom{n}{e} p^e (1-p)^{n-e},$$
$$\tfrac{n-e}{e+1}\, p < 1-p \quad \Rightarrow \quad p < \tfrac{e+1}{n+1}\,.$$

For the explicit values, it is $0.1 < \frac{2}{8} = 0.25$. The condition is satisfied for the specified values.

Solution of Problem 1.5

$n = 15, k = 11, m = 4,$

$$\mathbf{H} = \left(\begin{array}{ccccccccccc|cccc}
1&1&0&1&0&0&0&1&1&1&1&1&0&0&0\\
1&1&0&1&1&1&1&1&0&0&0&0&1&0&0\\
1&1&1&0&0&1&1&0&0&1&1&0&0&1&0\\
1&0&1&1&1&0&1&0&1&0&1&0&0&0&1
\end{array}\right),$$

$$\mathbf{G} = \left(\begin{array}{ccccccccccc|cccc}
1&0&0&0&0&0&0&0&0&0&0&1&1&1&1\\
0&1&0&0&0&0&0&0&0&0&0&1&1&1&0\\
0&0&1&0&0&0&0&0&0&0&0&0&0&1&1\\
0&0&0&1&0&0&0&0&0&0&0&1&1&0&1\\
0&0&0&0&1&0&0&0&0&0&0&0&1&0&1\\
0&0&0&0&0&1&0&0&0&0&0&0&1&1&0\\
0&0&0&0&0&0&1&0&0&0&0&0&1&1&1\\
0&0&0&0&0&0&0&1&0&0&0&1&1&0&0\\
0&0&0&0&0&0&0&0&1&0&0&1&0&0&1\\
0&0&0&0&0&0&0&0&0&1&0&1&0&1&0\\
0&0&0&0&0&0&0&0&0&0&1&1&0&1&1
\end{array}\right).$$

Solution of Problem 1.6

$$\begin{aligned}
E(n) &:= \sum_{i=0}^{n} i P_r(i) = \sum_{i=1}^{n} \binom{n}{i} p^i (1-p)^{n-i} i \\
&= \sum_{i=1}^{n} \tfrac{n}{i} \binom{n-1}{i-1} p^i (1-p)^{n-i} i \\
&= n \sum_{j=0}^{n-1} \binom{n-1}{j} p^{j+1} (1-p)^{n-j-1} \\
&= n p (p+1-p)^{n-1} = n p\,.
\end{aligned}$$

Solution of Problem 1.7

(a) The minimum distance $d = 5$ means that up to two errors can be corrected. Therefore the Hamming bound yields

$$2^7 (1+15+105) \leq 2^{15} \Leftrightarrow 121 \leq 2^8 = 256\,.$$

The Hamming bound is satisfied. The right side minus the left side is the number of vectors in $\mathbb{F}_2^{15}$ that are neither codewords, nor can be generated by addition of a codeword and a vector with weight 1, nor by a codeword and vector with weight 2. In the example, this value is $(256-121)\,2^7 = 135 \cdot 128 = 17280$. On the other hand, the number of the vectors inside the correction sphere is $121 \cdot 128 = 15488$. That means that the larger portion of vectors cannot be associated with a codeword, i.e. the larger portion of the vectors is not inside the correction sphere.

(b) We calculate the Hamming bound with $e = 3$:
$2^7 (1+15+105+455) \leq 2^{15}$, i.e. $576 \leq 256$: false, i.e. a code with the specified parameters cannot exist.

(c) We calculate the Hamming bound with $e = 3$:

$$2^{12} (1+23+23 \cdot 11 + 23 \cdot 11 \cdot 7) \leq 2^{23}, \quad 2048 = 2048\,.$$

The equal sign holds! $\Rightarrow$ Perfect code!

Solution of Problem 1.8

The $(4,1,4)$ code is a repetition code of length $n=4$ with the codewords (0000) and (1111). The standard array is

$$\begin{array}{ll}
\mathbf{b}=(0000):\ \{(0000),(1111)\}, & \mathbf{b}=(1000):\ \{(1000),(0111)\},\\
\mathbf{b}=(0001):\ \{(0001),(1110)\}, & \mathbf{b}=(1100):\ \{(1100),(0011)\},\\
\mathbf{b}=(0010):\ \{(0010),(1101)\}, & \mathbf{b}=(0110):\ \{(0110),(1001)\},\\
\mathbf{b}=(0100):\ \{(0100),(1011)\}, & \mathbf{b}=(1010):\ \{(1010),(0101)\}.
\end{array}$$

Solution of Problem 1.9

s/s-MAP:

$$P(x_k=0\,|\,\mathbf{y})=\underbrace{\frac{1}{P(\mathbf{y})}}_{\alpha}\sum_{\substack{\mathbf{x}\in C\\ x_k=0}}P(\mathbf{y}\,|\,\mathbf{x})P(\mathbf{x})=\alpha\sum_{\substack{\mathbf{x}\in C\\ x_k=0}}\prod_{i=1}^{n}P(y_i\,|\,x_i)P(x_i)$$

and, in analogy,

$$P(x_k=1\,|\,\mathbf{y})=\underbrace{\frac{1}{P(\mathbf{y})}}_{\alpha}\sum_{\substack{\mathbf{x}\in C\\ x_k=1}}P(\mathbf{y}\,|\,\mathbf{x})P(\mathbf{x})=\alpha\sum_{\substack{\mathbf{x}\in C\\ x_k=1}}\prod_{i=1}^{n}P(y_i\,|\,x_i)P(x_i)$$

where independence of the code bits is assumed.

Received codeword $\mathbf{y}=(1010)$, error probability of the first and fourth position is $p_1=0.3$, of the second and third position is $p_2=0.1$. Further, $P(x_i)=0.5\,\forall i$.
One obtains

$$\begin{array}{ll}
P(x_1=0\,|\,\mathbf{y})=\frac{\alpha}{2}0.0918, & P(x_1=1\,|\,\mathbf{y})=\frac{\alpha}{2}0.0238,\\
P(x_2=0\,|\,\mathbf{y})=\frac{\alpha}{2}0.0918, & P(x_2=1\,|\,\mathbf{y})=\frac{\alpha}{2}0.0238,\\
P(x_3=0\,|\,\mathbf{y})=\frac{\alpha}{2}0.0238, & P(x_3=1\,|\,\mathbf{y})=\frac{\alpha}{2}0.0918,\\
P(x_4=0\,|\,\mathbf{y})=\frac{\alpha}{2}0.0238, & P(x_4=1\,|\,\mathbf{y})=\frac{\alpha}{2}0.0918;
\end{array}$$

$\Rightarrow$ a MAP decoder would decode to (0011).

Solution of Problem 1.10

BMD decoding method: $P_{Block}=1-\sum_{j=0}^{e}\binom{n}{j}p^j(1-p)^{n-j}$.

Here $n=23$, $e=3$; therefore

$$P_{Block}=1-\left\{(1-p)^{23}+23p(1-p)^{22}+253p^2(1-p)^{21}+1771p^3(1-p)^{20}\right\}.$$

p (BSC)	0.05	0.02	0.01	0.005
P_{Block}	0.0258	0.001	$7.6\cdot 10^{-5}$	$5.1\cdot 10^{-6}$

Solution of Problem 1.11

One obtains all codewords by multiplication of all 2^k possible information vectors by the generator matrix $\mathbf{G}$. The weight distribution is $\mathbf{A}=(1,0,0,7,7,0,0,1)$, i.e. $A_0=A_7=1$ and $A_3=A_4=7$. Therefore

$$P_{FBlock}=\sum_{j=1}^{n}A_jp^j(1-p)^{n-j}=7p^3(1-p)^4+7p^4(1-p)^3+p^7.$$

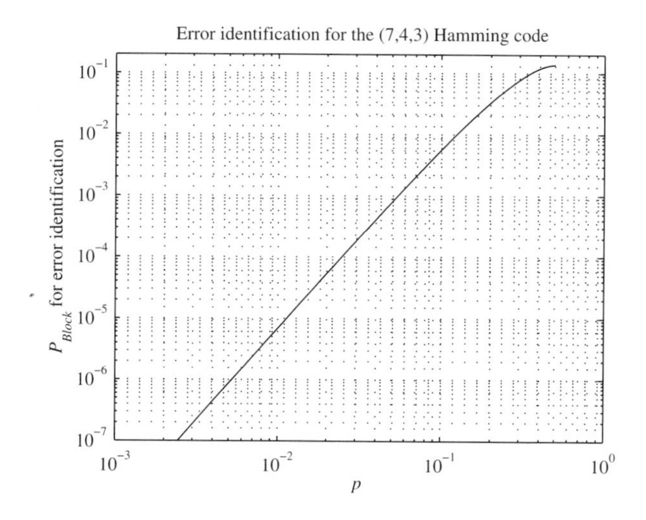

Solution of Problem 1.12

(a) $\mathcal{C}(n,k,d)$: linear block code, coset of a vector $\mathbf{b}$: $\mathbf{b}+\mathcal{C}=\{\mathbf{b}+\mathbf{a},\, \mathbf{a}\in\mathcal{C}\}$. Let $\mathbf{v}=\mathbf{b}+\mathbf{a}$; syndrome $\mathbf{s}=\mathbf{H}\cdot\mathbf{v}^T=\mathbf{H}\cdot(\mathbf{b}+\mathbf{a})^T=\mathbf{H}\cdot\mathbf{b}^T+\underbrace{\mathbf{H}\cdot\mathbf{a}^T}_{=0\,\forall\mathbf{a}\in\mathcal{C}}=\mathbf{H}\cdot\mathbf{b}$.

(b) It is possible to correct $\left\lfloor\frac{d-1}{2}\right\rfloor=1$ error.

correctable error patterns	syndrome	correctable error patterns	syndrome
0000001	001	0010000	110
0000010	010	0100000	101
0000100	100	1000000	011
0001000	111		

(c) Each of the 2^3-1 possible syndromes (not equal to $\mathbf{0}$) is uniquely associated with an error of weight 1, and can therefore be corrected. Errors with higher weight (linear combinations of error patterns) cannot be corrected, since the corresponding linear combinations of the syndromes are already associated with an error patterns.

Solution of Problem 1.13

Since we are dealing with a linear code, the zero codeword must already be included: (0000)

The property 'cyclic' denotes that all cyclic shifts of a codeword yield a codeword again: $(0011)\Rightarrow(0110)$, (1100), (1001) and $(0101)\Rightarrow(1010)$.

It is a parity check code with the parameters $n=4$, $k=\log_2 8=3$ and $d=\min\limits_{\mathbf{c}\in\mathcal{C}}(\mathrm{wt}(\mathbf{c}))=2$.

Solution of Problem 1.14

(a)

$$\mathbf{G}=\begin{pmatrix}1&0&0&0&0&1&1\\0&1&0&0&1&0&1\\0&0&1&0&1&1&0\\0&0&0&1&1&1&1\end{pmatrix}.$$

(b) 1010 will be mapped to 0101011; in given form, one obtains $(1010)\cdot\mathbf{G}=(1010101)$. The desired result can, for example, be obtained by switching the first and the second, the third and the fourth, and the fifth and sixth code coordinates, and thus the corresponding columns of the generator matrix.

$$\Rightarrow \hat{\mathbf{G}} = \begin{pmatrix} 0&1&0&0&1&0&1 \\ 1&0&0&0&0&1&1 \\ 0&0&0&1&1&1&0 \\ 0&0&1&0&1&1&1 \end{pmatrix} .$$

(c) In principle there are $16! \approx 2.0923 \cdot 10^{13}$ different possibilities for mapping information to codewords.

(d) There are $7! = 5040$ possible permutations.
$\binom{7}{2} = 21$ possibilities exist to exchange exactly two coordinates.

Solution of Problem 2.1

(a) Closure, the associative property holds, the neutral element is the zero vector, every vector $a \in \mathbb{F}_2^n$ is inverse to itself, i.e. the set $\mathbb{F}_2^n$ is a group.

(b) The set of integers is a group with respect to addition. With respect to multiplication, the set of integers does not constitute a group, since an inverse element does not always exist. For example, $5x = 1;\ x = \frac{1}{5}$ is not an element of the set of integers.

Solution of Problem 2.2
No, because $cb \neq bc \rightarrow$ is no commutative ring $\rightarrow$ no field.

Solution of Problem 2.3

$q = 2$:

+	0	1
0	0	1
1	1	0

·	0	1
0	0	0
1	0	1

$q = 3$:

+	0	1	2
0	0	1	2
1	1	2	0
2	2	0	1

·	0	1	2
0	0	0	0
1	0	1	2
2	0	2	1

$q = 4$:

+	0	1	2	3
0	0	1	2	3
1	1	2	3	0
2	2	3	0	1
3	3	0	1	2

·	0	1	2	3
0	0	0	0	0
1	0	1	2	3
2	0	2	0	2
3	0	3	2	1

$q = 5$:

+	0	1	2	3	4
0	0	1	2	3	4
1	1	2	3	4	0
2	2	3	4	0	1
3	3	4	0	1	2
4	4	0	1	2	3

·	0	1	2	3	4
0	0	0	0	0	0
1	0	1	2	3	4
2	0	2	4	1	3
3	0	3	1	4	2
4	0	4	3	2	1

$q = 6$:

+	0	1	2	3	4	5
0	0	1	2	3	4	5
1	1	2	3	4	5	0
2	2	3	4	5	0	1
3	3	4	5	0	1	2
4	4	5	0	1	2	3
5	5	0	1	2	3	4

·	0	1	2	3	4	5
0	0	0	0	0	0	0
1	0	1	2	3	4	5
2	0	2	4	0	2	4
3	0	3	0	3	0	3
4	0	4	2	0	4	2
5	0	5	4	3	2	1

Whenever q can be divided by some number a, $1 < a < q$, with zero remainder, multiplication is not always unique, and consequently no field is obtained. Only for the prime numbers $q = 2, 3, 5$ are fields obtained.

Solution of Problem 2.4
Definition For $m \in \mathbb{N} : \Phi(m) := |\{i|\gcd(i,m) = 1, 1 \leq i \leq m\}|$

Properties of Euler's Φ function:
1) $a, b \in \mathbb{Z}$, $\gcd(a,b) = 1 \Rightarrow \Phi(ab) = \Phi(a)\Phi(b)$
2) Prime numbers p: $\Phi(p) = p - 1$ and $\Phi(p^\alpha) = (p-1)p^{\alpha-1}$

$$\Rightarrow \Phi(70) = \Phi(2 \cdot 5 \cdot 7) = 1 \cdot 4 \cdot 6 = 24;$$
$$\Rightarrow \Phi(288) = \Phi(2^5 \cdot 3^2) = 1 \cdot 2^4 \cdot 2 \cdot 3 = 96.$$

Solution of Problem 2.5

Addition table for $(\mathbb{Z}_4, +)$

+	0	1	2	3
0	0	1	2	3
1	1	2	3	0
2	2	3	0	1
3	3	0	1	2

Multiplication table for $(\mathbb{Z}_4, \cdot)$

·	0	1	2	3
0	0	0	0	0
1	0	1	2	3
2	0	2	0	2
3	0	3	2	1

Addition table for $(\mathbb{Z}_{11}, +)$

+	0	1	2	3	4	5	6	7	8	9	10
0	0	1	2	3	4	5	6	7	8	9	10
1	1	2	3	4	5	6	7	8	9	10	0
2	2	3	4	5	6	7	8	9	10	0	1
3	3	4	5	6	7	8	9	10	0	1	2
4	4	5	6	7	8	9	10	0	1	2	3
5	5	6	7	8	9	10	0	1	2	3	4
6	6	7	8	9	10	0	1	2	3	4	5
7	7	8	9	10	0	1	2	3	4	5	6
8	8	9	10	0	1	2	3	4	5	6	7
9	9	10	0	1	2	3	4	5	6	7	8
10	10	0	1	2	3	4	5	6	7	8	9

Multiplication table for $(\mathbb{Z}_{11}, \cdot)$

·	0	1	2	3	4	5	6	7	8	9	10
0	0	0	0	0	0	0	0	0	0	0	0
1	0	1	2	3	4	5	6	7	8	9	10
2	0	2	4	6	8	10	1	3	5	7	9
3	0	3	6	9	1	4	7	10	2	5	8
4	0	4	8	1	5	9	2	6	10	3	7
5	0	5	10	4	9	3	8	2	7	1	6
6	0	6	1	7	2	8	3	9	4	10	5
7	0	7	3	10	6	2	9	5	1	8	4
8	0	8	5	2	10	7	4	1	9	6	3
9	0	9	7	5	3	1	10	8	6	4	2
10	0	10	9	8	7	6	5	4	3	2	1

(a)

	$(\mathbb{Z}_4, +)$	$(\mathbb{Z}_4 \backslash \{0\}, \cdot)$	$(\mathbb{Z}_{11}, +)$	$(\mathbb{Z}_{11} \backslash \{0\}, \cdot)$
closure	√	√	√	√
associativity	√	√	√	√
neutral element	$e = 0$	$e = 1$	$e = 0$	$e = 1$
inverse element	∃	2 not invertible	∃	∃
commutativity	√	√	√	√
algebraic structure	Abelian group	commutative semigroup	Abelian group	Abelian group

(b) $(\mathbb{Z}_4 \backslash \{0\}, \cdot)$ is commutative semigroup (no inverse element with respect to 2).
$(\mathbb{Z}_{11} \backslash \{0\}, \cdot)$ is Abelian group.

(c) $(\mathbb{Z}_4, +, \cdot)$: Abelian group with respect to addition
commutative semigroup with respect to multiplication
distributive property holds $\Rightarrow$ commutative ring

$(\mathbb{Z}_{11}, +, \cdot)$: Abelian group with respect to addition
Abelian group with respect to multiplication
distributive property holds $\Rightarrow$ field, Galois field $GF(11)$

(d) $7^1 = 7,\ 7^2 = 5,\ 7^3 = 2,\ 7^4 = 3,\ 7^5 = 10,\ 7^6 = 4,\ 7^7 = 6,\ 7^8 = 9,\ 7^9 = 8,\ 7^{10} = 1$
$\Rightarrow \operatorname{ord}_{11} 7 = 10 \Rightarrow$ 7 is primitive element of $GF(11)$.

(e) The number of relatively prime elements is $\Phi(10) = 4$.

Solution of Problem 2.6

$$\begin{aligned} 816 &= 2 \cdot 294 + 228, \\ 294 &= 1 \cdot 228 + 66, \\ 228 &= 3 \cdot 66 + 30, \\ 66 &= 2 \cdot 30 + 6, \\ 30 &= 5 \cdot 6 + 0, \end{aligned} \qquad \Rightarrow \gcd(294, 816) = 6.$$

Every remainder can be represented by 816 and 294:

$$\begin{aligned} 816 &= 1 \cdot 816 + 0 \cdot 294, \\ 294 &= 0 \cdot 816 + 1 \cdot 294, \\ 228 &= (1 - 2 \cdot 0)\, 816 + (0 - 2 \cdot 1)\, 294 = 1 \cdot 816 - 2 \cdot 294, \\ 66 &= (0 - 1 \cdot 1)\, 816 + (1 - 1(-2))\, 294 = -1 \cdot 816 + 3 \cdot 294, \\ 30 &= (1 - 3(-1))\, 816 + (-2 - 3 \cdot 3)\, 294 = 4 \cdot 816 - 11 \cdot 294, \end{aligned}$$

Solution of Problem 2.7

(a) The Euclidean algorithm yields

$$\begin{aligned} 1768 &= 3 \cdot 585 + 13, \\ 585 &= 45 \cdot 13 + 0 \quad \Rightarrow \quad \gcd(1768, 585) = 13\,. \end{aligned}$$

(b) The Euclidean algorithm applied to polynomials yields

$$\begin{aligned} x^{12}+x^{10}+x^7+x^4+x^3+x^2+x+1 &= x\,(x^{11}+x^9+x^7+x^6+x^5+x+1), \\ & \quad \text{remainder} \quad x^8+x^6+x^4+x^3+1; \\ x^{11}+x^9+x^7+x^6+x^5+x+1 &= x^3\,(x^8+x^6+x^4+x^3+1), \\ & \quad \text{remainder} \quad x^5+x^3+x+1; \\ x^8+x^6+x^4+x^3+1 &= x^3\,(x^5+x^3+x+1), \\ & \quad \text{remainder} \quad 1. \end{aligned}$$

$\gcd\big(u(x), v(x)\big) = 1$, which means that $u(x)$ and $v(x)$ are relatively prime.

Solution of Problem 2.8

(a) The $\gcd(6, 127)$ is calculated using the Euclidean algorithm. Because 127 is a prime number, the $\gcd(6, 127) = 1$:

$$\begin{aligned} \gcd(6, 127) &= a \cdot 6 + b \cdot 127, \\ 1 &= -21 \cdot 6 + 1 \cdot 127 \quad \Rightarrow 1 = -21 \cdot 6 \mod 127; \end{aligned}$$

$-21 = 106 \bmod 127$ is the inverse element to 6, i.e. $6^{-1} = 106 \bmod 127$.

Hence one obtains

$$\begin{aligned} 6^{-1} \cdot 6 \cdot x &= 6^{-1} \cdot 47 \mod 127, \\ x &= 106 \cdot 47 \mod 127 = 29 \mod 127\,. \end{aligned}$$

(b) **Definition** Let α be a primitive element of a Galois field $GF(p)$ and $a \in GF(p)$. Then the uniquely defined exponent e, $0 \le e \le \Phi(p)$, with

$$\alpha^e = a \bmod p,$$

is called the index or discrete logarithm modulo p of a to the base α:

$$e = \mathrm{ind}_{p,\alpha}(a)$$

After determination of a primitive element of $GF(17)$, e.g. $\alpha = 3$, one can rewrite, because $3^{11} = 7 \bmod 17$ and $3^5 = 5 \bmod 17$, the equation $7^x = 5 \bmod 17$ as:

$$3^{11x} = 3^5 \mod 17 \quad \Longrightarrow \quad 11x = 5 \mod 16,$$

$$x = \frac{5}{11} = \frac{5}{-5} = -1 = 15 \mod 16.$$

Solution of Problem 2.9

Check $p(x)$ for zeros from $GF(5)$: $p(1) = 3+4+2+1 = 10 = 0 \bmod 5$.
$\Rightarrow p(x)$ can be represented as product of two polynomials of smaller order.
$\Rightarrow p(x)$ is *not* irreducible over $GF(5)$, and $p(x)$ cannot be a primitive polynomial.

Solution of Problem 2.10

$$K_0 = \{0\}, \quad K_{11} = \{11,33,19,57\}, \quad K_{25} = \{25,75,65,35\},$$
$$K_1 = \{1,3,9,27\}, \quad K_{13} = \{13,39,37,31\}, \quad K_{26} = \{26,78,74,62\},$$
$$K_2 = \{2,6,18,54\}, \quad K_{14} = \{14,42,46,58\}, \quad K_{40} = \{40\},$$
$$K_4 = \{4,12,36,28\}, \quad K_{16} = \{16,48,64,32\}, \quad K_{41} = \{41,43,49,67\},$$
$$K_5 = \{5,15,45,55\}, \quad K_{17} = \{17,51,73,59\}, \quad K_{44} = \{44,52,76,68\},$$
$$K_7 = \{7,21,63,29\}, \quad K_{20} = \{20,60\}, \quad K_{50} = \{50,70\},$$
$$K_8 = \{8,24,72,56\}, \quad K_{22} = \{22,66,38,34\}, \quad K_{53} = \{53,79,77,71\},$$
$$K_{10} = \{10,30\}, \quad K_{23} = \{23,69,47,61\}.$$

Solution of Problem 2.11

(a)

+	00	01	02	10	11	12	20	21	22
00	00	01	02	10	11	12	20	21	22
01	01	02	00	11	12	10	21	22	20
02	02	00	01	12	10	11	22	20	21
10	10	11	12	20	21	22	00	01	02
11	11	12	10	21	22	20	01	02	00
12	12	10	11	22	20	21	02	00	01
20	20	21	22	00	01	02	10	11	12
21	21	22	20	01	02	00	11	12	10
22	22	20	21	02	00	01	12	10	11

·	00	01	02	10	11	12	20	21	22
00	00	00	00	00	00	00	00	00	00
01	00	01	02	10	11	12	20	21	22
02	00	02	01	20	22	21	10	12	11
10	00	10	20	11	21	01	22	02	12
11	00	11	22	21	02	10	12	20	01
12	00	12	21	01	10	22	02	11	20
20	00	20	10	22	12	02	11	01	21
21	00	21	12	02	20	11	01	22	10
22	00	22	11	12	01	20	21	10	02

(b) $11 \mathrel{\widehat{=}} \alpha^2 \neq 1;\quad (\alpha^2)^2 = \alpha^4 \neq 1;\quad (\alpha^2)^3 = \alpha^6 \neq 1;\quad (\alpha^2)^4 = \alpha^8 = \alpha^0 = 1 \qquad \Rightarrow$ (1,1) has order 4.

(c)

exponent	components	trace function
$-\infty$	00	
0	01	$\alpha^0 + \alpha^0 = 1 + 1 = 2$
1	10	$\alpha^1 + \alpha^3 = \alpha + 2\alpha + 1 = 1$
2	11	$\alpha^2 + \alpha^6 = \alpha + 1 + 2\alpha + 2 = 0$
3	21	$\alpha^3 + \alpha^1 = 1$
4	02	$\alpha^4 + \alpha^4 = 2 + 2 = 1$
5	20	$\alpha^5 + \alpha^7 = 2\alpha + \alpha + 2 = 2$
6	22	$\alpha^6 + \alpha^2 = \alpha + 1 + 2\alpha + 2 = 0$
7	12	$\alpha^7 + \alpha^5 = \alpha + 2 + 2\alpha = 2$

Solution of Problem 3.1

(a) Galois field $GF(7)$, primitive element $\alpha = 5$, minimum distance $d = 3$. Two consecutive positions have to be zero, according to the definition of RS codes, e.g. $A_4 = A_5 = 0$.

It follows that, for all codewords $a(x)$, $a(\alpha^{-4}) = 0$, $a(\alpha^{-5}) = 0$. Therefore the generator polynomial $g(x)$ is calculated as

$$\begin{aligned} g(x) &= (x-\alpha^{-4})\,(x-\alpha^{-5}) = (x-\alpha^2)\,(x-\alpha^1) \\ &= (x-4)\,(x-5) = (x+3)\,(x+2) = x^2+5x+6\,. \end{aligned}$$

(b) $h(x)\,g(x) = x^6 - 1$

$\Rightarrow h(x) = (x^6-1) : g(x) = x^6 - 1 : x^2 + 5x + 6 = x^4 + 2x^3 + 5x^2 + 5x + 1$

(c) There are three ways to check whether or not $c(x)$ is a codeword.

1. Division by the generator polynomial $g(x)$*:*

$$x^3+6x^2+4x+6 : x^2+5x+6 = x+1$$

The remainder is zero; therefore $c(x)$ is a codeword.

2. Multiplication by the parity check polynomial $h(x)$*:*

$$(x^3+6x^2+4x+6)(x^4+2x^3+5x^2+5x+1) = 0 \mod (x^6-1)$$

$c(x)h(x) = 0 \mod (x^n-1)$, i.e. $c(x)$ is a codeword.

3. Transform:

It has to hold that $A_4 = A_5 = 0$:
$A_4 = \frac{1}{n}c(\alpha^{-4})$, $\quad c(\alpha^{-4}) = c(\alpha^2) = c(4) = 4^3+6\cdot 4^2+4\cdot 4+6 = 1+5+2+6 = 0$,
$A_5 = \frac{1}{n}c(\alpha^{-5})$, $\quad c(\alpha^{-5}) = c(\alpha^1) = c(5) = 5^3+6\cdot 5^2+4\cdot 5+6 = 6+3+6+6 = 0$.
The condition for a codeword in the transformed domain is satisfied, i.e. $c(x)$ is a codeword.

(d) Encoding by multiplication by the generator polynomial $g(x) = x^2+5x+6$.

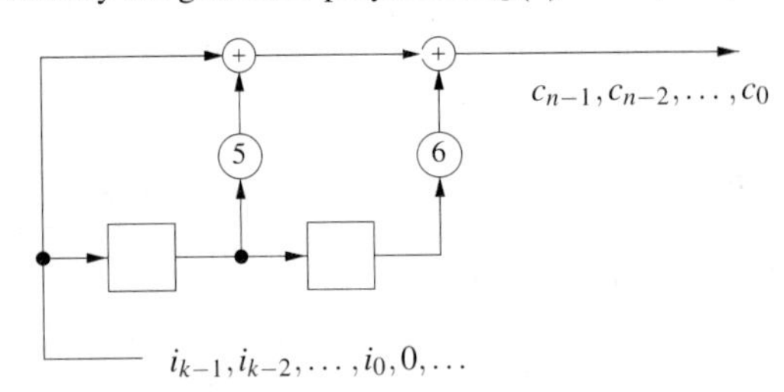

Systematic encoding with the generator polynomial: the k information symbols are contained in the codeword $a_6x^6+a_5x^5+\ldots+a_2x^2$.

$$g(x) = x^2+5x+6 \Rightarrow -g_0 = -6 = 1; -g_1 = -5 = 2$$

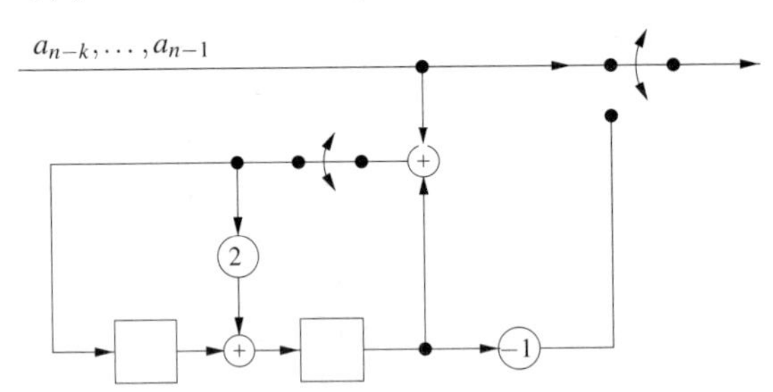

Encoding with the parity check polynomial $h(x) = x^4+2x^3+5x^2+5x+1$:

$$\begin{aligned} h_0 a_j &= -h_1 a_{j-1} - h_2 a_{j-2} - h_3 a_{j-3} - h_4 a_{j-4}, \\ a_j &= 2a_{j-1} + 2a_{j-2} + 5a_{j-3} + 6a_{j-4}. \end{aligned}$$

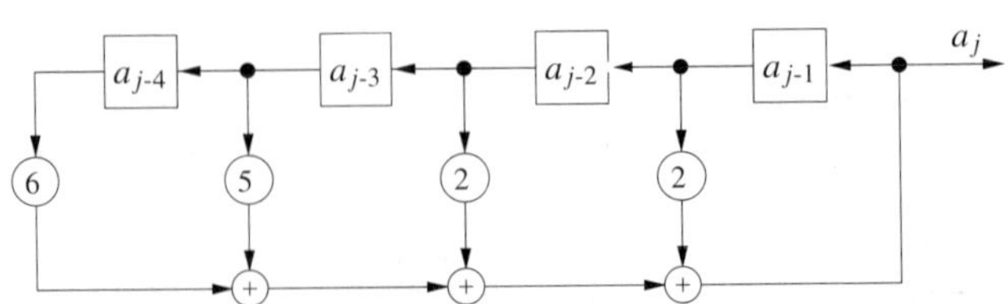

Solution of Problem 3.2

(a) Example: $a(x) = 5 + x + x^2$:

$$a(x)\,f(x) = 2x^4 + 3x^3 + 0x^2 + 1x + 1$$

input	register content		output
1	0	0	2
1	1	0	3 2
5	1	1	0 3 2
0	5	1	1 0 3 2
0	0	5	1 1 0 3 2
0	0	0	

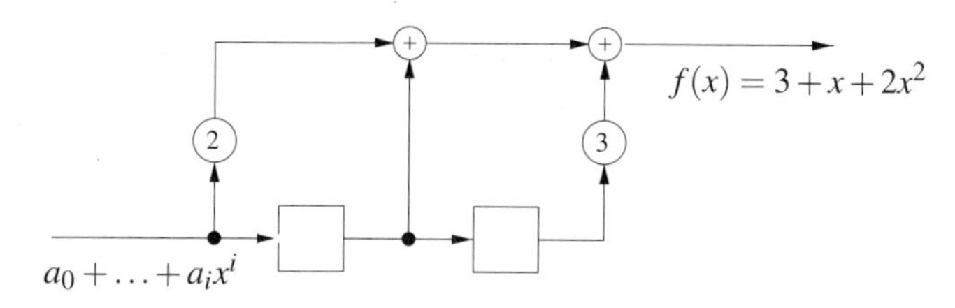

(b) $a(x) = x^6 + x^5$

$$x^6 + x^5 = 5x^3 + 4x^2 + 6x \mod (x^4 + 2x + 1)$$

input	register
1	0 0 0 0
1	1 0 0 0
0	1 1 0 0
0	0 1 1 0
0	0 0 1 1
0	6 5 0 1
0	6 4 5 0
0	0 6 4 5

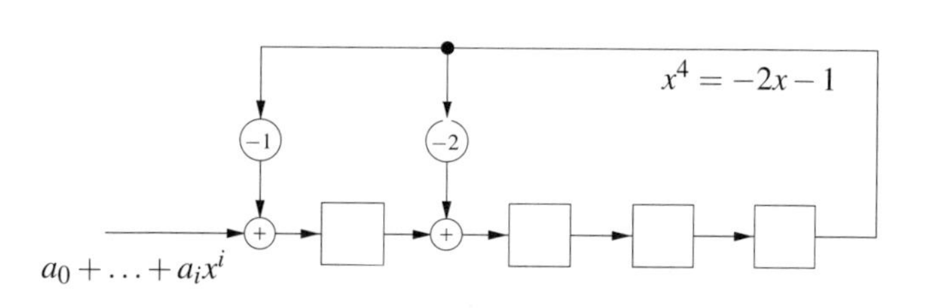

Solution of Problem 3.3

The following relation holds: $x^i\,c(x) \mod (x^n - 1) = x^i\,i(x)\,g(x) \mod (x^n - 1)$. Multiplication of a codeword by the parity check polynomial $h(x)$ must yield zero $\mod (x^n - 1)$:

$$x^i\,g(x)\,h(x) = x^i(x^n - 1) = 0 \mod (x^n - 1)$$

holds for arbitrary information polynomials. Every linear block code that can be generated using a generator polynomial is cyclic.

Solution of Problem 3.4

(a) The inverse transform yields $b_i = \sum B_j \alpha^{ij} = 2\,(5^4)^i + 5\,(5^5)^i$.

$$b_i = 2 \cdot 2^i + 5 \cdot 3^i \quad \Rightarrow \quad b = (0,5,4,4,3,5).$$

Using the convolutional theorem of the DFT: $B(x) = x^4 A(x) = C(x)A(x)$.

$$\begin{aligned} C(x) = x^4 \bullet\!\!-\!\!\circ\ c &= (1,2,4,1,2,4), \\ a &= (0,6,1,4,5,3), \\ b_i = c_i a_i \ \Rightarrow\ b &= (0,5,4,4,3,5)\,. \end{aligned}$$

(b) Syndrome calculation: $r = (3,5,3,4,3,5) \circ\!\!-\!\!\bullet R = (5,6,1,5,1,6)$

$$\Rightarrow S(x) = 5 + 6x + x^2 + 5x^3 .$$

Key equation: $C(x)\,S(x) = -T(x) \bmod x^4$. Because $S(x) \neq 0$, r is not a codeword. In the following we assume that an error has occurred, i.e. $e = 1$. The key equation then yields:

$$\begin{aligned} C_0 S_2 + S_1 &= 0, \qquad & C_0 + 6 &= 0, \\ C_0 S_3 + S_2 &= 0, \qquad & C_0 5 + 1 &= 0, \end{aligned}$$

$$\Rightarrow \left.\begin{aligned} C_0 &= 1, \\ C_0 &= 4, \end{aligned}\right\} \quad \text{contradiction}$$

Under the assumption that an error has occurred, the key equation cannot be solved explicitly. We have to assume that more than one error occurred, i.e. two errors ($e = 2$). Consequently, we obtain

$$\begin{aligned} C_0 S_3 + C_1 S_2 + S_1 &= 0, \qquad & C_0 5 + C_1 1 + 6 &= 0, \\ C_0 S_2 + C_1 S_1 + S_0 &= 0, \qquad & C_0 1 + C_1 6 + 5 &= 0, \end{aligned}$$

$$\Rightarrow \left.\begin{aligned} 5C_0 + \ \ C_1 &= 1 \\ C_0 + 6C_1 &= 2 \end{aligned}\right\} \Rightarrow \begin{aligned} C_1 &= 2, \\ C_0 &= 4 . \end{aligned}$$

The error location polynomial is calculated as $C(x) = 4 + 2x + x^2$.

Inverse transform:

i	1	5	4	c_i	
0	4	2	1	0	←
1	4	3	4	4	
2	4	1	2	0	←
3	4	5	1	3	
4	4	4	4	5	
5	4	6	2	5	

Thus errors have occurred at the 0th and 2nd positions.

Error value calculation:

$$f_i = n^{-1} \frac{\alpha^{-i} T(\alpha^i)}{C'(\alpha^i)}, \qquad \alpha = 5, \quad i: \text{ zeros of } C(x) .$$

The derivative of $C(x)$ is $C'(x) = 2 + 2x$.

We still have to calculate the error value polynomial:

$$T_j = -\sum_{i=0}^{j} C_i S_{j-i} \, ; \quad j = 0, \ldots, \underbrace{e-1}_{1} \quad \Rightarrow \quad \begin{aligned} T_0 &= -4 \cdot 5 &&= 1, \\ T_1 &= -4 \cdot 6 - 2 \cdot 5 &&= 1 . \end{aligned}$$

Thus the error value polynomial is $T(x) = 1 + x$.

Consequently, the error value calculation yields

$$f_0 = \frac{6 \cdot 5^0 T(5^0)}{C'(5^0)} = 3, \qquad f_2 = \frac{6 \cdot 5^{-2} T(5^2)}{C'(5^2)} = 6 .$$

Now we have calculated the error; it is $f(x) = 3 + 6x^2$.

The transmitted codeword was $b(x) = r(x) - f(x) = (0,5,4,4,3,5)$.

Solution of Problem 3.5

(a) x^4+x^3+1 is used as primitive polynomial. We obtain the following component representation:

$\Rightarrow \quad \alpha^4=\alpha^3+1$	component representation
$\alpha^0 = 1$	1 0 0 0
$\alpha^1 = \alpha$	0 1 0 0
$\alpha^2 = \alpha^2$	0 0 1 0
$\alpha^3 = \alpha^3$	0 0 0 1
$\alpha^4 = 1+\alpha^3$	1 0 0 1
$\alpha^5 = 1+\alpha+\alpha^3$	1 1 0 1
$\alpha^6 = 1+\alpha+\alpha^2+\alpha^3$	1 1 1 1
$\alpha^7 = 1+\alpha+\alpha^2$	1 1 1 0
$\alpha^8 = \alpha+\alpha^2+\alpha^3$	0 1 1 1
$\alpha^9 = 1+\alpha^2$	1 0 1 0
$\alpha^{10} = \alpha+\alpha^3$	0 1 0 1
$\alpha^{11} = 1+\alpha^2+\alpha^3$	1 0 1 1
$\alpha^{12} = 1+\alpha$	1 1 0 0
$\alpha^{13} = \alpha+\alpha^2$	0 1 1 0
$\alpha^{14} = \alpha^2+\alpha^3$	0 0 1 1

$\alpha^{15} = 1$

(b)

$$\begin{aligned} g(x) &= \prod_{i=1}^{6}(x+\alpha^i) \\ &= (x+\alpha)(x+\alpha^2)(x+\alpha^3)(x+\alpha^4)(x+\alpha^5)(x+\alpha^6) \\ &= (x^2+\alpha^{13}x+\alpha^3)(x^2+x+\alpha^7)(x^2+\alpha^2x+\alpha^{11}) \\ &= (x^4+\alpha^7x^3+\alpha^4x^2+\alpha^{12}x+\alpha^{10})(x^2+\alpha^2x+\alpha^{11}) \\ &= x^6+\alpha^{12}x^5+x^4+\alpha^2x^3+\alpha^7x^2+\alpha^{11}x+\alpha^6\ . \end{aligned}$$

(c) $r(x)=\alpha^6+\alpha^2x+\alpha^7x^2+\alpha^2x^3+\alpha^9x^4+\alpha^{12}x^5+x^6+\alpha^9x^{11}$

Syndrome calculation:

$$\begin{aligned} S_0 &= R_9 = r(\alpha^{-9}) = r(\alpha^6), & S_3 &= R_{12} = r(\alpha^{-12}) = r(\alpha^3), \\ S_1 &= R_{10} = r(\alpha^{-10}) = r(\alpha^5), & S_4 &= R_{13} = r(\alpha^{-13}) = r(\alpha^2), \\ S_2 &= R_{11} = r(\alpha^{-11}) = r(\alpha^4), & S_5 &= R_{14} = r(\alpha^{-14}) = r(\alpha)\ . \end{aligned}$$

$$\begin{aligned} S_0 = r(\alpha^6) &= 1+\alpha^3+\alpha^4+\alpha^5+\alpha^8+\alpha^{12} &&= \alpha^{13}, \\ S_1 = r(\alpha^5) &= 1+\alpha^4+\alpha^6+\alpha^{14} &&= \alpha^5, \\ S_2 = r(\alpha^4) &= 1+\alpha^2+\alpha^8+\alpha^9+\alpha^{10}+\alpha^{14} &&= \alpha^3, \\ S_3 = r(\alpha^3) &= \alpha^3+\alpha^5+\alpha^{11}+\alpha^{13} &&= \alpha^3, \\ S_4 = r(\alpha^2) &= \alpha+\alpha^2+\alpha^4+\alpha^6+\alpha^7+\alpha^8+\alpha^{11}+\alpha^{12} &&= \alpha^7, \\ S_5 = r(\alpha) &= \alpha^2+\alpha^3+\alpha^9+\alpha^{13} &&= \alpha^6, \end{aligned}$$

$$\Longrightarrow S = (S_0,S_1,S_2,S_3,S_4,S_5) = (\alpha^{13},\alpha^5,\alpha^3,\alpha^3,\alpha^7,\alpha^6)\ .$$

Berlekamp–Massey algorithm:

jlk	Δ_j	$C^{(j+1)}(x)$	$2l > j$	$B^{(j+1)}(x)$
		$C^{(0)}(x) = 1$		$B^{(0)}(x) = 0$
000	$S = \alpha^{13}$	$C^{(1)}(x) = 1$	no	$B^{(1)}(x) = \Delta_0^{-1}C^{(0)}(x) = \alpha^2$
111	$S_1 + C_1^{(1)}S_0 = \alpha^5$	$C^{(2)}(x) = 1 + \alpha^7 x$	yes	$B^{(2)}(x) = B^{(1)}(x) = \alpha^2$
212	$S_2 + C_1^{(2)}S_1$	$C^{(3)}(x) = 1 + \alpha^7 x + \alpha^7 x^2$	no	$B^{(3)}(x) = \Delta_2^{-1}C^{(2)}(x)$
	$= \alpha^3 + \alpha^7\alpha^5 = \alpha^5$			$= \alpha^{10} + \alpha^2 x$
321	$S_3 + C_1^{(3)}S_2 + C_2^{(3)}S_1$	$C^{(4)}(x) = 1 + \alpha^{11}x + \alpha^{12}x^2$	yes	$B^{(4)}(x) = B^{(3)}(x)$
	$= \alpha^3 + \alpha^7\alpha^3 + \alpha^7\alpha^5 = 1$			$= \alpha^{10} + \alpha^2 x$
422	$S_4 + C_1^{(4)}S_3 + C_2^{(4)}S_2$	$C^{(5)}(x) = 1 + \alpha^{11}x + \alpha^3 x^2$	no	$B^{(5)}(x) = \Delta_4^{-1}C^{(4)}(x)$
	$= \alpha^7 + \alpha^{11}\alpha^3 + \alpha^{12}\alpha^3 = \alpha^{10}$	$+\alpha^{12}x^3$		$= \alpha^5 + \alpha x + \alpha^2 x^2$
531	$S_5 + C_1^{(5)}S_4 + C_2^{(5)}S_3 + C_3^{(5)}S_2$	$C^{(6)}(x) = 1 + \alpha^3 x + \alpha^{12}x^2$	yes	$B^{(6)}(x) = B^{(5)}(x)$
	$= \alpha^6 + \alpha^{11}\alpha^7 + \alpha^3\alpha^3 + \alpha^{12}\alpha^3 = \alpha^4$	$+\alpha^{14}x^3$		

$\Rightarrow$ error location polynomial:

$$C(x) = 1 + \alpha^3 x + \alpha^{12}x^2 + \alpha^{14}x^3 = (1 + x\alpha^{-1})(1 + x\alpha^{-4})(1 + x\alpha^{-11}).$$

The zeros have been found using a Chien search. Because $C(x)$ has the roots α, α^4 and α^{11}, the components r_1, r_4 and r_{11} of the received word r are erroneous.

Error value polynomial $T_j^{(l)}$; here $l = -9 = 6$:

$$T_0^{(6)} = \alpha^{13}, \quad T_1^{(6)} = \alpha^4, \quad T_2^{(6)} = \alpha^{14},$$

$$\Rightarrow \begin{cases} T^{(6)}(x) &= \alpha^{13} + \alpha^4 x + \alpha^{14}x^2, \\ C'(x) &= \alpha^3 + \alpha^{14}x^2. \end{cases}$$

For the error value calculation, it holds that

$$f_i = x^{-l} n x^{-1} \frac{T^{(l)}(x)}{C'(x)} \quad \text{for } x = \alpha^i,$$

$$f_1 = \alpha^8 \frac{\alpha^6}{\alpha^{10}} = \alpha^4, \quad f_4 = \alpha^2 \frac{\alpha^6}{\alpha^6} = \alpha^2, \quad f_{11} = \alpha^{13} \frac{\alpha^3}{\alpha^7} = \alpha^9.$$

Consequently, the error polynomial is $f(x) = \alpha^4 x + \alpha^2 x^4 + \alpha^9 x^{11}$, and one calculates for the transmitted codeword $a(x)$

$$\begin{aligned} a(x) &= r(x) - f(x) \\ &= \alpha^6 + \alpha^2 x + \alpha^7 x^2 + \alpha^2 x^3 + \alpha^9 x^4 + \alpha^{12}x^5 + x^6 + \alpha^9 x^{11} \\ &\quad - \alpha^4 x - \alpha^2 x^4 - \alpha^9 x^{11} \\ &= \alpha^6 + \alpha^{11}x + \alpha^7 x^2 + \alpha^2 x^3 + x^4 + \alpha^{12}x^5 + x^6 = g(x)\,. \end{aligned}$$

Solution of Problem 3.6

(a) $\deg g(x) = 4 \quad \Rightarrow \quad k = 2^m - 1 - 4 = 2^m - 5.$

A symbol consists of m bits, i.e. the code has $m(2^m - 5)$ binary information coordinates.

(b) The code can correct two erroneous symbols. Two erroneous symbols can be obtained by two bits. In the case where all erroneous bits are in two symbols, $2m$ binary errors can be corrected.

Solution of Problem 3.7

(a) Coefficient a_{15} of the code extended by one coordinate:

$$a_{15} = \sum_{i=0}^{14} a_i = \alpha^6 + \alpha^{11} + \alpha^7 + \alpha^2 + 1 + \alpha^{12} + 1 = \alpha.$$

$$\Longrightarrow c_{\text{ext}}(x) = c(x) + \alpha x^{15}.$$

(b) A code of length $n = 17$ will be constructed over $GF(2^8)$. A primitive element $\eta \in GF(2^8)$ is necessary, because $2^8 - 1 = 255$, and 17 divides 255, and 15 divides 255.
The element $\alpha = \eta^{17}$ has order 15, and can therefore be used as primitive element of $GF(2^4)$. The element $\beta = \eta^{15}$ is 17. Unity root. Further,
$\alpha = \eta^{17}, \alpha^2 = \eta^{34}, \alpha^3 = \eta^{51}, \alpha^4 = \eta^{68}, \alpha^5 = \eta^{85}, \alpha^6 = \eta^{102}, \alpha^7 = \eta^{119}, \alpha^8 = \eta^{136}, \alpha^9 = \eta^{153}, \alpha^{10} = \eta^{170}, \alpha^{11} = \eta^{187}, \alpha^{12} = \eta^{204}, \alpha^{13} = \eta^{221}, \alpha^{14} = \eta^{238}, \alpha^{15} = \eta^{255}.$
It now follows that

$$\begin{aligned}
\beta+\beta^{-1} &= \eta^{15}+\eta^{240} = \eta^{204} = \alpha^{12}, & \beta^5+\beta^{-5} &= \eta^{75}+\eta^{180} = \eta^{187} = \alpha^{11},\\
\beta^2+\beta^{-2} &= \eta^{30}+\eta^{225} = \eta^{153} = \alpha^{9}, & \beta^6+\beta^{-6} &= \eta^{90}+\eta^{165} = \eta^{221} = \alpha^{13},\\
\beta^3+\beta^{-3} &= \eta^{45}+\eta^{210} = \eta^{238} = \alpha^{14}, & \beta^7+\beta^{-7} &= \eta^{105}+\eta^{150} = \eta^{119} = \alpha^{7},\\
\beta^4+\beta^{-4} &= \eta^{60}+\eta^{195} = \eta^{51} = \alpha^{3}, & \beta^8+\beta^{-8} &= \eta^{120}+\eta^{135} = \eta^{102} = \alpha^{6}.
\end{aligned}$$

Now $x^{17} - 1$ can be factorized with elements from $GF(2^4)$:

$$x^{17} - 1 = (x-1)(x^2+\alpha^{12}x+1)(x^2+\alpha^9x+1)(x^2+\alpha^{14}x+1)(x^2+\alpha^3x+1)$$
$$\cdot (x^2+\alpha^{11}x+1)(x^2+\alpha^{13}x+1)(x^2+\alpha^7x+1)(x^2+\alpha^6x+1).$$

We have constructed the following codes of length 17:

code	generator polynomial $g(x)$
$C(17,15,3)$	$x^2+\alpha^6x+1$
$C(17,14,4)$	$(x-1)(x^2+\alpha^{12}x+1)$
$C(17,13,5)$	$(x^2+\alpha^7x+1)(x^2+\alpha^6x+1)$
$C(17,12,6)$	$(x-1)(x^2+\alpha^{12}x+1)(x^2+\alpha^9x+1)$
$C(17,11,7)$	$(x^2+\alpha^{13}x+1)(x^2+\alpha^7x+1)(x^2+\alpha^6x+1)$
$C(17,10,8)$	$(x-1)(x^2+\alpha^{12}x+1)(x^2+\alpha^9x+1)(x^2+\alpha^{14}x+1)$
$C(17,9,9)$	$(x^2+\alpha^{11}x+1)(x^2+\alpha^{13}x+1)(x^2+\alpha^7x+1)(x^2+\alpha^6x+1)$
$C(17,8,10)$	$(x-1)(x^2+\alpha^{12}x+1)(x^2+\alpha^9x+1)(x^2+\alpha^{14}x+1)(x^2+\alpha^3x+1)$
$C(17,7,11)$	$(x^2+\alpha^3x+1)(x^2+\alpha^{11}x+1)(x^2+\alpha^{13}x+1)(x^2+\alpha^7x+1)(x^2+\alpha^6x+1)$
$C(17,6,12)$	$(x-1)(x^2+\alpha^{12}x+1)(x^2+\alpha^9x+1)(x^2+\alpha^{14}x+1)(x^2+\alpha^3x+1)(x^2+\alpha^{11}x+1)$
$C(17,5,13)$	$(x^2+\alpha^{14}x+1)(x^2+\alpha^3x+1)(x^2+\alpha^{11}x+1)(x^2+\alpha^{13}x+1)(x^2+\alpha^7x+1)(x^2+\alpha^6x+1)$
$C(17,4,14)$	$(x-1)(x^2+\alpha^{12}x+1)(x^2+\alpha^9x+1)(x^2+\alpha^{14}x+1)(x^2+\alpha^3x+1)(x^2+\alpha^{11}x+1)$ $(x^2+\alpha^{13}x+1)$
$C(17,3,15)$	$(x^2+\alpha^9x+1)(x^2+\alpha^{14}x+1)(x^2+\alpha^3x+1)(x^2+\alpha^{11}x+1)(x^2+\alpha^{13}x+1)(x^2+\alpha^7x+1)$ $(x^2+\alpha^6x+1)$
$C(17,2,16)$	$(x-1)(x^2+\alpha^{12}x+1)(x^2+\alpha^9x+1)(x^2+\alpha^{14}x+1)(x^2+\alpha^3x+1)(x^2+\alpha^{11}x+1)$ $(x^2+\alpha^{13}x+1)(x^2+\alpha^7x+1)$

Solution of Problem 4.1

The length of the code is $n = 31$. The cyclotomic coset K_1 with respect to the number 31 is $K_1 = \{1,2,4,8,16\}$.

Because of the two successive numbers, the BCH code has the planned minimum distance $d = 3$. The generator polynomial $g(x)$ is (α from table)

$$\begin{aligned}
g(x) :&= \prod_{i \in K_1} (x - \alpha^i),\\
&= (x-\alpha)(x-\alpha^2)(x-\alpha^4)(x-\alpha^8)(x-\alpha^{16})\\
&= x^5 + (\alpha^{30}+\alpha^{17})x^2 + 1 = x^5 + x^2 + 1
\end{aligned}$$

Solution of Problem 4.2

A Hamming code with the following parameters exists for every $m \geq 2$: length $n = 2^m - 1$, dimension $k = 2^m - m - 1 = n - m$, minimum distance $d = 3$. The Hamming code is a perfect code, capable of correcting one-error.

A binary primitive one-error correcting BCH code, constructed using the cyclotomic coset K_1 with respect to the number $n = 2^m - 1$ has the following parameters: $n = 2^m - 1$, $k = 2^m - m - 1 = n - m$, $d = 3$.

The cyclotomic coset $K_1 = \{1 = 1 \cdot 2^0, 2 = 1 \cdot 2^1, \ldots, 1 \cdot 2^{m-1}\}$ always has dimension m, since $2^m = 1 \bmod 2^m - 1$. The designed minimum distance is $d = 3$, because always two successive numbers, i.e. 1 and 2 are contained in K_1. For $\delta \geq 3$, any two columns of the parity check matrix are independent. Furthermore, all columns are different and $\neq \mathbf{0}$. A parity check matrix with a maximum of m rows has a maximum of $2^m - 1$ columns. Therefore every code with the parameters $(2^m - 1, 2^m - m - 1, d \geq 3)$ is a Hamming code (including permutations). In conclusion, every binary Hamming code has a cyclic representation.

Solution of Problem 4.3

First we determine the cyclotomic cosets with respect to the number 31. They are obtained as follows:

$$
\begin{aligned}
K_0 &= \{0\}, & K_7 &= \{7, 14, 19, 25, 28\}, \\
K_1 &= \{1, 2, 4, 8, 16\}, & K_{11} &= \{11, 13, 21, 22, 26\}, \\
K_3 &= \{3, 6, 12, 17, 24\}, & K_{15} &= \{15, 23, 27, 29, 30\}, \\
K_5 &= \{5, 9, 10, 18, 20\}.
\end{aligned}
$$

Consequently, we obtain

$$
\begin{aligned}
m_0(x) &= (x - 1) \\
m_1(x) &= (x - \alpha^1)(x - \alpha^2)(x - \alpha^4)(x - \alpha^8)(x - \alpha^{16}), \\
m_3(x) &= (x - \alpha^3)(x - \alpha^6)(x - \alpha^{12})(x - \alpha^{17})(x - \alpha^{24}), \\
m_5(x) &= (x - \alpha^5)(x - \alpha^9)(x - \alpha^{10})(x - \alpha^{18})(x - \alpha^{20}), \\
m_7(x) &= (x - \alpha^7)(x - \alpha^{14})(x - \alpha^{19})(x - \alpha^{25})(x - \alpha^{28}), \\
m_{11}(x) &= (x - \alpha^{11})(x - \alpha^{13})(x - \alpha^{21})(x - \alpha^{22})(x - \alpha^{26}), \\
m_{15}(x) &= (x - \alpha^{15})(x - \alpha^{23})(x - \alpha^{27})(x - \alpha^{29})(x - \alpha^{30}) .
\end{aligned}
$$

Generator polynomials for $2, 4, 6, 10, 14$ successive zeros can be calculated. The dimensions are

$$
\begin{aligned}
k_1 &= 31 - 5 &&= 26, & k_5 &= 11, \\
k_2 &= 31 - 2 \cdot 5 &&= 21, & k_7 &= 6, \\
k_3 &= 31 - 3 \cdot 5 &&= 16 .
\end{aligned}
$$

$n = 31$ is a prime number; therefore the cyclotomic cosets (except K_0) all have the dimension $m = 5$. We have $\prod_{i=0}^{15} m_i = x^{31} - 1$.

Solution of Problem 4.4

$g(x)$ has *successive* zeros $\alpha, \alpha^2, \alpha^3, \alpha^4$.

Consequently, the following syndromes can be calculated from the received word $r(x)$:

$$
\begin{aligned}
S_0 &= r(\alpha) &&= \alpha^{10} + \alpha^8 + \alpha^6 + \alpha^2 + 1 &&= \alpha^7, \\
S_1 &= r(\alpha^2) &&= \alpha^5 + \alpha + \alpha^{12} + \alpha^4 + 1 &&= \alpha^{14}, \\
S_2 &= r(\alpha^3) &&= 1 + \alpha^9 + \alpha^3 + \alpha^6 + 1 &&= \alpha^{11}, \\
S_3 &= r(\alpha^4) &&= \alpha^{10} + \alpha^2 + \alpha^9 + \alpha^8 + 1 &&= \alpha.
\end{aligned}
$$

The calculation of the error location polynomial using the Berlekamp–Massey algorithm yields $C(x) = 1+\alpha^7 x+\alpha^9 x^2$.

$C(x)$ has zeros α^0 and α^6. The error coordinates are obtained from the inverse zeros, i.e. $\frac{1}{\alpha^0} = \alpha^0$ and $\frac{1}{\alpha^6} = \alpha^9$. Therefore the 0th and 9th coordinate in $r(x)$ are false, i.e. the error polynomial is $e(x) = 1+x^9$.

The transmitted codeword is therefore

$$c(x) = r(x)+e(x) = x^{10}+x^9+x^8+x^6+x^2 = x^2\, g(x).$$

Solution of Problem 5.1

The quadratic residues $\mathcal{M}_Q$ with respect to the number 31 are

$$\mathcal{M}_Q = \{1,4,9,16,25,5,18,2,19,7,28,20,14,10,8\}\,.$$

The approximation of the minimum distance at 31 ($= 4\cdot 8-1$) yields

$$d^2-d+1>31 \quad\Rightarrow\quad d>6.$$

The set $\mathcal{M}_Q$ has as longest sequence the numbers 7, 8, 9, 10. This means that the designed minimum distance is $d=5$. Consequently, one can correct up to two errors with the Berlekamp–Massey algorithm.

Solution of Problem 5.2

$$\begin{array}{c|c}
+ & +-+++---+-- \\
+ & -+-+++---+- \\
+ & --+-+++---+ \\
+ & +--+-+++--- \\
+ & -+--+-+++-- \\
+ & --+--+-+++- \\
+ & ---+--+-+++ \\
+ & +---+--+-++ \\
+ & ++---+--+-+ \\
+ & +++---+--+- \\
+ & -+++---+--+ \\
\hline
+ & ++++++++++++
\end{array}$$

The cross-correlation of two arbitrary Legendre sequences of the matrix has the value -1. If the column $+1$ is added on the left then this value is increased to 0; hence the rows become orthogonal. Since no row has a DC component, one can add as additional row $++\ldots+$ at the bottom.
This way, a Hadamard matrix of order 12 is obtained.

Solution of Problem 5.3

$\mathcal{R}(0,2) \rightarrow (4,1,4) \rightarrow$ repetition code, $\mathbf{c}_1 = (0000), \mathbf{c}_2 = (1111)$.
$\mathcal{R}(1,2) \rightarrow (4,3,2) \rightarrow$ PC code, $2^3 = 8$ codewords.

$\mathbf{b} = (0000)$	(0000)	(1111)
$\mathbf{b} = (0011)$	(0011)	(1100)
$\mathbf{b} = (1001)$	(1001)	(0110)
$\mathbf{b} = (1010)$	(1010)	(0101)

The coset leaders form a $(4,2,2)$ code.

Solution of Problem 5.4

$\mathbf{G}$ has systematic form, i.e. $\mathbf{G} = (\mathbf{I}|-\mathbf{A}^T)$; hence the parity check matrix $\mathbf{H}$ can be calculated:

$$\mathbf{H} = (\mathbf{A}|\mathbf{I}) = \begin{pmatrix} 1&1&1&1&1&0 \\ 4&3&1&2&0&1 \end{pmatrix}.$$

For the syndrome, it holds that $\mathbf{s} = \mathbf{r} \cdot \mathbf{H}^T = (4\,4) = 4\,(1\,1)$. The syndrome corresponds to the 3rd column of the parity check matrix multiplied by 4. Consequently, the 3rd coordinate is false; the error has the value 4.

Correction of 3rd coordinate: $2 - 4 = -2 = 3 \bmod 5$.

The transmitted codeword is $\mathbf{c} = (1\,4\,3\,1\,1\,4)$.

Solution of Problem 5.5

Regarding BCH codes, the maximum number of errors per codeword is $20 \Rightarrow d_{\min} = 41$. From table 4.1 $\mathcal{C}_1(127, 29, 43)$, $\mathcal{C}_2(255, 115, 43)$.

Regarding the RS code, $n = 31 = 2^5 - 1$; each symbol can be represented by 5 bits. Hence 4 symbols are false in the best case and 5 in the worst case. Consequently, it must be capable of correcting 5 false *symbols* $\Rightarrow d_{\min} = 11$.

It holds that $k = n - d + 1 = 31 - 11 + 1 = 21$; therefore $\mathcal{C}_3(31, 21, 11)$.

The code rates are

$$R_1 = \tfrac{29}{127} \approx 0.228\,,\; R_2 = \tfrac{115}{255} \approx 0.451\,,\; R_3 = \tfrac{21}{31} = \tfrac{21 \cdot 5}{31 \cdot 5} \approx 0.677\,.$$

$\Rightarrow$ The binary interpretated RS code has the highest code rate.

Solution of Problem 5.6

(a) Weight$= 8$ $\surd$, distribution of runs $\surd \Rightarrow$ PN sequence, generated with primitive polynomial $p(x) = x^4 + x + 1 \Rightarrow$ can be codeword of a simplex code.

(b) Weight$= 8$ $\surd$, sequence contains a 5 run of ones $\Rightarrow$ no PN sequence $\Rightarrow$ cannot be a codeword of a simplex code.

(c) Weight$= 9$ $\Rightarrow$ cannot be a codeword of a simplex code.

Solution of Problem 5.7

The code parameters of $\mathcal{R}(5,8)$ can be calculated from the parameters of the codes $\mathcal{R}(4,7) = \mathcal{C}_1(128, 99, 8)$ and $\mathcal{R}(5,7) = \mathcal{C}_2(128, 120, 4)$: $k_{(5,8)} = k_{(4,7)} + k_{(5,7)} = 99 + 120 = 219$.

The code $\mathcal{R}(5,8)$ has $2^{219} \approx 8.425 \cdot 10^{65}$ codewords.

Solution of Problem 5.8

The most likely transmitted row can be determined by correlation of each row with the received vector:

$$\mathbf{H}_8 \cdot \mathbf{h}^T = \begin{pmatrix} 2 & 2 & -2 & 6 & -2 & -2 & 2 & 2 \end{pmatrix}^T$$

$\Rightarrow$ the maximum of the correlation is at the 4th coordinate according to the 4th row of $\mathbf{H}_8$

$\Rightarrow$ the 4th row was probably transmitted.

Solution of Problem 6.1

Let the number of branches e that end in a node ϑ be $m(\vartheta)$. To calculate the metric of a path that runs via a branch e to ϑ ($\vartheta' \xrightarrow{e} \vartheta$), one has to add the metric of the previous node ϑ' and the metric of the branch e. The number of additions is therefore $|\mathcal{E}|$.

To find the best out of $m(\vartheta)$ paths that lead to a node ϑ, $m(\vartheta) - 1$ binary comparisons have to be performed. The total amount of comparisons is consequently

$$\sum_{\vartheta \in \mathcal{V} \setminus \vartheta_A} m(\vartheta) - 1 = \sum_{\vartheta \in \mathcal{V} \setminus \vartheta_A} m(\vartheta) - \sum_{\vartheta \in \mathcal{V} \setminus \vartheta_A} 1 = |\mathcal{E}| - (|\mathcal{V}| - 1).$$

Solution of Problem 6.2

(a) The $(7,4,3)$ Hamming code will be used. A possible parity check matrix is

$$H = \begin{pmatrix} 0 & 0 & 0 & 1 & 1 & 1 & 1 \\ 0 & 1 & 1 & 0 & 0 & 1 & 1 \\ 1 & 0 & 1 & 0 & 1 & 0 & 1 \end{pmatrix}.$$

First step: construction of a trellis without minimization.

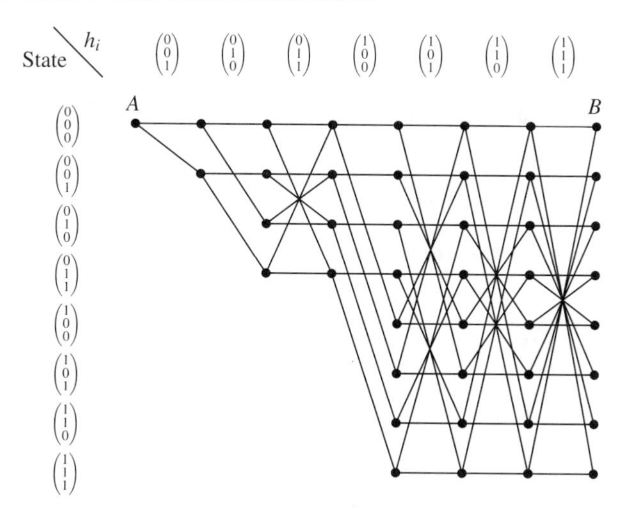

If the paths that do not lead to node B are removed then the following syndrome trellis is obtained:

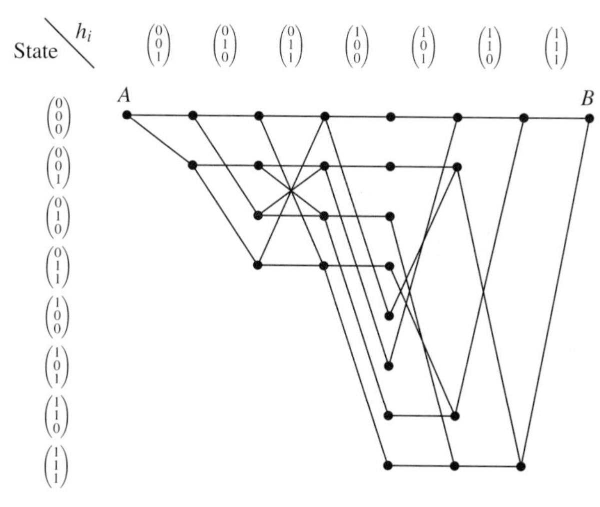

(b) A possible generator matrix of the $(7,4,3)$ Hamming code is

$$\mathbf{G} = \begin{pmatrix} 1 & 0 & 0 & 0 & 0 & 1 & 1 \\ 0 & 1 & 0 & 0 & 1 & 0 & 1 \\ 0 & 0 & 1 & 0 & 1 & 1 & 0 \\ 0 & 0 & 0 & 1 & 1 & 1 & 1 \end{pmatrix} = \begin{pmatrix} \mathbf{g}_1 \\ \mathbf{g}_2 \\ \mathbf{g}_3 \\ \mathbf{g}_4 \end{pmatrix}.$$

A minimal trellis is obtained if $\mathbf{G}$ has the LR property, i.e. $\mathbf{G}$ is first brought to LR format.

1. $\mathbf{g}_1 := \mathbf{g}_1 + \mathbf{g}_4$:
$$\mathbf{G}' = \begin{pmatrix} 1 & 0 & 0 & 1 & 1 & 0 & 0 \\ 0 & 1 & 0 & 0 & 1 & 0 & 1 \\ 0 & 0 & 1 & 0 & 1 & 1 & 0 \\ 0 & 0 & 0 & 1 & 1 & 1 & 1 \end{pmatrix};$$

2. $\mathbf{g}_3 := \mathbf{g}_2 + \mathbf{g}_3 + \mathbf{g}_4$:
$$\mathbf{G}'' = \begin{pmatrix} 1 & 0 & 0 & 1 & 1 & 0 & 0 \\ 0 & 1 & 1 & 1 & 1 & 0 & 0 \\ 0 & 0 & 1 & 0 & 1 & 1 & 0 \\ 0 & 0 & 0 & 1 & 1 & 1 & 1 \end{pmatrix};$$

3. $\mathbf{g}_1 := \mathbf{g}_1 + \mathbf{g}_2$:
$$\mathbf{G}_{LR} = \begin{pmatrix} 1 & 1 & 1 & 0 & 0 & 0 & 0 \\ 0 & 1 & 1 & 1 & 1 & 0 & 0 \\ 0 & 0 & 1 & 0 & 1 & 1 & 0 \\ 0 & 0 & 0 & 1 & 1 & 1 & 1 \end{pmatrix} = \begin{pmatrix} \mathbf{g}_1 \\ \mathbf{g}_2 \\ \mathbf{g}_3 \\ \mathbf{g}_4 \end{pmatrix}.$$

The following diagrams show how the minimal trellis $T(\mathcal{C})$ can be determined by applying the Shannon product to the basic trellises $T(\mathbf{g}_i)$,

$$T(\mathcal{C}) = T(\mathbf{g}_1) * T(\mathbf{g}_2) * T(\mathbf{g}_3) * T(\mathbf{g}_4)$$

(solid lines are labeled with 0, dashed lines with 1):

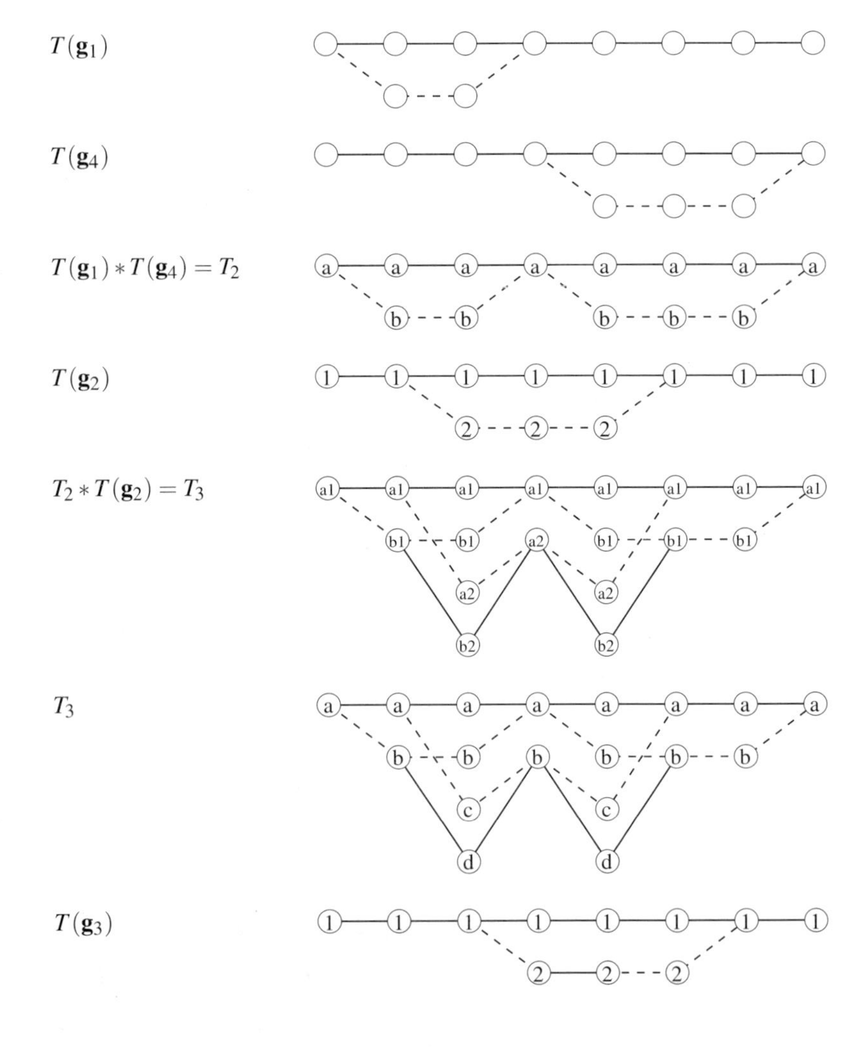

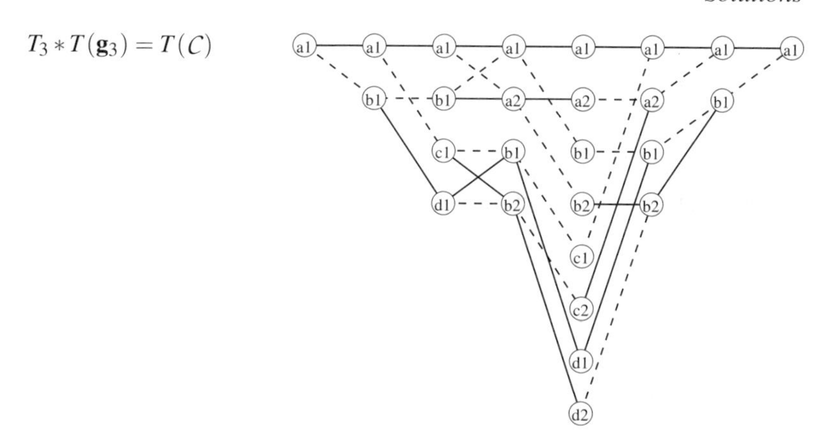

Solution of Problem 6.3
Let us consider the $(5,3,2)$ code with the parity check matrix

$$\mathbf{H} = \begin{pmatrix} 1 & 1 & 1 & 0 & 0 \\ 0 & 0 & 1 & 1 & 1 \end{pmatrix}$$

and the corresponding syndrome trellis

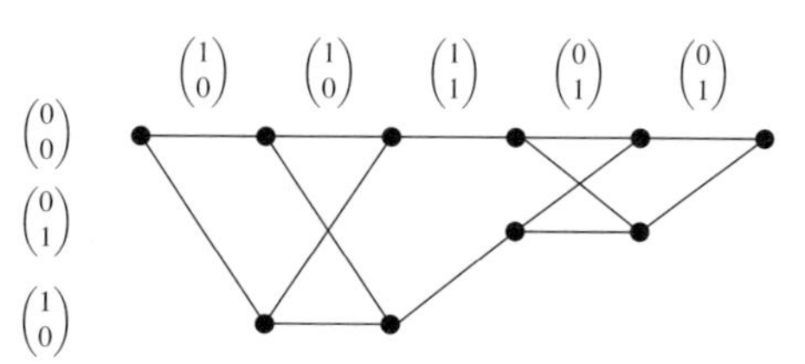

One can see that $|\mathcal{V}_t| \leq 2 < 2^{\min\{k,n-k\}} = 2^{\min\{3,2\}} = 4 \; \forall t$.

Solution of Problem 6.4
All minimal trellises of a code C are isomorphic and the syndrome trellis of C is minimal. Therefore $|\mathcal{V}|$ and $|\mathcal{V}^{\perp}|$ can be calculated using syndrome trellises. Let $\mathbf{G}$ and $\mathbf{H}$ be the generator matrix and the parity check matrix of a code C respectively. For a given t, the partition $\mathbf{G} = (\mathbf{G}_t^A | \mathbf{G}_t^B)$ and $\mathbf{H} = (\mathbf{H}_t^A | \mathbf{H}_t^B)$ is assumed. Every arbitrary codeword $\mathbf{c} = (c_1, \ldots, c_n)$ can be represented as $\mathbf{c} = \mathbf{u} \cdot \mathbf{G}$. At depth t, the syndrome trellis uses the following states:

$$\sigma_t(\mathbf{c}) = (c_1, \ldots, c_t) \cdot (\mathbf{H}_t^A)^T = \mathbf{u} \cdot \mathbf{G}_t^A \cdot (\mathbf{H}_t^A)^T.$$

This means that $|\mathcal{V}_t| = q^{\mathrm{rank}(\mathbf{G}_t^A (\mathbf{H}_t^A)^T)}$.

The parity check matrix and the generator matrix exchange roles for the dual code and therefore $|\mathcal{V}_t^{\perp}| = q^{\mathrm{rank}(\mathbf{H}_t^A \cdot (\mathbf{G}_t^A)^T)}$.

Since for arbitrary matrices $\mathbf{A}$ and $\mathbf{B}$, $(\mathbf{A} \cdot \mathbf{B})^T = \mathbf{B}^T \cdot \mathbf{A}^T$ and $\mathrm{rank}(\mathbf{A}) = \mathrm{rank}(\mathbf{A}^T)$, it follows that they are equal: $|\mathcal{V}_t^{\perp}| = |\mathcal{V}_t|$.

Solution of Problem 6.5
First a non-minimized syndrome trellis is constructed for the given (n,k) code. This trellis has an initial node ϑ_A and q^{n-k} final nodes, which are referred to by the $(n-k)$-tuples $\mathbf{s}_0 = 0, \mathbf{s}_1, \ldots, \mathbf{s}_{q^{n-k}-1}$.

A path $\mathbf{w}$ from ϑ_A to a final node $\mathbf{s}_i$ satisfies the equation

$$\mathbf{w} \cdot \mathbf{H}^T = \mathbf{s}_i.$$

Consequently, there is an unique correspondence between the coset that belongs to the syndrome $\mathbf{s}_i$, and paths from ϑ_A to the final nodes $\mathbf{s}_i$ in the trellis. If the Viterbi algorithm is applied to this trellis then the best path from ϑ_A to $\mathbf{s}_i$ $(i = 0, \ldots, q^{n-k} - 1)$ is obtained. This corresponds to a ML decoding of the coset $\mathbf{s}_i$.

As an example, consider the $(7,4,3)$ Hamming code. The complexity $\chi = 2|\mathcal{E}| - |\mathcal{V}| + 1$ of Viterbi decoding when using the minimal code trellis is $\chi = 47$ $(|\mathcal{V}| = 26, |\mathcal{E}| = 36)$. The complexity of decoding 8 cosets of the code is $\chi = 98$ $(|\mathcal{E}| = 70, |\mathcal{V}| = 43)$; thus only twice the number of computations are necessary to decode 8 cosets.

Solution of Problem 6.6

$$\mathbf{G} = \begin{pmatrix} 1&1&1&1&1&1&1&1 \\ 0&0&0&0&1&1&1&1 \\ 0&0&1&1&0&0&1&1 \\ 0&1&0&1&0&1&0&1 \end{pmatrix} \Longleftrightarrow \mathbf{G}_{LR} = \begin{pmatrix} 1&1&1&1&0&0&0&0 \\ 0&0&1&1&1&1&0&0 \\ 0&0&0&0&1&1&1&1 \\ 0&1&0&1&1&0&1&0 \end{pmatrix}.$$

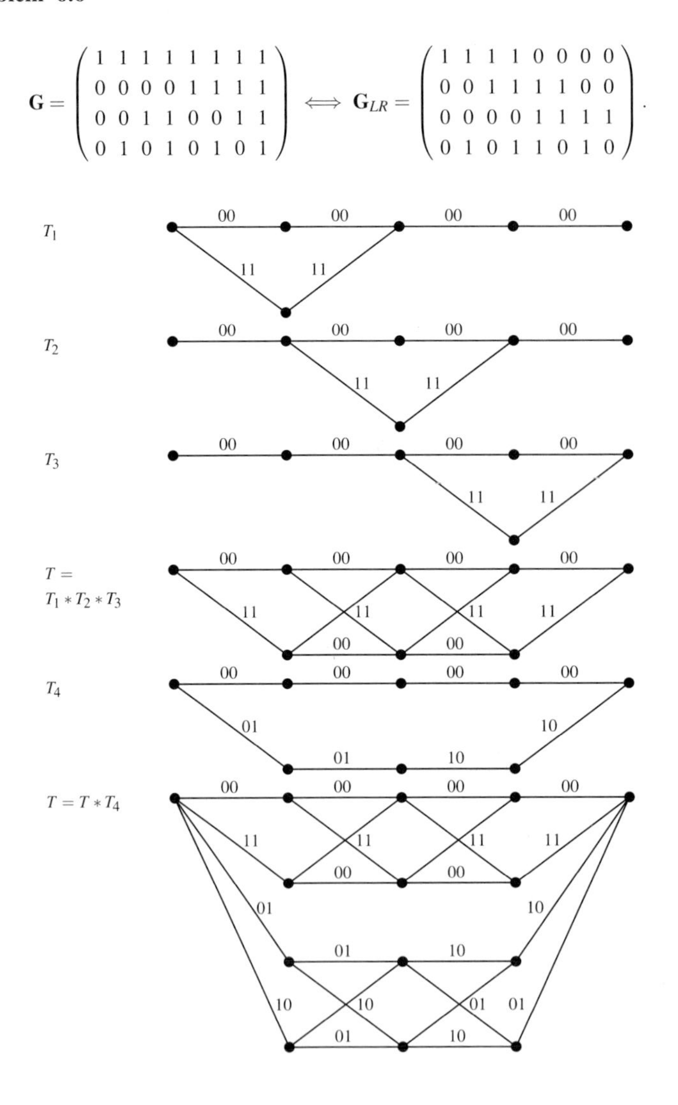

Solution of Problem 7.1

(a) The number of vectors of a coset corresponds to the number of codewords and the number of cosets corresponds to the number of (not necessarily unique) correctable error vectors. In our example the standard array is

$$\begin{array}{ll} \{(0000),(1111)\}, & \{(1000),(0111)\}, \\ \{(0001),(1110)\}, & \{(0011),(1100)\}, \\ \{(0010),(1101)\}, & \{(0110),(1001)\}, \\ \{(0100),(1011)\}, & \{(1010),(0101)\}. \end{array}$$

The array contains all 16 vectors of the field $\mathbb{F}_2^4$.

(b) The array must contain all vectors of $\mathbb{F}_2^7$, i.e. $2^7 = 128$ vectors. The number of columns corresponds to the number of codewords of the Hamming code, i.e. $k = 4 \Rightarrow 2^4 = 16$, and the number of cosets must therefore be $2^3 = 8$.

Because the Hamming code is a perfect one-error correcting code, the coset leaders must all be unique. The set of coset leaders corresponds to the set of all vectors with weight 1 and the zero vector:

$$\mathcal{M}_c := \left\{ \mathbf{a} \in \mathbb{F}_2^7 \mid \mathrm{wt}(\mathbf{a}) = 1,\ 0 \right\} .$$

Solution of Problem 7.2

(a) We need a set of permutations ϕ_i such that an arbitrary error location is mapped by at least one permutation to the redundancy portion. A possible four permutations are the following:

	redundancy				information										
	0	1	2	3	4	5	6	7	8	9	10	11	12	13	14
ϕ_1	r_0	r_1	r_2	r_3	r_4	r_5	r_6	r_7	r_8	r_9	r_{10}	r_{11}	r_{12}	r_{13}	r_{14}
ϕ_2	r_{11}	r_{12}	r_{13}	r_{14}	r_0	r_1	r_2	r_3	r_4	r_5	r_6	r_7	r_8	r_9	r_{10}
ϕ_3	r_7	r_8	r_9	r_{10}	r_{11}	r_{12}	r_{13}	r_{14}	r_0	r_1	r_2	r_3	r_4	r_5	r_6
ϕ_4	r_3	r_4	r_5	r_6	r_7	r_8	r_9	r_{10}	r_{11}	r_{12}	r_{13}	r_{14}	r_0	r_1	r_2

(b) First we need the parity check matrix of the code. Therefore we calculate the parity check polynomial $h(x)$ by dividing $x^n - 1$ by the generator polynomial $g(x)$:

$$x^{15} - 1 : x^4 + x + 1 = x^{11} + x^8 + x^7 + x^5 + x^3 + x^2 + x + 1 .$$

Using this parity check polynomial, we calculate the parity check matrix $\mathbf{H}$:

$$\begin{pmatrix} 1&0&0&0&1&0&0&1&1&0&1&0&1&1&1 \\ 0&1&0&0&1&1&0&1&0&1&1&1&1&0&0 \\ 0&0&1&0&0&1&1&0&1&0&1&1&1&1&0 \\ 0&0&0&1&0&0&1&1&0&1&0&1&1&1&1 \end{pmatrix}$$

Thereby rows 2–4 are the cyclically shifted parity check polynomial, and the first row is the cyclically shifted parity check polynomial plus the 4th row.

According to the permutation decoding algorithm in section 7.3.1, we multiply the received vector $\mathbf{r}$, after permutating it with the permutations ϕ_1 to ϕ_4, by the parity check matrix $\mathbf{H}$, and check whether or not the obtained syndrome has weight ≤ 1:

$$\begin{aligned} \mathbf{H} \cdot \phi_1(\mathbf{r})^T &= \begin{pmatrix} 1&0&0&0&1&0&0&1&1&0&1&0&1&1&1 \\ 0&1&0&0&1&1&0&1&0&1&1&1&1&0&0 \\ 0&0&1&0&0&1&1&0&1&0&1&1&1&1&0 \\ 0&0&0&1&0&0&1&1&0&1&0&1&1&1&1 \end{pmatrix} \cdot (1\ 1\ 0\ 0\ 0\ 0\ 0\ 1\ 0\ 1\ 0\ 0\ 1\ 0\ 0)^T \\ &= (1\ 0\ 1\ 1)^T . \end{aligned}$$

This syndrome has weight $3 > 1$. We try the next permutation:

$$\mathbf{H} \cdot \phi_2(\mathbf{r})^T = (0\ 0\ 1\ 0)^T .$$

This syndrome has weight 1, i.e. we have found the error. We now decode $\mathbf{r}$ in two steps:

1. $\mathbf{x}_1 = \phi_2(\mathbf{r}) - (S_0, \ldots, S_{n-k-1}, 0, \ldots, 0) = (0\,1\,1\,0\,1\,1\,0\,0\,0\,0\,0\,1\,0\,1\,0)$;
2. $\mathbf{x}_2 = \phi_2^{-1}(\mathbf{x}_1) = (1\,1\,0\,0\,0\,0\,0\,1\,0\,1\,0\,0\,1\,1\,0) = \hat{\mathbf{r}}$.

Solution of Problem 7.3

For the code $\mathcal{G}_{23}^{\perp}$, which is the dual code to the Golay code $\mathcal{G}_{23}$,

$$n = 23, \quad k^{\perp} = 23 - 12 = 11, \quad d^{\perp} = d + 1 = 8 \,.$$

At most, $\left\lfloor \frac{n-1}{d^{\perp}-1} \right\rfloor = \left\lfloor \frac{22}{7} \right\rfloor = 3$ parity check vectors can exist. Consequently, only one error can be decoded.

Solution of Problem 7.4

We have the following code parameters: $n = 15$, $k = 11$, $d = 5$.

(a) With each erasure, the order of the syndrome is decreased by one. The order of the syndrome is $4 - 2 = 2$, and therefore one error or two errasures can be corrected, according to $2e + t \leq 4 - 2$.

(b) We reduce the code by 7 symbols and obtain a code C^* with parameters $n^* = 8$, $k^* = 4$, $d^* = 5$. Because RS codes have the MDS property (*maximum distance separable*), arbitrary k^* coordinates determine the codeword $a^* \in C^*$ uniquely. We have 8 coordinates, which we sort into groups with 2 coordinates as follows:

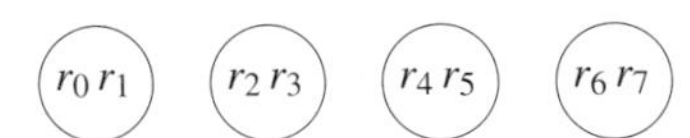

Two errors can corrupt at most two groups, i.e. at least two groups (i.e. 4 coordinates) are correct. Consequently, we can calculate the codeword from all $\binom{4}{2} = 6$ possible group combinations. The codeword that is calculated from two correct groups has distance $d_i \leq 2$ to the received word $r(x) = a(x) + f(x)$.

For a codeword $a(x)$,

$$a(x) = i(x) \cdot g(x)\,,\ \deg(g(x)) = 4\,,\ \deg(i(x)) < 4$$

$$\begin{aligned}
a_0 &= g_0 i_0, & a_4 &= g_1 i_3 + g_2 i_2 + g_3 i_1 + g_4 i_0,\\
a_1 &= g_0 i_1 + g_1 i_0, & a_5 &= g_2 i_3 + g_3 i_2 + g_4 i_1,\\
a_2 &= g_0 i_2 + g_1 i_1 + g_2 i_0, & a_6 &= g_3 i_3 + g_4 i_2,\\
a_3 &= g_0 i_3 + g_1 i_2 + g_2 i_1 + g_3 i_0, & a_7 &= g_4 i_3.
\end{aligned}$$

We choose, according to the groups, 4 equations, from which we can calculate the 4 unknowns i_j.

Solution of Problem 7.5

To perform step 1 of the decoding algorithm, we need to know the syndrome weight X. It is calculated as

$$X = \mathrm{WT}(\mathcal{B}, r): \begin{pmatrix}
0&0&0&0&0&0&0&1&0&0&0&1&0&1&1\\
0&0&0&0&0&0&1&0&0&0&1&0&1&1&0\\
0&0&0&0&0&1&0&0&0&1&0&1&1&0&0\\
0&0&0&0&1&0&0&0&1&0&1&1&0&0&0\\
0&0&0&1&0&0&0&1&0&1&1&0&0&0&0\\
0&0&1&0&0&0&1&0&1&1&0&0&0&0&0\\
0&1&0&0&0&1&0&1&1&0&0&0&0&0&0\\
1&0&0&0&1&0&1&1&0&0&0&0&0&0&0\\
0&0&0&1&0&1&1&0&0&0&0&0&0&0&1\\
0&0&1&0&1&1&0&0&0&0&0&0&0&1&0\\
0&1&0&1&1&0&0&0&0&0&0&0&1&0&0\\
1&0&1&1&0&0&0&0&0&0&0&1&0&0&0\\
0&1&1&0&0&0&0&0&0&0&1&0&0&0&1\\
1&1&0&0&0&0&0&0&0&1&0&0&0&1&0\\
1&0&0&0&0&0&0&0&1&0&0&0&1&0&1
\end{pmatrix} \cdot \begin{pmatrix}0\\0\\0\\0\\0\\1\\0\\0\\0\\0\\1\\0\\0\\0\\0\end{pmatrix} = \begin{pmatrix}0\\1\\1\\1\\1\\0\\1\\0\\1\\1\\0\\0\\1\\0\\0\end{pmatrix}.$$

This means $X = 8$.

Now we calculate the ε_i: $\varepsilon = (12, 8, 8, 8, 8, 4, 8, 8, 8, 8, 4, 8, 8, 8, 8)$.

We can now choose between correcting the 6th or the 11th coordinate. When choosing the 6th, a new ε is obtained: $\varepsilon = (8, 6, 6, 6, 6, 8, 6, 6, 6, 6, 0, 6, 6, 6, 6)$. The 11th coordinate is therefore in error. One also notes that, after correcting this coordinate, the syndrome has weight 0, and therefore the corrected vector represents a codeword.

Solution of Problem 8.1

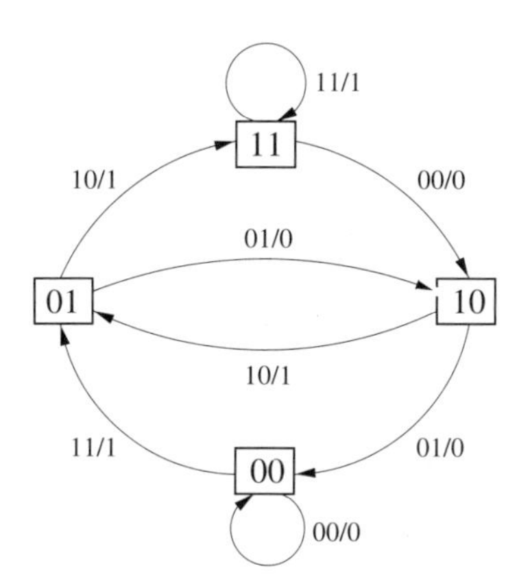

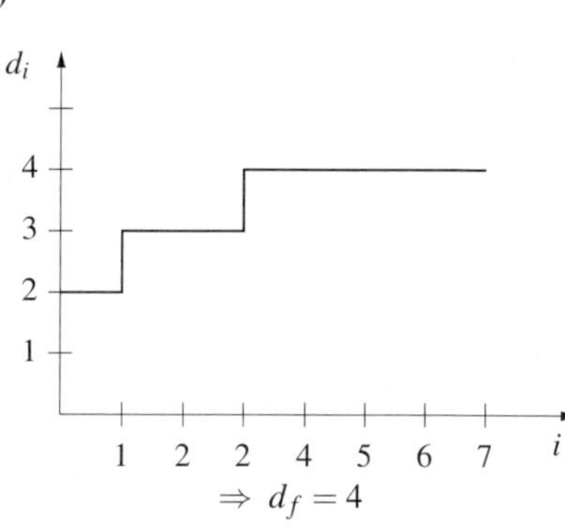

(c)

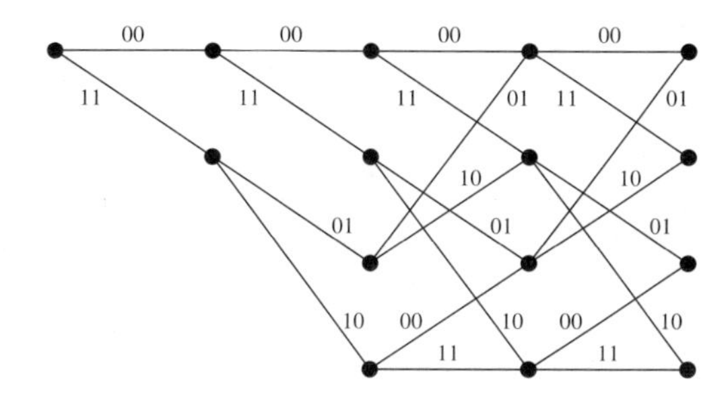

(d)

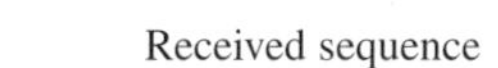

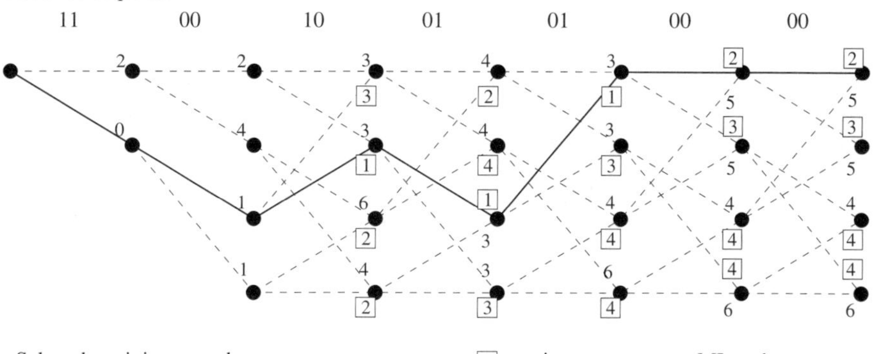

Solution of Problem 8.2

Received sequence: 10 10 11 11 ...; $p = 0.1$; $R = \frac{1}{2}$.

$$\text{Metric for } p = 0.1: \quad M(r_i|v_i) = \begin{cases} \log_2 2p - R & \text{for } r_i \neq v_i, \\ \log_2 2(1-p) - R & \text{for } r_i = v_i. \end{cases}$$

v_i/r_i	0	1
0	0.35	−2.8
1	−2.8	0.35

scaled $\Longrightarrow$

v_i/r_i	0	1
0	1	−8
1	−8	1

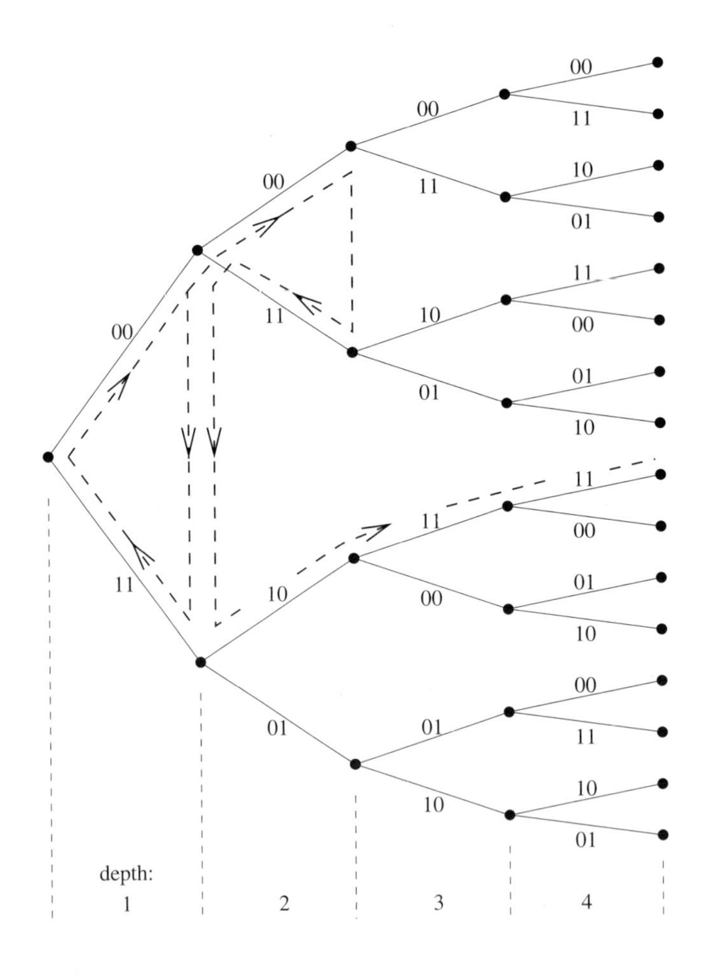

move	depth q	Λ_q	Λ_{q-1}	sequence	T	rule
—	0	0	$-\infty$	—	0	$1 \to$ V *
V	1	−7	0	00	0	$3 \to$ S
S	1	−7	0	11	0	$3 \to$ R
R	0	0	$-\infty$	—	−6	$4 \to$ V
V	1	−7	0	00	−6	$3 \to$ S
S	1	−7	0	11	−6	$3 \to$ R
R	0	0	$-\infty$	—	−12	$4 \to$ V
V	1	−7	0	00	−12	$2 \to$ V
V	2	−14	−7	00 00	−12	$3 \to$ S
S	2	−14	−7	00 11	−12	$3 \to$ R
R	1	−7	0	00	−12	$5 \to$ S
S	1	−7	0	11	−12	$2 \to$ V
V	2	−5	−7	11 10	−6	$1 \to$ V
V	3	−3	−5	11 10 11	−6	$1 \to$ V *
V	4	−1	−3	11 10 11 11	−6	$1 \to$ V *

* Change of threshold not possible, condition not satisfied

References

[AAS] J. B. Anderson, T. Aulin, C.-E. Sundberg: *Digital Phase Modulation.* Plenum, New York, 1986.

[AEZ85] L. Ahlin, T. Ericson, V. V. Zyablov: Performance of concatenated codes in a fading channel. Rep. LiTH-ISY-I0736, Linköping University, Sweden, 1985.

[AM63] E. F. Assmus, Jr., H. F. Mattson, Jr.: Error-correcting codes: An axiomatic approach. *Inf. and Control*, vol. 6, pp. 315–330, 1963.

[Art] E. Artin: *Geometric Algebra.* Wiley, New York, 1957.

[AS81a] T. Aulin, C.-E. Sundberg: Continuous phase modulation—part I: Full response signaling. *IEEE Trans. Commun.*, vol. COM-29, pp. 196–209, Mar. 1981.

[AS81b] T. Aulin, C.-E. Sundberg: Continuous phase modulation—part II: Partial response signaling. *IEEE Trans. Commun.*, vol. COM-29, pp. 210–225, Mar. 1981.

[BaCo] N. Balakrishnan, A. C. Cohen: *Order Statistics and Interference.* Academic Press, San Diego, 1991.

[BBLK98] M. Breitbach, M. Bossert, R. Lucas, C. Kempter: Soft-decison decoding of linear block codes as optimization problem. *Eur. Trans. Telecommun.*, vol. 9, no. 3, pp. 289–293, May/Jun. 1998.

[BBZS98] M. Breitbach, M. Bossert, V. V. Zyablov, V. R. Sidorenko: Array codes correcting a two-dimensional cluster of errors. *IEEE Trans. Inf. Theory*, vol. IT-44, pp. 2025–2031, Sep. 1998.

[BBZS99] M. Breitbach, M. Bossert, V. V. Zyablov, V. Sidorenko: Array codes correcting likely burst error patterns. *Eur. Trans. Telecommun.*, vol. 10, no. 1, Jan./Feb. 1999.

[BCJR74] L. R. Bahl, J. Cocke, F. Jelinek, J. Raviv: Optimal decoding of linear codes for minimizing symbol error rate. *IEEE Trans. Inf. Theory*, vol. IT-20, pp. 284–287, Mar. 1974.

[BDG79] Battail, M. C. Decouvelaere, P. Godlewski: Replication decoding. *IEEE Trans. Inf. Theory*, vol. 25, pp. 332–345, May 1979.

[BDS96a] M. Bossert, H. Dieterich, S. A. Shavgulidze: Generalized concatenation of convolutional codes. *Eur. Trans. Telecommun.*, vol. 7, no. 6, pp. 483–492, Nov./Dec. 1996.

[BDS96b] M. Bossert, H. Dieterich, S. A. Shavgulidze: Partitioning of convolutional codes using a convolutional scrambler. *Electron. Lett.*, vol. 32, no. 19, pp. 1758–1760, Sep. 1996.

[BDS97] M. Bossert, H. Dieterich, S. A. Shavgulidze: Methods for the partitioning of convolutional codes. Tech. Rep. ITUU-TR-1997/05, Univ. of Ulm, Dept. of Information Technology, Jun. 1997.

[BDS98] M. Bossert, H. Dieterich, S. A. Shavgulidze: Some general methods for the partitioning of convolutional codes. In *2. ITG Fachtagung, Codierung für Quelle Kanal und Übertragung*, pp. 219–224, Aachen, Germany, Mar. 1998.

[Ber] E. R. Berlekamp: *Algebraic Coding Theory*. Aegean Park Press, Laguna Hills, CA., 1984.

[BGT93] C. Berrou, A. Glavieux, P. Thitimajshima: Near Shannon limit error-correcting coding and decoding: Turbo-Codes (1). In *Proc. IEEE Int. Conf. on Communications '93*, pp. 1064–1070, Geneva, Switzerland, May 1993.

[BH86] M. Bossert, F. Hergert: Hard- and soft-decision decoding beyond the half minimum distance—An algorithm for linear codes. *IEEE Trans. Inf. Theory*, vol. IT-32, pp. 709–714, Sep. 1986.

[BHSD99] M. Bossert, A. Häutle, S. A. Shavgulidze, H. Dieterich: Generalized concatenation of encoded CPFSK modulation. To appear in *Eur. Trans. Telecommun.*

[BJ74] E. R. Berlekamp, J. Justesen: Some long cyclic linear binary codes are not so bad. *IEEE Trans. Inf. Theory*, vol. IT-20, pp. 351–356, May 1974.

[BKS92] M. Bossert, P. Klund, G. Schnabl: Coded modulation with generalized multiple concatenation on fading channels. In *Proc. of the 5th Tirrenia Int. Workshop on Digital Communications: Coded Modulation and Bandwidth-Efficient Transmission*, pp. 295–303, Elsevier, Amsterdam, 1992.

[Bla] R. E. Blahut: *Theory and Practice of Error Control Codes*. Addison-Wesley, Reading, MA, 1983.

[BM96] S. Benedetto, G. Montorsi: Unveiling turbo codes: Some results on parallel concatenated coding schemes. *IEEE Trans. Inf. Theory*, vol. IT-42, pp. 409–428, 1996.

[BM98] S. Benedetto, G. Montorsi: Serial concatenation of interleaved codes: Performance analysis, design, and iterative decoding. *IEEE Trans. Inf. Theory*, vol. IT-44, pp. 909–926, 1998.

[Bos80] A. Bos: Codes over groups with arbitrary metrics. T.H.-Rep. 80-WSK-06, Univ. of Techn., Eindhoven, Netherlands, 1980.

[Boss87] M. Bossert: On decoding binary quadratic residue codes. In *Proc. of AAECC-5*, vol. 356 of *Lecture Notes in Computer Science*, pp. 60–68, Springer-Verlag, Berlin, Heidelberg, 1987.

[BRC60a] R. C. Bose, D. K. Ray-Chaudhuri: On a class of error correcting binary group codes. *Inf. and Control*, vol. 3, pp. 68–79, 1960.

[BRC60b] R. C. Bose, D. K. Ray-Chaudhuri: Further results on error correcting binary group codes. *Inf. and Control*, vol. 3, pp. 279–290, 1960.

[Bre97] M. Breitbach: *On the Correction of Two-Dimensional Clusters of Errors*. No. 489 in Reihe 10, VDI Verlag Fortschrittberichte, 1997, Dissertation, Univ. of Ulm.

[BS86] Y. Be'ery, J. Snyders: Optimal soft-decision block decoders based on fast Hadamard transform. *IEEE Trans. Inf. Theory*, vol. IT-32, pp. 355–364, 1986.

[BS90] M. Bossert, G. Schnabl: Multiple concatenated codes in the euclidean space. In *Proc. Int. Workshop on Algebraic and Combinatorial Coding Theory*, Leningrad, USSR, 1990.

[BS96] M. Bossert, V. R. Sidorenko: Singleton-type bounds for blot-correcting codes. *IEEE Trans. Inf. Theory*, vol. IT-42, no. 3, pp. 1021–1023, May 1996.

[BSHD98] M. Bossert, S. Shavgulidze, A. Häutle, H. Dieterich: Generalized concatenation of encoded tamed frequency modulation. *IEEE Trans. Commun.*, vol. 46, no. 10, pp. 1337–1345, Oct. 1998.

[BSS90] E. Belitskaya, V. R. Sidorenko, P. Stenström: Testing of memory with defects of fixed configuration. In *Proc. 2nd Int. Workshop on Algebraic and Combinatorial Coding Theory*, pp. 24–28, Leningrad, USSR, Sep. 1990.

[BSU88] V. L. Banket, A. V. Salabai, N. A. Ugrelidze: Convolutional codes for channels with coherent CPFSK. *Proc. Moscow Radio Inst.*, pp. 10–17, Apr./Jun. 1988.

[BU88] V. L. Banket, N. A. Ugrelidze: Convolutional codes for channels with minimum shift keying. *Radiotechnika*, pp. 11–15, Jan. 1988.

[BZ74] E. L. Blokh, V. V. Zyablov: Coding of generalized concatenated codes. *Problemy Peredachi Informatsii*, vol. 10, no. 3, pp. 45–50, 1974.

[BZ95] M. Bossert, V. V. Zyablov: When and why erasures are good for Gaussian channel. In *Proc. 7th Joint Swedish–Russian Int. Workshop on Information Theory*, pp. 45–48, St. Petersburg, Russia, Jun. 1995.

[Cal89] A. R. Calderbank: Multilevel codes and multistage decoding. *IEEE Trans. Inf. Theory*, vol. IT-37, pp. 222–229, Mar. 1989.

[Cha72] D. Chase: A class of algorithms for decoding block codes with channel measurement information. *IEEE Trans. Inf. Theory*, vol. IT-18, pp. 170–182, 1972.

[Che96] J.-F. Cheng: On the decoding of certain generalized concatenated convolutional codes. In *IEEE Int. Commun. Conf.*, Dallas, TX, Jun. 1996.

[ClCa] G. C. Clark, J. B. Cain: *Error-Correction Coding for Digital Communications*. Plenum, New York, 1988.

[CMKH96] A. R. Calderbank, G. McGuire, P. V. Kumar, T. Helleseth: Cyclic codes over $\mathbb{Z}_4$, locator polynomials and Newton's identities. *IEEE Trans. Inf. Theory*, vol. IT-42, no. 1, pp. 217–226, Jan. 1996.

[Cos69] D. J. Costello, Jr.: A construction technique for random-error-correcting convolutional codes. *IEEE Trans. Inf. Theory*, vol. IT-19, pp. 631–636, 1969.

[CoSl] J. H. Conway, N. J. A. Sloane: *Sphere Packings, Lattices and Groups*. Springer-Verlag, Berlin, 1998.

[Cus84] E. L. Cusack: Error control codes for QAM signalling. *Electron. Lett.*, vol. 20, pp. 62–63, 1984.

[CW80] S. C. Chang, J. K. Wolf: A simple derivation of the MacWilliams identitiy for linear codes. *IEEE Trans. Inf. Theory*, vol. IT-26, 1980.

[DC87] R. H. Deng, D. J. Costello, Jr.: Reliability and throughput analysis on a concatenated coding scheme. *IEEE Trans. Commun.*, vol. COM-35, pp. 698–705, 1987.

[DN92] S. M. Dodunekov, J. E. M. Nilsson: Algebraic decoding of the Zetterberg codes. *IEEE Trans. Inf. Theory*, vol. 38, no. 5, pp. 1570–1573, Sep. 1992.

[Dor74] B. G. Dorsch: A decoding algorithm for binary block codes and *J*-ary output channels. *IEEE Trans. Inf. Theory*, vol. IT-20, pp. 391–394, May 1974.

[Dorn87] J. L. Dornstetter: On the equivalence between Berlekamp's and Euklid's algorithms. *IEEE Trans. Inf. Theory*, vol. IT-33, no. 3, pp. 428–431, May 1987.

[DRS93] U. Dettmar, R. Raschofer, U. Sorger: On the trellis complexity of block and convolutional codes. *Prob. Inf. Transmission*, vol. 32, no. 2, pp. 10–21, Jul.–Sep. 1993, translated from *Problemy Peredachi Informatsii*.

[DS92] U. Dettmar, S. A. Shavgulidze: New optimal partial unit memory codes. *Electron. Lett.*, vol. 28, no. 18, pp. 1748–1749, 1992.

[DS93] U. Dettmar, U. Sorger: New optimal partial unit memory codes based on extended BCH codes. *Electron. Lett.*, vol. 29, no. 23, pp. 2024, 1993.

[Dum96] I. Dumer: Suboptimal decoding of linear codes: Partition technique. *IEEE Trans. Inf. Theory*, vol. 42, no. 6, pp. 1971–1986, 1996.

[DZ97] J. Dahl, V. V. Zyablov: Interleaver design for turbo codes. In *Proc. IEEE Int. Symp. on Turbo Codes*, Brest, France, 1997.

[EBMS96] A. Engelhart, M. Bossert, J. Maucher, V. R. Sidorenko: Heuristic algorithms for ordering a linear block code to reduce the number of nodes of the minimal trellis. Tech. Rep. ITUU-TR-1996/03, Univ. of Ulm, Dept. of Information Technology, Nov. 1996.

[EEI+89] G. Einarsson, T. Ericson, I. Ingemarsson, R. Johannesson, K. Zigangirov, C.-E. Sundberg: *Topics in Coding Theory, In Honour of Lars H. Zetterberg*. Springer-Verlag, Berlin, 1989.

[Eli55] P. Elias: Coding for noisy channels. *IRE Conv. Rec. Part 4*, pp. 37–47, 1955.

[Enns87] V. I. Enns: New bounds of decoding domain for certain methods of error correction with decision. In *3rd Joint Soviet–Swedish Workshop on Information Theory*, pp. 347–350, Sochi, USSR, May 1987.

[Eri86] T. Ericson: Concatenated codes—principles and possibilities. In *AAECC-4*, Karlsruhe, 1986.

[Evs83] G. S. Evseev: On the complexity of decoding of linear block codes. *Problemy Peredachi Informatsii*, vol. 19, no. 1, pp. 3–8, 1983.

[EZ95] T. Ericson, V. A. Zinoviev: Spherical codes generated by binary partitions of symmetric pointsets. *IEEE Trans. Inf. Theory*, vol. IT-41, pp. 107–129, Jan. 1995.

[Fano63] R. M. Fano: A heuristic discussion of probabilistic decoding. *IEEE Trans. Inf. Theory*, vol. IT-9, pp. 64–74, Apr. 1963.

[Far92] P. G. Farrell: A survey of array error control codes. *Eur. Trans. Telecommun.*, vol. 3, no. 5, pp. 441–454, Sep./Oct. 1992.

[FB96] T. Frey, M. Bossert: A first approach to concatenation of coding and spreading for CDMA-Systems. In *Int. Symp. on Spread Spectrum Techniques and Applications*, pp. 667–671, Mainz, Germany, 22–25 Sep. 1996.

[FGL+84] G. D. Forney, Jr., R. G. Gallager, G. R. Lang, F. M. Longstaff, S. U. Qureshi: Efficient modulation for band-limited channels. *IEEE J. Select. Areas Commun.*, vol. 2, pp. 632–647, 1984.

[FL95] M. P. C. Fossorier, S. Lin: Soft-decision decoding of linear block codes based on ordered statistics. *IEEE Trans. Inf. Theory*, vol. IT-41, no. 5, pp. 1379–1396, Sep. 1995.

[FL96] M. P. C. Fossorier, S. Lin: Computationally efficient soft-decision decoding of linear block codes based on ordered statistics. *IEEE Trans. Inf. Theory*, vol. IT-42, pp. 738–751, May 1996.

[FL97] M. P. C. Fossorier, S. Lin: Complementary reliability-based decodings of binary linear block codes. *IEEE Trans. Inf. Theory*, vol. IT-43, no. 5, pp. 1667–1672, Sep. 1997.

[FLS98] M. P. C. Fossorier, S. Lin, J. Snyders: Reliability-based syndrome decoding of linear block codes. *IEEE Trans. Inf. Theory*, vol. IT-44, pp. 388–398, Jan. 1998.

[For66a] G. D. Forney, Jr.: *Concatenated Codes*. MIT Press, Cambridge, MA, 1966.

[For66b] G. D. Forney, Jr.: Generalized minimum distance decoding. *IEEE Trans. Inf. Theory*, vol. IT-12, no. 2, pp. 125–131, Apr. 1966.

[For70] G. D. Forney, Jr.: Convolutional codes I: Algebraic structure. *IEEE Trans. Inf. Theory*, vol. IT-16, pp. 720–738, Nov. 1970.

[For73] G. D. Forney, Jr.: The Viterbi algorithm. *Proc. IEEE*, vol. 61, pp. 268–278, Mar. 1973.

[For74] G. D. Forney, Jr.: Convolutional codes II: Maximum likelihood decoding. *Inf. and Control*, vol. 25, pp. 222–266, Jul. 1974.

[For88a] G. D. Forney, Jr.: Coset codes—part I: Introduction and geometrical classification. *IEEE Trans. Inf. Theory*, vol. IT-34, no. 5, pp. 1123–1151, Sep. 1988.

[For88b] G. D. Forney, Jr.: Coset codes—part II: binary lattices and related codes. *IEEE Trans. Inf. Theory*, vol. IT-34, no. 5, pp. 1152–1187, Sep. 1988.

[FT93] G. D. Forney, Jr., M. D. Trott: The dynamics of group codes: state spaces, trellis diagrams, and canonical encoders. *IEEE Trans. Inf. Theory*, vol. IT-39, pp. 1491–1513, Sep. 1993.

[Gab71] E. M. Gabidulin: Combinatorial metrics in coding theory. In *Proc. 2nd Int. Symp. on Information Theory*, pp. 169–176, Budapest, 1971.

[Gal] R. G. Gallager: *Information Theory and Reliable Communication*. Wiley, New York, 1968.

[Gal62] R. G. Gallager: Low-density parity-check codes. *IRE Trans. Inf. Theory*, vol. 8, no. 1, pp. 21–28, Jan. 1962.

[GB97] E. Gabidulin, M. Bossert: Phase rotation invariant block codes. In *4th Int. Symp. on Communication Theory and Applications*, Jul. 13–18 1997.

[GB98] E. Gabidulin, M. Bossert: Hard and soft-decision decoding of phase rotation invariant block codes. In *1998 Int. Zürich Seminar on Broadband Communications*, Feb. 17–19 1998.

[Gil52] E. Gilbert: A comparison of signalling alphabets. *Bell Syst. Tech. J.*, vol. 31, pp. 504–522, 1952.

[Gil60] E. N. Gilbert: Capacity of a burst-noise channel. *Bell Syst. Tech. J.*, vol. 39, pp. 1253–1265, Sep. 1960.

[Gol] S. W. Golomb: *Shift Register Sequences*. rev. edn., Aegean Park Press, Laguna Hills, CA, 1982.

[Gol49] M. J. E. Golay: Notes on digital coding. In *Proc. IEEE*, vol. 37, p. 657, 1949.

[Gol54] M. J. E. Golay: Binary coding. *IEEE Trans. Inf. Theory*, vol. 4, pp. 23–28, 1954.

[Gore70] W. C. Gore: Further results on product codes. *IEEE Trans. Inf. Theory*, vol. IT-16, pp. 446–451, 1970.

[GZ61] D. C. Gorenstein, N. Zierler: A class of error-correcting codes in p^m symbols. *J. Soc. Indust. Appl. Math.*, vol. 9, pp. 207–214, 1961.

[Hag88] J. Hagenauer: Rate compatible punctured convolutional codes (RCPC codes) and their applications. *IEEE Trans. Commun.*, vol. COM-36, pp. 389–400, Apr. 1988.

[Hag90] J. Hagenauer: Soft output Viterbi decoder. Tech. Rep., Deutsche Forschungsanstalt für Luft- und Raumfahrt (DLR), 1990.

[Hag95] J. Hagenauer: Source-controlled channel coding. *IEEE Trans. Commun.*, vol. COM-43, pp. 2449–2457, 1995.

[Ham50] R. Hamming: Error detecting and error correcting codes. *Bell Syst. Tech. J.*, vol. 29, pp. 147–160, 1950.

[HaWr] G. H. Hardy, E. M. Wright: *An Introduction to the Theory of Numbers*. 5th edn., Oxford University Press, Oxford, 1979.

[HBS93] H. Herzberg, Y. Be'ery, J. Snyders: Concatenated multilevel block coded modulation. *IEEE Trans. Commun.*, vol. COM-41, no. 1, pp. 41–49, Jan. 1993.

[HBSD98] A. Häutle, M. Bossert, S. A. Shavgulidze, H. Dieterich: Generalized concatenation of encoded tamed frequency modulation. In *2. ITG Fachtagung, Codierung für Quelle Kanal und Übertragung*, pp. 279–284, Aachen, Germany, Mar. 1998.

[HEK87] P. C. Hershey, A. Ephremides, R. K. Khatri: Performance of RS-BCH concatenated codes and BCH single stage codes on an interference satellite channel. *IEEE Trans. Commun.*, vol. COM-35, pp. 550–556, 1987.

[HH89] J. Hagenauer, P. Höher: A Viterbi algorithm with soft-decision outputs and its applications. In *Proc. GLOBECOM '89*, pp. 47.1.1–47.1.7, Dallas, TX, Nov. 1989.

[HHC93] Y. S. Han, C. R. P. Hartmann, C.-C. Chen: Efficient priority-first search maximum-likelihood soft-decision decoding of linear block codes. *IEEE Trans. Inf. Theory*, vol. IT-39, no. 5, pp. 1514–1523, Sep. 1993.

[HKC+94] A. R. Hammons, P. V. Kumar, A. R. Calderbank, N. J. A. Sloane, P. Sole: The $\mathbb{Z}_4$-linearity of Kerdock, Preparata, Goethals, and related code. *IEEE Trans. Inf. Theory*, vol. IT-40, pp. 301–319, 1994.

[Hoc59] A. Hocquenghem: Codes correcteurs d'erreurs. *Chiffres*, vol. 2, pp. 147–156, 1959.

[Hole88] K. J. Hole: New short constraint length rate $(N-1)/N$ punctured convolutional codes for soft-decision Viterbi decoding. *IEEE Trans. Inf. Theory*, vol. IT-34, pp. 1079–1081, Sep. 1988.

[HOP96] J. Hagenauer, E. Offer, L. Papke: Iterative decoding of binary block and convolutional codes. *IEEE Trans. Inf. Theory*, vol. IT-42, no. 2, pp. 429–445, Mar. 1996.

[HR76] C. R. P. Hartmann, L. D. Rudolph: An optimum symbol-by-symbol decoding rule for linear codes. *IEEE Trans. Inf. Theory*, vol. IT-22, pp. 514–517, Sep. 1976.

[HS73] H. J. Helgert, R. D. Stinaff: Minimum distance bounds for binary linear codes. *IEEE Trans. Inf. Theory*, vol. IT-19, pp. 344–356, May 1973.

[HTV82] H. Hoeve, J. Timmermanns, L. B. Vries: Error correction and concealment in the compact disc system. *Philips Tech. Rev.*, vol. 40, no. 6, pp. 166–172, 1982.

[Hub94] K. Huber: Codes over Gaussian integers. *IEEE Trans. Inf. Theory*, vol. IT-40, no. 1, pp. 207–216, Jan. 1994.

[IH77] H. Imai, S. H. Hirakawa: A new multilevel coding method using error-correcting codes. *IEEE Trans. Inf. Theory*, vol. IT-23, pp. 371–377, May 1977.

[Jak] W. C. Jakes: *Microwave Mobile Communications*. Wiley, New York, 1974.

[JD78] F. de Jager, C. B. Dekker: Tamed frequency modulation—a novel method to achieve spectrum economy in digital transmission. *IEEE Trans. Commun.*, vol. COM-26, pp. 534–542, May 1978.

[Jel69] F. Jelinek: A fast sequential decoding algorithm using a stack. *IBM J. Res. Dev.*, vol. 13, pp. 675–685, Nov. 1969.

[Jen85] J. M. Jensen: The concatenated structure of cyclic and abelian codes. *IEEE Trans. Inf. Theory*, vol. IT-31, pp. 788–793, 1985.

[Jen92] J. M. Jensen: Cyclic concatenated codes with constacyclic outer codes. *IEEE Trans. Inf. Theory*, vol. IT-38, pp. 950–959, May 1992.

[Jen96] J. M. Jensen: Cyclic concatenated codes. Personal communication.

[Joh75] R. Johannesson: Robustly-optional rate one-half binary convolutional codes. *IEEE Trans. Inf. Theory*, vol. IT-21, pp. 964–968, Jul. 1975.

[JoZi] R. Johannesson, K. S. Zigangirov: *Fundamentals of Convolutional Coding*. IEEE Press, Piscataway, NJ, 1999.

[JTZ88] J. Justesen, C. Thommesen, V. V. Zyablov: Concatenated codes with convolutional inner codes. *IEEE Trans. Inf. Theory*, vol. IT-34, part II, no. 5, pp. 1217–1225, Sep. 1988.

[Jus93] J. Justesen: Bounded distance decoding of unit memory codes. *IEEE Trans. Inf. Theory*, vol. IT-39, no. 5, pp. 1616–1627, Sep. 1993.

[JW93] R. Johannesson, Z. Wan: A linear algebra approach to minimal convolutional encoders. *IEEE Trans. Inf. Theory*, vol. IT-39, pp. 1219–1233, Jul. 1993.

[Kap] E. D. Kaplan (ed.): *Understanding GPS: Principles and Applications*. Artech House, Boston, 1996.

[Kar87] Y. D. Karyakin: Fast correlation decoding of Reed–Muller codes. *Problemy Peredachi Informatsii*, vol. 23, pp. 40–49, 1987.

[KKT+95] T. Kasami, T. Koumoto, T. Takata, T. Fujiwara, S. Lin: The least stringent sufficient condition on the optimality of suboptimally decoded codewords. In *Proc. of IEEE Int. Symp. on Information Theory*, p. 470, Whistler, Canada, Jun. 1995.

[KL87] T. Kasami, S. Lin: A cascade coding scheme for error control and its performance analysys. Tech. Rep., NASA-GSFC, 1987.

[KL93] A. D. Kot, C. Leung: On the construction and dimensionality of linear block code trellises. In *Proc. of IEEE Int. Symp. on Information Theory*, p. 291, 1993.

[KNIH94] T. Kaneko, T. Nishijima, H. Inazumi, S. Hirasawa: An efficient maximum-likelihood-decoding algorithm for linear block codes with algebraic decoder. *IEEE Trans. Inf. Theory*, vol. IT-40, no. 2, pp. 320–327, Mar. 1994.

[Kol96] E. Kolev: Binary mapped Reed–Solomon codes and their weight distribution. In *Proc. 5th Int. Workshop on Algebraic and Combinatorial Coding Theory*, pp. 161–169, Sozopol, Bulgaria, 1996.

[Kro89] E. A. Krouk: A bound on the decoding complexity of linear block codes. *Problemy Peredachi Informatsii*, vol. 25, no. 3, pp. 103–106, 1989.

[KS95] F. R. Kschischang, V. Sorokine: On the trellis structure of block codes. *IEEE Trans. Inf. Theory*, vol. IT-41, Part II, no. 6, pp. 1924–1937, Nov. 1995.

[Ksc96] F. Kschischang: The trellis structure of maximal fixed-cost codes. *IEEE Trans. Inf. Theory*, vol. IT-42, no. 6, pp. 1828–1838, Nov. 1996.

[KTFL93] T. Kasami, T. Takata, T. Fujiwara, S. Lin: On the optimum bit orders with respect to the state complexity of trellis diagrams for binary linear block codes. *IEEE Trans. Inf. Theory*, vol. IT-39, pp. 242–245, Jan. 1993.

[KTL86] T. Kasami, F. Tohru, S. Lin: A concatenated coding scheme for error control. *IEEE Trans. Commun.*, vol. COM-34, pp. 481–488, 1986.

[Lar73] K. J. Larsen: Short convolutional codes with maximal free distance for rates 1/2, 1/3 and 1/4. *IEEE Trans. Inf. Theory*, vol. IT-19, pp. 371–372, May 1973.

[LBB96] R. Lucas, M. Bossert, M. Breitbach: Iterative soft decision decoding of linear binary block codes. In *Proc. IEEE Int. Symp. on Information Theory and Its Applications*, Victoria, Canada, 1996.

[LBB98] R. Lucas, M. Bossert, M. Breitbach: On iterative soft-decision decoding of linear binary block codes and product codes. *IEEE J. Select. Areas Commun.*, vol. SAC-16, no. 2, pp. 276–296, Feb. 1998.

[LBBG96] R. Lucas, M. Bossert, M. Breitbach, H. Grießer: On iterative soft decision decoding of binary QR codes. In *Proc. 5th Int. Workshop on Algebraic and Combinatorial Coding Theory*, pp. 184–189, Sozopol, Bulgaria, 1996.

[LBD98a] R. Lucas, M. Bossert, A. Dammann: Improved list-decoding for Reed–Muller codes as generalized multiple concatenated (GMC) codes. In *Proc. of IEEE Int. Symp. on Information Theory*, p. 337, Cambridge, MA, Aug. 1998.

[LBD98b] R. Lucas, M. Bossert, A. Dammann: Improved soft-decision decoding of Reed–Muller codes as generalized multiple concatenated codes. In *2. ITG Fachtagung, Codierung für Quelle Kanal und Übertragung*, pp. 137–141, Aachen, Germany, Mar. 1998.

[LBT93] N. Lous, P. Bours, H. van Tilborg: On maximum likelihood soft-decision decoding of binary linear codes. *IEEE Trans. Inf. Theory*, vol. 39, no. 1, pp. 197–203, Jan. 1993.

[LCF] S. Lin, D. J. Costello, M. P. C. Fossorier: *Error Control Coding*. 2nd edn., Prentice-Hall, Englewood Cliffs, NJ, 1999.

[Lee76] L.-N. Lee: Short unit-memory byte-oriented binary convolutional codes having maximal free distance. *IEEE Trans. Inf. Theory*, vol. IT-22, pp. 349–352, May 1976.

[Lee94] L. H. C. Lee: New rate-compatible punctured convolutional codes for Viterbi decoding. *IEEE Trans. Commun.*, vol. COM-42, pp. 3073–3079, Dec. 1994.

[LiCo] S. Lin, D. J. Costello: *Error Control Coding, Fundamentals and Applications*. Prentice-Hall, Englewood Cliffs, NJ, 1983.

[LL87] M. Lin, S. Lin: On codes with multi-level error-correction capabilities. Tech. Rep., NASA-ECS, 1987.

[LNSM85] S. N. Litsyn, E. E. Nemirovsky, O. I. Shekhovtsov, L. G. Mikhailovskaya: The fast decoding of first order Reed–Muller codes in the Gaussian channel. *Prob. Control Inf. Theory*, vol. 14, pp. 189–201, 1985.

[LS83] S. N. Litsyn, O. I. Shekhovtsov: Fast decoding algorithm for first-order Reed–Muller codes. *Prob. Inf. Transmission*, vol. 19, pp. 87–91, 1983, translated from *Problemy Peredachi Informatsii*.

[Luc97] R. Lucas: *Iterative Decoding of Block Codes*. No. 511 in Reihe 10, VDI Verlag Fortschrittberichte, 1997, Dissertation, Univ. of Ulm.

[Man77] D. M. Mandelbaum: Method for decoding of generalized Goppa codes. *IEEE Trans. Inf. Theory*, vol. IT-23, pp. 137–140, Jan. 1977.

[Mas] J. L. Massey: *Threshold Decoding*. MIT Press, Cambridge, MA, 1963.

[Mas69] J. L. Massey: Shift register synthesis and BCH decoding. *IEEE Trans. Inf. Theory*, vol. IT-15, pp. 122–127, Jan. 1969.

[Mas78] J. L. Massey: Foundations and methods of channel coding. In *Proc. of Int. Conf. on Information Theory and Systems*, vol. 65, pp. 148–157, 1978, NTG-Fachberichte.

[Mas92] J. L. Massey: *Deep Space Communications and Coding: A Marriage Made in Heaven*, vol. 182 of *Lecture Notes in Control and Information Sciences*. Springer-Verlag, Berlin, 1992.

[Mau95] J. Maucher: A new construction of (partial) unit memory codes based on Reed–Muller codes. In *Proc. 7th Joint Swedish–Russian Int. Workshop on Information Theory*, pp. 180–184, St. Petersburg, Russia, Jun. 1995.

[McE96] R. J. McEliece: On the BCJR trellis for linear block codes. *IEEE Trans. Inf. Theory*, vol. IT-42, pp. 1072–1092, Jul. 1996.

[McE98] R. J. McEliece: The algebraic theory of convolutional codes. In *Handbook of Coding Theory*, vol. 1, North-Holland, Amsterdam, 1998.

[McWSl] F. J. MacWilliams, N. J. A. Sloane: *The Theory of Error-Correcting Codes*. North-Holland, Amsterdam, 1996.

[MH81] K. Murota, K. Hirade: GMSK modulation for digital mobile radio telphony. *IEEE Trans. Commun.*, vol. COM-29, pp. 1044–1050, Jul. 1981.

[MMHP94] F. Morales-Moreno, W. Holubowicz, S. Pasupathy: Optimization of trellis coded TFM via matched codes. *IEEE Trans. Commun.*, vol. COM-42, pp. 1586–1594, Feb./Mar./Apr. 1994.

[MMHP95] F. Morales-Moreno, W. Holubowicz, S. Pasupathy: Convolutional coding of binary CPM schemes with no increase in receiver complexity. *IEEE Trans. Commun.*, vol. COM-43, pp. 1221–1224, Feb./Mar./Apr. 1995.

[MMP88] F. Morales-Moreno, S. Pasupathy: Structure, optimization, and realization of FFSK trellis codes. *IEEE Trans. Inf. Theory*, vol. IT-34, pp. 730–751, Jul. 1988.

[MRRW77] R. J. McEliece, E. R. Rodemich, H. C. Rumsey, L. R. Welch: New upper bounds on the weight of a code via the Delsarte–MacWilliams inequalities. *IEEE Trans. Inf. Theory*, vol. IT-23, pp. 157–166, 1977.

[Mud88] D. J. Muder: Minimal trellises for block codes. *IEEE Trans. Inf. Theory*, vol. IT-34, no. 5, pp. 1049–1053, Sep. 1988.

[Mul54] D. E. Muller: Application of boolean algebra to switching circuit design and to error detection. *IEEE Trans. Comput.*, vol. 3, pp. 6–12, 1954.

[Nec91] A. A. Nechaev: Kerdock codes in a cyclic form. *Discrete Math. Appl.*, vol. 1, no. 4, pp. 365–384, 1991, (in Russian: *Discrete Math. (USSR)* 1989).

[NeWo] G. L. Nemhauser, L. A. Wolsey: *Integer and Combinatorial Optimization*. Wiley, New York, 1988.

[NK96] A. A. Nechaev, A. S. Kuzmin: Z-4-linearity, two approaches. In *Proc. 5th Int. Workshop on Algebraic and Combinatorial Coding Theory*, pp. 212–215, Sozopol, Bulgaria, 1996.

[Omu70] J. K. Omura: A probabilistic decoding algorithm for binary group codes (abstract). *IEEE Trans. Inf. Theory*, vol. 16, no. 1, pp. 123, Jan. 1970.

[OpWi] A. Oppenheim, A. S. Willsky: *Signals and Systems*. Prentice-Hall, Englewood Cliffs, NJ, 1983.

[Pal95] R. Palazzo, Jr.: A network flow approach to convolutional codes. *IEEE Trans. Commun.*, vol. COM-43, pp. 1429–1440, Feb./Mar./Apr. 1995.

[Pet60] W. W. Peterson: Encoding and error-correction procedures for the Bose–Chaudhuri codes. *IEEE Trans. Inf. Theory*, vol. 6, pp. 459–470, 1960.

[PeWe] W. W. Peterson, E. J. Weldon: *Error Correcting Codes*. MIT Press, Cambridge, MA, 1981.

[Plo60] M. Plotkin: Binary codes with specified minimum distances. *IEEE Trans. Inf. Theory*, vol. 6, pp. 445–450, 1960.

[Por85] S. L. Portnoy: Characteristics of coding and modulation systems from the standpoint of concatenated codes. *Problemy Peredachi Informatsii*, vol. 21, no. 3, pp. 14–27, 1985.

[Pra59] E. Prange: The use of coset equivalence in the analysis and decoding of group codes. Tech. Rep. AFCRC-TR-59-164, USAF Cambridge Research Center, Bedford, Mass., USA, 1959.

[Pro] J. G. Proakis: *Digital Communications*. 3rd edn., McGraw-Hill, New York, 1995.

[Reed54] I. S. Reed: A class of multiple-error-correcting codes and the decoding scheme. *IEEE Trans. Inf. Theory*, vol. 4, pp. 38–49, 1954.

[Rei60] S. H. Reiger: Codes for the correction of "clustered" errors. *IRE Trans. Inf. Theory*, vol. 6, no. 2, pp. 16–21, Mar. 1960.

[Rie98] S. Riedel: Symbol-by-symbol MAP decoding algorithm for high-rate convolutional codes that use reciprocal dual codes. *IEEE J. Select. Areas Commun.*, vol. SAC-16, no. 2, pp. 175–185, Feb. 1998.

[Rim88] B. Rimoldi: A decomposition approach to CPM. *IEEE Trans. Inf. Theory*, vol. IT-34, pp. 260–270, Mar. 1988.

[Rim89] B. Rimoldi: Design of coded CPFSK modulation systems for bandwidth and energy efficiency. *IEEE Trans. Commun.*, vol. COM-37, pp. 897–905, Sep. 1989.

[RL95] B. Rimoldi, Q. Li: Coded continuous phase modulation using ring convolutional codes. *IEEE Trans. Commun.*, vol. COM-43, pp. 2714–2720, Nov. 1995.

[RS60] I. S. Reed, G. Solomon: Polynomial codes over certain finite fields. *J. Soc. Indust. Appl. Math.*, vol. 8, pp. 300–304, 1960.

[RS95] S. Riedel, Y. V. Svirid: Iterative (turbo) decoding of threshold decodable codes. *Eur. Trans. Telecommun.*, vol. 6, pp. 527–534, 1995.

[RVH95] P. Robertson, E. Villebrun, P. Höher: A comparison of optimal and sub-optimal MAP decoding algorithms operating in the log domain. In *IEEE Int. Conf. on Communications*, pp. 1009–1013, Seattle, WA, Jun. 1995.

[RW95] P. Robertson, T. Wörz: Coded modulation scheme employing turbo codes. *Electron. Lett.*, vol. 31, pp. 1546–1547, 1995.

[Say86] S. I. Sayegh: A class of optimum block codes in signal space. *IEEE Trans. Commun.*, vol. COM-34, pp. 1043–1045, 1986.

[SB89] J. Snyders, Y. Be'ery: Maximum likelihood soft decoding of binary block codes and decoders for the Golay codes. *IEEE Trans. Inf. Theory*, vol. 35, pp. 963–975, Sep. 1989.

[SB90] G. Schnabl, M. Bossert: Coded modulation with generalized multiple concatenation of block codes. In *Proc. of AAECC-8*, Tokyo, Japan, 1990.

[SB94] G. Schnabl, M. Bossert: Reed–Muller codes as generalized multiple concatenated codes with soft-decision decoding. Tech. Rep. ITUU-TR-1994/01, Univ. of Ulm, Dept. of Information Technology, Germany, 1994.

[SB95] G. Schnabl, M. Bossert: Soft-decision decoding of Reed–Muller codes as generalized multiple concatenated codes. *IEEE Trans. Inf. Theory*, vol. IT-41, pp. 304–308, Jan. 1995.

[Sha48] C. E. Shannon: A mathematical theory of communication. *Bell Syst. Tech. J.*, vol. 27, pp. 379–423 and 623–656, 1948.

[Sid96] V. R. Sidorenko: The Viterbi decoding complexity of group and some nongroup codes. In *Proc. 5th Int. Workshop on Algebraic and Combinatorial Coding Theory*, pp. 259–265, Sozopol, Bulgaria, Jun. 1996.

[Sid97] V. R. Sidorenko: The Euler characteristic $|V| - |E|$ of the minimal code trellis is maximum. *Prob. Inf. Transmission*, vol. 32, no. 2, pp. 10–21, Jan. 1997, translated from *Problemy Peredachi Informatsii*.

[Sin64] R. C. Singleton: Maximum distance q-nary codes. *IEEE Trans. Inf. Theory*, vol. 10, pp. 116–118, 1964.

[SKHN75] Y. Sugiyama, M. Kasahara, S. Hirasawa, T. Namekawa: A method for solving key equation for decoding Goppa codes. *Inf. and Control*, vol. 27, pp. 87–99, 1975.

[SL83] G. Seroussi, A. Lempel: Maximum likelihood decoding of certain Reed-Muller codes. *IEEE Trans. Inf. Theory*, vol. IT-29, pp. 448–450, 1983.

[SMH96] V. R. Sidorenko, G. Markarian, B. Honary: Minimal trellis design for linear codes based on the Shannon product. *IEEE Trans. Inf. Theory*, vol. IT-42, no. 6, pp. 2048–2053, Nov. 1996.

[SMH97] V. R. Sidorenko, I. Martin, B. Honary: On separability of some known nonlinear block codes. In *Proc. of IEEE Int. Symp. on Information Theory*, p. 506, Jun. 1997.

[Sny91] J. Snyders: Reduced lists of error patterns for maximum likelihood soft decoding. *IEEE Trans. Inf. Theory*, vol. 37, no. 4, pp. 1194–1200, Jul. 1991.

[Sor93] U. K. Sorger: A new Reed–Solomon code decoding algorithm based on Newton's interpolation. *IEEE Trans. Inf. Theory*, vol. IT-39, no. 2, pp. 358–365, Mar. 1993.

[Sun86] C.-E. Sundberg: Continuous phase modulation. *IEEE Commun. Mag.*, vol. 24, no. 4, pp. 25–38, Apr. 1986.

[Svi95] Y. V. Svirid: Weight distributions and bounds for turbo codes. *Eur. Trans. Telecommun.*, vol. 6, pp. 543–555, 1995.

[TH95a] P. Tyczka, W. Holubowicz: GMSK modulation combined with convolutional codes under receiver complexity constraint. In *ISCTA*, pp. 114–121, Ambleside, UK, Jul. 1995.

[TH95b] P. Tyczka, W. Holubowicz: Trellis coding of Gaussian filtered MSK. In *Proc. of IEEE Int. Symp. on Information Theory*, p. 63, Whistler, Canada, Sep. 1995.

[TH97] P. Tyczka, W. Holubowicz: Comparison of several receiver structures for trellis-coded GMSK signals: Analytical and simulation results. In *Proc. of IEEE Int. Symp. on Information Theory*, p. 193, Ulm, Germany, Jun. 1997.

[TJ83] C. Thommesen, J. Justesen: Bounds on distances and error exponents of unit memory codes. *IEEE Trans. Inf. Theory*, vol. IT-29, pp. 637–649, Sep. 1983.

[TP91] D. J. Taipale, M. B. Pursley: An improvement to generalized minimum distance decoding. *IEEE Trans. Inf. Theory*, vol. IT-37, no. 1, pp. 167–172, Jan. 1991.

[TSKN82] K. Tokiwa, T. Sugimura, M. Kasahara, T. Namekawa: New decoding algorithm for Reed-Muller codes. *IEEE Trans. Inf. Theory*, vol. IT-28, pp. 779–787, 1982.

[Ulr57] W. Ulrich: Non-binary error-correcting codes. *Bell Syst. Tech. J.*, vol. 36, no. 6, pp. 1341–1387, 1957.

[Ung82] G. Ungerböck: Channel coding with multilevel/phase signals. *IEEE Trans. Inf. Theory*, vol. IT-28, pp. 55–67, Jan. 1982.

[Ung87a] G. Ungerböck: Trellis-coded modulation with redundant signal sets—part I. *IEEE Commun. Mag.*, vol. 25, pp. 5–11, 1987.

[Ung87b] G. Ungerböck: Trellis-coded modulation with redundant signal sets—part II: State of the art. *IEEE Commun. Mag.*, vol. 25, no. 2, pp. 12–21, Feb. 1987.

[US94] N. A. Ugrelidze, S. A. Shavgulidze: Convolutional codes over rings for CPFSK signalling. *Electron. Lett.*, vol. 30, pp. 832–834, May 1994.

[USA94] N. A. Ugrelidze, S. A. Shavgulidze, I. G. Asanidze: Simulated error performance of encoded MSK signals in Gaussian and Rician fading channels. *Electron. Lett.*, vol. 30, no. 12, pp. 932–933, Jun. 1994.

[Var57] R. Varshamov: Estimate of the number of signals in error correcting codes. Tech. Rep. 117, Dokl. Akad. Nauk, SSSR, 1957.

[VB91] A. Vardy, Y. Be'ery: Bit-level soft-decision decoding of Reed–Solomon codes. *IEEE Trans. Inf. Theory*, vol. IT-39, no. 3, Mar. 1991.

[Vit67] A. J. Viterbi: Error bounds for convolutional codes and an asymptotically optimum decoding algorithm. *IEEE Trans. Inf. Theory*, vol. IT-13, pp. 260–269, Apr. 1967.

[Vit71] A. J. Viterbi: Convolutional codes and their performance in communications systems. *IEEE Trans. Commun.*, vol. COM-19, pp. 751–772, 1971.

[VK96] A. Vardy, F. R. Kschischang: Proof of a conjecture of McEliece regarding the expansion index of the minimal trellis. *IEEE Trans. Inf. Theory*, vol. IT-42, no. 6, Nov. 1996.

[WE63] J. K. Wolf, B. Elspas: Error-locating codes—a new concept in error control. *IEEE Trans. Inf. Theory*, vol. 9, pp. 113–117, 1963.

[Wei84a] L. F. Wei: Rotationally invariant convolutional channel coding with expanded signal space—part I. *IEEE J. Select. Areas Commun.*, vol. SAC-2, pp. 659–672, 1984.

[Wei84b] L. F. Wei: Rotationally invariant convolutional channel coding with expanded signal space—part II. *IEEE J. Select. Areas Commun.*, vol. SAC-2, pp. 672–686, 1984.

[Wei87] L. F. Wei: Trellis-coded modulation with multidimensional constellations. *IEEE Trans. Inf. Theory*, vol. IT-33, no. 4, pp. 483–501, Jul. 1987.

[Wel71] E. J. Weldon, Jr.: Decoding binary block codes on q-ary output channels. *IEEE Trans. Inf. Theory*, vol. IT-17, pp. 713–718, 1971.

[WH93a] T. Wörz, J. Hagenauer: Decoding of M-PSK-multilevel codes. *Eur. Trans. Telecommun.*, vol. 4, no. 3, pp. 299–308, May/Jun. 1993.

[WH93b] T. Wörz, J. Hagenauer: Multistage decoding of coded modulation using soft output and source information. In *IEEE Information Theory Workshop*, pp. 43–44, Jun. 1993.

[WH95] U. Wachsmann, J. Huber: Power and bandwidth efficient digital communication using turbo codes in multilevel codes. *Eur. Trans. Telecommun.*, vol. 6, pp. 557–567, 1995.

[WHPH87] W. W. Wu, D. Haccoun, R. Peile, Y. Hirata: Coding for satellite communication. *IEEE J. Select. Areas Commun.*, vol. SAC-5, no. 4, pp. 724–748, May 1987.

[WLK$^+$94] J. Wu, S. Lin, T. Kasami, T. Fujiwara, T. Takata: An upper bound on the effective error coefficent of two-stage decoding, and good two-level decompositions of some Reed–Muller codes. *IEEE Trans. Commun.*, vol. COM-42, no. 2/3/4, pp. 813–818, Feb./Mar./Apr. 1994.

[WoJa] J. M. Wozencraft, I. M. Jacobs: *Principles of Communication Engineering*. Wiley, New York, 1965.

[Wolf65] J. K. Wolf: On codes derivable from the tensor product of check matrices. *IEEE Trans. Inf. Theory*, vol. 11, pp. 281–284, 1965.

[Wolf78] J. K. Wolf: Efficient maximum likelihood decoding of linear block codes using a trellis. *IEEE Trans. Inf. Theory*, vol. IT-24, no. 1, pp. 76–80, Jan. 1978.

[WoRe] J. M. Wozencraft, B. Reiffen: *Sequential Decoding*. MIT Press, Cambridge, MA, 1961.

[WS79] L. R. Welch, R. A. Scholtz: Continued fractions and Berlekamp's algorithm. *IEEE Trans. Inf. Theory*, vol. IT-25, pp. 19–27, Jan. 1979.

[YKH84] Y. Yasuda, K. Kashiki, Y. Hirata: High-rate punctured convolutional codes for soft decision Viterbi decoding. *IEEE Trans. Commun.*, vol. COM-32, pp. 315–319, Mar. 1984.

[YT94] R. H.-H. Yang, D. P. Taylor: Trellis-coded continuous-phase frequency-shift keying with ring convolutional codes. *IEEE Trans. Inf. Theory*, vol. IT-40, pp. 1057–1067, Jul. 1994.

[Zig66] K. Zigangirov: Some sequential decoding procedures. *Problemy Peredachi Informatsii*, vol. 2, pp. 13–25, 1966.

[Zin76] V. A. Zinoviev: Generalized cascade codes. *Problemy Peredachi Informatsii*, vol. 12, no. 1, pp. 5–15, 1976.

[Zin81] V. A. Zinoviev: Generalized concatenated codes for channels with error bursts and independent errors. *Problemy Peredachi Informatsii*, vol. 17, pp. 53–56, 1981.

[ZJTS96] V. V. Zyablov, J. Justesen, C. Thommesen, S. A. Shavgulidze: Bounds on distances for unit memory concatenated codes. *Problemy Peredachi Informatsii*, vol. 32, no. 1, pp. 58–69, 1996.

[ZMB98] V. V. Zyablov, J. Maucher, M. Bossert: On the equivalence of GCC and GEL codes. In *Proc. 6th Int. Workshop on Algebraic and Combinatorial Coding Theory*, pp. 255–259, Pskov, 1998.

[ZMB99] V. V. Zyablov, J. Maucher, M. Bossert: On the equivalence of generalized concatenated codes and generalized error location codes. To appear in *IEEE Trans. Inf. Theory*.

[ZP91] V. V. Zyablov, S. L. Portnoy: Construction of unit memory convolutional codes based on Reed-Muller codes. *Problemy Peredachi Informatsii*, vol. 27, no. 3, pp. 3–15, Sep. 1991.

[ZPS88] V. V. Zyablov, S. L. Portnoy, S. A. Shavgulidze: The construction and charakteristics of new systems of modulation and coding. *Problemy Peredachi Informatsii*, vol. 24, no. 4, pp. 17–28, 1988.

[ZPS93] V. V. Zyablov, V. G. Potapov, V. R. Sidorenko: Maximum-likelihood list decoding using trellises. *Problemy Peredachi Informatsii*, vol. 29, no. 4, pp. 3–10, 1993.

[ZS87] V. V. Zyablov, S. A. Shavgulidze: A bound on the distance for unit memory generalized convolutional concatenated codes. *Problemy Peredachi Informatsii*, vol. 23, no. 2, pp. 17–27, 1987.

[ZS94] V. V. Zyablov, V. R. Sidorenko: Bounds on complexity of trellis decoding of linear block codes. *Prob. Inf. Transmission*, vol. 29, no. 3, pp. 1–6, 1994, translated from *Problemy Peredachi Informatsii*.

[ZSJ95] V. V. Zyablov, S. A. Shavgulidze, J. Justesen: Some constructions of generalised concatenated codes based on unit memory codes. In *Cryptography and Coding*, vol. 1025 of *Lecture Notes in Computer Science*, pp. 237–256, Springer-Verlag, Berlin, 1995.

[Zya72] V. V. Zyablov: New interpretation of localization error codes, their error correcting capability and algorithms of decoding. In *Transmission of Discrete Information over Channels with Clustered Errors*, pp. 8–17, Nauka, Moscow, 1972, (in Russian).

[ZySha] V. V. Zyablov, S. A. Shavgulidze: *Generalized Concatenated Constructions on the Basis of Convolutional Codes*. Nauka, Moscow, 1991, (in Russian).

[ZZ79a] V. A. Zinoviev, V. V. Zyablov: Codes with unequal error protection of symbols. *Problemy Peredachi Informatsii*, vol. 15, pp. 50–58, 1979.

[ZZ79b] V. A. Zinoviev, V. V. Zyablov: Correction of error bursts and independent errors using generalized cascaded codes. *Problemy Peredachi Informatsii*, vol. 15, pp. 58–70, 1979.

Index

A

B

C

D

H

I

K

L

M